Lecture Notes in Physics

The Lecture Notes in Physics

The series Lecture Notes in Physics (LNP), founded in 1969, reports new developments in physics research and teaching – quickly and informally, but with a high quality and the explicit aim to summarize and communicate current knowledge in an accessible way. Books published in this series are conceived as bridging material between advanced graduate textbooks and the forefront of research and to serve three purposes:

- to be a compact and modern up-to-date source of reference on a well-defined topic
- to serve as an accessible introduction to the field to postgraduate students and nonspecialist researchers from related areas
- to be a source of advanced teaching material for specialized seminars, courses and schools

Both monographs and multi-author volumes will be considered for publication. Edited volumes should, however, consist of a very limited number of contributions only. Proceedings will not be considered for LNP.

Volumes published in LNP are disseminated both in print and in electronic formats, the electronic archive being available at springerlink.com. The series content is indexed, abstracted and referenced by many abstracting and information services, bibliographic networks, subscription agencies, library networks, and consortia.

Proposals should be sent to a member of the Editorial Board, or directly to the managing editor at Springer:

Christian Caron
Springer Heidelberg
Physics Editorial Department I
Tiergartenstrasse 17
69121 Heidelberg / Germany
christian.caron@springer.com

V.J. Martínez
E. Saar
E. Martínez-González
M.-J. Pons-Bordería (Eds.)

Data Analysis in Cosmology

Springer

Vicent J. Martínez
Observatori Astronòmic
Universitat de València
Edificio de Institutos de Investigación
Polígono la Coma
s/n, 46980 Paterna
Valencia Spain
vicent.martinez@uv.es

Enn Saar
Tartu Observatoorium
EE-61602
Tõravere
Estonia
saar@aai.ee

Enrique Martínez-González
Instituto de Física de Cantabria
Av. de los Castros
s/n, 39005 Santander
Spain
martinez@ifca.unican.es

María-Jesús Pons-Bordería
Escuela Universitaria de Estadística
Universidad Complutense de Madrid
Avda. Puerta de Hierro
s/n, 28040 Madrid
Spain
maria.pons@estad.ucm.es

Martínez V.J. et al. (Eds.), *Data Analysis in Cosmology*, Lect. Notes Phys. 665 (Springer, Berlin Heidelberg 2009), DOI 10.1007/978-3-540-44767-2

ISBN: 978-3-642-06305-3 e-ISBN: 978-3-540-44767-2

DOI 10.1007/978-3-540-44767-2

Lecture Notes in Physics ISSN: 0075-8450 e-ISSN: 1616-6361

Cover design: eStudio Calamar S.L., F. Steinen-Broo, Pau/Girona, Spain

Printed on acid-free paper

9 8 7 6 5 4 3 2 1

springer.com

Preface

Cosmology became a branch of physics during the twentieth century mainly thanks to important discoveries made with the use of big telescopes and radioantennas. Certainly cosmology is a science driven by the most modern technological achievements. Those discoveries were:

- The expansion of the universe proposed by Hubble in 1929 with his famous law showing the linear relationship between the radial velocities of galaxies and their distances.
- The relative abundance of the light chemical elements in the universe envisaged by Gamow and collaborators at the end of the 1940s.
- The discovery in 1965 of the cosmic background radiation by Penzias and Wilson in New Jersey, using the Holmdel antenna of the Bell Labs, while only a few miles from there a group of physicists at Princeton University (Dicke, Peebles, Roll and Wilkinson) had theoretically anticipated the radiation as a detectable fossil of the early universe.

The hot Big Bang model achieved then the range of a paradigm following Khunian terminology. The amount of cosmological data available has increased dramatically in the past decades due to an unprecedented development of telescopes, detectors and satellites.

The recent observations by the WMAP satellite of the cosmic microwave anisotropies and its polarization have become one of the pillars for the present standard cosmological model. The cosmological parameters are being fixed by this and other astronomical observations such as the wide and deep redshift catalogues of galaxies and clusters of galaxies, which provide a fundamental clue about how luminous matter is distributed through the Universe. Experiments on lensing make also possible to know the distribution of dark matter and to relate all of this, to end up with a satisfactory theory of structure formation.

High redshift supernovae showed only a few years ago that the Universe could now be following a phase of accelerated expansion, due to an unknown dark energy that should act as a repulsive gravity – astrophysicists then rescued the cosmological constant from Einstein's wastebasket.

In any case, the huge amount of data collected by the new instruments, comes now in order of terabytes per day and this needs not only big computers to be processed but also the development of intelligent algorithms and pipelines to analyze efficiently all this impressive amount of high-quality information. The idea of editing a volume on Data Analysis in Cosmology was born at an international school held in Valencia. As a follow-up project we undertook to collect tutorial lectures covering in a pedagogical way the best techniques to extract information from cosmological data, as well as reliable methods that should help us to improve the view of our Universe and to know more about the open questions that still arise in our model.

On the occasion of publishing this book we would like to take the opportunity to thank the Universidad Internacional Menéndez Pelayo for including this excellent course in its academic program and for all its financial support. We want also to acknowledge all institutions that have sponsored this school. We are grateful to the Universitat de València (Vicerrectorados de Extensión Universitaria y de Investigación), to the Universidad Politécnica de Cartagena and its Departamento de Matemática Aplicada y Estadística, to the Instituto de Física de Cantabria (Universidad de Cantabria and CSIC), to the Generalitat Valenciana (Conselleria de Empresa, Universidad y Ciencia), to the Spanish Ministerio de Industria, Turismo y Comercio and to the Spanish Ministerio de Educación y Ciencia, through the Acción Complementaria AYA2004-20067-E (Dirección General de Investigación).

Rector, University of Valencia *Francisco Tomás Vert*

Director, Astronomical Observatory, University of Valencia Chairman of the School *Vicent J. Martínez*

Contents

Part I Universal Tools

Part I

Universal Tools

The Sea of Wavelets

B.J.T. Jones

Kapteyn Institute, University of Groningen, 9700 AV Groningen, NL
jones@astro.rug.nl

Wavelet analysis has become a major tool in many aspects of data handling, whether it be statistical analysis, noise removal or image reconstruction. Wavelet analysis has worked its way into fields as diverse as economics, medicine, geophysics, music and cosmology.

Wavelets formalize the process of examining the characteristics of data on many scales through the use of special orthogonal function sets. At each scale we have a shrunken, less detailed, version of the original data. This comes with enough information on the residual details that were smoothed during the shrinking process to reconstruct the original data if required. Transformations using wavelets are reversible in the sense that the original data can be reconstructed from the transformed data.

The idea of looking at data on different scales provides an important approach to data analysis: this is what *multi-resolution analysis* is about. Not all multi-resolution analysis is done using wavelets, but using wavelets adds some important features to such analysis.

One of the main difficulties facing the would-be user of wavelet techniques is to extract what is needed from the vast and varied literature on the subject. In particular, there seems to be a gulf between users and uses of discrete and continuous transforms which this overview hopes to clarify. My task here is to provide an easily accessible introduction to this vast and complex subject.

1 Wavelets Everywhere

1.1 Origin of the Idea

Wavelet transforms were recognized long ago, notably the Haar transform dates from 1910 [21] and the Gabor's localized Fourier transform [18] dates from 1946. These had found specific engineering applications, but had not, at the time, been recognized as being specific examples of a far larger class of transforms. On the mathematical side we should note that Littlewood and

Jones, B.J.T.: *The Sea of Wavelets*. Lect. Notes Phys. **665**, 3–50 (2009)
DOI 10.1007/978-3-540-44767-2_1

Paley [24] provided an important description of a scaling decomposition as early as 1931. Later, the Calderón–Zygmund theory of singular integral operators was to play a major role in establishing the mathematical fundamentals of wavelet bases.

The recent explosion in the development and application of wavelets, as we now know them, started some 20 years ago. The foundations were laid by Grossmann and Morlet [20], who worked for an oil company analyzing seismic data, and by Mallat [26]. Grossmann and Morlet gave us the formal definition of the continuous wavelet transform. The crucial relationship with the theory of filters was established somewhat later: it is now more than 10 years since the seminal papers of Meyer [29] and Daubechies [10] and since Daubechies published her famous lecture notes *Ten Lectures on Wavelets* [9].

An important key development of the mid-1990s was the so-called Lifting mechanism for generating wavelets [39, 40, 41]. Lifting could be used to define wavelets and to improve their properties. All finite filters related to wavelets can be obtained via the lifting technique [12]. The lifting technique provides a starting point for defining wavelet transforms on irregular point sets or on a sphere.

Related ideas are the pyramidal and quad-tree decomposition of image data in which an image is described by examining it at various resolutions [5, 34]. The down-sampling required to traverse the pyramid could be achieved by any of a variety of methods: averaging groups of pixels, taking the median of a group of pixels or by some nonlinear transform that preserved features. Down-sampling using wavelets is an important case.

1.2 What Can Wavelets Do That Is Interesting?

Why bother with wavelets? Do they provide us with anything new? It is almost trendy to include the word "wavelet" in the title of a paper when all that has been done is to smooth or filter the data with a function or filter that has been dubbed a wavelet because it satisfies certain mathematical conditions. We say "smooth or filter" because the world of wavelets is divided into two apparently distinct parts: continuous wavelet transforms and data filters defined at discrete points. As we shall see, these are not really distinct: the situation has parallels in the Heisenberg or Schrodinger representations of states in quantum mechanics.

Functions or filters that are wavelets provide us with a *hierarchical* set of functions or filters that have important orthogonality properties. Mostly, but not always, these functions and filters are defined on a finite interval (compact support), so they act locally.

The word "hierarchical" is significant: we are simultaneously looking at the data on a range of scales. In the world of continuous wavelets all scales are looked at, whereas in the world of wavelet filters the scales differ by factors of two (frequency doubling). The word "local" is also important: it means that the function or filter acts hierarchically on a limited subset of the data, thereby making the data analysis local (as opposed to the Fourier transform which transforms the entire data set at one go). The local and hierarchical properties mean that we are effectively zooming into our data using a microscope. At low magnification we see the overall trends, and as we zoom in we see more and more details without being distracted by the surroundings.

Wavelets remove trends and highlight residuals in a hierarchical way. The residuals thus found can be subjected to scale-dependent statistical analysis with a view to separating out noise component and leaving behind de-noised data. This has an obvious role in image processing. The advantage of wavelets in this respect is that, in general, they are of finite extent and so can analyse local trends in a way that Fourier analysis cannot. This is one reason wavelets have made an impact on the analysis of non-stationary time series.

Wavelets also provide a way of representing a function or data in terms of a hierarchy of orthogonal functions. This is important in music, for example, in trying to separate out the sound from different instruments, or to identify the notes that make up a chord. Wavelets provide a hierarchical way of representing and compacting data. The wavelet representation is often very efficient and allows for high-performance lossy or lossless compression. The efficiency of the wavelet representation leads to high-performance algorithms for computation.

An interesting example is provided by the astronomically familiar problem of calculating potential interactions among particles. It is necessary to calculate sums of the form

$$V_j = \sum_{i=1}^{N} \frac{m_i m_j}{|x_i - x_j|} \ . \tag{1}$$

At first glance this involves calculating and summing N^2 terms. Over the past decades, techniques have been developed to reduce this to on the order of $N \log N$ computations by using hierarchical and multipole expansions of the potential distribution [2, 19]. The work of Rokhlin and collaborators on this problem [6, 4] has reduced the calculation to $O(N)$ by use of wavelets. Note, however, that $O(N)$ computation can be achieved without wavelets by a combination of tree-codes and fast multipole moment evaluations (Dehnen [13, 14]). For a different wavelet-based approach to this problem, see the recent papers by Romeo et al. [33]. There is no stopping progress!

Wavelets are indeed interesting and provide powerful new tools. However, there is almost an excess of goodies: the class of wavelets is enormous and the user is confronted by the question of which members of this class best suit the problem at hand. People often give up on this question and use the simplest continuous wavelet, the Mexican Hat, or the simplest discrete point filter, the Haar wavelet, without asking why they made that choice or seeking a better alternative. I hope this article will provide a relatively simple look into the rich fruits of the sea of wavelets.

1.3 What Will We Learn from All This?

Apart from the hope that this will provide, for some, a relatively straightforward introduction to a rapidly developing subject that is full of possibilities, there are some points, not usually stated, that I wish to emphasis.

First, if you are interested only in data analysis, but not in data reconstruction, multi-resolution analysis is a very powerful tool. You do not need wavelets for this: there are many very powerful alternatives. Wavelets can be used to construct the multi-resolution data pyramids and they may offer some, as yet largely unexplored, possibilities that will provide a different view of data.

Next, one has to distinguish clearly whether continuous wavelet analysis or discrete wavelet filters are the appropriate tool for a particular task. Having decided whether to go continuous or discrete, the burning question is which wavelet to use: there is, as we shall see, a whole ocean of wavelets. Different jobs require different tools. And remember that even continuous wavelets require discretization in order that the integrals can be evaluated.

Finally, the Lifting technique for creating designer wavelets may provide a powerful tool for attacking special problems. We have seen the importance of this with regard to defining wavelets on a sphere, or on non-regularly distributed point sets.

1.4 Decisions, Decisions

The development of wavelet theory has taken place on many fronts and can be compared with the nineteenth-century development of the theory of orthogonal polynomial sets that are solutions of ordinary differential equations. There are so many sets of orthogonal polynomials that the question frequently asked is "which should I use to study this data or solve this equation?". Since the polynomials are intimately linked with eigenvalue problems of differential equations the choice is most often motivated by the nature of the problem: problems on a sphere are invariably solved using the Legendre polynomials. However, when it comes to using orthogonal polynomial sets to analyse data, the situation rapidly becomes confused. Fitting a curve through a set of points is often done using a simple polynomial of the form $a + bx + cx^2 + \cdots + x^N$, but there are far better choices: Tchebycheff polynomials give guarantees for the nature of the residuals or any of a vast variety of spline polynomials give guarantees for the differentiability of the fitted curve.

The fact that wavelets are not, in general, associated with particular differential equations means that they are divorced from a direct relationship with physical problem. The generation and application of wavelets are thus led by interest in specific problems in particular disciplines.

1.5 Different Approaches

Contemporary wavelet analysis is the result of the convergence of developments in several quite distinct areas: function theory, engineering filters for signal processing, image processing and so on. As a consequence, the literature is quite diverse and often quite difficult to grasp. Engineering papers taking the filter theory approach frequently use z-transforms, mathematical papers use function theory and so on. One of the goals of this article is to present and inter-relate material from these approaches. Thus, we shall cover some elementary material on filters and data representation by orthogonal function sets and touch on the z-transforms that dominate the engineering literature.

Astronomical use of wavelets seems to be largely confined to continuous wavelet transforms based mainly on a few specific wavelet functions. The so-called Mexican Hat wavelet holds a predominant position in this respect, perhaps because it is among the simplest of wavelets even if it is possibly not the best choice.

One of the main issues is to grasp the relationship between the continuous and discrete wavelet transforms and to know which transform to use and when. Discussions of this are frequently shrouded in function theory (for good reasons), which can obfuscate what is actually going on.

Then there is the realization that the class of potentially useful wavelets is enormous. Why then should we fix on any particular choice? Indeed, in recent years an important general procedure for generating what might be termed "designer wavelets" has become of prime importance. The so-called "Lifting" process (Sect. 7) provides a mechanism which can be used to design wavelets having specific properties, and moreover to generalize the concept of wavelet analysis to wavelets on non-planar surfaces (e.g. spheres) and on non-uniformly distributed point sets.

1.6 Learning About Wavelets

There is now a significant number of fine texts on wavelets written from a wide variety of points of view. The World Wide Web provides an even greater repository of information, among which are many fine review articles. This article has been put together using a great number of these sources: I have taken what appears to me to be the best bits from among all I have read and put them into a (hopefully) coherent article. If in places you get a feeling of deja vu, you will have read an article that I also enjoyed and found particularly useful.

For pedagogical reasons I will start, as have other reviewers of this subject done, with the Recipe Approach: "this is what a simple so-and-so wavelet transform is" and "this is how to apply it". We shall then see this as part of a larger and deeper concept which we shall develop into a discussion of continuous and discrete wavelet transforms. I will present a list of the more famous wavelet transforms and then apply them to a couple of simple problems so that we can see both continuous and discrete wavelets in action. Then finally we shall get into some of the wider issues such as the different kinds of wavelet analysis that are available and their broader applications.

2 First Examples

2.1 The Simplest: The Haar Transform

We shall consider a one-dimensional data set: values of a function on a set of uniformly distributed points on a line. To keep things really simple we shall work with just four points: $S_0 : \{8, 6, 1, 5\}$. Figure 1 shows the data transformations.

At the first stage, we produce a new data set comprising the pairs $S_1 : \{7, 3\}$ and $D_1 : \{1, -2\}$. S_1 is obtained by adding neighbouring pairs of data values, D_1 is obtained by differencing neighbouring pairs of data values. At the second stage, we do to S_1 exactly what we did to S_0 and transform the pair $S_1 : \{7, 3\}$ into $S_2 : \{5\}$ and $D_2 : \{2\}$ by summing and differencing neighbouring data values in S_1.

This process results in the so-called "multi-resolution analysis":

$$S_0 : \{8, 6, 1, 5\} \to \{S_2, D_2, D_1\} = \{5, 2, 1, -2\} . \tag{2}$$

The process is reversible; we can go from $\{S_2, D_2, D_1\}$ back to $S_0 : \{8, 6, 1, 5\}$ just by adding or subtracting neighbouring pairs of numbers and using

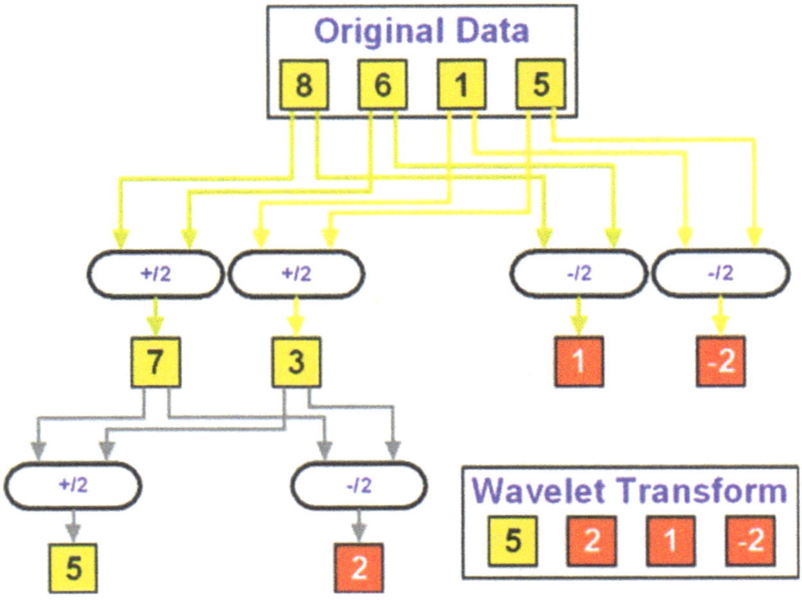

Fig. 1. Trivial wavelet transform of four data points

Table 1. Synonyms for transformed data components

S	D
Low pass	High pass
Sum	Difference
Smooth	Detail
Synthesis	Analysis
Trend	Fluctuation

appropriate normalizations. The sequence S_i is a set of views of the data at decreasing resolution (we averaged neighbouring pairs to generate the next member of the sequence). The sequence D_i provides, at each level, a view of the detail in the data that is not contained in the corresponding S_i. The S_i are sometimes called the "smooth" coefficients and the D_i are sometimes called the "detail" coefficients. The names are not unique: one of the confusing things is the variety of names that are used to describe the same thing. Table 1 lists some of the names used.

This shows "what" a wavelet transform is (albeit a rather trivial one) but not "why" we might have bothered. But where is the wavelet? What did we gain?

Part of the answer to this is that the wavelet representation of the data is in many ways more efficient than the original data array. The sequence of D_i allows new methods for filtering and compressing data. In this trivial example, the transform was executed by simply taking sums or differences of neighbouring elements of the data array. More sophisticated manipulation will yield different and possibly more useful transformed data sets.

2.2 More Formal Look

Let us recapitulate on and formalize the above procedure so that we can generalize it. The first step is to take an ordered data set $\{U\}$:

$$\{U : u_0, u_1, u_2, u_3, \dots\} = \{a, b, c, d, \dots\} . \tag{3}$$

Now create a new data set $\{S\}$ from $\{U\}$ by averaging neighbouring values:

$$\{S : s_0, s_1, \dots\} = \{\tfrac{1}{2}(a+b), \tfrac{1}{2}(c+d), \dots\} . \tag{4}$$

$\{S\}$ has half as many elements as $\{U\}$ and looks rather similar: it is a downsized version of $\{U\}$, having half the resolution. Some of the information in $\{U\}$ has been lost in the averaging process that created $\{U\}$; the missing information is contained in $\{D\}$.

Now construct the sequence $\{D\}$ of differences between neighbours:

$$\{D : d_0, d_1, \dots\} = \{\tfrac{1}{2}(a-b), \tfrac{1}{2}(c-d), \dots\} . \tag{5}$$

Note that $\{D\}$ has the same number of elements as does $\{S\}$.

We can use these differences to reconstruct the original data $\{U\}$ from the averaged sequence $\{S\}$ by means of the (trivial) formulae:

$$\begin{aligned} a &= s_0 + d_0 = \tfrac{1}{2}(a+b) + \tfrac{1}{2}(a-b) , \\ b &= s_0 - d_0 = \tfrac{1}{2}(a+b) - \tfrac{1}{2}(a-b) , \\ c &= s_1 + d_1 = \tfrac{1}{2}(c+d) + \tfrac{1}{2}(c-d) , \\ d &= s_1 - d_1 = \tfrac{1}{2}(c+d) - \tfrac{1}{2}(c-d) , \\ &\text{etc.} \end{aligned} \tag{6}$$

This is the simplest example of a wavelet transform: the Haar transform (we have suppressed normalization factors of $\sqrt{2}$ for the sake of clarity).

This, like other wavelet transforms, is a linear mapping of the data values $\{U\}$ into another set of data values $\{S, D\}$. The importance of this process is that the new set of values may possess some advantageous properties that were not apparent in the original data set. The quest is to select a wavelet transform that is best suited to the particular kind of data analysis that is required for $\{U\}$.

2.3 Alternative Strategies

In the preceding section we averaged pairs of neighbouring numbers. We could have chosen some more complex form of averaging involving four or more numbers. We could also have stored the data in a slightly different way. The following example is an important alternative strategy:

$$\begin{aligned} \{S' : s'_0, s'_1, \dots\} &= \{a, c, \dots\} , \\ \{D' : d'_0, d'_1, \dots\} &= \{b-a, d-c, \dots\} . \end{aligned} \tag{7}$$

Compare (4) and (5). The reconstruction from this sequence is trivial.

What is the difference? In (7) the datum b is only used once (when calculating the difference d'_0), and this calculation can be done in place since b is not used again. In the first method ((4) and (5)), a and b are both used twice. Thus, memory and cpu are used more efficiently and that may be important when transforming large amounts of data.

Yet another strategy is as follows. We use values $a, c, e, \dots$ at the even-numbered data points to make a prediction of what the values $b, d, \dots$ at the odd-numbered points would be on the basis of linear interpolation and then calculate and store a correction for this rather naive prediction:

$$\begin{aligned} \mu_0 &= \tfrac{1}{2}(a+c) , \\ \delta_1 &= b - \mu_0 , \\ \mu_1 &= \tfrac{1}{2}(c+e) , \end{aligned}$$

$$\begin{aligned} \delta_2 &= d - \mu_1 , \\ \{S' : s'_0, s'_1, \dots\} &= \{a, c, e, \dots\} , \\ \{D' : d'_0, d'_1, \dots\} &= \{\delta_0, \delta_1, \dots\} . \end{aligned} \tag{8}$$

Here μ_0 is the naive prediction, based on linear interpolation, of what the value at the intermediate point should be. δ_0 corrects this prediction so that $b = \mu_0 + \delta_0$. We might note at this point that the mean of the coefficients in the original signal $\{a, b, c, \dots\}$ is not equal to the mean of the downsized coefficients $\{a, c, e, \dots\}$: the "energy" of the signal has not been preserved. This is a general property of this kind of interpolating wavelet. We shall show later how to apply a further correction to the values $s'_0, s'_0, \dots$ so that the desirable property of energy conservation can be realized.

This seemingly trivial point lies at the basis of an important technique in wavelet analysis: the so-called Lifting Operation, by which we can refine a given wavelet transform in such a way that there is the possibility of satisfying additional constraints. The above (trivial) analysis can usefully be viewed in a slightly different way. The $\{S' : s'_i\}$ can be interpreted as *predictors* of the values of the data. The $\{D' : \delta_i\}$ are the *corrector* numbers required for the correction of the coarser prediction. The final step will be to update the prediction so that additional constraints (such as energy conservation) are satisfied.

The lifting process is certainly one of the most significant development in wavelet theory. It allows us, for example, to define wavelets on a sphere, to deal with irregularly distributed data points and to define wavelet bases for solving partial differential equations.

3 Some Formal Definitions

The language of wavelets is replete with definitions and notational complexities. We have "Mother" and "Father" wavelets, scaling functions and so on. Table 2 summarizes some of these terms and give pointers to where in this article they are defined.

Table 2. Some of the wavelet jargon

Function	Name(s)	Equation
$\phi(x)$	Scaling function Father wavelet	(15)
$\phi_{n,k}(x)$	Rescaled translate of $\phi(x)$	(16)
h_k	Refinement coefficients	(15)
$\psi(x)$	Mother wavelet	(26)
$\psi_{n,k}(x)$	Rescaled translate of $\psi(x)$	(27)
g_k	Wavelet numbers	(32)

3.1 Multi-resolution Analysis – Heuristics

Before starting on this subject, look at Fig. 2, which is simply a more general version of Fig. 1. In this figure, 16 points are depicted as being transformed, and two levels of the transform are being shown. Look at the end result, the line labelled "MRA". The first four points S_2 are the level 2 shrunken version of the level 1 data S_1. The next four are the detail data D_2 for level 2. This goes with S_2 and is what as left over when S_1 was shrunk to S_2. The last eight data points of MRA are the detail data D_1 that went with S_1 when S_0 was transformed. We could write this level 2 transformed data set as

$$MRA_L(k) = \{z_{L,1}, \ldots, z_{L,4}, e_{L,1}, \ldots, e_{L,4}, d_{L-1,1}, \ldots, d_{L-1,8}, \ldots\}, \quad L = 2\,. \tag{9}$$

Figure 2 depicts this transformed data array.

What is important is that the shrunk data sets $S_0, S_1, S_2, \ldots$ are transformed, "downsized", for want of a better term, versions of the original data, using rescaled versions of the same scaling functions which we can denote by $\phi_L(x)$. Likewise, the detail data sets $D_1, D_2, \ldots$ are transformed versions of the original data, using rescaled versions of the same detail, or analysis, functions which we can denote by $\psi_L(x)$. It would be ideal if the components S_L and D_L were orthogonal. This would be achieved if the functions $\phi_L(x)$ and $\psi_L(x)$ were orthogonal.

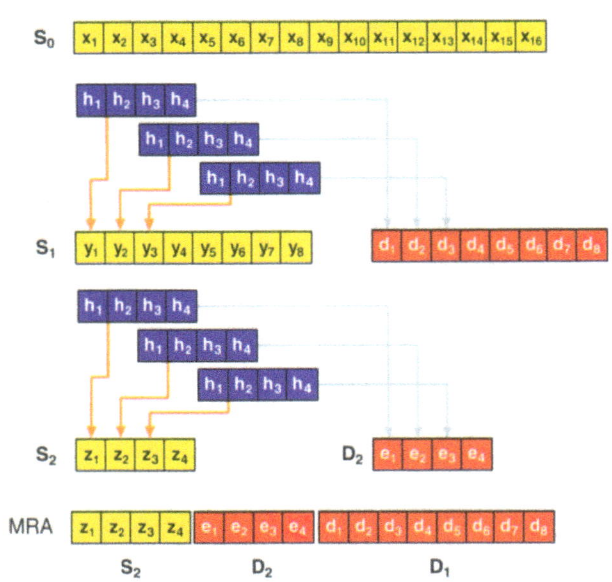

Fig. 2. Multi-resolution analysis

Focus on the Lth level and let the corresponding scaling and detail basis functions be $\phi_L(x)$ and $\psi_L(x)$. We might hazard a guess then that the original data $f(x)$ can be written in terms of what we have at the Lth level as a sum

$$\underbrace{f(x)}_{\text{data}} = \underbrace{\textstyle\sum_k s_L \phi_{L,k}(x)}_{\text{downsized}} + \underbrace{\textstyle\sum_k}_{\text{levels}} \underbrace{\textstyle\sum_{j\le L} \overbrace{d_{j,k}}^{\text{wavelet coefficients}} \psi_{j,k}(x)}_{\text{details}} . \tag{10}$$

The detail data involves all levels $0, \ldots, L$, so basis functions $\psi_k(x)$ at all levels are involved. Only one shrunk set of data is present in $MRA_L(x)$, that is S_L and that is described only by the scaling functions $\phi_L(x)$ appropriate to that level.

If the number of data points is a power of 2, going to as many levels as possible reduces the first sum to a single term, a constant. This is the full wavelet transform of the original data. It should be noted in passing that (10) may not be the best way of computing the transform since the double sum may be inefficient to calculate as it stands. A large part of wavelet technology is devoted to making these transforms more efficient: as in Fourier analysis there is a fast Fourier transform; in wavelet analysis there is a fast wavelet transform.

At this point we need to know what are suitable sets of scaling functions $\phi_L(x)$ and wavelet functions $\psi_L(x)$.

3.2 Scaling Function

Formally, the *scaling function* $\phi(x)$ is a, possibly complex, square-integrable function of a real argument x that is normalized so that

$$\int \phi(x)dx = 1 , \tag{11}$$

$$\int |\phi(x)|^2 dx = 1 . \tag{12}$$

It has the additional property that its translates

$$\phi_k = \{\phi(x-k)\} \tag{13}$$

form an orthonormal set:

$$< \phi_k, \phi_m >= \int_{-\infty}^{\infty} \phi(x-k)\phi(x-m)dx = \delta_{k,m} . \tag{14}$$

We use $< \phi_k, \phi_m >$ to denote the scalar product.

Orthogonality makes the functions useful since they can be used to represent any function on the interval where they live. In particular the function $\phi(x)$ itself has representation of the form

$$\phi(x) = \sqrt{2} \sum_{k=0}^{k=L-1} h_k \phi(2x - k) \tag{15}$$

for some L and some set of L coefficients h_k. The value of L is related to the number of vanishing moments of the function $\phi(x)$.

The h_k are referred to as the *refinement coefficients.* The $\sqrt{2}$ factor is conventional in theoretical work, though in computational work it may be dropped in order to speed up calculations by removing non-integer factors. This equation, called the *dilation equation,* expresses the original function $\phi(x)$ in terms of the same function scaled by a factor two. The scaling function is occasionally referred to as the *Father wavelet.*

Functions satisfying this kind of equation are called m-refinable functions.

For notational convenience we will occasionally use the symbol $\phi_{n,k}(x)$ to denote the rescaled and shifted version of $\phi(x)$:

$$\phi_{n,k}(x) = \phi(2^n x - k), \tag{16}$$

and if the value of k is not pertinent we may shorten this to $\phi_n(x)$.

From (15) we can express the refinement coefficients h_k in terms of the scaling function using orthogonality of the $\phi(2x - k)$:

$$h_k = \sqrt{2} \int \phi(x)\phi(2x - k)dx, \quad k = 0, \dots, L - 1\,. \tag{17}$$

There are L of these coefficients. If $\phi(x)$ were a complex valued function, we would use its complex conjugate $\bar{\phi}(x)$ in the integrand.

The condition (11) taken with the dilation equation (15) imposes a condition on the sum of h_k:

$$\sum_k^{L-1} h_k = \sqrt{2}\,. \tag{18}$$

Multiplying (15) by $\phi(x+p)$, integrating and using (17) yield the very important orthogonality relationship for the refinement coefficients h_k:

$$\sum_k^{L-1} h_k h_{k+2p} = \delta_{0,p}\,. \tag{19}$$

When $p = 0$, this shows that the sum of the squares of the h_k is unity, hence the reason for the $\sqrt{2}$ normalization in the dilation equation (15). This relationship will play an important role later on: we shall see how it works on data sets.

The refinement coefficients h_k provide the all-important link between the wavelet scaling function $\phi(x)$ and the discrete version of the wavelet transform. Indeed, it is often the case that we know the h_k but have no analytic expression for the function $\phi(x)$. We can, in effect, do wavelet analysis without ever knowing what the wavelets are.

3.3 Mother Wavelet

We can now construct sets of orthogonal functions $\psi_k(x)$ that are the translations $\psi_k(x) = \psi(x-k)$ of some function $\psi(x)$, and which are orthogonal to the translates $\phi_k(x) = \phi(x-k)$ of the scaling function $\phi(x)$:

$$< \phi_k(x), \psi_n(x) >= \int \phi(x-k)\psi(x-n)dx \ . \tag{20}$$

We shall insist on a few conditions. First that the function $\psi(x)$ has mean zero:

$$\int_{-\infty}^{\infty} \psi(x)dx = 0 \ . \tag{21}$$

In other words, $\psi(x)$ must be oscillatory with mean value zero. It is the oscillatory nature of these functions that led to the name "wavelet". Furthermore, $\psi(x)$ should be square integrable:

$$\int |\psi(x)|^2 dx = 1 \tag{22}$$

(this assures us that the function is not everywhere zero!) and its translates

$$\{\psi(x-k)\} \tag{23}$$

should form an orthonormal basis. This latter condition means that the function $\psi(mx)$ can be expressed as a linear sum of $\psi(x-k)$. Both these last conditions exclude the Fourier transform basis $e^{i\omega x}$ from the class of wavelets.

This definition is rather general and it is usual to add further conditions that make the functions useful in terms of data analysis. One important such condition is the "admissibility condition" which imposes a constraint on the Fourier transform of $\psi(x)$: the Fourier transform of $\psi(x)$

$$\Psi(k) = \int_{-\infty}^{\infty} \psi(x) e^{-2\pi \mathrm{i}\, kx} dx \tag{24}$$

should satisfy Calderón's admissibility condition

$$0 < \int_0^{\infty} \frac{|\Psi(k)|^2}{k} dk < \infty \ . \tag{25}$$

This ensures that the wavelet transform of a function should be reversible: we can reconstruct the function from its transform.

Since $\phi(x)$ are an orthonormal set, there is a set of coefficients g_k such that

$$\psi(x) = \sqrt{2} \sum_{k=0}^{L-1} g_k \phi(2x-k) \tag{26}$$

for the same L as in the dilation equation (15). The set $\{g_k\}$ is referred to as the *scaling filter*. By analogy with (16) we can introduce the notation

$$\psi_{n,k}(x) = \psi(2^n x - k) , \tag{27}$$

and if the value of k is not pertinent we may shorten this to $\psi_n(x)$.

Using the orthogonality of $\phi(x)$ we find that

$$g_k = \sqrt{2} \int \phi(x)\psi(2x - k)dx, \quad k = 0, \ldots, L - 1 . \tag{28}$$

As with h_k, we may know g_k without explicitly knowing the function $\psi(x)$.

Substituting (15) and (26) into (20) yields

$$< \phi(x), \psi(x) >= \sum_{k=0}^{L-1} h_k g_k = 0 . \tag{29}$$

Likewise, we can use the orthogonality of the translates $\psi_k(x) = \psi(x - k)$ to obtain another condition on the g_k:

$$< \psi(x - k), \psi(x - m) >= \sum_i g_i g_{i-2(k-m)} = \delta_{k,m} . \tag{30}$$

Equation (21) which shows us that the mean of the mother wavelet function is zero implies a condition on the g_k:

$$\sum_i g_i = 0 . \tag{31}$$

This follows from (26) and (11).

Equations (29) and (30) have an important solution, referred to as the *quadrature mirror relationship*:

$$g_k = (-1)^k h_{L-k-1} . \tag{32}$$

This is discussed in Sect. 10.2 (see (128)). So we see that the refinement coefficients h_k completely define the transforms. The scaling function $\phi(x)$ and the mother wavelet $\psi(x)$ allow continuous wavelet transforms, while the refinement coefficients h_k allow discrete wavelet transforms.

3.4 Continuous Wavelet Transform

The wavelet transform of a function $f(x)$ is just the convolution of the function with the wavelet. Given a function $f(x)$ and a wavelet $\psi(x)$ we may be able to construct, for any a and b, the localized scaling transforms $F(a, b)$ of $f(x)$ by evaluating the integrals

$$F(a,b) = \frac{1}{\sqrt{a}} \int_0^\infty f(x) \psi \left(\frac{x-b}{a} \right) dx \; . \tag{33}$$

We would like this transform to be invertible so that $f(x)$ can be reconstructed from $F(a,b)$ by using the same function $\psi(x)$ as follows:

$$f(x) = \frac{1}{C} \int_{-\infty}^\infty \int_0^\infty F(a,b) \frac{1}{\sqrt{a}} \psi \left(\frac{x-b}{a} \right) \frac{dadb}{a^2}, \tag{34}$$

for some finite normalization constant C. The requirement that C be finite is equivalent to constraint (25).

The continuous wavelet transform maps a function of one variable into a function of two variables, scale and location: it contains a lot of redundant information. We can see a as providing a microscope that allows us to zoom in on what is happening in the vicinity of a point b. It is the statement "... in the vicinity of a point b" that is important since this is quite unlike the Fourier transform which merely describes an average behaviour on all scales.

If at some point b, we see the value of $F(a,b)$ rise as we make a smaller, we would surmise that we were looking at some kind of spike or even a singularity in the function $f(x)$. This might be construed as evidence of noise. On the other hand, if we saw scaling behaviour for ever larger values of a we might surmise that there was structure in the signal at this point.

What we are saying is that the structure of the data is reflected in the behaviour of its singularities and that the continuous wavelet transform is a good tool for getting at this. The fact that the continuous wavelet transform leads to a description of local singularities in the signal structure further suggests that the continuous wavelet transform is somehow related to the multifractal description of the data (see Sect. 11.10).

3.5 Vanishing Moments

The mth moment of the wavelet function $\psi(x)$ is defined to be

$$\mathcal{M}_m = \int_{-\infty}^\infty x^m \psi(x) dx \; . \tag{35}$$

Daubechies ([11]) argued that good filters should have vanishing moments, i.e. $\mathcal{M}_m = 0$ for $m = 0, \ldots, r-1$ for some positive integer r:

$$\int_{-\infty}^\infty x^m \psi(x) dx = 0, \quad \forall m : 0 \leq m < r \; . \tag{36}$$

Under these circumstances $\psi(x)$ is said to be orthogonal to all polynomials of order up to degree r. Let me emphasize that (36) is an option.

If $\psi(x)$ and $r-1$ of its derivatives are continuous everywhere, then it has r vanishing moments (the converse is not true). Continuity of $\psi(x)$ and its derivatives is one ingredient in minimizing artefacts in signal analysis that might be due to $\psi(x)$ itself rather than the data. It is always a good idea to have the first moment (the mean) vanish since that conserves the energy in the signal. We have seen (Sect. 2.3) that simple interpolating filters do not have this property: we shall see how to correct this by using the lifting process.

However, there is no free lunch: the rate of convergence of the wavelet approximation as a function of levels used is lessened as the number of vanishing moments increases. It is then pertinent to ask what is the best wavelet to use from the point of view of vanishing moments and speed of convergence. The answer, provided by Daubechies, is the wavelet system known as "Coiflets" (Sect. 5.3).

Imposing the zero-moment condition of (36) leads to additional constraints on the values of the h_k:

$$\sum_{k=0}^{r-1} h_k k^m = 0, \quad 0 < m < r \, . \tag{37}$$

Imposing this condition improves the match between the original data values and the smoothed, transformed, values by fitting a higher order polynomial to sets of filtered points. This leads to a family of wavelets called Coiflets.

3.6 Representing Data

We now turn to the all-important issue of how to represent data in terms of wavelets and how to invert the transform so as to recover the data. This is technically quite complex, but it is central to the issue of generating and using wavelet transforms and their inverse.

Consider a function $f(x)$. It can be viewed at two neighbouring resolutions and represented in the following two ways:

$$f(x) = \sum_k s_j[k]\phi(2^j x - k) \tag{38}$$

$$= \sum_k s_{j-1}[k]\phi(2^{j-1}x - k) + \sum_k d_{j-1}[k]\psi(2^{j-1}x - k) \, . \tag{39}$$

The first of these is the view from what might be called "level j" and the second is the view from "level $j-1$". At level j we are seeing the function in all its detail: it is described by the coefficient set $s_j[k]$. At level $j-1$ we have split it into a smooth part described by the coefficient set $s_{j-1}[k]$ and a detail part described by the coefficient set $d_{j-1}[k]$.

Multiply these equations by $\phi(2^{j-1}x - k)$ and integrate, using the orthogonality relations (14) and the definitions of the refinement coefficients (17):

$$s_{j-1}[m] = \sum_k s_j[k]h_{k-2m} \, . \tag{40}$$

Likewise, integrating with $\psi(2^{j-1}x - k)$ and using (28) gives

$$d_{j-1}[m] = \sum_k s_j[k] g_{k-2m} \; . \tag{41}$$

We can rewrite the indices on the right-hand sides to give

$$s_{j-1}[m] = \sum_n h_n s_j[n + 2m] \; , \tag{42}$$

$$d_{j-1}[m] = \sum_n g_n s_j[n + 2m] \; . \tag{43}$$

So $s_{j-1}[m]$ is derived from $s_j[2m]$ and its neighbours, with weightings given by the h_n. (Note that some authors replace n by $-n$ and redefine h_n and g_n accordingly so that these equations look more like a convolution.)

The fact that the $s_j[2m]$ and its neighbours are used introduces a subtlety into the inversion of these equations; we shall come to that below.

With these equations, the coefficient sets $s_j[k]$ and $d_j[k]$ can be calculated recursively. It is the essence of the fast wavelet transform. The recursion is depicted in Fig. 2. The starting point for the recursion is the original data $f(x)$ sampled at discrete points:

$$s_0[k] = f(x)\delta(x - k) \; . \tag{44}$$

Since data is usually presented that way, this does not pose a problem. However, if the analytic form of the function $f(x)$ is given, this choice for the initial samples $s_0[k]$ would not be unique: it would depend on the binning. Likewise, if $f(x)$ were defined on some irregular point set, the values of the $s_0[k]$ would depend on the interpolation used to create a regularly spaced data set.

The recursive sums (42) and (43) are generally far easier to compute than integrals like

$$s_j[k] \simeq \int f(x)\phi(2^j x - k)dx \tag{45}$$

$$d_j[k] \simeq \int f(x)\psi(2^j x - k)dx \tag{46}$$

(normalization factors have been omitted). Moreover, recall that we might not even know what the function $\phi(x)$ is, though we may know the coefficient set $\{h_k\}$!

3.7 Orthogonality Revisited – Reverse Transforms

It is worth recapitulating on the various relationships that have been used in performing wavelet transforms and their inverses. We introduced a scaling function $\phi(x)$ which was associated with a set of refinement coefficients $\{h_k\}$

(the low-pass filter). The corresponding wavelet $\psi(x)$ was associated with a set of filter coefficients $\{g_k\}$ (the high-pass filter) related to $\{h_k\}$ via the quadrature mirror relationship (32).

We have seen how to go from data to a representation in terms of wavelet coefficients ((42) and (43)). We have not said how to go backwards from these coefficients and recover $f(x)$ from $s_j[k]$ and $d_j[k]$. Equation (39) does this, but it is expressed in terms of the functions $\phi(x)$ and $\psi(x)$, which we might not know explicitly. We would rather have an expression explicitly involving filters like $\{h_k\}$ and $\{g_k\}$.

The analogue of (39) that suggests itself is

$$s_j[k] = \sum_n \bar{h}_{k-2n} s_{j-1}[n] + \sum_n \bar{g}_{k-2n} d_{j-1}[n] \tag{47}$$

for some coefficient sets $\{\bar{h}_k\}$ and $\{\bar{g}_k\}$ as yet to be determined. This can be done by imposing orthogonality conditions on the coefficient sets $\{h_k\}$ and $\{g_k\}$, $\{\bar{h}_k\}$ and $\{\bar{g}_k\}$, and between the barred and unbarred coefficient sets. Such coefficient sets are referred to as *bi-orthogonal.*

An important example of such coefficient sets is the 5-3 wavelet that we shall come to in a different context later:

$$\{h_k\} = \tfrac{\sqrt{2}}{8}\{-1, 2, 6, 2, -1\}\,, \quad \{g_k\} = \tfrac{\sqrt{2}}{4}\{-1, -2, -1\}\,, \tag{48}$$

$$\{\bar{h}_k\} = \tfrac{\sqrt{2}}{4}\{1, 2, 1\}\,, \quad \{\bar{g}_k\} = \tfrac{\sqrt{2}}{8}\{-1, -2, 6, -2, -1\}\,. \tag{49}$$

The way in which the coefficients for the inverse transform are used is not at first sight obvious since each of the forward transformations reduces the size of the data by a factor two. We see this in (47) where the summation on the right-hand side involves $\bar{h}_{2n}$ coupling with the data values $s[n]$. We can show this if we write out the inverse transformation for the coefficient set (48) for even and odd k:

$$s_j[2n] = s_{j-1}[n] - \tfrac{1}{4}\left\{d_{j-1}[n-1] + d_{j-1}[n]\right\}\,, \tag{50}$$

$$\begin{aligned} s_j[2n+1] = {} & \tfrac{1}{2}\left\{s_{j-1}[n] + s_{j-1}[n+1]\right\} \\ & + \tfrac{1}{8}\left\{-d_{j-1}[n-1] + 6d_{j-1}[n] - d_{j-1}[n+1]\right\}\,. \end{aligned} \tag{51}$$

We see how the various terms from (47) pick out different components of $\{\bar{h}_k\}$ and $\{\bar{g}_k\}$ in (48) and (49) for even or odd k.

We can impose a further, more restrictive, condition on the coefficients:

$$\bar{h}_k = g_{-k}\,, \tag{52}$$

$$\bar{g}_k = h_{-k}\,. \tag{53}$$

Wavelet coefficient sets satisfying these conditions are referred to as *orthogonal.* They must have an even number of coefficients. This symmetry means that the forward and backward transforms involve the same coefficients.

4 Famous Scaling Functions and Wavelets

4.1 Haar

The simplest wavelet is the Haar wavelet whose scaling function is

$$\phi_{Haar}(x) = \begin{cases} 1 & \text{if } 0 \le x < 1 \, , \\ 0 & \text{otherwise} \, . \end{cases} \tag{54}$$

It can be readily verified that it satisfies the dilation equation

$$\phi_{Haar}(x) = \phi_{Haar}(2x) + \phi_{Haar}(2x-1) \, . \tag{55}$$

There are just two terms (i.e. $L = 2$ in (15)) and the coefficients h_i are

$$\begin{aligned} h_0 &= 1/\sqrt{2} \, , \\ h_1 &= 1/\sqrt{2} \, . \end{aligned} \tag{56}$$

The Haar scaling function has compact support but is discontinuous. This is the wavelet used in the simple example of Sect. 2.1.

The Haar Mother wavelet is

$$\psi(x) = \begin{cases} 1 & \text{if } 0 \le x < \frac{1}{2} \, , \\ -1 & \text{if } \frac{1}{2} \le x < 1 \, , \\ 0 & \text{otherwise} \, . \end{cases} \tag{57}$$

This can be derived from (15) and (32).

4.2 Hat, or Tent, Scaling Function

The scaling function is very simple

$$\phi_{Tent}(x) = \begin{cases} x & 0 \le x < 1 \\ 2 - x & 1 \le x < 2 \\ 0 & \text{otherwise} \end{cases} \tag{58}$$

and is a member of an important family of wavelets ("CDF(2,2)"). The dilation equation can be derived by expressing the tent function in terms of the Haar function:

$$\phi_{Tent}(x) = \int_0^x \left[\phi_{Haar}(u) - \phi_{Haar}(u-1)\right] du \, , \tag{59}$$

and using the Haar dilation equation (55) to replace u in the integrand with $2u$. It is then easily found that

$$\phi_{Tent}(x) = \tfrac{1}{2}\left[\phi_{Tent}(2x) + 2\phi_{Tent}(2x-1) + \phi_{Tent}(2x-2)\right] \, . \tag{60}$$

The coefficient set for this hat function follows from (15):

$$\tfrac{1}{\sqrt{2}}\{\tfrac{1}{2}, 1, \tfrac{1}{2}\}\{-\tfrac{1}{4}, \tfrac{1}{2}, 1, \tfrac{1}{2}, -\tfrac{1}{4}\} \, . \tag{61}$$

This has compact support and is symmetric. It is the coefficient set $\{\bar{h}_k\}$ of (49).

4.3 Quadratic Battle–Lemarié Wavelet

These wavelets are based on B-splines. The scaling function is bell-shaped with compact support:

$$\phi_{BM2}(x) = \begin{cases} \frac{1}{2}x^2 & 0 \le x < 1\,, \\ -x^2 + 3x - \frac{3}{2} & 1 \le x < 2\,, \\ \frac{1}{2}(x-3)^2 & 2 \le x < 3\,, \\ 0 & \text{otherwise}\,, \end{cases} \tag{62}$$

and satisfies the dilation equation:

$$\phi_{BM2}(x) = \frac{1}{4}\phi_{BM2}(2x) + \frac{3}{4}\phi_{BM2}(2x-1) + \frac{3}{4}\phi_{BM2}(2x-2) + \frac{1}{4}\phi_{BM2}(2x-3)\,. \tag{63}$$

This equation can be derived by expressing $\phi_{BM2}(x)$ in terms of $\phi_{Tent}(x)$:

$$\phi_{BM2}(x) = \int_0^x [\phi_{Tent}(u) - \phi_{Tent}(u-1)]\, du\,, \tag{64}$$

and using scaling relationships for $\phi_{Tent}(u)$ to express the integrand in terms of $\phi_{Tent}(2u)$. A simple change of integration variable $2u \to v$ then yields (63).

The cubic Battle–Lemarié wavelet can be similarly derived and has scaling function

$$\begin{aligned} \phi_{BM3}(x) &= \frac{1}{8}\phi_{BM3}(2x) + \frac{1}{2}\phi_{BM3}(2x-1) \\ &\quad + \frac{3}{4}\phi_{BM3}(2x-2) + \frac{1}{2}\phi_{BM3}(2x-3) + \frac{1}{8}\phi_{BM3}(2x-4)\,. \end{aligned} \tag{65}$$

4.4 Shannon Scaling Function

This scaling function does not have compact support, but it is smooth: all derivatives exist and are continuous:

$$\phi(x) = \begin{cases} \dfrac{\sin(\pi t)}{\pi t} & t \neq 0\,, \\ 1 & t = 0\,. \end{cases} \tag{66}$$

The corresponding wavelet function $\psi(x)$, sometimes called a *sinclet*, can be shown to be

$$\psi(x) = \begin{cases} \dfrac{\sin 2\pi t - \sin(\pi t)}{\pi t} & t \neq 0\,, \\ 1 & t = 0\,, \end{cases} \tag{67}$$

and the refinement coefficients are

$$\begin{aligned} h_0 &= g_0 = \frac{\sqrt{2}}{2} , \\ h_{2n} &= g_{2n} = 0 , \\ h_{2n+1} &= -g_{2n+1} = (-1)^n \frac{\sqrt{2}}{(2n+1)\pi} . \end{aligned} \tag{68}$$

Since the wavelet does not have finite support there is an infinite number of these coefficients.

A signal $s(t)$ is referred to as being "band-limited" if its Fourier transform has compact support. In practical terms this means that the Fourier transform of the signal is zero for frequencies ω higher than some cut-off frequency: $|\omega| > \sigma$. A band-limited signal can be properly sampled at the *Nyquist interval* $T = \pi/\sigma$ by giving its values at the points nT $(-\infty < n < \infty)$. This is the essence of the Shannon sampling theorem.

Such a band-limited signal has a representation of the form

$$s(t+\tau) = \sum_{n=-\infty}^{\infty} s(t+nT) \frac{\sin \sigma(\tau - nT)}{\sigma(\tau - nT)} , \quad T = \frac{\pi}{\sigma} . \tag{69}$$

Putting $T = 2^{-(m+1)}$, this can be re-expressed in terms of the Shannon scaling function (66) as

$$s(t+\tau) = \sum_{n=-\infty}^{\infty} s\left(t + \frac{n}{2^{m+1}}\right) \phi_{Shannon}(2^{m+1}\tau - n) \tag{70}$$

for all t and τ. In practical terms we can never know the entire past and future of the signal, as is implicit in this representation. See Papoulis [30] Sect. 11.5 for a thorough discussion of this.

4.5 Mexican Hat, or Marr, Wavelet

The Mexican Hat wavelet, so called because of its appearance, is perhaps the most popularly used wavelet among the set of continuous wavelet transforms (though there is no particular scientific reason why it should be so popular). Marr [27] recognized the physical significance of this for the human visual system. There is no simple form for the scaling function, but the mother wavelet is

$$\psi(x) = \frac{2}{\pi^{1/4}\sqrt{3}\sigma}(1 - \frac{x^2}{\sigma^2})\, \mathrm{e}^{-x^2/2\sigma^2} . \tag{71}$$

This function is just the second derivative of a Gaussian. Note that this function does not have compact support.

The Mexican Hat function can usefully be approximated using

$$\begin{aligned} G(x;\sigma) &= \frac{1}{\sqrt{2\pi}\sigma}\, \mathrm{e}^{-x^2/2\sigma^2} , \\ H(x;\sigma) &= G''(x;\sigma) \approx \alpha G(\alpha x;\sigma) - G(x;\sigma) , \quad \alpha \simeq 1.6 . \end{aligned} \tag{72}$$

This shows that the wavelet is approximately the difference between Gaussians of different widths.

4.6 Morlet Wavelet

The Morlet wavelet is a complex exponential whose amplitude is modulated by a Gaussian:

$$\psi(x) = \pi^{-1/4}\,\mathrm{e}^{-\mathrm{i}\,\omega x}\,\mathrm{e}^{-x^2} \,. \tag{73}$$

Since this function does not integrate to zero it is not admissable. Nonetheless, for frequencies $\omega > 10$ or so, the integral is so small that approximate reconstruction is satisfactory.

There are two scales in the Morlet wavelet: the width of the wave packet and the frequency. This can be made more explicit by rewriting (73) as

$$\psi(x) = \sigma^{-1/2}\,\mathrm{e}^{-\mathrm{i}\,\omega x \sigma}\,\mathrm{e}^{-(x/\sigma)^2} \tag{74}$$

(up to a proportionality constant). The width parameter is then σ and the frequency parameter is ω. In this form the wavelet is sometimes referred to as the Gabor Wavelet since it follows the spirit of the Gabor transformation (122).

A modification of the Morlet wavelet that is admissable is

$$\psi(x) = C\,\mathrm{e}^{-\mathrm{i}\,\omega x}\,\mathrm{e}^{-x^2}(1 - \sqrt{2}\,\mathrm{e}^{-\omega^2/4}) \,, \tag{75}$$

where the normalization constant C depends on ω.

5 Famous Wavelet Filters

5.1 Daubechies 4-Point

After the Haar wavelet, the most famous wavelet is arguably the first member of a huge class of wavelets discovered by Daubechies. For $L = 4$, conditions (18) and (19) are simply

$$\begin{aligned} h_1 + h_2 + h_3 + h_4 &= \sqrt{2} \,, \\ h_1^2 + h_2^2 + h_3^2 + h_4^2 &= 1 \,, \\ h_1 h_3 + h_2 h_4 &= 0 \,. \end{aligned} \tag{76}$$

There are only three equations for the four h_ks, so we can impose a further condition.

$$0g_1 + 1g_2 + 2g_3 + 3g_4 = 0 \,, \tag{77}$$

which expresses a zero-first-moment relationship among the wavelet coefficients g_k. Using $g_k = (-1)^k h_{L-k-1}$ (see (32)) we get

$$0h_4 - 1h_3 + 2h_2 - 3h_1 = 0\,. \tag{78}$$

Solving these four equations for h_k, we find that this four-point transform has coefficients

$$h_1 = \frac{1+\sqrt{3}}{4\sqrt{2}}\,, \quad h_1 = \frac{3+\sqrt{3}}{4\sqrt{2}}\,, \quad h_1 = \frac{3-\sqrt{3}}{4\sqrt{2}}\,, \quad h_1 = \frac{1-\sqrt{3}}{4\sqrt{2}}\,. \tag{79}$$

It is perhaps surprising that these equations have a closed form expression.

5.2 Daubechies 6-Point

For $L = 6$, conditions (18) and (19) are simply

$$\begin{aligned} h_1 + h_2 + h_3 + h_4 + h_5 + h_6 &= \sqrt{2}\,, \\ h_1^2 + h_2^2 + h_3^2 + h_4^2 + h_5^2 + h_6^2 &= 1\,, \\ h_1 h_3 + h_2 h_4 + h_3 h_5 + h_4 h_6 &= 0\,, \\ g_1 + g_2 + g_3 + g_4 + g_5 + g_6 &= 0\,. \end{aligned} \tag{80}$$

We can use $g_k = (-1)^k h_{L-k-1}$ to replace g_k with h_ks in the last of these. There are only four equations for the six h_ks, so we can impose two further conditions on the moments of g_k:

$$\begin{aligned} 0g_1 + 1g_2 + 2g_3 + 3g_4 + 4g_5 + 5g_6 &= 0\,, \\ 0^2 g_1 + 1^2 g_2 + 2^2 g_3 + 3^2 g_4 + 4^2 g_5 + 5^2 g_6 &= 0\,, \end{aligned} \tag{81}$$

and again we can replace g_ks with h_ks.

Amazingly, these too have a solution that is expressible in terms of surds.[1] It can be found in several places (including Numerical Recipes [31]), but for completeness it is given:

$$\begin{aligned} h_1 &= (1 + \sqrt{10} + \sqrt{5 + 2\sqrt{10}})/14\sqrt{2}\,, \\ h_2 &= (5 + \sqrt{10} + 3\sqrt{5 + 2\sqrt{10}})/14\sqrt{2}\,, \\ h_3 &= (10 - 2\sqrt{10} + 2\sqrt{5 + 2\sqrt{10}})/14\sqrt{2}\,, \\ h_4 &= (10 - 2\sqrt{10} - 2\sqrt{5 + 2\sqrt{10}})/14\sqrt{2}\,, \\ h_5 &= (5 + \sqrt{10} - 3\sqrt{5 + 2\sqrt{10}})/14\sqrt{2}\,, \\ h_6 &= (1 + \sqrt{10} - \sqrt{5 + 2\sqrt{10}})/14\sqrt{2}\,. \end{aligned}$$

[1] Editor's note: For those who do not know – surds are roots (fractional powers). I did not.

5.3 Coiflet 6-Point

The Coiflets are wavelets designed so that all moments vanish, so there are additional constraints. As before it is necessary to satisfy the relationships

$$\begin{aligned} h_1 + h_2 + h_3 + h_4 + h_5 + h_6 &= \sqrt{2} , \\ h_1^2 + h_2^2 + h_3^2 + h_4^2 + h_5^2 + h_6^2 &= 1 , \\ g_1 + g_2 + g_3 + g_4 + g_5 + g_6 &= 0 \end{aligned} \tag{82}$$

(compare (80)). We can use $g_k = (-1)^k h_{L-k-1}$ to replace g_k with h_ks in the last of these. The additional constraints are

$$\begin{aligned} 0g_1 + 1g_2 + 2g_3 + 3g_4 + 4g_5 + 5g_6 &= 0 , \\ -2h_1 - 1h_2 + 0h_3 + 1h_4 + 2h_5 + 3h_6 &= 0 , \\ (-2)^2 h_1 + (-1)^2 h_2 + 0^2 h_3 + 1^2 h_4 + 2^2 h_5 + 3^2 h_6 &= 0 \end{aligned} \tag{83}$$

(compare (81)). Solving these leads to the 6-point Coiflet coefficients

$$\begin{aligned} h_1 = \frac{1-\sqrt{7}}{16\sqrt{2}} , \quad h_2 = \frac{5+\sqrt{7}}{16\sqrt{2}} , \quad h_3 = \frac{14+2\sqrt{7}}{16\sqrt{2}} , \\ h_4 = \frac{14-2\sqrt{7}}{16\sqrt{2}} , \quad h_5 = \frac{1-\sqrt{7}}{16\sqrt{2}} , \quad h_6 = \frac{-3+\sqrt{7}}{16\sqrt{2}} . \end{aligned} \tag{84}$$

Again, it is remarkable that these are expressible in closed form.

5.4 Deslauriers–Dubuc Wavelet

Deslauriers and Dubuc provided an important method for interpolating functions defined at integer points by a process of successive refinement at binary rational points [15, 17]. The interpolating function is continuous at the rational interpolating points. Given values of the function at integer points, $f(x-nh)$, we can interpolate the value of $f(x)$ from its neighbours using the simple interpolation

$$f(x) = \tfrac{9}{16}[f(x-h) + f(x+h)] - \tfrac{1}{16}[f(x-3h) + f(x+3h)] . \tag{85}$$

This is just Lagrange four-point interpolation to the mid-point.

Successive refinement at binary rational points gives rise to a series of wavelet transforms. This four-point interpolation leads to the transformation

$$\begin{aligned} d_{j-1}[k] &= x_j[2k+1] \\ &\quad + \tfrac{1}{16}(-x_j[2k-2] + 9x_j[2k] + 9x_j[2k+2] - x_j[2k+4]) , \\ s_{j-1}[k] &= x_j[2k] \\ &\quad - \tfrac{1}{32}(-d_{j-1}[k-2] + 9d_{j-1}[k-1] + 9d_{j-1}[k] - d_{j-1}[k+1]) . \end{aligned} \tag{86}$$

This equation for the wavelet transform is not expressed directly in terms of the coefficients h_k as were the previous examples. Rather, it has been written in the form that would be used in computation of the transform. The calculation of $s_{j-1}[k]$ involves the previously calculated $d_{j-1}[k]$ instead of the function values at the data points. This, as we shall see, is because (86) has been derived from (85) via the Lifting process.

This is an interpolating wavelet (by construction!). The manner of its definition at binary rational points means that this is somewhat different from usual wavelets in that it is not related to a square-integrable function. However, that does not make it any less useful.

The important generalization of (85) introduced by Deslauriers and Dubuc [15] was to write

$$f(x) = \sum_{k \in Z} F\left(\frac{k}{2}\right) f(x + kh) , \tag{87}$$

where the weighting function $F(x)$ satisfies the relationship

$$F\left(\frac{x}{2}\right) = \sum_{k \in Z} F\left(\frac{k}{2}\right) F(x - k) . \tag{88}$$

With this the interpolated version of the original function $f(x)$ can be written as the convolution

$$f(x) = \sum_{k \in Z} f(x) F(x - k) \quad x \in R . \tag{89}$$

For $F(x)$ Deslauriers and Dubuc selected the family $\mathcal{L}^L_{2m-1}(x)$ of Lagrange polynomials of degree $L - 1$ with $M = 2L$ zeros at $-L+1, -L+3, \ldots, L-1$. The general Lagrange polynomials of degree N having zeros on a set of points $\{x_k\}$ can be written as

$$\mathcal{L}^N_i(x) = \frac{(x - x_0)\ldots(x - x_{i-1})(x - x_{i+1})\ldots(x - x_N)}{(x_i - x_0)\ldots(x_i - x_{i-1})(x_i - x_{i+1})\ldots(x_i - x_N)} , \quad i = 0, \ldots, N , \tag{90}$$

where the factor $(x - x_i)$ is missing from the numerator. So in the Deslauriers and Dubuc case we have

$$F(x) : \quad \mathcal{L}^L_{2m-1}(x) = \prod_{l=-M+1, l \neq m}^{M} \frac{x - (2l - 1)}{(2m - 1) - (2l - 1)} , \quad L = 2M , \tag{91}$$

and then

$$f(x) = \sum_{k=-M+1}^{M} \mathcal{L}^L_{2k-1}(0) f(x + (2k - 1)h) , \quad L = 2M , \tag{92}$$

which, considering what went before, is remarkably simple. The case $L = 4$ is commonly used: this is the transform given in (86).

6 Matrix View

It is useful to express the wavelet transform as a matrix operating on a data vector. This can provide some useful insights.

6.1 4-Point Orthogonal Transform

We can illustrate the use of these coefficients as a simple matrix multiplying a column vector of $2N$ data points. We have the following relationships among the four refinement coefficients h_k:

$$\begin{aligned} h_1 + h_2 + h_3 + h_4 &= \sqrt{2} , \\ h_1^2 + h_2^2 + h_3^2 + h_4^2 &= 1 , \\ h_1 h_3 + h_2 h_4 &= 0 . \end{aligned} \tag{93}$$

The first of these comes from the normalization, while the last two come from (19). Note that we have three equations constraining the four coefficients h_i: the four-point wavelets are a one-parameter set. This gives rise to the notion of "tunable" wavelets. The corresponding wavelet numbers (32) are

$$g_1 = h_4 , \quad g_2 = -h_3 , \quad g_3 = -h_2 , \quad g_4 = h_1 . \tag{94}$$

For $2N$ data points the transformation matrix is

$$W^{(2N)} = \begin{bmatrix} h_1 & h_2 & h_3 & h_4 & 0 & 0 & 0 \ldots 0 & 0 & 0 \\ g_1 & g_2 & g_3 & g_4 & 0 & 0 & 0 \ldots 0 & 0 & 0 \\ 0 & 0 & h_1 & h_2 & h_3 & h_4 & 0 \ldots 0 & 0 & 0 \\ 0 & 0 & g_1 & g_2 & g_3 & g_4 & 0 \ldots 0 & 0 & 0 \\ \cdots & & & & & & & & \\ h_3 & h_4 & 0 & 0 & 0 & 0 & 0 \ldots 0 & h_1 & h_2 \\ g_3 & g_4 & 0 & 0 & 0 & 0 & 0 \ldots 0 & g_1 & g_2 \end{bmatrix} . \tag{95}$$

There are as many columns and rows as there are data points. Note the wrap-around for the last pair of data points. This is a generally undesirable feature unless the data is periodic. There are a number of available strategies for avoiding this problem.

It is easy to verify from (93) and (94) that

$$W^T W = I . \tag{96}$$

The matrix W is orthogonal: its transpose is its inverse and so doing a wavelet transform of a wavelet transform brings us back to where we started.

If we denote the data by $U^{(2N)}$ and its transform by $V^{(2N)}$ we have

$$V_i^{(2N)} = \sum_{i=1}^{N} W_{ij}^{(2N)} U_j^{(2N)} , \quad i = 1, \ldots, 2N . \tag{97}$$

Because of the orthogonality of W, the inverse transform recovering the original data is just the wavelet transform of V.

We note that the entries of the data vector V are an alternating mixture of sum and difference values: all odd-numbered entries are created using the h_k coefficients, while all the even-numbered entries are generated using the g_k coefficients. It is customary to split the odd and even, (i.e. the sum and difference) parts into two separate vectors, one consisting entirely of entries created by h_k and the other of entries created by g_k.

In practice this is achieved by defining two $2N \times N$ matrices $\mathbf{G}$ and $\mathbf{H}$:

$$\mathbf{H}^{(2N\times N)} = \begin{bmatrix} h_1 & h_2 & h_3 & h_4 & 0 & 0 & 0 \ldots 0 & 0 & 0 \\ 0 & 0 & h_1 & h_2 & h_3 & h_4 & 0 \ldots 0 & 0 & 0 \\ \cdots & \cdots & \cdots & \cdots & \cdots & \cdots & \cdots & \cdots & \cdots \\ h_3 & h_4 & 0 & 0 & 0 & 0 & 0 \ldots 0 & h_1 & h_2 \end{bmatrix} , \tag{98}$$

$$\mathbf{G}^{(2N\times N))} = \begin{bmatrix} g_1 & g_2 & g_3 & g_4 & 0 & 0 & 0 \ldots 0 & 0 & 0 \\ 0 & 0 & g_1 & g_2 & g_3 & g_4 & 0 \ldots 0 & 0 & 0 \\ \cdots & \cdots & \cdots & \cdots & \cdots & \cdots & \cdots & \cdots & \cdots \\ g_3 & g_4 & 0 & 0 & 0 & 0 & 0 \ldots 0 & g_1 & g_2 \end{bmatrix} . \tag{99}$$

Note that by virtue of (94) the rows of $\mathbf{G}$ are orthogonal to one another and by virtue of (93) the rows of $\mathbf{H}$ are orthogonal to one another. The rows of $\mathbf{G}$ are also orthogonal to the rows of $\mathbf{H}$. We have orthogonality everywhere and such wavelet transforms are referred to a "bi-orthogonal".

We can usefully block the two matrices $\mathbf{G}$ and $\mathbf{H}$ together to form a version of W whose rows have been rearranged:

$$\mathbf{W}^{(2N\times 2N)} = \begin{bmatrix} \mathbf{H} \\ \mathbf{G} \end{bmatrix} . \tag{100}$$

The wavelet transform of the $2N$ component data vector $U^{(2N)}$ can then be written as

$$\begin{bmatrix} S^{(N)} \\ D^{(N)} \end{bmatrix} = \begin{bmatrix} \mathbf{H} \\ \mathbf{G} \end{bmatrix} \left[U^{(2N)} \right] . \tag{101}$$

This differs from (97) only in the organization of the transformed data coefficients. We have split these coefficients into two sets: a sum $S^{(N)}$ and a difference $D^{(N)}$. The sum is a reduced version of the original data and the differences are what is needed to reconstruct the original data from this sum.

6.2 Multi-resolution View

Having noted that S is a data array of size N, half the size of the original data array, we can proceed to do an N-point wavelet transform of S, producing a yet smaller (lower resolution) representation of the original data:

$$\begin{bmatrix} S_{j-2}^{(N/2)} \\ D_{j-2}^{(N/2)} \end{bmatrix} = \begin{bmatrix} \mathbf{H}^{N\times N/2} \\ \mathbf{G}^{N\times N/2} \end{bmatrix} \left[S_{j-1}^{(N)} \right] . \tag{102}$$

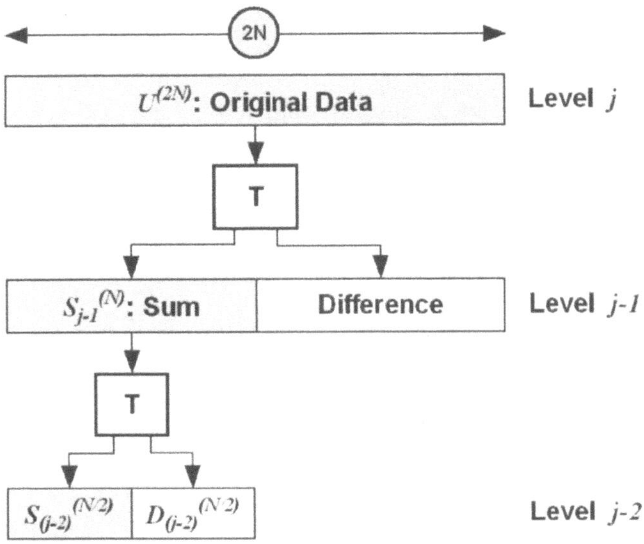

Fig. 3. Discrete wavelet transform produces a multi-resolution view of the data by splitting data in a hierarchy of sums and differences

We have explicitly written in the dimensions of the various arrays to emphasize that this is not the same transform that was executed at the first pass (Fig. 3).

The end result of this process is a hierarchy of copies of the original data, together with a hierarchy of the difference data that is required to restore one level from the previous (smaller) levels. Each level has half the resolution of its predecessor.

6.3 5-3 Wavelet Again

We have met the 5-3 wavelet in Sect. 3.7 and will meet it yet again in Sect. 7.2 from the point of view of the Lifting process. It is straightforward to write down the matrix versions of this wavelet transform from (48) for the forward transform and (50) and (51) for the backward transform.

The forward transform is

$$\begin{bmatrix} \cdots \\ S[n] \\ D[n] \\ S[n+1] \\ D[n+1] \\ \cdots \end{bmatrix} = \begin{bmatrix} \cdots & \cdots & \cdots & \cdots & \cdots & \cdots & \cdots & \cdots \\ -\frac{1}{8} & \frac{1}{4} & \mathbf{\frac{3}{4}} & \frac{1}{4} & -\frac{1}{8} & 0 & 0 & 0\ldots \\ 0 & 0 & -\frac{1}{2} & 1 & -\frac{1}{2} & 0 & 0 & 0\ldots \\ 0 & 0 & -\frac{1}{8} & \frac{1}{4} & \frac{3}{4} & \frac{1}{4} & -\frac{1}{8} & 0\ldots \\ 0 & 0 & 0 & 0 & -\frac{1}{2} & 1 & -\frac{1}{2} & 0\ldots \\ \cdots & \cdots & \cdots & \cdots & \cdots & \cdots & \cdots & \cdots \end{bmatrix} \begin{bmatrix} \cdots \\ S[2n-2] \\ S[2n-1] \\ \mathbf{S[2n]} \\ S[2n+1] \\ S[2n+2] \\ \cdots \end{bmatrix} . \qquad (103)$$

Notice the alignment of the elements in the rows and the alignment of the rows and columns. The $\mathbf{\frac{3}{4}}$ has been emphasized to show that its horizontal position corresponds with the $\mathbf{S[2n]}$ in the data column.

The backward transform is

$$\begin{bmatrix} \cdots \\ S[2n] \\ S[2n+1] \\ S[2n+2] \\ S[2n+3] \\ \cdots \end{bmatrix} = \begin{bmatrix} \cdots & \cdots & \cdots & \cdots & \cdots & \cdots & \cdots & \cdots & \cdots & \cdots \\ 0 & -\frac{1}{4} & \mathbf{1} & -\frac{1}{4} & 0 & 0 & 0 & 0 \ldots 0 & 0 \\ 0 & -\frac{1}{8} & \frac{1}{2} & \frac{3}{4} & \frac{1}{2} & -\frac{1}{8} & 0 & 0 \ldots 0 & 0 \\ 0 & 0 & 0 & -\frac{1}{4} & 1 & -\frac{1}{4} & 0 & 0 \ldots 0 & 0 \\ 0 & 0 & 0 & -\frac{1}{8} & \frac{1}{2} & \frac{3}{4} & \frac{1}{2} & -\frac{1}{8} \ldots 0 & 0 \\ \cdots & \cdots & \cdots & \cdots & \cdots & \cdots & \cdots & \cdots & \cdots & \cdots \end{bmatrix} \begin{bmatrix} \cdots \\ S[n-1] \\ D[n-1] \\ \mathbf{S[n]} \\ D[n] \\ S[n+1] \\ D[n+1] \\ \cdots \end{bmatrix} . \quad (104)$$

Again, the alignment is important: the **1** in the first row of the matrix must line up with the **S[n]** of the data. In that way the $\frac{3}{4}$ of the following row multiplies the $D[n]$.

Notice that, because of the interleaving of the smooth and detail values, this matrix organization is not computationally efficient if the transform is to be done to several levels. The rows and columns of the transform matrices have to be re-organized to handle that.

7 Lifting

We alluded to the so-called lifting process in our preamble on wavelet transforms (Sect. 2.3, (8)). There we saw that we could take the values at odd-numbered points as the smoothed (downsized) sample values and use the average of these values as a prediction of what the intermediate values would be. The detail values would then be the difference between the prediction and the actual value.

This is the first part of a three-step process that involves splitting the data into odd and even points, making a prediction on the basis of the even points about the values at the odd points, and then updating the prediction.

In this section we shall change the notation slightly so that, for example, $s[k]$ might refer to the data value at the point k, as opposed to using the functional form $s(x)$.

7.1 Split, Predict and Update

If the data points are denoted by $s_j[k]$ at level j, the data $s_{j-1}[k]$ at the next level is

$$s'_{j-1}[k] = s_j[2k] \, . \quad (105)$$

The prediction for the point $2k+1$ is based on $s_j[2k]$ and $s_j[2k+2]$:

$$\mu_j[2k+1] = (s_j[2k] + s_j[2k+2])/2 \, , \quad (106)$$

so that the detail coefficient at level $j-1$ is

$$d_{j-1}[k] = s_j[2k+1] - \mu_j[2k+1] = s_j[2k+1] - \tfrac{1}{2}(s_j[2k] + s_j[2k+2]) \ . \quad (107)$$

This recalls what was done in Sect. 2.3. However, we note that with this prescription

$$\sum_k s'_{j-1}[k] \neq \tfrac{1}{2} \sum_k s_j[2k] \ . \quad (108)$$

The average value of the signal (the energy) is not preserved.

It would have been nicer if the average value of the signal had been preserved. We can fix this by adjusting the $s_{j-1}[k]$ in (105) using the values of the $d_{j-1}[k]$ we have just calculated:

$$s_{j-1}[k] = s_j[2k] - \tfrac{1}{4}(d_{j-1}[k-1] + d_{j-1}[k]) \ . \quad (109)$$

It is easy to verify that now

$$\sum_k s_{j-1}[k] = \tfrac{1}{2} \sum_k s_j[2k] \ . \quad (110)$$

This adjustment or updating step is an important part of the process.

Although we have used a specific predictor and update, the three-stage process described is quite generic. The data values are split into values at the odd and even points. The even points are used to make a prediction of the value at an intermediate odd point. That is compared with the actual value and the comparison is used to update the even point data values. Figure 4 summarizes the three important steps.

Inverting the transform is simply a matter of running the diagram backwards. Exactly how this is done will be made clearer in the next section.

7.2 5-3 Wavelet Again

The above process is described by the following equations

$$\begin{aligned} d_{j-1}[k] &= s_j[2k+1] - \tfrac{1}{2}(s_j[2k] + s_j[2k+2]) \\ s_{j-1}[k] &= s_j[2k] + \tfrac{1}{4}(d_{j-1}[k-1] + d_{j-1}[k]) \end{aligned} \quad (111)$$

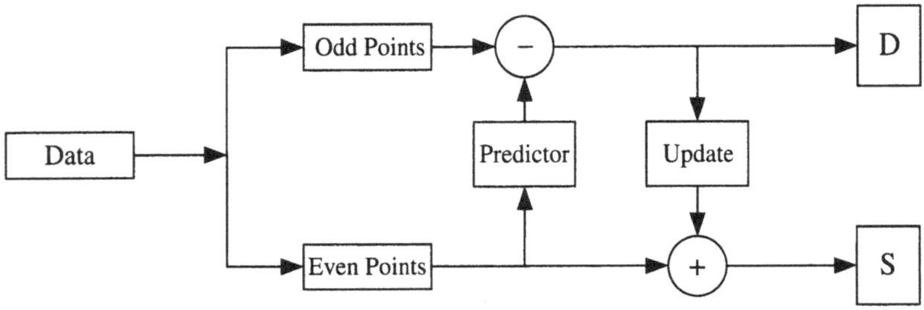

Fig. 4. Lifting strategy

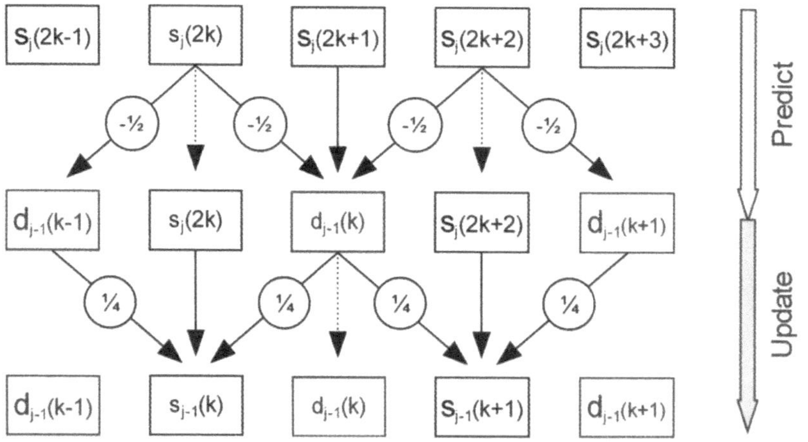

Fig. 5. Wavelet construction by Lifting. The inverse transform is derived by reversing the arrows and changing the signs of the multipliers

and is shown in Fig. 5. The backward transform going from the set of values $s_{j-1}[k]$ and $d_{j-1}[k]$ to recover $s_j[k]$ is obtained by reversing the arrows and changing the signs of the coefficients:

$$\begin{aligned} s_j[2k] &= s_{j-1}[k] - \tfrac{1}{4}(d_{j-1}[k-1] + d_{j-1}[k]) \,, \\ s_j[2k+1] &= d_{j-1}[k] + \tfrac{1}{2}(s_j[2k] + s_j[2k+2]) \,. \end{aligned} \tag{112}$$

This is formally referred to as "CDF(2,2)".

If we unravel the forward transform recursion relation (111), and change the notation slightly for convenience, we can show that

$$\begin{aligned} S[1] &= -\tfrac{1}{8}s[0] + \tfrac{1}{4}s[1] + \tfrac{3}{4}s[2] + \tfrac{1}{4}s[3] - \tfrac{1}{8}s[4] \,, \\ D[1] &= -\tfrac{1}{2}s[2] + s[3] - \tfrac{1}{2}s[4] \,. \end{aligned} \tag{113}$$

In other words, the wavelet transform is described by filter coefficients

$$\begin{aligned} \{h\} &= \{-\tfrac{1}{8}, \tfrac{1}{4}, \tfrac{3}{4}, \tfrac{1}{4}, -\tfrac{1}{8}\} \,, \\ \{g\} &= \{-\tfrac{1}{2}, 1, -\tfrac{1}{2}\} \,. \end{aligned} \tag{114}$$

Up to normalization factors, this is the 5-3 coefficient set encountered in Sect. 3.7. The lifting version (112) of the inverse transform is somewhat easier to appreciate than the direct version of (50) and (51).

7.3 9-7 Wavelet

This is the wavelet used in the JPEG2000 update of the JPEG image storage format. It is defined through its lifting steps as follows:

$$\begin{aligned}
d^{(1)}_{j-1}[k] &= s_j[2k+1] + [\alpha(s_j[2k] + s_j[2k+2]) + \tfrac{1}{2}] \,, \\
s^{(1)}_{j-1}[k] &= s_j[2k] + [\beta(d^{(1)}_{j-1}[k] + d^{(1)}_{j-1}[k-1]) + \tfrac{1}{2}] \,, \\
d_{j-1}[k] &= d^{(1)}_{j-1}[k] + [\gamma(s^{(1)}_{j-1}[k] + s^{(1)}_{j-1}[k+1]) + \tfrac{1}{2}] \,, \\
s_{j-l}[k] &= s^{(1)}_{j-1}[k] + [\delta(d_{j-l}[k] + d_{j-1}[k-1]) + \tfrac{1}{2}] \,,
\end{aligned} \tag{115}$$

where the coefficients α, β, γ and δ are

$$\begin{aligned}
\alpha = -1.586134342 \,, \quad & \beta = -0.05298011854 \,, \\
\gamma = 0.8829110762 \,, \quad & \delta = 0.4435068522 \,.
\end{aligned} \tag{116}$$

The intermediate values $d^{(1)}_l, s^{(1)}_l$ are calculated first and then updated. There are nine coefficients in the smoothing part of the wavelet and seven in the analysis part, and both parts have four vanishing moments.

The explicit values of the coefficients are given in Table 3.

Table 3. Coefficients for the Daubechies 9-7 wavelet

k	Low pass	High pass
0	+0.6029490182363579	+1.115087052456994
±1	+0.2668641184428723	−0.5912717631142470
±2	−0.07822326652898785	−0.05754352622849957
±3	−0.01686411844287495	+0.09127176311424948
±4	+0.026748741080976	

7.4 Wavelets on Irregularly Distributed Point Sets

In one dimension, the cheap way out is to ignore the fact that the data points are irregularly distributed and just calculate as if they were on a regular mesh. This can work, but it is difficult to deal properly and consistently with large gaps in the point distribution.

It is better to use a systematic interpolation method, such as Lagrangian interpolation or spline curves, to fit the data as given, though we are still left wondering whether this actually solves the large gap problem.

The lifting procedure would then suggest that we fit such a curve to all even-numbered points, predict the values at the odd points and calculate and store the residuals. If the residuals are small, data compression and reconstruction will be possible. An alternative is to use points for prediction where the sampling frequency is highest rather than every other point regardless of sampling density.

In two or more dimensions this is not so easy since there is no natural extension of the concept of "odd" and "even" points. However, we do have a concept of "neighbouring" points via constructs like the Delaunay tessellation and this provides the possibility of "predicting" values at a point from the Delaunay neighbours or some related set (such as the "natural neighbours"). The question of which predictor to use is more complex.

8 Transforms in Two Dimensions: Image Data

So far it has been assumed that the data is a one-dimensional set of values (like a time series or the value of a function on an axis). Wavelet transforms of two-dimensional data are straightforward: just transform all the rows or columns of the data as if they were independent one-dimensional data sets. However, to make the transformation process useful is not quite as straightforward and there are several possible approaches.

8.1 Schemes

Image data is a two-dimensional array of pixels, usually rectangular though occasionally triangular sampling is used. The general idea behind performing a wavelet transform on rectangular sampled image data is to treat all rows and columns of the image as independent data strings. However, that leaves several available strategies for handling the rows and columns.

We start by transforming all the rows, keeping the smoothed data of the rows on the left and the detail data on the right. Then there is a choice. The first choice is to treat the columns as though they were data arrays and regardless of whether they contain smoothed or detailed row data. This produces four regions which we label **SS**, **SD**, **DS** and **DD** depending on what kind of data is there (see left panel of Fig. 6).

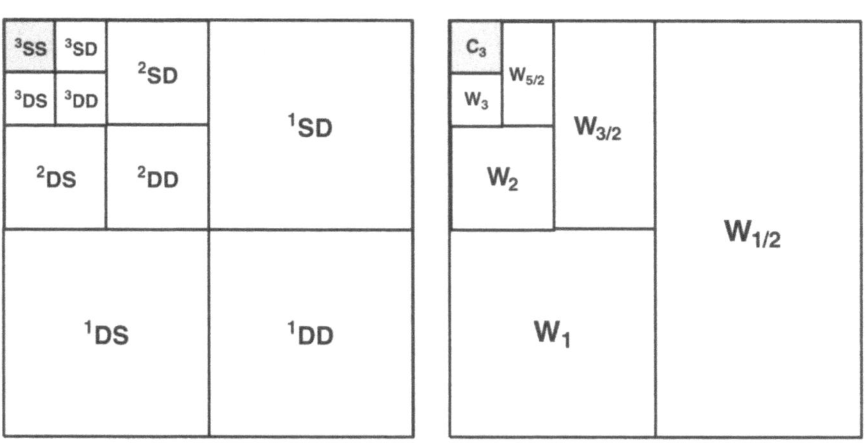

Fig. 6. *Left*: Mallat partition. *Right*: Feauveau partition

The **DD** part of the array is horizontal detail data that itself has been transformed to give a vertical detail of the horizontal detail (!). The **SS** part of the array is just a half-size image. We can repeat the process dividing the **SS** part into four components as before. This is referred to as Mallat's scheme.

Alternatively, having done the first horizontal row transform, we could transform only the columns of horizontal smoothed data, leaving the horizontal details alone. This again leaves a half-size shrunken image in the top left of the array, and this can in turn be subdivided. This is referred to as Feauveau's scheme.

Although the Feauveau scheme requires only half the computational effort of the Mallat scheme, the latter is most commonly used since it treats the horizontal and vertical directions equally. From here on we shall stick to discussing the Mallat partition.

8.2 Interpretations

The four sections labelled **SS**, **SD**, **DS** and **DD** each reveal a different aspect of the image data. The **SS** part is simply a shrunken version of the image using the chosen scaling function. The **DD** part is detailed data that arises from computing details in both horizontal and vertical directions. The **SD** and **DS** parts reveal, respectively, the vertical and horizontal structures within the image.

The detail part of the wavelet transform is effectively a derivative of the data. Hence noise in the original data is strongly manifest in the details. If one can recognize in the detail components what is noise and what is image, then cleaning the image becomes possible. The methods that are available for noise handling are numerous and even look vaguely familiar: there is a wavelet analogue of the Wiener Filter, for example.

The main difficulty is to annihilate the noise without causing damage to the data that represents features. Most of that damage is damage to edges that delineate fine image structure: some noise filters may simply smooth the image, thereby wiping out faint structures. Context-sensitive noise filtering involving feature recognition through nonlinear filters can be very effective.

It is often the case when compressing image data that the **DD** can be set to zero without suffering too much visible damage to the reconstructed image.

8.3 Example

As a specific example we take an image of buildings in Valencia designed by the architect Calatrava.[2] Figure 7 shows the original while the left panel of Fig. 8 shows two levels of the wavelet transform. The image of the wavelet transform

[2] I am grateful to Rien van de Weygaert for this fine picture. Moreover, it makes a change not to use the image of Lena.

Fig. 7. Calatrava architecture in Valencia

has been enhanced so as to reveal fine detail in the wavelet transform. One is struck by the large amount of noisy (uncorrelated) data in the **SD**, **DS** and **DD** components. If the noise component is filtered out the figure on the right results. We see a lot of black spaces: it is the encoding of these spaces that leads to strong compression.

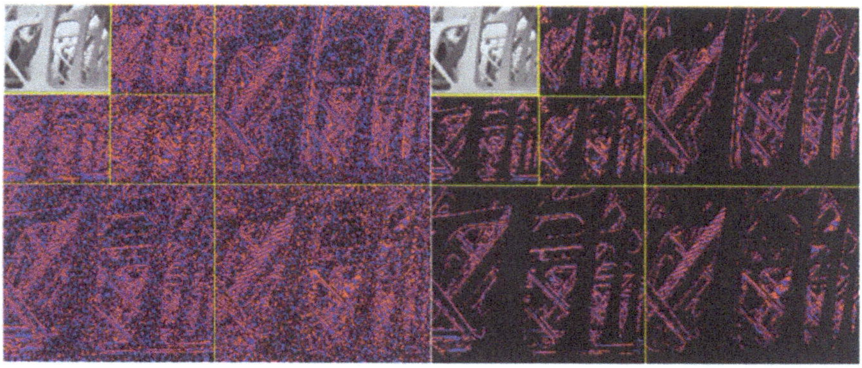

Fig. 8. *Left*: Wavelet transform of the Calatrava architecture image. *Right*: Censored version of wavelet transform of Calatrava image. The gamma has been changed for ease of viewing

9 Specific Example

The following example of the use of the continuous wavelet transform comes from James Walker's magnificent "Primer on Wavelets and their Scientific Applications" [42]. It illustrates what difference the choice of wavelet makes to the analysis. I have somewhat amplified that discussion by considering the effect of noise on the data.

The example starts with a one-dimensional signal consisting of three wave packets. The wave packets are themselves built from two sinusoidal oscillations having different frequencies, localized by applying Gaussian window (see Fig. 9). Two of the packets are single frequency and the third is a mixture of both. The question is whether we can use wavelet analysis to recognize how the mixture is built.

The figures show the data and scalograms generated by applying a Mexican Hat wavelet transforms and complex Gabor wavelet transforms. Both these transforms have a width parameter w: the wavelet transforms are computed for a series of values of w and stacked to form the scalogram with small values of w at the top and larger ones at the bottom.

The data consists of regularly spaced samples of the function

$$\begin{aligned} s(x) = {} & \sin(40\pi x)\exp\left(-100\pi(x-0.2)^2\right) \\ & + [\sin(40\pi x) + 2\cos(160\pi x)]\exp\left(-50\pi(x-0.5)^2\right) \\ & + 2\sin(160\pi x)\exp\left(-100\pi(x-0.8)^2\right). \end{aligned} \tag{117}$$

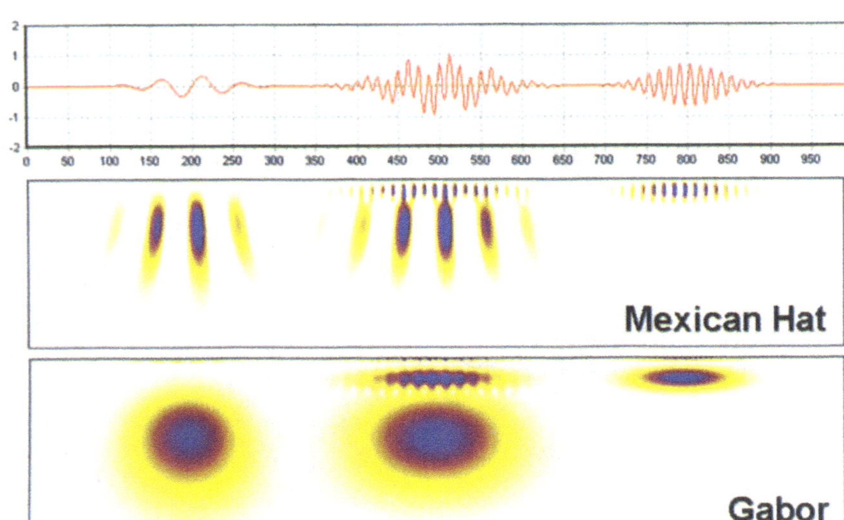

Fig. 9. *Top panel*: the original data consisting of three wave packets, the central one being a mixture of the other two. *Middle panel*: Mexican Hat scalogram. *Bottom panel*: the Gabor scalogram

The analysing wavelets are the Mexican Hat wavelet and the complex Gabor wavelet. The Mexican Hat used here is

$$\psi(x) = \frac{2\pi}{w^{1/2}} \left[1 - 2\pi \left(\frac{x}{w}\right)^2\right] \exp\left(-\pi(x/w)^2\right), \tag{118}$$

where w is a parameter that is varied to generate the scalogram. The data is transformed for a series of values of w.

The Gabor wavelet has real and imaginary parts that can be written as

$$\begin{aligned} \psi_R(x) &= w^{-1/2} \exp\left(-\pi(x/w)^2\right) \cos(2\pi\nu x/w)\,, \\ \psi_I(x) &= w^{-1/2} \exp\left(-\pi(x/w)^2\right) \sin(2\pi\nu x/w)\,. \end{aligned} \tag{119}$$

The Gabor wavelet has two parameters w and ν. The width parameter w is just like the width parameter w in the Mexican Hat wavelet and is used to generate the scalogram. The frequency ν is an important extra parameter that can be used to tune the analysis. The figures shown here use a value of ν that is optimized for the analysis of the noise-free data and the same value of ν is used for the noisy data. In fact the noisy data could be shown to better advantage by using different values of ν, but we shall not go into that here.

A cursory glance at the clean data analysis in Fig. 9 shows the difference between the Mexican Hat and Gabor wavelet analyses. While both analyses are capable of revealing the components, even when they are mixed, the differentiation is far clearer in the case of the Gabor analysis where we have used the extra parameter ν to advantage to resolve the mixture into two quite separate islands. What is important is that the choice of wavelet effects how easily and unambiguously the result can be interpreted.

This is even more so when looking at the same data when it has been polluted by additive noise (Fig. 10) having amplitude comparable with the amplitude of the original signal (can you see the signal in the noise, even though you know what you are looking for?). The noise that has been added here has a slightly red spectrum in order to make the signal discrimination more difficult.

The resulting scalograms have been distorted by the noise, but the situation is nonetheless still clear in the case of the Gabor wavelet analysis. The message from this simple example is that you should carefully formulate the questions that are to be asked of the data and choose the best wavelet to provide the answers. There is no lack of available transforms! The question that remains unanswered is how to best choose that wavelet and at the moment the only answer is "trial and error".

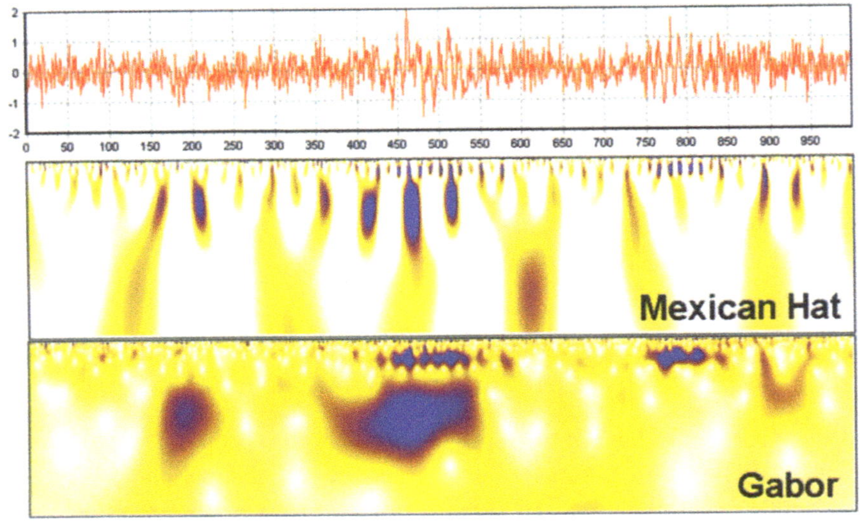

Fig. 10. *Top panel*: the noise-added data consisting of three wave packets, the central one being a mixture of the other two. *Middle panel*: Mexican Hat scalogram. *Bottom panel*: the Gabor scalogram

10 Other Data Transforms

10.1 Windowed Fourier Transform

The familiar Fourier transform of a function $f(x)$ is

$$F(\omega) = \int_{-\infty}^{\infty} f(x)\, \mathrm{e}^{-\mathrm{i}\omega x} dx\ , \tag{120}$$

which has inverse

$$f(x) = \frac{1}{2\pi} \int_{-\infty}^{\infty} F(\omega)\, \mathrm{e}^{\mathrm{i}\omega x} dx\ . \tag{121}$$

(There are restrictions on $f(x)$ in order that these integrals exist.) This looks like a special case of the continuous wavelet transform, but such appearances are misleading. $G(\omega)$ describes the frequency content of the function $f(x)$ over the entire domain of $f(x)$: the frequency information, while accurate, is not localized.

Localization of frequency information can be achieved by using a windowed version of the Fourier transform (also known as the Gabor transform):

$$F(\omega) = \int_{-\infty}^{\infty} f(x) W(x)\, \mathrm{e}^{-\mathrm{i}\,i\omega x} dx \tag{122}$$

for some window function $W(x)$. The problem with this form of the transform is that $W(x)$ does not itself depend on the frequency ω.

10.2 Fourier Transform

Wavelets made their way into the second edition of *Numerical Recipes* [31]. That short review opens up a perspective into the world of filters by defining a function $H(\omega)$ from the filter coefficients $\{h_i\}$:

$$H(\omega) = \sum_j h_j \, \mathrm{e}^{\mathrm{i}\, j\omega} . \tag{123}$$

$H(\omega)$ is periodic with period 2π. From the orthogonality relations (14) and the dilation equation (15) it can be shown that $H(\omega)$ satisfies the equations

$$\frac{1}{2}|H(0)|^2 = 1 , \tag{124}$$

$$\frac{1}{2}\left[|H(\omega)|^2 + |H(\omega+\pi)|^2\right] = 1 . \tag{125}$$

Any further conditions on the h_j impose further conditions on $H(\omega)$.

What is interesting is that it is simple to concoct functions satisfying all these conditions and so by inverting (123) come up with a set of N wavelet coefficients:

$$h_j = \frac{1}{N}\sum_{k=0}^{N-1} H\left(\frac{2\pi k}{N}\right) \mathrm{e}^{-2\pi \mathrm{i}\, jk/N} . \tag{126}$$

However, not all $H(\omega)$ will produce wavelets having compact support.

Equation (26) expressing the mother wavelet $\psi(x)$ in terms of the scaling functions $\phi(x)$ gave rise to the set of scaling filter coefficients g_k that are related to h_k via the quadrature mirror relationship $g_k = (-1)^k h_{L-k-1}$ (see (32)). If we write

$$G(\omega) = \sum_j g_j \, \mathrm{e}^{\mathrm{i}\, j\omega} , \tag{127}$$

then it can be shown that

$$G(\omega) = \mathrm{e}^{-\mathrm{i}\,\omega} H^*(\omega+\pi) . \tag{128}$$

$H(\omega)$ and $G(\omega)$ are called "quadrature mirror filters" [36]. The derivation of any of these equations in nontrivial.

So given $H(\omega)$ (say from (126)) we can determine both the scaling function $\phi(x)$ and the mother wavelet $\psi(x)$, and hence the complete multi-resolution analysis. Owing to the recursive nature of the wavelet bases, all computations should be performed using the quadrature mirror filters $H(\omega)$ and $G(\omega)$.

We can define the correlation functions of $\phi(x)$ and $\psi(x)$ as

$$\Phi(x) = \int_{-\infty}^{\infty} \phi(t)\phi(t-x)dt , \tag{129}$$

$$\Psi(x) = \int_{-\infty}^{\infty} \psi(t)\psi(t-x)dt . \tag{130}$$

It can be shown, using the properties of Fourier transforms, that

$$\Psi(x) = 2\Phi(2x) - \Phi(x) \, . \tag{131}$$

This should be compared with the approximation of a Mexican Hat wavelet by the difference of two Gaussians in (72).

10.3 z-Transform

In the engineering literature we encounter an important variant of the Fourier transform called the z-transform. The Fourier series, or frequency domain, representation of a one-dimensional data set $\{x(i)\}$ is given by

$$X(\omega) = \sum_{k \in \mathbf{Z}} x(k) \, \mathrm{e}^{\mathrm{i}\, k\omega} = X(\mathrm{e}^{\mathrm{i}\,\omega}) \, , \quad \omega \in \mathbf{R} \tag{132}$$

(when this sum exists). The last equality has been written so as to emphasize that the dependence is in fact on the variable $e^{i\omega}$ and that X is periodic with period 2π when viewed as a function of ω . With this we can substitute

$$z = \mathrm{e}^{\mathrm{i}\,\omega} \tag{133}$$

and so rewrite the Fourier series as

$$X(z) = \sum_{k \in \mathbf{Z}} x(k) z^k \, . \tag{134}$$

The difference now is that we can analytically extend the z-dependence of $X(z)$ away from the circle on which $X(\omega)$ was defined. There is of course a singularity at $z = 0$. In the particular case when the original data $x(k)$ is defined on a finite interval, $X(z)$ is simply a polynomial and can be analytically extended over the entire complex plane, with the exception of the origin.

The z-transform is linear; we have an entire battery of tools from complex variable theory that can be used, and it retains many nice properties of the Fourier representation from which we derived it. An example of the last is the convolution theorem. If

$$w(n) = \sum_{k \in \mathbf{Z}} x(k) y(n-k) \, , \tag{135}$$

then

$$W(z) = X(z) Y(z) \, . \tag{136}$$

A further result is that if, from the sequence $\mathbf{x} = \{x(i)\}$, we define a second sequence $\mathbf{x}_{left} = \{x(i+1)\}$, then

$$X_{left}(z) = zX(z) \, , \tag{137}$$

which is trivial to verify. There is a similar result for the right-shifted sequence $\mathbf{x}_{right} = \{x(i-1)\}$:

$$X_{right}(z) = z^{-1}X(z) . \tag{138}$$

Two of the most important operations concern the "2-down-sampled" and "2-up-sampled" sequences derived from the sequence $\{x(i)\}$. By 2-down-sampling we mean the sequence $\{x_{2\downarrow}(i)\}$ created from $\{x(i)\}$ by deleting all odd-indexed entries from the sequence and re-indexing:

$$x_{2\downarrow}(i) = x(2i), \quad i \in \mathbf{Z} . \tag{139}$$

By 2-up-sampling we mean the sequence $\{x_{2\uparrow}(i)\}$ created from $\{x(i)\}$ by adding a new zero valued term in between every pair of terms $x(m)$ and $x(m+1)$, and then re-indexing the sequence:

$$x_{2\uparrow}(i) = \begin{cases} 0 & \text{if } i \in \mathbf{Z} \text{ is odd ,} \\ x([i/2]) & \text{if } i \in \mathbf{Z} \text{ is even .} \end{cases} \tag{140}$$

These two sequences have z-transforms

$$X_{2\downarrow}(z) = \frac{1}{2}\left(X(z^{1/2}) + X(-z^{1/2})\right) , \tag{141}$$

$$X_{2\uparrow}(z) = X(z^2) . \tag{142}$$

The shift and up- and down-sampling operators are the basic tools for the design of very complex wavelet filters. This is an entire subject in its own right.

The z-transform is not used to compute the transform of a data set, rather it is used as a theoretical tool to construct and analyse wavelet filters. It provides, for example, a tool for classifying wavelets through the properties of the zeroes of the various wavelet generators.

11 Further Issues – In Brief

11.1 Wavelet Packet Transforms

There is no reason why the difference data D resulting from a wavelet transform $\{S, D\}$ of a data set $\{X\}$ should not itself be transformed: doing so might, for example, give an improved compression of the data or provide for better noise handling. The skill here is to decide how many levels of the D-data transform are needed and to choose an appropriate wavelet at each level.

This procedure can be particularly effective when dealing with two-dimensional image data. We have noted that the **SD** and **DS** parts of the transform show predominately vertical and horizontal image structures. There is some advantage to be gained in doing a vertical transform on the **SD** component and a horizontal transform on the **DS** part. Again the choice of wavelet and the number of levels to transform are key issues. In practice, it is possible to test several strategies and choose the most effective.

11.2 Period Doubling

Period doubling is implicit in everything we have done so far (specifically because the scaling functions satisfy the dilation equation (15)). This does not normally pose any problems in using wavelets for analysis. For some problems, particularly problems where the time dependence of a variable is the subject of investigation, it may be useful to have more levels. Chiu et al. [8] have sought to generalize the basic scaling equations to

$$\phi_k^{j,n}(x) = (2^j \kappa_n)^{1/2} \phi(2^j \kappa_n x - k) \tag{143}$$

$$\psi_k^{j,n}(x) = (2^j \kappa_n)^{1/2} \psi(2^j \kappa_n x - k) \tag{144}$$

$$\kappa_n = \frac{2^N}{2^N + n} \tag{145}$$

in order to provide $2^N - 1$ additional levels of analysis.

An obvious alternative is to use the continuous wavelet transform if the scaling function is known explicitly.

11.3 Shift Invariance

The usual wavelet transforms are not invariant under a linear shift in the data. Thus, the coefficients of the wavelet transform of the shifted data are not the same as the coefficients of the wavelet transform of the original data: there may be a significant redistribution of energy among the wavelet components. This may or may not be a problem; it depends on what you are doing with your wavelet transform. Matching images, such as stereo pairs, benefits from using shift-invariant wavelets.

Wavelets having shift invariance can be designed by going to complex wavelets, but one of the costs of this is that some redundancy is introduced and there is a considerable computational overhead. The subject is reviewed by Kingsbury [23] (this paper uses the z-transform).

11.4 Horizontal–Vertical Bias

Two-dimensional wavelet transforms are often regarded as a horizontal transform followed by a vertical transform: this inevitably produces a bias towards vertically or horizontally aligned structures. We can see this in the two-dimension wavelet map where the **DD** component that describes diagonal structure is always of lower order than the **SD** and **DS** components (one more differentiation is implied in **DD**).

11.5 Locality

Wavelets are localized (compact support) and so, in one dimension, can deal with discontinuities more effectively than Fourier representations. One reason

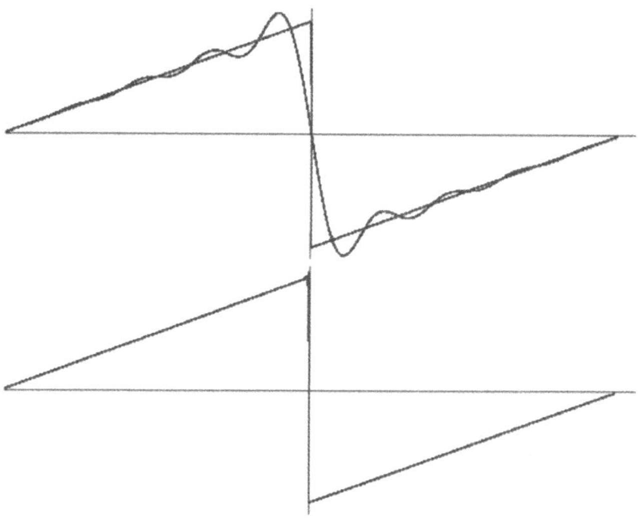

Fig. 11. *Top*: sawtooth approximated by 17 Fourier coefficients. *Bottom*: sawtooth approximated by 17 wavelet coefficients

wavelets are thought to be useful in economics (time series analysis) is they can deal with sudden change.

This is nicely illustrated in Fig. 11 where we compare the Fourier and wavelet representations of a simple sawtooth function using 17 terms in the respective expansions. The Fourier representation shows a strong "Gibbs" effect that is entirely absent in the wavelet representation.

However, two-dimensional wavelets suffer problems with handling linear features and as a consequence produce artefacts. One solution that has been proposed that deals with this is the generalization to "Ridglets".

11.6 Orthogonality

The wavelets we have been dealing with have various orthogonality properties. The orthogonality is only necessary if we are to reconstruct the original data from the transformed data. If we have no interest in the reconstruction, there is no need to insist on orthogonality.

11.7 Continuous Versus Discrete Transforms

We see two types of wavelet transforms: the continuous wave transform (CWT) as exemplified by the Mexican Hat wavelet and the discrete wavelet transform (DWT) as exemplified by Dauechies' wavelets. In order to use the continuous wavelet transform on a data set whose values are given on a discrete set of points it is inevitably necessary to discretize the wavelet function in order to do the integrals, so in a practical sense the continuous wavelet transform reduces to a discrete transform. However, there is more to it than that.

11.8 Discretizing the Continuous Wavelet Transform

If we consider the wavelet transform

$$F(a,b) = \frac{1}{\sqrt{a}} \int_0^\infty f(x)\psi\left(\frac{x-b}{a}\right) dx \tag{146}$$

and choose

$$a = 2^{-j}, \quad b = 2^{-j}k \tag{147}$$

for some k, we get what looks like a discrete wavelet transform with scales parametrized by the level number j. This certainly allows us to construct a hierarchical representation of the data $f(x)$. However, this transform is not in general invertible.

11.9 Hybrid Approaches

There is some advantage to be gained by working on image data in the Fourier domain and doing the wavelet operations there before transforming back to the real data space.

11.10 Multifractals

It was remarked that the wavelet transform of $f(x)$, $F(a,b)$, is dominated by cusps that represent the scaling behaviour of $f(x)$ in local neighbourhoods. Given a wavelet that has vanishing moment of all orders up to some N, this cuspy behaviour has asymptotic behaviour

$$F(a,b) \sim |a|^{h(x_0)} \tag{148}$$

for some function $h(x)$ describing the scaling of the singularity behaviour of $f(x)$. If we look in real space, as opposed to wavelet space, the local behaviour of the function $f(x)$ in the vicinity of spikes will have the form

$$|f(x) - f(x_0)| \sim |x - x_0|^{\alpha_0} . \tag{149}$$

Here α_0 is called the Hölder or singularity exponent. In other disciplines, notably in the area of finance modelling, $H = 1 - \alpha_0/2$ is referred to as the Hurst exponent, after H.E. Hurst who studied the structure of the river Nile. The Hurst exponent is calculated by comparing the range of the data with its standard deviation: the so-called R/S statistic. This exponent α_0 will be an erratic function of position x_0, and so we may prefer to write α_0 explicitly as a function of position, $\alpha_0(x)$, in order to emphasize this.

12 Further Reading

I have used a great number of references from papers, the World Wide Web and books in compiling this review, too many to mention individually. Here I list some of the main sources which will perhaps aid others wishing to plunge further into this wavelet sea.

12.1 General Introduction

James Walker's *Primer on Wavelets and their Scientific Applications* [42] was one of my starting points. It is full of instructive examples, one of which I have borrowed here (Sect. 9). *Discovering Wavelets* by Aboufadel and Schlicker [1] is perhaps more formal, but it makes for an excellent course book since everything is worked out; it has problems and answers and there are sample routines written in *Maple*.

There are excellent review articles available on the World Wide Web by Dremin, Ivanov and Nechitailo [16], and by Bultheel which are available from `http://arxiv.org/abs/hep-ph/0101182` , `http://www.cs.kuleuven.ac.be/~ade/WWW/WAVE/contents.html` , respectively.

12.2 Applications

The monograph *Ripples in Mathematics: The Discrete Wavelet Transform* [22] by Jensen and Cour-Harbo focuses on the discrete wavelet transform as might be applied to signal processing. The focus is on the Lifting process and filters. Many practical problems are addressed, such as what to do at the boundaries of finite stretches of data. A magnificent book from which I have learnt, and am still learning, a lot. Examples are in *MATLAB* format.

The proceedings of the Royal Society discussion meeting *Wavelets: The Key to Intermittent Information?* edited by B.W. Silverman and J.C. Vassilicos [35] is a fine collection of articles spanning a large range of applications of wavelets. There are papers on the analysis of economic and financial data, intermittency in fluid motion, statistics and other diverse topics. A fascinating insight into what wavelets can do for us, showing how work across different fields stimulates advances in the subject.

12.3 Astronomical

The only kid on the block is the monograph *Image Processing and Data Analysis: The Multiscale Approach* by Starck, Murtagh and Bijaoui [37] who were perhaps the pioneers in exploiting wavelet and multiscale analysis into astronomy. The book is perhaps more about multiscale image analysis than about wavelets as such, but it is also a rich source of information on advanced imaging techniques in general.

There are many papers using wavelets to analyse various all-sky microwave background data sets (COBE, WMAP and Planck) to cite but a few: Barreiro et al. [3], Cayon et al. [7], Starck et al. [38], Rocha et al. [32]. The paper by Wiaux et al. [43] is particularly interesting, in that they generate a wavelet on a sphere via inverse stereographic projection. McEwen et al. [28] use Morlet wavelets on the sphere.

The paper of Maisinger et al. [25] looks at using maximum entropy reconstruction techniques in wavelet space to de-noise and deconvolve microwave background maps and images in general.

12.4 Web Sites

The World Wide Web is a vast, almost impenetrable, source of information: the amount of wavelet-related content is truly awesome. However, there are a few sites that might be described as primary resources.
`http://www.wavelet.org`,
`http://www.amara.com/current/wavelet.html`,
`http://www.ondelette.com/indexen.html`,
`http://www-stat.stanford.edu/~wavelab/`.
The first of these issues an e-zine with valuable up-to-date information on preprints, books and meetings. The second is a fine general resource: almost anything can be found from here. The third is a messageboard (in English) with some useful resources such as programs. Wavelab is a collection of MatLab functions that do all kinds of things with wavelets.

Acknowledgements

It was a great pleasure participating in this school and I have Vicent Martinez and Enn Saar to thank for that. Together with Sylvestre Paredes they were responsible for my getting into this subject and related areas such as the multifractal analysis of structure. Rien van de Weygaert has been an invaluable source of discussion about such things. He also provided me with a library of fine pictures of the Valencia architecture of Calatrava.

References

1. Aboufadel, E., Schlicker, S.: Discovering Wavelets. John Wiley, New York (1999)
2. Barnes J., Hut, P.: Nature **324**, 446 (1986)
3. Barreiro, R.B., Martinez-Gonzalez E., Sanz, J.L.: Mon. Not. Roy. Astr. Soc. **322**, 411 (2001)
4. Beylkin, G., Coifman, R.R., Rokhlin, V.: Comm. Pure Appl. Math. **44**, 141 (1991)

5. Burt, P.J., Adelson, A.E.: IEEE Trans. Commun. **31** (1983)
6. Carrier, Greengard L., Rokhlin, V.: SIAM J. Sci. Stat. Comput. **9**, 4 (1988)
7. Cayon, L., Sanz, J.L., Martinez-Gonzalez, E., Banday, A.J., Argueso, F., Gallegos, J.E., Gorski, K.M., Hinshaw, G.: Mon. Not. Roy. Astr. Soc. **326**, 1243 (2001)
8. Chiu, C.K., Goswami, J.C., Chan, A.K.: Numer. Math. **70**, 283 (1995)
9. Daubechies, I.: Ten Lectures on Wavelets. SIAM, Philadelphia (1992)
10. Daubechies, I.: Commun. Pure Appl. Math. **41**, 909 (1988)
11. Daubechies, I.: SIAM J. Math. Anal. **24**, 499 (1993)
12. Daubechies, I., Sweldens, W.: J. Fourier Anal. Appl. **4**, 245 (1998)
13. Dehnen, W.: J. Comp. Phys. **179**, 27 (2002)
14. Dehnen, W.: Mon. Not. Roy. Astron. Soc. **324**, 273 (2001)
15. Deslauriers, G., Dubuc, S.: Construct. Approx. **5**, 49 (1989)
16. Dremin, I.M., Ivanov, O.V., Nechitailo, V.A.: Phys. Usp. **44**, 447 (2001)
17. Dubuc, S.: J. Math. Anal. Appl. **114**, 185 (1986)
18. Gabor, D.: J. IEE: **93**, 429 (1946)
19. Greengard, L., Rohklin, V.: Acta Numer. **6**, 229 (1997)
20. Grossmann, A., Morlet, J.: SIAM J. Math. Anal. **15**, 723 (1984)
21. Haar, A.: Math. Ann. **69**, 331 (1910)
22. Jensen, A., la Cour-Harbo, A.: Ripples in Mathematics: The Discrete Wavelet Transform. Springer, New York (2001)
23. Kingsbury, N.G.: In: Silverman, B.W., Vassilicos, J.C. (eds.) Wavelets: the Key to Intermittent Information? p. 166 Oxford University Press, Oxford (1999)
24. Littlewood, J., Paley, R.: J. London Math. Soc. **6**, 20 (1931)
25. Maisinger, K., Hobson, M.P., Lasenby, A.N.: Mon. Not. Roy. Astr. Soc. **347**, 339 (2004)
26. Mallat, S.: IEEE Trans. Pattern Anal. Mach. Intell. **11**, 674 (1989)
27. Marr, S.: Vision. Freeman, W.H. (1983)
28. McEwen, J.D., Hobson, M.P., Lasenby A.N., Mortlock, D.J.: Mon. Not. Roy. Astr. Soc. **359**, 1583 (2005)
29. Meyer, Y., Jaffard, S., Rioul, O.: Pour la Sci. **119**, 28 (1987)
30. Papoulis, A.: Probability, Random Variables and Stochastic Processes 3rd edn. McGraw Hill, New York (1991)
31. Press, W.H., Teukolsky, S.A., Vetterling, W.T., Flannery, B.P.: Numerical Recipes: the Art of Scientific Computing. Cambridge University Press, Cambridge (2002)
32. Rocha, G., Cayon, L., Bowen, R., Canavezes, A., Silk, J., Banday, A.J., Gorski, K.M.: Mon. Not. Roy. Astron. Soc. **351**, 769 (2004)
33. Romeo, A.B., Horellou, C., Bergh, J.: Mon. Not. Roy. Astron. Soc. **354**, 1280 (2004)
34. Samet, H.: ACM Comput. Surv. **16**, 187 (1984)
35. Silverman, B.W., Vassilicos, J.C.: Wavelets: the Key to Intermittent Information?. Oxford University Press, Oxford (1999)
36. Smith, M.J., Barnwell, T.P.: IEEE Trans. ASSP **34**, 434 (1986)
37. Starck, J.-L., Murtagh, F., Bijaoui, A.: Image Processing and Data Analysis: the Multiscale Approach. Cambridge University Press, Cambridge (1998)
38. Starck, J.L., Aghanim, N., Forni, O.: Astron. Astrophys. **416**, 9 (2004)
39. Sweldens, W.: Appl. Comput. Harmon. Anal. **3**, 186 (1996)
40. Sweldens, W.: SIAM J. Math. Anal. **29**, 511 (1997)

41. Sweldens, W., Schroder, P.: Wavelets in Computer Graphics, p. 15. ACM SIGGRAPH course notes (1996)
42. Walker, J.S.: Wavelets and their Scientific Applications. Chapman and Hall, Boca Raton (1999)
43. Wiaux, Y., Jacques, L., Vandergheynst, P.: Astrophys. J. **632**, 15 (2005)

Fisher Matrices and All That: Experimental Design and Data Compression

A. Heavens

Institute for Astronomy, University of Edinburgh, Blackford Hill, Edinburgh EH9 3HJ
afh@roe.ac.uk

Abstract. We consider the general problem of estimating parameters from data, and consider how one can use Fisher matrices to analyse survey designs before any data are taken, to see whether the survey will actually do what is required. We also take a look at data compression methods, such as Karhunen–Loève and MOPED, which can be extremely valuable for speeding up analysis of large data sets, and making the analysis more stable.

1 Introduction: Data Analysis

Most data analysis problems are *inverse problems.* You have a set of data $\{\boldsymbol{x}\}$, and you wish to interpret the data in terms of a model. The model will typically have some parameters $\{\boldsymbol{\Theta}\}$ in it, which you want to determine. Clearly, what one wants is the probability distribution of $\boldsymbol{\Theta}$, given the data $\boldsymbol{x}$, i.e.

$$p(\boldsymbol{\Theta}|\boldsymbol{x}) \,. \tag{1}$$

From this one can calculate the expectation values of the parameters, and their errors. What may be readily calculable is not this, rather the opposite, $p(\boldsymbol{x}|\boldsymbol{\Theta})$. For example, consider a model which is a Gaussian with mean μ and variance σ^2. The probability of a single variable x given the parameters $\boldsymbol{\Theta} = (\mu, \sigma)$ is

$$p(x|\boldsymbol{\Theta}) = \frac{1}{\sqrt{2\pi}\sigma} \exp\left[-\frac{(x-\mu)^2}{2\sigma^2}\right] , \tag{2}$$

but this is not what we actually want. However, we can relate this to $p(\boldsymbol{\Theta}|\boldsymbol{x})$ using Bayes' Theorem:

$$p(\boldsymbol{\Theta}|\boldsymbol{x}) = \frac{p(\boldsymbol{\Theta}, \boldsymbol{x})}{p(\boldsymbol{x})} = \frac{p(\boldsymbol{x}|\boldsymbol{\Theta})p(\boldsymbol{\Theta})}{p(\boldsymbol{x})} \,. \tag{3}$$

- $p(\boldsymbol{x}|\boldsymbol{\Theta})$ is called the *likelihood* $L(\boldsymbol{x}; \boldsymbol{\Theta})$.

Heavens, A.: *Fisher Matrices and All That: Experimental Design and Data Compression.* Lect. Notes Phys. **665**, 51–72 (2009)
DOI 10.1007/978-3-540-44767-2_2

- $p(\boldsymbol{\Theta})$ is called the *prior*, and expresses what we know about the parameters prior to the experiment being done. This may be the result of previous experiments, or theory (e.g. some parameters, such as the age of the Universe, may have to be positive). In the absence of any previous information, the prior is often assumed to be a constant (a "flat prior"), but other choices can sometimes be justified.
- $p(\boldsymbol{x})$ is the *evidence*. In the end, we are interested in the relative probabilities of the parameters, and this term does not depend on them, so is often ignored. It can play a role when more than one theoretical model is being considered, and one wants to choose which model is most likely, whatever the parameters are. (A different model should be interpreted loosely, e.g. one may consider a flat CDM Universe as a model, or a CDM Universe with non-zero curvature. The curvature is another parameter, but the "model" in this context differs, as it has one more parameter.)

Thus for flat priors, we have simply

$$p(\boldsymbol{\Theta}|\boldsymbol{x}) \propto L(\boldsymbol{x};\boldsymbol{\Theta}) . \tag{4}$$

Although we may have the full probability distribution for the parameters, often one simply uses the peak of the distribution as the estimate of the parameters. This is then a *maximum-likelihood* estimate. Note that if the priors are not flat, the peak in $p(\boldsymbol{x}|\boldsymbol{\Theta})$ is not the maximum-likelihood estimate.

My rule of thumb is that if the priors are assigned theoretically, and they influence the result significantly, the data are usually not good enough.

2 Fisher Matrix Analysis

This has been adapted from [14] (hereafter TTH).

How accurately can we estimate model parameters from a given data set? This question was basically answered 60 years ago [1], and we will now summarize the results, which are both simple and useful.

Suppose for definiteness that our data set consists of N real numbers $x_1, x_2, \ldots, x_N$, which we arrange in an n-dimensional vector $\boldsymbol{x}$. These numbers could for instance denote the measured temperatures in the N pixels of a CMB sky map, the counts-in-cells of a galaxy redshift survey, the first N coefficients of a Fourier–Bessel expansion of an observed galaxy density field, or the number of gamma-ray bursts observed in N different flux bins. Before collecting the data, we think of $\boldsymbol{x}$ as a random variable with some probability distribution $L(\boldsymbol{x};\boldsymbol{\Theta})$, which depends in some known way on a vector of M model parameters $\boldsymbol{\Theta} = (\theta_1, \theta_2, \ldots, \theta_M)$.

Such model parameters might for instance be the spectral index of density fluctuations, the Hubble constant h, the cosmic density parameter Ω, or the mean redshift of gamma-ray bursts. We will let $\boldsymbol{\Theta}_0$ denote the true parameter values and let $\boldsymbol{\Theta}$ refer to our estimate of $\boldsymbol{\Theta}$. Since $\boldsymbol{\Theta}$ is some function of the data vector $\boldsymbol{x}$, it too is a random variable. For it to be a good estimate, we would of course like it to be unbiased, i.e.

$$\langle \boldsymbol{\Theta} \rangle = \boldsymbol{\Theta}_0 \, , \tag{5}$$

and give as small error bars as possible, i.e. minimize the standard deviations

$$\Delta\theta_i \equiv \left(\langle \theta_i^2 \rangle - \langle \theta_i \rangle^2 \right)^{1/2} \, . \tag{6}$$

In statistics jargon, we want the BUE θ_i, which stands for the "best unbiased estimator".

A key quantity in this context is the so-called *Fisher information matrix*, defined as

$$\mathsf{F}_{ij} \equiv \left\langle \frac{\partial^2 \mathcal{L}}{\partial\theta_i \partial\theta_j} \right\rangle \, , \tag{7}$$

where $\mathcal{L} \equiv -\ln L$. Another key quantity is the *maximum-likelihood estimator*, or *ML-estimator* for brevity —indexML-estimator defined as the parameter vector $\boldsymbol{\Theta}_{\rm ML}$ that maximizes the likelihood function $L(\boldsymbol{x}; \boldsymbol{\Theta})$.

Using this notation, a number of powerful theorems have been proved (see, e.g. [8, 7]):

1. For any unbiased estimator, $\Delta\theta_i \geq 1/\sqrt{\mathsf{F}_{ii}}$.
2. If there is a BUE $\boldsymbol{\Theta}$, then it is the ML-estimator or a function thereof.
3. The ML-estimator is asymptotically BUE.

The first of these theorems, known as the Cramér–Rao inequality, thus places a firm lower limit on the error bars that one can attain, regardless of which method one is using to estimate the parameters from the data. This is called the *conditional error*, and is the minimum error bar attainable on θ_i if all the other parameters are known. *It is rarely relevant and should almost never be quoted.*

The normal case is that the other parameters are estimated from the data as well, in which case the minimum standard deviation rises to

$$\Delta\theta_i \geq \left(\mathsf{F}^{-1} \right)_{ii}^{1/2} \, . \tag{8}$$

This is called the *marginal error*, and is normally the relevant error to quote.

The second theorem shows that maximum-likelihood (ML) estimates have quite a special status: if there is a best method, then the ML-method is the one. Finally, the third result basically tells us that in the limit of a very large data set, the ML-estimate for all practical purposes is the best estimate, the one that for which the Cramér–Rao inequality becomes an equality. It is these nice properties that have made ML-estimators so popular.

Note that conditional and marginal errors coincide if F is diagonal. If it is not, then the *estimates* of the parameters are correlated (even if the parameters themselves are uncorrelated), e.g. in the example shown in Fig. 1, estimates of the baryon density parameter Ω_b and the dark energy equation

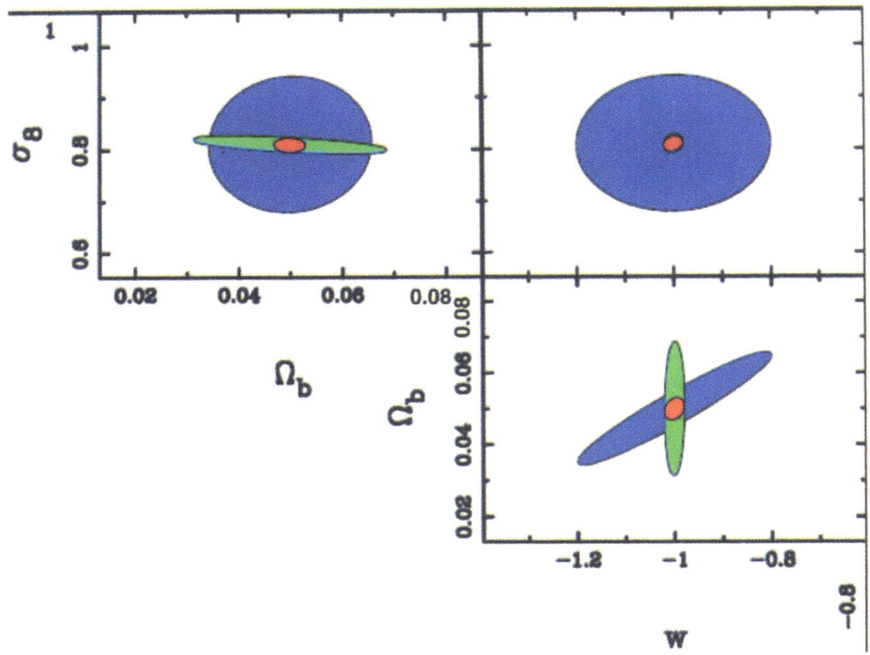

Fig. 1. The expected error ellipses for cosmological parameters (σ_8, baryon density parameter Ω_b, and dark energy equation of state $w \equiv p/\rho c^2$) from a 3D weak lensing survey of 1000 square degrees, with a median redshift of 1 and a photometric redshift error of 0.15. Probabilities are marginalized over all other parameters, except that $n = 1$ and a flat Universe are assumed. Dark ellipses represent a prior from WMAP, pale represents the 3D lensing survey alone, and the central ellipses show the combination (Figure courtesy Tom Kitching)

of state $w \equiv p/\rho c^2$ are strongly correlated with WMAP data alone, so we will tend to overestimate both Ω_b and w, or underestimate both. However, the value of Ω_b in the Universe has nothing obvious to do with w - they are independent.

2.1 Fisher Information Matrix

Although the proof of the Cramér–Rao inequality is rather lengthy, it is quite easy to acquire some intuition for why the Fisher information matrix has the form that it does. This is the purpose of the present section.

Let us Taylor expand $\mathcal{L}$ around the ML-estimate $\mathbf{\Theta}$. By definition, all the first derivatives $\partial\mathcal{L}/\partial\theta_i$ will vanish at the ML-point, since the likelihood function has its maximum there, so the local behaviour will be dominated by the quadratic terms. Since $L = \exp[-\mathcal{L}]$, we thus see that the likelihood function will be approximately Gaussian near the ML-point. If the error bars are quite small, L usually drops sharply before third-order terms have become important, so that this Gaussian is a good approximation to L everywhere. Interpreting L as a Bayesian probability distribution in parameter space, the covariance matrix T is thus given simply by the second derivatives at the ML-point, as the inverse of the Hessian matrix:

$$(\mathsf{T}^{-1})_{ij} \equiv \frac{\partial^2 \mathcal{L}}{\partial\theta_i \partial\theta_j} . \tag{9}$$

Note that the Fisher information matrix F is simply the expectation value of this quantity at the point $\boldsymbol{\Theta} = \boldsymbol{\Theta}_0$ (which coincides with the ML-point on average if the ML-estimate is unbiased). It is basically a measure of how fast (on average) the likelihood function falls off around the ML-point, i.e. a measure of the width and shape of the peak. From this discussion, it is also clear that we can use its inverse, F^{-1}, as an estimate of the covariance matrix

$$\mathsf{T} \equiv \langle \boldsymbol{\Theta}\boldsymbol{\Theta}^t \rangle - \langle \boldsymbol{\Theta} \rangle \langle \boldsymbol{\Theta} \rangle^t \tag{10}$$

of our parameter estimates when we use the ML-method.

2.2 Gaussian Case

Let us now explicitly compute the Fisher information matrix for the case when the probability distribution is Gaussian, i.e. where (dropping an irrelevant additive constant $n \ln[2\pi]$)

$$2\mathcal{L} = \ln\det\mathsf{C} + (\boldsymbol{x} - \boldsymbol{\mu})\,\mathsf{C}^{-1}(\boldsymbol{x} - \boldsymbol{\mu})^t , \tag{11}$$

where in general both the mean vector $\boldsymbol{\mu}$ and the covariance matrix

$$\mathsf{C} = \langle (\boldsymbol{x} - \boldsymbol{\mu})(\boldsymbol{x} - \boldsymbol{\mu})^t \rangle \tag{12}$$

depend on the model parameters $\boldsymbol{\Theta}$. Although vastly simpler than the most general situation, the Gaussian case is nonetheless general enough to be applicable to a wide variety of problems in cosmology. Defining the data matrix

$$\mathsf{D} \equiv (\boldsymbol{x} - \boldsymbol{\mu})(\boldsymbol{x} - \boldsymbol{\mu})^t \tag{13}$$

and using the matrix identity (see exercises) $\ln\det\mathsf{C} = \mathrm{Trace}\ln\mathsf{C}$, we can re-write (11) as

$$2\mathcal{L} = \mathrm{Trace}\left[\ln\mathsf{C} + \mathsf{C}^{-1}\mathsf{D}\right] . \tag{14}$$

We will use the standard comma notation for derivatives, where for instance

$$\mathsf{C}_{,i} \equiv \frac{\partial}{\partial\theta_i}\mathsf{C} . \tag{15}$$

Since C is a symmetric matrix for all values of the parameters, it is easy to see that all the derivatives $\mathsf{C}_{,i}$, $\mathsf{C}_{,ij}$, will also be symmetric matrices. Using the matrix identities $(\mathsf{C}^{-1})_{,i} = -\mathsf{C}^{-1}\mathsf{C}_{,i}\,\mathsf{C}^{-1}$ and $(\ln \mathsf{C})_{,i} = \mathsf{C}^{-1}\mathsf{C}_{,i}$ (see exercises), we find

$$2\mathcal{L}_{,i} = \mathrm{Trace}\left[\mathsf{C}^{-1}\mathsf{C}_{,i} - \mathsf{C}^{-1}\mathsf{C}_{,i}\,\mathsf{C}^{-1}\mathsf{D} + \mathsf{C}^{-1}\mathsf{D}_{,i}\right] . \tag{16}$$

When evaluating C and $\boldsymbol{\mu}$ at the true parameter values, we have $\langle \boldsymbol{x} \rangle = \boldsymbol{\mu}$ and $\langle \boldsymbol{x}\boldsymbol{x}^t \rangle = \mathsf{C} + \boldsymbol{\mu}\boldsymbol{\mu}^t$, which gives

$$\begin{cases} \langle \mathsf{D} \rangle & = \mathsf{C}\,, \\ \langle \mathsf{D}_{,i} \rangle & = 0\,, \\ \langle \mathsf{D}_{,ij} \rangle & = \boldsymbol{\mu}_{,i}\,\boldsymbol{\mu}_{,j}^t + \boldsymbol{\mu}_{,j}\,\boldsymbol{\mu}_{,i}^t\,. \end{cases} \tag{17}$$

Using this and (16), we obtain $\langle \mathcal{L}_{,i} \rangle = 0$. In other words, the ML-estimate is correct on average in the sense that the average slope of the likelihood function is zero at the point corresponding to the true parameter values. Applying the chain rule to (16), we obtain

$$\begin{aligned} 2\mathcal{L}_{,ij} = \mathrm{Trace}[& - \mathsf{C}^{-1}\mathsf{C}_{,i}\,\mathsf{C}^{-1}\mathsf{C}_{,j} + \mathsf{C}^{-1}\mathsf{C}_{,ij} \\ & + \mathsf{C}^{-1}(\mathsf{C}_{,i}\,\mathsf{C}^{-1}\mathsf{C}_{,j} + \mathsf{C}_{,j}\,\mathsf{C}^{-1}\mathsf{C}_{,i})\mathsf{C}^{-1}\mathsf{D} \\ & - \mathsf{C}^{-1}(\mathsf{C}_{,i}\,\mathsf{C}^{-1}\mathsf{D}_{,j} + \mathsf{C}_{,j}\,\mathsf{C}^{-1}\mathsf{D}_{,i}) \\ & - \mathsf{C}^{-1}\mathsf{C}_{,ij}\,\mathsf{C}^{-1}\mathsf{D} + \mathsf{C}^{-1}\mathsf{D}_{,ij}] \,. \end{aligned} \tag{18}$$

Substituting this and (17) into (7) and using the trace identity Trace[AB] = Trace[BA], many terms drop out and the Fisher information matrix reduces to simply

$$\mathsf{F}_{ij} = \langle \mathcal{L}_{,ij} \rangle = \frac{1}{2}\mathrm{Trace}[\mathsf{C}^{-1}\mathsf{C}_{,i}\mathsf{C}^{-1}\mathsf{C}_{,j} + \mathsf{C}^{-1}\mathsf{M}_{ij}]\,, \tag{19}$$

where we have defined the matrix $\mathsf{M}_{ij} \equiv \langle \mathsf{D}_{,ij} \rangle = \boldsymbol{\mu}_{,i}\,\boldsymbol{\mu}_{,j}^t + \boldsymbol{\mu}_{,j}\,\boldsymbol{\mu}_{,i}^t$.

This result is extremely powerful. If the data have a (multivariate) Gaussian distribution (and the errors can be correlated; C need not be diagonal), and you know how the means $\boldsymbol{\mu}$ and the covariance matrix C depend on the parameters, you can calculate the Fisher matrix *before you do the experiment.* The Fisher matrix gives you the expected errors, so you know how well you can expect to do if you do a particular experiment, and you can then design an experiment to give you, for example, the best (marginal) error on the parameter you are most interested in.

Note that if the prior is not uniform, then you can simply add a "prior matrix" to the Fisher matrix before inversion. Figure 1 shows an example, where a prior from CMB experimental results has been added to a hypothetical 3D weak lensing survey.

For a good discussion of how to interpret likelihood plots in more than one dimension, see the Numerical Recipes books (e.g. [12]; available online at `www.nr.com`).

3 Data Compression

In some cases, one can throw away a lot of data without losing much information. This does not necessarily mean discarding actual data points themselves, but might mean binning the data and keeping the average in the bins but throwing away the individual measurements. This can reduce the size of the data set, but might not increase the error bars on the parameters significantly. Reducing the size of the data set is called *data compression.*

Why might one want to do this? Either for storage reasons or (more likely) for speed of analysis. This is especially true if the data are correlated, for which a maximum-likelihood solution needs the inverse of matrices. Matrix inversion is an N^3 process in the general case, for N data points, so data compression by a modest factor of two can lead to almost an order of magnitude speed-up in processing time.

When might one be able to do this? If the data are *noisy* and/or *correlated.* An example of data which are both noisy and correlated are the COBE microwave background data. The individual pixels had signal-to-noise of around unity, and because of the beam of the telescope, nearby pixels are correlated. For these reasons you rarely see an original map – they are usually presented at lower resolution.

To make these point clearer, if one has a pair of data points which are supposed to measure the same thing, and one point is much noisier than the other, one will lose very little information if one simply throws away the very noisy point, thus achieving a factor of two data compression. In fact one can do better than this, by reducing the data set size by two by *inverse-variance weighting* of the data:

$$y = \frac{\frac{x_1}{\sigma_1^2} + \frac{x_2}{\sigma_2^2}}{\frac{1}{\sigma_1^2} + \frac{1}{\sigma_2^2}} . \tag{20}$$

If we want the mean of the distribution from which x_1 and x_2 are drawn, then it turns out that the estimate of the mean (given by y) is as accurate as if we would used both x_1 and x_2, i.e. for this problem we lose nothing in accuracy if we compress the data like this. It is a simple exercise to prove this.

Similarly it is easy to convince oneself that if two points are highly correlated, then the second point tells us very little more than the first. In this case, a linear combination of the data proportional to $\mathsf{C}^{-1}\boldsymbol{x}$ is optimal.

So how do we do data compression more generally? The key is that if we restrict ourselves to *linear* data compression methods, then the Fisher matrix for the linear combinations is easily calculable. To see how this works, consider a data set $\boldsymbol{x}$ and a set of linear combinations of the data

$$\boldsymbol{y} = \mathsf{B}\boldsymbol{x} . \tag{21}$$

It is clear that if B is square and invertible, then $\boldsymbol{y}$ must tell us just as much about the parameters as $\boldsymbol{x}$. Why is this obvious? Simply because one can take

$\boldsymbol{y}$ and re-create the original data by $\boldsymbol{x}=\mathsf{B}^{-1}\boldsymbol{y}$. What is more interesting is if B is not square. If $\boldsymbol{y}$ has fewer elements than $\boldsymbol{x}$, then the data have been compressed, and B is not square (and clearly not invertible). How does this affect what we can learn about the parameters? In order to answer this, we calculate the Fisher matrix (19) for the new data set $\boldsymbol{y}$. For this we need the mean and covariance matrix of the new data.

The covariance matrix of $\boldsymbol{y}$ is readily calculated:

$$\begin{aligned} \mathsf{C}_{ij}^{(y)} &= \langle y_i y_j\rangle - \langle y_i\rangle\langle y_j\rangle \\ &= \langle B_{ik}x_k B_{jm}x_m\rangle - \langle B_{ik}x_k\rangle\langle B_{jm}x_m\rangle \\ \Rightarrow \mathsf{C}^{(y)} &= \mathsf{BCB}^t \,. \end{aligned} \tag{22}$$

The mean is simply $\langle\boldsymbol{y}\rangle = \mathsf{B}\langle\boldsymbol{x}\rangle$. Substitution of these into (19) gives the Fisher matrix for the $\boldsymbol{y}$ data set:

$$\mathsf{F}_{ij}^{(y)} = \frac{1}{2}\mathrm{Trace}[(\mathsf{BCB}^t)^{-1}(\mathsf{BC}_{,i}\mathsf{B}^t)(\mathsf{BCB}^t)^{-1}\mathsf{BC}_{,j}\mathsf{B}^t + (\mathsf{BCB}^t)^{-1}\mathsf{M}_{ij}^{(y)}]\,, \tag{23}$$

and $\mathsf{M}_{ij}^{(y)} \equiv \mathsf{B}\boldsymbol{\mu}_i\mathsf{B}\boldsymbol{\mu}_j$.

Now, if B is square and invertible, then the B matrices cancel out, and the Fisher matrix is unchanged. This makes perfect sense, since, if B is invertible, then we can always recover the original data from the $\boldsymbol{y}$ values, so we should not be losing any information in doing so.

The more interesting situation is when we make the length of the $\boldsymbol{y}$ vector smaller than the length of $\boldsymbol{x}$. In this case, we have genuine data compression. Each row vector of B specifies one of the numbers in the new data set (as some linear combination of the original data $\boldsymbol{x}$), and we wish to know which such row vectors capture the most information about the parameters $\boldsymbol{\Theta}$. If B has merely a single row, then we can write $\mathsf{B} = \mathbf{b}^t$ for some vector $\mathbf{b}$, and the Fisher matrix for the compressed datum is

$$\mathsf{F}_{ij}^{(y)} = \frac{1}{2}\left[\frac{(\mathbf{b}^t\mathsf{C}_{,i}\,\mathbf{b})(\mathbf{b}^t C_{,j}\,\mathbf{b})}{(\mathbf{b}^t\mathsf{C}\mathbf{b})^2}\right] + \frac{(\mathbf{b}^t\boldsymbol{\mu}_{,i}\,)(\mu_j\mathbf{b})}{(\mathbf{b}^t\mathsf{C}\mathbf{b})}\,. \tag{24}$$

The trick, of course, is to choose a vector $\mathbf{b}$ which preserves as much information about the parameter(s) of interest as possible. Really, what one wants is to minimize the *marginal* error on a parameter, i.e. assuming that all of the parameters are to be determined simultaneously from the data. However, this involves maximizing a component of the *inverse* of F, and this problem may be difficult and possible impossible analytically. What one can do is at least to minimize the *conditional* error on a parameter. This means maximizing the diagonal ($i = j$) component of the Fisher matrix:

$$\mathsf{F}_{ii}^{(y)} = \frac{1}{2}\left(\frac{\mathbf{b}^t\mathsf{C}_{,i}\,\mathbf{b}}{\mathbf{b}^t\mathsf{C}\mathbf{b}}\right)^2 + \frac{(\mathbf{b}^t\boldsymbol{\mu}_{,i}\,)^2}{(\mathbf{b}^t\mathsf{C}\mathbf{b})}\,. \tag{25}$$

The most general problem has both the mean $\boldsymbol{\mu}$ and the covariance matrix C depending on the parameters of the spectrum, and the resulting maximization leads to an eigenvalue problem which is nonlinear in $\boldsymbol{b}$.

There are, however, two situations where this maximization can be readily done:

- The mean $\boldsymbol{\mu}$ is known
- The covariance matrix C is known

These may appear to be strong restrictions, but cover many cases, at least approximately. For example, if the data set consists of overdensity measurements in a galaxy distribution, or $\Delta T/T$ measurements in the CMB, or Fourier/spherical harmonic transforms of these data, then the means are known (often they are just zero; partial sky coverage may mean the transform coefficients are not zero on average, but the mean is fixed and calculable). In these cases, all of the information comes from the scatter about the mean, i.e. from the (co)variances of the data, rather than the mean of the data (Table 1).

Another set of examples arises when the main information comes from the mean of the signal, rather than the covariance. Examples here include the CMB power spectrum [3] and the stellar flux from a galaxy (e.g. [4, 13, 10, 5, 6]). In these cases, the main information comes from the mean. In both cases, the covariance does include a small amount of additional information. In the CMB case, this is because the cosmic variance component of the scatter in the power spectrum estimates depends on the power spectrum itself, and hence is dependent on the cosmological parameters. In the stellar flux from a galaxy, one wishes to recover the star formation history from the spectrum. One part of the noise comes from photon noise in the source, which is again

Table 1. Example cosmological data sets, characterized by the mean signal, source(s) of covariance, and parameters on which they depend. Bold font is used where the mean or covariance depends on the parameters strongly; italic font is used if the dependence is slight, and roman font is used if there is no dependence on the parameters. Note that the first two examples have zero mean, whereas the last two have covariance matrices which depend only weakly on the parameters

	Large-scale structure	CMB	CMB power spectrum	Galaxy spectrum
Data	Fourier coefficients	$\Delta T/T$	C_ℓ	f_ν
Mean $\boldsymbol{\mu}$	0	0	$\boldsymbol{C_\ell(\Omega_m, h, \ldots)}$	**Flux (star formation history, ...)**
Cov. matrix C	**P(k)**, shot noise	**Correlation function** $(\Omega_m, h, \ldots)$	*Cosmic variance*, detector noise	Background, detector, *galaxy photon noise*

dependent on the mean signal, and hence on the star formation history of the galaxy. In this case, this component of the noise is almost always very small in comparison with other sources of noise, such as sky background subtraction.

3.1 Case I: Mean Is Known (Karhunen–Loève Decomposition)

When $\boldsymbol{\mu}$ is independent of θ_i, the second term in (24) vanishes, and we simply wish to maximize the quantity

$$\left(2\mathsf{F}_{ii}^{(y)}\right)^{1/2} = \frac{|\mathbf{b}^t\mathsf{C}_{,i}\,\mathbf{b}|}{\mathbf{b}^t\mathsf{C}\mathbf{b}} \,. \tag{26}$$

Since the mean $\boldsymbol{\mu}$ does not depend on the parameters (i.e. it is known), let us for simplicity redefine our data by $\boldsymbol{x} \mapsto \boldsymbol{x} - \boldsymbol{\mu}$, so that $\langle \boldsymbol{x} \rangle = 0$. Note that whereas the denominator in (24) is positive (since C is a covariance matrix and thus positive definite), the expression $\mathbf{b}^t\mathsf{C}_{,i}\,\mathbf{b}$ in the numerator might be negative, since $\mathsf{C}_{,i}$ is not necessarily positive definite. We therefore want to make $(\mathbf{b}^t\mathsf{C}_{,i}\,\mathbf{b})/(\mathbf{b}^t\mathsf{C}\mathbf{b})$ either as large as possible or as small as possible, depending on its sign. Regardless of the sign, we thus seek the $\mathbf{b}$ for which this ratio takes an extremum, i.e. we want the derivatives with respect to the components of $\mathbf{b}$ to vanish. Since this ratio is clearly unchanged if we multiply $\mathbf{b}$ by a constant, we can without loss of generality choose to normalize $\mathbf{b}$ so that the denominator equals unity. We thus seek an extremum of $\mathbf{b}^t\mathsf{C}_{,i}\,\mathbf{b}$ subject to the constraint that

$$\mathbf{b}^t\mathsf{C}\mathbf{b} = 1. \tag{27}$$

This optimization problem is readily solved by introducing a Lagrange multiplier λ and extremizing the Lagrangian

$$\mathbf{b}^t\mathsf{C}_{,i}\,\mathbf{b} - \lambda\mathbf{b}^t\mathsf{C}\mathbf{b} \,. \tag{28}$$

Varying $\mathbf{b}$ and setting the result equal to zero, we obtain the generalized eigenvalue problem

$$\mathsf{C}_{,i}\,\mathbf{b} = \lambda\mathsf{C}\mathbf{b} \,. \tag{29}$$

Since C is symmetric and positive definite, we can Cholesky decompose it (e.g. [12]) as $\mathsf{C} = \mathsf{L}\mathsf{L}^t$ for some invertible matrix L. Multiplying (29) by L^{-1} from the left, we can re-write it as

$$(\mathsf{L}^{-1}\mathsf{C}_{,i}\,\mathsf{L}^{-t})(\mathsf{L}^t\mathbf{b}) = \lambda(\mathsf{L}^t\mathbf{b}) \,, \tag{30}$$

which we recognize as an ordinary eigenvalue problem for the symmetric matrix $(\mathsf{L}^{-1}\mathsf{C}_{,i}\,\mathsf{L}^{-t})$, whose solution will be n orthogonal eigenvectors $(\mathsf{L}^t\mathbf{b}_k)$ and corresponding eigenvalues λ_k, $k = 1, 2, \ldots, n$. Let us sort them so that

$$|\lambda_1| \geq |\lambda_2| \geq \ldots, \mathrm{geq}|\lambda_n| \,. \tag{31}$$

Let us choose the kth row of B to be the row vector $\mathbf{b}_k^t$, so that the compressed data set is given by $y_k = \mathbf{b}_k^t \boldsymbol{x}$. The orthogonality property $(\mathsf{L}^t\mathbf{b}_k)\cdot(\mathsf{L}^t\mathbf{b}_{k'}) \propto \delta_{kk'}$ in combination with our chosen normalization of (27) then tells us that our compressed data satisfies

$$\langle y_k y_{k'}\rangle = \langle(\mathbf{b}_k^t\boldsymbol{x})(\boldsymbol{x}^t\mathbf{b}_{k'})\rangle = \mathbf{b}_k^t\mathsf{C}\mathbf{b}_{k'} = \mathbf{b}_k^t\mathsf{L}\mathsf{L}^t\mathbf{b}_{k'} = \delta_{kk'} , \tag{32}$$

i.e. $\langle \boldsymbol{y}\boldsymbol{y}^t\rangle = \mathsf{I}$. The compressed data values $\boldsymbol{y}$ thus have the nice property that they are what is known as *statistically orthonormal*, i.e. they are all uncorrelated and all have unit variance. Since we are assuming that everything is Gaussian, this also implies that they are statistically independent — in fact, $\boldsymbol{y}$ is merely a vector of independent unit Gaussian random variables. In other words, knowledge of y_k gives us no information at all about the other y values. This means that the entire information content of the initial data set $\boldsymbol{x}$ has been portioned out in disjoint pieces in the new compressed data set, where y_1 contains the most information about θ_i, y_2 is the runner up (once the information content of y_1 has been removed, y_2 contains the most information about θ_i), and so on, i.e. we order the linear combinations in *increasing order of uselessness.*

If we use all the N vectors $\mathbf{b}_k$ as rows in B, then B will be invertible, and we clearly have not achieved any compression at all. However, once this matrix B has been found, the compression prescription is obvious: if we want a compressed data set of size $N' < N$, then we simply throw away all but the first N' rows of B. It is straightforward to show (see, e.g. [15, 17]) that if we fix an integer N' and then attempt to minimize $\Delta\theta_i$, we will arrive at exactly the same B as with our prescription above (or this B multiplied by some invertible matrix, which as we saw will leave the information content unchanged).

How does the error bar $\Delta\theta_i$ depend on the choice of N'? Equation (29) implies that

$$\mathsf{C}_{,i}\,\mathsf{B}^t = \mathsf{C}\mathsf{B}^t\mathbf{\Lambda} , \tag{33}$$

where $\Lambda_{ij} \equiv \delta_{ij}\lambda_i$, i.e. a diagonal matrix containing all the eigenvalues. Since (32) implies that

$$\mathsf{B}\mathsf{C}\mathsf{B}^t = \mathsf{I} , \tag{34}$$

it follows from (33) that

$$\mathsf{B}\mathsf{C}_{,i}\,\mathsf{B}^t = \mathbf{\Lambda} . \tag{35}$$

Therefore the diagonal part of (23) reduces to simply

$$2\mathsf{F}_{ii} = \mathrm{Trace}[\{(\mathsf{B}\mathsf{C}\mathsf{B}^t)^{-1}(\mathsf{B}\mathsf{C}_{,i}\,\mathsf{B}^t)\}^2] = \mathrm{Trace}\mathbf{\Lambda}^2 = \sum_{k=1}^{N'}\lambda_k^2 . \tag{36}$$

It is thus convenient to plot λ_k^2 as a function of the mode number k, as we have done in Fig. 2 for the early example, measuring the redshift distortion parameter from the 1.2 Jy redshift survey, to see at what k the information

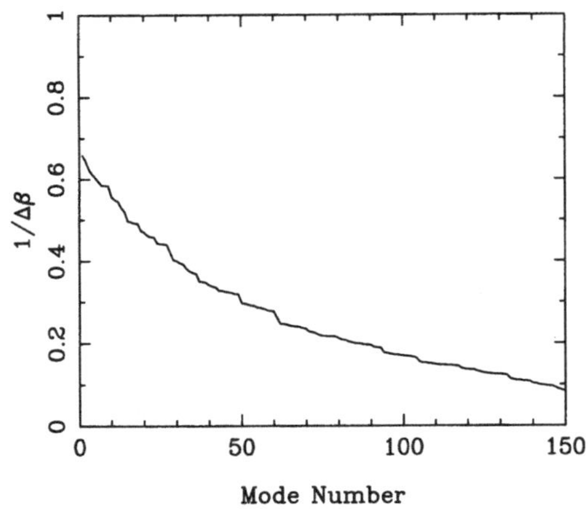

Fig. 2. KL eigenvalues $\lambda = 1/\Delta\beta$

content starts petering out. The data here are 150 coefficients of a spherical harmonic transform of the galaxy positions. As an alternative, one can plot $\Delta\theta_i$ as a function of n' as in Fig. 3 and see when the curve flattens out, i.e. when the computational cost of including more modes begins to yield diminishing returns.

If one feels uncomfortable about choosing N' by "eyeballing", one can obtain a more quantitative criterion by noting that using only a single mode $\mathbf{b}_k$ would give $\Delta\theta_i = 1/|\lambda_k|$. Thus the quantity $\theta_i|\lambda_k|$ can be thought of as the signal-to-noise ratio for the kth mode, and one can choose to keep only those modes where this ratio exceed unity (or some other constant such as 0.2). An important point to note here is that, although the information in the high mode numbers is rather small, there is still *some* information there. So we do lose something in the data compression process, but in many cases,

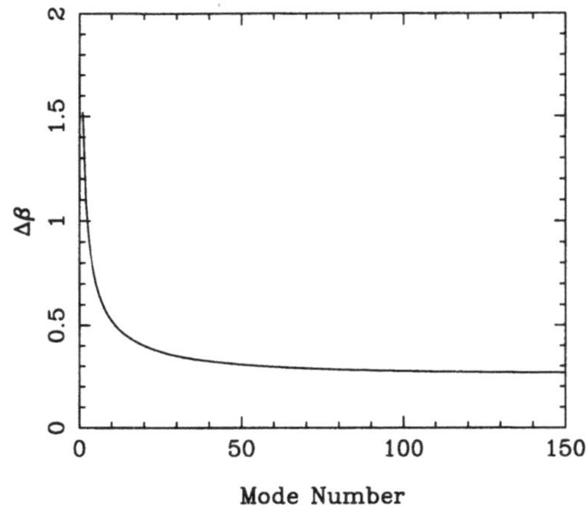

Fig. 3. Error bar on β as a function of N', the number of eigenmodes used

we lose very little indeed. This decomposition is called Karhunen–Loève (or KL) decomposition, although evidence that they actually solved this general problem has so far eluded me.

So far we have discussed how to measure a single parameter as accurately as possible from a compressed data set. In most cases of interest, there is more than one parameter to estimate, and here the formal problem is much harder to solve. Some suggestions have been made, such as finding the modes which contain most information about the first parameter and combining these with modes which contain the most about the second parameter, and so on. However, one has to do this carefully. If two parameter estimates are highly correlated, the contours of likelihood are long and thin, and oriented at some angle to the parameter axes. This is illustrated in Fig. 4, which shows estimates of the amplitude of fluctuations and the redshift distortion parameter from the 2dF galaxy redshift survey [11]. The parameter estimates are clearly highly correlated. If one performs two data compression steps on the full data, first maximizing F_{11} and then F_{22}, one effectively tries to prevent the width of the likelihood surface growing *along the directions of the parameter axes*. The data compression will do this, but it will not necessarily stop the likelihood surface getting longer along the "ridge". In practice this tends to happen, which is a disaster for the *marginal* errors on the parameters – both of them are determined by the length of the ridge, not its width, so both can grow substantially. An effective strategy in these cases is to rotate the parameter axes to lie along and across the ridge, and then perform the data compression using the new parameter which lies along the ridge. Keeping the likelihood surface as narrow as possible in this direction does what one desires, and the data compression can be very effective.

There is a further computational advantage to data compression; with large data sets, the covariance matrix can become numerically singular (even

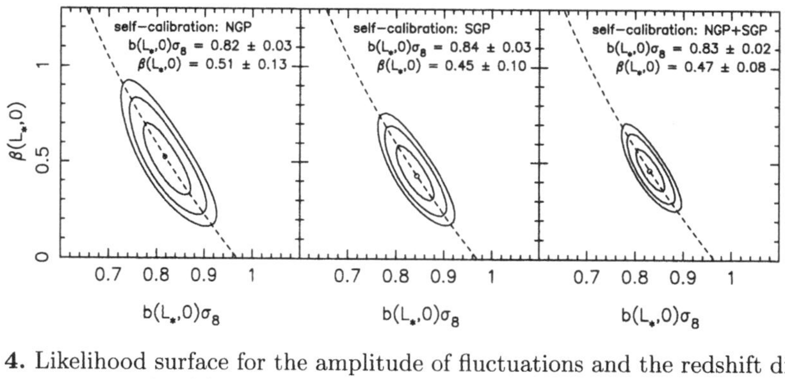

Fig. 4. Likelihood surface for the amplitude of fluctuations and the redshift distortion parameter in the AAT 2dF Galaxy Redshift Survey (from [11]), showing highly correlated parameter estimates (from two halves of the survey, and combined). The marginal errors on *both* parameters are largely determined by the length of the ridge and are largely unaffected by the width of the ridge

if physically it cannot be). Data compression, keeping the modes which contain most of the information, tends to lead to a well-behaved covariance matrix with few numerical problems.

3.2 Case II: Covariance Matrix Is Known: MOPED Algorithm

The second tractable data compression case occurs when the information comes from the mean signal $\boldsymbol{\mu}$, and no information comes from the covariance matrix C. Remarkably, this case has the potential for absolutely massive data compression, without increasing the error bars at all. The data compression is achieved with the MOPED algorithm (multiple optimized parameter estimation and data compression). It appears in [4], on which this section is based.

Examples of Case II include determining the star formation history, metallicity history, and dust content of galaxies, using their spectrum as the data set. In this case the noise is dominated by effects which are nothing to do with the galaxy, especially uncertainties in sky background subtraction and detector noise. There is a source of error which is dependent on the galaxy parameters, source photon shot noise, but this is negligible for faint sources.

As in Case I, we start with a single vector and maximize F subject to the constraint that $\boldsymbol{b}^t\mathsf{C}\boldsymbol{b} = 1$. As before

$$\mathsf{F}_{ij} = \frac{1}{2}\left(\frac{\boldsymbol{b}^t\mathsf{C}_{,i}\boldsymbol{b}}{\boldsymbol{b}^t\mathsf{C}\boldsymbol{b}}\right)\left(\frac{\boldsymbol{b}^t\mathsf{C}_{,j}\boldsymbol{b}}{\boldsymbol{b}^t\mathsf{C}\boldsymbol{b}}\right) + \frac{\left(\boldsymbol{b}^t\boldsymbol{\mu}_{,i}\right)\left(\boldsymbol{b}^t\boldsymbol{\mu}_{,j}\right)}{\left(\boldsymbol{b}^t\mathsf{C}\boldsymbol{b}\right)}, \tag{37}$$

which now simplifies because the first term disappears. Note that the denominators are simply numbers. It is clear from this expression that if we multiply $\boldsymbol{b}$ by a constant, we get the same F. This makes sense: multiplying the data by a constant factor does not change the information content. We can therefore fix the normalization of $\boldsymbol{b}$ at our convenience.

TTH showed how to solve for the case of estimation of a *single* parameter when C is known (i.e. does not depend on the parameters). MOPED generalizes this to the problem of estimating many parameters at once. For a single parameter, TTH showed that the entire data set could be reduced to a single number, with no loss of information about the parameter. We show below that if we have M parameters to estimate, then we can reduce the data set to M linear combinations, i.e. M numbers. These M numbers contain just as much information as the original data set, i.e. the data compression is lossless.

It is worth reflecting on this for a moment; it is far from obvious that this level of lossless data compression can be achieved. It is evident that the data can be *described* by M numbers (there are M parameters, after all), but it is not at all clear that one can form M linear combinations of the data which contain essentially all of the information about the parameters – as much as the entire data set. Note the use of "essentially" here – MOPED ensures that the expectation value of the likelihood surface is the same for the compressed data as for the full data, but only locally at the maximum-likelihood solution. It does not ensure that the likelihood surface is identical everywhere in parameter space. If this were so, the compressed data would be what is called "sufficient statistics" for the problem. MOPED coefficients are not, so the process is sometimes described as "locally lossless".

Note that the degree of locally lossless data compression can be very large – determined only by the ratio of data points to parameters. The CMB primary power spectrum has ~ 2500 numbers in it, yet is determined by ~ 12 cosmological parameters. The data compression factor is around 200 or so in this case.

We consider the parameters in turn. With C independent of the parameters, F simplifies, and maximizing F_{11} subject to the constraint requires

$$\frac{\partial}{\partial b_i}\left(b_j \mu_{,1\,j} b_k \mu_{,1\,k} - \lambda b_j C_{jk} b_k\right) = 0 \;, \tag{38}$$

where λ is a Lagrange multiplier, and we assume the summation convention ($j, k \in [1, N]$). This leads to

$$\boldsymbol{\mu}_{,1}(\boldsymbol{b}^t \boldsymbol{\mu}_{,1}) = \lambda \mathsf{C}\boldsymbol{b} \;, \tag{39}$$

with solution properly normalized

$$\boldsymbol{b}_1 = \frac{\mathsf{C}^{-1}\boldsymbol{\mu}_{,1}}{\sqrt{\boldsymbol{\mu}_{,1}^t \mathsf{C}^{-1}\boldsymbol{\mu}_{,1}}} \;, \tag{40}$$

and our compressed datum is the single number $y_1 = \boldsymbol{b}_1^t \boldsymbol{x}$. This solution makes sense – ignoring the unimportant denominator, the method weights high those data which are parameter sensitive and low those data which are noisy.

To see whether the compression is lossless, we compare the Fisher matrix element before and after the compression. Substitution of $\boldsymbol{b}_1$ into (37) gives

$$\mathsf{F}_{11} = \boldsymbol{\mu}_{,1}^t \mathsf{C}^{-1}\boldsymbol{\mu}_{,1} \;, \tag{41}$$

which is identical to the Fisher matrix element using the full data (19) if C is independent of θ_1. Hence, as claimed by TTH, the compression from the *entire* data set to the single number y_1 loses no information about θ_1. For example, if $\boldsymbol{\mu} \propto \theta$, then $y_1 = \sum_i \boldsymbol{x}_i / \sum_i \boldsymbol{\mu}_i$ and is simply an estimate of the parameter itself.

3.3 Estimation of Many Parameters

The problem of estimating a single parameter from a set of data is unusual in practice. Normally one has several parameters to estimate simultaneously, and this introduces substantial complications into the analysis. How can we generalize the single-parameter estimate above to the case of many parameters? We proceed by finding a second number $y_2 \equiv \boldsymbol{b}_2^t \boldsymbol{x}$ by the following requirements:

- y_2 is uncorrelated with y_1. This demands that $\boldsymbol{b}_2^t \mathsf{C} \boldsymbol{b}_1 = 0$.
- y_2 captures as much information as possible about the second parameter θ_2.

This requires two Lagrange multipliers (we normalize $\boldsymbol{b}_2$ by demanding that $\boldsymbol{b}_2^t \mathsf{C} \boldsymbol{b}_2 = 1$ as before). Maximizing and applying the constraints gives the solution

$$\boldsymbol{b}_2 = \frac{\mathsf{C}^{-1}\boldsymbol{\mu}_{,2} - (\boldsymbol{\mu}_{,2}^t \boldsymbol{b}_1)\boldsymbol{b}_1}{\sqrt{\boldsymbol{\mu}_{,2}\mathsf{C}^{-1}\boldsymbol{\mu}_{,2} - (\boldsymbol{\mu}_{,2}^t \boldsymbol{b}_1)^2}} . \tag{42}$$

This is readily generalized to any number M of parameters. There are then M orthogonal vectors $\boldsymbol{b}_m$, $m = 1, \ldots M$, each y_m capturing as much information about parameter α_m which is not already contained in y_q; $q < m$. The constrained maximization gives

$$\boldsymbol{b}_m = \frac{\mathsf{C}^{-1}\boldsymbol{\mu}_{,m} - \sum_{q=1}^{m-1}(\boldsymbol{\mu}_{,m}^t \boldsymbol{b}_q)\boldsymbol{b}_q}{\sqrt{\boldsymbol{\mu}_{,m}^t \mathsf{C}^{-1}\boldsymbol{\mu}_{,m} - \sum_{q=1}^{m-1}(\boldsymbol{\mu}_{,m}^t \boldsymbol{b}_q)^2}} . \tag{43}$$

This procedure is analogous to Gram–Schmidt orthogonalization with a curved metric, with C playing the role of the metric tensor. Note that the procedure gives precisely M eigenvectors and hence M numbers, so the data set has been compressed from the original N data down to the number of parameters M.

Since, by construction, the numbers y_m are uncorrelated, the likelihood of the parameters is obtained by multiplication of the likelihoods obtained from each statistic y_m. The y_m have mean $\langle y_m \rangle = \boldsymbol{b}_m^t \boldsymbol{\mu}$ and unit variance, so the likelihood from the compressed data is simply

$$\ln \mathcal{L}(\theta_\alpha) = \text{constant} - \sum_{m=1}^{M} \frac{(y_m - \langle y_m \rangle)^2}{2} \tag{44}$$

and the Fisher matrix of the combined numbers is just the sum of the individual Fisher matrices. Note once again the role of the fiducial model in setting the weightings $\boldsymbol{b}_m$: the orthonormality of the new numbers only holds if the fiducial model is correct. Multiplication of the likelihoods is thus only approximately correct, but iteration could be used if desired.

For our CMB example, the data are estimates of the CMB power spectrum, which for partial sky coverage are correlated. C is therefore non-diagonal, and inversion is an N^3 process. Compression by a factor of 200 then leads to a speed-up by a factor of about 10^7 in calculating the likelihood.

Figure 5 shows some of the $\boldsymbol{b}$ vectors for the study of galaxy spectra (courtesy Ben Panter). The parameters for the modelling in this case are the numbers of stars made in each of 11 ranges of lookback time, 11 metallicities of gas, and a single dust parameter. The $\boldsymbol{b}$ vectors show some features which will be recognizable. In practice, emission lines are masked out and not used, as these are not modelled in the stellar spectra.

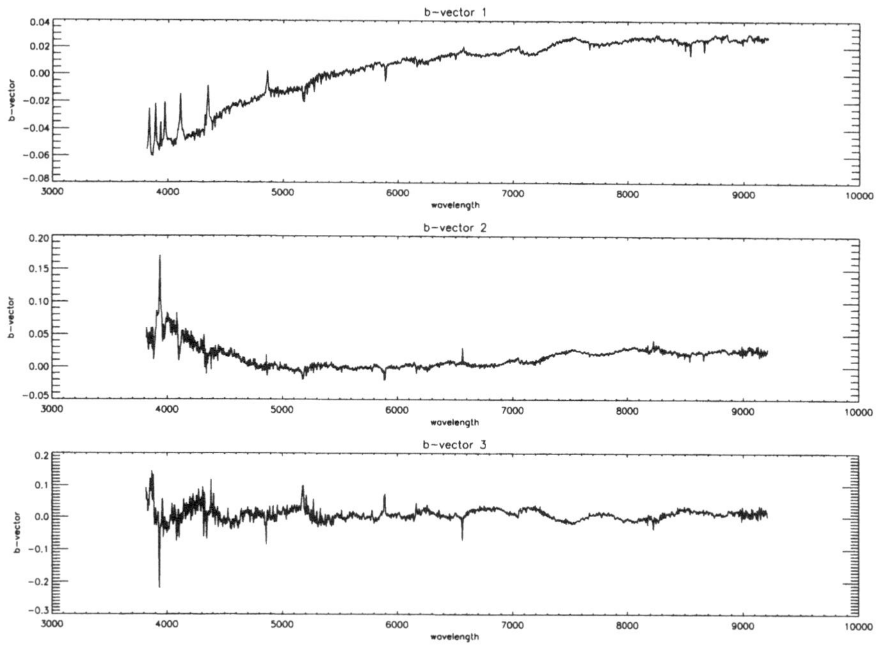

Fig. 5. The first three $\boldsymbol{b}$ vectors for galaxy spectra, for the three most recent star formation fractions (Figure courtesy Ben Panter)

Proof that the Method Can Be Lossless for Many Parameters

Under the assumption that the covariance matrix is independent of the parameters, reduction of the original data to the M numbers y_m results in no loss of information about the M parameters at all. In fact the set $\{y_m\}$ produces, on average, a likelihood surface which is locally identical to that from the entire data set – no information about the parameters is lost in the compression process. With the restriction that the information is defined locally by the Fisher matrix, the set $\{y_m\}$ is a set of sufficient statistics for the parameters $\{\theta_\alpha\}$. A proof of this for an arbitrary number of parameters is given in Appendix.

Fiducial Model

Since the MOPED vectors $\boldsymbol{b}$ depend on C and the derivatives $\boldsymbol{\mu}_{,i}$, one has to choose a set of parameters as a first guess. This "fiducial" model is in principle important, as the $\boldsymbol{y}$ vector contains as much information about the parameters only if the initial guess at the parameters is correct. In practice, one does not know beforehand what the true solution is, so one has to make an initial guess for the parameters. We compute the covariance matrix C and the gradient of the mean $(\boldsymbol{\mu}_{,i})$ for this fiducial model, to construct the $\boldsymbol{b}_i$. The Fisher matrix for the compressed data is the sum of components (37), but with the fiducial values inserted. In general this is not the same as Fisher matrix at the true solution. In practice one can iterate: choose a fiducial model, use it to estimate the parameters, and then repeat, using the estimate as the estimated parameters as the fiducial model. However, in all the practical cases investigated so far, iteration has been completely unnecessary.

4 Discussion

We have looked at how the Fisher matrix can be used to determine the best errors which a given experiment can provide on a set of theoretical model parameters. This can be very useful in designing an observational strategy – should you go wide or deep? For given observing time, you may not have both. The answer depends on what you want to measure. A simple example is set as one of the exercises.

We have also seen how the Fisher matrix may be used in data compression: by taking carefully chosen linear combinations of the data, one can reduce the size of the data set with little or (in the case of MOPED) no change to the Fisher matrix, and hence little or no increase in the error bars. A few cautionary remarks are in order. First, the analysis assumes that the errors are Gaussian; secondly, the error analysis assumes that a quadratic expansion of the $\ln \mathcal{L}$ surface is sufficient to describe the likelihood surface. This is good in the limit of small errors, but if they are large such a description may be inadequate. The next point is that the Fisher analysis only tells us about the statistical errors; in practice, systematic errors may dominate, so the Fisher errors should be used with caution. However, what is certain is that if the Fisher analysis tells you that the survey you propose is going to lead to uselessly large error bars at best, then there is no point starting.

Note that the data compression here is all linear: there are situations where nonlinear data compression can be highly effective. The obvious example here is in compressing a CMB map into a power spectrum. This is a quadratic compression and is completely lossless if the map is Gaussian – all of the statistical information is in the power spectrum.

Finally, I draw your attention to the Markov chain Monte Carlo (MCMC) technique, which is being increasingly common in astronomy. It is a quick way to explore a multidimensional likelihood surface, which works by constructing a chain of points in parameter space, where each point is randomly perturbed to give the next point. The next point is rejected or accepted, according to a criterion which ensures that in the long term, the points sample the likelihood surface with a density proportional to the likelihood. Marginalization is then trivial – one just ignores all the parameter coordinates of the points except for the parameter of interest. Priors can be easily included by giving each point in the chain a weight from the prior. For more details, see [2, 9, 16].

5 Exercises

1. Prove that $(\mathsf{C}^{-1})_{,i} = -\mathsf{C}^{-1}\mathsf{C}_{,i}\,\mathsf{C}^{-1}$.
2. Prove that $(\ln \mathsf{C})_{,i} = \mathsf{C}^{-1}\mathsf{C}_{,i}$.
3. Prove that $\ln \det \mathsf{C} = \mathrm{Trace}\ \ln \mathsf{C}$.
4. A set of N data is drawn from a population with mean μ, but with variable Gaussian errors σ_i; $i = 1 \dots N$. The mean is to be estimated from the data. Show that MOPED recovers the standard result that the data should be weighted with the inverse of the variance, and that the error on the mean is the same whether the whole data set is used or the single compressed MOPED datum.
5. A survey is proposed to determine the mean number density of a certain type of astronomical object, whose positions are random. The survey measures the number of objects in each of N independent cells, such that the mean number per cell, $\bar{n}$, is $\gg 1$. If the cells are observed to have n_i objects in them ($i = 1, \dots N$), show that the Fisher matrix (a scalar in this case) for $\bar{n}$ is

$$\mathsf{F} = \frac{N}{2\bar{n}^2} + \bar{n}N \tag{45}$$

and hence that the best survey strategy is the one which observes the most cells in the given observing time. Where is most of the information coming from, the dependence of $\boldsymbol{\mu}$ on $\bar{n}$ or C?

Hints:

1. What is $\mathsf{C}\mathsf{C}^{-1}$?
2. For matrices, $\exp(\mathsf{A}) \equiv \mathsf{I} + \mathsf{A} + \mathsf{A}^2/2! + \dots$. ln is the inverse of exp.
3. If you rotate coordinate axes (in parameter space in this case), the determinant and trace do not change.

Appendix 1: Linear Compression Algorithm Is Lossless

In this appendix, we prove that the linear compression algorithm for estimation of an arbitrary number M of parameters is lossless, provided the noise is independent of the parameters, $\mathsf{C}_{,\alpha} = 0$. Specifically, loss-free means the Fisher matrix for the set of M numbers $y_m = \boldsymbol{b}_m^t\boldsymbol{x}$ is identical to the Fisher matrix of the original data set $\boldsymbol{x}$:

$$F^O_{\alpha\beta} = \langle\alpha|\beta\rangle \,. \tag{46}$$

By construction, the y_m are uncorrelated, so the likelihoods multiply and the Fisher matrix for the set $\{y_m\}$ is the sum of the derivatives of the log-likelihoods from the individual y_m:

$$F_{\alpha\beta} = \sum_m F_{\alpha\beta}(m) \,. \tag{47}$$

From (37),

$$F_{\alpha\beta}(m) = (\boldsymbol{b}^t_m \boldsymbol{\mu}_{,\alpha})(\boldsymbol{b}^t_m \boldsymbol{\mu}_{,\beta}) \,. \tag{48}$$

With (43), we can write

$$\boldsymbol{b}^t_m = \frac{\boldsymbol{\mu}^t_{,m}\mathsf{C}^{-1} - \sum_{q=1}^{m-1}\left(\boldsymbol{b}^t_q \boldsymbol{\mu}^t_{,m}\right)\boldsymbol{b}_q}{\sqrt{\langle m|m\rangle - \sum_{q=1}^{m-1}\left(\boldsymbol{b}^t_q \boldsymbol{\mu}^t_{,m}\right)^2}} \,. \tag{49}$$

Hence

$$F_{\alpha\beta}(m) = \left[\langle\alpha|m\rangle - \sum_{q=1}^{m-1} F_{\alpha m}(q)\right] \times \frac{\left[\langle\beta|m\rangle - \sum_{q=1}^{m-1} F_{\beta m}(q)\right]}{\left[\langle m|m\rangle - \sum_{q=1}^{m-1} F_{mm}(q)\right]} \,. \tag{50}$$

Consider first $\beta = m$:

$$F_{\alpha m}(m) = \langle\alpha|m\rangle - \sum_{q=1}^{m-1} F_{\alpha m}(q)$$
$$\Rightarrow F_{\alpha M} = \sum_{q=1}^{M} F_{\alpha M}(q) = \langle\alpha|M\rangle = F^O_{\alpha M} \,, \tag{51}$$

proving that these terms are unchanged after compression. We therefore need to consider $F_{\alpha\beta}(m)$ for α or $\beta < m$. First we note that

$$F_{\alpha\beta}(m) = \frac{F_{\alpha m}(m)F_{m\beta}(m)}{F_{mm}(m)} \tag{52}$$

and, from (51),

$$\sum_{q=1}^{\beta} F_{\alpha\beta}(q) = \langle\alpha|\beta\rangle \,. \tag{53}$$

We want the sum to extend to M. However, the terms from $\beta + 1$ to M are all zero. This can be shown as follows: (52) shows that it is sufficient to show that $F_{\alpha m}(m) = 0$ if $m > \alpha$. Setting $\beta = m$ in (53), and reversing α and m, we get

$$\sum_{\alpha+1}^{m} F_{\alpha m}(q) = 0 \,. \tag{54}$$

Now, the contribution from q does not depend on derivatives with higher numbered parameters, so we can evaluate $F_{\alpha m}(\alpha + 1)$ by setting $m = \alpha + 1$.

The sum (54) implies that this term is zero. Increasing m successively by one up to M, and using (54), proves that all the terms are zero, proving that the compression is lossless.

Appendix 2: Solutions to Exercises

1. Prove that $(\mathsf{C}^{-1})_{,i} = -\mathsf{C}^{-1}\mathsf{C}_{,i}\,\mathsf{C}^{-1}$
   Since $\mathsf{C}\mathsf{C}^{-1} = \mathsf{I}$, its derivative is zero. Hence $\mathsf{C}(\mathsf{C}^{-1})_{,i} + \mathsf{C}_{,i}\mathsf{C}^{-1} = 0$. Result follows after premultiplication by C^{-1}.
2. Prove that $(\ln \mathsf{C})_{,i} = \mathsf{C}^{-1}\mathsf{C}_{,i}$.
   Let $\mathsf{A} = \ln \mathsf{C}$, so
   $$\mathsf{C} = \exp \mathsf{A} = \mathsf{I} + \mathsf{A} + \frac{\mathsf{A}^2}{2!} + \ldots = \sum_{n=0}^{\infty} \frac{\mathsf{A}^n}{n!} .$$
   Hence
   $$\mathsf{C}_{,i} = \sum_{n=1}^{\infty} \frac{\mathsf{A}^{n-1}}{(n-1)!}\,\mathsf{A}_{,i} = \sum_{m=0}^{\infty} \frac{\mathsf{A}^m}{m!}(\ln \mathsf{C})_{,i} = \mathsf{C}(\ln \mathsf{C})_{,i}$$
   and the result follows.
3. Prove that $\ln \det \mathsf{C} = \mathrm{Trace}\, \ln \mathsf{C}$.
   C is a symmetric matrix and therefore be diagonalized. In the new basis (where we have a new set of parameters which are linear combinations of the old ones), the diagonal components of C are strictly positive (they are variances of the new parameters).
   Since the trace and determinant are unchanged by the diagonalization, we can prove the result in the rotated system. If C is diagonal, then $\ln \mathsf{C}$ is diagonal, with components $\ln \mathsf{C}_{11},\ \ln \mathsf{C}_{22}, \ldots$. So $\mathrm{Tr} \ln \mathsf{C} = \sum_n \mathsf{C}_{nn}$. Since $\det \mathsf{C} = \prod_n \mathsf{C}_{nn}$,
   $$\ln \det \mathsf{C} = \sum_n \ln \mathsf{C}_{nn} = \mathrm{Tr} \ln \mathsf{C} .$$
   This proof is rigorous, but there may be a neater solution without diagonalization.
4. For a single parameter, the single MOPED vector is given by (40): $\boldsymbol{b} = \frac{\mathsf{C}^{-1}\boldsymbol{\mu}_{,1}}{\sqrt{\boldsymbol{\mu}_{,1}^t\mathsf{C}^{-1}\boldsymbol{\mu}_{,1}}}$. $\boldsymbol{\mu}_{,1}$ is just a vector of ones, and the inverse of C is $\mathrm{diag}(1/\sigma_1^2, 1/\sigma_2^2, \ldots, 1/\sigma_N^2)$, so
   $$\boldsymbol{b} = \frac{1}{\sqrt{\sum \frac{1}{\sigma_i^2}}} \begin{pmatrix} 1/\sigma_1^2 \\ \vdots \end{pmatrix} . \qquad (55)$$
   The multiplier is of no great importance, but we see that each datum is inverse-variance weighted. The Fisher matrix for the full data set is

obtained from (19), where $\mathsf{C}_{,1} = 0$, and M_{ij} is a square matrix full of ones. This gives $\mathsf{F} = \sum(1/\sigma_i^2)$. For the compressed data set,

$$\mathsf{F}^{(y)} = (\boldsymbol{b}^t \boldsymbol{\mu}_{,1})^2 = \frac{\left[(1/\sigma_1^2, \ldots) \begin{pmatrix} 1 \\ \vdots \end{pmatrix}\right]^2}{\sum(1/\sigma_i^2)} = \sum(1/\sigma_i^2) \,, \tag{56}$$

which is the same as for the full data set. This is of course a standard example which does not need MOPED, which can be used for much more general problems, involving arbitrarily many parameters and arbitrary dependence of the data on the parameters.

5. For $\bar{n} \gg 1$, we can approximate the Poisson distribution by a Gaussian, with $\langle n_i \rangle = \bar{n}$, and $\sigma_i^2 = \bar{n}$. Hence $\boldsymbol{\mu}^t = \bar{n}(1, 1, \ldots 1)$ and $\mathsf{C} = \bar{n}^{-1}\mathrm{diag}(1, 1, \ldots, 1)$. Hence $\mathsf{C}_{,1} = (-1/\bar{n}^2 a \mathrm{diag}(1, 1, \ldots, 1))$, and M_{ij} is an $N \times N$ matrix of 2 s. The result

$$\mathsf{F} = \frac{N}{2\bar{n}^2} + \bar{n}N \tag{57}$$

follows. The first term arises from $\mathsf{C}_{,1}$, and the second from $\boldsymbol{\mu}_{,1}$. The second dominates since $\bar{n} \gg 1$.

References

1. Fisher, R.A.: J. Roy. Stat. Soc. **98**, 39 (1935)
2. Gilks, W.R., Richardson, S., Spiegelhalter, D.J.: Markov Chain Monte Carlo in Practice. Chapman & Hall, London (1996)
3. Gupta, S., Heavens, A.F.: MNRAS **334**, 167 (2002)
4. Heavens, A.F., Jimenez, R., Lahav, O.: MNRAS **317**, 965 (2000)
5. Heavens, A.F., Panter, B.D., Jimenez, R., Dunlop, J.S.: Nature **428**, 625 (2004)
6. Jimenez, R., Panter, B.D., Heavens, A.F., Verde, L.: MNRAS **356**, 495 (2005)
7. Kendall, M.G., Stuart, A.: The Advanced Theory of Statistics, vol. II. Griffin, London (1969)
8. Kenney, J.F., Keeping, E.S.: Mathematics of Statistics, Part II, 2nd edn. Van Nostrand, New York (1951)
9. Lewis, A., Bridle, S.: PRD **66**, 103511 (2002)
10. Panter, B.D., Heavens, A.F., Jimenez, R.: MNRAS **355**, 764 (2004)
11. Percival, W.J., et al.: MNRAS **353**, 1201 (2004)
12. Press, W., et al.: Numerical Recipes in Fortran. CUP, Cambridge, U.K. (1992)
13. Reichardt, C., Jimenez, R., Heavens, A.F.: **327**, 849 (2001)
14. Tegmark, M., Taylor, A.N., Heavens, A.F.: ApJ **480**, 22 (1997) (TTH)
15. Therrien, C.W.: Discrete Random Signals and Statistical Signal Processing. Englewood Cliffs, Prentice Hall (1992)
16. Verde, L., et al.: ApJS **148**, 195 (2003)
17. Vogeley, M., Szalay, A.: ApJ **465**, 34 (1996)

Data Compression, Classification and Parameter Estimation. Methods: Examples from Astronomy

O. Lahav

Department of Physics and Astronomy, University College London, Gower Street, London WC1E 6BT, UK
lahav@star.ucl.ac.uk

Abstract. This is a brief summary of review talks given at the Valencia Summer School on *Data Analysis in Cosmology* and the 2005 UK National Astronomical Meeting. We describe the following examples: galaxy classification with principle component analysis, photometric redshift estimation with artificial neural networks and multi-parameter estimation with Markov chain Monte Carlo in a Bayesian framework.

1 Introduction

The dramatic increase in data in astronomy has renewed the interest in the principles and applications of statistical methods. The statistical methods can be viewed as a bridge between the data and the models, i.e. the same statistic can be applied to both data and models, as an effective way of comparing the two. It is well known that there is an ongoing debate between the "Frequentist" approach, where the probability is interpreted as the frequency of the outcome of a repeatable experiment, and the "Bayesian" approach where the interpretation of probability is more general and includes a degree of belief. In the Bayesian approach, the choice of *priors* might affect quite strongly the inference, but it is an "honest" approach in the sense that all the assumptions are spelled out in a logical manner.

Conceptually, it is helpful to distinguish between three goals commonly addressed in astronomy and other physical sciences:

- Data compression
- Classification
- Parameter estimation

The three might of course be related. For example, estimation of cosmological parameters from the cosmic microwave background or galaxy redshift surveys is commonly deduced from the compressed information, usually in the form of

Lahav, O.,: *Data Compression, Classification and Parameter Estimation. Methods: Examples from Astronomy*. Lect. Notes Phys. **665**, 73–76 (2009)
DOI 10.1007/978-3-540-44767-2_3

the angular and three-dimensional power spectra, respectively. As another example, galaxy classification can be done in a compressed space of the spectra, or in the space of astrophysical parameters estimated from the spectra.

This summary is in the form of an extended abstract rather than a comprehensive review. We describe and provide references to examples of applications of principle component analysis, artificial neural networks and Markov chain Monte Carlo parameter estimation.

2 Principal Component Analysis (PCA) of Galaxy Spectra

In machine learning a distinction is made between *unsupervised* methods (where the data "speak for themselves", in a model-independent way) and *supervised* methods (based on training sets of models, or other data sets).

Principal Component Analysis (PCA) is an example of an unsupervised method. PCA has previously been applied to data compression and classification of spectral data of stars (e.g. Murtagh & Heck [11]), QSO (e.g. Francis et al. [6]) and galaxies (e.g. Connolly et al. [2]; Lahav [7]; Madgwick et al.[9] and references therein). A galaxy spectrum, like any other vector, can be thought of as a point in a high-dimensional parameter space. One may wish for a more compact description of the data. By identifying the *linear* combination of input parameters with maximum variance, PCA finds the principal components that can be most effectively used to characterize the spectra.

The eigenvectors onto which we project the spectra can be viewed as optimal filters of the spectra, in analogy with other spectral diagnostics such as colour filter or spectral index. For example, by comparison with spectra from models, Madgwick et al. [10] have shown that a combination of the first and second principal components could be interpreted in terms of the galaxy star formation history.

3 Artificial Neural Networks for Photometric Redshifts

We describe now an example of supervised learning. In the absence of spectroscopic data, redshifts of galaxies may be estimated using multi-band photometry, which may be thought of as very low-resolution spectroscopy. Photometric redshift methods rely on measuring the signal in the photometric data arising from prominent "break" features present in galaxy spectra, e.g. the 4000 Å break in red, early-type galaxies or the Lyman break at 912 Å in blue, star-forming galaxies. There are two basic approaches to measuring galaxy photometric redshifts (e.g. Csabai et al. [3]; Padmanabhan et al. [12] and references therein). The first, template matching, relies on fitting model galaxy spectral energy distributions (SEDs) to the photometric data, where the models span a range of expected galaxy redshifts and spectral types. The second approach utilizes an existing spectroscopic redshift sample as a training set to derive an empirical photometric redshift fitting relation. There are advantages and disadvantages to each approach, as well as a number of variants and hybrids of these basic techniques (e.g. Csabai [3]). Photo-z methods proved very useful, e.g. in the recent analysis of the GOODS and COMBO-17 surveys.

The training-based method ANNz which utilizes artificial neural networks was developed by Firth, Lahav & Somerville [5] and Collister & Lahav [1]. ANNz was made freely available by the authors[1] and it will soon be incorporated as a standard tool in Astro-Grid. When applied to SDSS, the rms error using ANNz is $\delta z = 0.02$ compared with $\delta z = 0.07$ using a template method. For reference we note that some authors quote the photo-z error as $\sigma_z = \delta z/(1+z)$. The photo-z methods are crucial for the analysis of future deep imaging surveys such as the Dark Energy Survey and VISTA. While the redshift error per galaxy is relatively large, having a large number of galaxies could reduce the errors on measures of the galaxy clustering. Photometric surveys over large areas of the sky may compete well with spectroscopic surveys.

4 Parameter Estimation of Markov Chain Monte Carlo

Astronomers have faced a growing number of free parameters in modelling astrophysical systems, e.g. cosmological parameters or extra-solar planet orbital parameters. In the case of a model with N free parameters marginalizing over $N-1$ parameters is computationally expensive if the parameter space is mapped into a grid. An alternative method, the Markov chain Monte Carlo (MCMC), has been known since the 1950s and a wide range of methods exist in the literature to implement it, e.g. the Metropolis–Hasting algorithm.

The key idea is to turn a probability distribution function in N dimensions into a cloud of points which represents the probability distribution function. The probability distribution function could incorporate the probabilities for the priors, in the Bayesian spirit. The MCMC algorithm constructs a random walk in the model parameter space with steps drawn from a multi-dimensional proposal distribution (e.g. a Gaussian). It is crucial to verify convergence, i.e. that the parameter space is fully sampled, in particular if there are several peaks in a highly dimensional space.

MCMC algorithms have been applied recently to parameter estimation from the cosmic microwave background and other cosmological data sets (e.g. Lewis & Bridle [8]; Verde et al. [13]) and to both detecting and characterizing orbits of extrasolar planets (e.g. Ford [4]).

[1] `http://www.star.ucl.ac.uk/~lahav/annz.html`

References

1. Collister, A., Lahav, O.: PASP **116**, 345 (2004)
2. Connolly A., Szalay, A., Bershady, M., Kinney, A., Calzetti, D.: AJ **110**, 1071 (1995)
3. Csabai, I., et al.: AJ **125**, 580 (2003)
4. Ford, E.B.: ApJ **642**, 505 (2006)
5. Firth, A.E., Lahav, O., Somerville, R.S.: MNRAS **339**, 1195 (2003)
6. Francis, P.J., Hewett, P.C., Foltz, G.B., Chaffee, F.H.: ApJ **398**, 476 (1992)
7. Lahav, O.: In: Banday, A. J., Zaroubi, S., Bartelmann, M. (ed.) Mining the Sky, Proceedings of the MPA/ESO/MPE Workshop Held at Garching, Germany, P. 33. Springer-Verlag, Heidelberg (2001).
8. Lewis, A., Bridle, S.L.: PRD **66**, 103511 (2002)
9. Madgwick, D.S., Lahav, O., the 2dFGRS team: MNRAS **333**, 133 (2002)
10. Madgwick, D.S., Somerville, R., Lahav, O., Ellis, R.S.: MNRAS **343**, 871 (2003)
11. Murtagh, F., Heck, A.: Multivariate Data Analysis. Astrophysics and Space Science Library, Reidel, Dordrecht (1987)
12. Padmanabhan, N., et al.: MNRAS **359**, 237 (2005)
13. Verde, L., the WMAP Team: ApJS **148**, 196 (2003)

Part II

Statistics of Cosmic Background Radiation

Cosmic Microwave Background Anisotropies: The Power Spectrum and Beyond

E. Martínez-González

Instituto de Física de Cantabria, CSIC-Universidad de Cantabria, Avenida Los Castros s/n, 39005 Santander, Spain
martinez@ifca.unican.es

Most of the cosmological information extracted from the CMB has been obtained through the power spectrum; however, there is much more to be learnt from the statistical distribution of the temperature random field. We review some recent developments in the study of the cosmic microwave background (CMB) anisotropies and present a description of the novel tools developed to analyse the properties of the CMB anisotropies beyond the power spectrum.[1]

1 Introduction

The standard scenario of the Universe includes an early inflationary phase during which the Universe experienced a drastic expansion. As a consequence of this inflationary period the homogeneity, isotropy and flatness of the Universe that we observe today can be understood as a natural outcome, without a fine tuning of the initial conditions. Besides, quantum fluctuations of the fields dominating the dynamics at that early phase constitute the seeds which, after gravitational growing, will form the large-scale structure (LSS) of the Universe at the present time (for a detailed description of inflation and its consequences see, e.g. [126]). The statistical distribution of the quantum fluctuations of the field responsible for inflation, known as inflaton, in the vacuum state is Gaussian. In addition, the energy density fluctuations generated from them are also Gaussian since they are related by linear theory. Moreover, the

[1] This review was almost complete before the second release of WMAP data which was made publicly available on March 2006. The major improvement has been made in the polarization data which are now much better understood regarding both systematics and foregrounds. This has implied a significant change in the value of the optical depth ($\tau \approx 0.09$). Another difference with respect to the values of the cosmological parameters estimated from the first-year data is the possible significant deviation of the spectral index from unity as derived from the second release of data. Apart from these changes the rest of the parameters as well as the deviations from Gaussianity ("anomalies") remain very much unaltered.

Martínez-González, E.: *Cosmic Microwave Background Anisotropies: The Power Spectrum and Beyond*. Lect. Notes Phys. **665**, 79–120 (2009)
DOI 10.1007/978-3-540-44767-2_4

anisotropies of the CMB are related to the energy density fluctuations by the linearized Einstein–Boltzmann equations, being therefore Gaussian distributed as well. This is a very important result since, for this particular statistical distribution, all the properties of the temperature random field are included in the second-order moment. In harmonic space this is called the radiation power spectrum, usually denoted by C_ℓ.

The cosmological parameters which characterize the Universe leave their imprint in the radiation power spectrum. In the standard inflationary scenario we only need to measure this second-order moment to obtain all the possible information carried out by the CMB. This is in fact what many experiments have been doing with increasing precision in recent years. The BOOMERANG [57] and MAXIMA [80] experiments onboard stratospheric balloons were among the first to establish the spatial flatness of the Universe. Most notably, the NASA Wilkinson Microwave Anisotropy Probe (WMAP) has measured the C_ℓ with high precision up to the multipole $\ell \approx 600$ [21]. This measurement has implied a high improvement in the determination of the cosmological parameters, reducing the uncertainties below 10% [199]. The uncertainties are further reduced when other cosmological data sets, like the LSS distribution of galaxies and SN Ia magnitude-redshift data, are combined with the CMB anisotropies [205]. Moreover, the degeneracy among some parameters present when only a single data set is considered is broken. This *concordance model* has been further confirmed by the first detection of polarization fluctuations by DASI [108] and the further determination of the polarization power spectra by WMAP [107], DASI [121], CBI [172] and BOOMERANG [148, 164]. An independent piece of evidence comes from the cross-correlation between CMB anisotropies and fluctuations in the galaxy number density. Several authors have found positive cross-correlations between the WMAP data and the NVSS radio galaxy survey [46], implying that the model of the Universe is different from Einstein–de Sitter and, in the case of a flat Universe, that the Universe is dominated by a dark energy component [8, 28, 69, 142, 154, 216].

Measurements of the statistical distribution of the CMB temperature and polarization random field represent a unique probe of the early history of the Universe. Thus, if the distribution is consistent with Gaussianity at a high precision then it would imply a nice confirmation of the standard inflationary model. On the contrary, if some significant deviations from normality are found then this result would contradict the standard single-field model and motivate the search for alternative scenarios of the early history of the Universe. There have already been many suggestions of the latter in the past based on many different physical phenomena, most notably, non-standard inflation models based on extensions of the standard one in many different ways, -e.g., multifield scenarios as the curvaton mechanism [130], inhomogeneous reheating scenario, etc. (see [20] for a review)- and cosmic defects including global and local strings, monopoles, domain walls and textures [217].

The structure of the paper is as follows. In Sect. 2 we summarize the most relevant physical effects which generate the CMB temperature and polarization anisotropies. Section 3 deals with the constraints imposed by recent CMB data sets on the cosmological parameters as well as other cosmological data sets combined with the CMB. In Sect. 4 we describe the properties of the CMB anisotropies beyond the power spectrum and discuss possible sources of non-Gaussianity, inhomogeneity and anisotropy in their distribution on the celestial sphere. In Sect. 5 we discuss the methods developed for studies of the CMB temperature distribution and of its global isotropy whereas in Sect. 6 we review the analyses performed so far on CMB data sets measured by different experiments. Finally the conclusions are given in Sect. 7.

2 Theory

The CMB anisotropies are produced by different physical effects acting on this radiation before the last-scattering surface, when the Universe was around 380000 years old. These *primary anisotropies* contain very valuable information on the early history of the Universe and can be observed today almost unaffected. After the temperature of the photons drop below some 3000 degrees they are not able to ionize the hydrogen atoms and propagate freely. The physical effects producing the primary anisotropies at and after recombination can be summarized by the following equation [134, 184]:

$$\frac{\Delta T}{T}(\boldsymbol{n}) \approx \frac{\phi_e(\boldsymbol{n})}{3} + 2\int_e^o \frac{\partial \phi}{\partial t} dt + \boldsymbol{n} \cdot (\boldsymbol{v}_o - \boldsymbol{v}_e) + \left(\frac{\Delta T}{T}(\boldsymbol{n})\right)_e . \qquad (1)$$

The gravitational redshift suffered by the photons in their travel from the last-scattering surface to the observer is given by the first two terms in the r.h.s. of (1). They are known as *Sachs–Wolfe* and *late integrated Sachs–Wolfe* effect, respectively [182]. The velocity of the baryon–photon fluid at recombination generates a Doppler effect. The intrinsic temperature of the photons at recombination, represented by the fourth term in the r.h.s. of (1), also contributes to the total anisotropy. Equation (1) accounts for the anisotropies at recombination. Before recombination, changes in the gravitational potential, due to the imperfect baryon–photon coupling, can also produce a gravitational redshift (the *early integrated Sachs–Wolfe effect*). The accurate computation of all the contributions requires to solve the linearized coupled Einstein–Boltzmann equations. Several codes have been developed to numerically solve those equations as a function of the cosmological parameters (about 10, see below), being some of them publicly available (e.g. CMBFAST [192], CAMB [123]). A comparison among the results of different codes showed that the accuracy achieved is very good, reaching $\approx 0.1\%$ up to $\ell = 3000$ [191].

As will be explained below, a very relevant statistical quantity to compute is the 2-point correlation function of the temperature anisotropies or equivalently the correlation of the spherical harmonic coefficients:

$$< a_{\ell m} a^*_{\ell' m'} >= C_\ell \delta_{\ell\ell'} \delta_{mm'} \ , \tag{2}$$

where C_ℓ is the anisotropy power spectrum and the a_{lm} are the coefficients of the spherical harmonic expansion

$$\frac{\Delta T}{T}(\boldsymbol{n}) = \sum_{\ell,m} a_{\ell m} Y_{\ell m}(\boldsymbol{n}). \tag{3}$$

The homogeneity and isotropy of the Universe have been assumed in (2), i.e. the a_{lm} are not correlated for different l or m. In Fig. 1 the scales at which the different effects dominate the power spectrum C_ℓ are marked. As we can see, the gravitational effects dominate at the lower multipoles (large angular scales). At intermediate scales, $100 < \ell < 1000$, the spectrum is dominated by several oscillations, usually called *acoustic peaks.* These peaks appear as a consequence of the balance between the gravitational force and the radiation pressure. At the smallest scales, $\ell > 1000$, the C_ℓ are damped because of the width of recombination and the imperfections in the coupling of the photon–baryon fluid (*Silk effect*, [194]).

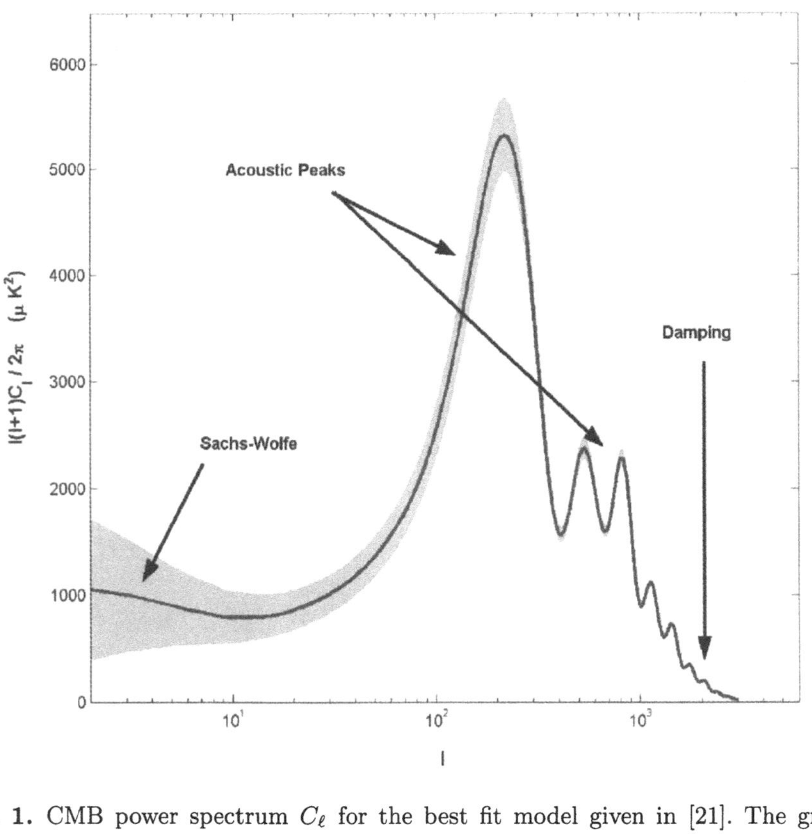

Fig. 1. CMB power spectrum C_ℓ for the best fit model given in [21]. The grey band represents the cosmic variance. The spectrum has been computed with the CMBFAST code [192]

When the small scales in the matter distribution become non-linear and their collapse gives rise to the formation of the first stars and quasars they start to reionize the surrounding matter. This ionized matter can interact again with the microwave photons and produce new anisotropies. *Secondary anisotropies* can be produced either by the scattering with free electrons or by the gravitational effect of the matter density evolution. The scattering can leave a very clear imprint when the photons happen to cross the core of rich galaxy clusters where the electron density and temperature are very high (of several $10^{-3}\mathrm{cm}^{-3}$ and $\sim 10^8$K, respectively). In this case the electrons inject energy to the photons through inverse Compton scattering, producing a distortion of the black-body spectrum. This effect is called *the Sunyaev–Zeldovich* effect [200].

The evolving gravitational wells in the large-scale structure produce a gravitational redshift in the photons. This is known as the *late integrated Sachs–Wolfe* effect [182] or *the Rees–Sciama* effect when the evolution is non-linear [180]. Besides, the trajectory of the photons is lensed by the same gravitational wells producing a noticeable effect on the CMB anisotropies at arcmin angular scales. One important issue in the observation of the CMB anisotropies is therefore to distinguish between primary and secondary anisotropies. In the case of the SZ effect, as the frequency dependence is different from the Planckian one, it is possible to separate the effect from the intrinsic CMB anisotropies by using multifrequency observations. However, for the lensing effect this is not possible and other properties, such as high-order correlations in the CMB temperature and polarization fields, have to be used (see, e.g. [88, 74, 193]).

In the same manner that anisotropies in the temperature are expected in the gravitational instability scenario for structure formation, anisotropies in the polarization of the CMB are also expected. The physical effects giving rise to these anisotropies are, however, different. Linear polarization is generated by Thompson scattering at the end of recombination when the growth of the mean free path of the photons allows temperature anisotropies to grow. In particular a quadrupole is formed in the reference frame of each electron producing polarization after the scattering (see, e.g. [41]). The expected level of polarization is, however, small $\approx 5\%$. As shown by [230, 99] two rotationally invariant quantities E, B can be constructed from the Stokes parameters Q, U. The different behavior of E, B under parity transformations implies that only three spectra are needed to characterize CMB polarization: $C_\ell^E, C_\ell^B, C_\ell^{TE}$.

In addition to the anisotropies generated by physical effects associated to the energy–matter density perturbations (*scalar perturbations*), new anisotropies can be generated by a background of primordial gravitational waves (*tensor perturbations*). This background is also a generic prediction of inflation. The primordial power spectra for both scalar and tensor perturbations are usually characterized by a scale-free law of the form:

$$P_s(k) = A_s(k/k_0)^{n_s}, \quad P_t(k) = A_t(k/k_0)^{n_t}, \quad k_0 = 0.05Mpc^{-1} \ . \tag{4}$$

There is no observational evidence of the primordial background of gravitational waves yet. Upper limits have been imposed by the combination of WMAP and other high-resolution CMB data with large-scale structure data implying a tensor-to-scalar amplitude ratio $r \equiv A_t/A_s < 0.9$ [21], see next sections. In any case, the amplitude of both temperature and polarization power spectra produced by tensor perturbations are believed to be several orders of magnitude below the ones corresponding to the scalar perturbations. However, the B-mode can only be generated by gravitational waves and thus its detection represents a unique evidence of the existence of this primordial background. This is the reason why there is now so much interest in planning new more sensitive CMB experiments to detect polarization (see, e.g. [27]). The detection of the B-mode is expected to be much harder than the already detected E-mode because of its intrinsically smaller amplitude and the relatively larger foreground emissions (expected from galactic synchrotron and thermal dust and extragalactic radio sources; for a recent discussion of these issues see [208]). An additional complication comes from the lensing conversion of the E-mode to the B-mode [231] which is expected to dominate over the primordial B-mode at multipoles $\ell \gtrsim 100$ [104, 101]. Several methods have been developed to reconstruct the gravitational lensing potential from the E and B-mode polarization correlations and remove the lensing contamination [91, 102, 190]. A more technical problem is the separation of the mixing of E and B-modes for observations with partial sky coverage for which other methods have been proposed [124, 32, 42].

Outside the inflationary scenario there are other possibilities to generate B-modes. For instance, cosmic strings or primordial magnetic fields would generate a vector and a vector plus tensor components of metric perturbations, respectively, contributing to the B-mode polarization (see, e.g. [166, 125]). It is important to remark, however, that in the case of cosmic strings and other topological defects their role as seeds for the large-scale structure formation is already very much constrained by the observed C_ℓ, see below.

In Fig. 2 the different temperature and polarization power spectra corresponding to the primary anisotropies for the concordance model are plotted for the scalar and tensor perturbations assuming a value of $r = 0.01$. The power of the tensor perturbations is a few orders of magnitude below the corresponding scalar power spectra for $\ell \lesssim 100$ and decays strongly for larger multipoles.

Many parameters are needed to characterize the standard cosmological model. Their variation changes the amplitude and shape of the temperature and polarization power spectra in many different ways (see Fig. 3 for their effect on the temperature power spectrum). The cosmological parameters can be classified depending on whether they characterize the background Universe or the primordial power spectrum.

The background Friedman–Robertson–Walker Universe and its matter and energy content are determined by the following parameters: *physical baryonic density*, $w_b = \Omega_b h^2$; *physical matter density*, $w_m = \Omega_m h^2$, where Ω_m is the matter density parameter including the contributions from baryons, cold

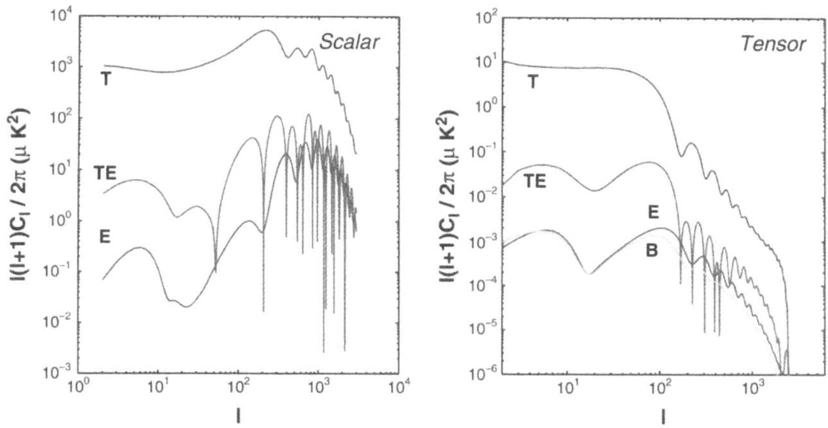

Fig. 2. Angular power spectra for temperature and polarization. The C_ℓs are plotted for scalar perturbations on the left and for tensor perturbations on the right, where a tensor/scalar ratio of $r = 0.01$ has been chosen. The rest of the parameters have been fixed to the best fit model of [21]. All the spectra have been computed with the CMBFAST code

dark matter and neutrinos, $\Omega_m = \Omega_b + \Omega_{CDM} + \Omega_\nu$; *physical neutrino density*, $w_\nu = \Omega_\nu h^2$; *dark energy equation of state parameter*, $w \equiv p_{DE}/\rho_{DE}$; *dark energy density*, Ω_{DE} (in case $w = -1$ the dark energy takes the form of a cosmological constant and its energy contribution is denoted by Ω_Λ) and *the Hubble constant*, $h \equiv H_0/100\text{km s}^{-1}\ \text{Mpc}^{-1}$.

The scalar and tensor primordial power spectra are characterized by the following parameters: *amplitude of the primordial scalar power spectrum*, A_s, defined by $P_s(k) = A_s(k/k_0)^{n_s}$, where $k_0 = 0.05\text{Mpc}^{-1}$; *scalar spectral index*, n_s; *running index*, $\alpha = dn_s/d\ln k$, normally determined at the scale $k_0 = 0.05\text{Mpc}^{-1}$; *tensor-to-scalar ratio*, $r = A_t/A_s$; *tensor spectral index*, n_t, which is normally assumed to be $n_t = -r/8$ from the consistency relation of inflation (see, e.g. [126]).

There is an extra parameter that accounts for the reionization history of the Universe: *the optical depth*, $\tau = \sigma_T \int_{t_r}^{t_0} n_e(t)dt$, where σ_T is the Thompson cross-section and $n_e(t)$ is the electron number density as a function of time. Besides, there are two possible types of matter density fluctuations: *adiabatic* which preserves the entropy per particle and *isocurvature* which preserves the total energy density. The standard inflationary scenario predicts fluctuations of the adiabatic type.

It is important to note that the manner in which some of the parameters enter in the calculation of the C_ℓ produces degeneracies. In particular, there is the well-known geometrical degeneracy involving Ω_m, Ω_Λ and Ω_k, where Ω_k is the curvature density parameter, $\Omega_k \equiv 1 - \Omega_m - \Omega_\Lambda$ (an example of this degeneracy can be found in Fig. 5 of [138]). This fact makes clear the need to combine different cosmological data sets to break those degeneracies, as will be shown below.

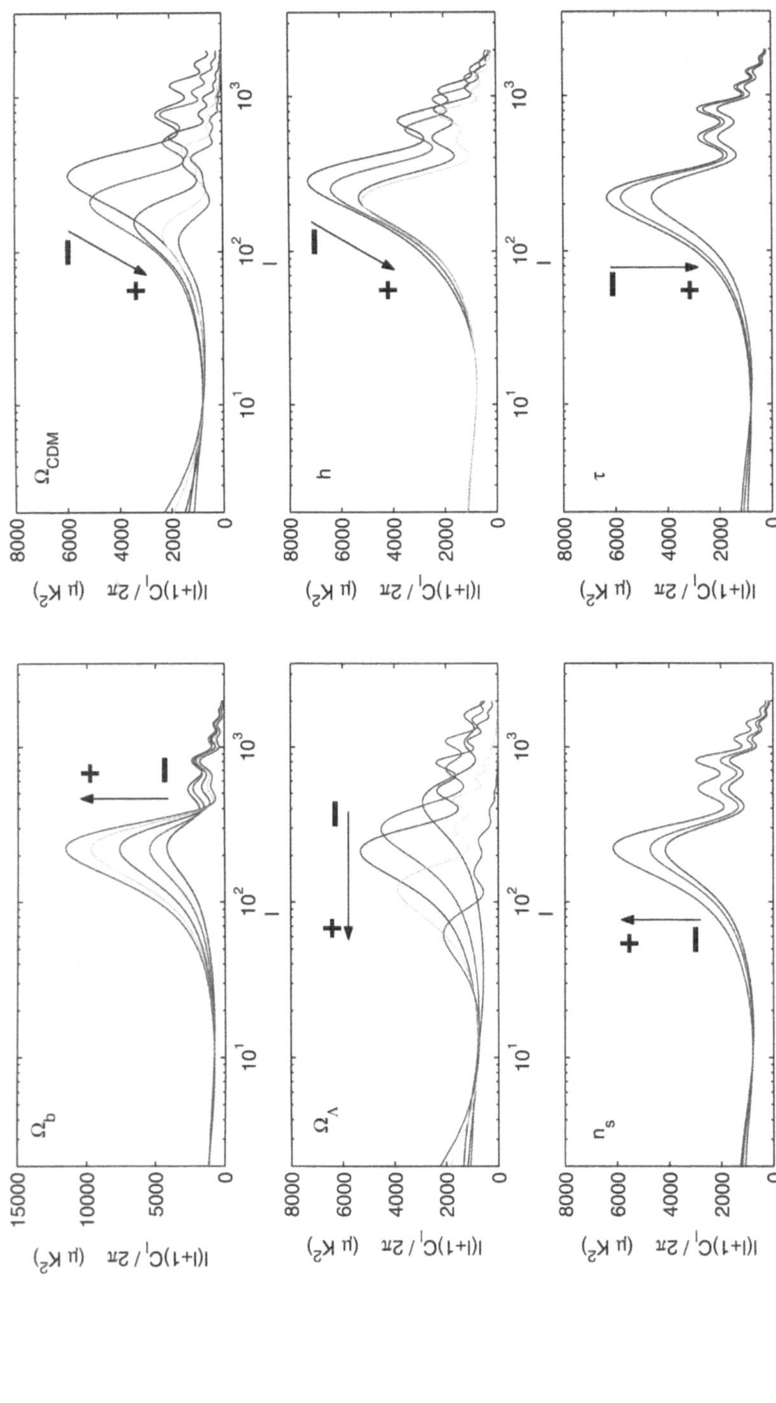

Fig. 3. Dependence of the temperature power spectrum C_ℓ on some relevant cosmological parameters (the spectra have been computed with the CMBFAST code)

3 Cosmological Constraints

Several issues are important to bear in mind in the analysis of the CMB anisotropy data. First of all, the microwave sky is interpreted as a realization of the temperature anisotropy random field whose origin is in the inflationary era during the first moments in the history of the Universe. The quantum fluctuations in the field dominating the dynamics at that epoch, called the inflaton field, generate energy perturbations which produce anisotropies in the CMB through the physical effects discussed in the previous section. The anisotropy field so generated is not ergodic on the sphere and so averages on the field do not coincide with averages on the sphere. This fact implies an intrinsic uncertainty in the statistical quantities extracted from the data, in particular in the power spectrum C_ℓ, called *the cosmic variance*. For the standard model this uncertainty is easy to compute since the random field is expected to be Gaussian, $\Delta C_\ell = C_\ell/\sqrt{\ell + 0.5}$ (this is valid for the whole sky, however, when only a fraction of the sky f_{sky} is observed then the error is increased by approximately $f_{sky}^{-1/2}$). The value of this uncertainty in plotted in Fig. 1 for the standard model.

Second, at the millimeter wavelengths where the intensity of the black-body spectrum with temperature $\approx 2.725\,\mathrm{K}$ is maximum there are other astronomical *foregrounds* which are also bright. These are the galactic foregrounds: synchrotron, free-free and thermal dust, and also the extragalactic sources whose emission peaks in the radio or the infrared bands. The problem of separating the foreground emission from the cosmic signal is crucial and requires multifrequency observations and a good understanding of the frequency and spatial properties of the foregrounds. In recent years a strong effort in this direction is being made by the CMB community, devising new instruments optimized to observe at frequencies around 100 GHz where the CMB emission dominates over the foregrounds (for both temperature and polarization, see, e.g. [208]) and developing sophisticated methods of component separation (see, e.g. [60, 16] in this volume). In any case, there are still regions of the sky (near the galactic plane and others dominated by extragalactic sources) where the CMB is unavoidably hampered by the foregrounds. These contaminated regions are usually masked and therefore removed from the data analyses.

In the next subsections we will discuss the results obtained on the model of the Universe based on the C_ℓ. Following the standard model and some observational evidences, the anisotropies are assumed to be Gaussian and thus all the statistical information is included in the second-order moments. It is for this reason that most of the analyses performed to constrain the cosmological parameters are based on the C_ℓ. Even the likelihood methods to derive those parameters are based on the Gaussian assumption (see for instance the review by [90]). In Sect. 5 and after we will also review the works made to test the Gaussian assumption.

3.1 Cosmological Parameters from the C_ℓ

In the last decade there has been a strong technological development in CMB devices allowing very sensitive measurements of the temperature of the microwave sky. Due to the atmospheric emission at microwave frequencies the experiments have to be located at very dry places on Earth or onboard of balloons or satellites. In particular the space mission NASA WMAP has made possible to cover the whole sky from a privileged position (the Lagrangian point 2 of the Sun–Earth system), saved from contaminating emissions coming from the Earth, Moon and Sun. In Table 1 the values of the cosmological parameters determined with the first-year data of that mission are shown. An interesting result that can be derived from this table is that the Einstein–de Sitter model (i.e. flat geometry with null dark energy density) is many standard deviations away from the best fit model (assuming a flat Universe and an optical depth $\tau < 0.3$) [199].

The resolution of the WMAP experiment allows a good precision in the determination of the C_ℓ up to multipoles $\ell \lesssim 600$. Other experiments based on the ground like ACBAR [113], CBI [171], VSA [61] and balloon-borne like BOOMERANG [98] are able to reach higher resolutions covering small patches on the sky with good sensitivity. The data obtained with these experiments therefore complement very well the WMAP ones. In Fig. 4 the temperature anisotropy results from WMAP and the other high-resolution experiments as a function of the multipole are shown. We can see that the concordance model follows quite well the data up to $\ell \approx 2000$. For higher multipoles only ACBAR and CBI experiments have enough resolution and the data seem to indicate an excess of power as compared to the model prediction. This excess can be interpreted as a contribution from secondary anisotropies coming from the SZ effect [26] and/or the emission of radio sources [209]. However, new multifrequency data at arcmin resolution are necessary to confirm the excess and discriminate among possible causes.

Table 1. Cosmological parameters using only WMAP first-year data. In the fit the Universe is assumed to be spatially flat and the value of the optical depth is constrained to $\tau < 0.3$ (from [199]). The physical baryonic density parameter w_b is defined by $w_b = \Omega_b h^2$ and similarly for the physical matter density w_m which includes the contribution from all the matter species: baryons, cold dark matter (CDM) and neutrinos

Parameter	Values (68% CL)
w_b	0.024±0.001
w_m	0.14±0.02
h	0.72±0.05
A_s	0.9±0.1
τ	$0.166^{+0.076}_{-0.071}$
n_s	0.99±0.04

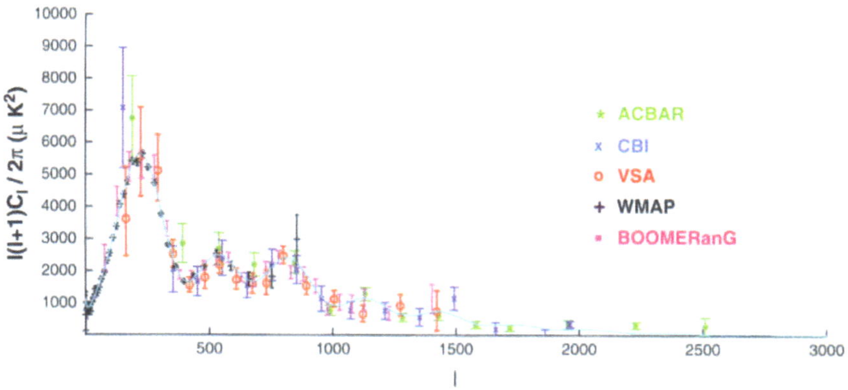

Fig. 4. The temperature power spectrum C_ℓ measured by WMAP, ACBAR, CBI, VSA and BOOMERANG compared to the best fit model given by [21]

Since the recent detection of polarization by the DASI experiment [108] several experiments have measured the TE cross-power spectrum C_ℓ^{TE} (WMAP [107], CBI [172] and BOOMERANG [164]) and the EE one C_ℓ^E (CBI [172] and BOOMERANG [148]). The sensitivity of the polarization data is not comparable to the temperature data. However, it is already sufficient to confirm the main features of the concordance model and, more specifically, the adiabatic nature of the primordial matter density fluctuations, namely, that the peaks of the polarization power spectrum, C_ℓ^E, are out of phase with the temperature C_ℓ ones. The TE cross-power spectrum, C_ℓ^{TE}, and the E-mode polarization power spectrum, C_ℓ^E, are shown in Fig. 5.

The next challenge is the detection of the B-mode polarization. As commented in the previous section this detection would unambiguously indicate the existence of a background of primordial gravitational waves. Moreover, it is the only chance that we have to know about its possible existence in the next decade since experiments aimed to directly detect gravitational waves are still not sensitive enough. The amplitude of this background, A_t, or equivalently the tensor-to-scalar ratio, r, is proportional to the energy scale of inflation. Whereas the all-sky ESA Planck space mission[2] [165] will be limited by instrument sensitivity, being able to detect values down to $r \approx 0.05$ [208], planned very sensitive ground-based experiments covering small patches of the sky and carrying arrays of thousands of detectors like PolarBear [167], Clover[131], BRAIN [139] or QUIET [169] are expected to be limited by the ability to remove the foregrounds and are expected to reach values of $r \approx 0.01$ in the best case. In order to significantly improve this limit we will have to wait for the next generation of space missions with $10^3 - 10^4$ detectors, now under discussion by ESA, NASA and other space agencies [222, 27]. The combination of complete sky coverage, very sensitive multifrequency observations and an

[2] `http://www.rssd.esa.int/Planck`

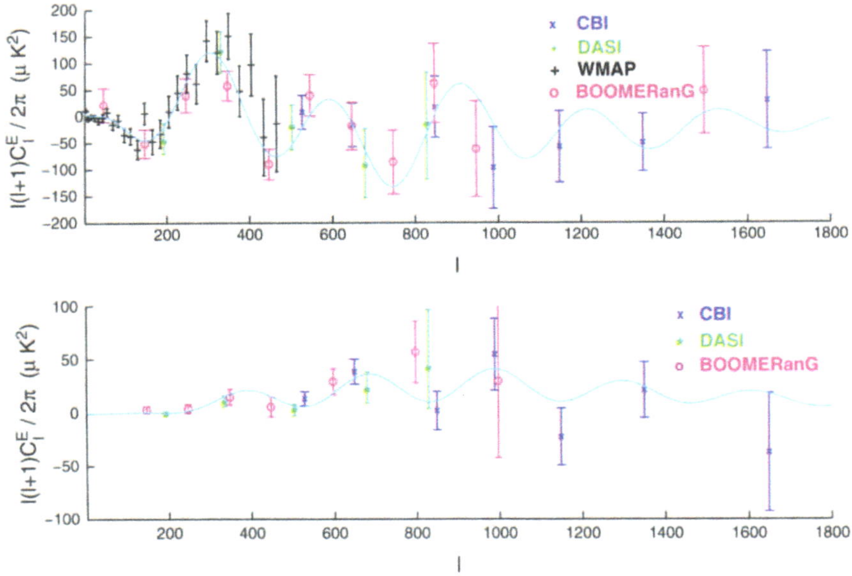

Fig. 5. Polarization power spectra measured by DASI, WMAP, CBI and BOOMERANG (TE) and DASI, CBI and BOOMERANG (E). Also plotted is the best fit model given by [21]

optimal control of systematics makes the space missions the ultimate experiments to go down to values $r \sim 10^{-3} - 10^{-4}$ allowing to probe energy scales for inflation down to $\approx 5 \times 10^{15}$GeV [208].

3.2 Combination with Other Cosmological Data Sets

The first observations showing the existence of a dark energy dominating the dynamics of the Universe were those based on the *luminosity distance-redshift diagram* determined with supernovae SN Ia [174, 162]. Those results were, however, taken with certain caution because of the assumption made that the low and high redshift SN Ia had the same light curve behaviour (in addition, there were many other possible systematics which raised some concern). In any case, SN Ia data complements very well the CMB anisotropies since for the former the dependence of the model prediction on the dark matter and dark energy density parameters enters approximately as $\Omega_m - \Omega_\Lambda$, and as $\Omega_m + \Omega_\Lambda$ for the latter (see Fig. 6).

The large galaxy surveys, *2-degree Field Galaxy Redshift Survey* 2dFGRS [161] and *Sloan Digital Sky Survey* SDSS [204] have recently reached ≈ 200000 measured redshifts on large fractions of the sky. A 3D power spectrum $P(k)$ of density perturbations is estimated with each data set. The cosmological parameters enter in the prediction of $P(k)$ through the initial power spectrum (A_s, n_s) and the transfer function which linearly connects the initial

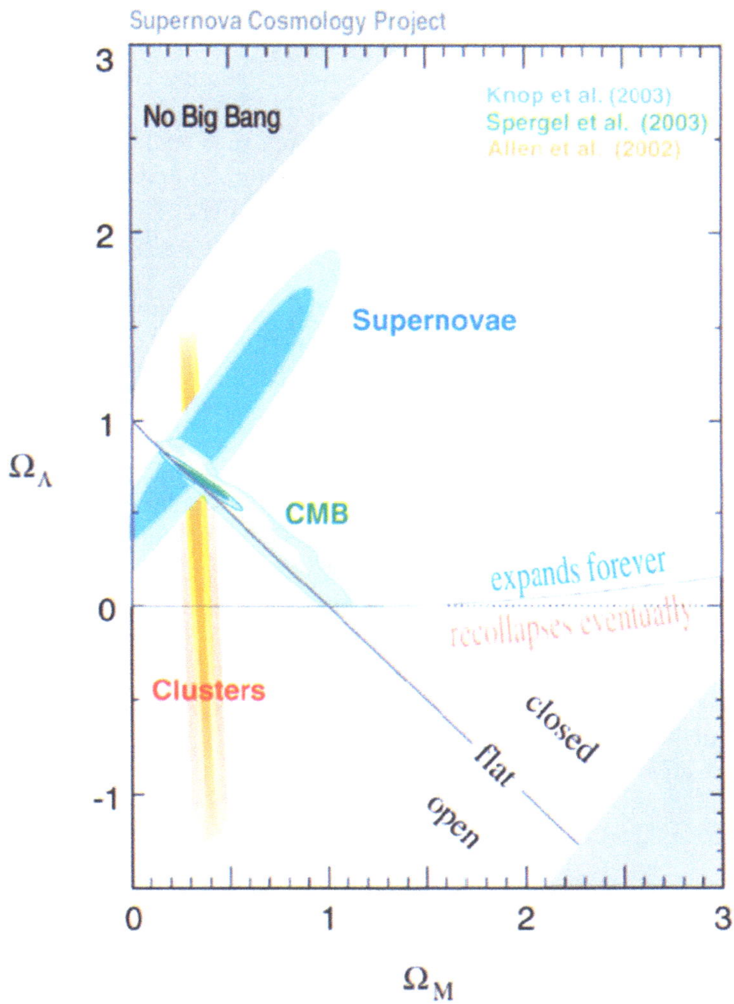

Fig. 6. Confidence contours for the pair $(\Omega_m, \Omega_\Lambda)$ combining SN Ia, CMB and cluster density data (taken from the Supernova Cosmology Project)

and present spectra. Besides, an additional parameter, *the bias b*, is required to link the galaxy power spectrum to the matter one $P(k)$. Although this parameter will in principle depend on the scale, however, it is found that at large scales b is scale independent (see, e.g. [205]). There are still other problematics that need to be corrected before using the data for accurate parameter determination such as the redshift-space distortions due to galaxy peculiar velocities, survey geometry effects, etc. Once those effects are corrected the galaxy redshift surveys play a key role in breaking the CMB degeneracies. For instance, the degeneracies of the model predictions on both data sets on the plane formed by the spectral shape parameter $h\Omega_m$ and the baryon fraction Ω_b/Ω_m are almost orthogonal.

Results on 11 free cosmological parameters combining CMB data (WMAP [86], CBI [171], ACBAR [113]) with the 2dFGRS Galaxy Redshift Survey [161] are given in Table 2. As can be seen from that table, the combination of different cosmological data sets has allowed for the first time an accuracy $\lesssim 10\%$ in the determination of most parameters. Results combining WMAP data with the SDSS Galaxy Redshift Survey [204] have also produced similar values for the parameters [205].

In addition to CMB and galaxy surveys, other combinations including the HST key project value for the Hubble parameter [72], SN Ia magnitude-redshift data [175, 207], Lyα forest power spectrum [54, 140] or abundances of reach clusters of galaxies can help to improve the results [199, 189, 170, 96]. Confidence contours in the plane $(\Omega_m, \Omega_\Lambda)$ for the combination CMB+SN Ia+cluster abundances are given in Fig. 6. The complementarity of the three data sets to break degeneracies is clearly shown. A similar effect in the plane (Ω_m, w) is obtained by combining CMB+2dFGRS+SN Ia (see Fig. 7).

Recently, an independent piece of evidence that the Universe is not Einstein–de Sitter (flat geometry with null dark energy density) has been found by cross-correlating the CMB map with galaxy survey maps. The same evolving gravitational potential wells which generate the large-scale structure of the galaxy distribution also produce the gravitational redshift in the CMB photons at late times (the *late ISW* effect, see Sect. 2). The amplitude and sign of the cross-correlation depend on three parameters Ω_{DE}, Ω_k and h. For a flat Universe, as indicated by many different observations as discussed above, a positive signal will unambiguously imply the presence of dark energy. This is

Table 2. Cosmological parameters from WMAP, CBI, ACBAR and 2dFGRS combined data (from [21])

Parameter	Values (68% CL)
w_b	0.0224 ± 0.0009
w_m	$0.135^{+0.008}_{-0.009}$
w_ν	< 0.0076 (95% CL)
w	< -0.78 (95% CL)
Ω_{DE}	0.73 ± 0.04
h	$0.71^{+0.04}_{-0.03}$
τ	0.17 ± 0.04
A_s	$0.833^{+0.086}_{-0.083}$
n_s	0.93 ± 0.03
α	$-0.031^{+0.016}_{-0.018}$
r	< 0.90 (95% CL)

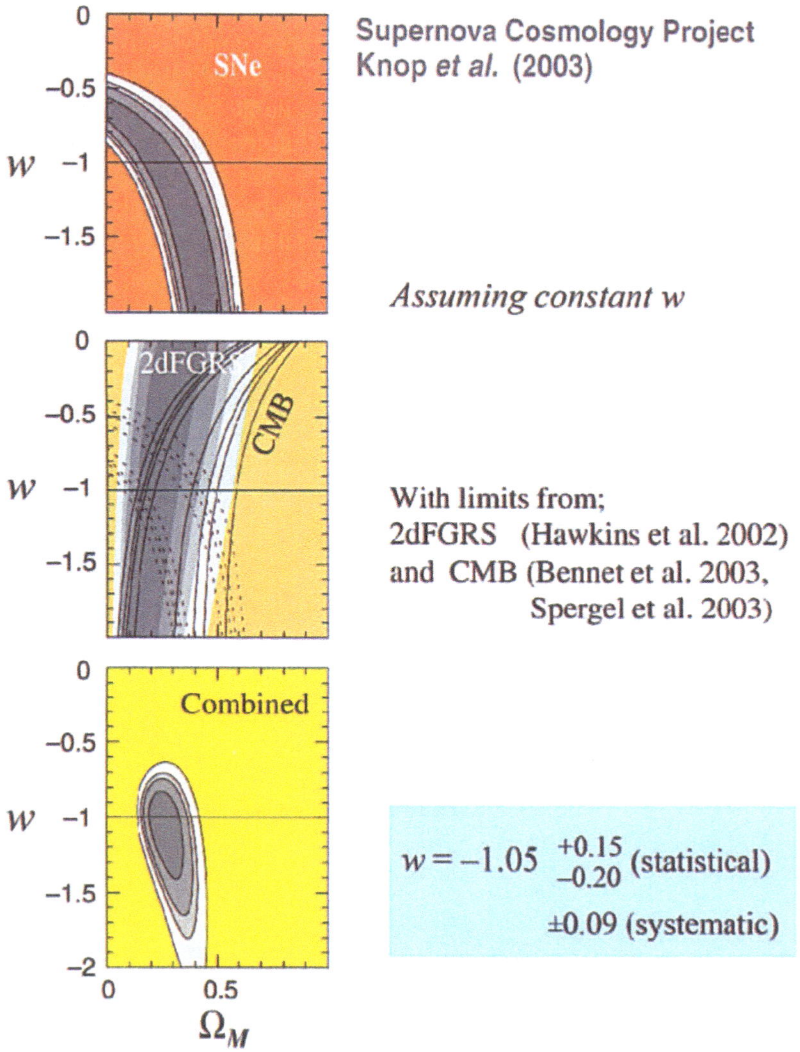

Fig. 7. Confidence contours for the pair (Ω_m, w) using either SN Ia, 2dFGRS and CMB or a combination of the three data sets (taken from the Supernova Cosmology Project)

the case found recently when cross-correlating the WMAP map with different large-scale structure surveys (the radio source survey NVSS [46], the X-ray survey HEAO-1 [24], the optical SDSS [1], the near-infrared survey 2MASS [195]) [28, 69, 154, 8, 216]. In [216] three different methods were used to estimate the *WMAP–NVSS cross-correlation*: direct temperature anisotropy galaxy number density, cross-power spectrum and covariance wavelet coefficients. A clear positive signal was found in the three cases using a maximum likelihood analysis. The significance of a non-null dark energy reaches 3.5σ for the cross-power spectrum. This is also the maximum significance detection of the ISW effect up to date (see also [142] for a recent analysis using directional wavelets). The reason for the different detection levels obtained with the three methods lies in the incomplete sky coverage available for the analysis. Otherwise, for a whole sky coverage the same detection level will be guaranteed by the maximum likelihood method since the likelihood is invariant under linear transformations of the data. In Fig. 8 we plot the two maps together with the combined mask needed to avoid the regions contaminated by foregrounds in

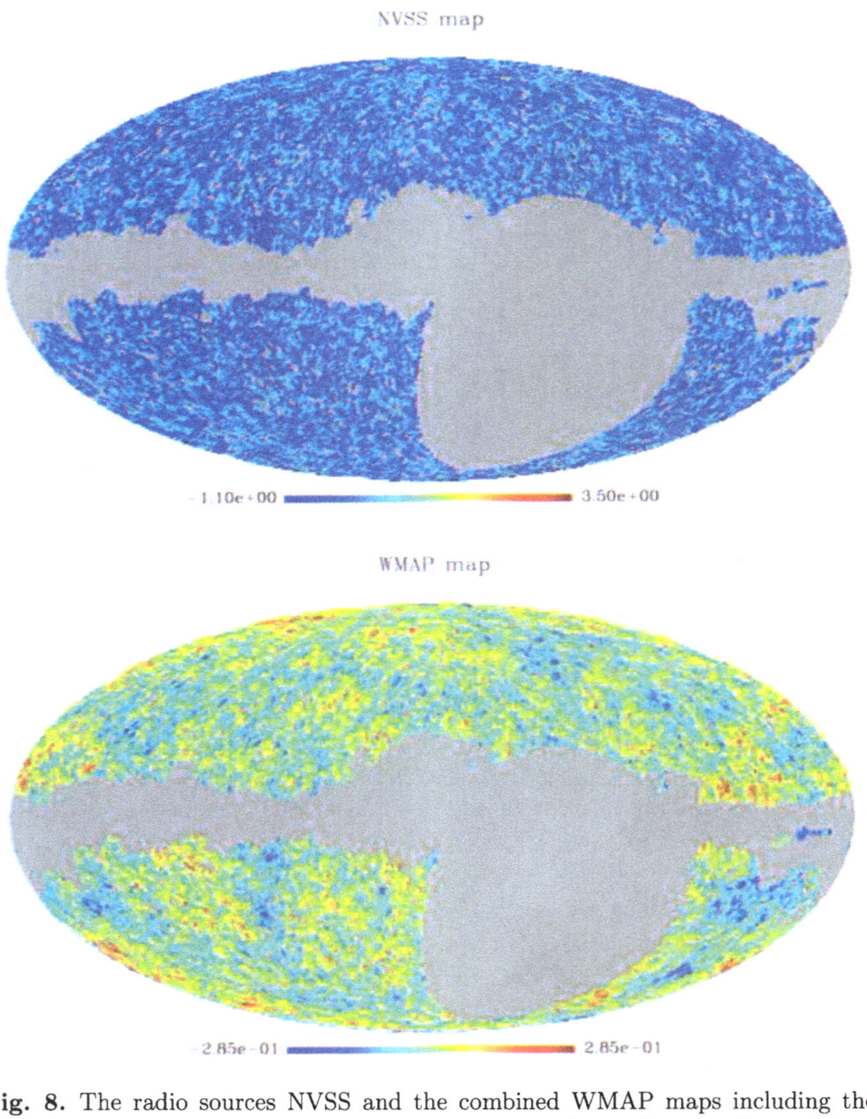

Fig. 8. The radio sources NVSS and the combined WMAP maps including the common masks used in the cross-correlation analyses (taken from [216])

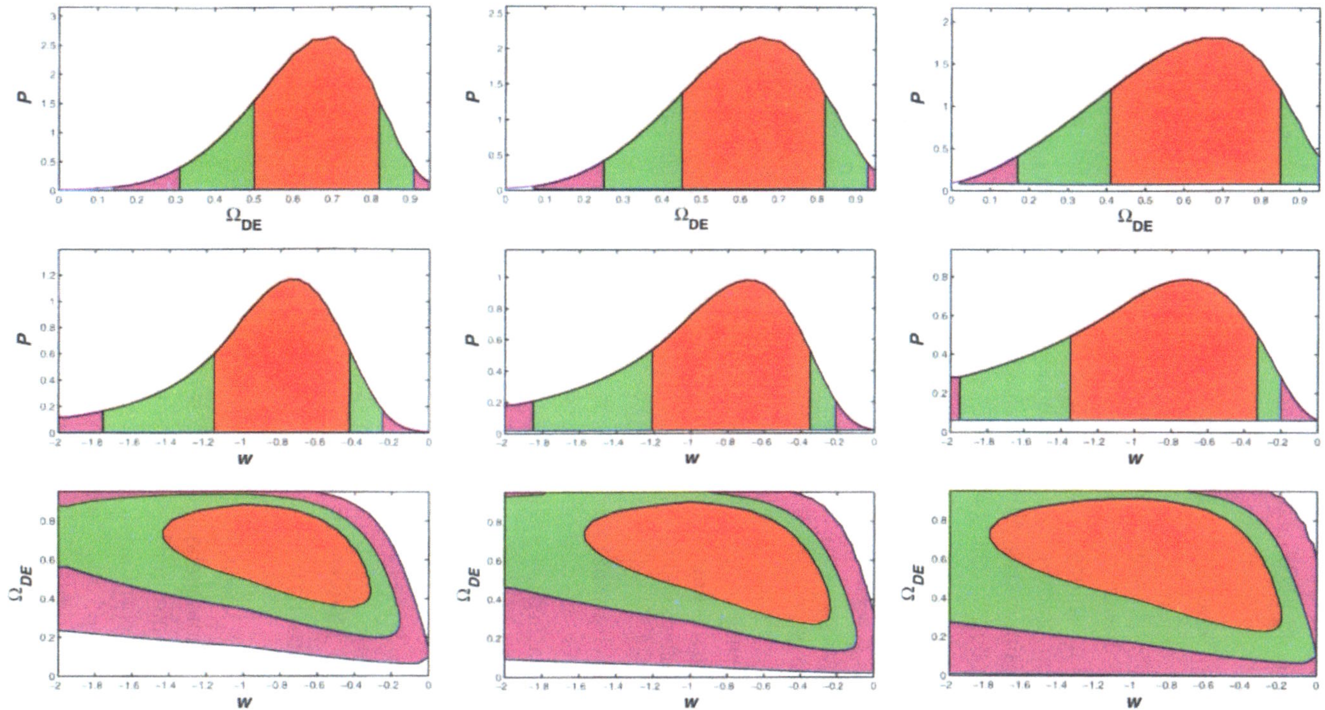

Fig. 9. The marginalized pdf for Ω_{DE} (*upper*) and w (*middle*). Contours of the 2D likelihood are given in the lower panel (taken from [216]). From left to right the three columns correspond to the results obtained with the WMAP–NVSS cross-power spectrum, covariance wavelet coefficients and temperature anisotropy–galaxy number density cross-correlation, respectively

the WMAP map and unobserved regions of the NVSS survey. An analysis allowing variations in the dark energy equation of state parameter w also shows a preferred value close to -1. The 2D confidence contours for the pair Ω_{DE}, w and the marginalized likelihoods are plotted in Fig. 9.

3.3 Summary of the Main Results

From the previous discussion about the cosmological implications of the large amounts of data already collected on the CMB anisotropies, galaxy redshift surveys, etc. the most important result is the convergence of all those cosmological data sets towards the same model of the Universe, *the concordance model*. Below we summarize the main characteristics of this model and other consequences implied by the data:

- The geometry of the Universe is very close to flat.
- The dynamics is dominated by a dark energy whose equation of state is almost that of a cosmological constant.
- Most of the matter content is in the form of cold dark matter.
- The large-scale structure of the Universe was seeded by quantum fluctuations in the very early Universe that evolved via gravitational instability (the first evidence of this came from COBE-DMR [198]).
- The initial matter–energy density fluctuations were of the adiabatic type.
- Topological defects did not play a dominant role in the structure formation of the Universe.
- Primordial gravitational waves do not appreciably contribute to neither the temperature nor the E-mode polarization anisotropies (up to the present sensitivities reached).
- The reionization happened at relatively early times, $z \gtrsim 10$.[3]
- The initial fluctuations were close to a homogeneous and isotropic Gaussian random field (see, however, the next sections).

4 Beyond the Power Spectrum: Sources of Non-Gaussianity

As it was said before, single-field inflationary models predict Gaussian temperature fluctuations homogeneously and isotropically distributed on the sphere. In this case all the information is contained in the power spectrum or, equivalently, the 2-point correlation function. In the previous sections all the discussion has been based on such statistical quantity. However, this picture can be modified if the origin of the primordial fluctuations is different (e.g. the presence of topological defects or non-standard inflation). If this is the case, the generated fluctuations will show non-Gaussian features (e.g. non-random

[3] Considering the second release of WMAP data, z is indeed around 10.

phases) and the statistical information is not only contained in the second-order moment. Even if the primordial fluctuations generated in the early Universe are Gaussian, non-Gaussianities can appear after recombination when the non-linear matter density evolution takes place and new anisotropies are generated. These secondary anisotropies, as already discussed in Sect. 2, can be produced by the gravitational lensing effect, the interaction of CMB photons with reionized matter or the gravitational effect of the non-linear evolution of the large-scale structure. On the other hand, the Universe is believed to be a Friedman–Robertson–Walker (FRW) model with a simply connected topology. Departures from this scenario, either its geometry or/and topology, will induce inhomogeneity or/and anisotropy in the statistical nature of the temperature distribution on the sphere (e.g. hot spots, rings, multiple images). Systematics arising from instruments and data processing as well as residues left after foreground removal to get the intrinsic CMB signal will inevitably introduce artefacts which will contaminate the primordial signal. It is therefore crucial to disentangle these artefacts from primary and secondary non-Gaussianity (see [17] for a recent study of the residues left by different component separation methods).

4.1 Models of the Early Universe

Even in the standard inflationary scenario small deviations from normality can appear as a consequence of second-order effects during the evolution of density fluctuations after inflation finishes (see [20] for a detailed discussion). The non-Gaussianity produced by those second-order effects must be therefore present in any inflationary model. In the case of non-standard inflation, as the curvaton scenario or inhomogeneous reheating, the primordial non-Gaussianity should be added to that term. The level of non-Gaussianity in the cosmological perturbations can be phenomenologically characterized by the *non-linear parameter* f_{NL} in the gravitational potential

$$\Phi = \Phi_L + f_{NL}{\Phi_L}^2 , \tag{5}$$

where Φ_L is the gravitational potential at linear order. Although f_{NL} will in general depend on the scale, from a practical data analysis point of view it is useful to consider it as a constant [20]. Recently, a complete second-order calculation has allowed an estimate of an effective f_{NL}^{eff} for some inflationary scenarios and considering only a limited number of multipoles $\ell_{max} \leq 500$ due to computational cost [127]. The result is a value $f_{NL}^{eff}(\ell_{max}) \simeq 4$ with an increasing trend towards higher multipoles. This level of non-Gaussianity might be marginally detected with future experiments like Planck.

Topological defects in the Universe are expected to appear as a natural consequence of phase transitions in the early Universe (for a detailed description see [217]). They are associated to symmetry breaking of fields responsible for the different particle interactions. According to their dimensionality they can be classified in the following types: monopoles, cosmic strings, domain walls and textures. We already know from the mismatch between the observed oscillations of the C_ℓ and the generic featureless spectra predicted by cosmic defects that their role in the generation of the cosmological fluctuations has to be sub-dominant. The contribution of both global and local defects to the C_ℓ is constrained to be $\lesssim$ 10–20% [22, 70, 118, 228, 227]. However, their intrinsically active role in seeding structure formation in the Universe produces a characteristic non-Gaussian CMB temperature field. On large angular scales the superposition of many defects tends to Gaussianity by the central limit theorem. On scales smaller than the projected inter-defect separation non-Gaussianity can be seen as line-like discontinuities in the case of strings, and as hot/cold spots in the case of monopoles and textures. Cosmic strings, the best studied defects, have recently received renewed attention because they are produced in a wide class of string theory models where inflation arises from the collisions of brains.

4.2 Geometry and Topology

The angular distribution of the CMB temperature and polarization fluctuations on the sky is very sensitive to the spacetime metric of the Universe. On the largest scales it is assumed that the metric corresponds to the Friedman–Robertson–Walker one. Deviations from this metric, even if small, can lead to noticeable signatures in the CMB anisotropy. In particular for homogeneous and anisotropic models (classified as Bianchi models), it is well known that global shear or rotation can produce a spiral pattern or hot spots on the CMB sky [18, 94]. Since those signatures are on large angular scales, strong limits on those models were already derived from the COBE-DMR experiment [106, 136]. More recently, asymmetries/non-Gaussianities found in the WMAP first-year data have been nicely fitted with a Bianchi VII_h model [94]. However, even considering extensions of the Bianchi VII_h models in which a Ω_Λ term is included, the values of the parameters which fit the large-scale asymmetries are ruled out by current observations at high significance [29, 95].

Due to the local character of the General Theory of Relativity the global topology of the Universe is not theoretically constrained (for different topologies discussed in the literature see, e.g. [114, 122, 220, 213]). Non-trivial topologies leave imprints in the CMB anisotropy at large angular scales which can be observed as anisotropic patterns, matched circles or, more generically, deviations from a Gaussian random field (see, e.g. [52, 92, 177]). The observability of those signatures depends on the topology scale as compared to the horizon scale [186]. Detailed CMB simulations in non-trivial topologies are required to extract the maximum information from the data (see [87, 173]). Precision all-sky surveys are necessary to obtain strong limits on global topology. The COBE-DMR data already constrained several topological models (see, e.g. [58, 178, 177]). More recently, the WMAP data were also analysed to further constrain topology finding no evidence of non-trivial topologies

[53, 112]. Moreover, theoretical limits are derived for the size of the topologies that can be detected [112]. Besides, [10, 129, 179] have claimed a dodecahedral topology for the Universe to explain the anomalies found at the lowest multipoles of the WMAP data. The next all-sky experiment, the Planck mission, will obtain more sensitive data in a wider frequency range allowing a better separation of the CMB from galactic emissions and therefore a better determination of the low-order multipoles. It will thus help to clarify the present situation and to determine whether the present anomalies are due to systematics, foreground residues or are indeed true anisotropies of the CMB.

4.3 Secondary Anisotropies

As already commented in Sect. 2, the temperature anisotropies of the CMB are usually divided into primary and secondary. The former are generated prior to recombination and are directly related to the initial density fluctuations. The latter are generated after matter–radiation decoupling and arise from the interaction of the CMB photons with the matter. These interactions can be of gravitational type (e.g. *Rees–Sciama* effect, [180, 134]) or of scattering type when the matter is ionized (e.g. *Sunyaev–Zel'dovich (SZ)* effect of galaxy clusters [200] or *Ostriker–Vishniac* effect for a homogeneous reionization [156, 218] or, on the contrary, *inhomogeneous reionization* of the Universe [3, 77, 103]). Secondary fluctuations will in general induce non-Gaussian signatures that will add to the intrinsic statistical properties of the CMB anisotropies. Their contributions to the bispectrum (third-order moment in harmonic space, see next section) during re-ionization have been studied in [48]. More recently, it has been shown that the imprint of the Ostriker–Vishniac effect can be characterized by an undetectable bispectrum and a significant trispectrum (fourth-order moment in harmonic space) [35]. The Rees–Sciama effect due to non-linear evolution of matter is expected to produce 3-point correlations much below the cosmic variance [144].

The SZ effect due to galaxy clusters is one of the most important sources of secondary anisotropies. Two types of SZ effect can be distinguished: the *thermal* effect, induced by the CMB photon scattering off free electrons in the hot intra-cluster gas, and the *kinetic SZ* effect due to Doppler shift of the photons when the clusters move with respect to the CMB rest frame. Owing to its peculiar spectral signature the thermal effect will be separated to a good accuracy and thus removed from the cosmological signal. However, the contribution from kinetic SZ effect, which is spectrally indistinguishable from the primary anisotropies, will not be easily subtracted, and will thus remain a potentially significant non-Gaussian foreground contribution. The SZ effect of galaxy clusters is by nature a non-Gaussian process. Several studies have aimed at characterizing the non-Gaussian signature, either in wavelet space (see, e.g. [4]) or in the real one ([49, 229]), in order to address its detectability and extract the maximum information from it.

On its way from the last-scattering surface to us, the CMB radiation passes mass inhomogeneities which deflect its paths by the gravitational lensing effect. This process can approximately be described as a random walk of light through a continuous field of mass inhomogeneities and thus leads to a diffusion process. The diffusion tends to broaden structures in the CMB on angular scales smaller than $\sim 10'$ (see [19] for a review).

As long as the density inhomogeneities can be described as a Gaussian random field, the light is slightly deflected and its distribution is a Gaussian random process. A purely Gaussian CMB would therefore remain approximately Gaussian despite the deflection at least for $l \lesssim 1000$. However, at smaller scales non-Gaussianity is imposed on an originally Gaussian CMB by producing correlations between temperature fluctuations at large scales (through the *late ISW* effect) and light deflections which generate noticeable effects at several arcmin resolution [88, 74, 193]. Those correlations induce high-order correlations in the CMB temperature and polarization fields. In particular, a bispectrum is generated whose signal to cosmic variance noise is higher for polarization and polarization–temperature bispectra than for those involving the temperature alone [88].

Gravitational lensing by galaxy clusters can also lead to non-Gaussian features in the CMB through their strong lensing effect. While their origin is not physically distinct from lensing by large-scale structures, galaxy clusters can be individually identified through their thermal Sunyaev–Zel'dovich effect and, for the brightest cases, extracted from CMB maps. Methods to remove non-Gaussianity from CMB maps therefore have to distinguish between lensing by clusters and by large-scale structures.

5 Methods to Detect and Characterize Deviations from Gaussianity, Standard Geometry and Trivial Topology

There are infinite ways in which a random field can depart from Gaussianity. For instance, the cumulants above order 2 should be zero for a Gaussian distribution and thus a non-null value taken by any of them would represent a deviation from normality. In Fig. 10 it is shown a Gaussian map corresponding to the concordance model together with other non-Gaussian ones with the same power spectrum but with small amounts of skewness or kurtosis. Hence, there is not a unique way to detect and characterize deviations from Gaussianity. Depending on the kind of features that are investigated some specific methods will prove to be more efficient than others. A variety of methods have been already proposed to search for non-Gaussianity in the temperature or polarization maps. Acting in different spaces (real, harmonic, wavelet, eigenmode, etc.) they are able to extract relevant information which is otherwise hidden in the fluctuation maps. Improved methods for the combined temperature and polarization maps are expected to be developed in the coming years

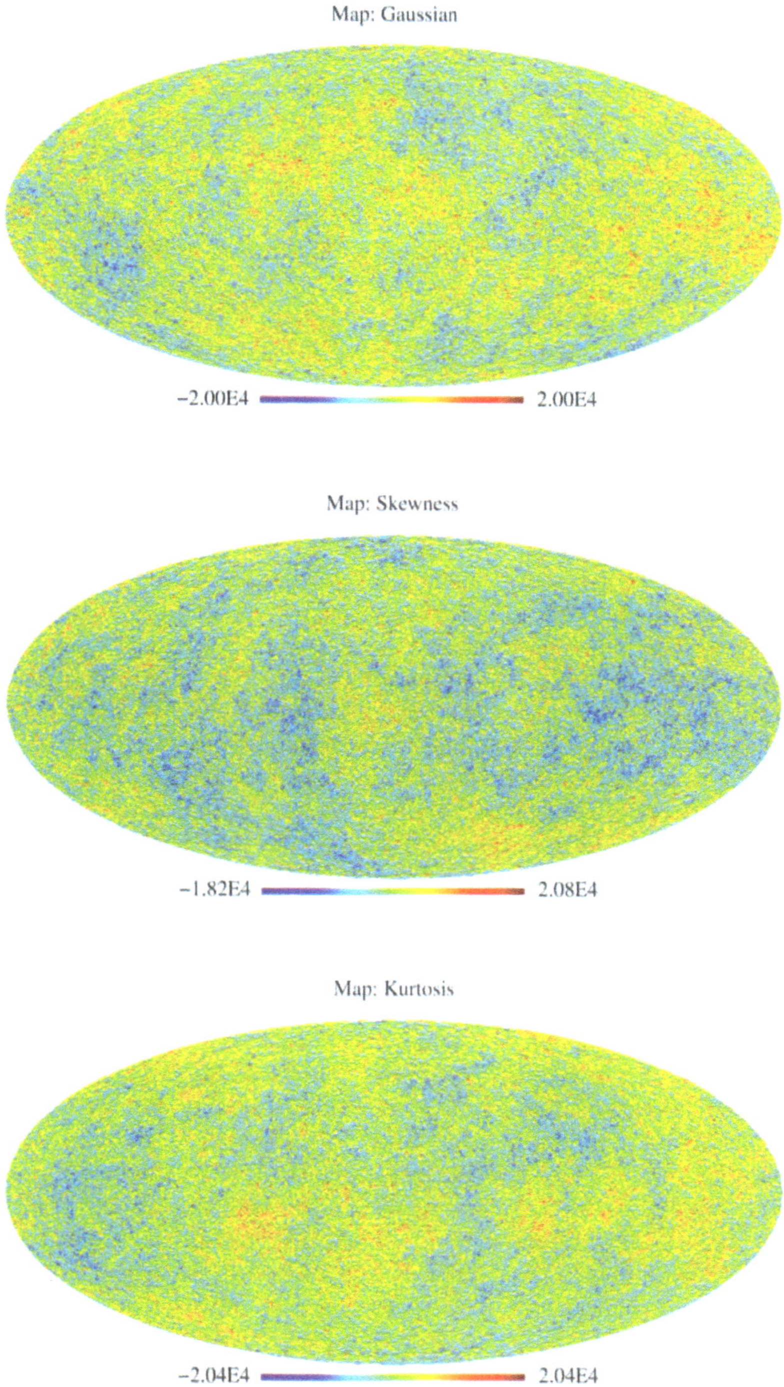

Fig. 10. Map simulations for Planck LFI 33 GHz ($N_{\rm side} = 256$, FWHM = 33 arcmin). Non-Gaussian features have been introduced in the map using the Edgeworth expansion (from [137])

motivated by the large amounts of new data expected to be collected by the new experiments (specially the all-sky WMAP and Planck satellites).

An important difference among the methods is whether they are guided or blind. The former are intended to test specific models of non-Gaussianity. The latter do not assume any form of non-Gaussianity. Guided methods would be more powerful in constraining non-Gaussianity than the blind ones at the cost of being model dependent. Both methods have been used in the literature with different results: while guided methods have imposed upper limits on the non-linear parameter f_{NL} some blind methods have obtained significant deviations of Gaussianity (see Sect. 6.2).

Below we summarize different methods which can be used to detect and characterize non-Gaussianity in temperature/polarization maps.

Methods based on *real space* are very useful for picking out statistical features. However, they may be insensitive to any localized features. Standard statistics like *moments* and *cumulants*, obtained from the moment-generating function and its logarithm or cumulant-generating function, are included in the 1-point probability distribution function (pdf). Any non-zero amplitude for any cumulant of order higher than 2 would be an indication of the presence of non-Gaussianity. It is more common to use the skewness and kurtosis, the third and fourth standardized moments, which are defined by the ratio of the 3 and 4-order cumulants normalized to the square root of the 2-order cumulant to the power of the corresponding order, respectively. All the statistical information is, however, included in the n-point pdf. For the Gaussian case, it is sufficient to specify the 2-point correlations. Any positive detection of reduced correlations above order 2 would imply deviations from Gaussianity in the data. Tests of normality based on the *Edgeworth expansion* for estimating the skewness of the COBE maps have been developed by [47]. The Edgeworth expansion allows to describe perturbations of the Gaussian distribution in terms of a series of cumulants (see [100]). An alternative expansion derived from the Hilbert space of a linear harmonic oscillator, producing a proper normalized distribution under any truncation, has been proposed by [176]. Other approaches based on *goodness-of-fit tests* of normality have recently been developed and applied to the MAXIMA data [37].

A different approach to test Gaussianity is based on *scalar quantities* on the sphere. These quantities are constructed with the first and second covariant derivatives of the temperature field. A well-known example is the local curvature (see [64, 82]) but many others can be defined as the modulus of the gradient, shape index, determinant of the Hessian matrix, etc. (see [146] for a detailed study of those quantities). These scalar quantities have been recently shown to be good detectors of small deviations from Gaussianity as given by the Edgeworth expansion [147]. Analyses based on *statistics of extrema* and *excursion sets* (e.g. number, correlations, eccentricities) provide very robust results since extrema are often many sigmas above the noise level. Those quantities can be computed (semi-)analytically for a Gaussian field as was shown in the pioneering work [25]. Also, it has been shown that the *Gaussian curva-*

ture of peaks is an efficient discriminator of non-Gaussianity [15]. The *extrema correlation function* for a Gaussian field is accurately computed in [84, 85] whereas [12] computed the *correlation of excursion sets* for Gaussian and some non-Gaussian temperature distributions. *Minkowski functionals* describe the morphology of the excursion sets which are defined by thresholding the temperature field. They are three for the 2D sphere: total contour of the excursion set, total area and the total curvature (or the genus). Analytical expressions can be obtained for the Gaussian field. These morphological descriptors have been implemented on the HEALPix pixelization [75] of the sphere by [219].

Alternative methods based on *multifractal analysis* and *surface roughness*, widely used in many branches of physics, have also been proposed to detect non-Gaussianity in CMB maps [62, 145].

Spherical harmonics space statistics are attractive as the predictions for Gaussian fields can in some cases be computed analytically, and their covariance properties may also be computed in principle. In practice, we may not be able to benefit from these advantages due to the complexities and size of the experimental data sets. As for the other statistics, sensitivity to artefacts such as those arising from the observing strategy and, in particular, incomplete foreground subtraction, need to be assessed. The *bispectrum*, $B_{\ell_1\ell_2\ell_3}$, is the rotationally invariant third-order moment of spherical harmonic coefficients and has been very often used to test different data sets. It is given by the following expression (see [89, 20] for a derivation of this and higher order moments):

$$B_{\ell_1\ell_2\ell_3} = \sum_{m_1 m_2 m_3} \begin{pmatrix} \ell_1 & \ell_2 & \ell_3 \\ m_1 & m_2 & m_3 \end{pmatrix} a_{\ell_1 m_1} a_{\ell_2 m_2} a_{\ell_3 m_3} , \tag{6}$$

where (...) is the Wigner-3j symbol. The bispectrum must satisfy the selection rules, requiring that $\ell_1 + \ell_2 + \ell_3$ be even, the triangular rule $|\ell_1 - \ell_2| \leq \ell_3 \leq \ell_1 + \ell_2$ and that $m_3 = m_1 + m_2$. For simplicity reasons often the case that all multipoles are equal, $\ell_1 = \ell_2 = \ell_3$, has been considered (see, e.g. [68] for an application to COBE-DMR). Inter-scale correlations of the form $B_{\ell-1\ell\ell+1}$ have also been used (e.g. [132]). They are generically expected in models with scaling seeds like cosmic defects. Other components with more distant multipoles may be considered but they are often dominated by noise. A fast algorithm to compute the bispectrum from HEALPix-formatted data has been developed by [109] and applied to the HEALPix-formatted COBE-DMR data. Kunz et al. [111] have recently implemented an algorithm to compute the *trispectrum* (fourth-order moment in harmonic space) in the HEALPix pixelization and have applied it to the COBE-DMR data.

Novel methods have been recently developed based on other spaces where the data are transformed by either *filters, wavelets or eigenvectors*. Since the operations involved are linear the new coefficients should form a set of Gaussian variables if the underlying map is Gaussian. *Wavelets* are compensated filters which allow one to extract information which is localized in both real and harmonic space and for this reason they can be more sensitive than classical methods (for a detailed review of the properties of wavelets see [97]). In recent years they have been applied to the non-Gaussian analysis of the CMB. Many of the statistics and techniques used on real or harmonic space can be translated and applied to wavelet space (e.g. moments, scale–scale correlations, 1-point pdf). On small patches of the sky several *planar wavelet families* have been successfully applied to detect simulated cosmic strings embedded in a Gaussian signal [14] and search for discontinuities due to secondary anisotropies [4]. The *Haar wavelet* (the simplest wavelet that one can construct) [13, 206] and the *Mexican Hat wavelet* (given by the Laplacian of a Gaussian) [36, 137] have been now implemented on the sphere (see Fig. 11 for an application of the latter wavelet to COBE). Martínez-González et al. [137] has shown that the spherical Mexican Hat wavelet can be more sensitive than the spherical Haar wavelet for detecting types of non-Gaussianity expected in some non-standard scenarios. Moreover, the former wavelet has been found to be more sensitive than the bispectrum for constraining the non-linear coupling parameter appearing in non-standard inflation using the COBE/DMR data [38]. Recently, *directional wavelets* having a preferred direction have been implemented on the sphere – e.g. *elliptical Mexican Hat* and *Morlet* – and applied to the CMB analyses [141]. The computational cost of exploring all possible directions can be greatly reduced by using *steerable wavelets* where the coefficients at any direction can be obtained as linear combinations of the ones of a basis with a limited number of elements [223, 224].

Filters can also help to extract non-Gaussian features from a given data set. This can be done in different ways, e.g. by removing the non-cosmological information (*Wiener*), equalizing the CMB signal (*signal-whitening*) or the noise (*noise-whitening*) [225] or taking *field derivatives* to study discontinuities (for an application to cosmic strings see [76]). In the noise-whitening case the corresponding signal-to-noise eigenvalues allow also a compression of the data (see [226]).

In addition to the CMB temperature, the determination of the statistical nature of polarization is crucial to test the Gaussianity of the CMB. Most of the tests proposed above for the temperature can be implemented for the CMB polarization. There are, however, other properties which are specific of the pseudo-vector nature of polarization. Up to now only a few tests have been developed based on *geometric characteristics of polarization*: peculiar points, up-crossing and down-crossing points of the modulus of polarization (see for instance [151]). In any case, it would be very desirable to device methods combining temperature and polarization maps in an optimal way.

Some of the methods already discussed can as well be used to search for possible features present in the CMB maps arising from *global anisotropy, inhomogeneity* or *topology*. Particularly interesting are those based on wavelets, optimal filters and pattern recognition. If the Universe is *anisotropic/inhomogeneous* (as in the case of, e.g. Bianchi, Tolman or Swiss-Cheese models) then the CMB temperature fluctuation field will be anisotropic/inhomogeneous on

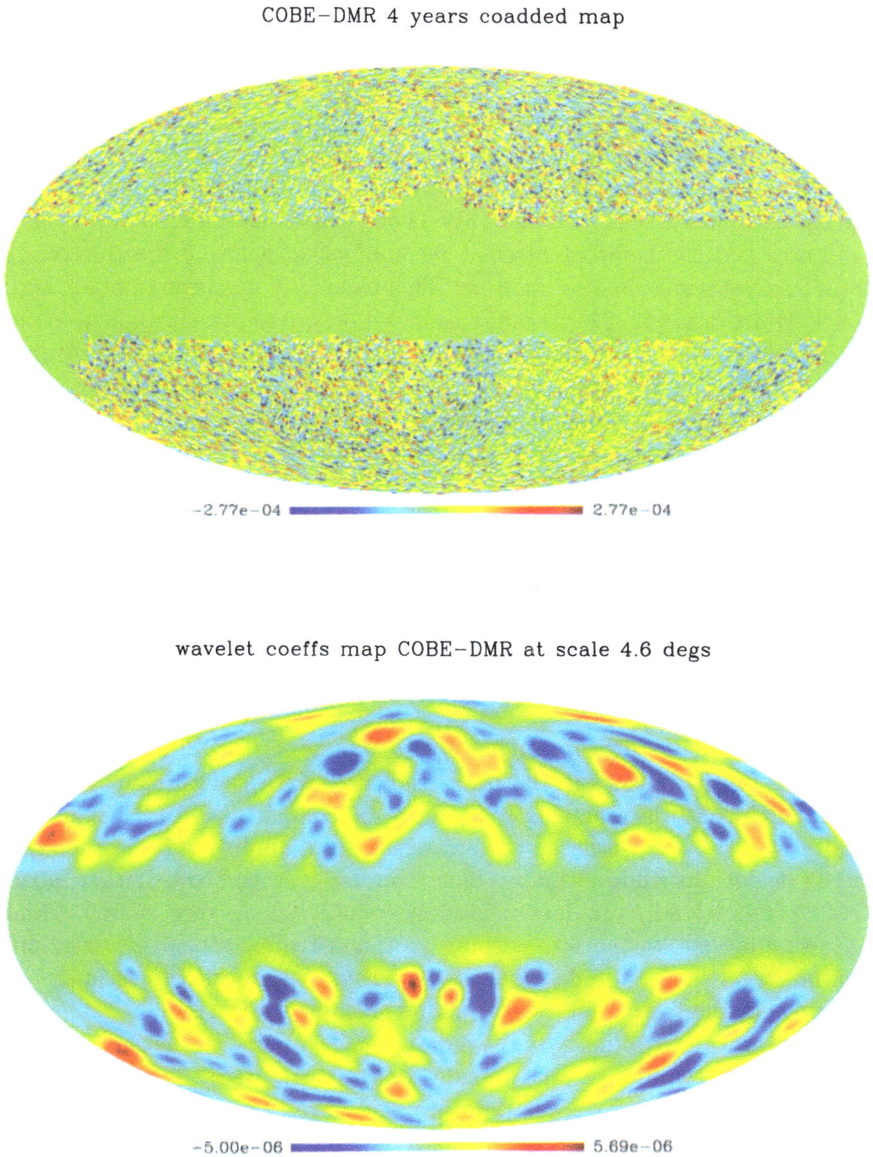

Fig. 11. Spherical Mexican Hat wavelet coefficients for COBE-DMR (from [36])

the sphere. The space–time geometry of the Universe is expected to affect mainly the low-order multipoles (i.e. large angular scales). However, focusing of geodesics can also produce a large variety of features at smaller scales (e.g. hot spots, rings of fire, spiral patterns, North–South asymmetries, etc. see [18, 95]. For recent applications to detect those features in WMAP data see Sect. 6.2. If the *topology* of the Universe is not trivial then anisotropic features will naturally appear in the CMB maps. Also symmetries are expected in some topologies (e.g. toroidal one). A method to probe non-trivial topology is based on the S-statistic which measures the symmetry of a map [202, 59]. Another technique consists of the *identified circles* principle. Circular patterns of temperature fluctuations are generated in compact topologies. A given circle should be seen in different positions of the CMB sky since it can travel to the observer following different geodesics. Notice, however, that it will not look exactly the same because of the ISW effect caused by different intervening structures and the Doppler effect at recombination whose projection on different geodesics will be also different. This technique has been applied to the COBE-DMR data [52, 51]. Techniques based on wavelets that allow to detect the patterns of spots imprinted by the different topologies have been implemented in [177]. The results show the potential of scale–scale correlations for distinguishing among different topologies. Recently, other techniques involving *phase correlations* have been applied to the WMAP data to constrain the topology of the Universe [63].

6 Data Analysis Beyond the Power Spectrum

6.1 The Search for Non-Gaussianity Before WMAP

The first systematic analyses that have been done on CMB data in search for non-Gaussianity have been performed on the 4-year COBE-DMR data set. This was for a decade the only whole-sky data set publicly available. The COBE-DMR data have been found to be compatible with the Gaussian hypothesis by the majority of the statistical tests applied. The few analyses finding deviations from normality have been later proved to be either incomplete or interpreted as undocumented systematic effects (see below). Monte Carlo simulations, taking into account all the instrumental and observational constraints of the data under analysis, are usually performed to estimate distributions of the testing statistic as well as confidence levels.

The degree of asymmetry of the data was measured with the *3-point correlation function* (equilateral and pseudo-collapsed) by [105]. *Extrema distribution and correlations* of the temperature field have been tested by [105, 163]. Several works have relied on morphological characteristics of the observed field to detect non-Gaussianity, *the Minkowski functionals* [187, 155] and in particular the *genus* (number of peaks above the threshold minus the number of holes [105, 163]) to study the COBE-DMR data. The *partition function*, based on the combination of information at several scales and the contribution of different moments of the measure (defined in this case as the sum of absolute temperatures in a defined box), and the roughness of the last-scattering surface were proposed by [62] and [145], respectively, as powerful statistics to detect deviations from Gaussianity. All the analyses performed in those works were done in real space. No evidence of departure from Gaussianity of the COBE-DMR anisotropies was found in all the cases discussed above.

Tests were also carried out in spaces defined by eigenmodes and wavelets (see previous section). A *principal component analysis* was done by [30] to extract the eigenmodes from the COBE-DMR data. χ^2, Kolmogorov–Smirnoff, cumulants up to fourth order and significance of the top few outliers were calculated from the eigenmodes and tested against Gaussianity. None of these tests rejected the Gaussian hypothesis at the 95% confidence level. Skewness, kurtosis and scale–scale correlations of the COBE-DMR *wavelet* coefficients have been studied in the north and south faces of the quad-cube COBE pixelization, using different wavelet bases [157, 149, 5]. Spherical wavelets have also been convolved with the COBE data (HEALPix pixelization) to test the distributions of wavelet coefficients at several scales against Monte Carlo simulations assuming Gaussianity [13, 36, 38]. A detection of non-Gaussianity was claimed by [157] finding the value of the scale–scale correlation (between scales 11° and 22°) for the COBE data outside the 99% confidence level. However, [149] pointed out the orientation dependence of orthogonal wavelet basis. A rotation of the data by 180° makes the COBE data set compatible with Gaussianity.

Several works have studied the *bispectrum* of the COBE-DMR data in a simplified form that ignores correlations between different multipoles [68, 83, 11]. A rejection of Gaussianity was claimed by [68] at the 98% confidence level and appeared at $l = 16$. In a subsequent paper [11] analysed the bispectrum as a function of frequency, channel and time interval. The non-Gaussian signal was found to come only from the 53 GHz map, implying a non-cosmological origin. Removal of data corresponding to a period of time dominated by a systematic effect (data collected during an eclipse) resulted in a COBE-DMR bispectrum compatible with Gaussianity.

An *extended bispectrum* including components with correlations among different multipoles was studied by [132]. Gaussianity was rejected by the COBE-DMR data at a confidence level larger than 98%. An exhaustive study of the possible non-cosmological origin of this detection is included in that work giving negative results. With the aid of the WMAP data, this detection was later proved to be due to pixelization artefacts of the Quad-Cube tessellation of the sky and the eclipse data.

All the previous works calculated the COBE-DMR bispectrum for a limited number of modes. More recently, [109] have measured all the modes available from those data, 466 in total. They conclude that the data are consistent with Gaussianity. A recent study using the *trispectrum* also shows consistency with Gaussianity (having previously removed the data affected by the systematic effect, [111]).

While all the works mentioned above performed blind analyses, some works tried to detect deviations from Gaussianity in the form given by the *non-linear parameter* f_{NL} defined in Sect. 4.1. Only upper limits were obtained at least two orders of magnitude above the values expected in the most optimistic non-standard inflationary models (see, e.g. [109, 38] for analyses involving the bispectrum and spherical wavelets, respectively).

Uncertainty remained about whether the CMB anisotropies are Gaussian distributed or not. Scales observed by the COBE-DMR experiment are above the causal horizon at the last-scattering surface. It could well be that even if the anisotropies are non-Gaussian distributed at smaller scales they appear Gaussian above degree scales due to the central limit theorem. Therefore small angular scale observations are needed to answer this question. At present many of these new small scale observations have been tested against Gaussianity. The *genus* of the data of the balloon-borne QMAP, ground-based Saskatoon and QMASK (a combination of QMAP and Saskatoon) experiments was compared to Gaussian predictions by [158]. Only in the case of QMASK was an asymmetry in the genus curve found. However, this deviation from the Gaussian genus curve was not statistically significant. Shandarin et al. [188] calculated several *morphological quantities* from the QMASK data and concluded that they were consistent with the values expected for a Gaussian field. Several statistics have been computed to test the Gaussianity of the balloon-borne MAXIMA-1 data [226, 183, 37, 6]. Wu et al. [226] apply *moments, cumulants, the Kolmogorov–Smirnoff test, the* χ^2 *test* and *Minkowski functionals* in eigen, real, Wiener-filtered and signal-whitened spaces; Santos et al. [183] estimate the *bispectrum*; The authors of [37, 6] apply *goodness-of-fit* tests to the data. Consistency with Gaussianity was found in all cases. The balloon observations of BOOMERANG maps were also searched for deviations from Gaussianity. Polenta et al. [168] perform a real-space analysis, computing *skewness, kurtosis* and *Minkowski functionals* of those data. No significant deviation from Gaussianity was found.

More recently and contemporary to the WMAP analyses, the statistical distribution of interferometric observations taken with the VSA experiment have been also examined. Savage et al. [185] apply some tests in real and harmonic spaces, the former to the maximum-entropy reconstruction of the observed regions. Smith et al. [197] focus on the *bispectrum* of the data. Rubiño-Martín et al. [181] apply *smooth tests of goodness-of-fit* developed by Aliaga et al. [7]. The results of the analyses seem to be in relatively good agreement with the Gaussian assumption (except for maybe one or two fields in the analysis of [181]).

6.2 WMAP and Non-Gaussianity

The quality of the first-year all-sky data provided by the WMAP satellite has motivated many works dealing with the Gaussianity and isotropy of the CMB signal (also in relation to the geometry and topology of the Universe, see Sect. 4.2). Clean maps are constructed by the WMAP team and made publicly available to the community, together with noise templates and beam responses for all the channels. Also a mask is provided to ignore those pixels in the analysis where the foreground contamination is believed to be unavoidable. All this information is essential to perform "realistic" simulations of the data needed to compare the values of the chosen statistics with the ones obtained from

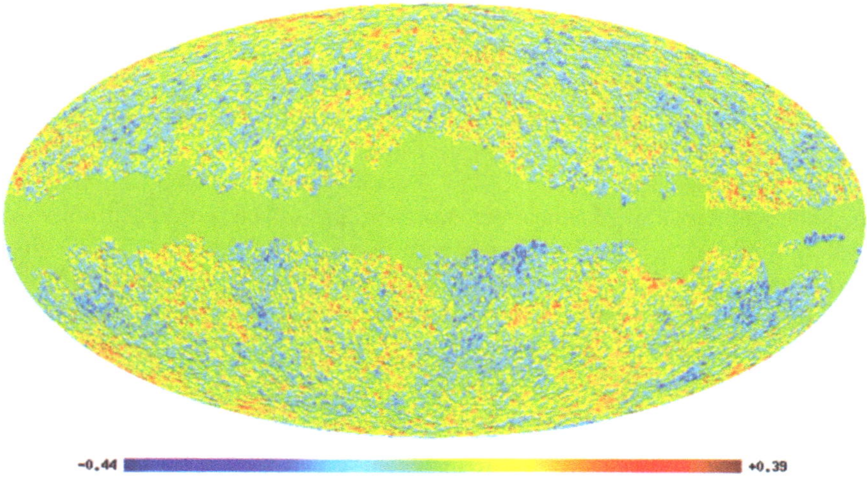

Fig. 12. The first-year WMAP combined, foreground-cleaned Q-V-W map together with the kp0 mask

the true data. Figure 12 shows the first-year WMAP map obtained after a noise-weighted combination of the foreground-cleaned Q-V-W channels (hereafter WCM). Although other foreground-cleaned maps have been produced based on different techniques [21, 67, 160, 203], for non-Gaussianity studies the WCM (together with the most conservative kp0 mask) is preferred since the noise properties are well defined for this map.

There have been a wide range of analyses of the temperature distribution using different quantities. *Minkowski functionals* have been computed by [45, 65, 110, 133, 159]. The results obtained from those works are diverse: whereas compatibility with Gaussianity is found in [45, 110], others find significant deviations from it [65, 133, 159]. The main reason for the discrepancy is that, although based on the same quantities, they construct different measures from them as the asymmetry in the genus above/below a given threshold or in different hemispheres (these latter asymmetries have also been found in the power spectrum or the n-point correlation functions for different regions of the sky, see below).

N-point correlation functions have been calculated by [66, 73] for the temperature. [66] find North-South asymmetries on large scales in ecliptic coordinates. Statistics based on the *alignment of low-multipoles* have been computed by [23, 50, 59, 115, 196, 201, 221]. Uncomfortably low probability for the observed quadrupole and octopole alignment has been obtained in those works (the so-called axis of evil). The *bispectrum* have been calculated for different subsets of all its components up to the WMAP resolution [34, 110, 117]. Significant non-null results have been obtained in several of those analyzes which agree well with the emission expected from undetected point sources [9, 110]. In addition those results can be translated to limits on the non-linear parameter f_{NL} [110, 34]. By using another subset of multipole configurations, namely the closest inter-ℓ components, [117] have also found North–South asymmetries in agreement with previous results [66, 81].

Phase correlations have been used to test the hypothesis that the CMB temperature fluctuations on the celestial sphere constitute a homogeneous and isotropic Gaussian random field. This hypothesis implies that the phases of the spherical harmonic coefficients corresponding to the temperature fluctuations are independent and uniformly distributed in the interval $[0, 2\pi]$. Phase analyses of different derived maps of the WMAP data have been performed by [43, 44, 152, 153]. These works find departures from uniformity and/or independency which in general reflect galactic contamination or noise gradients rather than primordial non-Gaussianity.

Several statistics based on the *spherical wavelet coefficients* of the CMB data have been used to test the Gaussian hypothesis [39, 55, 128, 141, 150, 215]. Highly significant deviations have been found in the *kurtosis* of the WMAP data for both *symmetric* [150, 215] and *directional wavelets* [141]. An analysis of *extrema of the wavelet coefficients* identifies a very *cold spot* as the cause of the large kurtosis found with the *spherical Mexican Hat wavelet* [55] (see Fig. 13). This cold spot has been also revealed by the *higher criticism* test applied to the wavelet coefficients [39]. Moreover, [56] finds a nearly spherical shape for the cold spot and also that it cannot be explained with foreground emissions. A test of global isotropy has been recently performed by [224] using *steerable wavelets*. Convolving the CMB map with the steerable wavelet given by the second derivative of the Gaussian at a fixed scale, a preferred orientation is signalled by the maximum of the coefficients corresponding to

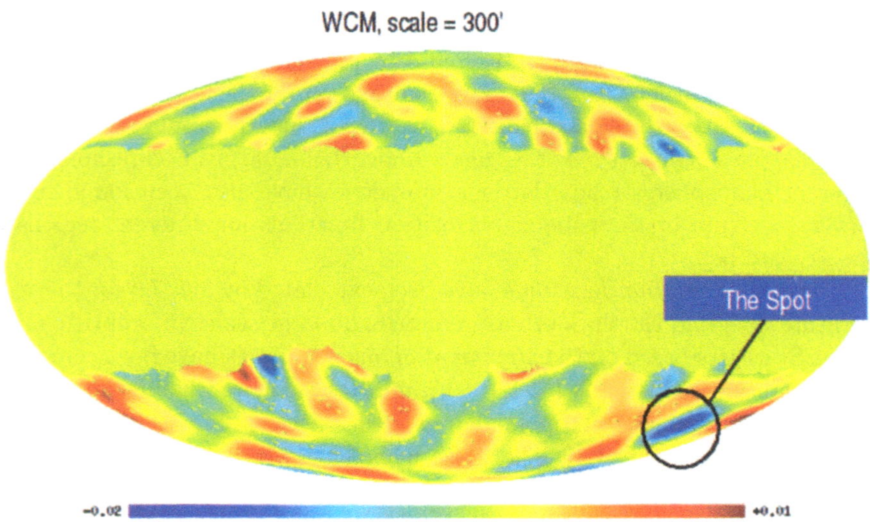

Fig. 13. The cold Spot found in the WMAP combined map after convolving with a spherical Mexican Hat wavelet with scale 300′ (see [215, 55, 56])

the different orientations and assigned to each pixel. Each pixel is thus looking at all the pixels intersected by the great circle formed by extrapolating the preferred orientation. Then the number of times each pixel is seen by any other pixel is counted. In the case of a homogeneous and isotropic Gaussian random field all the pixels are seen the same number of times on average. Performing this test on the WMAP first-year data, a number of anomalous directions were found at a very high significance level. Their mean direction is very close to the ecliptic poles, being $\approx 10°$ from it (see Fig. 14). Moreover, the anomalous directions are located along a great circle whose perpendicular is pointing towards a direction close to the dipole and to the axis of evil. This result synthesizes for the first time the previously reported anomalies commented above on the North–South asymmetry and the alignment of low multipoles.

Contrary to these results, analyses based on the bipolar spectra do not find violation of the global isotropy of the Universe [78, 79].

Number and *correlation of extrema* are computed by [119, 120, 210]. Larson and Wandelt [119, 120] find that the maxima and minima in the WMAP data are not hot and cold enough. Tojeiro et al. [210] find evidence for non-Gaussianity on large scales which could be associated with residual foregrounds.

Local curvature has been investigated by [82, 33]. They classify the points in the data as hills, lakes or saddles according to the sign of the eigenvalues of the Hessian matrix, and consider the proportions of each of them above a certain threshold. A clear asymmetry is found which is maximum for hemispheres centred near the ecliptic poles, consistent with the results by [66].

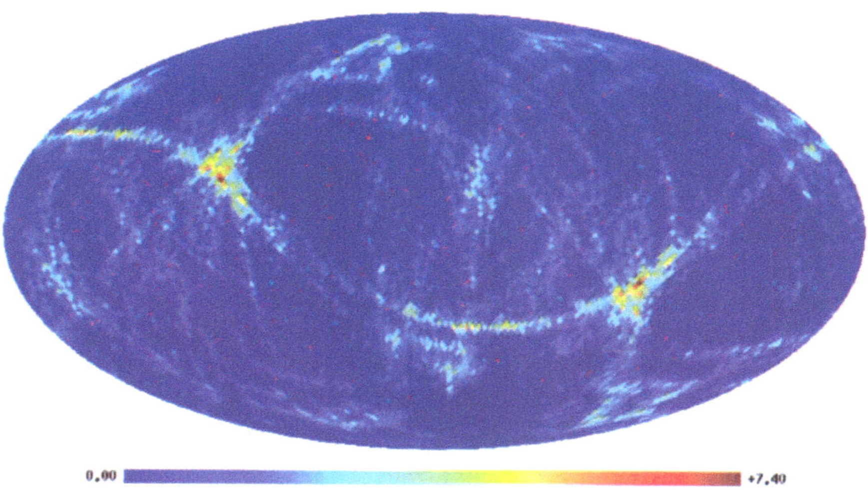

Fig. 14. Excess of the number of times each pixel is seen by any other in σ units, where σ is the dispersion corresponding to Gaussian simulations (see [224])

Several suggestions have been made to explain the anomalies found in the first-year WMAP data at the large angular scales, namely the power deficit and alignment of the low-order multipoles, the asymmetries between different hemispheres and the cold spot.[4]

A suggestion to explain the low-multipole anomaly invokes a non-trivial topology which due to the finite size of the Universe tends to cut power at the lowest multipoles and in some cases to align them [129, 179, 10]. However, apart from this nice consequence, there is no evidence for a possible non-trivial topology. Another possibility proposed is the lensing of the dipole by local large-scale structures which can potentially produce a quadrupole and octopole similar to those found in the WMAP data [214]. It remains to be demonstrated that the appropriate physical characteristics of those structures (in terms of mass, size and position) are in fact present in the local Universe.

Possible explanations for the cold spot are the RS effect produced by the most prominent voids or massive structures in the Universe [135, 211, 93], topological defects in the form or textures [212] or, a more radical one, based on an inhomogeneous model of the Universe given by a finite sphere of dust and dark energy [2].

Anisotropic and inhomogeneous geometries leaving spiral or stripy patterns in the CMB sky have been proposed to correct the large-angle anomalies found (asymmetries, the cold spot and the low-multipole power deficit and alignment) [94, 116, 40, 29] (see, however, [143] for further non-Gaussian detections in the Bianchi VII_h-corrected WMAP first-year data). Although a good fit can be found for most of the anomalies with the anisotropic geometries (Bianchi VII_h), however, the values required for the matter content are too high [95, 29].

Small changes in the magnitude of the WMAP dipole vector [71], together with possible systematics or foreground residuals associated with the ecliptic (as might be inferred from the recent results by [66, 81, 224]), might be considered as a possibility to explain the large-scale anomalies. However, much more work is still needed in order to find a convincing case for them.

7 Conclusions

Cosmology is passing through a period of abundance of new data which imply a significant improvement on the precision in the determination of the parameters characterizing the Universe. The CMB field is probably the major contributor to that improvement and is expected to be so in the next decade. With only the first year of WMAP data, the power spectrum of the temperature anisotropies has been determined with error bars below cosmic variance up to multipoles $\ell \approx 600$. Information on larger multipoles is already provided by several high-resolution experiments allowing a consistent determination of

[4] These anomalies remained almost unaltered in the 3-year WMAP data.

the third acoustic oscillation and providing an incomplete coverage of the multipoles up to $\ell \approx 2500$. In the next few years, the ESA Planck mission is expected to provide accurate and high-resolution all-sky images of the CMB temperature anisotropies in a wide frequency range which will allow an optimal removal of foregrounds and a very good control of systematics. From those high-quality images a complete coverage of multipoles at the cosmic variance limit will be obtained up to $\ell \approx 2500$. The C_ℓ so obtained will be the ultimate temperature power spectrum of the primary CMB anisotropies since secondary anisotropies are expected to start to be dominant beyond those multipoles.

The large-scale galaxy surveys as well as the high-z SN Ia light curves have also played a crucial role in establishing the concordance model. The strong complementarity of these two cosmological data sets with the CMB anisotropies has implied a significant improvement on the parameter determination compared to each data set considered individually. And, what is more important is a consistency check of the best fit model. In the next years, that combination of data sets is expected to be strengthened with weak lensing observations of background galaxies in large areas of the sky.

In addition to the power spectrum, the recent studies on the statistical distribution of the temperature anisotropies provide independent tests of the standard inflationary model. Significant deviations from a homogeneous and isotropic Gaussian random field have already been found in the first-year WMAP data. However, the origin of those deviations, whether cosmological, residual foregrounds or systematics, is still a matter of debate. More precised maps from next years of WMAP data and specially the high-quality maps expected from the Planck mission will certainly help in the clarification of the possible causes of the anomalies found.

Finally, high-sensitive polarization maps to search for the B-mode will be the next major quest after Planck. The B-mode polarization is at present the best way to study the primordial background of gravitational waves expected in the standard model of inflation. It is therefore not surprising that both space agencies, NASA and ESA, have considered polarization missions in their future programs "Beyond Einstein" and "Cosmic Vision", respectively.

Acknowledgements

I thank J.L. Sanz and P. Vielva for useful comments on the manuscript. I acknowledge financial support from the Spanish MEC project ESP2004-07067-C03-01. I also acknowledge the use of LAMBDA, support for which is provided by the NASA Office of Space Science. I acknowledge the use of the software package CMBFAST (`http://www.cmbfast.org`) developed by Seljak and Zaldarriaga. The work has also used the software package HEALPix (`http://www.eso.org/science/healpix`) developed by K.M. Gorski, E.F. Hivon, B.D. Wandelt, J. Banday, F.K. Hansen and M. Barthelmann.

References

1. Abazajian, K., et al.: AJ **128**, 502 (2004)
2. Adler, R.J., Bjorken, J.D., Overduin, J.M.: gr-qc/0602102 (2006)
3. Aghanim, N., Desert, F.X., Puget, J.L., Gispert, R.: A&A **311**, 1 (1996)
4. Aghanim, N., Forni, O.: A&A **347**, 409 (1999)
5. Aghanim, N., Forni, O., Bouchet, F.R.: A&A **365**, 341–346 (2001)
6. Aliaga, A.M., Martínez-González, E., Cayón, L., Argüeso, F., Sanz, J.L., Barreiro, R.B.: New Astron. Rev. **47**, 821 (2003)
7. Aliaga, A.M., Rubiño-Martín, J.A., Martínez-González, E., Barreiro, R.B., Sanz, J.L.: MNRAS **356**, 1559 (2005)
8. Afshordi, N., Loh, Y.-S., Strauss, M.A.: Phys. Rev. D **69**, 3524 (2004)
9. Argüeso, F., González-Nuevo, J., Toffolatti, L.: ApJ **598**, 86–96 (2003)
10. Aurich, R., Lustig, S., Steiner, F.: MNRAS **369**, 240, astro-ph/0510847 (2006)
11. Banday, A.J., Zaroubi, S., Gorski, K.M.: ApJ **533**, 575–587 (2000)
12. Barreiro, R.B., Sanz, J.L., Martínez-González, E., Silk, J.: MNRAS **296**, 693 (1998)
13. Barreiro, R.B., Hobson, M.P., Lasenby, A.N., Banday, A.J., Gorski, K.M., Hinshaw, G.: MNRAS **318**, 475–481 (2000)
14. Barreiro, R.B., Hobson, M.P.: MNRAS **327**, 813, astro-ph/0104300 (2001)
15. Barreiro, R.B., Martínez-González, E., Sanz, J.L.: MNRAS **322**, 411 (2001)
16. Barreiro, R.B.: Data Analysis in Cosmology. Springer, Heidelbery, astro-ph/0512538 (2006)
17. Barreiro, R.B., Martínez-González, E., Vielva, P., Hobson, M.P.: MNRAS **368**, 226 (2006)
18. Barrow, J.D., Juszkiewicz, R., Sonoda, D.H.: MNRAS **213**, 917 (1985)
19. Bartelmann, M., Schneider, P.: Phys. Rep. **340** 291 (2001)
20. Bartolo, N., Komatsu, E., Matarrese, S., Riotto, A.: Phys. Rep. **402**, 103 (2004)
21. Bennett, C.L., et al.: ApJS **148**, 1 (2003)
22. Bevis, N., Hindmarsh, M., Kunz, M.: Phys. Rev. D **70**, 043508 (2004)
23. Bielewicz, P., Eriksen, H.K., Banday, A.J., Gòrski, K.M., Lilje, P.B.: ApJ **635**, 750 (2005)
24. Boldt, E.: Phys. Rev. **146**, 215 (1987)
25. Bond, J.R., Efstathiou, G.: MNRAS **226**, 655 (1987)
26. Bond, J.R., et al.: ApJ **626**, 12 (2005)
27. Bouchet, F.R., Benoît, A., Camus, Ph., Désert, F.X., Piat, M., Ponthieu, N.: In: *SF2A-2005*, Casoli F., et al. (eds.) Charting the New Frontier of the Cosmic Microwave Background Polarization vol. 675. EDP Sciences, France (2005) astro-ph/0510423
28. Boughn, S.P., Crittenden, R.G.: Nature **427**, 45 (2004)
29. Bridges, M., McEwen, J.D., Lasenby, A.N., Hobson, M.P.: MNRAS **377**, 1473 (2007)
30. Bromley, B.C., Tegmark, M.: ApJ **524**, L79–L82 (1999)
31. Bunn, E.F., Ferreira, P., Silk, J.: Phys. Rev. Lett. **77**, 2883 (1996)
32. Bunn, E.F., Zaldarriaga, M., Tegmark, M., de Oliveira-Costa, A.: Phys. Rev. D **67**, 023501 (2003)
33. Cabella, P., Liguori, M., Hansen, F.K., Marinucci, D., Matarrese, S., Moscardini, L., Vittorio, N.: MNRAS **358**, 684 (2005)

34. Cabella, P., Hansen, F.K., Liguori, M., Marinucci, D., Matarrese, S., Moscardini, L., Vittorio, N.: MNRAS **369**, 819, astro-ph/0512112 (2006)
35. Castro, P.G.: Phys. Rev. D **67**, 123001, astro-ph/0212500 (2003)
36. Cayón, L., Sanz, J.L., Martínez-González, E., Banday, A.J., Argüeso, F., Gallegos, J.E., Gorski, K.M., Hinshaw, G.: MNRAS **326**, 1243 (2001)
37. Cayón, L., Argüeso, F., Martínez-González, E., Sanz: MNRAS **344**, 917 (2003)
38. Cayón, L., Martínez-González, E., Argüeso, F., Banday, A.J., Gorski, K.M.: MNRAS **339**, 1189 (2003)
39. Cayón, L., Jin, L., Treaster, A.: MNRAS **362**, 826 (2005)
40. Cayón, L., Banday, A.J., Jaffe, T., Eriksen, H.K., Hansen, F.K., Gorski, K.M., Jin, J.: MNRAS **369**, 598 (2006)
41. Challinor, A.: Data Analysis in Cosmology. Springer, Heidelberg, astro-ph/0502093 (2006)
42. Challinor, A., Chon, G.: MNRAS **360**, 509 (2005)
43. Chiang, L.-Y., Naselsky, P.D., Verkhodanov, O.V., Way, M.J.: ApJL **590**, 65 (2003)
44. Coles, P., Dineen, P., Earl, J., Wright, D.: MNRAS **350**, 989 (2004)
45. Colley, W.N., Gott, J.R.: MNRAS **344**, 686 (2003)
46. Condon, J.J., et al.: AJ **115**, 1693 (1998)
47. Contaldi, C.R., Ferreira, P.G., Magueijo, J.: ApJ **534**, 25 (2000)
48. Cooray, A.R., Hu, W: ApJ **534**, 533 (2000)
49. Cooray, A: Phys. Rev. **D64**, 063514, astro-ph/0105063 (2001)
50. Copi, C.J., Huterer, D., Starkman, G.D.: Phys. Rev. D **70**, 043515, astro-ph/0310511 (2004)
51. Cornish, N.J., Spergel, D.N., Starkman, G.D.: Phys. Rev. Lett. **77**, 215 (1996)
52. Cornish, N.J., Spergel, D.N., Starkman, G.D.: Class. Quant. Grav. **15**, 2657–2670 (1998)
53. Cresswell, J.G., Liddle, A.R., Mukherjee, P., Riazuelo, A.: Phys. Rev. D **73**, 041302, astro-ph/0512017 (2006)
54. Croft, R.A.C., et al.: ApJ **581**, 20 (2002)
55. Cruz, M., Martínez-González, E., Vielva, P., Cayón, L.: MNRAS **356**, 29 (2005)
56. Cruz, M., Tucci, M., Martínez-González, E., Vielva, P.: MNRAS **369**, 57 (2006)
57. de Bernardis, P., et al.: Nature **404**, 955 (2000)
58. de Oliveira-Costa, A., Smoot, G.F., Starobinsky, A.A.: ApJ **468**, 457 (1996)
59. de Oliveira-Costa, A., Tegmark, M., Zaldarriaga, M., Hamilton, A.: Phys. Rev. D **69**, 063516 (2004)
60. Delabrouille, R.B.: Data Analysis in Cosmology. Springer, Heidelberg (2006)
61. Dickinson, C., et al.: MNRAS **353**, 732 (2004)
62. Diego, J.M., Martínez-González, E., Sanz, J.L., Mollerach, S., Martínez, V.J.: MNRAS **306**, 427–436 (1999)
63. Dineen, P., Rocha, G., Coles, P.: MNRAS **358**, 1285 (2005)
64. Doré, O., Colombi, S., Bouchet, F.R.: MNRAS **344**, 905 (2003)
65. Eriksen, H.K., Novikov, D.I., Lilje, P.B., Banday, A.J., Gorski, K.M.: ApJ **612**, 64 (2004)
66. Eriksen, H.K., Hansen, F.K., Banday, A.J., Gorski, K.M., Lilje, P.B.: ApJ **605**, 14 (2004)
67. Eriksen, H.K., Banday, A.J., Gorski, K.M., Lilje, P.B.: ApJ **612**, 633 (2004)

68. Ferreira, P.G., Magueijo, J., Gorski, K.M.: ApJ **503**, L1–L4 (1998)
69. Fosalba, P., Gaztañaga, E., Castander, F.: ApJL, **597**, 89 (2004)
70. Fraisse, A.A.: Phys. Rev. Lett. submitted, astro-ph/0503402 (2005)
71. Freeman, P.E., Genovese, C.R., Miller, C.J., Nichol, R.C., Wasserman, L.: ApJ **638**, 1 (2006)
72. Freedman, W.L., et al.: ApJ **553**, 47 (2001)
73. Gaztañaga, E., Wagg, J.: Phys. Rev. D **68**, 021302 (2003)
74. Goldberg, D.M., Spergel, D.N.: Phys. Rev. D **59**, 103002 (1999)
75. Gorski, K.M., Hivon, E., Banday, A.J., Wandelt, B.D., Hansen, F.K., Reinecke, M., Bartelman, M.: ApJ **622**, 759 (2005)
76. Gott, J.R. III, Park, C., Juszkiewicz, R., Bies, W. E., Bennett, D.P., Bouchet, F.R., Stebbins, A.: ApJ **352**, 1 (1990)
77. Gruzinov, A., Hu, W.: ApJ **508**, 435 (1998)
78. Hajian, A., Souradeep, T.: ApJ **597**, 5 (2005)
79. Hajian, A., Souradeep, T., Cornish, N.: ApJ **618**, 63 (2005)
80. Hanany, S., et al.: ApJ **545**, L5 (2000)
81. Hansen, F.K., Banday, A.J., Gorski, K.M.: MNRAS **354**, 641 (2004)
82. Hansen, F.K., Cabella, P., Marinucci, D., Vittorio, N.: ApJ **607**, L67 (2004)
83. Heavens, A.F.: MNRAS **299**, 805–808 (1998)
84. Heavens, A.F., Sheth, R.K.: MNRAS **310**, 1062 (1999)
85. Heavens, A.F., Gupta, S.: MNRAS **324**, 960, astro-ph/0010126 (2000)
86. Hinshaw, G., et al.: ApJ **148**, 135 (2003)
87. Hipólito-Ricaldi, W.S., Gomero, G.I.: Phys. Rev. D **72**, 103008 (2005)
88. Hu, W.: Phys. Rev. D **62**, 043007 (2000)
89. Hu, W.: Phys. Rev. D **64**, 083005 (2001)
90. Hu, W., Dodelson, S.: ARAA **40**, 171 (2002)
91. Hu, W., Okamoto, T.: ApJ **574**, 566 (2002)
92. Inoue, K.T.: Prog. Theor. Phys. **106**, 39, astro-ph/0102222 (2001)
93. Inoue, K.T., Silk, J.: ApJ **648**, 23 (2006)
94. Jaffe, T.R., Banday, A.J., Eriksen, H.K., Gorski, K.M., Hansen, F.K.: ApJL **629**, L1, astro-ph/0503213 (2005)
95. Jaffe, T.R., Hervik, S., Banday, A.J., Gorski, K.M.: ApJ **644**, 701, astro-ph/0512433 (2006)
96. Jassal, H.K., Bagla, J.S., Padmanabhan, T.: Phys. Rev. D **72**, 103503, astro-ph/0506748 (2005)
97. Jones, B.J.T.: Data Analysis in Cosmology. Springer, Heidelberg (2006)
98. Jones, W.C., et al.: ApJ **647**, 823, astro-ph/0507494 (2006)
99. Kamionkowski, M., Kosowsky, A., Sttebins, A.: Phys. Rev. D **55**, 7368 (1997)
100. Kendall, M.G., Stuart, A.: The Advance Theory of Statistics. Charles Griffin London: (1977)
101. Kesden, M., Cooray, A., Kamionkowski, M.: Phys. Rev. Lett. **89**, 011304 (2002)
102. Kesden, M., Cooray, A., Kamionkowski, M.: Phys. Rev. D **67**, 123507 (2003)
103. Knox, L., Scoccimarro, R., Dodelson, S.: Phys. Rev. Lett. **81**, 2004 (1998)
104. Knox, L., Song, Y.-S.: Phys. Rev. Lett. **89**, 011303 (2002)
105. Kogut, A., Banday, A.J., Bennett, C.L., Gorski, K.M., Smoot, G.F., Wright, E.L.: ApJ **464**, L29–L33 (1996)
106. Kogut, A., Hinshaw, G., Banday, A.J.: Phys. Rev. D **55**, 1901 (1997)
107. Kogut, A., et al.: ApJS **148**, 161 (2003)

108. Kovac, J.M., et al.: Nature **420**, 772 (2002)
109. Komatsu, E., Wandelt, B.D., Spergel, D.N., Banday, A.J., Gorski, K.M.: ApJ. **566**, 19 (2002)
110. Komatsu, E., et al.: ApJS **148**, 119 (2003)
111. Kunz, M., Banday, A.J., Castro, P.G., Ferreira, P.G., Gorski, K.M.: ApJ **563**, 99 (2001)
112. Kunz, M., Aghanim, N., Cayón, L., Forni, O., Riazuelo, A.: Phys. Rev. D **73**, 023511, astro-ph/0510164 (2006)
113. Kuo, C.L., et al.: ApJ **600**, 32 (2004)
114. Lachièze-Rey, M., Luminet, J.-P.: Phys. Rep. **254**, 135 (1995)
115. Land, K., Magueijo, J.: Phys. Rev. Lett. **95**, 071301 (2005)
116. Land, K., Magueijo, J.: MNRAS **367**, 1714, astro-ph/0509752 (2005)
117. Land, K., Magueijo, J.: MNRAS **357**, 994 (2005)
118. Landriau, M., Shellard, E.P.S.: Phys. Rev. D **69**, 023003 (2004)
119. Larson, D.L., Wandelt, B.D.: ApJ **613**, 85 (2004)
120. Larson, D.L., Wandelt, B.D.: Phys. Rev. D, submitted, astro-ph/0505046 (2005)
121. Leitch, E.M., Kovac, J.M., Halverson, N.W., Carlstrom, J.E., Pryke, C., Smith, M.W.E.: ApJ **624**, 10, astro-ph/0409357 (2005)
122. Levin, J.: Phys. Rep. **365**, 251–333 (2002)
123. Lewis, A., Challinor, A., Lasenby, A.: ApJ **538**, 473, `http://camb.info` (2000)
124. Lewis, A., Challinor, A., Turok, N.: Phys. Rev. D **65**, 023505 (2002)
125. Lewis, A.: Phys. Rev. D **70**, 043011 (2004)
126. Liddle, A.R., Lyth, D.H.: Cosmological Inflation and Large-Scale Structure. Cambridge University Press, Cambridge (2000)
127. Liguori, M., Hansen, F.K., Komatsu, E., Matarresse, S., Riotto, A.: Phys. Rev. D **73**, 043505 (2006)
128. Liu, X., Zhang, S.N.: ApJ **633**, 542 (2005)
129. Luminet, J.-P., Weeks, J.R., Riazuelo, A., Lehoucq, R., Uzan, J.-P.: Nature **425**, 593 (2003)
130. Lyth, D.H., Wands, D.: Phys. Lett. B **524**, 5 (2002)
131. Maffei, B., et al.: EAS Publications Series **14**, 251–256 (2005)
132. Magueijo, J.: ApJ **528**, L57–L60 (2000)
133. Magueijo, J., Medeiros, J.: MNRAS **351**, L1 (2004)
134. Martínez-González, E., Sanz, J.L., Silk, J.: ApJL **335**, L5 (1990)
135. Martínez-González, E., Sanz, J.L.: MNRAS **247**, 473 (1990)
136. Martínez-González, E., Sanz, J.L.: A&A **300**, 346 (1995)
137. Martínez-González, E., Gallegos, J.E., Argüeso, F., Cayón, L., Sanz, J.L.: MNRAS **336**, 22 (2002)
138. Martínez-González, E., Vielva, P.: The cosmic microwave background anisotropies: open problems. In: del Toro, C., et al. (eds.) JENAM Workshop: The Many Scales in the Universe, vol. 1, Kluwer, Dordrect astro-ph/0510003 (2005)
139. Masi, S., et al.: EAS Publications Series **14**, 87–92 (2005)
140. McDonald, P., et al.: ApJ **635**, 761, astro-ph/0407377 (2005)
141. McEwen, J.D., Hobson, M.P., Lasenby, A.N. Mortlock, D.J., MNRAS **359**, 1583 (2005)
142. McEwen, J.D., Vielva, P., Hobson, M.P., Martínez-González, E., Lasenby, A.N. MNRAS **373**, 1211, astro-ph/0602398 (2007)

143. McEwen, J.D., Hobson, M.P., Lasenby, A.N., Mortlock, D.J.: MNRAS **359**, 1583 (2005)
144. Mollerach, S., Gangui, A., Luchin, F., Matarrese, S.: ApJ **453**, 1 (1995)
145. Mollerach, S., Martínez, V.J., Diego, J.M., Martínez-González, E., Sanz, J.L., Paredes, S.: ApJ **525**, 17 (1999)
146. Monteserín, C., Barreiro, R.B., Sanz, J.L., Martínez-González, E.: MNRAS **360**, 9 (2005)
147. Monteserín, C., Barreiro, R.B., Martínez-González, E., Sanz, J.L.: MNRAS **371**, 312 (2006)
148. Montroy, T.E., et al.: ApJ **647**, 813, astro-ph/0507514 (2006)
149. Mukherjee, P., Hobson, M.P., Lasenby, A.N.: MNRAS **318**, 1157–1163 (2000)
150. Mukherjee, P., Wang, Y.: ApJ **613**, 51 (2004)
151. Naselsky, P.D., Novikov, D.I.: ApJ **507**, 31 (1998)
152. Naselsky, P.D., Doroshkevich, A.G., Verkhodanov, O.V.: ApJL **599**, L53 (2003)
153. Naselsky, P., Chiang, L.-Y., Olesen, P., Novikov, I.: Phys. Rev. D **72**, 063512 (2005)
154. Nolta, M.R., et al.: ApJ **608**, 10 (2004)
155. Novikov, D., Schmalzing, J., Mukhanov, V.F.: A&A **364**, 17–25 (2000)
156. Ostriker, J.P., Vishniac, E.T.: ApJ **306**, 510 (1986)
157. Pando, J., Valls-Gabaud, D., Fang, L.-Z.: Phys. Rev. Lett. **81**, 4568–4571 (1998)
158. Park, C.-G., Park, C., Ratra, B., Tegmark, M.: ApJ **556**, 852, astro-ph/0102406 (2001)
159. Park, C.-G.: MNRAS **349**, 313 (2004)
160. Patanchon, G., Cardoso, J.-F., Delabrouille, J., Vielva, P.: MNRAS **364**, 1185 (2005)
161. Percival, W.J., et al.: MNRAS **327**, 1297 (2001)
162. Perlmutter, S.: ApJ **517**, 565 (1999)
163. Phillips, N.G., Kogut, A.: ApJ **548**, 540–549 (2001)
164. Piacentini, F., et al.: ApJ **647**, 823, astro-ph/0507507 (2005)
165. The Plank Consortia, 2005, in Planck: The Scientific Programme, European Space Agency, ESA-SCI(2005)1
166. Pogosian, L., Tye, S.-H.H., Wasserman, I., Wyman, M.: Phys. Rev. D **68**, 0235506 (2003)
167. POLARBEAR, `http://bolo.berkeley.edu/polarbear`
168. Polenta, G., et al.: ApJ **572**, L27, astro-ph/0201133 (2002)
169. QUIET, `http://quiet.uchicago.edu`
170. Rapetti, D., Steven, S.W., Weller, J.: MNRAS **360**, 555 (2005)
171. Readhead, A.C.S., et al.: ApJ **609**, 498 (2004)
172. Readhead, A.C.S., et al.: Science **306**, 836 (2004)
173. Riazuelo, A., Uzan, J.-P., Lehoucq, R., Weeks, J.: Phys. Rev. D **69**, 103514, astro-ph/0212223 (2004)
174. Riess, et al.: ApJ **116**, 1009 (1998)
175. Riess, et al.: ApJ **560**, 49 (2001)
176. Rocha, G., Magueijo, J., Hobson, M., Lasenby, A.: Phys. Rev. D **64**, 063512, astro-ph/0008070 (2001)
177. Rocha, G., Cayón, L., Bowen, R., Canavezes, A., Silk, J., Banday, A.J., Gorski, K.M.: MNRAS **351**, 769, astro-ph/0205155 (2004)

178. Roukema, B.: MNRAS **312**, 712 (2000)
179. Roukema, B.F., Lew, B., Cechowska, M., Marecki, A., Bajtlik, S.: A&A **423**, 821, astro-ph/0402608 (2004)
180. Rees, M.J., Sciama, D.W.: Nature **517**, 611 (1968)
181. Rubiño-Martín, et al.: MNRAS **369**, 909 (2006)
182. Sachs, R.K., Wolfe, A.M.: ApJ **147**, 73 (1967)
183. Santos, M.G., et al.: Phys. Rev. Lett. **88**, 241302, astro-ph/0107588 (2002)
184. Sanz, J.L.: In: Lineweaver, C.H., et al. (eds.) The Cosmic Microwave Background, p. 33. Kluwer Academic Publishers, Dordrect (1997)
185. Savage, R., et al.: MNRAS **349**, 973 (2004)
186. Scannapieco, E., Levin, J., Silk, J: MNRAS **303**, 797 (1999)
187. Schmalzing, J., Gorski, K.M.: MNRAS **297**, 355 (1998)
188. Shandarin, S.F., Feldman, H.A., Xu, Y., Tegmark, M.: ApJS **141**,1, astro-ph/0107136 (2002)
189. Seljak, U., et al.: Phys. Rev. D **71**, 103515 (2005)
190. Seljak, U., Hirata, C.M.: Phys. Rev. D **69**, 043005 (2004)
191. Seljak, U., Sugiyama, N., White, M., Zaldarriaga, M.: Phys. Rev. D **68**, 3507 (2003)
192. Seljak, U., Zaldarriaga, M.: ApJ **469**, 437 (`http://www.cmbfast.org`) (1996)
193. Seljak, U., Zaldarriaga, M.: Phys. Rev. D **60**, 043504 (1999)
194. Silk, J.: ApJ **151**, 459 (1968)
195. Skrutskie, M.F., et al.: In: Gazón, F., et al. (eds.) The Impact of Large Scale Near-IR Sky Survey, p. 187. Kluwer, Dordrecht (1997)
196. Slosar, A., Seljak, U.: Phys. Rev. D **70**, 083002 (2004)
197. Smith, S., et al.: MNRAS **352**, 887 (2004)
198. Smoot, G.F., et al.: ApJL **396**, L1 (1992)
199. Spergel, D.N., et al.: ApJS **148**, 175 (2003)
200. Sunyaev, R.A., Zeldovich, Y.B.: Comm. Astrophys. Space Phys. **4**, 173 (1972)
201. Schwartz, D.J., Starkman, G.D., Huterer, D., Copi, C.J.: Phys. Rev. Lett. **93**, 1301 (2004)
202. Starobinsky, A.A.: JETP Lett. **57**, 622 (1993)
203. Tegmark, M., de Oliveria-Costa, A., Hamilton, A.J.: Phys. Rev. D **68**, 123523 (2003)
204. Tegmark, M., et al.: ApJ **606**, 702 (2004)
205. Tegmark, M. et al.: Phys. Rev. D **69**, 103501 (2004)
206. Tenorio, L., Jaffe, A.H., Hanany, S., Lineweaver, C.H.: MNRAS **310**, 823 (1999)
207. Tonry, J.L., et al.: ApJ **594**, 1 (2003)
208. Tucci, M., Martínez-González, E., Vielva, P., Delabrouille, J.: MNRAS **360**, 935 (2005)
209. Toffolatti, L., Negrello, M., González-Nuevo, J., de Zotti, G., Silva, L., Granato, G.L., Argüeso, F.: A&A **438**, 475 astro-ph/0410605 (2005)
210. Tojeiro, R., Castro, P.G., Heavens, A.F., Gupta, S.: MNRAS **365**, 265 (2006)
211. Tomita, K.: Phys. Rev. D **72**, 103506 (2005)
212. Turok, N., Spergel, D.: Phys. Rev. D **64**, 2736 (1990)
213. Uzan, J.-P., Lehoucq, R., Luminet, J.-P.: In: Aubourg, E., Montmerle, T., Paul, J., Peter, P. (eds.) Proceedings of the XIXth Texas Symposium, Paris 14–18 December 1998 North Holland, The Netherlands, article-no: 04/25 (1999)
214. Vale, C.: ApJL, submitted, astro-ph/0509039 (2005)

215. Vielva, P., Martínez-González, E., Barreiro, R.B., Sanz, J.L., Cayón, L.: ApJ **609**, 22 (2004)
216. Vielva, P., Martínez-González, E., Tucci, M.: MNRAS **365**, 891, astro-ph/0408252 (2006)
217. Vilenkin, A., Shellard, E.P.S.: Cosmic Strings and Other Topological Defects, Cambridge University Press, Cambridge (1994)
218. Vishniac, E.T.: ApJ **322**, 597 (1986)
219. Wandelt, B.D., Hivon, E., Gorski, K.M.: Topological analysis of high-Resolution CMB maps. In: Tran Thanh Van (eds.) Fundamental Parameters in Cosmology, proceedings of the XXXIIIrd Rencontres de Moriond, Frontières 1998, astro-ph/9803317 (1998)
220. Weeks, J.: Class. Quant. Grav. **15**, 2599 (1998)
221. Weeks, J.R.: astro-ph/0412231 (2004)
222. White, N.E.: Adv. Space Res. **35**, 96 (2005)
223. Wiaux, Y., Jaques, L., Vandergheynst, P.: ApJ **652**, 820, astro-ph/0508516 (2005)
224. Wiaux, Y., Vielva, P., Martínez-González, E., Vandergheynst, P.: Phys. Rev. Lett. **96**, 151303 (2006)
225. Wu, J.H.P.: New method of extracting non-gaussian signals in the CMB. In: Wheeler, J.C., Martel, H. (eds.) 20th Texas Symposium on relativistic astrophysics, Austin, Texas, 10–15 December, vol. 586, p. 211, AIP Conference Proceedings, New York (2000)
226. Wu, J.H.P., Balbi, A., Borril, J., Ferreira, P.G., Hanany, S., Jaffe, A.H., Lee, A.T., Rabii, B., Richards, P.L., Smoot, G.F., Stompor, R., Winant, C.D.: Phys. Rev. Lett. **87**, 251303, astro-ph/0104248 (2001)
227. Wu, J.-H.P.: astro-ph/0501239 (2005)
228. Wyman, M., Pogosian, L., Wasserman, I.: Phys. Rev. D **72**, 023513 (2005)
229. Yoshida, N., Sheth, R.K., Diaferio, A.: MNRAS **328**, 669, astro-ph/0104332 (2001)
230. Zaldarriaga, M., Seljak, U.: Phys. Rev. D **55**, 1830 (1997)
231. Zaldarriaga, M., Seljak, U.: Phys. Rev. D **58**, 023003 (1998)

Cosmic Microwave Background Polarization Analysis

A. Challinor

Astrophysics Group, Cavendish Laboratory, Madingley Road,
Cambridge, CB3 0HE, U.K.
a.d.challinor@mrao.cam.ac.uk

Abstract. With polarization of the cosmic microwave background (CMB) now detected, and confirmed by several independent experiments, the next goal is to characterize accurately its statistical properties. In these lecture notes we review the physical motivation for pursuing CMB polarization and the basic statistical properties of the polarization fields. We then discuss some of the key aspects of the analysis of CMB polarization data, focusing on the additional complications that arise compared to temperature data due to the tensor character of the polarization field.

1 Introduction

The observed statistical properties of the temperature anisotropies in the cosmic microwave background (CMB) have played a key role in shaping our understanding of the large-scale properties of the Universe. The angular power spectrum of the anisotropies, quantifying the contribution to the variance of the anisotropies as a function of angular scale, has now been measured exquisitely over two decades of scale by the Wilkinson Microwave Anisotropy Probe (WMAP) satellite [1]. A number of ground and balloon-borne experiments extend the power spectrum measurements by a further decade in scale but with somewhat less precision than has been attained on the larger scales (see [2] for a recent review). Current CMB data is remarkably consistent with predictions based on the passive evolution of density perturbations, with adiabatic initial conditions drawn from a Gaussian distribution with an almost scale-invariant power spectrum, in a spatially flat Universe. Such primordial fluctuations, and the observed flat geometry of space, are natural outcomes of the simplest models of inflation in the early Universe, e.g. [3].

In addition to having angular variations in the total intensity, with r.m.s. $\sim 120\,\mu\mathrm{K}$, the CMB is also partially linearly polarized with r.m.s. $\sim 6\,\mu\mathrm{K}$. Polarization is generated by Thomson scattering of CMB photons around the time of recombination, once scattering becomes sufficiently rare to allow the

Challinor, A.: *Cosmic Microwave Background Polarization Analysis*. Lect. Notes Phys. **665**, 121–158 (2009)
DOI 10.1007/978-3-540-44767-2_5

development of anisotropies in the CMB intensity. Polarization is further generated at late times by re-scattering once the Universe reionizes. Despite being predicted shortly after the discovery of the CMB [4], linear polarization was not detected until 2002 by the Degree Angular Scale Interferometer (DASI) team [5]. It has now been measured by two further groups [6, 7] and also detected via its correlation with the temperature anisotropies in the first-year WMAP data [8]. The direct measurements are currently much less precise than those for the temperature anisotropies, but they are fully consistent with predictions in models favoured by the temperature data. Future, precise measurements of CMB polarization will be very valuable, allowing accurate determination of several cosmological parameters that are almost degenerate in their effect on the temperature anisotropies [9]. The biggest winners will be the reionization history [10], with the temperature–polarization correlation observed by WMAP already constraining the optical depth back to reionization to be ~ 0.15 [8], and constraints on any gravitational-wave background [11, 12] and sub-dominant isocurvature modes [13].

As the accuracy and size of CMB data sets have grown, so have the demands on the techniques used to process the data. A standard set of compression and cleaning steps have emerged that ultimately distill $O(10^{10})$ time-ordered observations to several hundred power spectrum measurements, and only around 10 cosmological parameters and their errors (see, e.g. [14] for a review). Although the detail of some of these steps are instrument specific, most analyses can be regarded as (approximate) variants on a small number of generic algorithms. Three of the most important steps are reviewed in detail elsewhere in this volume: for map making and power spectrum estimation see the contribution by Borrill; for (astrophysical) component separation and foreground removal see that by Delabrouille. Most of the steps in the analysis of CMB polarization data are straightforward generalizations of those for the temperature, but some new issues do arise mainly due to the tensorial nature of the polarization field. Furthermore, the small amplitude of the polarization signal demands that even more careful attention be paid to potential systematic effects in the analysis of polarization data than for the temperature anisotropies. Inevitably, this makes polarization data processing more instrument specific and was the reason that the WMAP team did not release polarization maps with their first-year data.

The purpose of these lecture notes is to describe, in outline, the scientific motivation for pursuing CMB polarization and discuss some of the key steps in the analysis of polarization data. We shall focus on those parts of the analysis that are complicated by the different geometric character of linear polarization compared to temperature anisotropies. To keep the discussion general, we will not have much to say on detailed instrument-specific parts of the analysis, although these are clearly crucial to the success of any given observation. We start in Sect. 2 with a discussion of the statistics of the CMB polarization fields and their representation in terms of orthonormal basis functions on the sphere. Section 3 reviews the basic physics of CMB polarization, and the current detections and best upper limits on CMB polarization are discussed in Sect. 4. What we have already learnt from current polarization measurements, and what we may learn in the future, is described in Sect. 5. The analysis of polarization data is then treated in Sect. 6, where discussions of map making, astrophysical component separation, E- and B-mode separation, power spectrum estimation and non-Gaussian lensing issues can be found.

2 Statistics of CMB Polarization

The polarization on the sky along some line of sight $\hat{\boldsymbol{n}}$ is conveniently described in terms of Stokes-specific brightness parameters $I(\hat{\boldsymbol{n}})$, $Q(\hat{\boldsymbol{n}})$, $U(\hat{\boldsymbol{n}})$ and $V(\hat{\boldsymbol{n}})$. For the CMB the total intensity $I(\hat{\boldsymbol{n}})$ has a Planck spectrum along any line of sight, but with angular temperature variations $\Delta T(\hat{\boldsymbol{n}})$ at the 10^{-5} level on a 2.725-K background. In linear theory the anisotropic contribution to $I(\hat{\boldsymbol{n}})$ is thus proportional to $\Delta T(\hat{\boldsymbol{n}})$ and has a spectrum given by the derivative of the Planck function (with respect to temperature). The parameters Q and U describe linear polarization, and inherit the spectrum of the anisotropic part of $I(\hat{\boldsymbol{n}})$ since CMB polarization is produced by frequency-independent Thomson scattering. The parameter V describes circular polarization and is expected to vanish for the CMB.

For a line of sight $\hat{\boldsymbol{n}}$, the radiation is propagating along $-\hat{\boldsymbol{n}}$. If we introduce a pair of orthogonal directions in the surface of the sphere, and call these x and y, such that x, y and $-\hat{\boldsymbol{n}}$ form a right-handed basis, then the linear Stokes parameter Q can be defined operationally as the difference between the intensities transmitted by perfect polarizers oriented along x and y. Similarly, U is the difference in intensities if the polarizers are rotated by 45°. We shall adopt the convention that in a spherical-polar coordinate system, the local x-axis is along the direction of decreasing polar angle θ (i.e. north) and the y-axis is along the direction of increasing azimuthal angle ϕ (i.e. east). This is in accordance with the IAU recommendations, but differs from, e.g. [11] and the latest version (1.2) of the HEALPix package.[1] If the local x- and y-axes are rotated through an angle ψ in a right-handed sense about the propagation direction, Q and U transform as

$$\begin{aligned} Q &\to Q\cos 2\psi + U\sin 2\psi \, , \\ U &\to U\cos 2\psi - Q\sin 2\psi . \end{aligned} \tag{1}$$

These show that Q and U properly form the components (in an orthonormal basis) of a rank-2, symmetric trace-free tensor $\mathcal{P}^{ab}(\hat{\boldsymbol{n}})$ that is transverse to the line of sight:

$$\mathcal{P}^{ab}(\hat{\boldsymbol{n}}) = \frac{1}{2}[Q(\hat{\boldsymbol{\theta}}\otimes\hat{\boldsymbol{\theta}} - \hat{\boldsymbol{\phi}}\otimes\hat{\boldsymbol{\phi}}) - U(\hat{\boldsymbol{\theta}}\otimes\hat{\boldsymbol{\phi}} + \hat{\boldsymbol{\phi}}\otimes\hat{\boldsymbol{\theta}})]. \tag{2}$$

[1] `http://www.eso.org/science/healpix/`

This linear polarization tensor is proportional to the correlation tensor (at zero lag) of the electric field components for the radiation propagating along $-\hat{\boldsymbol{n}}$.

Any two-dimensional tensor with the symmetries of $\mathcal{P}_{ab}$ can be derived from two scalar fields, denoted P_E and P_B, via [11, 12]:

$$\mathcal{P}_{ab} = \nabla_{\langle a}\nabla_{b\rangle}P_E + \epsilon^c{}_{\langle a}\nabla_{b\rangle}\nabla_c P_B , \tag{3}$$

where angle brackets denote the symmetric, trace-free part of the enclosed indices, ∇_a is the covariant derivative on the sphere and ϵ_{ab} is the alternating tensor. The part derived from P_E is the *electric* (or gradient) part and that from P_B the *magnetic* (or curl) part of $\mathcal{P}_{ab}$. The divergence $\nabla^a\mathcal{P}_{ab}$ is a pure gradient if $P_B = 0$ and a curl if $P_E = 0$. The decomposition (3) is analogous to writing a vector field as a gradient and a divergence-free vector, since in two dimensions the latter is always of the form $\epsilon_a{}^b\nabla_b\chi$. To gain some intuition for E modes and B modes, consider the case where P_E and P_B behave locally like a plane wave across a small patch of the sky (that can accurately be treated as flat). The electric and magnetic contributions to the polarization are depicted in Fig. 1. Quite generally, for a given potential P, the magnetic polarization $\mathcal{P}^B_{ab} \equiv \epsilon^c{}_{\langle a}\nabla_{b\rangle}\nabla_c P$ generated from P is related to the electric polarization $\mathcal{P}^E_{ab} \equiv \nabla_{\langle a}\nabla_{b\rangle}P$ by $\mathcal{P}^B_{ab} = \epsilon^c{}_{(a}\mathcal{P}^E_{b)c}$ which is a right-handed rotation of $\mathcal{P}^E_{ab}$ by 45° about the line of sight.

The fields P_E and P_B are simple scalars that can be expanded in spherical harmonics with $l \geq 2$ as

$$P_E(\hat{\boldsymbol{n}}) = \sum_{lm}\sqrt{\frac{(l-2)!}{(l+2)!}}E_{lm}Y_{lm}(\hat{\boldsymbol{n}}) , \quad P_B(\hat{\boldsymbol{n}}) = \sum_{lm}\sqrt{\frac{(l-2)!}{(l+2)!}}B_{lm}Y_{lm}(\hat{\boldsymbol{n}}) . \tag{4}$$

(The normalization is conventional.) The polarization tensor can then be written as

$$\mathcal{P}_{ab}(\hat{\boldsymbol{n}}) = \frac{1}{\sqrt{2}}\sum_{lm} E_{lm}Y^E_{(lm)ab} + B_{lm}Y^B_{(lm)ab} , \tag{5}$$

where the trace-free, symmetric tensors

Fig. 1. Polarization patterns for a pure electric mode (*left*) and pure magnetic mode (*right*) on a small patch of the sky for potentials that are locally Fourier modes. The shading denotes the amplitude of the potential. For the electric pattern the polarization is aligned with or perpendicular to the Fourier wavevector depending on the sign of the potential; for the magnetic pattern the polarization is at 45°

$$Y^{E}_{(lm)ab} \equiv \sqrt{\frac{2(l-2)!}{(l+2)!}} \nabla_{\langle a}\nabla_{b\rangle} Y_{lm} ,$$

$$Y^{B}_{(lm)ab} \equiv \sqrt{\frac{2(l-2)!}{(l+2)!}} \epsilon^{c}{}_{(a}\nabla_{b)}\nabla_{c} Y_{lm} , \tag{6}$$

form orthonormal tensor bases for electric and magnetic polarization, respectively. The electric and magnetic multipoles, E_{lm} and B_{lm}, can be extracted directly from the polarization tensor by contraction with the appropriate tensor harmonic followed by integration over the sphere. For example,

$$E_{lm} = \sqrt{2} \int \mathcal{P}^{ab} Y^{E*}_{(lm)ab} \, \mathrm{d}\hat{\boldsymbol{n}} . \tag{7}$$

Methods of recovering E and B modes when observations cover only a limited part of the sky are discussed later in Sect. 6. Reality of P_E and P_B demands that $E^*_{lm} = (-1)^m E_{l\,-m}$, and similarly for B_{lm}.

We end this subsection by noting an alternative formalism for expressing symmetric, trace-free tensors on the sphere: the spin-weighted formalism [15], first introduced to CMB physics in [11]. The complex polarization $P \equiv Q + \mathrm{i}U$ is scaled by a phase factor $e^{-2\mathrm{i}\psi}$ under a right-handed rotation of the local x- and y-directions about the propagation direction, and so is defined to have spin -2. The appropriate basis functions for expanding functions of definite spin on the sphere are the spin-weighted spherical harmonics, ${}_{\pm 2}Y_{lm}(\hat{\boldsymbol{n}})$, which are related to the components of the tensor harmonics, $Y^{E}_{(lm)ab}$ and $Y^{B}_{(lm)ab}$, in a null dyad:

$$Y^{E}_{(lm)ab} = \frac{1}{\sqrt{2}} \left({}_{-2}Y_{lm} m_a m_b + {}_{2}Y_{lm} m^*_a m^*_b \right) ,$$

$$Y^{B}_{(lm)ab} = \frac{1}{\mathrm{i}\sqrt{2}} \left({}_{-2}Y_{lm} m_a m_b - {}_{2}Y_{lm} m^*_a m^*_b \right) . \tag{8}$$

Here the null (complex) vector $\boldsymbol{m} \equiv (\hat{\boldsymbol{\theta}} + \mathrm{i}\hat{\boldsymbol{\phi}})/\sqrt{2}$ satisfies $\boldsymbol{m} \cdot \boldsymbol{m} = 0$ and $\boldsymbol{m} \cdot \boldsymbol{m}^* = 1$. With Q and U expressed on the north–east basis of a spherical-polar coordinate system, the multipole expansion of the complex polarization is

$$(Q \pm \mathrm{i}U)(\hat{\boldsymbol{n}}) = \sum_{lm} (E_{lm} \mp \mathrm{i}B_{lm}) {}_{\mp 2}Y_{lm}(\hat{\boldsymbol{n}}) . \tag{9}$$

The electric and magnetic multipoles can be extracted using the orthonormality of the spin-weighted harmonics, e.g.

$$E_{lm} \mp \mathrm{i}B_{lm} = \int (Q \pm \mathrm{i}U) {}_{\mp 2}Y^*_{lm} \, \mathrm{d}\hat{\boldsymbol{n}} . \tag{10}$$

2.1 Polarization Power Spectra

The decomposition of the polarization field into electric and magnetic parts is invariant under rotations, and the electric and magnetic multipoles at fixed

l transform irreducibly under rotations, i.e. $E_{lm} \to D^l_{mm'} E_{lm'}$ where $D^l_{mm'}$ are the Wigner functions (see, e.g. [16]). Under the operation of parity, $(Q \pm \mathrm{i}U)(\hat{\boldsymbol{n}}) \to (Q \mp \mathrm{i}U)(-\hat{\boldsymbol{n}})$ so that $E_{lm} \to (-1)^l E_{lm}$ (electric parity) while $B_{lm} \to (-1)^{l+1} B_{lm}$ (magnetic parity). These transformations ensure that the potential P_E is a scalar under parity, $P_E(\hat{\boldsymbol{n}}) \to P_E(-\hat{\boldsymbol{n}})$, but P_B is a pseudo-scalar, $P_B(\hat{\boldsymbol{n}}) \to -P_B(-\hat{\boldsymbol{n}})$. The temperature anisotropies $\Delta T(\hat{\boldsymbol{n}})$ are a scalar function and so can be expanded in spherical harmonics in the usual way: $\Delta T(\hat{\boldsymbol{n}}) = \sum_{lm} T_{lm} Y_{lm}$. The multipoles T_{lm} have electric parity.

In the absence of parity-violating interactions, the homogeneous and isotropic background Universe on which cosmological perturbations are assumed to propagate can support an ensemble of fluctuations that is invariant under parity and rotations. What this means is that in the ensemble any fluctuation is just as likely as its parity-reversed, or rotated, counterpart. The implication of this for the statistics of the CMB is that any expectation value must be preserved if we rotate or invert the fields involved. For the two-point statistics, this limits the non-zero correlations between the observable multipoles to

$$\begin{aligned} \langle T_{lm} T^*_{l'm'} \rangle &= \delta_{ll'} \delta_{mm'} C_l^T \,, \qquad \langle E_{lm} E^*_{l'm'} \rangle = \delta_{ll'} \delta_{mm'} C_l^E \,, \\ \langle B_{lm} B^*_{l'm'} \rangle &= \delta_{ll'} \delta_{mm'} C_l^B \,, \qquad \langle T_{lm} E^*_{l'm'} \rangle = \delta_{ll'} \delta_{mm'} C_l^{TE} \,. \end{aligned} \tag{11}$$

For Gaussian fluctuations, such as those predicted by the simplest models of inflation, these power spectra C_l^T, C_l^E, C_l^B and C_l^{TE} contain all of the statistical information in the CMB. There have been several claims of violations of Gaussianity/rotational invariance in the 1-year WMAP temperature data, e.g. [17, 18, 19, 20, 21], but the issue of whether these are of primordial origin, or rather artefacts of the instrument or local Universe, is unresolved [22]. Example power spectra of the CMB observables for adiabatic density perturbations in a flat, ΛCDM model are shown in Fig. 2. The physics that gives rise to these spectra is reviewed briefly in Sect. 3.

It is also interesting to consider the two-point correlations between the Stokes parameters (and also the temperature anisotropies) in real space [23]. For polarization, the correlation will be manifestly rotationally invariant only if we work with Stokes parameters defined on a basis intrinsic to the two points, $\hat{\boldsymbol{n}}_1$ and $\hat{\boldsymbol{n}}_2$, under consideration. If we define the local x-direction by the geodesic connecting the two points and denote the polarization on these bases by an overbar, the correlations are [12, 24, 25]

$$\begin{aligned} \langle \bar{Q}(\hat{\boldsymbol{n}}_1) \bar{Q}(\hat{\boldsymbol{n}}_2) \rangle &= \frac{1}{2} \sum_l \frac{2l+1}{4\pi} [C_l^E (d^l_{2\,2} + d^l_{2\,-2})(\beta) + C_l^B (d^l_{2\,2} - d^l_{2\,-2})(\beta)] \,, \\ \langle \bar{U}(\hat{\boldsymbol{n}}_1) \bar{U}(\hat{\boldsymbol{n}}_2) \rangle &= \frac{1}{2} \sum_l \frac{2l+1}{4\pi} [C_l^E (d^l_{2\,2} - d^l_{2\,-2})(\beta) + C_l^B (d^l_{2\,2} + d^l_{2\,-2})(\beta)] \,, \end{aligned}$$

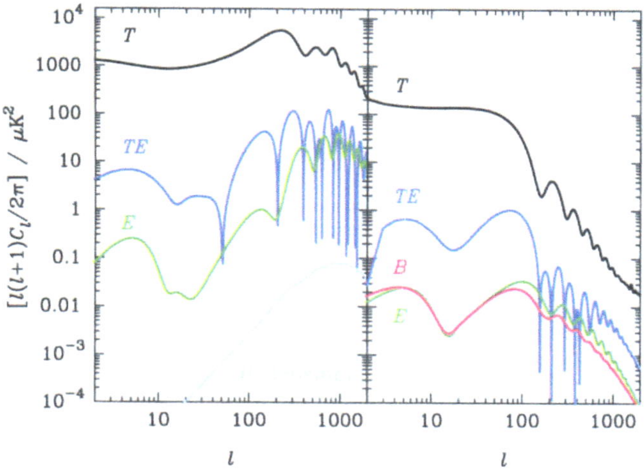

Fig. 2. Power spectra produced by adiabatic scalar perturbations (*left*) and tensor perturbations (*right*) for a tensor-to-scalar ratio $r = 0.38$

$$\langle T(\hat{\boldsymbol{n}}_1)\bar{Q}(\hat{\boldsymbol{n}}_2)\rangle = \sum_l \frac{2l+1}{4\pi} C_l^{TE} d^l_{2\,0}(\beta) \,,$$

$$\langle T(\hat{\boldsymbol{n}}_1)T(\hat{\boldsymbol{n}}_2)\rangle = \sum_l \frac{2l+1}{4\pi} C_l^{T} d^l_{00}(\beta) \,, \qquad (12)$$

where β is the angle between the two points ($0 \leq \beta \leq \pi$), and $d^l_{mn}(\beta)$ are the reduced Wigner functions. The correlations $\langle \bar{Q}\bar{U}\rangle$ and $\langle T\bar{U}\rangle$ vanish due to parity and rotational invariance. Examples of the non-zero polarization correlation functions are plotted in Fig. 3, along with a simulated Gaussian realization of the temperature anisotropies on a flat patch of the sky with the *correlated part* of the polarization overlaid. As first noted in [23], the negative tail to the temperature–polarization correlation function on large scales is generic, arising from the infall of the photon–baryon plasma into potential wells around recombination, and gives a tangential pattern of correlated polarization around large-scale temperature hot spots. On smaller scales the sign of the cross-correlation oscillates. For adiabatic perturbations on scales smaller than $\sim 0.3°$, the cross-correlation is positive and this gives rise to the tangential polarization pattern around small-scale cold spots that can be seen in Fig. 3.

Cosmic Variance

Since for Gaussian CMB fluctuations all cosmological information is encoded in the various power spectra, the precision with which the CMB can constrain cosmology is determined, in part, by the accuracy to which these spectra can

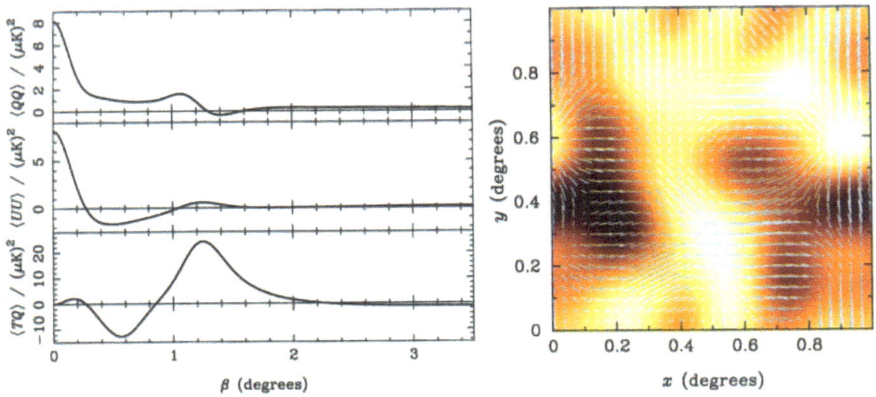

Fig. 3. Polarization correlation functions for adiabatic density perturbations (*left*). The right-hand plot overlays a simulation of the *correlated part* of the polarization on the CMB temperature anisotropies

be determined. In the absence of instrumental noise, and with full-sky coverage, cosmic variance provides a fundamental limit to this accuracy. Cosmic variance arises from having a single realization of a non-ergodic process (the sphere is compact) from which to estimate the power spectra. In this case, the obvious estimators of the power spectra, e.g.

$$\hat{C}_l^E = \frac{1}{2l+1} \sum_m |E_{lm}|^2 , \tag{13}$$

are optimal in that they have the minimum variance of all unbiased estimators. At each l we have only $2l+1$ independent real numbers from which to estimate a variance, so the cosmic variance in our estimator $\hat{C}_l^E$ for Gaussian fields is

$$\mathrm{var}(\hat{C}_l^E) = \frac{2}{2l+1}(C_l^E)^2 . \tag{14}$$

Equivalent results apply to the two other auto-spectra C_l^T and C_l^B. For the cross-spectrum, the cosmic variance is only slightly more complicated [12, 26]:

$$\mathrm{var}(\hat{C}_l^{TE}) = \frac{1}{2l+1}\left[(C_l^{TE})^2 + C_l^T C_l^E\right] . \tag{15}$$

The second term arises from chance correlations in the single realization that we have available, and would fundamentally limit the accuracy of searches for parity violations via, e.g. the cross-correlation between the temperature anisotropies and magnetic polarization. Due to the presence of correlations between the temperature and the electric polarization, estimates of C_l^T and C_l^E have non-vanishing covariance, and each is also correlated with the estimator for C_l^{TE}:

$$\mathrm{cov}(\hat{C}_l^T, \hat{C}_l^E) = \frac{2}{2l+1}(C_l^{TE})^2\,, \qquad \mathrm{cov}(\hat{C}_l^{T/E}, \hat{C}_l^{TE}) = \frac{2}{2l+1}C_l^{T/E}C_l^{TE}\,. \tag{16}$$

Of course, in the realistic case of non-zero instrument noise and limited sky coverage the precision of power spectrum estimates falls short of the cosmic variance limit. The effect of only observing a fraction f_{sky} of the sky is not only to increase the variance [27], but also to correlate estimates across Δl given roughly by the inverse linear size of the retained portion of the sky [28, 29]. The increase in variance can be crudely accounted for by reducing the number of independent modes to $(2l+1)f_{\mathrm{sky}}$. For polarization, this mode-counting argument ignores the loss of ambiguous modes that cannot be properly classified as electric or magnetic with partial sky coverage (see Sect. 6). This effect is minimized for surveys covering large connected regions of the sky.

3 Physics of CMB Polarization

When the mean free path was small compared to the spatial scale (i.e. wavelength for a plane wave) of the perturbations, Thomson scattering kept the CMB radiation isotropic in the rest frame of the electron–baryon plasma. As protons and electrons started to recombine to form neutral hydrogen, the mean free path increased and anisotropies started to develop. Subsequent Thomson scattering of the quadrupole component of the anisotropy generated linear polarization; this was then preserved for the free-streaming CMB until the Universe was reionized by the ionizing radiation from the first non-linear structures. The photon visibility function – the probability that a photon last scattered around recombination as a function of cosmic time – peaks around 370 kyr after the big bang and has a width ~115 kyr [30]. Since polarization is only generated by scattering it provides a very clean probe of conditions around the time of recombination and reionization.

Consider unpolarized radiation with a temperature distribution $\Theta(\boldsymbol{e})$, where $\boldsymbol{e}$ is the radiation propagation direction. If the radiation field possesses a quadrupole anisotropy $\Theta_2(\boldsymbol{e})$, Thomson scattering generates linear polarization with a polarization tensor [31, 32]

$$\mathcal{P}_{ab}(\boldsymbol{e}) = -\frac{1}{20}\mathrm{d}\tau\nabla_{\langle a}\nabla_{b\rangle}\Theta_2\,, \tag{17}$$

where $\mathrm{d}\tau$ is the differential optical depth, and the covariant derivatives are in the surface $\boldsymbol{e}^2 = 1$. Polarization is only generated from the quadrupole component of the incident radiation and can be seen to be purely electric quadrupole in character. If the incident radiation is polarized, there is an additional contribution to the polarization produced due to the electric quadrupole of the incident polarization (Θ_2 should be replaced by $\Theta_2 - 12P_{E,2}$ [31, 32]). For the CMB, the linear polarization we observe (at $\boldsymbol{x} = 0$) along a line of sight $\hat{\boldsymbol{n}}$ is that produced at last scattering along the radiation direction $\boldsymbol{e} = -\hat{\boldsymbol{n}}$, but at a direction-dependent comoving position $\boldsymbol{x} = \Delta\eta\hat{\boldsymbol{n}}$. (We are assuming the Universe is flat for simplicity, and $\Delta\eta$ is the conformal time between last scattering and the present.) The character of the *observed* polarization thus depends on the spatial distribution of the polarization at last scattering which, in turn, depends on the type of perturbations (e.g. density or gravitational waves) we consider.

3.1 Density Perturbations

If we consider a single plane-wave density perturbation of comoving wavenumber k, the peculiar velocity $\boldsymbol{v}_{\mathrm{b}}$ of the electron–baryon plasma around recombination is in the direction of the wavevector $\hat{\boldsymbol{k}}$. Over the mean free time since the previous scattering, a temperature quadrupole

$$\Theta_2(\boldsymbol{e}) \sim -\ell_p T_{\mathrm{CMB}} \nabla_{\langle i} v_{\mathrm{b},j\rangle} e^i e^j \propto T_{\mathrm{CMB}} \ell_p k P_2(\hat{\boldsymbol{k}} \cdot \boldsymbol{e}) \tag{18}$$

is produced to leading order in $\ell_p k$ where ℓ_p is the mean free path. The quadrupole is azimuthally symmetric about the wavevector and Thomson scatters to generate linear polarization parallel to the projection of $\hat{\boldsymbol{k}}$ on the sky (see Fig. 4). The polarization tensor at last scattering has a spatial dependence $\exp(\mathrm{i}\boldsymbol{k}\cdot\boldsymbol{x})$ appropriate to the plane wave, and this leads to a further modulation of the polarization observed today by $\exp(\mathrm{i}\Delta\eta\boldsymbol{k}\cdot\hat{\boldsymbol{n}})$. This modulation transfers polarization from the $l = 2$ to higher multipoles (with most appreciable power appearing at $l = k\Delta\eta$), but preserves the electric character of the polarization [11, 12]. To see that this is reasonable, note that the modulated polarization field has its polarization direction either parallel or

Fig. 4. Generation of linear polarization by Thomson scattering for a single Fourier mode of the density perturbations, following the treatment of [33]. The temperature anisotropies are azimuthally symmetric about the wavevector; the quadrupole part is shown on the left with cold lobes along the $\pm z$-directions. The linear polarization that is produced is then a pure $l = 2$, $m = 0$ electric polarization pattern locally (*right*). To get the observed polarization after free streaming from last scattering, the local polarization is modulated by the plane wave $\exp(i\boldsymbol{k}\cdot\hat{\boldsymbol{n}}\Delta\eta)$

perpendicular to the direction in which the polarization amplitude is changing [33]. Comparison with Fig. 1 shows that such a pattern is pure electric polarization. This important observation, that linear density perturbations do not produce magnetic polarization, can also be understood from the fact that the spatial distribution of $\nabla_{\langle i} v_{\mathrm{b},j\rangle}$ is curl free [32].

An example of the polarization power spectra produced by adiabatic density perturbations is plotted in Fig. 2. They were computed with the Boltzmann code CAMB [34]. The E-mode power peaks around $l \sim 1000$, corresponding to the angle subtended by the width of the visibility function at recombination. On larger scales the polarization probes the electron–baryon velocity at last scattering, as described above. Acoustic oscillations of the plasma prior to last scattering imprint an oscillatory structure on the angular power spectra of the CMB observables. Fourier modes of the photon density that are caught at the extrema of their oscillation at last scattering produce the peaks in the temperature-anisotropy power spectrum, but troughs in the polarization since $\boldsymbol{v}_{\mathrm{b}}$ then vanishes for such modes. Modes caught at either the extrema or the midpoint of their oscillation give zeroes in the temperature–polarization cross-correlation. Large-angle polarization from the last scattering surface is very small: the optimal wavelength for generating power at scale l is $l/\Delta\eta$, but this is too small to allow significant generation of quadrupole temperature anisotropies over a mean free path before last scattering. (Large-angle polarization is further suppressed since $\boldsymbol{v}_{\mathrm{b}}$ is small on super-Hubble scales.) The increase in polarization on large scales in Fig. 2 is due to reionization [10]. Re-scattering of the temperature quadrupole generates linear polarization peaking at multipoles roughly twice the ratio of the conformal look-back time to reionization to the conformal time elapsed between recombination and reionization. The amplitude of the polarization produced at reionization is proportional to the number of photons that scatter (i.e. the optical depth to reionization, taken to be $\tau = 0.15$ in Fig. 2).

Although linear density perturbations do not produce B-mode polarization, this is produced by a number of second-order effects. The most notable is weak gravitational lensing by large-scale structure [35], which effectively re-maps the polarization on the sky according to the lensing deflection field. This can distort a curl-free pattern into one with curl and so generate magnetic polarization. The power spectrum of this effect is also included in Fig. 2. On large scales the spectrum is almost white with $C_l^B \sim (1.3\,\mathrm{nK})^2$. The lens-induced B-mode spectrum peaks at $l \sim 1000$ corresponding to the peak in C_l^E.

3.2 Gravitational Waves

A stochastic background of gravitational waves, as predicted in all inflationary models, can also influence the CMB. Locally, a gravitational wave causes a quadrupole anisotropy in the expansion of space, so photons arriving at a scatterer from different lines of sight will suffer different redshifts since their previous scattering event. The longest wavelength gravitational waves are geometrically most efficient at generating a quadrupole moment. For wavelengths long compared to the mean free path, the quadrupole moment $\Theta_2(\boldsymbol{e}) \propto -\ell_p T_{\mathrm{CMB}} \dot{h}_{ij} e^i e^j / 2$, where h_{ij} is the transverse, trace-free perturbation to the Robertson–Walker metric,

$$\mathrm{d}s^2 = a^2(\eta)[\mathrm{d}\eta^2 - (\delta_{ij} + h_{ij})\mathrm{d}x^i \mathrm{d}x^j] , \tag{19}$$

and an overdot denotes differentiation with respect to proper time. However, the shear of the gravitational wave $\dot{h}_{ij}$ is zero for wavelengths well outside the horizon and undergoes damped oscillation inside the horizon. It follows that the temperature quadrupole at last scattering is dominated by modes with wavelength of the order of the horizon size, and this sets the characteristic scale ($l \sim 100$) on which the polarization power spectra peak (see Fig. 2). Thomson scattering of the temperature quadrupole from gravitational waves generates a polarization tensor at last scattering $\mathcal{P}_{ab} \propto [\dot{h}_{ab}]^{\mathrm{TT}}$, i.e. proportional to the trace-free projection of the shear onto the sphere $\boldsymbol{e}^2 = 1$. For a single plane wave with wavevector along the z-axis, the quadrupole moment and local polarization from the plane wave have only $m = \pm 2$ modes (see Fig. 5 and [31] for further details). The observed polarization from a plane wave is further modulated by $\exp(\mathrm{i}\Delta\eta \boldsymbol{k} \cdot \hat{\boldsymbol{n}})$ in free streaming from last scattering. Modulating the polarization in Fig. 5 gives a polarization pattern that locally looks electric along the lobes of the temperature quadrupole, and magnetic in between [33]. It follows that both electric and magnetic polarization are

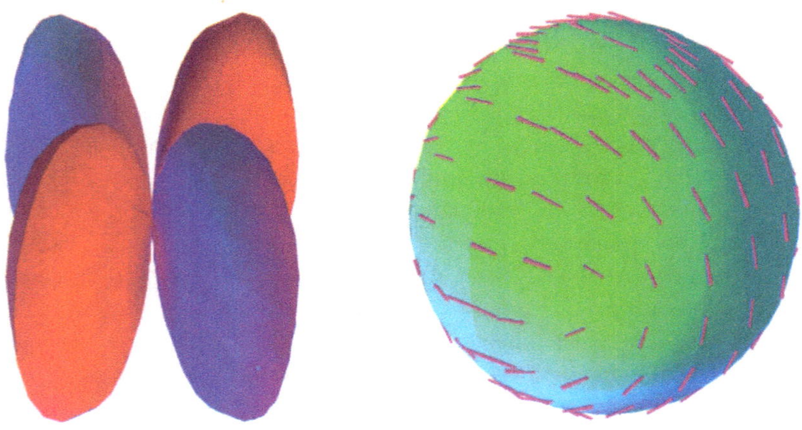

Fig. 5. As in Fig. 4 but for a Fourier mode of the gravitational wave perturbations. The quadrupole part of the temperature anisotropy now has $m = \pm 2$ modes (*left*); the hot lobes correspond to the directions in which space is locally being contracted by the wave. The linear polarization produced by Thomson scattering (*right*) is again electric and quadrupole in character at the last scattering surface. Unlike density perturbations, magnetic polarization is now generated by free streaming since the polarization field modulated by the plane wave has a non-zero curl

generated by gravitational waves [11, 12], and with roughly equal powers [31]. The generation of B-mode polarization from gravitational waves can also be understood as a consequence of the non-vanishing (spatial) curl of the shear $\dot{h}_{ij}$ at last scattering [32].

The power spectra of the CMB observables from a scale-invariant background of gravitational waves is shown in Fig. 2. As for density perturbations, large-angle polarization is generated by reionization. The amplitude of any gravitational wave background is currently only weakly constrained by CMB observations. A recent analysis, combining CMB observations and data from the Sloan Digital Sky Survey (SDSS), constrained the tensor-to-scalar ratio r – a measure of the ratio of the primordial power in gravitational waves to curvature (density) perturbations – to $r < 0.36$ at 95% confidence in a seven-parameter flat model. This is similar to the value adopted in Fig. 2, and shows that the gravitational wave contribution to the temperature anisotropies and electric polarization is sub-dominant.

4 Current Status of CMB Polarization Measurements

The first detection of polarization of the CMB was announced in September 2002 [5]. The measurements were made with DASI, a compact interferometric array operating at 30 GHz, deployed at the South Pole. The DASI team have now analyzed 3 years of data; they report a detection of E-mode polarization at high significance (6.3σ) and at a level perfectly consistent with predictions based on temperature-anisotropy data [36]. The only other reported direct detections to date are from the Cosmic Anisotropy Polarization MAPper (CAPMAP; [7]), a heterodyne correlation polarimeter operating at 90 GHz from New Jersey, and from the Cosmic Background Imager (CBI; [6]), an interferometer operating in the range 26–36 GHz from Chile. The results on C_l^E from maximum-likelihood power spectrum analyses by the DASI, CBI and CAPMAP teams are shown in Fig. 6, along with the prediction based on the best-fit to all the first-year WMAP data. Although the data are very noisy and broad-band (due to the limited sky coverage), the agreement with the prediction is striking. In addition, the TE cross-correlation has been measured with good precision by WMAP [8], and has also been detected by DASI (see Fig. 6).

The B-mode polarization of the CMB is currently subject only to weak upper limits. The best limit on large scales, $l \sim 10$, is from the POLAR experiment [37], while the best limits on smaller scales are from DASI [36]. These (95%) upper limits are shown in Fig. 6, along with the predicted B-mode spectrum in a model with tensor-to-scalar ratio $r = 0.38$, close to the current 95% limit from temperature anisotropies and SDSS [38]. Current direct constraints on the amplitude of gravitational waves from B-mode data are clearly much weaker than this. Furthermore, upper limits on the B-mode power on

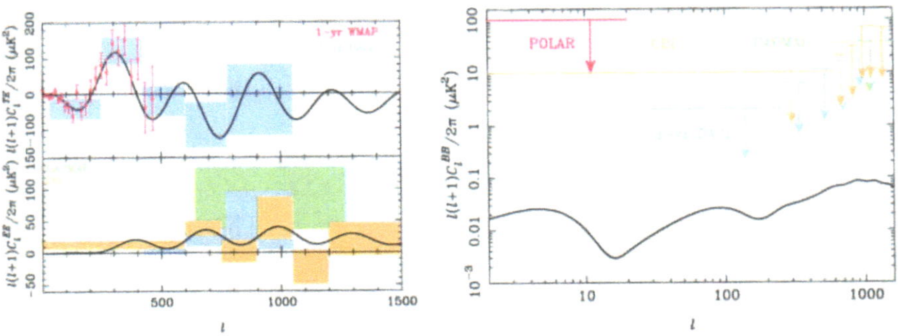

Fig. 6. Current polarization detections and best upper limits (as of January 2005). The left plot shows C_l^{TE} measurements. The points with $1-\sigma$ errors are from the first 1-year data release from WMAP. The error boxes are the flat bandpower results from DASI, CBI and CAPMAP data, centred on the maximum-likelihood bandpower and spanning the 68% intervals. The solid lines are the predicted power from the best-fit model to all the WMAP data. The right plot shows the upper limits on C_l^B from DASI, CBI, CAPMAP and POLAR. The solid line assumes a tensor-to-scalar ratio $r = 0.38$

small scales are at least two orders of magnitude above the level expected from gravitational lensing.

5 What Do We Learn from CMB Polarization?

Electric polarization primarily traces the velocity of the electron–baryon plasma at recombination due to density perturbations. It thus provides complementary information to the temperature anisotropies which are sourced mainly by the photon density variations at last scattering on sub-degree scales. Polarization observations can thus be expected to bring significant improvements on some of the 'acoustic' cosmological parameters [9], such as the physical density of baryons and cold dark matter, although current polarization data is not yet of sufficient quality relative to the temperature data to see this effect [6]. Furthermore, the *relative* phase of the acoustic peaks in the temperature and electric-polarization power spectra (and the cross-correlation) can be used for a largely model-independent test of the physics of acoustic oscillations. For both adiabatic and isocurvature perturbations the acoustic peaks in C_l^T and C_l^E should be in antiphase. (Admixtures of these perturbations can, however, partially destroy this coherence.) A recent analysis by the CBI team [6] found the phase of the oscillations in the E-mode spectrum to be fully consistent with expectations for adiabatic models. The spectrum C_l^{TE} from the 1-year WMAP data [8] has also been used to improve constraints on the amplitude of any isocurvature fluctuations, e.g. [39, 40, 41].

Polarization on scales of tens of degrees ($l < 10$) can be used to probe reionization. Significantly, it breaks the degeneracy between the amplitude of density perturbations, A_s, and the optical depth to reionization, τ, that leaves only $A_s e^{-2\tau}$ well determined from the temperature anisotropies. (The power spectra of the CMB observables are reduced by $e^{-2\tau}$ on scales corresponding to Fourier modes that are sub-horizon at reionization.) The 1-year WMAP measurement of C_l^{TE} shows a significant excess power on large scales that is attributed to early reionization ($11 < z_{\rm re} < 30$) giving a significant optical depth $\tau = 0.17 \pm 0.04$ (68% interval) [8]. The amplitude of curvature (density) perturbations (at 0.05 Mpc) is then constrained to $A_s = (2.7 \pm 0.3) \times 10^{-9}$ from the CMB alone [30]. Looking further ahead, the fine details of the large-angle polarization power can, in principle, distinguish different ionization histories with the same optical depth, although this is hampered by the large cosmic variance at low l [42, 43].

Perhaps the greatest ambition for CMB polarization observations is the detection of the B-mode signal from gravitational waves. Although gravitational waves do generate temperature anisotropies and electric polarization, they do so only on large scales where the cosmic variance of the dominant density perturbations is large. Indeed, a perfect temperature-anisotropy experiment can only detect $r > 0.07$, even assuming all other parameters are known. Supplementing the temperature anisotropies with electric polarization makes only a modest further improvement: $r > 0.02$. Although the current upper limit on r from B-mode polarization is much weaker than that inferred from the CMB temperature anisotropies (see Fig. 6), the former route ultimately has the potential to probe much lower values as thermal noise levels improve. Assuming that astrophysical foregrounds can be removed adequately, the ultimate limit to a detection of the gravitational wave signal is set by the B-modes generated by gravitational lensing [44, 45, 46].

A detection of a stochastic background of gravitational waves would have significant implications for fundamental physics. In inflation models, the power produced in gravitational waves as a function of wavelength is a direct measure of the expansion rate of the Universe at the time that wavelength was stretched beyond the horizon. In slow-roll inflation in a potential V, the power in gravitational waves thus constrains the 'energy scale of inflation', $V^{1/4}$, via

$$V^{1/4} = 3.33 \times 10^{16} r^{1/4} (A_s / 2.3 \times 10^{-9})^{1/4}\,\mathrm{GeV}\,. \tag{20}$$

The current 95% limit $r < 0.36$ [38] constrains $V^{1/4} < 2.3 \times 10^{16}$ GeV. By way of contrast, the recently introduced cyclic model [47] would produce negligible gravitational waves as the fluctuations are produced during a phase of very slow contraction [48]. A detection of cosmological gravitational waves would effectively rule out such models. In addition to constraining the energy scale of inflation, measuring r is important for constraining the *shape* of the inflaton potential. For the simplest models of slow-roll inflation, the observables r and n_s (the spectral index of curvature perturbations) are both required to specify the first two derivatives of the inflaton potential. Without the tight limits on r from future B-mode searches, constraints on inflation models will be largely degenerate along the direction of constant n_s. The most recent constraints on

r and n_s from the CMB and galaxy surveys already rule out some popular models such as $V \propto \phi^4$ [38].

There are several other potential sources of B-mode polarization that could dominate over any signal of primordial gravitational waves from inflation. A sub-dominant background of cosmic strings – currently enjoying a resurgence of interest following the realization that they are naturally produced in some models of brane inflation [49] – would provide a seed for gravitational waves (and vortex, or vector, modes) on the last scattering surface. Both give rise to B-mode polarization, and the resulting power spectrum is expected to peak on sub-degree scales [50]. It has been argued that a statistically isotropic, but highly non-Gaussian, pattern of B-mode polarization with power peaking on small scales would be a smoking-gun signature for a background of cosmic strings [51]. In addition, primordial magnetic fields have several interesting effects on the CMB and its polarization, although the theory is still rather underdeveloped. The anisotropic stress of a stochastic field also excites gravitational waves and vector modes (e.g. [52]) leading to (non-Gaussian) B-mode polarization. Faraday rotation by either large-scale coherent [53, 54] or stochastic fields [55] can also rotate electric polarization into magnetic. The hope is that these effects could be separated on the basis of their non-Gaussian statistical properties, and, in the case of Faraday rotation, by its frequency dependence.

Finally, we note that weak gravitational lensing effects on the CMB are a valuable source of additional cosmological information, most notably on the properties of dark energy and neutrino masses [56]. As we describe further in Sect. 6.6, CMB polarization is particularly useful here since intrinsically it has more power on small scales than the temperature and so is more sensitive to the small-scale power in the lensing deflection.

6 Polarization Data Analysis

The CMB polarization signal is small. Current measurements are consistent with the expected r.m.s. signal $\sim 6\,\mu\mathrm{K}$, which should be compared with the $\sim 120\,\mu\mathrm{K}$ r.m.s. of the temperature anisotropies. The low signal level requires high thermal sensitivity and demands exquisite control of systematic effects. The impact on the final science products of any systematics that cannot be designed out of the instrument must be carefully assessed (usually with a large number of simulations), and any that are found to be significant must be accounted for in the data analysis. As noted in the Introduction, this inevitably makes the analysis of polarization data more instrument specific than for temperature-anisotropy experiments. Despite this, there are a number of common steps in the analysis pipeline of any polarization experiment: map making from time-series data; astrophysical foreground removal; E–B mode separation; power spectrum estimation; and parameter estimation. (Separation of E and B modes can also be performed statistically during the estimation of power spectra, and this has been the case for all analyses of real data to date.) In the following subsections we briefly summarize most of these key steps, but focus on E–B separation and power spectrum estimation in particular. While the other steps are, of course, crucially important, the analysis methods employed there are relatively straightforward generalizations of those developed for the temperature anisotropies, for which excellent reviews already exist (see, e.g. [57]).

6.1 Modelling the Polarimeter Response

We begin by looking at how to model the response of a polarimeter, i.e. the relation between the observed data in the time domain to the fields on the sky. The most general (instantaneous) linear response of a polarimeter to the polarized sky brightness at each frequency is of the form

$$s \propto \int (\tilde{I}I + \tilde{Q}Q + \tilde{U}U - \tilde{V}V)\,\mathrm{d}\hat{\boldsymbol{n}}\,. \tag{21}$$

Here, $I(\hat{\boldsymbol{n}},\nu)$, $Q(\hat{\boldsymbol{n}},\nu)$, $U(\hat{\boldsymbol{n}},\nu)$ and $V(\hat{\boldsymbol{n}},\nu)$ are the Stokes brightness parameters at frequency ν, and the equivalent quantities with tildes refer to the instrument response. The latter parameters are determined by the instrument optics and receiver. They inherit a time-dependence not only from the scanning of the instrument on the sky – $\tilde{I}(\hat{\boldsymbol{n}},\nu)$ and $\tilde{V}(\hat{\boldsymbol{n}},\nu)$ rotate with the instrument as scalar functions and $\tilde{Q}(\hat{\boldsymbol{n}},\nu)$ and $\tilde{U}(\hat{\boldsymbol{n}},\nu)$ as components of a spin-2 function – but also from any modulation scheme employed to ameliorate the effects of low-frequency noise in the instrument. The signal contribution to the data is given by further integrating (21) over the spectral response of the instrument.

A useful idealization is an instrument with ideal optics and receiver, and a narrow spectral bandpass. For a dual-polarization system, such as a pair of polarization-sensitive bolometers on the high-frequency instrument of the Planck satellite, the beam patterns on the sky for the two polarization modes should have field directions that are orthogonal and constant, and the amplitude of the response should be azimuthally symmetric. In the case of the bolometric system, the ideal instrument response is then fully determined by $\tilde{I}(\hat{\boldsymbol{n}})$, which is a function of θ only. The response to circular polarization vanishes and $\tilde{Q} \pm \mathrm{i}\tilde{U} = -\tilde{I}(\theta)\mathrm{e}^{\pm 2i\phi}$ for the y-polarization, and minus this for the x-polarization. If we now rotate the instrument so the optic axis is along the direction $\hat{\boldsymbol{n}}$ and the local y-polarization direction is at an angle ψ to the ϕ-direction there, the signal received in the y-polarization simplifies to [58]

$$\int (\tilde{I}I + \tilde{Q}Q + \tilde{U}U - \tilde{V}V)\,\mathrm{d}\hat{\boldsymbol{n}} = I_\mathrm{s}(\hat{\boldsymbol{n}}) - Q_\mathrm{s}(\hat{\boldsymbol{n}})\cos 2\psi + U_\mathrm{s}(\hat{\boldsymbol{n}})\sin 2\psi\,, \tag{22}$$

where the beam-smoothed fields are

$$I_{\mathrm{s}}(\hat{\boldsymbol{n}}) = \sum_{lm} W_l T_{lm} Y_{lm}(\hat{\boldsymbol{n}}) , \tag{23}$$

$$(Q_{\mathrm{s}} \pm \mathrm{i}U_{\mathrm{s}})(\hat{\boldsymbol{n}}) = \sum_{lm} {}_2W_l (E \mp \mathrm{i}B)_{lm\,\mp 2}Y_{lm}(\hat{\boldsymbol{n}}) . \tag{24}$$

Here, the W_l is proportional to the $m = 0$ multipoles of $\tilde{I}(\theta)$ and ${}_2W_l$ to the $m = 2$ electric multipoles of $\tilde{Q} \pm \mathrm{i}\tilde{U}$. For a Gaussian beam of dispersion $\sigma \ll 1$, we have beam functions

$$W_l \approx \exp\left(-l(l+1)\sigma^2/2\right) \quad \text{and} \quad {}_2W_l \approx \exp\left(-[l(l+1)-4]\sigma^2/2\right) . \tag{25}$$

For $l \gg 1$ the polarized and total intensity beam functions are very nearly equal. The form of (22) is equivalent to a point sampling of the smoothed fields along the optic axis with a polarizer in orientation ψ. The signal in the x-polarization is got by replacing ψ by $\psi - \pi/2$. The generalization of (22) to other ideal polarimeters is straightforward: in each case the signal remains a linear combination of the beam-smoothed fields.

6.2 Polarized Map Making

The aim of map making is twofold: (i) to produce pixelized images of the temperature and polarization fields on the sky from the observed data; and (ii) to compress the time-stream data to a more manageable size, while preserving as much of the cosmological information present as possible. We shall only consider the ideal instrument model described above; for a case study of some of the real-world problems that must be solved in the map-making process for total intensity data see [59]. For the ideal model, the signal contribution to the data is linear in the beam-smoothed fields. Instrument noise is assumed additive, in which case our simple model for the data from the ith detector at time t becomes

$$d_t^i = A_{tp}^i (I_{\mathrm{s},p} - Q_{\mathrm{s},p} \cos 2\psi_t^i + U_{\mathrm{s},p} \sin 2\psi_t^i) + n_t^i . \tag{26}$$

Here, A_{tp}^i is the detector-pointing matrix and is unity if the beam centre of detector i points towards pixel p at time t, and is zero otherwise. The ith detector at time t is sensitive to polarization along a direction at angle ψ_t^i to the ϕ-direction on the sky, and has instrument noise n_t^i. We assume that all detectors have the same narrow spectral bandpass, and the same angular resolution, so that we can combine the signals of all detectors into a single vector $\boldsymbol{d}$ and the instrument noise into $\boldsymbol{n}$. We attempt to reconstruct the pixelized beam-smoothed fields, which we represent by the vector $\boldsymbol{s}$, by a regularized inversion of $\boldsymbol{d} = \mathsf{A}\boldsymbol{s} + \boldsymbol{n}$, where the $(it)p$ element of A is A_{tp}^i.

If the noise is Gaussian, with known covariance $\langle \boldsymbol{n}\boldsymbol{n}^T \rangle = \mathsf{N}$, the maximum-likelihood solution.

$$\hat{\boldsymbol{s}} = (\mathsf{A}^T\mathsf{N}^{-1}\mathsf{A})^{-1}\mathsf{A}^T\mathsf{N}^{-1}\boldsymbol{d}\,, \tag{27}$$

is optimal in the sense of being the minimum-variance unbiased estimate. The covariance of the errors is $\langle(\hat{\boldsymbol{s}} - \boldsymbol{s})(\hat{\boldsymbol{s}} - \boldsymbol{s})^T\rangle = (\mathsf{A}^T\mathsf{N}^{-1}\mathsf{A})^{-1}$. If the CMB fields are Gaussian, the maximum-likelihood map contains all of the cosmological information present in the time-stream data. In the language of statistics, the maximum-likelihood map is a sufficient statistic for parameter estimation, i.e. the data $\boldsymbol{d}$ only enters the posterior probability of the chosen (cosmological) parameters $\boldsymbol{\alpha}$ given the data, $\Pr(\boldsymbol{\alpha}|\boldsymbol{d})$, through $\hat{\boldsymbol{s}}$. Forming the maximum-likelihood map can thus also be viewed as lossless compression of the time-stream data. The question of the feasibility of computing the maximum-likelihood map is discussed in detail elsewhere in this volume by Borrill. Here we make just a few general remarks:

- Any matrix can be used in place of the noise covariance matrix N in (27) and the solution is still unbiased. However, if the matrix used does not properly capture the noise correlations described by off-diagonal elements of N, the resulting maps will have correlations along the scan directions ('stripes').
- Brute force evaluation of $\hat{\boldsymbol{s}}$ is prohibitive for large data sets so approximate methods must be used. For stationary noise, N is circulant and can be applied efficiently in the Fourier domain. Furthermore, direct inversion of $\mathsf{A}^T\mathsf{N}^{-1}\mathsf{A}$ can be avoided by solving $\mathsf{A}^T\mathsf{N}^{-1}\mathsf{A}\hat{\boldsymbol{s}} = \mathsf{A}^T\mathsf{N}^{-1}\boldsymbol{d}$ with iterative techniques.
- For stationary noise, the power spectrum (the diagonal elements of N in Fourier space) can be estimated from the data if it is not known a priori. Errors in the noise estimation increase the errors in the map.
- Maximum-likelihood map making is easily generalized to include template fitting and marginalization. For example, the unknown amplitude of some additive systematic effect whose 'shape' is known can be marginalized over during the map-making stage.
- Fast, instrument-specific alternatives to maximum-likelihood map making have been proposed, e.g. the 'destriping' technique for Planck [60].

6.3 Astrophysical Foreground Removal

The dominant diffuse polarized foregrounds in the frequency range 20–400 GHz are expected to be galactic synchrotron and vibrational (thermal) dust emission. Synchrotron radiation in a uniform magnetic field is linearly polarized. The polarization fraction depends on the energy spectrum of the electrons, but can plausibly be as high as 70%. Thermal dust emission will also be linearly polarized if non-spherical grains are aligned in a magnetic field. Any variation of the galactic magnetic field along the line of sight will tend to depolarize the radiation reducing the polarization fraction. This depolarization tends to cancel the increased emission along the plane making the polarization dependence on galactic latitude weaker than in total intensity.

Synchrotron and dust emission can be expected to produce both electric and magnetic polarization in similar amounts [61].

Synchrotron radiation is expected to be the dominant polarized foreground at frequencies below ~100 GHz. At the time of writing, the only well-surveyed part of the sky for which polarized data is available is within the galactic plane, and then only at frequencies below 2.7 GHz (e.g. [62]). The only high-latitude observations available are very patchy [63], and most are not useful for assessing foreground contamination in clean regions of the sky that will likely be the target of future deep polarization observations. (But see [64] for a useful exception.) Analysis of the 2.7 GHz data of the Southern Galactic plane from [62] gives E- and B-mode power spectra going as $l^{-1.5}$ [65], but care should be taken in extrapolating this to higher latitudes. A power law index of -1.5 is shallower than the large-angle polarization for both density perturbations and gravitational waves. Modelling at higher latitudes, based on total intensity observations (e.g. [65]), suggests that at frequencies around 100 GHz, synchrotron emission will not be a major foreground for E-mode polarization, but that its removal will be essential for large-angle B-mode searches. We eagerly await the release of polarization maps from the 2-year WMAP data; these should greatly improve our knowledge of high-latitude synchrotron polarization.

The power spectra from diffuse galactic polarized dust emission have been measured recently at 353 GHz by the Archeops team [66] over 20% of the sky. Extrapolating their measurements to 100 GHz, they found for $3 \leq l \leq 70$ that $(l+1)C_l^{E/B}/2\pi < 0.2\,\mu\text{K}^2$, excluding data with 5° of the galactic plane. Removal of the dust contamination is thus also likely to be necessary for future large-angle B-mode searches.

A variety of methods have been developed to remove diffuse foreground emission from CMB maps using multi-frequency data; see the contribution by Delabrouille in this volume for a review. Many of these were developed for total intensity data (temperature), but they generalize to polarization straightforwardly when applied directly to Stokes maps. Of these, some, such as Wiener filtering [67], require knowledge of the frequency spectra of all the components present and their angular power spectra. Others attempt to derive this information from the data by assuming statistical independence of the components [68, 69]. Finally, linear techniques, based on template fitting [1], or internal spectral combinations that preserve the CMB signal while minimizing the r.m.s. of the combined field can also be applied very effectively to polarization data provided that the spectral properties of the contaminants do not vary too greatly over the survey area. The results of applying these techniques to simulated polarization data are very encouraging (e.g. [70]), but we must await further observations to assess the true extent to which foregrounds will compromise the scientific returns of CMB polarization surveys.

Extra-galactic radio sources are a significant polarized contaminant at small angular scales. They contribute equally to E- and B-mode power and, for Poisson-distributed sources, have white angular power spectra. An effective way to deal with these is simply to exclude pixels within an observing beam of known sources above some flux cut. Since the degree of polarization is typically rather low ($< 10\%$), source contamination for E-mode polarization is relatively less troublesome than for temperature anisotropies. Having Stokes maps with a large number of excluded pixels does complicate the problem of separating E and B modes, and this is worse for low-resolution observations since the fraction of the area that is removed is then larger.

6.4 Map-Level E- and B-Mode Separation

Separating the recovered polarization field into its electric and magnetic parts amounts to solving for the potentials P_E and P_B in (3). This can be done trivially if the observation covers the full sky by using the orthogonality of the electric and magnetic tensor harmonics to recover the multipoles E_{lm} and B_{lm} as in (7). However, problems arise for observations over only part of the sky since the decomposition $\mathcal{P}_{ab} = \nabla_{\langle a}\nabla_{b\rangle}P_E + \epsilon^c{}_{\langle a}\nabla_{b\rangle}\nabla_c P_B$ is then not unique [71, 72]. Even if full sky observations were available, we would want to remove certain regions (e.g. the galactic plane) from the subsequent analysis, and this is most naturally done before the non-local decomposition into E and B modes.

To see how the ambiguity comes about, consider attempting to solve for P_E and P_B by recalling that the electric contribution to the divergence $\nabla^b\mathcal{P}_{ab}$ is curl-free and the magnetic part is divergence free, i.e.

$$2\nabla^a\nabla^b\mathcal{P}_{ab} = \nabla^2(\nabla^2+2)P_E \,, \tag{28}$$

$$-2\epsilon_{ac}\nabla^c\nabla^b\mathcal{P}_{ab} = \nabla^2(\nabla^2+2)P_B \,. \tag{29}$$

Solving these equations over part of the sphere requires one to specify boundary data, namely the value of the potentials and their normal derivatives on the boundary [71]. But this information is not known since it is the potentials that we are attempting to solve for. We can, of course, impose boundary conditions by force, in which case we should ensure that we do not produce spurious E or B modes, i.e. we should recover $P_B = 0$ if there are no B-modes present. This will be true if we solve (28) and (29) with the boundary conditions that both the potentials and their gradients vanish on the boundary [71, 72]. In this manner, we obtain a unique solution for P_E and P_B, but they are systematically wrong by ΔP_E and ΔP_B which are the unique solutions of $\nabla^2(\nabla^2+2)\Delta P_E = 0$ with $\Delta P_E = P_E$ and $\nabla_a\Delta P_E = \nabla_a P_E$ on the boundary (similar equations hold for ΔP_B). Reconstructing the polarization field from the recovered potentials, we see that this differs from the original field by $\Delta\mathcal{P}_{ab} \equiv \nabla_{\langle a}\nabla_{b\rangle}\Delta P_E + \epsilon^c{}_{\langle a}\nabla_{b\rangle}\nabla_c\Delta P_B$. This part of the polarization field, which satisfies $\nabla^a\nabla^b\mathcal{P}_{ab} = 0$ and $\epsilon_{ac}\nabla^c\nabla^b\Delta\mathcal{P}_{ab} = 0$, constitutes the ambiguous part since it cannot be classified in terms of electric and magnetic modes. Some examples of these ambiguous modes are illustrated in [71, 73]. A useful vector analogy to keep in mind is that the ambiguous modes are like the electromagnetic fields of (static) charges and currents that are localized outside the observed region.

By excluding the ambiguous modes from subsequent analysis, we are throwing away some information. As we might expect, the ambiguous modes tend to be localized near the boundary, and a rough accounting of the fractional loss of modes as a function of scale is given by the ratio of the product of the boundary length and the coherence length to the observed area.

Practical methods for performing the separation described above are given in [71, 72, 73]. The basic idea is to avoid differentiating the data by constructing instead of that tensor bases that can be used to project out the unambiguous E and B modes. Consider, for example, isolating magnetic polarization. If we construct a symmetric, trace-free tensor $W_{B\,ab} \equiv \sqrt{2}\epsilon^c{}_{(a}\nabla_{b)}\nabla_c W$ from a scalar field W that vanishes, along with its gradient, on the boundary but is otherwise arbitrary, contracting with the polarization tensor and integrating by parts gives [72]

$$\int W_B^{ab*}\mathcal{P}_{ab}\,\mathrm{d}\hat{\boldsymbol{n}} = -2\int W^*\epsilon^b{}_c\nabla^c\nabla^a\mathcal{P}_{ab}\,\mathrm{d}\hat{\boldsymbol{n}}\,, \tag{30}$$

where there are no boundary terms on account of the conditions imposed on W. It follows that the integral over the observed area is a non-local measure of the B-mode component. Similarly, the tensor $W_{E\,ab} \equiv \sqrt{2}\nabla_{\langle a}\nabla_{b\rangle}W$, with the same scalar field W, extracts a measure of the E-mode component. The construction of a complete set of such tensors can be performed in spherical-harmonic space [72, 73] (see also Sect. 6.5), or directly in pixel space [71]. An orthonormal tensor basis can be constructed by taking the W to be eigenfunctions of $\nabla^2(\nabla^2+2)$ (with the boundary conditions given above). Eigenfunctions with different eigenvalues are then orthogonal over the observed area, and the same is true of the tensors derived from them. Examples of $W_{B\,ab}$ basis functions for a circular patch of the sky are shown in Fig. 7. In this case the tensor bases can be constructed with W, an eigenfunction of ∂/∂_ϕ. For $m \geq 2$ exactly two B modes are lost per m; for $m = 1$ only one is lost; and for $m = 0$ none is lost. The basis functions in the figure are constructed to maximize the expected power in that mode.

The loss of information due to the ambiguous modes has interesting implications for survey design. For a fixed integration time, the 'best' strategy for measuring the variance of a Gaussian, statistically isotropic field, in the limit of low signal-to-noise, is to survey only a small field since the error on the variance scales as the square-root of the survey area. (Of course, this is not a wise strategy as you may be unlucky and choose a field where the signal happens to be very low.) Applying this to searches for gravitational waves with B-mode polarization, the implication is that a small survey is best. However, the coherence length of the desired signal is around a degree so the loss of information to ambiguous modes becomes significant for small fields. The optimum field size on this basis turns out to be a circular patch with radius around 10° [72]. This argument ignores the effect of weak gravitational lensing.

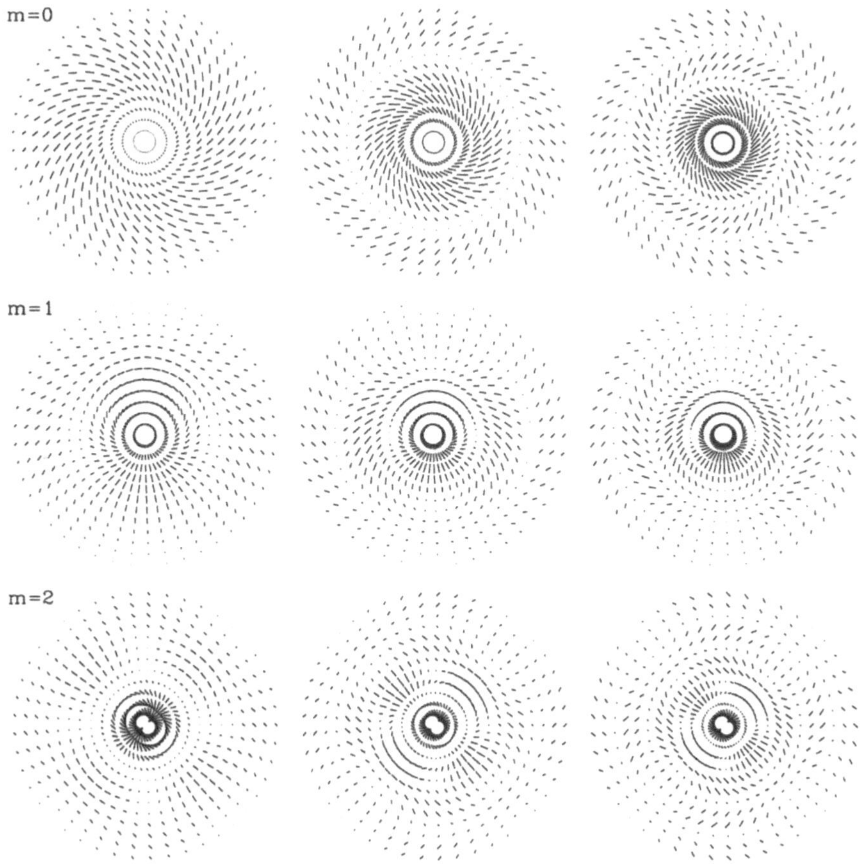

Fig. 7. Magnetic tensor basis functions $W_{B\,ab}$ for projecting onto pure magnetic-polarization modes for an azimuthally symmetric sky patch of radius 10° [72]. Only $m = 0$, 1 and 2 modes are shown. In each case only the three modes with the largest expected power are shown (with the left-most having greatest power). Figure from [72]

If we include the lens-induced B modes as an additional Gaussian noise term,[2] i.e. we make no attempt to clean out the lensing contribution as described in Sect. 6.6, the optimum size is increased further such that the contributions to the error on the variance from thermal noise and lensing sample variance are roughly equal.

We end this subsection by noting that coherent E–B separation is not strictly necessary for constraining cosmology. A statistical separation can be attempted during power spectrum estimation (see Sect. 6.5), and for the

[2] The B modes produced by lensing are not Gaussian. Their (connected) four-point function does have a significant effect on the variance and covariance of the estimated B-mode power spectrum [74].

optimal maximum-likelihood method this should be very effective. However, there are good reasons to perform a coherent, map-level separation as well. For example, visualizing the B modes will likely be a useful diagnostic of unknown systematic effects and foreground residuals.

6.5 Power Spectrum Estimation

The statistical properties of Gaussian random fields are fully specified by their power spectra. For this reason, power spectrum estimation is an important step in the analysis of CMB data. Ideally, this involves mapping out the likelihood $\Pr(\boldsymbol{s}|C_l^R)$, where $\boldsymbol{s}$ are the observed CMB fields, as a function of the power spectra C_l^R ($R = T$, E, B or TE). In practice, what is normally done is to approximate the likelihood as a Gaussian function of the power spectra, in which case all of the information contained in the original CMB maps is distilled into the position of maximum likelihood, and the curvature of the likelihood there [75]. In this process, we achieve a very significant compression of the CMB maps into, at most, a few thousand maximum-likelihood C_ls, and their associated covariances. Even this programme is not achievable exactly for large data sets, so a number of faster techniques based on estimators have been suggested. Here, the idea is to construct (unbiased) estimators of the power spectra, and to use their covariances – or, better, their full sampling distributions [76] – to constrain cosmology. A particularly fast class of estimators are the heuristically weighted, quadratic or 'pseudo-C_l' estimators (e.g. [77, 78]). For an interesting critique of pseudo-C_l and maximum-likelihood methods for temperature anisotropies, see [79]. For polarization, both methods can be applied directly to maps of the Stokes parameters, in which case one must rely on a statistical separation of the E and B modes, or after map-level E–B separation.

An important exception that demands a treatment beyond the power spectrum is the effect of weak gravitational lensing. Special techniques have been developed for this problem which allow reconstruction of the underlying Gaussian CMB and deflection fields (see Sect. 6.6). The power spectra of these reconstructed fields then contain all of the statistical (hence cosmological) information present in the non-Gaussian, lensed fields.

Maximum-Likelihood Methods

Maximum-likelihood methods, and the problems one faces in their practical implementation, are reviewed extensively elsewhere in this volume by Borrill. Here, we briefly summarize the technique. Maximum-likelihood methods have been used extensively in the analysis of real polarization data, e.g. from DASI [5] and CBI [6].

For the direct analysis of Stokes maps (i.e. assuming no E–B separation has been performed) we combine the observed Stokes parameters in a vector $\boldsymbol{s} = (I_p, Q_p, U_p)^T$. The covariance matrix of this vector, $\mathsf{C} = \mathsf{S} + \mathsf{N}$, has contributions S from the signal and N from the noise. The former is given essentially by the temperature and polarization correlation functions (12), but with the following modifications: (i) the power spectra are multiplied by the appropriate product of the beam functions W_l and ${}_2W_l$ to account for convolution with the instrument beam; and (ii) there are additional factors to account for the rotation of the Stokes parameters from the polar basis to the geodesic basis defined by the points p and p'. The noise contribution N includes the projection of the instrument noise onto the map, which will generally be modified by the foreground removal process, and an estimate of the covariance of any residual foreground errors. Both S and N have $(3N_{\rm pix})^2$ elements; S is dense in the pixel domain but is sparse in spherical-harmonic space, while N is typically sparse in the pixel domain. If map-level E–B separation has already been performed, the data $\boldsymbol{s}$ can be taken as the unambiguous E and B modes, and these can be analysed independently (although the noise may correlate the errors in the estimated E- and B-mode power spectra). Assuming Gaussian signals and errors, the likelihood function is

$$\Pr(\boldsymbol{s}|C_l^R) = \frac{1}{\sqrt{\det(2\pi\mathsf{C})}} \exp\left(-\boldsymbol{s}^T\mathsf{C}^{-1}\boldsymbol{s}/2\right) . \tag{31}$$

Maximizing the likelihood is equivalent to maximizing the posterior probability $\Pr(C_l^R|\boldsymbol{s})$ if a uniform prior on the power spectra, C_l^R, is adopted.

We now make some general remarks on this procedure.

- For full-sky coverage, and no noise, the log-likelihood is (up to an irrelevant constant)

$$-2\ln\Pr(\boldsymbol{s}|C_l^R) = \sum_l (2l+1)\left[\frac{1}{\Delta_l}(\hat{C}_l^T C_l^E + \hat{C}_l^E C_l^T - 2\hat{C}_l^{TE} C_l^{TE}) + \ln\Delta_l + \frac{\hat{C}_l^B}{C_l^B} + \ln C_l^B\right] , \tag{32}$$

  where $\Delta_l \equiv C_l^T C_l^E - (C_l^{TE})^2$. The likelihood is approximately a Gaussian function of the C_l^R at high l. The maximum-likelihood point coincides with the simple estimators, e.g. $\hat{C}_l^E = \sum_m |E_{lm}|^2/(2l+1)$ at low l.
- Generally, the maximum of the likelihood can be located by solving $\partial\ln\Pr(\boldsymbol{s}|C_l^R)/\partial C_l^R = 0$, with, e.g. the Newton–Raphson method. The data enters the gradient of the log-likelihood, and hence the maximum-likelihood point, in the form $\boldsymbol{s}^T\mathsf{C}^{-1}(\partial\mathsf{S}/\partial C_l^R)\mathsf{C}^{-1}\boldsymbol{s}$. This is proportional to the simple estimate $\hat{C}_l^R$, but with the spherical harmonic transforms taken only over the observed region, and with the data first weighted non-locally by C^{-1}.
- Approximating the likelihood as Gaussian requires the computation of the curvature of the likelihood. At any C_l^R, the expectation value of the second derivatives of (minus) the log-likelihood there, under the assumption that

these are the true power spectra, defines the Fisher matrix. The inverse of the Fisher matrix, evaluated at a smoothed version of the maximum-likelihood power spectra, is often quoted as the error on the power spectra. For the ideal case, where the log-likelihood is given by (32), the inverse of the Fisher matrix has elements given by the cosmic variance formulae of Sect. 2.1.

- A brute-force implementation to locate the maximum likelihood, and evaluate the Fisher matrix, is not feasible for mega-pixel data sets. Both the operation count, $O(N_{\text{pix}}^3)$, and the storage requirements are prohibitive.
- Additional regularization is required for small surveys since the problem is then degenerate if we attempt to solve for every multipole l. A common fix is to solve for bandpower amplitudes [75]. For narrow bands, approximating the $l(l+1)C_l$s as piecewise constant is common. As the polarization power spectra are steeper functions of l than C_l^T, particularly on large scales, shaped bandpowers are often employed in polarization analyses (e.g. [5]).

Maximum-likelihood estimation is closely related to the optimal quadratic estimators [80, 81]. These take the form

$$\hat{C}_l^R = \frac{1}{2}\sum_{l'R'} F^{-1}_{(lR)(l'R')}\text{trace}[(\boldsymbol{s}\boldsymbol{s}^T - \mathsf{N})\mathsf{C}^{-1}\partial\mathsf{S}/\partial C_{l'}^{R'}\mathsf{C}^{-1}], \tag{33}$$

where the Fisher matrix is $F_{(lR)(l'R')} \equiv \text{trace}[\mathsf{C}^{-1}\partial\mathsf{S}/\partial C_l^R\mathsf{C}^{-1}\partial\mathsf{S}/\partial C_{l'}^{R'}]/2$. The quadratic estimator is unbiased for any (invertible) matrix C, but the variance is minimized if C is truly the covariance of the data. Of course, this is not known a priori so we instead construct an approximation to C from some reasonable guess for the power spectra. It can be shown that iterating the quadratic estimator (i.e. constructing C for the next estimate from the current power spectrum estimates $\hat{C}_l^R$) is equivalent to performing a Newton–Raphson maximization of the likelihood, with the curvature approximated by the Fisher matrix [75]. The optimal quadratic estimator has the same computational problems as maximum-likelihood methods.

Heuristically Weighted, Quadratic Estimates

The computational complexity of the optimal quadratic estimator (33) arises from the non-local weighting of the data by the inverse covariance matrix C^{-1}. Heuristically weighted, quadratic methods [25, 76, 77, 78, 82] circumvent this problem by adopting a local weighting scheme. The use of fast-spherical transforms, as implemented in, e.g. the HEALPIX package [83], reduce the operations count to $O(N_{\text{pix}}^{3/2})$ which is fast enough to be used in high-volume simulations.

We shall concentrate here on the polarization auto-correlations only. The cross-correlation between temperature and polarization is straightforward

[25, 82]. Assuming we are analysing Stokes maps, we adopt a real weighting $w(\hat{\boldsymbol{n}})$ that is zero where we have no data. For the case of temperature anisotropies, weighting with the inverse noise variance on scales where the noise dominates the signal, and uniformly where the signal dominates, can be shown to be close to optimal [79]. Care should be taken in generalizing this reasoning to low-amplitude B-mode polarization because of the additional complication of E–B mixing. We extract the electric and magnetic multipoles of the weighted polarization field with spherical transforms,

$$\begin{aligned}\tilde{E}_{lm} \pm \mathrm{i}\tilde{B}_{lm} &\equiv \int w(\hat{\boldsymbol{n}})(Q \mp \mathrm{i}U)(\hat{\boldsymbol{n}})_{\pm 2}Y^*_{lm}(\hat{\boldsymbol{n}})\,\mathrm{d}\hat{\boldsymbol{n}}\,, \\ &= \sum_{(lm)'} {}_{\pm 2}I_{(lm)(lm)'}(E_{(lm)'} \pm \mathrm{i}B_{(lm)'})\,,\end{aligned} \tag{34}$$

where the Hermitian coupling matrices are

$${}_{\pm 2}I_{(lm)(lm)'} \equiv \int w(\hat{\boldsymbol{n}})_{\pm 2}Y_{(lm)'}(\hat{\boldsymbol{n}})_{\pm 2}Y^*_{lm}(\hat{\boldsymbol{n}})\,\mathrm{d}\hat{\boldsymbol{n}}\,, \tag{35}$$

and can be expressed in terms of the (scalar) multipoles, w_{lm}, of the weight function and the Wigner $3j$ symbols [82]. We have left the effect of smoothing by the instrument beam implicit in (34); the factors ${}_2W_l$ can be absorbed into the multipoles E_{lm} and B_{lm}, and, later, ${}_2W_l^2$ into the power spectra C_l^E and C_l^B. We compress the pseudo-multipoles into *pseudo-*C_l*s* according to

$$\tilde{C}_l^E \equiv \frac{1}{2l+1}\sum_m |\tilde{E}_{lm}|^2 \quad , \quad \tilde{C}_l^B \equiv \frac{1}{2l+1}\sum_m |\tilde{B}_{lm}|^2\,. \tag{36}$$

These quantities are equivalent to the $\boldsymbol{s}^T\mathsf{C}^{-1}(\partial\mathsf{S}/\partial C_l^R)\mathsf{C}^{-1}\boldsymbol{s}$ that appear in the optimal quadratic estimator (33) if we replace C^{-1} by heuristic weights.

E–*B* Mixing Revisited

We make a brief aside to note the relation between the pseudo-multipoles and the issue of separating E and B modes with incomplete sky coverage. The pseudo-multipoles mix E and B modes according to

$$\tilde{E}_{lm} = \sum_{(lm)'} \left({}_{+}I_{(lm)(lm)'}E_{(lm)'} + \mathrm{i}\,{}_{-}I_{(lm)(lm)'}B_{(lm)'}\right), \tag{37}$$

$$\tilde{B}_{lm} = \sum_{(lm)'} \left({}_{+}I_{(lm)(lm)'}B_{(lm)'} - \mathrm{i}\,{}_{-}I_{(lm)(lm)'}E_{(lm)'}\right), \tag{38}$$

where we have defined the Hermitian ${}_{\pm}I_{(lm)(lm)'}$ to be

$${}_{+}I_{(lm)(lm)'} \equiv \frac{1}{2}({}_{+2}I_{(lm)(lm)'} + {}_{-2}I_{(lm)(lm)'}) = \int wY^{E\,*}_{(lm)\,ab}Y^{E\,ab}_{(lm)'}\,\mathrm{d}\hat{\boldsymbol{n}}\,, \tag{39}$$

$${}_{-}I_{(lm)(lm)'} \equiv \frac{1}{2}({}_{+2}I_{(lm)(lm)'} - {}_{-2}I_{lm(lm)'}) = -i\int wY^{E\,*}_{(lm)\,ab}Y^{B\,ab}_{(lm)'}\,\mathrm{d}\hat{\boldsymbol{n}}. \tag{40}$$

The mixing matrix ${}_-I_{(lm)(lm)'}$ controls the mixing of E modes into $\tilde{B}_{lm}$, and vice versa. For the case of uniform weighting over the full sky, the mixing matrix vanishes and ${}_+I_{(lm)(lm)'} \propto \delta_{ll'}\delta_{mm'}$. For partial sky coverage, if $w(\hat{\boldsymbol{n}})$ is unity inside the observed region but zero elsewhere, integrating by parts in (40) shows that ${}_-I_{(lm)(lm)'}$ reduces to a line integral around the boundary [72]:

$$i\,{}_-I_{(lm)(lm)'} = \sqrt{\frac{2(l-2)!}{(l+2)!}} \oint \left(\nabla_b Y^*_{lm} Y^{B\,ab}_{(lm)'} - Y^*_{lm} \nabla_b Y^{B\,ab}_{(lm)'} \right) \mathrm{d}l_a \,. \qquad (41)$$

If we contract the $\tilde{B}_{lm}$ with the multipoles W^*_{lm} of any scalar function $W(\hat{\boldsymbol{n}}) = \sum_{lm} W_{lm} Y_{lm}(\hat{\boldsymbol{n}})$, such that W and $\nabla_a W$ vanish on the boundary of the observed region, we see that $\sum_{lm} W^*_{lm}\,{}_-I_{(lm)(lm)'} = 0$ and we will have projected out the E-mode contamination from the $\tilde{B}_{lm}$. Since the construction $\sum_{lm} W^*_{lm} \tilde{B}_{lm}$ is equivalent to integrating the contraction of $W_B^{ab\,*} = \sqrt{2}\epsilon^c{}_{(a}\nabla_{b)}\nabla_c W^*$ and $\mathcal{P}_{ab}$ over the observed region, isolating those functions in harmonic space that are orthogonal to the range of ${}_-I_{(lm)(lm)'}$ is equivalent to finding elements of the tensor basis of the unambiguous E and B modes over the observed region (see Sect. 6.4). The harmonic-space method for E–B separation developed in [72] uses singular value techniques to project out the range of the mixing matrix ${}_-I_{(lm)(lm)'}$.

For a more general weight function $w(\hat{\boldsymbol{n}})$, if we assume that it smoothly apodizes the edges of a connected region of linear dimension $\sim R$, the relative sizes of ${}_-I$ and ${}_+I$ at scale l^{-1} are in the ratio $1/(lR)$ [84]. It follows that the geometric effect of mode mixing is suppressed on scales small compared to the survey size.

Mean Values of the Pseudo-C_ls

Returning to the pseudo-C_ls, their mean values contain two contributions, one from the covariance of the instrument noise (and foreground residuals) and one from the CMB power spectra. The former can be estimated by using a large number of simulated observations: $\sim 10^4$ observations should give $\sim 1\%$ accuracy. The mean level in the simulations can then be subtracted from the observed $\tilde{C}_l$ to remove the noise bias in the mean. From now on we shall assume that the pseudo-C_ls have been corrected in this way. Alternatively, if only correlations between different instrument channels are used in the construction of the pseudo-$C_l s$, the noise bias will vanish if the noise is uncorrelated between channels [85]. The means of the pseudo-C_ls are then linearly related to the true power spectra:

$$\langle \tilde{C}^E_l \rangle = \sum_{l'} P_{ll'} C^E_{l'} + M_{ll'} C^B_{l'} \,, \qquad \langle \tilde{C}^B_l \rangle = \sum_{l'} M_{ll'} C^E_{l'} + P_{ll'} C^B_{l'} \,. \qquad (42)$$

Here, the matrices are $P_{ll'} \equiv \sum_{mm'} |{}_+I_{(lm)(lm)'}|^2/(2l+1)$, with a similar expression for $M_{ll'}$. They depend on the weight function $w(\hat{\boldsymbol{n}})$ only through the rotational-invariant $w_l \equiv \sum_m |w_{lm}|^2/(2l+1)$, and can be expressed in terms of $3j$ symbols as

$$(P/M)_{ll'} = \frac{2l'+1}{8\pi} \sum_L (2L+1) w_L [1 \pm (-1)^{l+l'+L}] \begin{pmatrix} l & l' & L \\ -2 & 2 & 0 \end{pmatrix}^2 . \quad (43)$$

For uniform weighting over the full sky, only w_0 is non-zero and so $M_{ll'}$ vanishes as it must (the $3j$ symbol forces $l = l'$ and hence $l + l' + L$ to be even). For weight functions that smoothly apodize observations over a region of linear size $\sim R$, the weight function is almost band-limited to $l_{\max} \sim 1/R$. In this case the $P_{ll'}$ and $M_{ll'}$ matrices are close to band diagonal, and, for spectra that are smooth on the scale of R^{-1}, the means of the pseudo-C_ls are approximately

$$\langle \tilde{C}_l^E \rangle \approx C_l^E \sum_{l'} P_{ll'} + C_l^B \sum_{l'} M_{ll'} , \qquad \langle \tilde{C}_l^B \rangle \approx C_l^E \sum_{l'} M_{ll'} + C_l^B \sum_{l'} P_{ll'} . \quad (44)$$

The relative size of the mixing of E- and B-mode power in the pseudo-C_ls is thus set by the ratio of the normalizations $\sum_{l'} M_{ll'}$ to $\sum_{l'} P_{ll'}$. For $l \gg 1/R$, it can be shown that [84]

$$\sum_{l'} P_{ll'} \approx \frac{1}{4\pi} \int w^2(\hat{\boldsymbol{n}}) \, \mathrm{d}\hat{\boldsymbol{n}} , \quad (45)$$

$$\sum_{l'} M_{ll'} \approx \frac{1}{2\pi} \frac{1}{l(l+1)} \int (\nabla w)^2(\hat{\boldsymbol{n}}) \, \mathrm{d}\hat{\boldsymbol{n}} , \quad (46)$$

so that mixing of power is suppressed by $1/(lR)^2$. Given that it is expected that $C_l^E \gg C_l^B$ on all scales, $\langle \tilde{C}_l^E \rangle$ is essentially a scaled version of the true spectrum, but this need not be the case for $\tilde{C}_l^B$. These ideas are illustrated in Fig. 8 which show typical rows of the matrices $P_{ll'}$ and $M_{ll'}$ and the mean pseudo-C_ls that they give rise to, for observations covering a circle of radius 15° with the last 5° apodized with a cosine function. The mean of $\tilde{C}_l^E$ is seen to be a scaled version of the true power spectrum except on large scales, but this is not the case for $\tilde{C}_l^B$.

'Unbiased' Estimators

The pseudo-C_ls can be used directly for parameter estimation [76], but it is often desirable to deconvolve the geometric effects of the weight function so we have unbiased estimates of the underlying power spectra. For example, for presentation purposes it is useful if we can plot the data in a form where they can be easily compared with true power spectra.

For observations covering a significant fraction of the sky, deconvolution is straightforward. The matrices $P_{ll'} \pm M_{ll'}$ will then be invertible and we can form unbiased estimates of the power spectra according to

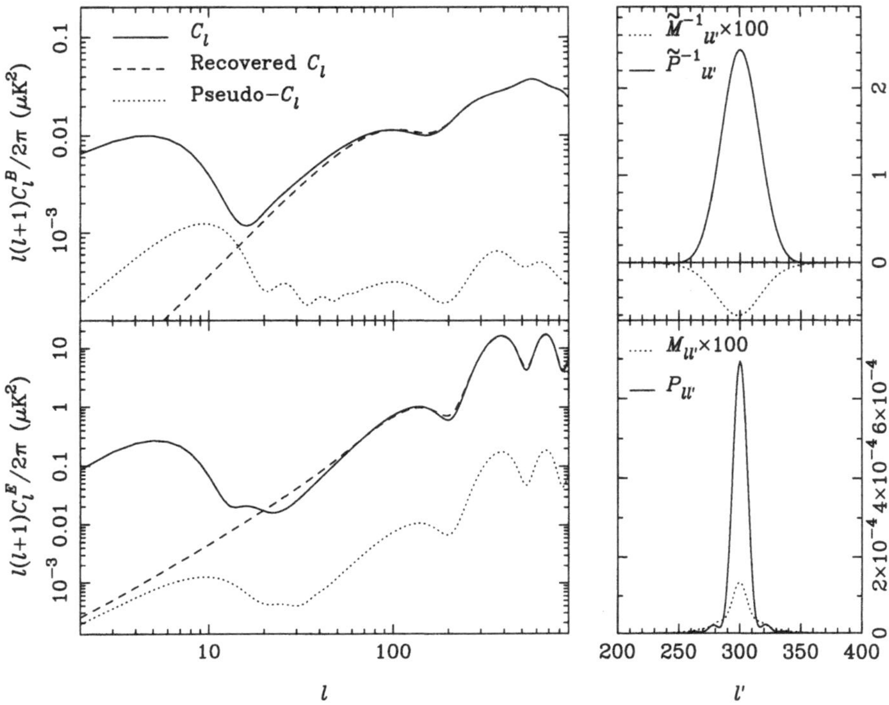

Fig. 8. *Left*: power spectra for B (*top*) and E (*bottom; solid lines*) compared to the mean pseudo-C_ls (dotted lines) and the recovered power spectra (*dashed lines*). *Right*: the bottom panel shows representative matrices $P_{ll'}$ (*solid lines*) and $M_{ll'}$ (*dotted lines*); the top panel shows the pseudo-inverses $\tilde{P}^{-1}_{ll'}$ (*solid lines*) and $\tilde{M}^{-1}_{ll'}$ (*dotted lines*). Note that $M_{ll'}$ and $\tilde{M}^{-1}_{ll'}$ have been multiplied by a factor of 100 for clarity. The weight function applied to the map is uniform inside a circle of 10° radius, with cosine apodization out to 15°. To obtain the pseudo-inverses, a Gaussian apodization of 4° HWHM is applied to the correlation functions. Figure from [84]

$$\hat{C}_l^E \pm \hat{C}_l^B = \sum_{l'} (P \pm M)^{-1}_{ll'} (\tilde{C}_{l'}^E \pm \tilde{C}_{l'}^B) . \tag{47}$$

An equivalent formulation of this approach can be made in terms of correlation functions [25, 78]. Consider estimating the correlation functions (11) by averaging over pairs of pixels separated by β, with pairs weighted by $w(\hat{\boldsymbol{n}}_1)w(\hat{\boldsymbol{n}}_2)$. It turns out that the data enters the estimators only through its pseudo-C_ls. The estimators $\hat{\xi}_\pm(\beta)$ for $\langle(\bar{Q} \mp \mathrm{i}\bar{U})(\hat{\boldsymbol{n}}_1)(\bar{Q} + \mathrm{i}\bar{U})(\hat{\boldsymbol{n}}_2)\rangle$, where $\hat{\boldsymbol{n}}_1 \cdot \hat{\boldsymbol{n}}_2 = \cos\beta$, can be expressed as [25]

$$\hat{\xi}_\pm(\beta) = \frac{\sum_l (2l+1)(\tilde{C}_l^E \pm \tilde{C}_l^B) d^l_{2\,\pm 2}(\beta)}{\sum_l w_l P_l(\cos\beta)} , \tag{48}$$

which also provides an efficient $O(N_{\mathrm{pix}}^{3/2})$ way of estimating the correlation functions. If enough sky is available to estimate the correlation functions for all β, then we can recover unbiased estimates of the power spectra by integral transforms. Using the orthogonality of the reduced Wigner functions, we have

$$\hat{C}_l^E \pm \hat{C}_l^B = 2\pi \int_{-1}^{+1} \hat{\xi}_\pm(\beta) d^l_{2\,\pm 2}(\beta)\,\mathrm{d}\cos\beta\,. \tag{49}$$

The integration can be performed essentially exactly with Gauss–Legendre integration, and a significant feature of estimating the correlation functions via the pseudo-C_ls is that they can be evaluated for any β up to the maximum pixel separation in the map. This avoids the need to re-sample the correlation functions onto the points required for the quadrature. This method has recently been applied to estimate the power spectra from the polarized channel of Archeops [66]. Aside from discretization issues, this correlation function method is equivalent to performing the direct inversion in (47).

For small surveys we will not have pixels separated by all angles $\leq 180°$, in which case we cannot estimate the correlation functions for all β. The inversion in (49) is then not possible. This is equivalent to saying that the survey does not have the spectral resolution to perform the matrix inversions in (47). Various schemes can be adopted to regularize the inversion, such as the use of bandpowers [77]. Alternatively, we can settle for pseudo-inverses $\tilde{P}^{-1}_{ll'}$ and $\tilde{M}^{-1}_{ll'}$, constructing estimators, e.g.

$$\hat{C}_l^B = \sum_{l'} \tilde{P}^{-1}_{ll'} \tilde{C}^B_{l'} + \tilde{M}^{-1}_{ll'} \tilde{C}^E_{l'}\,, \tag{50}$$

that have a simpler relation to the true power spectra than do the pseudo-C_ls. One desirable property we might enforce on the pseudo-inverses is that E–B mixing is removed *in the mean*, i.e. $\langle \hat{C}_l^B \rangle$ has no contribution from C_l^E. A convenient way to find such a pseudo-inverse is given in [25], which builds on earlier work in the context of cosmic shear analysis [86]. We make use of the correlation function estimates (48) in the angular range $(0, \beta_{\max})$, where $\beta_{\max}$ is the maximum separation of pixels in the Stokes maps. It is shown in [25] that the estimators $\hat{\xi}(\beta) \pm \hat{\xi}_-(\beta)$, which are linear in the pseudo-C_ls, contain only E- and B-mode power, respectively, in the mean. The function $\hat{\xi}(\beta)$ is obtained in the range $(0, \beta_{\max})$ by quadrature of $\hat{\xi}_+$ in the same range:

$$\begin{aligned}\hat{\xi}(\beta) = \hat{\xi}_+(\beta) &+ \frac{1}{\sin^2(\beta/2)} \int_{\cos\beta}^{1} \xi_+(\beta') \sec^4(\beta'/2)\,\mathrm{d}\cos\beta' \\ &- \frac{2(2+\cos\beta)}{\sin^4(\beta/2)} \int_{\cos\beta}^{1} \xi_+(\beta') \frac{\tan^3(\beta'/2)}{\sin\beta'}\,\mathrm{d}\cos\beta'\,,\end{aligned} \tag{51}$$

and satisfies

$$\frac{1}{2}\langle \hat{\xi}(\beta) \pm \hat{\xi}_-(\beta)\rangle = \sum_l \frac{2l+1}{4\pi} C_l^{E/B} d^l_{2\,-2}(\beta)\,. \tag{52}$$

We can recover estimates $\hat{C}_l^E$ and $\hat{C}_l^B$ that are linear combinations of the true power spectra, but with no mixing of E and B in the mean, by an apodized integral transform:

$$\hat{C}_l^{E/B} = 2\pi \int_{\cos\beta_{\max}}^{1} \frac{1}{2}[\hat{\xi}(\beta) \pm \hat{\xi}_-(\beta)] d^l_{2\,-2}(\beta) f(\beta)\, \mathrm{d}\cos\beta \,, \tag{53}$$

where $f(\beta)$ is chosen to apodize the discontinuity at $\beta_{\max}$. In this manner, we construct pseudo-inverses that deconvolve the pseudo-C_ls up to a window function ${}_{-2}K_{ll'}$, i.e. $\langle \hat{C}_l^E \rangle = \sum_{l'} {}_{-2}K_{ll'} C_{l'}^E$ and similarly for C_l^B. The window function is determined solely by the apodizing function $f(\beta)$ and is given by

$$_{-2}K_{ll'} = \frac{2l'+1}{2} \int_{\cos\beta_{\max}}^{1} f(\beta) d^l_{2\,-2}(\beta) d^{l'}_{2\,-2}(\beta)\, \mathrm{d}\cos\beta \,. \tag{54}$$

Oscillations in the window function can be suppressed with a careful choice of apodizing function, but the minimum achievable width is determined by the survey size $1/\beta_{\max}$. An example of the construction of 'unbiased estimates' with this route is given in Fig. 8.

Covariance of the Pseudo-C_ls

The covariance matrix of the pseudo-C_ls can be used to quantify the errors on any heuristically weighted quadratic estimate of the power spectra. Having accurate error information is clearly essential to establish reliable constraints on cosmological models. For Gaussian fields, computing the pseudo-C_l covariance directly is simple in principle, but the computational costs are prohibitive at high l. This is particularly problematic given that the sample covariance (i.e. that due to the random nature of the CMB fields) depends on the power spectra, and so should ideally be recomputed at each point in parameter space when constraining models. This model dependence also makes computing the sample variance 'blindly' from a large-suite of simulations unattractive. Fortunately, for the temperature anisotropies the sample covariance can be accurately approximated with analytic methods for l large compared to the inverse of the survey size [79]. Extending this calculation to polarization is difficult because the effect of E–B mixing can be the dominant source of variance in $\tilde{C}_l^B$ on a wide range of scales for small surveys. However, useful analytic approximations do now exist [84]. For an accurate calculation on all scales, one can combine the analytical calculations at high l (possibly calibrated from a small number of simulations if required) with a direct calculation at reduced resolution at low l [79]. (See also [87] for an alternative prescription.) The noise contribution (and any residual errors from foreground removal) is generally best handled with simulations.

An important issue for heuristically weighted quadratic estimators is that the weighting scheme adopted generally makes no attempt to separate E

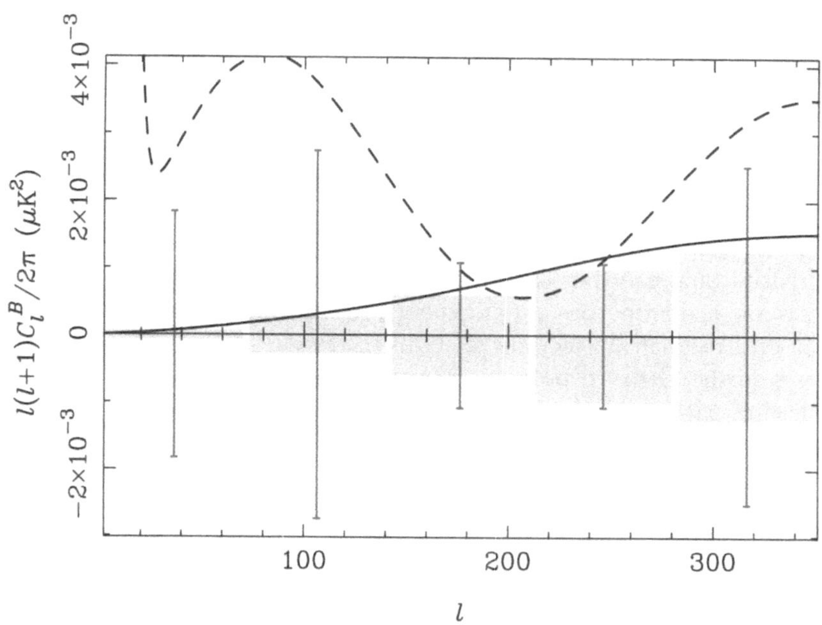

Fig. 9. Sample variance errors on the recovered $\hat{C}_l^B$ using the estimator (53) in the null hypothesis $r = 0$. The survey area and weight function are the same as in Fig. 8. The error boxes are the contribution purely from C_l^B (i.e. the lens-induced B modes) to the one-sigma errors on flat bandpower estimates with a $\Delta l = 70$. These are thus representative of the errors that would be obtained with the pseudo-C_l method if it were applied to maps from which the E modes had been removed (see Sect. 6.4, but here we have ignored the loss of information to the ambiguous modes). The error bars are the contribution purely from C_l^E and arise from E–B mixing. The dashed and solid lines are crude 'rule-of-thumb' approximations to the errors; see [84] for full details. Figure from [84]

and B modes coherently. While we can remove E–B mixing from the power spectra in the mean, in any realization $\hat{C}_l^B$ need not vanish even if there are no B modes present. This cross-contribution to the sample variance of pseudo-C_l estimators makes them inaccurate for high-sensitivity, small-area B-mode surveys. This is illustrated in Fig. 9, where it is shown that the contribution to the sample variance of $\hat{C}_l^B$ due to leakage of E modes is well in excess of that due to lens-induced B modes. (We have assumed there are no gravitational waves present, so lensing is the only source of B modes.) For the 15°-radius survey considered, with no instrument noise, this excess variance increases the error on the tensor-to-scalar ratio r in the null hypothesis to $\Delta r = 0.15$ (3σ). One way to remove this problem is first to isolate the B modes with the methods described in Sect. 6.4. Ignoring the loss in this process, we find that the limit on r is then reduced to 3.0×10^{-4}; this arises solely from the sample variance of the lens-induced B modes.

6.6 Non-Gaussianity and Lensing Reconstruction

The effect of lensing on the CMB will probably be seen first in the angular power spectra (e.g. [35, 88] and references therein), or, indirectly, via cross-correlation with large-scale structure [89]. As with gravitational wave searches, the most promising place to look from the standpoint of cosmic variance is the B-mode spectrum since, on small scales, lensing may be the dominant signal. Lensing is sensitive to the evolution of the gravitational potential at low redshift and can thus can be used to probe parameters, such as neutrino masses and the properties of dark energy, to which the primary CMB spectra are insensitive [56]. At the sensitivity of the Planck experiment, the effect of lensing on the observed power spectra must be included to avoid bias when estimating parameters such as the baryon density. As thermal noise levels improve, and the sample variance of lens-induced B-modes comes to dominate the error budget on C_l^B, corrections due to the non-Gaussianity from lensing will need to be included in the C_l^B errors and their covariance [74].

Looking further ahead, a number of authors have proposed reconstructing the lensing deflection field directly from the CMB by exploiting the non-Gaussianity of the lensing effect (see, e.g. [90, 91] and references therein). This allows one to extract more information from the observed (lensed) CMB fields than is possible with an analysis of their power spectra alone, as well as providing a coherent reconstruction of the projected mass distribution. In essence, the reconstruction methods work by using the locally anisotropic effect of lensing shear on the statistics of the CMB fluctuations within a region over which the deflections are coherent ($\sim 60°$). The fact that we do not know the unlensed CMB fields, but only their statistical properties, introduces a noise in the reconstruction analogous to the effect of the scatter in intrinsic ellipticities of background galaxies in cosmic shear analyses. CMB polarization is particularly useful here [92, 93, 94]: it has more power on small scales than the temperature anisotropies and so allows reconstruction of the deflection to smaller scales. It was shown in [93] that *local* correlations between E and the lens-induced B modes, estimated with simple quadratic estimators, reconstruct the deflection field with the highest signal-to-noise on all scales once thermal noise permits imaging of the lens-induced B modes on small scales. This requires both high sensitivity ($\sim 1\,\mu$K in an arcmin pixel) and high angular resolution (a few arcmin). The E–B estimator is particularly powerful since there is no contribution to its sample variance from chance correlations in the unlensed fields since there are no unlensed small-scale B modes.

Recently, it has been argued that the relatively larger non-Gaussianity of the small-scale (lensed) B-modes, compared to the temperature anisotropies and E modes, allows further improvement in the lensing reconstruction if more optimal techniques are employed [94]. Indeed, a simple counting argument suggests that, in the absence of thermal noise, the assumption that lensing is the only source of small-scale B modes allows perfect reconstruction of the deflection field (except possibly for a few degenerate configurations). In practice, errors in the theoretical modelling of lensing ultimately limit the reconstruction, but this effect is unimportant until thermal noise levels fall well below the very optimistic level of 0.25 μK-arcmin.

An important application of lensing reconstruction is to 'clean out' the large-scale B modes where any gravitational wave background is expected to contribute. If the lens-induced modes are treated simply as an additional source of Gaussian confusion, the limit on the detectability imposed by lensing (i.e. with no thermal noise or errors in foreground removal) is $r > 4 \times 10^{-5}$ at 3σ for a full-sky survey [44]. This can be improved by more than a factor of 40 with optimal reconstruction methods [46], corresponding to an energy scale of inflation $V^{1/4} \approx 1 \times 10^{15}$ GeV.

7 Conclusion

Detections of CMB polarization are still at an early stage, but already we are beginning to see the promise of polarization data for constraining cosmological models being realized. The electric-polarization power spectrum measurements from DASI [36] and CBI [6] reveal acoustic oscillations with an amplitude and phase perfectly consistent with the best-fit adiabatic models to the temperature anisotropies. This is a non-trivial test on the dynamics of the photons and baryons around the time of recombination. Furthermore, the large-angle measurements of the temperature–polarization cross-correlation from WMAP [8] indicate a large optical depth to reionization and hence a complex ionization history. We can expect further rapid progress observationally, with more accurate measurements of the power spectra of E-mode polarization, and its correlation with the temperature anisotropies, expected shortly from a number of ground and balloon-borne experiments. These experiments will also greatly increase our knowledge of polarized astrophysical foregrounds at CMB frequencies. Instruments are currently being commissioned that should have the sensitivity to detect the power of B-mode polarization induced by weak gravitational lensing [95], and, already, several groups are working on a new generation of polarimeters with the ambition of detecting gravitational waves and reconstructing the projected mass distribution from CMB polarization observations.

The success of these programmes will depend critically on many complex data analysis steps. We have attempted to summarize here some of the generic parts of a polarization data analysis pipeline, but, inevitably, have had to leave out many topics that are more instrument specific. Important omissions include calibration of the instrument, cleaning and other low-level reductions of time-stream data, noise estimation, propagation of errors and the broad topic of statistical accounting for non-ideal instrument effects. Given the exquisite control of systematic effects – both instrumental and astrophysical – that searching for sub-μK signals demands [96], these omissions will almost certainly prove to be the most critical steps.

Acknowledgements

AC acknowledges a Royal Society University Research Fellowship and thanks the organizers for the invitation to participate in an interesting and productive summer school.

References

1. Bennett, C.L., et al.: Astrophysical. J. Suppl. **148**, 97 (2003)
2. Bond, J.R., Contaldi, C.R., Pogoysan, D.: Phil. Trans. R. Soc. Lond. A **361**, 2435 (2003)
3. Guth, A.H.: Phys. Rev. D **23**, 347 (1981)
4. Rees, M.J.: Astrophysical. J. Lett. **153**, 1 (1968)
5. Kovac, J.M., et al.: Nature **420**, 772 (2002)
6. Readhead, A.C.S., et al.: Science **306**, 836 (2004)
7. Barkats, D., et al.: Astrophysical. J. Lett. **619**, 127 (2005)
8. Kogut, A., Spergel, D.N., Barnes, C.: Astrophysical. J. Suppl. **148**, 161 (2003)
9. Zaldarriaga, M., Spergel, D.N., Seljak, U.: Astrophysical. J. **488**, 1 (1997)
10. Zaldarriaga, M.: Phys. Rev. D **55**, 1822 (1997)
11. Seljak, U., Zaldarriaga, M.: Phys. Rev. Lett. **78**, 2054 (1997)
12. Kamionkowski, M., Kosowsky, A., Stebbins, A.: Phys. Rev. Lett. **78**, 2058 (1997)
13. Bucher, M., Moodley, K., Turok, N.: Phys. Rev. Lett. **87**, 191301 (2001)
14. Bond, J.R., et al.: Comput. Sci. Eng. **1**, 21 (1999)
15. Goldberg, J.N., et al.: J. Math. Phys. **8**, 2155 (1967)
16. Varshalovich, D.A., Moskalev, A.N., Khersonskii, V.K.: Quantum Theory of Angular Momentum, pp. 72–117. World Scientific, Singapore (1988)
17. de Oliveira-Costa, A., et al.: Phys. Rev. D **69**, 063516 (2004)
18. Vielva, P., et al.: Astrophysical. J. **609**, 22 (2004)
19. Copi, C.J., Huterer, D., Starkman, G.D.: Phys. Rev. D **70**, 043515 (2004)
20. Eriksen, H.K., et al.: Astrophysical. J. **605**, 14 (2004)
21. Hansen, F.K., et al.: Astrophysical. J. Lett. **607**, 67 (2004)
22. Schwarz, D.J., et al.: Phys. Rev. Lett. **93**, 221301 (2004)
23. Coulson, D., Crittenden, R.G., Turok, N.G.: Phys. Rev. Lett. **73**, 2390 (1994)
24. Ng, K., Liu, G.: Int. J. Mod. Phys. D **8**, 61 (1999)
25. Chon, G., et al.: Mon. Not. R. Astron. Soc. **350**, 914 (2004)
26. Zaldarriaga, M., Seljak, U.: Phys. Rev. D **55**, 1830 (1997)
27. Scott, D., Srednicki, M., White, M.: Astrophysical. J. Lett. **421**, 5 (1994)
28. Hobson, M.P., Magueijo, J.: Mon. Not. R. Astron. Soc. **283**, 1133 (1996)
29. Tegmark, M.: Mon. Not. R. Astron. Soc. **280**, 299 (1996)
30. Spergel, D.N., et al.: Astrophysical. J. Suppl. **148**, 175 (2003)
31. Hu, W., White, M.: Phys. Rev. D **56**, 596 (1997)
32. Challinor, A.: Phys. Rev. D **62**, 043004 (2000)
33. Hu, W., White, M.: New Astronomy **2**, 323 (1997)
34. Lewis, A., Challinor, A., Lasenby, A.: Astrophysical. J. **538**, 473 (2000)
35. Zaldarriaga, M., Seljak, U.: Phys. Rev. D **58**, 023003 (1998)
36. Leitch, E.M., et al.: Astrophysical. J. **624**, 10 (2005)
37. Keating, B.G., et al.: Astrophysical. J. Lett. **560**, 1 (2001)

38. Seljak, U., et al.: Phys. Rev. D **71**, 103515 (2005)
39. Peiris, H.V., et al.: Astrophysical. J. Suppl. **148**, 213 (2003)
40. Gordon, C., Lewis, A.: Phys. Rev. D **67**, 123513 (2003)
41. Bucher, M., et al.: Phys. Rev. Lett. **93**, 081301 (2004)
42. Holder, G.P., et al.: Astrophysical. J. **595**, 13 (2003)
43. Hu, W., Holder, G.P.: Phys. Rev. D **68**, 023001 (2003)
44. Knox, L., Song, Y.: Phys. Rev. Lett. **89**, 011303 (2002)
45. Kesden, M., Cooray, A., Kamionkowski, M.: Phys. Rev. Lett. **89**, 011304 (2002)
46. Seljak, U., Hirata, C.M.: Phys. Rev. D **69**, 043005 (2004)
47. Steinhardt, P.J., Turok, N.: Science **296**, 1436 (2002)
48. Boyle, L.A., Steinhardt, P.J., Turok, N.: Phys. Rev. D **69**, 127302 (2004)
49. Jones, N., Stoica, H., Tye, S.-H.H.: J. High Energy Phys. **7**, 51 (2002)
50. Seljak, U., Pen, U., Turok, N.: Phys. Rev. Lett. **79**, 1615 (1997)
51. Pogosian, L., et al.: Phys. Rev. D **68**, 023506 (2003)
52. Lewis, A.: Phys. Rev. D **70**, 043011 (2004)
53. Kosowsky, A., Loeb, A.: Astrophysical. J. **469**, 1 (1996)
54. Scannapieco, E.S., Ferreira, P.G.: Phys. Rev. D **56**, 7493 (1997)
55. Kosowsky, A., et al.: Phys. Rev. D **71**, 043006 (2005)
56. Hu, W.: Phys. Rev. D **65**, 023003 (2002)
57. Bond, J.R., Crittenden, R.: CMB analysis. In: Crittenden R., Turok N.G. (eds.) NATO ASIC Proc. 565: Structure Formation in the Universe, pp. 241–280. Kluwer Academic Publishers, Dordrecht (2001)
58. Challinor, A., et al.: Phys. Rev. D **62**, 123002 (2000)
59. Stompor, R., et al.: Phys. Rev. D **65**, 022003 (2002)
60. Delabrouille, J.: Astron. Astrophys. Suppl. **127**, 555 (1998)
61. Zaldarriaga, M.: Phys. Rev. D **64**, 103001 (2001)
62. Duncan, A.R., et al.: Mon. Not. R. Astron. Soc. **291**, 279 (1997)
63. Brouw, W.N., Spoelstra, T.A.T.: Astron. Astrophys. Suppl. **26**, 129 (1976)
64. Bernardi, G., et al.: Astrophysical. J. Lett. **594**, 5 (2003)
65. Giardino, G., et al.: Astron. Astrophys. **387**, 82 (2002)
66. Ponthieu, N., et al.: Astron. Astrophys. **444**, 327 (2005)
67. Bouchet, F.R., Prunet, S., Sethi, S.K.: Mon. Not. R. Astron. Soc. **302**, 663 (1999)
68. Baccigalupi, C., et al.: Mon. Not. R. Astron. Soc. **354**, 55 (2002)
69. Delabrouille, J., Cardoso, J.-F., Patanchon, G.: Mon. Not. R. Astron. Soc. **346**, 1089 (2003)
70. Tucci, M., et al.: Mon. Not. R. Astron. Soc. **360**, 926 (2005)
71. Bunn, E.F., et al.: Phys. Rev. D **67**, 023501 (2003)
72. Lewis, A., Challinor, A., Turok, N.: Phys. Rev. D **65**, 023505 (2002)
73. Lewis, A.: Phys. Rev. D **68**, 083509 (2003)
74. Smith, K.M., Hu, W., Kaplinghat, M.: Phys. Rev. D **70**, 043002 (2004)
75. Bond, J.R., Jaffe, A.H., Knox, L.: Phys. Rev. D **57**, 2117 (1998)
76. Wandelt, B.D., Hivon, E., Górski, K.M.: Phys. Rev. D **64**, 083003 (2001)
77. Hivon, E., et al.: Astrophysical. J. **567**, 2 (2002)
78. Szapudi, I., Prunet, S., Colombi, S.: Astrophysical. J. Lett. **561**, 11 (2001)
79. Efstathiou, G.: Mon. Not. R. Astron. Soc. **349**, 603 (2004)
80. Tegmark, M.: Phys. Rev. D **55**, 5895 (1997)
81. Tegmark, M., de Oliveira-Costa, A.: Phys. Rev. D **64**, 063001 (2001)
82. Hansen, F.K., Górski, K.M.: Mon. Not. R. Astron. Soc. **343**, 559 (2003)

83. Gorski, K.M., et al.: Astrophysical. J. **622**, 759 (2005)
84. Challinor, A., Chon, G.: Mon. Not. R. Astron. Soc. **360**, 509 (2005)
85. Hinshaw, G., et al.: Astrophysical. J. Suppl. **148**, 135 (2003)
86. Crittenden, R.G., et al.: Astrophysical. J. **568**, 20 (2002)
87. Brown, M.L., Castro, P.G., Taylor, A.N.: Mon. Not. R. Astron. Soc. **360**, 1262 (2005)
88. Seljak, U.: Astrophysical. J. **463**, 1 (1996)
89. Hirata, C.M., et al.: Phys. Rev. D **70**, 103501 (2004)
90. Seljak, U., Zaldarriaga, M.: Phys. Rev. Lett. **82**, 2636 (1999)
91. Hu, W.: Astrophysical. J. Lett. **557**, 79 (2001)
92. Benabed, K., Bernardeau, F., van Waerbeke, L.: Phys. Rev. D **63**, 043501 (2001)
93. Hu, W., Okamoto, T.: Astrophysical. J. **574**, 566 (2002)
94. Hirata, C.M., Seljak, U.: Phys. Rev. D **68**, 083002 (2003)
95. Bowden, M., et al.: Mon. Not. R. Astron. Soc. **349**, 321 (2004)
96. Hu, W., Hedmān, M.M., Zaldarriaga, M.: Phys. Rev. D **67**, 043004 (2003)

Diffuse Source Separation in CMB Observations

J. Delabrouille[1] and J.-F. Cardoso[2]

[1] APC, 11 place Marcelin Berthelot, F75231 Paris Cedex 05
delabrouille@apc.univ-paris7.fr
[2] LTCI, 46 rue Barrault, F75634 Paris Cedex 13 and
APC, 11 place Marcelin Berthelot, F75231 Paris Cedex 05
cardoso@enst.fr

1 Introduction

Spectacular advances in the understanding of the Big-Bang model of cosmology have been due to increasingly accurate observations of the properties of the cosmic microwave background (CMB). The detector sensitivities of modern experiments have permitted to measure fluctuations of the CMB temperature with such a sensitivity that the contamination of the data by astrophysical foreground radiations, rather than by instrumental noise, is becoming the major source of limitation. This will be the case, in particular, for the upcoming observations by the Planck mission, to be launched by ESA in 2008 (Lamarre et al. [36], Mandolesi et al. [41], Lamarre et al. [35]), as well as for next-generation instruments dedicated to the observation of CMB polarization.

In this context, the development of data analysis methods dedicated to identifying and separating foreground contamination from CMB observations is of the utmost importance for future CMB experiments. In many astrophysical observations indeed, and in particular in the context of CMB experiments, signals and images contain contributions from several components or *sources*. Some of these sources may be of particular interest (CMB or other astrophysical emission), some may be unwanted (noise).

Obviously, components cannot be properly studied in data sets in which they appear only as a mixture. Component separation consists, for each of them, in isolating the emission from all the other components present in the data, in the best possible way.

It should be noted that what "best" means depends on what the isolated data will be used for. Very often, one tries to obtain, for each component, an estimated map (or a set of maps at different frequencies) minimizing the total error variance, i.e. minimizing

Delabrouille, J., Cardoso, J.-F.: *Diffuse Source Separation in CMB Observations*. Lect. Notes Phys. **665**, 159–205 (2009)
DOI 10.1007/978-3-540-44767-2_6

$$\chi^2 = \sum_p |\widehat{s}(p) - s(p)|^2 \ , \tag{1}$$

where $s(p)$ is the true component emission and $\widehat{s}(p)$ its estimated value. p indexes the space of interest for the component, typically a set of pixels (θ_p, ϕ_p), or modes (ℓ, m) of a spherical harmonic decomposition of a full sky map, or a set of Fourier modes (k_x, k_y).

More generally, the objective of component separation is to estimate a *set of parameters* which describes the component of interest. In the simplest case, this set of parameters may be emission in pixels, but it may also be instead parameters describing statistical properties such as power spectra, spectral indices, etc. Since the set of parameters depends on the model assumed for the components, this model is of the utmost importance for efficient component separation. In the following, a significant part of the discussion will thus be dedicated to a summary of existing knowledge and of component modelling issues.

In the following, it is assumed that we are given a set of observations $y_i(p)$, where i, ranging from 1 to N_{chann}, indexes the observation frequency. The observed emission in each of the frequency bands is supposed to result from a mixture of several astrophysical components, with additional instrumental noise.

In this chapter, we discuss in some detail the problem of diffuse component separation. The paper is organized as follows: in the next section, we review the principles and implementation of the ILC, a very simple method to average the measurements obtained at different frequencies; Sect. 3 reviews the known properties of diffuse sky emissions, useful to model the observations and put priors on component parameters; Sect. 4 discusses observation and data reduction strategies to minimize the impact of foregrounds on CMB measurements, based on physical assumptions about the various emissions; Sect. 5 discusses the model of a linear mixture, and various options for its linear inversion to separate astrophysical components; Sect. 6 discusses a nonlinear solution for inverting a linear mixture, based on a maximum entropy method; Sect. 7 presents the general ideas behind blind source separation (BSS) and independent component analysis (ICA); Sect. 8 discusses a particular method, the spectral matching ICA (SMICA); Sect. 10 concludes with a summary, hints about recent and future developments, open issues, and a prospective.

Let the reader be warned beforehand that this review may not do full justice to much of the work which has been done on this very exciting topic. The discussion may be somewhat partial, although not intentionally. It has not been possible to the authors to review completely and compare all of the relevant work, partly for lack of time and partly for lack of details in the published papers. As much as possible, we have nonetheless tried to mention all of the existing work, to comment the ideas behind the methods, and to quote most of the interesting and classical papers.

2 ILC: Internal Linear Combination

The internal linear combination (ILC) component separation method assumes very little about the components. One of them (e.g. the CMB) is considered to be the only emission of interest, all the other being unwanted foregrounds.

It is assumed that the template of emission of the component of interest is the same at all frequencies of observation, and that the observations are calibrated with respect to this component, so that for each frequency channel i we have

$$y_i(p) = s(p) + f_i(p) + n_i(p) , \tag{2}$$

where $f_i(p)$ and $n_i(p)$ are foregrounds and noise contributions, respectively, in channel i.

A very natural idea, since all the observations actually measure $s(p)$ with some error $f_i(p) + n_i(p)$, consists in averaging all these measurements, giving a specific weight w_i to each of them. Then, we look for a solution of the form

$$\widehat{s}(p) = \sum_i w_i(p) y_i(p) , \tag{3}$$

where the weights $w_i(p)$ are chosen to maximize some criterion about the reconstructed estimate $\widehat{s}(p)$ of $s(p)$, while keeping the component of interest unchanged. This requires in particular that for all p, the sum of the coefficients $w_i(p)$ should be equal to 1.

2.1 Simple ILC

The simplest version of the ILC consists in minimizing the variance σ^2 of the map $\widehat{s}(p)$ using weights independent of p (so that $w_i(p) = w_i$ independent of p), with $\sum w_i = 1$. In this case, the estimated component is

$$\begin{aligned}\widehat{s}(p) &= \sum_i w_i y_i(p) \\ &= s(p) + \sum_i w_i f_i(p) + \sum_i w_i n_i(p) .\end{aligned} \tag{4}$$

Hence, under the assumption of decorrelation between $s(p)$ and all foregrounds, and between $s(p)$ and all noises, the variance of the error is minimum when the variance of the ILC map itself is minimum.

2.2 ILC Implementation

We now outline a practical implementation of the ILC method. For definiteness (and simplicity), we will assume here that the data is in the form of harmonic coefficients $s(\ell, m)$. The variance of the ILC map is

$$\sigma^2 = \sum_{\ell \geq 1} \sum_{m=-\ell}^{\ell} \boldsymbol{w}^\dagger \, \mathbf{C}(\ell, m) \, \boldsymbol{w} = \boldsymbol{w}^\dagger \, \mathbf{C} \, \boldsymbol{w} \, , \tag{5}$$

where $\mathbf{C}(\ell, m) = \langle \boldsymbol{y}(\ell, m) \boldsymbol{y}^\dagger(\ell, m) \rangle$ is the covariance matrix of the observations in mode (ℓ, m), and $\mathbf{C}$ is the covariance summed over all modes except $\ell = 0$. $\boldsymbol{y}(\ell, m)$ and $\boldsymbol{w}$ stand for the vectors of generic element $y_i(\ell, m)$ and w_i, respectively. The minimum, under the constraint of $\sum w_i = 1$, is obtained, using the Lagrange multiplier method, by solving the linear system

$$\begin{aligned} \forall i \, , \qquad \frac{\partial}{\partial w_i} \left[\sigma^2 + \lambda \left(1 - \sum w_i \right) \right] &= 0 \, , \\ \sum_i w_i &= 1 \, . \end{aligned} \tag{6}$$

Straightforward linear algebra gives the solution

$$w_i = \frac{\sum_j \left[\mathbf{C}^{-1} \right]_{ij}}{\sum_{ij} \left[\mathbf{C}^{-1} \right]_{ij}} \, . \tag{7}$$

Note that if the template of emission of the component of interest is the same at all frequencies of observation, but the observations are *not* calibrated with respect to this component, (2) becomes

$$y_i(p) = a_i s(p) + f_i(p) + n_i(p) \, . \tag{8}$$

In this case, it is still possible to implement an ILC. The solution is

$$\boldsymbol{w} = \frac{\boldsymbol{A}^T \mathbf{C}^{-1}}{\boldsymbol{A}^T \mathbf{C}^{-1} \boldsymbol{A}} \, , \tag{9}$$

where $\boldsymbol{A}$ is the vector of recalibration coefficients a_i. This solution of (9) is equivalent to first changing the units in all the observations to make the response 1 in all channels, and then implementing the solution of (7).

2.3 Examples of ILC Separation: Particular Cases

This idea of ILC is quite natural. It has, however, several unpleasant features, which makes it non-optimal in most real-case situations. Before discussing this, let us examine now what happens in two simple particular cases.

Case 1: Noisy Observations with No Foreground

If there are no foregrounds, and the observations are simply noisy maps of $s(p)$, with independent noise for all channels of observation, the ILC solution should lead to a noise-weighted average of the measurements.

Let us assume for simplicity that we have two noisy observations, y_1 and y_2, with $y_i = s + n_i$. In the limit of very large maps, so that cross-correlations between s, n_1 and n_2 vanish, the covariance matrix of the observations takes the form

$$\mathbf{C} = \begin{bmatrix} S + N_1 & S \\ S & S + N_2 \end{bmatrix},$$

where S is the variance of the signal (map) of interest, and N_1 and N_2 the noise variances for the two channels. The inverse of $\mathbf{C}$ is

$$\mathbf{C}^{-1} = \frac{1}{\det(\mathbf{C})} \begin{bmatrix} S + N_2 & -S \\ -S & S + N_1 \end{bmatrix}$$

and applying (7), we get $w_1 = N_2/(N_1 + N_2)$ and $w_2 = N_1/(N_1 + N_2)$. This is the same solution as weighting each map i proportionally to $1/N_i$.

Case 2: Noiseless Observations with Foregrounds

Let us now examine the opposite extreme, where observations are noiseless linear mixtures of several astrophysical components. Consider the case of two components, with two observations. We can write the observations as $\boldsymbol{y} = \mathbf{A}\boldsymbol{s}$, where $\mathbf{A}$ is the so-called "mixing matrix" and $\boldsymbol{s} = (s_1, s_2)^\dagger$ the vector of sources.

The covariance of the observations is

$$\mathbf{C} = \boldsymbol{y}\boldsymbol{y}^\dagger = \mathbf{A}\,\boldsymbol{s}\boldsymbol{s}^\dagger\mathbf{A}^\dagger$$

and its inverse is

$$\mathbf{C}^{-1} = [\mathbf{A}^\dagger]^{-1}\,[\boldsymbol{s}\boldsymbol{s}^\dagger]^{-1}\mathbf{A}^{-1} \tag{10}$$

Let us assume that we are interested in the first source. The data are then calibrated so that the mixing matrix $\mathbf{A}$ and its inverse are of the form

$$\mathbf{A} = \begin{bmatrix} 1 & a_{12} \\ 1 & a_{22} \end{bmatrix} \quad \text{and} \quad \mathbf{A}^{-1} = \frac{1}{\det(\mathbf{A})} \begin{bmatrix} a_{22} & -a_{12} \\ -1 & 1 \end{bmatrix}.$$

Then, if we assume that components 1 and 2 are uncorrelated, (10) yields

$$\mathbf{C}^{-1} = \frac{1}{(\det(\mathbf{A}))^2} \begin{bmatrix} a_{22} & -1 \\ -a_{12} & 1 \end{bmatrix} \begin{bmatrix} S_1^{-1} & 0 \\ 0 & S_2^{-1} \end{bmatrix} \begin{bmatrix} a_{22} & -a_{12} \\ -1 & 1 \end{bmatrix}, \tag{11}$$

where S_1 and S_2 are the variances of components 1 and 2, respectively. After expansion of the matrix product, we get

$$\mathbf{C}^{-1} = \frac{1}{(\det(\mathbf{A}))^2} \begin{bmatrix} (a_{22}^2 S_1^{-1} + S_2^{-1}) & (-a_{22}a_{12}S_1^{-1} - S_2^{-1}) \\ (-a_{22}a_{12}S_1^{-1} - S_2^{-1}) & (a_{12}^2 S_1^{-1} + S_2^{-1}) \end{bmatrix} \tag{12}$$

and using (7), we get

$$w = \frac{1}{(a_{22} - a_{12})} \begin{bmatrix} a_{22} \\ -a_{12} \end{bmatrix} , \tag{13}$$

which is the transpose of the first line of matrix $\mathbf{A}^{-1}$, so that $\widehat{s}_1 = \boldsymbol{w}.\boldsymbol{y} = s_1$. As expected, *if the covariance of the two components vanishes*, the ILC solution is equivalent, for the component of interest, to what is obtained by inversion of the mixing matrix.

What happens now if the two components are correlated? Instead of the diagonal form $\text{diag}(S_1, S_2)$, the covariance matrix of the sources contains an off-diagonal term S_{12}, so that (11) becomes

$$\mathbf{C}^{-1} = \frac{1}{(\det(\mathbf{A}))^2} \frac{1}{(S_1 S_2 - S_{12}^2)} \begin{bmatrix} a_{22} & -1 \\ -a_{12} & 1 \end{bmatrix} \begin{bmatrix} S_2 & S_{12} \\ S_{12} & S_1 \end{bmatrix} \begin{bmatrix} a_{22} & -a_{12} \\ -1 & 1 \end{bmatrix} , \tag{14}$$

which yields the solution

$$w = \frac{1}{(a_{22} - a_{12})} \begin{bmatrix} a_{22} + S_{12}/S_2 \\ -a_{12} - S_{12}/S_2 \end{bmatrix} . \tag{15}$$

The ILC is not equivalent anymore to the inversion of the mixing matrix $\mathbf{A}$. Instead, the estimate $\widehat{s}_1$ of s_1 is

$$\widehat{s}_1 = \boldsymbol{w}.\boldsymbol{y} = s_1 - \frac{S_{12}}{S_2} s_2 . \tag{16}$$

The ILC result is biased, giving a solution in which a fraction of s_2 is subtracted erroneously, in proportion to the correlation between s_1 and s_2, to lower as much as possible the variance of the output map. The implication of this is discussed in Sect. 2.6.

2.4 Improving the ILC Method

With the exception of the CMB, diffuse sky emissions are known to be very non-stationary (e.g. galactic foregrounds are strongly concentrated in the galactic plane). In addition, most of the power is concentrated on large scales (the emissions are strongly correlated spatially). As the ILC method minimizes the total variance of the ILC map (the integrated power from all scales, as can be seen in (5)), the weights w_i are strongly constrained essentially by regions of the sky close to the galactic plane, where the emission is strong, and by large scales, which contain most of the power. In addition, the ILC method finds weights resulting from a compromise between reducing astrophysical foreground contamination and reducing the noise contribution. In other words, for a smaller variance of the output map, it pays off more to reduce the galactic contamination in the galactic plane and on large scales, where it is strong, rather than at high galactic latitude and on small scales, where there is little power anyway. This particularity of the ILC when implemented globally is quite annoying for CMB studies, for which all scales are interesting, and essentially the high galactic latitude data is useful.

Away from the galactic plane and on small scales, the best linear combination for cleaning the CMB from foregrounds and noise may be very different from when it is close to the galactic plane and on large scales. A very natural idea to improve on the ILC is to decompose sky maps in several regions and/or scales and apply an ILC independently to all these maps. The final map is obtained by adding-up all the ILC maps obtained independently in various regions and at different scales. Applications of these ideas are discussed in the next paragraph.

2.5 ILC-Based Foreground-Cleaned CMB Map from WMAP Data

A map of CMB anisotropies has been obtained using the ILC method (Bennett et al. [7]) on the first year data from the WMAP mission, and has been released to the scientific community as part of the first year WMAP data products.

The input data is the set of five all sky, band-averaged maps for the K, Ka, Q, V and W frequency bands, all of which smoothed to the same 1 degree resolution for convenience. The ILC is performed independently in 12 regions, 11 of which being in the WMAP kp2 mask at low galactic latitudes, designed to mask out regions of the sky highly contaminated by galactic foregrounds. This division into 12 regions is justified by the poor performance of the ILC on the full sky, interpreted as due to varying spectral indices of the astrophysical foregrounds. Discontinuities between the regions are reduced by using smooth transitions between the regions.

Little detail is provided on the actual implementation of the ILC by the WMAP team. Apparently, a non-linear iterative minimization algorithm was used, instead of the linear solution outlined in Sect. 2.2. Although there does not seem to be any particular reason for this choice, in principle the particular method chosen to minimize the variance does not matter, as long as it finds the minimum efficiently. There seem to be, however, indications that the convergence was not perfect, as discussed by Eriksen and collaborators in a paper discussing the ILC and comparing the results of the several of its implementations on WMAP data (Eriksen et al. [20]). Caution should probably be taken when using the WMAP ILC map for any purpose other than a visual impression of the CMB.

Tegmark and collaborators have improved the ILC method in several respects, and provide an independent CMB map obtained from WMAP data by ILC (Tegmark et al. [60]). Their implementation allows the weights to depend not only on the region of the sky, but also on angular scale, as discussed in Sect. 2.4. In addition, they partially deconvolve the WMAP maps in harmonic space to put them all to the angular resolution of the channel with the smallest beam, rather than smoothing all maps to put them all to the angular resolution of the channel with the largest beam. As a result, their ILC

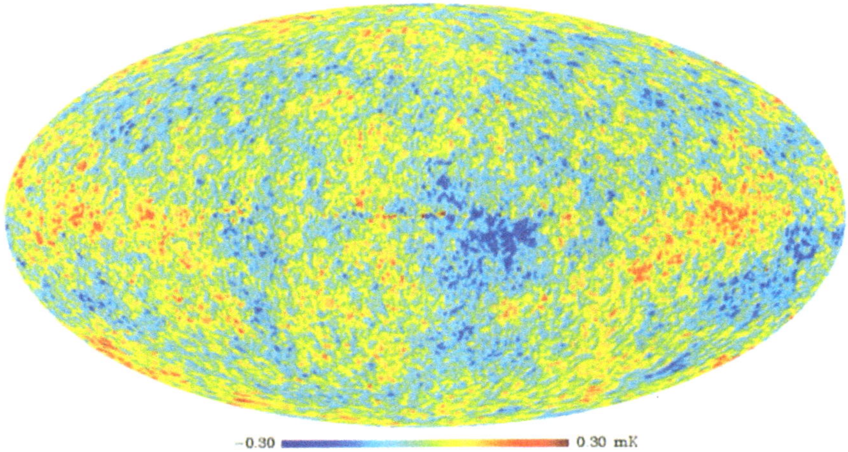

Fig. 1. The ILC map of the CMB obtained by the WMAP team (1 year data). Residuals of galactic emission are clearly visible in the centre of the map. The colour scale spans a range of −300 to +300 μ K thermodynamic, although localized residuals exceed these values. Figure from NASA LAMBDA web site

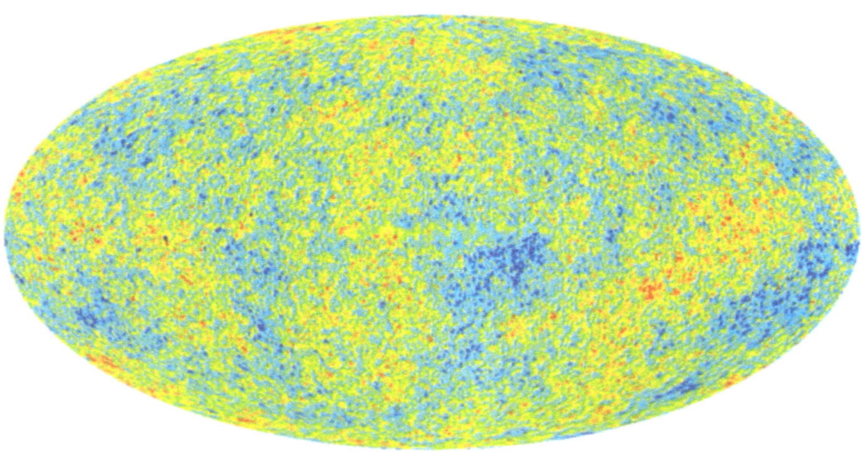

Fig. 2. The foreground-cleaned CMB map of Tegmark et al., obtained by the ILC method described in Tegmark et al. [60], after Wiener filtering. The effect of the region and scale-dependent weighting can be seen in the centre of the map (galactic centre) where the map looks smoother and flatter than elsewhere. The colour scale spans a range of −300 to +300 μ K, although localized residuals exceed these values, as in Fig. 1. The superior angular resolution can clearly be seen

map has better angular resolution, but higher total noise. The high-resolution map, however, can be filtered using a Wiener filter for minimal variance of the error. The Wiener-filtered map is obtained by multiplying each $a_{\ell m}$ mode of the map by a factor

$$W(\ell, m) = C_\ell / S_\ell ,$$

where C_ℓ is the estimated CMB power spectrum (computed for the cosmological model fitting best the WMAP data estimate) and S_ℓ is the estimated power spectrum of the noisy CMB map obtained by the authors using their ILC method.

The CMB map obtained by the WMAP team from first year data is shown in Fig. 1. For comparison, the map obtained by Tegmark et al. is shown in Fig. 2. Both give a good visual perception of what the CMB field looks like.

2.6 Comments About the ILC

The ILC has been used essentially to obtain a clean map of CMB emission. In principle, nothing prevents using it also for obtaining cleaned maps of other emissions, with the caveats that

- The data must be calibrated with respect to the emission of interest, so that the data takes the form of (2). This implies that the template of emission of the component of interest should not change significantly with the frequency band of observation. This is the case for the CMB (temperature and polarization) or for the SZ effect (to first order at least... more on this later).
- The component of interest should not be correlated with other components. Galactic components, being all strongly concentrated in the galactic plane, can thus not be recovered reliably with the ILC.

This issue of decorrelation of the component of interest $s(p)$ and the foregrounds can also generate problems in cases where the empirical correlation between the components does not vanish. As demonstrated in Sect. 2.3, the ILC method will not work properly (biasing the result) if the components of interest $s(p)$ are correlated, for whatever reason. In particular, small data sets, even if they are realizations of actually uncorrelated random processes, are always empirically correlated to some level. For this reason, the ILC should not be implemented independently on too small subsets of the original data (very small regions, very few modes).

Finally, whereas the ILC is a powerful tool when nothing is known about the data, it is certainly non-optimal when prior information is available. Foreground emissions are discussed in some detail in the following section.

3 Sky Emission Model: Components

"Know your enemy". This statement, borrowed from elementary military wisdom, applies equally well in the fight against foreground contamination. Prior knowledge about astrophysical components indeed has been widely used in all practical CMB data analyses. Methods can then be specifically tailored to remove foregrounds based on their physical properties, in particular their morphology, their localization, and their frequency scaling based on the physical understanding of their emission mechanisms.

In addition to knowledge about the unwanted foregrounds, prior knowledge about the component of interest is of the utmost importance for its identification and separation in observations. In the ILC method discussed above, for instance, the prior knowledge of the emission law of the CMB (derivative of a blackbody) is specifically used.

3.1 Various Astrophysical Emissions

Astrophysical emissions relevant to the framework of CMB observations can be classified into three large categories (in addition to the CMB itself) – diffuse galactic emission, extragalactic emission, and solar system emission.

Diffuse galactic emissions originate from the local interstellar medium (ISM) in our own galaxy. The ISM is constituted of cold clouds of molecular or atomic gas, of an intercloud medium which can be partly ionized, and of hot ionized regions presumably formed by supernovae. These different media are strongly concentrated in the galactic plane. The intensity of corresponding emissions decreases with galactic latitude with a cosecant law behaviour (the optical depth of the emitting material scales proportionally to $1/\sin b$). Energetic free electrons spiralling in the galactic magnetic field generate synchrotron emission, which is the major foreground at low frequencies (below a few tens of GHz). Warm ionized material emits free–free (Bremstrahlung) emission, due to the interaction of free electrons with positively charged nuclei. Small particles of matter (dust grains and macromolecules) emit radiation as well, through thermal grey-body emission, and possibly through other mechanisms.

Extragalactic emissions arise from a large background of resolved and unresolved radio and infrared galaxies, as well as clusters of galaxies. The thermal and kinetic Sunyaev–Zel'dovich effects, due to the inverse Compton scattering of CMB photons off hot electron gas in ionized media, are of special interest for cosmology. These effects occur, in particular, towards clusters of galaxies, which are known to comprise a hot (few keV) electron gas. Infrared and radio-galaxies emit also significant radiation in the frequency domain of interest for CMB observations, and contribute both point source emission from nearby bright objects and a diffuse background due to the integrated emission of a large number of unresolved sources, too faint to be detected individually, but which contribute sky background inhomogeneities which may pollute CMB observations.

Solar system emission comprises emissions from the planets, their satellites, and a large number of small objects (asteroids). In addition to those, there is diffuse emission due to dust particles and grains in the ecliptic plane (zodiacal light). The latter is significant essentially at the highest frequencies of an instrument like the Planck HFI (Maris et al. [42]).

In the rest of this section, we briefly outline the general properties of these components and the modelling of their emission in the centimetre to submillimetre wavelength range.

3.2 Cosmic Microwave Background

The cosmic microwave background, relic radiation from the hot big bang emitted at the time of decoupling when the Universe was about 370,000 years old, is usually thought of (by cosmologists) as the component of interest in the sky emission mixture. Millimetre and submillimetre wave observations, however, sometimes aim at measuring not only CMB anisotropies, but also other emissions. In this case, the CMB becomes a noxious background which has to be subtracted out of the observations, just as any other.

The CMB emission is relatively well known already. The main theoretical framework of CMB emission can be found in any modern textbook on cosmology, as well as in several reviews (Hu & Dodelson [29], White & Cohn [63]). The achievement of local thermal equilibrium in the primordial plasma before decoupling, together with the very low level of the perturbations, guarantees that CMB anisotropies are properly described as the product of a spatial template $\Delta T(p) = T_{\rm CMB}(p) - \overline{T}_{\rm CMB}$ and a function of ν (frequency scaling) which is the derivative of a blackbody with respect to temperature:

$$\Delta I_\nu(p) = \Delta T_{\rm CMB}(p) \left[\frac{\partial B_\nu(T)}{\partial T} \right]_{T=\overline{T}_{\rm CMB} \simeq 2.726\,{\rm K}} . \qquad (17)$$

In the standard cosmological model, the CMB temperature fluctuation map $\Delta T(p)$ is expected to be a realization of a stationary Gaussian random field, with a power spectrum C_ℓ displaying a series of peaks and troughs (the acoustic peaks), the location and relative size of which are determined by a few free parameters of the cosmological model.[3]

Good maps of sky emission at a resolution of about 15 arcmin, obtained from WMAP data in the frequency range 20–90 GHz, clearly comprise at high galactic latitude an astrophysical component compatible with all these predictions. The power spectrum is measured with excellent accuracy by WMAP up to the second Doppler peak, while complementary balloon-borne and ground-based experiments yield additional measurements at higher ℓ (smaller scales).

Efficient diffuse component separation methods should make use of this current status of knowledge about the CMB:

[3] The power spectrum C_ℓ is defined as the set of variances of the coefficients $a_{\ell m}$ of the expansion of the random field representing CMB-relative temperature fluctuations $\Delta T(p)/\overline{T}_{\rm CMB}$ onto the basis of spherical harmonics on the sphere $Y_{\ell m}(\theta, \phi)$. The stationarity and isotropy of the random field guarantees that the variance of $a_{\ell m}$ (coefficients C_ℓ) is independent of m.

- Law of emission, known to a high level of precision to be the derivative of a blackbody with respect to temperature, as expected theoretically and checked experimentally with the Boomerang (de Bernardis et al. [16]) and Archeops (Tristram et al. [61]) multifrequency data sets, as well as with the WMAP data (Bennett et al. [6], Patanchon et al. [47])
- Stationarity and Gaussianity to a high level of accuracy, as expected theoretically and checked on WMAP data (Komatsu et al. [33])
- Good cosmological prior on the power spectrum of the fluctuations, validated experimentally with several data sets (Netterfield et al. [46], Hinshaw et al. [26])

A good visual impression of all-sky CMB emission is given in Figs. 1 and 2. The present status of knowledge of the power spectrum C_ℓ is shown in Fig. 3.

The extraction of CMB emission from a set of multifrequency observations may be done with the following objectives in mind (at least):

- Get the best possible map of the CMB (in terms of total least square error, from noise and foregrounds together)
- Get the CMB map with the least possible foreground contamination

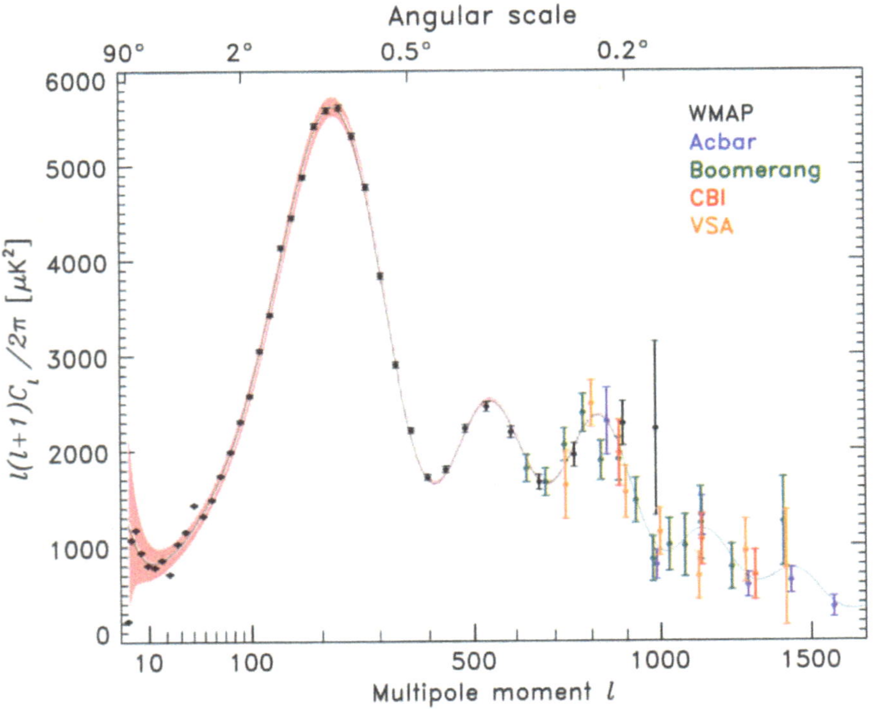

Fig. 3. Present-day best constraints of the CMB temperature power spectrum (from Hinshaw et al. [26]). Data sets in addition to WMAP 3-year data are from Jones et al. [31], Kuo et al. [34], Readhead et al. [50], Dickinson et al. [17]

- Get the CMB map for which spurious non-Gaussianity from foregrounds, noise and systematic effects is minimal
- Get the best possible estimate of the CMB angular power spectrum

Obviously, the best component separation method for extracting the CMB will depend on which of the above is the primary objective of component separation.

3.3 Emissions from the Diffuse Interstellar Medium

Synchrotron Emission

Synchrotron emission arises from energetic charged particles spiralling in a magnetic field. In our galaxy, such magnetic fields extend outside the galactic plane. Energetic electrons originating from supernovae shocks, spiralling in this field, can depart the galactic plane and generate emission even at high galactic latitudes. For this reason, synchrotron emission is less concentrated in the galactic plane than free–free and dust.

The frequency scaling of synchrotron emission is a function of the distribution of the energies of the radiating electrons. For number density distributions $N(E) \propto E^{-\gamma}$, the flux emission is also in the form of a power law, $I_\nu \propto \nu^{-\alpha}$, with $\alpha = (\gamma - 1)/2$. In Rayleigh–Jeans (RJ) temperature $\Delta T \propto \nu^{-\beta}$ with $\beta = \alpha + 2$. Typically, β ranges from 2.5 to 3.1, and is somewhat variable across the sky.

In spite of a moderate sensitivity for current standards, the 408 MHz all-sky map (Haslam et al. [25]), dominated by synchrotron emission, gives a good visual impression of the distribution of synchrotron over the sky.

In principle, synchrotron emission can be highly polarized, up to 50–70%

Free–Free Emission

Free–free emission is the least well known observationally of the three major emissions originating from the galactic interstellar medium in the millimetre and centimetre wavelength range. This emission arises from the interaction of free electrons with ions in ionized media, and is called "free–free" because of the unbound state of the incoming and outgoing electrons. Alternatively, free–free is called "Bremsstrahlung" emission ("braking radiation" in German), because photons are emitted while electrons lose energy by interaction with the heavy ions.

Theoretical calculations of free–free emission in an electrically neutral medium consisting of ions and electrons give an estimate of the brightness temperature at frequency ν for free–free emission of the form

$$T_{\rm ff} \simeq 0.08235\, T_e^{-0.35}\, \nu^{-\beta} \int_{\rm l.o.s.} N_e N_i dl \,, \tag{18}$$

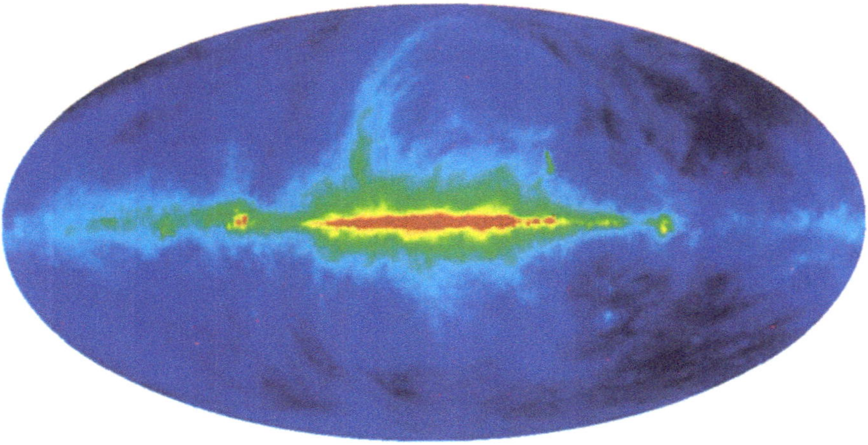

Fig. 4. The 408-MHz all-sky synchrotron map (Haslam et al. [25]). Data and images are available on the NASA Lambda web site

where T_e is in Kelvin, ν is in GHz and the line of sight integral of electron and ion density in cm^{-6}pc (Smoot [54]). Theoretical estimates of the spectral index, β, range from about 2.1 to 2.15, with errors of ± 0.03 (Fig. 4).

While free–free emission is not observed directly, as it never dominates over other astrophysical emissions, the source of its emission (mainly ionized hydrogen clouds) can be traced with hydrogen recombination emission lines, and particularly Hα emission. The connection between Hα and free–free has been discussed extensively by a number of authors (Smoot [54], Valls-Gabaud [62], McCullough et al. [44]). We have

$$\frac{T_{\rm ff}[\mathrm{mK}]}{I_\alpha[\mathrm{R}]} \simeq 10.4\, \nu^{-2.14}\, T_4^{0.527}\, 10^{0.029/T_4}\, (1+0.08)\ , \tag{19}$$

where $T_{\rm ff}$[mK] is the free–free brightness temperature in mK, I_α[R] the Hα surface brightness in Rayleigh, ν the frequency, and T_4 the temperature of the ionized medium in units of 10^4 K. The Rayleigh (R) is defined as 1 R = $(10^6/4\pi)$ photons/cm^2/s/sr.

Free–free emission, being due to incoherent emissions from individual electrons scattered by nuclei in a partially ionized medium, is not polarized (to first order at least).

Thermal Emission of Galactic Dust

The present knowledge of interstellar dust is based on extinction observations from the near infrared to the UV domain and on observations of its emission from radio frequencies to the infrared domain.

Dust consists in small particles of various materials, essentially silicate and carbonaceous grains of various sizes and shapes, in amorphous or crystalline form, sometimes in aggregates or composites. Dust is thought to comprise also large molecules of polycyclic aromatic hydrocarbon (PAH). The sizes of the grains range from few nanometres for the smallest, to micrometres for the largest. They can emit through a variety of mechanisms. The most important for CMB observations is grey-body emission in the far infrared, at wavelengths ranging from few hundreds of microns to few millimetres. The grey-body emission is typically characterized by a temperature $T_{\rm dust}$ and by an emissivity proportional to a power of the frequency ν:

$$I_\nu \propto \nu^\beta B_\nu(T_{\rm dust}) \,, \tag{20}$$

where $B_\nu(T)$ is the usual blackbody emission

$$B_\nu(T) = \frac{2h\nu^3}{c^2}\frac{1}{e^{h\nu/kT}-1} \,. \tag{21}$$

This law is essentially empirical. In practice, dust clouds along the line of sight can have different temperatures and different compositions: bigger or smaller grains, different materials. They can thus have different emissivities as well. Temperatures for interstellar dust are expected to range from about 5 K to more than 30 K, depending on the heating of the medium by radiation from nearby stars, with typical values of 16–18 K for emissivity indices $\beta \simeq 2$.

In principle, thermal emission from galactic dust should not be strongly polarized, unless dust particles are significantly asymmetric (oblate or prolate), and there exists an efficient process for aligning the dust grains in order to create a significant statistical asymmetry. Preliminary dust observations with the Archeops instrument (Benoît et al. [8], Ponthieu et al. [48]) seem to indicate polarization levels of the order of few per cent, and as high as 15–20% in a few specific regions.

Spinning Dust or Anomalous Dust Emission

In the last years, increasing evidence for dust-correlated emissions at frequencies below 30 GHz, in excess to expectations from synchrotron and free–free, has aroused interest in possible non-thermal emissions from galactic dust (Kogut et al. [32], Leitch et al. [37]). Among the possible non-thermal emission mechanisms, spinning dust grains offer an interesting option (Draine & Lazarian [19]).

At present, there still is some controversy on whether the evidence for non-thermal dust emission is robust enough for an unambiguous statement. Observations of different sky regions, indeed, yield somewhat different results (Dickinson et al. [18], Fernández-Cerezo et al. [22]), which may be due either to varying dust cloud properties or to differences in the analyses and interpretations, or both. Certainly, investigating this question is an objective of primary interest for diffuse component separation methods (especially blind ones) in the near future.

3.4 SZ Effect

The Sunyaev–Zel'dovich (SZ) effect (Sunyaev & Zeldovich [56]) is the inverse Compton scattering of CMB photons on free electrons in ionized media. In this process, the electron gives a fraction of its energy to the scattered CMB photon. There are, in fact, several SZ effects: The thermal SZ effect is due to the scattering of photons on a high-temperature electron gas, such as can be found in clusters of galaxies. The kinetic SZ effect is due to the scattering on a number of electrons with a global radial bulk motion with respect to the cosmic background. Finally, the polarized SZ effect is a second-order effect due to the kinematic quadrupole of the CMB in the frame of an ensemble of electrons with a global transverse bulk motion with respect to the CMB.

SZ effects are not necessarily linked to clusters of galaxies. Any large body with hot ionized gas can generate significant effects. It has been proposed that signatures of inhomogeneous reionization can be found via the kinetic and thermal SZ effect (Aghanim et al. [1], Gruzinov & Hu [23], Yamada et al. [65]). However, the largest expected SZ signatures originate from ionized intra-cluster medium.

Clusters of Galaxies

Clusters of galaxies, the largest known massive structures in the Universe, form by gravitational collapse of matter density inhomogeneities on large scales (comoving scales of few Mpc). They can be detected either optically from concentrations of galaxies at the same redshift or in the submillimetre by their thermal SZ emission or by the effect of their gravitational mass in weak shear maps or in X-ray. The hot intra-cluster baryonic gas can be observed through its X-ray emission due to Bremsstrahlung (free–free) emission of the electrons on the nuclei, which permits to measure the electron temperature (typically a few keV). On the sky, typical cluster angular sizes range from about 1 arcmin to about 1 degree. Clusters are scattered over the whole sky, although this distribution follows the repartition of structure on the largest scales in the Universe. Large-scale SZ effect observations may also be used to survey the distribution of hot gas on these very large scales, although such SZ emission, from filaments and pancakes in the distribution, is expected to be at least an order of magnitude lower in intensity than thermal SZ emission from the clusters themselves.

Each cluster of galaxies has its own thermal, kinetic, and polarized SZ emission. These various emissions and their impact on CMB observations and for cosmology have been studied by a variety of authors. Useful reviews have been made by Birkinshaw [9] and Rephaeli [51], for instance.

Thermal SZ

The thermal SZ effect generated by a gas of electrons at temperature T_e is, in fact, a spectral distortion of the CMB emission law. It is common to consider as the effect the difference $\Delta I_\nu = I_\nu - B_\nu(T_{\rm CMB})$ between the distorted CMB photon distribution I_ν and the original one $B_\nu(T_{\rm CMB})$. In the non-relativistic limit (when T_e is lower than about 5 keV, which is the case for most clusters), the shape of the spectral distortion does not depend on the temperature. The change in intensity due to the thermal SZ effect is

$$\Delta I_\nu = y \frac{x e^x}{(e^x - 1)} \left[\frac{x(e^x + 1)}{(e^x - 1)} - 4 \right] B_\nu(T_{\rm CMB}) \,, \qquad (22)$$

where $B_\nu(T_{\rm CMB})$ is the Planck blackbody emission law at CMB temperature

$$B_\nu(T_{\rm CMB}) = \frac{2h\nu^3}{c^2} \frac{1}{e^x - 1}$$

and $x = h\nu / kT_{\rm CMB}$. The dimensionless parameter y (Comptonization parameter) is proportional to the integral of the electron pressure along the line of sight:

$$y = \int_{\rm los} \frac{kT_e}{m_e c^2} n_e \sigma_{\rm thomson} dl \,,$$

where T_e is the electron temperature, m_e the electron mass, c the speed of light, n_e the electron density, and $\sigma_{\rm thomson}$ the Thomson cross-section.

Kinetic SZ

The kinetic SZ effect is generated by the scattering of CMB photons off an electron gas in motion with respect to the CMB. This motion generates spectral distortions with the same frequency scaling as CMB temperature fluctuations, and are directly proportional to the velocity of the electrons along the line of sight. As the effect has the same frequency scaling as CMB temperature fluctuations, it is, in principle, indistinguishable from primordial CMB. However, since the effect arises in clusters of galaxies with typical sizes of 1 arcmin, it can be distinguished to some level from the primordial CMB by spatial filtering, especially if the location of the clusters most likely to generate the effect is known from other information (e.g. the detection of the clusters through the thermal SZ effect).

Polarized SZ

The polarized SZ effect arises from the polarization dependence of the Thomson cross-section:

$$\sigma_T \propto |\boldsymbol{e_1} . \boldsymbol{e_2}|^2 \,,$$

where $\boldsymbol{e_1}$ and $\boldsymbol{e_2}$ are the polarization states of the incoming and outgoing photon, respectively. A quadrupole moment in the CMB radiation illuminating the cluster electron gas generates a net polarization, at a level typically two orders of magnitude lower than the kinetic SZ effect (Sunyaev & Zeldovich [57],

Audit & Simmons [2], Sazonov & Sunyaev [52]). Therefore, the kinetic SZ effect has been proposed as a probe to investigate the dependence of the CMB quadrupole with position in space. Cluster transverse motions at relativistic speed, however, generate also such an effect from the kinematic quadrupole induced by the motion. Multiple scattering of CMB photons also generates a low-level polarization signal towards clusters.

The polarized SZ effects have a distinctive frequency scaling, independent (to first order) of cluster parameters and of the amplitude of the effect. Amplitudes are proportional

- to τ for the intrinsic CMB quadrupole effect
- to $(v_t/c)^2\tau$ for the kinematic quadrupole effect
- to $(kT_e/m_ec^2)\tau^2$ and $(v_t/c)\tau^2$ for polarization effects due to double scattering

Here τ is the optical depth, v_t the transverse velocity, c the speed of light, k the Boltzmann constant, and T_e and m_e the electron temperature and mass.

As polarized effects arise essentially in galaxy clusters, they can be sought essentially in places where the much stronger thermal effect is detected, which will permit to improve the detection capability significantly. Polarized SZ emission, however, is weak enough that it is not expected to impact significantly the observation of any of the main polarized emissions.

Diffuse Component or Point Source Methods for SZ Effect Separation?

The SZ effect is particular in several respects. As most of the emission comes from compact regions towards clusters of galaxies (at arcminute scales), most of the present-day CMB experiments do not resolve clusters individually (apart from a few known extended clusters). For this reason, it seems natural to use point source detection methods for cluster detection (see review by Barreiro [4]). However, the very specific spectral signature, the presence of a possibly large background of clusters with emission too weak for individual cluster detection, and the interesting possibility to detect larger scale diffuse SZ emission make looking for SZ effect with diffuse component separation methods an interesting option.

3.5 Infrared Background of Unresolved Sources

The added emissions from numerous unresolved infrared sources at high redshift make a diffuse infrared background, detected originally in the FIRAS and DIRBE data (Puget et al. [49]). Because each source has its specific emission law, and because this emission law is redshifted by the cosmological expansion, the background does not have a very well-defined frequency scaling. It appears thus, in the observations at various frequencies, as an excess emission correlated between channels. The fluctuations of this background are expected to be significant at high galactic latitudes (where not masked by much stronger emissions from our own galaxy), and essentially at high frequencies (in the highest frequency channels of the Planck HFI).

3.6 Point Sources

The "point sources" component comprises all emissions from astrophysical objects such as radio galaxies, infrared galaxies, quasars, which are not resolved by the instruments used in CMB observations. For such sources, the issues are both their detection and the estimation of parameters describing them (flux at various frequencies, location, polarization...), and specific methods are devised for this purpose. For diffuse component separation, they constitute a source of trouble. Usually, pixels contaminated by significant point source emission are blanked for diffuse component separation.

4 Reduction of Foreground Contamination

The simplest way of avoiding foreground contamination consists in using prior information on emissions to reduce their impact on the data: by adequate selection of the region of observation, by masking some directions in the sky, by choosing the frequency bands of the instrument, or, finally, by subtracting an estimate of the contamination. All of these methods have been used widely in the context of CMB experiments.

4.1 Selection of the Region of Observation

Perhaps the most obvious solution to avoid contamination by foregrounds is to design the observations in such a way that the contamination is minimal. This sensible strategy has been adopted by ground-based and balloon-borne experiments observing only a part of the sky. In this case, CMB observations are made away from the galactic plane, in regions where foreground contamination from the galactic emissions is known to be small. The actual choice of regions of observation may be based on existing observations of dust and synchrotron emission at higher and lower frequencies, picking those regions where the emission of these foregrounds is known to be the lowest.

The drawback of this strategy is that the observations do not permit to estimate very well the level of contamination, nor the properties of the foregrounds.

4.2 Masking

For all-sky experiments, a strategy for keeping the contamination of CMB observations by foregrounds consists in masking regions suspected to comprise significant foreground emissions and deriving CMB properties (in particular the CMB power spectrum) in the "clean" region. The drawback of this strategy is that sky maps are incomplete.

Typically, for CMB observations, pixels contaminated by strong point sources (radio and infrared galaxies) are blanked, as well as a region containing the galactic plane. Such masks have been used in the analysis of WMAP data.

4.3 Selection of the Frequency Bands of the Instrument

Of course, the selection of the frequency of observation to minimize the overall foreground contamination is a sensible option. For this reason, many CMB experiments aim at observing the sky around 70–100 GHz. Ground-based observations, however, need to take into account the additional foreground of atmospheric emission, which leaves as best windows frequency bands around 30, 90, 150, and 240 GHz.

Figure 5 shows the expected typical frequency scaling for the major diffuse emission astrophysical components, including the CMB. For efficient

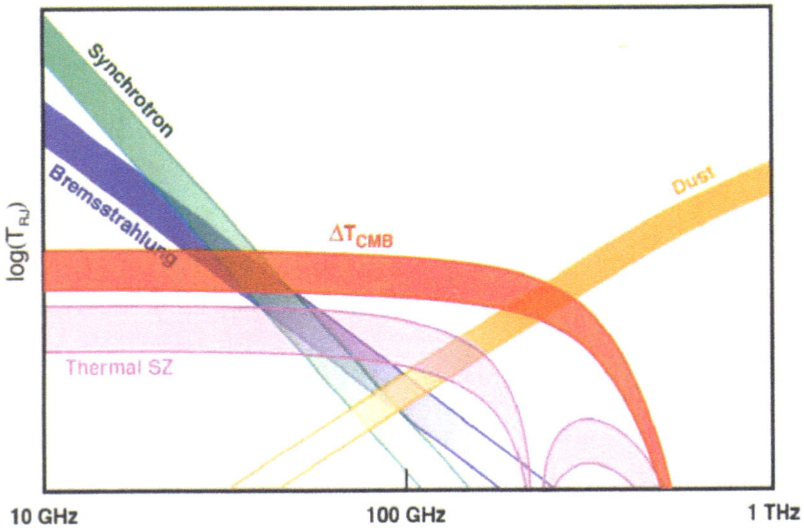

Fig. 5. The frequency scaling of CMB and most relevant diffuse foregrounds, in Rayleigh–Jeans temperature, between 10 GHz and 1 THz. Depending on the relative amplitude of synchrotron, bremsstrahlung and dust emissions, the minimum of galactic foregrounds is somewhat below 100 GHz. Free–free emission decreases roughly as $\nu^{-2.1}$ and synchrotron as ν^{-3}, while dust increases as ν^2. The SZ effect is the major emission towards rich clusters, but is very localized. The thickness of the bands illustrates uncertainties as to the level of foregrounds, as well as uncertainties in the frequency scaling for synchrotron, free–free, and dust emissions. Anomalous dust emission is not represented, due to our present lack of knowledge of the existence and nature of such a component

component separation, CMB experiment, ideally, should comprise two or three channels around 70–100 GHz where CMB dominates, one channel around 217 GHz (the zero of the SZ effect), two channels at higher frequencies to monitor dust emission, and three to four channels at lower frequencies to monitor low-frequency foregrounds.

Below 100 GHz, the present state of the art technology suggests the use of radiometers with high electron mobility transistor (HEMT) amplifiers, whereas above 100 GHz, low temperature bolometers provide a significantly better sensitivity that any other techniques. Typically, a single experiment uses one technology only. For Planck specifically, two different instruments have been designed to cover all the frequency range from 30 to 850 GHz.

4.4 Foreground Cleaning

As a refinement to the above simple observational strategies, a first-order estimate of foreground contamination, based on observations made at low and high frequencies, can be subtracted from the observations. Depending on the accuracy of the model, the overall level of contamination can be reduced by a factor of a few at least, which permits to reduce the amount of cut sky. This strategy, in particular, has been used by the WMAP team for the analysis of first-year WMAP data (Bennett et al. [7]).

Observations at low frequencies (10–40 GHz) can be used to map synchrotron emission and model its contribution in the 70–100 GHz range. Similar strategies can be used towards the high-frequency side to model dust emission and subtract its contribution from CMB channels. For this purpose, models of emission as good as possible are needed, and the cleaning can be no better than the model used. There is always, therefore, a trade-off between a sophisticated model with simple correction methods (subtraction of an interpolation, simple decorrelation) and a simple model with sophisticated statistical treatments (multifrequency filtering, independent component analysis). Which approach is best depends on a number of issues, and the answer is not completely clear yet.

5 Linear Model and System Inversion

The most popular model of the observations for source separation in the context of CMB observations probably is the linear mixture.

In this model, all components are assumed to have an emission which can be decomposed as the product of a spatial template independent of the frequency of observation and of a spectral emission law which does not depend on the pixel. The total emission at frequency ν, in pixel p, of a particular emission process j is written as

$$x_j(\nu, p) = a(\nu)s_j(p)$$

or alternatively, in spherical harmonics space,

$$x_j(\nu, \ell m) = a(\nu) s_j(\ell m) .$$

Forgetting for the moment annoying details concerning the response of the instrument (beams, frequency bands, etc.) the observation with a detector is then

$$y_i(p) = \sum_j x_j(\nu_i, p) + n_i(p) ,$$

where $n_i(p)$ is the contribution of noise for detector i. For a set of detectors, this can be recast in a matrix-vector form as

$$\boldsymbol{y}(p) = \mathbf{A}\boldsymbol{s}(p) + \boldsymbol{n}(p) . \tag{23}$$

Here, $\boldsymbol{y}(p)$ represent the set of maps observed with all detectors and $\boldsymbol{s}(p)$ are the unobserved components (one template map per astrophysical component). The mixing matrix $\mathbf{A}$ which does not depend on the pixel for a simple linear mixture has one column per astrophysical component and one line per detector.

If the observations are given in CMB temperature for all detectors and if the detectors are properly calibrated, each element of the column of the mixing matrix corresponding to CMB is equal to 1.

The problem of component separation consists in inverting the linear system of (23). Here we first concentrate on linear inversion, which consists in finding the "best" possible matrix $\mathbf{W}$ (such that $\widehat{\boldsymbol{s}} = \mathbf{W}\boldsymbol{y}$ is "as good an estimator of $\boldsymbol{s}$ as possible").

Covariances and Multivariate Power Spectra

In the following, a lot of use will be made of second-order statistics of various sorts of data. In general, for a collection of maps $\boldsymbol{x}(p) = \{x_i(p)\}$, the covariance will be noted as $\mathbf{R_x}(p, p')$, the elements of which are

$$R_{ij}(p, p') = \mathrm{cov}(x_i(p), x_j(p')) .$$

Alternatively, in harmonic space, we denote as $\mathbf{R_x}(\ell)$ the multivariate power spectrum of $\boldsymbol{x}$, i.e. the collection of matrices

$$\mathbf{R_x}(\ell) = \left\langle \boldsymbol{x}(\ell, m)\boldsymbol{x}^\dagger(\ell, m)\right\rangle ,$$

where the brackets $\langle . \rangle$ denote ensemble average and the dagger † denotes the transpose of the complex conjugate. Such a power spectrum is well defined only for stationary/isotropic random fields on the sphere for which $\langle \boldsymbol{x}(\ell, m)\boldsymbol{x}^\dagger(\ell, m)\rangle$ does not depend on m.

5.1 Simple Inversion

If $\mathbf{A}$ is square and non-singular, in absence of any additional information, then the inversion is obtained by

$$\mathbf{W} = \mathbf{A}^{-1} \tag{24}$$

and we have

$$\widehat{\boldsymbol{s}} = \mathbf{A}^{-1}\boldsymbol{y} = \boldsymbol{s} + \mathbf{A}^{-1}\boldsymbol{n} \,. \tag{25}$$

Note that because of the remaining noise term, this inversion is not always the best solution in terms of residual error, in particular in the poor signal-to-noise regimes. For instance, if we have two measurements of a mixture of CMB + thermal dust in a clean region of the sky (low foregrounds), one of which, at 150 GHz, is relatively clean and the other, at 350 GHz, quite poor because of high-level noise, then it may be better to use the 150 GHz as the CMB template (even with some dust contamination), rather than to invert the system, subtracting little dust and adding a large amount of noise.

In terms of residual foreground contamination, however (if the criterion is to reject astrophysical signals, whatever the price to pay in terms of noise), the only solution here is matrix inversion. The solution is unbiased, but may be noisy.

Note that an ILC method would produce a different solution, possibly slightly biased (as discussed in Sect. 2.6), but possibly better in terms of signal-to-noise ratio of the end product.

This solution can be applied if the full matrix $\mathbf{A}$ is known (not only the column of the component of interest, i.e. the CMB), without further prior knowledge of the data.

5.2 Inversion of a Redundant System Using the Pseudo-inverse

If there are more observations than components, but nothing is known about noise and signal levels, then the inversion is obtained by

$$\mathbf{W} = \left[\mathbf{A}^{\dagger}\mathbf{A}\right]^{-1}\mathbf{A}^{\dagger} \tag{26}$$

and we have

$$\widehat{\boldsymbol{s}} = \left[\mathbf{A}^{\dagger}\mathbf{A}\right]^{-1}\mathbf{A}^{\dagger}\boldsymbol{y} = \boldsymbol{s} + \left[\mathbf{A}^{\dagger}\mathbf{A}\right]^{-1}\mathbf{A}^{\dagger}\boldsymbol{n} \,. \tag{27}$$

Again, this estimator is unbiased, but may contain a large amount of noise and may not be optimal in terms of signal-to-noise ratio. All the comments made in the previous paragraph hold as well for this solution.

Note that there is no noise weighting here, so that one single very bad channel may contaminate significantly all the data after inversion. It is therefore not a very good idea to apply this estimator with no further thoughts.

Note that, again, this solution can be implemented without any further knowledge about signal and noise – only the entries of the mixing matrix for all components are needed.

5.3 Noise-Weighted Scheme: Generalized Least Square Solution

Let us now assume that we know something additional about the noise, namely, its second-order statistics. These are described by noise correlation matrices in real space or alternatively by noise power spectra in Fourier (for small maps) or in harmonic (for all-sky maps) space.

We denote as $\mathbf{R_n}$ the noise correlation matrix and assume, for the time being, that the noise for each detector i is a realization of a random Gaussian field, the generalized (or global) least square (GLS) solution of the system of (23) is

$$\mathbf{W} = \left[\mathbf{A}^{\dagger}\mathbf{R_n}^{-1}\mathbf{A}\right]^{-1}\mathbf{A}^{\dagger}\mathbf{R_n}^{-1} \tag{28}$$

and we have

$$\widehat{\boldsymbol{s}} = \left[\mathbf{A}^{\dagger}\mathbf{R_n}^{-1}\mathbf{A}\right]^{-1}\mathbf{A}^{\dagger}\mathbf{R_n}^{-1}\boldsymbol{y} = \boldsymbol{s} + \left[\mathbf{A}^{\dagger}\mathbf{R_n}^{-1}\mathbf{A}\right]^{-1}\mathbf{A}^{\dagger}\mathbf{R_n}^{-1}\boldsymbol{n}\,. \tag{29}$$

Again, the solution is unbiased. Although there remains a noise contribution, this is the solution yielding the minimum variance error map for a deterministic signal (in contrast with the Wiener solution below, which optimizes the variance of the error when the signal is stochastic, i.e. assumed to be a random field). It is also the best linear solution in the limit of large signal-to-noise ratio.

This solution is also theoretically better than the ILC when the model holds, but the price to pay is the need for more prior knowledge about the data (knowledge of the mixing matrix and of noise covariance matrices or power spectra). If that knowledge is insufficient, one has to design methods to get it from the data itself. Such "model learning" methods will be discussed in Sect. 7.

5.4 Wiener Solution

The Wiener filter (Wiener [64]) has originally been designed to filter time series in order to suppress noise, but has been extended to a large variety of applications since then. Wiener's solution requires additional information regarding the spectral content of the original signal and the noise. Wiener filters are characterized by the following:

- Both the noise and the signal are considered as stochastic processes with known spectral statistics (or correlation properties) – contrarily to the GLS method which considers the noise only to be stochastic, the signal being deterministic
- The optimization criterion is the minimum least square error
- The solution is linear

In signal processing, a data stream $y(t) = s(t) + n(t)$ assumed to be a noisy measurement of a signal s can be filtered for denoising as follows: in Fourier space, each mode $y(f)$ of the data stream is weighted by a coefficient

$$W(f) = \frac{S(f)}{S(f) + N(f)} ,$$

where $S(f) = \langle |s(f)|^2 \rangle$ and $N(f) = \langle |n(f)|^2 \rangle$ are ensemble averages of the square moduli of the Fourier coefficients of the stochastic processes s and n.

In the limit of very small noise level $N(f) \ll S(f)$, the Wiener filter value is $W(f) = 1$, and the filter does not change the data. In the limit of very poor signal to noise $S(f) \ll N(f)$, the filter suppresses the data completely, because that mode adds noise to the total data stream and nothing else.

It can be shown straightforwardly that the Wiener filter minimizes the variance of the error of the signal estimator $\widehat{s}(f) = W(f)y(f)$ (so that $\langle \int_f |\widehat{s}(f) - s(f)|^2 \mathrm{d}f \rangle$ is minimal).

The Wiener solution can be adapted for solving our component separation problem, provided the mixing matrix $\mathbf{A}$ and the second-order statistics of the components and of the noise are known (Tegmark & Efstathiou [58], Bouchet & Gispert [12]) as

$$\mathbf{W}^{(1)} = \left[\mathbf{A}^{\dagger}\mathbf{R_n}^{-1}\mathbf{A} + \mathbf{R_s}^{-1}\right]^{-1} \mathbf{A}^{\dagger}\mathbf{R_n}^{-1} , \tag{30}$$

where $\mathbf{R_s}$ is the correlation matrix of the sources (or power spectra of the sources, in the Fourier or harmonic space) and $\mathbf{R_n}$ the correlation matrix of the noise. The superscript (1) is used to distinguish two forms of the Wiener filter (the second is given later in this section).

An interesting aspect of the Wiener filter is that

$$\begin{aligned} \widehat{\boldsymbol{s}} &= \left[\mathbf{A}^{\dagger}\mathbf{R_n}^{-1}\mathbf{A} + \mathbf{R_s}^{-1}\right]^{-1} \mathbf{A}^{\dagger}\mathbf{R_n}^{-1}\boldsymbol{y} \\ &= \left[\mathbf{A}^{\dagger}\mathbf{R_n}^{-1}\mathbf{A} + \mathbf{R_s}^{-1}\right]^{-1} \mathbf{A}^{\dagger}\mathbf{R_n}^{-1}\mathbf{A}\boldsymbol{s} \\ &\quad + \left[\mathbf{A}^{\dagger}\mathbf{R_n}^{-1}\mathbf{A} + \mathbf{R_s}^{-1}\right]^{-1} \mathbf{A}^{\dagger}\mathbf{R_n}^{-1}\boldsymbol{n} . \end{aligned} \tag{31}$$

The matrix in front of $\boldsymbol{s}$ is not the identity, and thus the Wiener filter does not give an unbiased estimate of the signals of interest. Diagonal terms can be different from unity. In addition, non-diagonal terms may be non-zero, which means that the Wiener filter allows some residual foregrounds to be present in the final CMB map – the objective being to minimize the variance of the residuals, irrespective of whether these residuals originate from instrumental noise or from astrophysical foregrounds.

As noted in Tegmark & Efstathiou [58], the Wiener solution can be "debiased" by multiplying the Wiener matrix by a diagonal matrix removing the impact of the filtering. The authors argue that for the CMB this debiasing is desirable for subsequent power spectrum estimation on the reconstructed CMB map. Each mode of a given component is divided by the diagonal element of the Wiener matrix for that component and that mode. This, however, destroys the minimal variance property of the Wiener solution and can increase the noise very considerably. There is an incompatibility between the objective of obtaining a minimum variance map and the objective of obtaining an unbiased map which can be used directly to measure the power spectrum of the CMB. There is no unique method for both.

Before moving on, it is interesting to check that the matrix form of the Wiener filter given here reduces to the usual form when there is one signal only and when the matrix **A** reduces to a scalar equal to unity. In that case, the Wiener matrix **W** of (30) reduces to

$$W(f) = [1/S(f) + 1/N(f)]^{-1}/N(f) = S(f)/[N(f) + S(f)] ,$$

where S and N are the signal and noise power spectra, and we recover the classical Wiener formula.

Two Forms of the Wiener Filter

In the literature, another form can be found for the Wiener filter matrix:

$$\mathbf{W}^{(2)} = \mathbf{R_s A}^{\dagger} \left[\mathbf{R_n} + \mathbf{A R_s A}^{\dagger}\right]^{-1} . \tag{32}$$

It can be shown straightforwardly that if the matrices

$$\mathbf{M}_1 = \left[\mathbf{A}^{\dagger}\mathbf{R_n}^{-1}\mathbf{A} + \mathbf{R_s}^{-1}\right]$$

and

$$\mathbf{M}_2 = \left[\mathbf{R_n} + \mathbf{A R_s A}^{\dagger}\right]$$

are regular, then the forms of (30) and (32) are equivalent (simply multiply both forms by $\mathbf{M}_1$ on the left and $\mathbf{M}_2$ on the right, and expand).

It may seem that the form of (32) is more convenient, as it requires only one matrix inversion instead of three. Each form, however, presents specific advantages or drawbacks, which appear clearly in the high signal-to-noise ratio (SNR) limit and if power spectra of all signals are not known.

The High SNR Limit

The above two forms of the Wiener filter are not equivalent in the high SNR limit. In this regime, (30) yields in the limit

$$\mathbf{W}^{(1)}_{\text{limit}} = \left[\mathbf{A}^{\dagger}\mathbf{R_n}^{-1}\mathbf{A}\right]^{-1} \mathbf{A}^{\dagger}\mathbf{R_n}^{-1} ,$$

which is the GLS solution of (28), and depends only on the noise covariance matrix, whereas (32) tends to

$$\mathbf{W}^{(2)}_{\text{limit}} = \mathbf{R_s A}^{\dagger} \left[\mathbf{A R_s A}^{\dagger}\right]^{-1} ,$$

which depends only on the covariance of the signal. Therefore, some care should be taken when applying the Wiener filter in the high SNR ratio regime, when numerical roundup errors may cause problems.

Note that if $\left[\mathbf{A}^\dagger\mathbf{A}\right]$ is regular, then

$$\begin{aligned}\mathbf{W}^{(2)}_{\text{limit}} &= \mathbf{R_s}\mathbf{A}^\dagger\left[\mathbf{A}\mathbf{R_s}\mathbf{A}^\dagger\right]^{-1} \\ &= \left[\mathbf{A}^\dagger\mathbf{A}\right]^{-1}\left[\mathbf{A}^\dagger\mathbf{A}\right]\mathbf{R_s}\mathbf{A}^\dagger\left[\mathbf{A}\mathbf{R_s}\mathbf{A}^\dagger\right]^{-1} \qquad (33)\\ &= \left[\mathbf{A}^\dagger\mathbf{A}\right]^{-1}\mathbf{A}^\dagger \qquad (34)\end{aligned}$$

and the limit is simply the pseudo-inverse of matrix $\mathbf{A}$, without any noise weighting. Of course, when there is no noise at all, $\mathbf{W}^{(1)}$ cannot be implemented at all, and the Wiener solution is pointless anyway.

What if Some Covariances Are Not Known?

It is interesting to note that even if the covariance matrix (or equivalently multivariate power spectrum) $\mathbf{R_s}$ of all sources is not known, it is still possible to implement an approximate Wiener solution if the maps of observations are large enough to allow a good estimate of the covariance matrix of the observations.

If $\boldsymbol{y} = \mathbf{A}\boldsymbol{s} + \boldsymbol{n}$ and if the noise and the components are independent, the covariance $\mathbf{R_y}$ of the observations is of the form

$$\mathbf{R_y} = \mathbf{R_n} + \mathbf{A}\mathbf{R_s}\mathbf{A}^\dagger \,.$$

Therefore, form 2 of the Wiener filter can be recast as

$$\mathbf{W}^{(2)} = \mathbf{R_s}\mathbf{A}^\dagger\left[\mathbf{R_y}\right]^{-1} \,. \qquad (35)$$

If all components are decorrelated, the matrix $\mathbf{R_s}$ is diagonal. For the implementation of a Wiener solution for one single component (e.g. CMB), only the diagonal element corresponding to the CMB (i.e. the power spectrum C_ℓ of the CMB) is needed, in addition to the multivariate power spectrum of the observations $\mathbf{R_y}$. The latter can be estimated directly using the observations.

5.5 Comment on the Various Linear Inversion Solutions

The above four linear solutions to the inversion of the linear system of (23) have been presented by order of increasing generality, increasing complexity, and increasing necessary prior knowledge. The various solutions are summarized in Table 1. Three comments are necessary.

First, we note that the Wiener solutions require the prior knowledge of the covariance matrices (or equivalently power spectra) of both the noise and the signal. For CMB studies, however, the measurement of the power spectrum of the CMB field is precisely the objective of the observations. Then, the question of whether the choice of the prior on the CMB power spectrum biases the final result or not is certainly of much relevance. For instance, the prior

Table 1. Summary of linear solutions to component separation when the mixing matrix **A** is known

Solution	$\mathbf{W} =$	Required prior knowledge	Comments
Inverse	$\mathbf{A}^{-1}$	$\mathbf{A}$	When there are as many channels of observation as components. Unbiased, contamination free
Pseudo-inverse	$[\mathbf{A}^\dagger \mathbf{A}]^{-1} \mathbf{A}^\dagger$	$\mathbf{A}$	When there are more channels of observation than components. Unbiased, contamination free
GLS	$[\mathbf{A}^\dagger \mathbf{R_n}^{-1} \mathbf{A}]^{-1} \mathbf{A}^\dagger \mathbf{R_n}^{-1}$	$\mathbf{A}$ and $\mathbf{R_n}$	Minimizes the variance of the error for deterministic signals. Unbiased, contamination free
Wiener 1	$[\mathbf{A}^\dagger \mathbf{R_n}^{-1} \mathbf{A} + \mathbf{R_s}^{-1}]^{-1} \mathbf{A}^\dagger \mathbf{R_n}^{-1}$	$\mathbf{A}$, $\mathbf{R_n}$, and $\mathbf{R_s}$	Minimizes the variance of the error for stochastic signals. Biased, not free of contamination. Tends to the GLS solution in the limit of high SNR
Wiener 2	$\mathbf{R_s} \mathbf{A}^\dagger [\mathbf{R_n} + \mathbf{A} \mathbf{R_s} \mathbf{A}^\dagger]^{-1}$	$\mathbf{A}$, $\mathbf{R_n}$, and $\mathbf{R_s}$	Equivalent to Wiener 1. Tends to the pseudo-inverse in the limit of high SNR
Debiased Wiener	$\Lambda \mathbf{R_s} \mathbf{A}^\dagger [\mathbf{R_n} + \mathbf{A} \mathbf{R_s} \mathbf{A}^\dagger]^{-1}$	$\mathbf{A}$, $\mathbf{R_n}$, and $\mathbf{R_s}$	The diagonal matrix Λ, inverse of the diagonal of **WA** where **W** is the Wiener solution, removes for each mode the filtering effect of the Wiener filter. Unbiased, but not contamination free

assumption that the power spectrum of the CMB is small in some ℓ range will result in filtering the corresponding modes, and the visual impression of the recovered CMB will be that indeed there is little power at the corresponding scales. For power spectrum estimation on the maps, however, this effect can be (and should be) corrected, which is always possible as the effective filter induced by the Wiener solution is known (for an implementation in harmonic space, it is equal for each mode ℓm, for each component, to the corresponding term of the diagonal of $\mathbf{W}_{\ell m}\mathbf{A}$). In Sect. 8, a solution will be proposed for estimating first, on the data themselves, all the relevant statistical information (covariance matrices and frequency scaling) and then using this information for recovering maps of the different components.

Second, we should emphasize that the choice of a linear solution should be made with a particular objective in mind. If the objective is to get the best possible map in terms of least square error, then the Wiener solution is the best solution if the components are Gaussian. The debiased Wiener is not really adapted to any objective in particular. The GLS solution is the best solution if the objective is an unbiased reconstruction with no filtering and no contamination. In practice, it should be noted that small uncertainties on $\mathbf{A}$ result in errors (biases and contamination) even for the GLS solution.

As a third point, we note that it can be shown straightforwardly that for Gaussian sources and noise, the Wiener solution maximizes the posterior probability $P(\boldsymbol{s}|\boldsymbol{y})$ of the recovered sources given the data. From Bayes theorem, the posterior probability is the product of the likelihood $P(\boldsymbol{y}|\boldsymbol{s})$ and the prior $P(\boldsymbol{s})$, normalized by the evidence $P(\boldsymbol{y})$. The normalizing factor does not depend on s. We can write

$$\begin{aligned} P(\boldsymbol{s}|\boldsymbol{y}) &\propto \exp\left[-(\boldsymbol{y}-\mathbf{A}\boldsymbol{s})^\dagger \mathbf{R_n}^{-1}(\boldsymbol{y}-\mathbf{A}\boldsymbol{s})\right] \exp\left[-\boldsymbol{s}^\dagger \mathbf{R_s}^{-1}\boldsymbol{s}\right] , \\ \log(P(\boldsymbol{s}|\boldsymbol{y})) &= -\left[(\boldsymbol{y}-\mathbf{A}\boldsymbol{s})^\dagger \mathbf{R_n}^{-1}(\boldsymbol{y}-\mathbf{A}\boldsymbol{s})\right] - \left[\boldsymbol{s}^\dagger \mathbf{R_s}^{-1}\boldsymbol{s}\right] + \text{const} , \end{aligned} \quad (36)$$

where $\exp\left[-\boldsymbol{s}^\dagger \mathbf{R_s}^{-1}\boldsymbol{s}\right]$ is the Gaussian prior for s. The requirement that

$$\frac{\partial}{\partial \boldsymbol{s}}\log(P(\boldsymbol{s}|\boldsymbol{y})) = 0$$

implies

$$\begin{aligned} -\mathbf{A}^\dagger \mathbf{R_n}^{-1}(\boldsymbol{y}-\mathbf{A}\boldsymbol{s}) + \mathbf{R_s}^{-1}\boldsymbol{s} &= 0 , \\ \mathbf{A}^\dagger \mathbf{R_n}^{-1}\mathbf{A}\boldsymbol{s} + \mathbf{R_s}^{-1}\boldsymbol{s} &= \mathbf{A}^\dagger \mathbf{R_n}^{-1}\boldsymbol{y} , \\ \left[\mathbf{A}^\dagger \mathbf{R_n}^{-1}\mathbf{A} + \mathbf{R_s}^{-1}\right]\boldsymbol{s} &= \mathbf{A}^\dagger \mathbf{R_n}^{-1}\boldsymbol{y} . \end{aligned} \quad (37)$$

Thus we get the solution $\mathbf{W}^{(1)}$. In Sect. 6, we will discuss the case where the Gaussian prior is replaced by an entropic prior, yielding yet another solution for s.

5.6 Pixels, Harmonic Space, or Wavelets?

The simple inversion of $\mathbf{A}$ using the inverse or pseudo-inverse can be implemented equivalently with any representations of the maps, in pixel domain, harmonic space, or on any decomposition of the observations on a set of functions as, e.g. wavelet decompositions (Moudden et al. [45]). The result in terms of separation is independent of this choice, as far as the representation arises from a linear transformation.

If all sources and signals are Gaussian random fields, the same is true for GLS or Wiener inversions, provided all the second-order statistics are properly described by the covariance matrices $\mathbf{R_n}$ and $\mathbf{R_s}$.

These covariance matrices, in pixel space, take the form of the set of covariances:

$$\mathbf{R_n} = \left\{R_{n_i n_j}\left(p_i, p_j\right)\right\} ,$$

where

$$R_{n_i n_j}\left(p_i, p_j\right) = \langle n_i(p_i) n_j(p_j)\rangle .$$

Similarly, in harmonic space, we have

$$\mathbf{R_n} = \left\{R_{n_i n_j}\left(\ell_i, m_i, \ell_j, m_j\right)\right\} ,$$

where

$$R_{n_i n_j}\left(\ell_i, m_i, \ell_j, m_j\right) = \langle n_i\left(\ell_i, m_i\right) n_j\left(\ell_j, m_j\right)\rangle .$$

If the number of pixels is large, if we deal with several sources and many channels at the same time (tens today, thousands in a few years), the implementation of the GLS or Wiener solution may be quite demanding in terms of computing. For this reason, it is desirable to implement the solution in the space where matrices are the most easy to invert.

For stationary Gaussian random fields, harmonic space implementations are much easier than direct space implementations, because the covariance between distinct modes vanish, so that

$$R_{n_i n_j}\left(\ell_i, m_i, \ell_j, m_j\right) = \langle n_i(\ell_i, m_i) n_j(\ell_i, m_i)\rangle \, \delta_{\ell_i \ell_j} \delta_{m_i m_j} .$$

The full covariance matrix consists in a set of independent covariance matrices (one for each mode), each of which is a small matrix of size $N_{\text{channels}} \times N_{\text{channels}}$ for $\mathbf{R_n}$, and of size $N_{\text{sources}} \times N_{\text{sources}}$ for $\mathbf{R_s}$.

5.7 Annoying Details

Under the assumption that the response of each detector i in the instrument can itself be decomposed into the product of a spectral response $h_i(\nu)$ and a frequency-independent symmetrical beam B_i, the contribution of component j to the observation obtained with detector i is

$$y_{ij}(\ell m) = \left[\int_{\nu} h_i(\nu) a(\nu) \mathrm{d}\nu\right] B_{i,\ell}\, s_j(\ell m) ,$$

where $B_{i,\ell}$ are the coefficients of the expansion of the symmetric beam of detector i on Legendre polynomials.

The mixing matrix of this new linear model is seen to include a band integration, assumed to first order to be independent of ℓ, and the effect of a beam, which depends on ℓ. Both can be taken into account in a linear inversion, if known a priori.

6 Maximum Entropy Method

The Wiener filter provides the best (in terms of minimum variance or maximum likelihood) estimate of the component maps if two main assumptions hold. First, the observations should be a linear mixture of distinct emissions. Second, the components and the noise should be (possibly correlated) Gaussian stationary random processes.

Unfortunately, the sky is known to be neither Gaussian nor stationary, with the possible exception of the CMB itself. Is this critical?

The maximum entropy method (MEM) of component separation (Hobson et al. [28]) is a method which inverts the same linear system of component mixtures, but assumes non-Gaussian probability distributions.

6.1 Maximum Entropy

The concept of entropy in information theory has been introduced by Shannon in 1948 [53]. The entropy of a discrete random variable X on a finite set of possible values $\{x_i\}$ with probability distribution function $p(x_i) = p(X = x_i)$, is defined as

$$H(X) = -\sum_{i=1}^{N} p(x_i) \log p(x_i) \; . \tag{38}$$

The principle of maximum entropy is based of the idea that whenever there is some choice to be made about the distribution function of the random variable X, one should choose the least informative option possible. Entropy measures the amount of information in a probability distribution, and entropy maximization is a way of achieving this.

For instance, in the absence of any prior information, the probability distribution which maximizes the entropy of (38) is a distribution with uniform probability, $p(x_i) = 1/N$, i.e. the least informative choice of a probability distribution on the finite set $\{x_i\}$, where all outcomes are equally likely. This is the most natural choice if nothing more is said about the probability distribution.

In the opposite, a most informative choice would be a probability which gives a certain result (for instance always $X = x_1$). This is a probability distribution which minimizes entropy.

In the continuous case where X can achieve any real value x with probability density $p(x)$, entropy can be defined as

$$H(X) = -\int_{-\infty}^{\infty} p(x) \log p(x)\, dx \, . \tag{39}$$

Of course, maximum entropy becomes really useful when there is also additional information available. In this case, the entropy must be maximized within the constraints given by additional information.

For instance, the maximum entropy distribution of a real random variable of mean μ and variance σ^2 is the normal (Gaussian) distribution :

$$p(x) = \frac{1}{2\pi\sigma} \exp\left[-\frac{(x-\mu)^2}{2\sigma^2} \right] \, .$$

For this reason, in the absence of additional information about the probability distribution of a random variable of known mean and variance, it is quite natural, according to the maximum entropy principle, to assume a Gaussian distribution – which maximizes the entropy, and hence corresponds to the least informative choice possible.

Another useful example is the maximum entropy distribution of a real positive random variable of mean μ, which is the exponential distribution

$$p(x) = \frac{1}{\mu} \exp(-x/\mu) \, .$$

6.2 Relative Entropy

In fact, the differential entropy of (39) has an unpleasant property. It is not invariant under coordinate transformations (on spaces with more than one dimension).

The definition of *relative entropy* (or Kullback–Leibler divergence) between two distributions solves the issue. It can be interpreted as a measure of the amount of additional information one gets from knowing the actual (true) probability distribution $p(x)$, instead of an imperfect model $m(x)$, and is given by

$$D_{\mathrm{KL}}(p||m) = \int_{-\infty}^{\infty} p(x) \log \frac{p(x)}{m(x)}\, dx \, . \tag{40}$$

Later in this chapter (in Sect. 8), we will make use of the Kullback–Leibler divergence for measuring the "mismatch" between two positive matrices $\mathbf{R_1}$ and $\mathbf{R_2}$. It will actually correspond to the KL divergence between two Gaussian distributions with covariance matrices $\mathbf{R_1}$ and $\mathbf{R_2}$.

The relative entropy is invariant under coordinate transformations (because both the ratio $p(x)/m(x)$ and $p(x)dx$ are invariant under coordinate transformations).

6.3 Component Separation with the MEM

In principle, replacing the Gaussian prior by some other prior is perfectly legitimate. In practice, the choice of such a prior is not obvious, as the full statistical description of a complex astrophysical component is difficult to apprehend.

Following the maximum entropy principle, one may decide to use as a prior the distribution which maximizes the entropy given a set of constraints. If the constraints are the value of the mean and the variance, then the maximum entropy prior is the Gaussian prior.

Hobson and collaborators, in their MEM paper (Hobson et al. [28]), argue that based on the maximum entropy principle, an appropriate prior for astrophysical components s is

$$p(s) = \exp\left[-\alpha S_c(\boldsymbol{s}, \boldsymbol{m}_u, \boldsymbol{m}_v)\right] \tag{41}$$

with

$$S_c(\boldsymbol{s}, \boldsymbol{m}_u, \boldsymbol{m}_v) = \sum_{j=1}^{L} \left\{ \psi_j - m_{uj} - m_{vj} - s_j \ln\left[\frac{\psi_j + s_j}{2m_{uj}}\right]\right\} ,$$

where $\psi_j = [{s_j}^2 + 4m_{uj}m_{vj}]^{1/2}$, and where $\boldsymbol{m}_u$ and $\boldsymbol{m}_v$ are models of two positive additive distributions (which are not clearly specified) used to represent the astrophysical components.

A derivation for this is given by Hobson & Lasenby [27], but the connection to entropy is not direct. In particular, the definition of entropy does not require the *values* of the random variables to be positive, but their *probability densities*, which makes the discussion unconvincing.

Pragmatically, the choice for the prior of (41) seems to be validated a posteriori by the performance of the separation, which is not worse (and actually better for some of the components) than that obtained with the Wiener filter. It is not likely to be optimal, however, because the non-stationarity of components implies correlations in the harmonic domain, which are not fully taken into account in the MEM implementation.

The maximization of the posterior probability (and hence of the product of the likelihood and the prior) is done with a dedicated fast maximization algorithm. We refer the reader to the relevant papers for additional details (Hobson & Lasenby [27], Stolyarov et al. [55]).

This method has been applied to the separation of components in the COBE data (Barreiro et al. [5]).

6.4 Comments About the MEM

Although entropy has a clear meaning in terms of information content in the discrete case (e.g. it defines the minimum number of bits necessary to represent a sequence), there is no such interpretation in the continuous case. Entropy maximization, understood as minimizing the amount of arbitrary information in the assumed distribution, hence, is not very clearly founded for continuous images.

The "principle" of maximum entropy, as the name indicates, is not a theorem, but a reasonable recipe which seems to work in practice. In the context of the CMB, there is no guarantee that it is optimal, among all non-linear solutions of the mixing system. MEM outperforms the Wiener filter solution for some components in particular because the entropic prior of Hobson and Lasenby allows heavier tails than the Gaussian prior. Other priors, however, based on a physical model of the emissions, might well perform even better in some cases. This question remains as an open problem in the field.

7 ICA and Blind Source Separation

7.1 About Blind Separation

The term "blind separation" refers to a fascinating possibility: if the components of a linear mixture are *statistically independent*, they can be recovered even if the mixing matrix $\mathbf{A}$ is unknown *a priori*. In essence, this is possible because statistical independence is, at the same time, a strong mathematical property and, quite often, a physically plausible one.

There is an obvious and strong motivation for attempting blind component separation: allowing underlying components to be recovered blindly makes it possible to analyse multi-detector data with limited, imperfect, or even outright missing knowledge about the emission laws of the components. Even better, one can process data without knowing in advance which components might be "out there". Hence, the blind approach is particularly well suited for exploratory data analysis.

In the last 15 years, blind component separation has been a very active area of research in the signal processing community where it goes by the names of "blind source separation" (BSS) and "independent component analysis" (ICA). This section outlines the principles underlying some of the best known approaches to blind source separation. There is not a single best approach because there is not a unique way in which to express statistical independence on the basis of a finite number of samples.

7.2 Statistical Independence

This section explains why blind component separation is possible in the first place. For the sake of exposition, the main ideas are discussed in the simplest case: there is no observation noise and there are as many "channels" as underlying components. Thus the model reduces to

$$\mathbf{y}(p) = \mathbf{A}\mathbf{s}(p) ,$$

where $\mathbf{A}$ is an $n \times n$ matrix and we are looking for an $n \times n$ matrix "separating matrix" $\mathbf{W}$. Of course, if the mixing matrix $\mathbf{A}$ is known, there is little mystery about separation: one should take $\mathbf{W} = \mathbf{A}^{-1}$ and be done with it.

If nothing is known about $\mathbf{A}$ but the components are known (or assumed) to be statistically independent, the idea is to determine $\mathbf{W}$ in such a way that the entries of vector $\mathbf{Wy}$ are independent (or as independent as possible). In other words, the hope is that by restoring independence, one would restore the components themselves. Amazingly enough, this line of attack works. Even better, under various circumstances, it can be shown to correspond to maximum likelihood estimation and there is therefore some statistical optimality to it, provided the hypothesis of statistical independence is expressed vehemently enough.

Note, however, that no matter the amount of statistical ingenuity thrown at blind component separation, there is no hope to recover completely the mixing matrix (or equivalently: the components). This is because a scalar factor can always be exchanged between each entry of $\mathbf{s}$ and the corresponding column of $\mathbf{A}$ without changing what the model predicts (i.e. the value of the product $\mathbf{As}$) and without destroying the (hypothetical) independence between the entries of $\mathbf{s}$. The same is true of a renumbering of the columns of $\mathbf{A}$ and of the entries of $\mathbf{s}$. In other words, blind recovery is possible only up to rescaling and permutation of the components. In many applications, this will be "good enough". If these indeterminancies have to be fixed, it can be done only by imposing additional constraints or resorting to side information.

For any possible choice $\mathbf{W}$ of a candidate-separating matrix, denote

$$\mathbf{x}(p) = \mathbf{Wy}(p)$$

the corresponding vector of candidate components. If $\mathbf{W} = \mathbf{A}^{-1}$ then the entries of $\mathbf{x}$ are independent (since, in this case $\mathbf{x}(p) = \mathbf{s}(p)$). Under which circumstances would the *converse* be true? Whenever the converse is true, it will be possible to recover the sources by looking for the linear transform $\mathbf{W}$ which makes them independent. Hence, we have a blind separation principle: to separate components, make them independent.

7.3 Correlations

The main difficulty in blind source separation is to define a measure of independence. The problem is that the simple decorrelation condition[4] between any two candidate components,

$$\frac{1}{P}\sum_{p=1}^{P} x_i(p)x_j(p) = 0 \quad \text{for} \quad 1 \leq i \neq j \leq n, \tag{42}$$

does not cut it. This is in fact obvious from the fact that this decorrelation condition between x_i and x_j is *symmetric*. Hence decorrelation provides only $n(n-1)/2$ constraints while n^2 constraints are needed to determine $\mathbf{W}$

[4] Here, as in the rest of this section, all signals are assumed to have zero mean.

uniquely. Therefore, more expressive forms of independence must be used. Two main avenues are possible: non-linear correlations and localized correlations, as described next.

Non-linear Correlations

The "historical approach" to blind separation has been to determine a separating matrix $\mathbf{W}$ in order to obtain "non-linear decorrelations", i.e.

$$\frac{1}{P}\sum_{p=1}^{P}\psi_i(x_i(p))\,x_j(p)=0 \quad \text{for} \quad 1\leq i\neq j\leq n\,, \tag{43}$$

where functions $\psi_1,\ldots,\psi_n : R\mapsto R$ are *non-linear* functions (more about choosing them below). By using non-linear functions, symmetry is broken and the required number of constraints is obtained, namely $n(n-1)$ (with n additional arbitrary constraints, needed for fixing the scale of each component).

Localized Correlations

Another approach is to look for "localized decorrelation" in the sense of solving

$$\frac{1}{P}\sum_{p=1}^{P}\frac{x_i(p)}{\sigma_{ip}^2}x_j(p)=0 \quad \text{for} \quad 1\leq i\neq j\leq n\,, \tag{44}$$

where for each component i, a sequence $\{\sigma_{ip}\}_{p=1}^{P}$ of positive number must be defined (more about this soon). Again, blind identification is possible because symmetry is broken, provided no two sequences of σs are proportional.

Maximum Likelihood

Why using the particular proposals (43) or (44) as extended decorrelation conditions rather than any other form, possibly more complicated? One reason is that reasonable algorithms exist for computing the $\mathbf{W}$ such that $\mathbf{x}=\mathbf{Wy}$ is a solution of (43) or (44). Another, more important reason is that these two conditions actually characterize the maximum likelihood estimate of $\mathbf{W}$ in simple and well-understood models. Because of this, we can understand what the algorithm does and we have guidance for choosing the non-linear functions ψ_i in condition (43) or the varying variance profiles σ_{iq}^2 in condition (44) as stated next.

Non-linear correlations. Assume that each component $\{s_i(p)\}$ is modelled as having all pixels independently and identically distributed according to some probability density p_i. In this model, the most likely value of $\mathbf{A}$ given the observations has for inverse a matrix $\mathbf{W}$ such that condition (43) holds with $\psi_i = -p_i'/p_i$. Hence, if the model is true (or approximately true), the non-linear function appearing in condition (43) should be taken as minus the derivative of the log-density of $s_i(p)$. For a Gaussian distribution p_i, the corresponding function ψ_i is linear: here, the necessary non-linearity of ψ_i corresponds to the non-Gaussianity of the corresponding component.

Localized correlations. Alternatively, one may model each component $\{s_i(p)\}$ as having all pixels independently and *normally* distributed with zero-mean and "local" variance σ_{ip}^2. Then, in this model, the likeliest value of A given the observations has for inverse a matrix $\mathbf{W}$ such that $\mathbf{x} = \mathbf{W}\mathbf{y}$ satisfies condition (44).

7.4 ICA in Practice

For the simple noise-free setting under consideration (the noisy case is addressed in next section), the algorithmic solutions depend on the type of decorrelation one decides to use.

Non-linear Decorrelation

Two popular ICA algorithms based on non-linear decorrelation (hence exploiting non-Gaussianity) are JADE (Cardoso [14]) and FastICA (Hyvärinen [30]). In practice, however, these algorithms do not exactly solve an equation in the form (43). Rather, for algorithmic efficiency, they try to solve it under the additional constraint that the components are uncorrelated, i.e. that condition (42) is satisfied exactly. The underlying optimization engine is a joint diagonalization algorithm for JADE and a fixed point technique for FastICA.

Localized Decorrelation

Efficient algorithms for solving the localized decorrelation conditions (44) are based on assuming some regularity in the variance profiles: the sequences $\{\sigma_{ip}^2\}$ are approximated as being constant over small domains. Hence, the global set $[1, P]$ is partitioned into Q subsets $\mathcal{I}_1, \ldots, \mathcal{I}_Q$, each containing a number P_q of points (so that $P = \sum_{q=1}^{Q} P_q$). In practice, these pixel subsets are (well-chosen) spatial regions. With a slight abuse of notation, we write $\sigma_{ip}^2 = \sigma_{iq}^2$ if $p \in \mathcal{I}_q$. Then, a small amount of maths turns the decorrelation conditions (44) into

$$\left[\sum_{q=1}^{Q} P_q \mathbf{\Sigma}_q^{-1} \mathbf{W} \widehat{\mathbf{R}}_q \mathbf{W}^\dagger\right]_{ij} = 0 \quad \text{for} \quad 1 \leq i \neq j \leq n\,, \tag{45}$$

where $\widehat{\mathbf{R}}_q$ is a localized covariance matrix

$$\widehat{\mathbf{R}}_q = \frac{1}{P_q} \sum_{p \in q} \mathbf{y}(p)\mathbf{y}(p)^\dagger \quad \text{and} \quad \mathbf{\Sigma}_q = \operatorname{diag}\left(\sigma_{1q}^2, \ldots, \sigma_{nq}^2\right)\,. \tag{46}$$

An important point here is that by assuming piecewise constant variance profiles, the localized decorrelation condition can be expressed entirely in terms of the localized covariance matrices $\widehat{\mathbf{R}}_q$. Hence the localized covariance matrices appear as *sufficient statistics* in this model. Even better, the likelihood of $\mathbf{A}$ can be understood as a mismatch between these statistics and their predicted form, namely $\mathbf{R}_q = \mathbf{A}\mathbf{\Sigma}_q\mathbf{A}^\dagger$. Specifically, in this model the probability $p(\mathbf{y}(1), \dots \mathbf{y}(P)|\mathbf{A}, \mathbf{\Sigma})$ of the data given $\mathbf{A}$ and the set $\mathbf{\Sigma} = \{\mathbf{\Sigma}_1, \dots, \mathbf{\Sigma}_Q\}$ of covariance matrices is given by

$$\log p(\mathbf{y}(1), \dots \mathbf{y}(P)|\mathbf{A}, \mathbf{\Sigma}) = -\phi(\mathbf{A}, \mathbf{\Sigma}) + \mathrm{cst} ,$$

where function ϕ is defined as

$$\phi\left(\mathbf{A}, \mathbf{\Sigma}\right) = \sum_q P_q K\left(\widehat{\mathbf{R}}_q, \mathbf{A}\mathbf{\Sigma}_q\mathbf{A}^\dagger\right) \tag{47}$$

and where $K(\cdot, \cdot)$ is a measure of divergence between two matrices defined as

$$K(\mathbf{R}_1, \mathbf{R}_2) = \frac{1}{2}\left(\mathrm{Tr}\left(\mathbf{R}_1\mathbf{R}_2^{-1}\right) - \log\det\left(\mathbf{R}_1\mathbf{R}_2^{-1}\right) - n\right) . \tag{48}$$

This shows that maximum likelihood estimation of $\mathbf{A}$ amounts to the minimization of the weighted mismatch (47) between the set of localized covariance matrices $\widehat{\mathbf{R}}_q$ (computed from the data) and their expected value $\mathbf{R}_q = \mathbf{A}\mathbf{\Sigma}_q\mathbf{A}^\dagger$ (predicted by the model).

In the noise-free case considered here, it turns out that there is a simple and very efficient algorithm (due to D.T. Pham) for minimizing the spectral mismatch.

8 SMICA

We have developed a component separation technique dubbed SMICA for 'spectral matching ICA' which is based on the ideas sketched in previous section but improves on them in several ways.

In its simplest form, SMICA is based on *spectral* statistics, that is, on statistics which are localized not in space but in frequency. These statistics are binned auto- and cross-spectra of the channels. More specifically, for a given set of N_{chann} multi-channel maps $\{y_i(p)\}$, we form for each (ℓ, m) the $N_{\mathrm{chann}} \times 1$ vector $\mathbf{y}(\ell, m)$ of their harmonic coefficients and define

$$\widehat{\mathbf{R}}(\ell) = \frac{1}{2\ell + 1} \sum_{m=-\ell}^{m=+\ell} \mathbf{y}(\ell, m)\mathbf{y}(\ell, m)^\dagger .$$

These empirical spectral covariance matrices are then binned. In the simplest case, we define Q top-hat bins, with the qth frequency bin containing all frequencies ℓ between $\ell_q^{\min}$ and $\ell_q^{\max}$. We consider the binned spectra:

$$\widehat{\mathbf{R}}_q = \frac{1}{P_q} \sum_{\ell=\ell_q^{\min}}^{\ell=\ell_q^{\max}} (2\ell+1)\widehat{\mathbf{R}}(\ell) , \quad \text{where} \quad P_q = \sum_{\ell=\ell_q^{\min}}^{\ell=\ell_q^{\max}} (2\ell+1) .$$

Here P_q is the number of Fourier modes summed together in a single estimate $\widehat{\mathbf{R}}_q$.

The mixture model $\mathbf{y} = \mathbf{A}\mathbf{s} + \mathbf{n}$ predicts that the empirical spectra $\mathbf{R}_q$ have an expected value

$$\mathbf{R}_q = \left\langle \widehat{\mathbf{R}}_q \right\rangle = \mathbf{A}\mathbf{\Sigma}_q\mathbf{A}^\dagger + \mathbf{N}_q ,$$

where $\mathbf{\Sigma}_q$ are the binned spectral covariance matrix for the components in bin q and $\mathbf{N}_q$ is the same for noise, assumed to be uncorrelated from the components. The unknown parameters can be collected in a big vector θ:

$$\theta = \{\mathbf{A}, \{\mathbf{\Sigma}_q\}, \{\mathbf{N}_q\}\} ,$$

but in practice we will not fit such a large model. Many constraints can be imposed on θ. A typical choice is to assume that the components are uncorrelated between themselves and that the noise also is uncorrelated between channels. Such a choice would result in a smaller parameter set

$$\theta = \{\mathbf{A}, \{\mathrm{diag}\mathbf{\Sigma}_q\}, \{\mathrm{diag}\mathbf{N}_q\}\} ,$$

but infinitely many other options are possible, both more stringent (like assuming that the noise in each channel is a smooth function of the bin index q) or less stringent (like assuming that some components may not be uncorrelated). In the following, we do not assume a specific parametrization of the binned spectral covariance matrices. Rather, we denote where θ is some parameter set which uniquely determines the values of $\mathbf{A}$ and each $\mathbf{R}_q$ and $\mathbf{N}_q$:

$$\{\mathbf{R}_q\} = \{\mathbf{R}_q(\theta)\} = \left\{\mathbf{A}(\theta)\mathbf{\Sigma}_q(\theta)\mathbf{A}(\theta)^\dagger + \mathbf{N}_q(\theta)\right\} .$$

SMICA determines the set θ of unknown parameters by fitting the empirical spectral covariance matrices to whichever structure is predicted by the model. Specifically, the unknown parameters are found by minimizing the "spectral mismatch"

$$\phi(\theta) = \sum_q P_q K\left(\widehat{\mathbf{R}}_q, \mathbf{R}_q(\theta)\right) , \tag{49}$$

averaged across bins. Some comments are in order regarding the matching criterion, the issue of non-stationarity and practical implementation.

Matching Criterion

The reason for choosing this particular form of mismatch between data and model is that minimizing (49) is identical to maximizing the likelihood of the data in a model where all components are (1) Gaussian, (2) stationary, and (3) have harmonic spectra which are constant over bins. Of course, these assumptions are not met in practice so one could choose a different criterion for matching $\widehat{\mathbf{R}}_q$ to $\mathbf{R}_q(\theta)$ but we have little statistical guidance for picking up an alternate matching measure. Furthermore, the assumptions (1) and (2) are met by the CMB and (3) is approximately correct for narrow bins. In addition, the failure of stationarity can be alleviated by using localized statistics (see below).

Non-stationarity and Localization

The spectral approach to building a likelihood function has some benefits, in particular the fact that it is perfectly suited to describing the statistical properties of the CMB. Another beneficial side effect is that it makes it easy to deal with varying resolution from channel to channel as long as the beam can be considered to be symmetrical.

However, going straight away to harmonic space seems unreasonable to deal with highly non-stationary components such as the galactic components. This issue can be addressed to some extent by resorting to localized spectral statistics. It is a simple matter to use spatial window functions to partition the sky into spatial domains (Cardoso et al. [15]). Although not a perfect solution, it certainly allows to capture a good deal of the non-stationary features of the galactic sky.

Implementation

The definition of the spectral matching criterion (49) encapsulates all of the statistical modelling but leaves open the separate and possibly tricky issue of minimizing $\phi(\theta)$.[5] Because the criterion is a likelihood in disguise, it is possible to use the EM algorithm for its minimization, with the components taken as latent variables. However, EM is often not fast enough and also is not able to deal with arbitrary parametrization of $\mathbf{\Sigma}_q(\theta)$ and $\mathbf{N}_q(\theta)$. It has been found necessary to use general optimization techniques. A conjugate gradient algorithm can be implemented because a reasonably tractable expression for the gradient of the criterion is available as

$$\frac{\partial \phi}{\partial \theta} = \sum_q P_q \operatorname{Tr}\left(\mathbf{R}_q(\theta)^{-1}\left(\mathbf{R}_q(\theta) - \widehat{\mathbf{R}}_q\right)\mathbf{R}_q(\theta)^{-1}\frac{\partial \mathbf{R}_q(\theta)}{\partial \theta}\right).$$

[5] We note in passing that some authors seem to make a confusion between the objective function (the criterion which has to be minimized, which derives from a statistical model) and the algorithm used for minimization. For instance, some authors use the terms "EM method" or "MCMC method" to design a method in which they use the EM algorithm or Monte Carlo Markov chains. This is rather unfortunate and contributes to a certain level of confusion.

However, in our context, the conjugate gradient algorithm also requires preconditioning. A preconditioner can be classically obtained as the inverse of the Fisher information matrix $\mathrm{FIM}(\theta)$ which is taken as an approximation to the Hessian of $\phi(\theta)$:

$$\frac{\partial^2 \phi}{\partial \theta^2} \approx \mathrm{FIM}(\theta) = \sum_q P_q \operatorname{Tr}\left(\mathbf{R}_q(\theta)^{-1} \frac{\partial \mathbf{R}_q(\theta)}{\partial \theta} \mathbf{R}_q(\theta)^{-1} \frac{\partial \mathbf{R}_q(\theta)}{\partial \theta}\right) .$$

Mismatch Control and Error Bars

A benefit of the SMICA approach is that it comes with a built-in measure of the quality of the model. Indeed, if we properly fit all the auto–cross-spectra, then $\phi(\theta)$ should be 'statistically small'. Visual control of the quality of the spectral matching is obtained by plotting $\phi_q = P_q K(\widehat{\mathbf{R}}_q, \mathbf{R}_q(\hat{\theta}))$ versus q where $\hat{\theta}$ is the minimizer of $\phi(\theta)$. This quantity should be understood as a χ^2. If the model holds (Gaussian stationary components and noise) and when all spectral parameters are freely estimated ϕ_q behaves approximately as a χ^2 with a number of degrees of freedom equal to $N_1 - N_2$, where $N_1 = N_{\mathrm{chann}}(N_{\mathrm{chann}}+1)/2$ is the number of degrees of freedom for a sample covariance matrix of size N_{chann} and where $N_2 = N_{\mathrm{comp}} + N_{\mathrm{chann}}$ is the number of adjustable spectral parameters (the variances of each components and noise levels in a given frequency bin).

9 Other Blind, Semi-blind, or Model Learning Methods

This paper would not be complete without a quick review of some of the recent work. We quote here a few papers which we think deserve reading for further exploration of component separation issues and methods. Although unevenly mature, these methods provide complementary approaches, with advantages and drawbacks which deserve to be investigated.

9.1 FastICA

A blind component separation based on the FastICA method has been developed for CMB data reduction by Baccigalupi et al. [3], with an extension to the full sky by Maino et al. [38]. This blind approach uses, as "engine" for component separation, a measure of independence based on non-Gaussianity. Therefore, it is essentially equivalent to finding components which cancel non-linear correlations in the sense of (43).

For CMB applications, characterizing independence via non-linear correlations of the form 43 has some limitations. First, theory shows that this characterization allows for the separation of at most one Gaussian component [13]. The Gaussian component is somehow found "by default", as the particular component which is orthogonal to (uncorrelated with) all other non-Gaussian components. This is a concern for component separation performed with the CMB as the main target. Second, the non-linear decorrelation conditions do not take the noise into account. Even though this can be fixed in some ad hoc fashion, it is computationally demanding to do it in maximum likelihood sense. Finally, pixel space implementations cannot easily handle channel-dependent beams (unless explicit beam deconvolution is performed). If, to circumvent this problem, one considers harmonic space implementation, performance suffers from the fact that Fourier tend to be more Gaussian than the original, pixel-domain maps.

FastICA, however, can outperform other component separation methods for some applications. Spectral-based methods (like SMICA) cannot blindly separate two components if their angular power spectra are proportional. FastICA does not suffer from this limitation and therefore has an edge for separating galactic components. If all galactic components have similar power spectra (say, proportional to ℓ^{-3}) then SMICA is expected to perform poorly without prior information.

Although both FastICA and SMICA are blind methods entering in the general class of "independent component analysis", it should thus be stressed that they are conceptually very different. Performance, therefore, is expected to be very different also and to depend on the actual properties of the data sets.

FastICA has been used on COBE and on WMAP data [39, 40].

9.2 Other Recent Developments

A "semi-blind" approach to component separation has been proposed by Tegmark and collaborators in a work where they model the foreground emissions using a number of physical parameters, which they estimate directly in the data sets (Tegmark et al. [59]). They estimate the impact of estimating these extra parameters in terms of accuracy loss on parameters of interest for CMB science. This paper was the first to address seriously the problem of component spectral indices varying over the sky.

Martínez-González and collaborators have proposed a method for the extraction of the CMB specifically and for the estimation of its power spectrum (Martínez-González et al. [43]). The EM algorithm is the main tool of the implementation.

Eriksen and collaborators have developed a method based on a refined modelling of the astrophysical components and fitted this model to the data to obtain estimates of foreground parameters (Eriksen et al. [21]). The fit of the parameters is made pixel by pixel at low resolution using a MCMC technique for exploring the likelihood. After this first "model learning" step, the parameters obtained are used to estimate high-resolution component maps.

Recently, Hansen and collaborators have proposed a CMB cleaning method based on a wavelet fit of component emissions obtained by differencing observations in different channels and subtraction of the fit from observations made at frequencies where the CMB dominates (Hansen et al. [24]).

Bonaldi and collaborators have recently published a paper for estimating parameters of emission of astrophysical components (emission laws, described by spectral indices). The statistics used are based on estimations of the correlations of the observations using a subset of points on the sphere (Bonaldi et al. [11]).

An alternate way of performing component separation has been proposed by Bobin and collaborators based on sparse representations of the various emissions (Bobin et al. [10]). The basic principle of this method consists in decomposing the observations in a set of (redundant) dictionaries chosen so that each component can be represented sparsely in one of the dictionaries. Separation is achieved by minimizing the number of coefficients required to represent the data set.

A comparison of these different methods on a common data set for investigating their strengths and weaknesses and evaluating their relative performance for various objectives would be an interesting work to improve the quality of component separation with the data set of upcoming space missions.

10 Conclusion and Prospects

With improving data quality and increasingly demanding performance in component characterization, component separation will play an important role in the analysis of CMB data sets in the next decade.

In this paper, we have reviewed the main issues for component separation, concentrating on diffuse components specifically.

Although substantial work has been performed, open questions remain. Polarization, for instance, is one of the next major objectives of CMB science, for which much better sensitivities are required and for which foreground emission is poorly known. Time varying sources, as the emission due to zodiacal light (modulated by the trajectory of the instrument in the ecliptic), as solar system objects in general, and as intrinsically time-varying radio sources, require specific methods tailored for their extraction.

The upcoming Planck data set, expected to become available to the Planck consortium in 2008, will provide a fantastic and challenging data set for extracting the emission from all astrophysical processes emitting in the millimetre range.

References

1. Aghanim, N., Desert, F.X., Puget, J.L., Gispert, R.: Ionization by early quasars and cosmic microwave background anisotropies. Astron. Astrophys. **311**, 1 (1996)

2. Audit, E., Simmons, J.F.L.: The kinematic Sunyaev-Zel'dovich effect and transverse cluster velocities. MNRAS **305**, L27 (1999)
3. Baccigalupi, C., et al.: Neural networks and the separation of cosmic microwave background and astrophysical signals in sky maps. MNRAS **318**, 769 (2000)
4. Barreiro, R.B.: *Techniques for Compact Source Extraction on CMB Maps, to appear in "Data Analysis in Cosmology"*. Lect. Notes Phys. Springer-Verlag, Valencia, 6–10 September (2004) astro-ph/0512538
5. Barreiro, R.B., Hobson, M.P., Banday, A.J., Lasenby, A.N., Stolyarov, V., Vielva, P., Górski, K.M.: Foreground separation using a flexible maximum-entropy algorithm: an application to COBE data. MNRAS **351**, 515 (2004)
6. Bennett, C.L., et al.: First-Year Wilkinson Microwave Anisotropy Probe (WMAP) Observations: Preliminary Maps and Basic Results. ApJ Supplement Series **148**, 1 (2003)
7. Bennett, C.L., et al.: First year Wilkinson Microwave Anisotropy Probe (WMAP) observations: foreground emission. ApJ Supplement Series **148**, 97 (2003)
8. Benoît, A., et al.: First detection of polarization of the submillimetre diffuse galactic dust emission by Archeops. Astron. Astrophys. **424**, 571 (2004)
9. Birkinshaw, M.: The Sunyaev Zel'dovich effect. Phys. Rep. **310**, 97 (1999)
10. Bobin, J., Moudden, Y., Starck, J.-L., Elad, M.: Morphological Diversity and Source Separation IEEE Transaction on Signal Processing, (2006, in press)
11. Bonaldi, A., Bedini, L., Salerno, E., Baccigalupi, C., de Zotti, G.: Estimating the spectral indices of correlated astrophysical foregrounds by a second-order statistical approach. MNRAS **373**, 271 (2006)
12. Bouchet, F. R., Gispert, R.: Foregrounds and CMB experiments I. Semi-analytical estimates of contamination. New Astron. 4, 443 (1999)
13. Cardoso, J.-F.: Blind signal separation: statistical principles. In: "Proceedings of the IEEE. Special issue on blind identification and estimation", 9, 2009 (1998)
14. Cardoso, J.-F.: High-order contrasts for independent component analysis. Neural Comput. **11** (1) 157–192 (1999)
15. Cardoso, J.-F., et al.: Statistiques direction-multipôle pour la séparation de composantes dans le fonds de rayonnement cosmologique, Actes du GRETSI. Louvain-la-Neuve, Belgique (2005)
16. de Bernardis, P., et al.: A flat Universe from high-resolution maps of the cosmic microwave background radiation. Nature **404**, 955 (2000)
17. Dickinson, C., et al.: High-sensitivity measurements of the cosmic microwave background power spectrum with the extended Very Small Array. MNRAS **353**, 732 (2004)
18. Dickinson, C., Casassus, S., Pineda, J.L., Pearson, T.J., Readhead, A.C.S., Davies, R. D.: An upper limit on anomalous dust emission at 31 GHz in the diffuse cloud 201.663+1.643. Astrophys. J. Lett. 643, L111 (2006)
19. Draine, B. T., Lazarian, A.: Diffuse Galactic Emission from Spinning Dust Grains. Astrophys. J. Lett. **494**, L19 (1998)
20. Eriksen, H.K., Banday, A.J., Górski, K.M., Lilje, P.B.: On foreground removal from the WMAP data by an ILC method: limitations and implications. ApJ **612**, 633 (2004)
21. Eriksen, H. K., et al.: Cosmic microwave background component separation by parameter estimation. ApJ **641**, 665 (2006)

22. Fernández-Cerezo, S., et al.: Observations of the cosmic microwave background and galactic foregrounds at 12–17 GHz with the COSMOSOMAS experiment. MNRAS **370**, 15 (2006)
23. Gruzinov, A., Hu, W.: Secondary cosmic microwave background anisotropies in a universe reionized in patches. Astrophys. J. **508**, 435 (1998)
24. Hansen, F.K., Banday, A.J., Eriksen, H.K., Górski, K.M., Lilje, P.B.: Foreground subtraction of cosmic microwave background maps using WI-FIT (Wavelet-based high-resolution fitting of internal templates). ApJ **648**, 784 (2006)
25. Haslam, C.G.T., Klein, U., Salter, C.J., Stoffel, H., Wilson, W.E., Cleary, M.N., Cooke, D.J., Thomasson, P.: A 408 MHz all-sky continuum survey. I - Observations at southern declinations and for the North Polar region. Astron. Astrophys. **100**, 209 (1981)
26. Hinshaw, G., et al.: Three-year Wilkinson Microwave Anisotropy Probe (WMAP) observations: temperature analysis. ApJS **170**, 288 (2007)
27. Hobson, M.P., Lasenby, A.N.: The entropic prior for distributions with positive and negative values. MNRAS **298**, 905 (1998)
28. Hobson, M.P., Jones, A.W., Lasenby, A.N., Bouchet, F.R.: Foreground separation methods for satellite observations of the cosmic microwave background. MNRAS 300, 1 (1998)
29. Hu, W., Dodelson, S.: Cosmic microwave background anisotropies. Annu. Rev. Astron. Astrophys. **40**, 171 (2002)
30. Hyvvärinen, A.: Fast and robust fixed-point algorithms for independent component analysis. IEEE Trans. Neural Networ. **10**(3) 626–634 (1999)
31. Jones, W.C., et al.: A measurement of the angular power spectrum of the CMB temperature anisotropy from the 2003 flight of boomerang. ApJ **647**, 823 (2006)
32. Kogut, A., Banday, A.J., Bennett, C.L., Gorski, K.M., Hinshaw, G., Reach, W.T.: High-latitude galactic emission in the COBE differential microwave radiometer 2 year sky maps. Astrophys. J. **460**, 1 (1996)
33. Komatsu, E., et al.: First-year Wilkinson Microwave Anisotropy Probe (WMAP) observations: tests of gaussianity. ApJ Supplement Series **148**, 119 (2003)
34. Kuo, C. L., et al.: High-resolution observations of the cosmic microwave background power spectrum with ACBAR. Astrophys. J. **600**, 32 (2004)
35. Lamarre, J. M., et al.: The high frequency instrument of planck: design and Performances. Astrophys. Lett. Commun. **37**, 161 (2000)
36. Lamarre, J. M., et al.: The planck high frequency instrument, a third generation CMB experiment, and a full sky submillimeter survey. New Astron. Rev. **47**, 1017 (2003)
37. Leitch, E.M., Readhead, A.C.S., Pearson, T.J., Myers, S.T.: An anomalous component of galactic emission. Astrophys. J. Lett. **486**, L23 (1997)
38. Maino, D., et al.: All-sky astrophysical component separation with Fast Independent Component Analysis (FASTICA). MNRAS **334**, 53 (2002)
39. Maino, D., Banday, A.J., Baccigalupi, C., Perrotta, F., Górski, K.M.: Astrophysical component separation of COBE-DMR 4-yr data with FASTICA. MNRAS **344**, 544 (2003)
40. Maino, D., Donzelli, S., Banday, A.J., Stivoli, F., Baccigalupi, C.: Cosmic microwave background signal in Wilkinson Microwave Anisotropy Probe three-year data with FASTICA. MNRAS **374**, 1207 (2007)

41. Mandolesi, N., Bersanelli, M., Burigana, C., Villa, F.: The planck low frequency instrument. Astrophys. Lett. Commun. **37**, 151 (2000)
42. Maris, M., Burigana, C., Fogliani, S., Astron. Astrophys. **452**, 685 (2006)
43. Martínez-González, E., Diego, J.M., Vielva, P., Silk, J.: Cosmic microwave background power spectrum estimation and map reconstruction with the expectation-maximization algorithm. MNRAS **345**, 1101 (2003)
44. McCullough, P.R., et al.: Implications of halpha observations for studies of the CMB. ASP Conf. Ser. 181: Microwave Foregrounds **181**, 253 (1999)
45. Moudden, Y., Cardoso, J.-F., Starck, J.-L., Delabrouille, J.: Blind Component Separation in Wavelet Space: Application to CMB Analysis. Eurasip J. Appl. Signal Process. 15 2437–2454 (2005)
46. Netterfield, C.B., et al.: A measurement by BOOMERANG of multiple peaks in the angular power spectrum of the cosmic microwave background. Astrophys. J. **571**, 604 (2002)
47. Patanchon, G., Cardoso, J.-F., Delabrouille, J., Vielva, P.: Cosmic microwave background and foregrounds in Wilkinson Microwave Anisotropy Probe first-year data. MNRAS **364**, 1185 (2005)
48. Ponthieu, N., et al.: Temperature and polarization angular power spectra of Galactic dust radiation at 353 GHz as measured by Archeops. Astron. Astrophys. **444**, 327 (2005)
49. Puget, J.-L., Abergel, A., Bernard, J.-P., Boulanger, F., Burton, W. B., Desert, F.-X., Hartmann, D.: Tentative detection of a cosmic far-infrared background with COBE. Astron. Astrophys. **308**, L5 (1996)
50. Readhead, A.C.S., et al.: Extended mosaic observations with the cosmic background imager. Astrophys. J. **609**, 498 (2004)
51. Rephaeli, Y.: The Sunyaev-Zeldovich effect: Recent progress and future prospects. AIP Conf. Proc. 616: Experimental Cosmology at Millimetre Wavelengths **616**, 309 (2002)
52. Sazonov, S. Y., Sunyaev, R. A.: Microwave polarization in the direction of galaxy clusters induced by the CMB quadrupole anisotropy. MNRAS **310**, 765 (1999)
53. Shannon,: A mathematical theory of communication Bell Sys. Tech. J. **27**, 379–423, 623–656, July, October. (1948)
54. Smoot, G. F.: Galactic Free-Free and H-alpha emission (1998) astro-ph/9801121
55. Stolyarov, V., Hobson, M.P., Ashdown, M.A.J., Lasenby, A.N.: All-sky component separation for the Planck mission. MNRAS **336**, 97 (2002)
56. Sunyaev, R.A., Zeldovich, Y. B.: The observations of relic radiation as a test of the nature of X-ray radiation from the clusters of galaxies. Comment Astrophys. Space Phys. **4**, 173 (1972)
57. Sunyaev, R. A., Zeldovich, I.B.: The velocity of clusters of galaxies relative to the microwave background – The possibility of its measurement. MNRAS **190**, 413 (1980)
58. Tegmark, M., Efstathiou, G.: A method for subtracting foregrounds from multifrequency CMB sky maps. MNRAS **281**, 1297 (1996)
59. Tegmark, M., Eisenstein, D.J., Hu, W., de Oliveira-Costa, A.: Foregrounds and forecasts for the cosmic microwave background. Astrophys. J. **530**, 133 (2000)
60. Tegmark, M., de Oliveira-Costa, A., Hamilton, A.J.: High resolution foreground cleaned CMB map from WMAP. Phys. Rev. D. **68**, 123523 (2003)
61. Tristram, M., et al.: The CMB temperature power spectrum from an improved analysis of the Archeops data. Astron. Astrophys. **436**, 785 (2005)

62. Valls-Gabaud, D.: Cosmological applications of H-alpha surveys. Publ. Astron. Soc. Aust. **15**, 111 (1998)
63. White, M., Cohn, J.D.: The theory of anisotropies in the cosmic microwave background. Am. J. Phys. **70**, 106 (2002)
64. Wiener, N.: Extrapolation, Interpolation, and Smoothing of Stationary Time Series. Wiley, New York (1949) ISBN 0262730057
65. Yamada, M., Sugiyama, N., Silk, J.: The Sunyaev-Zeldovich effect by cocoons of radio galaxies. Astrophys. J. **522**, 66 (1999)

Techniques for Compact Source Extraction in CMB Maps

R.B. Barreiro

Instituto de Física de Cantabria, CSIC – Universidad de Cantabria, Avda. de los Castros s/n, 39005 Santander, Spain
barreiro@ifca.unican.es

Abstract. The detection of compact sources embedded in a background is a very common problem in many fields of astronomy. In these lecture notes, we present a review of different techniques developed for the detection and extraction of compact sources, with an especial focus on their application to the field of the cosmic microwave background radiation. In particular, we will consider the detection of extragalactic point sources and the thermal and kinematic Sunyaev–Zeldovich effects from clusters of galaxies.

1 Introduction

Observations of an astrophysical signal in the sky are usually corrupted by some level of contamination (called *noise* or *background*) due to other astrophysical emissions and/or the detector itself. A common situation is that the signals of interest are spatially well localized, i.e. each of them covers only a small fraction of the image, but we do not know a priori its position and/or its amplitude. Some examples are the detection of extragalactic sources in cosmic microwave background (CMB) observations (see Fig. 1), the identification of local features (emission or absorption lines) in noisy one-dimensional spectra or the detection of objects in X-ray images. It is clear that our ability to extract all the useful information from the image will critically depend on our capacity to disentangle the signal(s) of interest from the background.

The process to detect a localized signal in a given image usually involves three different steps, which are not necessarily independent:

1. Processing: some processing of the data (commonly linear filtering) is usually performed in order to amplify the searched signal over the background. This is an important step because in many cases the signals are relatively weak with respect to the background and it becomes very difficult to detect them in the original image. This is illustrated in Fig. 2: the top panel shows a simulation of white noise where a source with a Gaussian profile has been added in the centre of the map; the bottom panel gives the same simulation

Barreiro, R.B.: *Techniques for Compact Source Extraction in CMB Maps*. Lect. Notes Phys. **665**, 207–235 (2009)
DOI 10.1007/978-3-540-44767-2_7

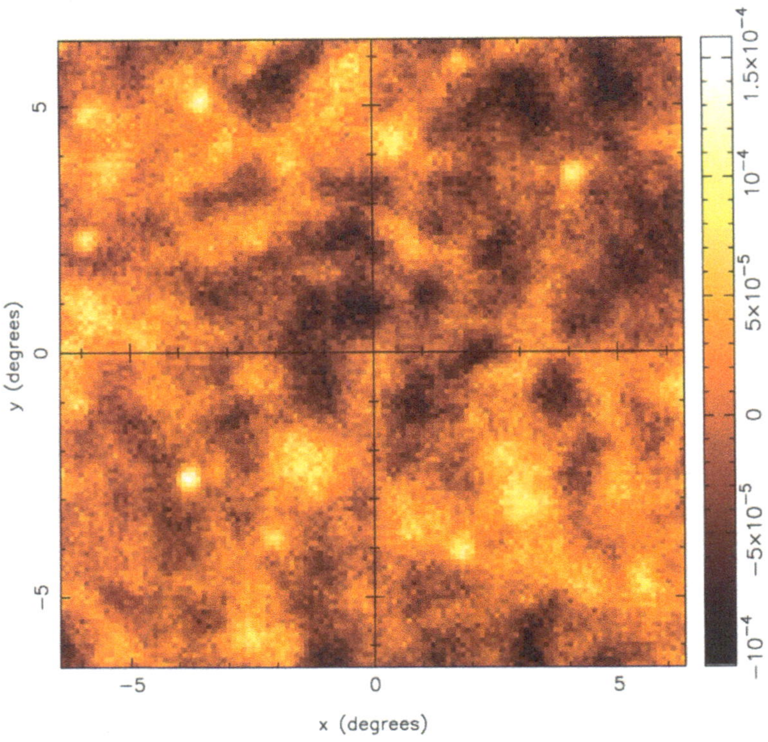

Fig. 1. Simulation of the 44 GHz Planck frequency channel in a small patch of the sky, containing CMB, galactic foregrounds, extragalactic point sources and instrumental noise. The point sources can be seen as localized objects embedded in the background. The Planck Mission [69] is a satellite of the European Space Agency to be launched in 2008 that will provide with multifrequency observations of the whole sky with unprecedented resolution and sensitivity

after filtering with the so-called matched filter. It becomes apparent that the source was hidden in the original image whereas it has been enhanced in the filtered image.

2. Detection: we need a detection criterion, the *detector*, to decide if some structure in the image is actually a real signal or if it is due to the background. A very simple and widely used detector in astronomy is *thresholding*: if the intensity of the image is above a given value (e.g. 5σ, where σ is the dispersion of the map), a detection of the signal is accepted, otherwise one assumes that only background is present. In the example of Fig. 2, we see that several peaks appear in the filtered image (right panel) but only one of them is above the considered threshold $\nu\sigma$. Therefore, in this case, we would accept only the highest peak as a true signal. Note that thresholding uses only the intensity of the data to make the decision; however, other useful information could also be included in order to improve the detector (e.g. curvature, size).

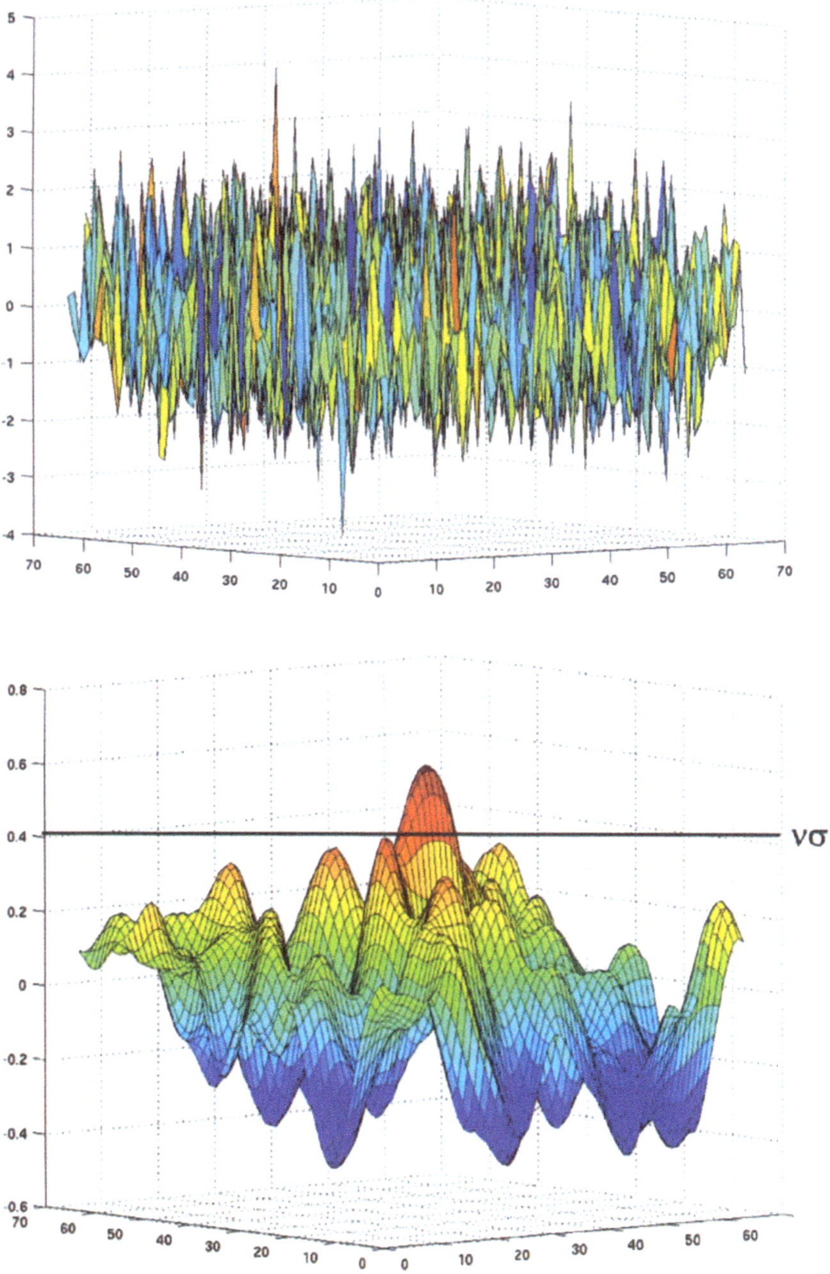

Fig. 2. This example illustrates the importance of filtering. A source with a Gaussian profile has been placed in the centre of the image in a background of white noise. The source cannot be distinguished in the original image (*top*); however, after filtering (*bottom*), the source is enhanced over the background fluctuations

3. Estimation: a procedure must be established to estimate the parameters (amplitude, size, position, etc.) characterizing the detected signal. For instance, a simple possibility is to estimate the required parameters by fitting the signal to its theoretical profile.

The aim of these lecture notes is to present the problem of the extraction of localized signals (compact sources) in the context of CMB astronomy and to review some of the methods developed to deal with it. In Sect. 2, we outline the problem of component separation in CMB observations. Section 3 reviews some of the techniques developed for extraction of point sources, including, among others, the matched filter and the Mexican Hat wavelet. Sections 4 and 5 deal with the extraction of the thermal and kinematic Sunyaev–Zeldovich effects in multifrequency microwave observations, respectively. Section 6 briefly discusses some techniques for the extraction of statistical information from undetected sources. Finally, in Sect. 7 we present our conclusions.

2 Microwave Sky and the Component Separation Problem

Microwave observations contain not only the cosmological signal but also galactic foregrounds, thermal and kinetic Sunyaev–Zeldovich (SZ) effects from clusters of galaxies and emission from extragalactic point sources [5, 11, 15, 83]. In addition, they also contain some level of noise coming from the detector itself. In order to recover all the wealth of information encoded in the CMB anisotropies, it is crucial to separate the cosmological signal from the rest of the components of the sky. Moreover, the foregrounds themselves contain very valuable information about astrophysical phenomena [28]. Therefore, the development of tools to reconstruct the different components of the microwave sky is of great interest not only to clean the CMB signal but also to recover all the useful information present in the foregrounds.

The main galactic foregrounds are the synchrotron, free–free and thermal dust emissions. The synchrotron emission is due to relativistic electrons accelerated in the galactic magnetic field. The free–free emission is the thermal bremsstrahlung from hot electrons when accelerated by ions in the interstellar gas. The observed dust emission is the sum over the emission from each dust grain along the line of sight (dust grains in our galaxy are heated by the interstellar radiation field, absorbing UV and optical photons and re-emitting the energy in the far infrared). In addition, there is some controversy about the presence of an anomalous foreground at microwave frequencies [26, 27, 37, 56, 66] that could be due to the emission of spinning dust grains [33, 34].

The thermal SZ effect [78, 79] is a spectral distortion of the blackbody spectrum of the CMB produced by inverse Compton scattering of microwave photons by hot electrons in the intracluster gas of a cluster of galaxies. In addition, the radial peculiar velocities of clusters also produce secondary anisotropies in the CMB via the Doppler effect, known as the kinetic SZ effect [80]. For a review on the SZ effect, see [13].

The thermal SZ effect has a distinct spectral signature. It produces a temperature decrement below 217 GHz and an increment above that frequency. The change in intensity (see Fig. 3) is given by

$$\Delta I = \frac{2(kT_o)^3}{(hc)^2}\frac{x^4\,\mathrm{e}^x}{(\mathrm{e}^x-1)^2}y_c\left[x\coth\frac{x}{2}-4\right]\,,\quad x=\frac{h\nu}{kT_o}\,, \tag{1}$$

where $y_c \equiv (k\sigma_T/m_e)\int dl\ T_e n_e$ is the Compton parameter and is a function of the electron density n_e and temperature T_e. This distinct frequency dependence can be used in multifrequency observations to separate the thermal SZ effect from the rest of the components of the microwave sky and, in particular, from the CMB.

The Doppler shift induced by the kinetic SZ effect in the CMB temperature fluctuations is given by

$$\frac{\Delta T}{T} = -\tau\frac{v_r}{c}\,, \tag{2}$$

where v_r is the radial velocity of the cluster and $\tau = \sigma_T\int n_e dl$ is the optical depth.

The thermal and kinetic SZ effects imprint anisotropies in the CMB at scales below a few arcminutes. Therefore, they are compact sources whose shape is given by the convolution of the beam of the experiment with the

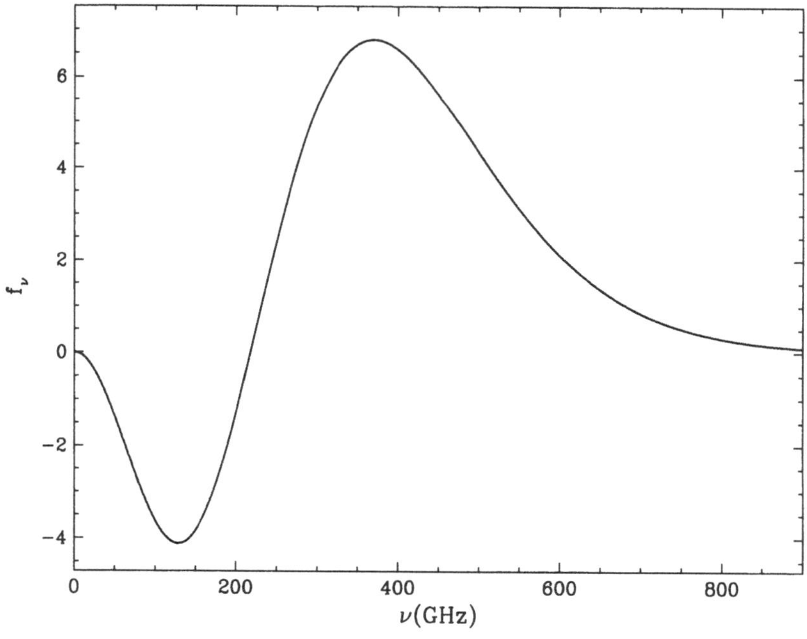

Fig. 3. Frequency dependence of the thermal SZ effect (in arbitrary units)

cluster profile. It is anticipated that it will be very difficult to detect the kinetic SZ effect and to separate it from the cosmological signal. This is due to the fact that it has the same frequency dependence as the CMB (since it is just a Doppler shift). Moreover, it is a very weak effect, around one order of magnitude lower than the thermal effect. The SZ effect is a very useful cosmological probe. Future SZ surveys will allow one to obtain very valuable information about some of the cosmological parameters, such as H_0, Ω_m, Ω_Λ and σ_8 (for a review, see [17]).

Emission from extragalactic point sources is an important contaminant for high-resolution CMB experiments. The term *point source* means that the typical angular size of these objects is much smaller than the resolution of the experiment (which is usually the case in CMB observations) and therefore they appear in the data as point-like objects convolved with the beam of the instrument. There are two main source populations: radio sources, which dominate at lower frequencies ($\lesssim 300$ GHz), and far-IR sources which give the main contribution at higher frequencies ($\gtrsim 300$ GHz). These populations consist mainly of compact AGN, blazars and radio-loud QSOs in the radio and of inactive spiral galaxies in the far IR. Different models for the radio [29, 39, 85, 86] and infrared [32, 40, 41] point source populations have been proposed. However, there are still many uncertainties with regard to the number of counts and the spectral behaviour of these objects due to a lack of data at the frequency range explored by CMB experiments. Therefore, experiments such as Planck will provide with unique information to understand the astrophysical processes taking place in these populations of sources. An additional problem is the heterogeneous nature of extragalactic point sources since, among other complications, each source has its own frequency dependence. Therefore, they cannot be treated as a single foreground to be separated from the other components by means of multifrequency observations.

There are basically two different approaches to perform component separation. The first one tries to reconstruct simultaneously all the components of the microwave sky whereas the second one focuses on just one single component. The first type of methods includes the Wiener filter [14, 81], maximum-entropy method [8, 11, 49, 50, 76, 77] and blind source separation [4, 10, 25, 62, 63, 68]. These methods usually assume that the components to be reconstructed can be factorized in a spatial template times a frequency dependence (but see [36] for a recent work where this assumption is not necessary). This assumption is correct for the CMB and the SZ effects but it is only an approximation for the galactic foregrounds. In addition, point sources cannot be factorized in this way, and therefore, these techniques are not well suited for extracting this contaminant. Regarding the second approach, it consists of methods designed to extract a particular component of the sky. For instance, the blind EM algorithm of [65] or the internal linear combination of [11, 35, 84] tries to recover only the CMB component. Moreover, this type of approach is especially useful for the detection of localized objects such as extragalactic point sources or the SZ effects. In these lectures we will describe some of these methods that have been developed with the aim of extracting compact sources from microwave observations.

3 Techniques for Extraction of Point Sources

The most common approach to detect point sources embedded in a background is probably linear filtering. A linearly filtered image $w(x)$ is obtained as the convolution[1] of the data $y(x)$ with the filter $\psi(x)$:

$$w(x) = \int y(u)\psi(x-u)du \; . \tag{3}$$

Note that those parts of the data that resemble the shape of the filter will be enhanced in the filtered map. Therefore, the filter should have a similar profile to that of the sought signal. Equivalently, we can work in Fourier space:

$$w(x) = \int y(q)\psi(q)\, \mathrm{e}^{-\mathrm{i}\, qx} dq \; , \tag{4}$$

where $f(q)$ denotes the Fourier transform of f. From the previous equation, we see that the filter favours certain Fourier modes of the data.

In principle, it is equivalent to perform the filtering in real or Fourier space. However, from the practical point of view, direct convolution is a very CPU time-consuming operation. Therefore, working in Fourier space, where a simple product is performed, is preferred.

Different linear filters have been proposed in the literature to detect point sources in CMB maps, including the matched filter [82], the Mexican Hat wavelet [18, 87, 89], the scale-adaptive filter [44, 73], the biparametric scale-adaptive filter [59] or the adaptive top hat filter [20]. In addition, non-linear techniques have also been proposed, such as the Bayesian method of [51] or the non-linear fusion of [60, 61]. In the next sections we give an overview of some of these techniques, including applications to CMB simulated data.

3.1 Matched Filter

Let us assume that we have a signal of amplitude A at position x_0 embedded in a background of dispersion σ. The amplification $\mathcal{A}$ of the signal obtained with a filter is given by

$$\mathcal{A} = \frac{w(x_0)/\sigma_w}{A/\sigma} \; , \tag{5}$$

where $w(x_0)$ is the value of the filtered map at the position of the source and σ_w is the dispersion of the filtered map. Therefore, if the amplification is

[1] Strictly speaking, the filtered image can be written as a convolution, provided the filter is linear and spatially homogeneous (see, e.g., [46]).

greater than one, the contrast between the signal and the background has been increased in the filtered map, improving the chances of detecting the source with respect to the original data. This is the main idea behind filtering: it puts you in a better position to detect the sources.

The *matched filter* (MF) is defined as the linear filter that gives maximum amplification of the signal. As an example, we will outline how to construct the MF for a source $s(x)$ with spherical symmetry (a more detailed derivation can be found, e.g., in [46]). Let us consider a set of two-dimensional data $y(\boldsymbol{x})$:

$$\begin{aligned} y(\boldsymbol{x}) &= s(x) + n(\boldsymbol{x}) \ , \\ s(x) &= A\tau(x) \ , \end{aligned} \tag{6}$$

where $\boldsymbol{x}$ is a two-dimensional vector of position and $x = |\boldsymbol{x}|$. The source is characterized by a (spherically symmetric) profile $\tau(x)$ and an amplitude $A = s(0)$. $n(\boldsymbol{x})$ is the noise (or background) contribution which, for simplicity, is assumed to be a homogeneous and isotropic random field with zero mean and characterized by a power spectrum $P(q)$ ($q = |\boldsymbol{q}|$), i.e.

$$\langle n(\boldsymbol{q}) n^*(\boldsymbol{q}') \rangle = P(q) \delta^2(\boldsymbol{q} - \boldsymbol{q}') \ , \tag{7}$$

where $n(\boldsymbol{q})$ is the two-dimensional Fourier transform.

Let us introduce a filter ψ with spherical symmetry. The filtered field w is given by

$$w(\boldsymbol{x}) = \int y(\boldsymbol{q}) \psi(q) \, \mathrm{e}^{-\mathrm{i}\, \boldsymbol{q}\boldsymbol{x}} d\boldsymbol{q} \ . \tag{8}$$

It can be shown that the filtered field at the position of the source (for simplicity we will assume that the source is at the origin) is given by

$$w(\mathbf{0}) = 2\pi \int q s(q) \psi(q) dq \ ,$$

whereas the variance of the filtered field is obtained as

$$\sigma_w^2 = 2\pi \int q P(q) \psi^2(q) dq \ . \tag{9}$$

We want to find the filter that satisfies the following two conditions:

1. $\langle w(\mathbf{0}) \rangle = A$,
2. σ_w^2 is a minimum with respect to the filter ψ.

The first condition means that the filter is an unbiased estimator of the amplitude of the source and gives straightforwardly the constraint

$$\int q\tau(q) \psi(q) dq = \frac{1}{2\pi} \ . \tag{10}$$

In order to minimize the variance of the filtered map (condition 2) including the previous constraint, we introduce a Lagrange multiplier λ:

$$\mathcal{L}(\psi) = \sigma_w^2(\psi) + \lambda \left[\int q\tau(q)\psi(q)dq - \frac{1}{2\pi} \right] . \tag{11}$$

Taking variations with respect to ψ and setting the result equal to zero, we find the matched filter:

$$\psi(q) = k \frac{\tau(q)}{P(q)} , \tag{12}$$

$$k = \left[2\pi \int q \frac{\tau^2(q)}{P(q)} dq \right]^{-1} . \tag{13}$$

Note that the matched filter is favouring those modes where the contribution of the signal (τ) is large and that of the noise (P) is small.

Assuming simple models for the galactic foregrounds, [82] has given an estimation of the catalogue of point sources that Planck will produce. According to this work, the number of sources detected by Planck above a 5σ level in a sky area of 8 sr will range from around 650 for the 70 GHz channel to around 38,000 at 857 GHz.

The matched filter has also been applied for the detection of point sources in the first-year WMAP data [11]. Using this technique, a catalogue of 208 extragalactic point sources has been provided.

3.2 Mexican Hat Wavelet

Wavelet techniques are very versatile tools that only recently have been applied to the analysis of CMB maps. The main property that makes wavelet transforms so useful is that they retain simultaneously information about the scale and position of the image. This means that we can study the structure of an image at different scales without loosing all the spatial information (as it occurs in the case of the Fourier transform). There is no unique way to construct a wavelet transform (see, e.g., [16, 24]). In this section we will focus on one particular wavelet, the Mexican Hat wavelet (MHW), that has been successfully implemented for the detection of point sources with a Gaussian profile in CMB simulated observations [18, 87, 89]. The MHW is the second derivative of the Gaussian function (see Fig. 4):

$$\psi(x) = \frac{1}{\sqrt{2\pi}} \left[2 - \left(\frac{x}{R} \right)^2 \right] \exp\left(-\frac{x^2}{2R^2} \right) \tag{14}$$

and in Fourier space it is given by

$$\psi(q) \propto (qR)^2 \exp\left(-\frac{(qR)^2}{2} \right) , \tag{15}$$

where R is the scale of the MHW, a parameter that determines the width of the wavelet. Note that $\psi(x)$ – and wavelet functions in general – are compensated,

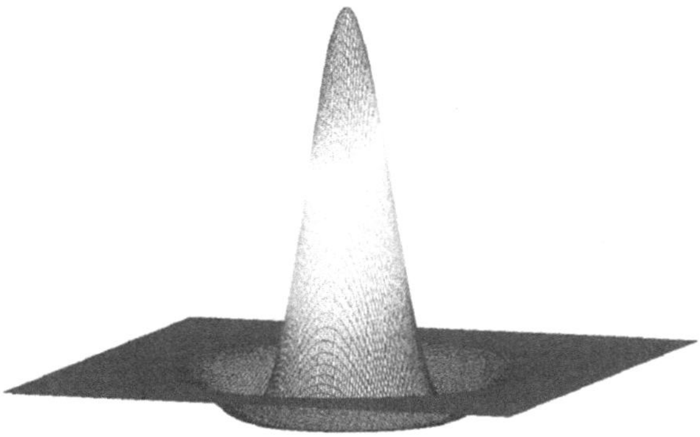

Fig. 4. The Mexican Hat wavelet in two dimensions

i.e. the integral below the curve is zero. When filtering the data with the MHW, this property helps to remove contributions of the background with a scale of variation larger than that of the wavelet.

The method to detect point sources is based on the study of the wavelet coefficients map (i.e. the image convolved with the MHW) at a given scale. Those wavelet coefficients above a fixed threshold are identified as point source candidates. The reason why this works well is because point sources are amplified in wavelet space. This can be easily seen in Fig. 5, which shows a graphical example of the performance of the MHW for a simulation of the Planck 857 GHz channel. Three of the panels correspond to the dust emission (top left), which is the dominant contaminant at this frequency, the emission of the extragalactic point sources (top right) and the total emission at this frequency (bottom left) including the galactic foregrounds, CMB, SZ effect, extragalactic point sources and instrumental noise. The last panel (bottom right) shows the total emission map after convolution with a MHW at a certain scale R_0. It is clear that in the wavelet coefficients map, a large fraction of the background has been removed and that the signal of the point sources has been enhanced. Therefore, the detection level in wavelet space is greater than the detection level in real space:

$$\frac{\omega(R)}{\sigma_\omega(R)} > \frac{A}{\sigma} , \tag{16}$$

where $\omega(R)$ is the wavelet coefficient at scale R at the position of the source, σ_ω is the dispersion of the wavelet coefficients map, A is the amplitude of the source and σ is the dispersion of the real map.

Note that the amplification (defined as the ratio between the detection level in wavelet space and the detection level in real space) depends on the wavelet scale R. In fact, for a given image, there exists an optimal scale R_0 that gives maximum amplification for the point sources and that can be determined from the data. For a point source convolved with a Gaussian beam

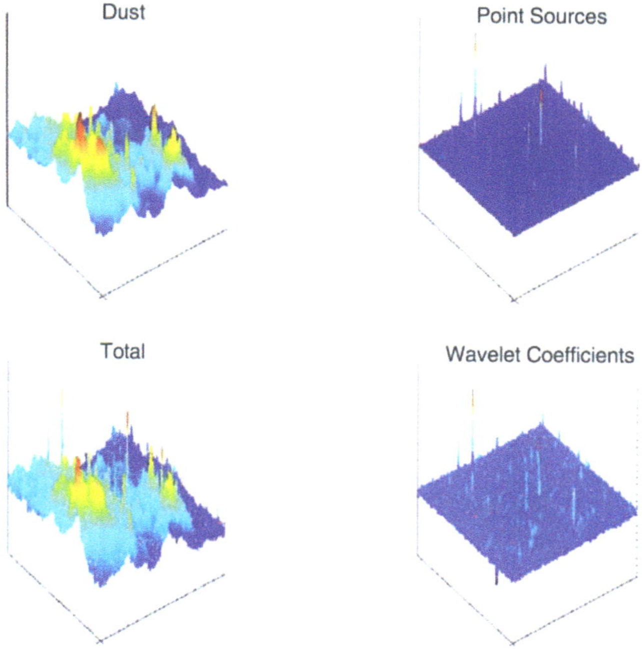

Fig. 5. An example of the performance of the MHW for a simulation of the Planck 857 GHz channel (see text for details)

of dispersion σ_b, the value of the wavelet coefficient $\omega(R)$ at the position of the source is given by

$$\omega(R) = 2R\sqrt{2\pi}A\frac{(R/\sigma_b)^2}{[1+(R/\sigma_b)^2]^2} , \tag{17}$$

whereas the dispersion of the wavelet coefficients map at scale R is

$$\sigma_\omega^2(R) = 2\pi R^2 \int P(q)|\psi(qR)|^2 q dq , \tag{18}$$

where $P(q)$ is the power spectrum of the background. Taking the previous expressions into account, one can obtain the optimal scale R_0 by maximizing the amplification $\mathcal{A}$ versus R. Figure 6 shows the amplification of the signal versus the scale for simulated observations of the Planck Low and High Frequency Instruments (LFI and HFI). Note that the optimal scale is close to σ_b. This is expected since this is the scale that characterizes the source, but the value of R_0 will also depend on the background. For instance, if the noise contribution is more important at scales smaller than that of the source, this will tend to move the optimal scale to values greater than σ_b and vice versa. Although the MHW will produce, in general, slightly less amplification than

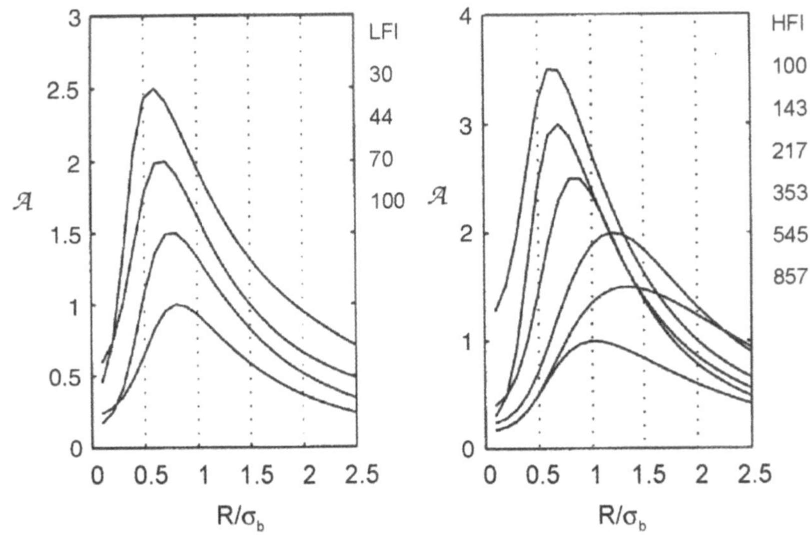

Fig. 6. Amplification (in arbitrary units) versus wavelet scale R (in units of the beam dispersion) for a typical region of the sky (the 100 GHz channel of the Low Frequency Instrument has been withdrawn from the final Planck payload due to financial shortage)

the MF, it has the advantage of being an analytical function, which greatly simplifies the use of this technique.

The procedure to detect point sources is as follows. First, the optimal scale R_0 is obtained and the data are filtered with a MHW of scale R_0. Those pixels k above a given threshold are identified as point sources. The amplitude of the detected sources is then estimated using a multiscale χ^2 fit for each pixel k:

$$\chi_k^2 = \sum_{i,j} \left(\omega_{i,k}^{th} - \omega_{i,k}^{o}\right)^t V_{ij}^{-1} \left(\omega_{j,k}^{th} - \omega_{j,k}^{o}\right) , \tag{19}$$

where V is the covariance matrix between the different scales i, j and $\omega_{i,k}^{th}$, $\omega_{i,k}^{o}$ correspond, respectively, to the theoretical (given by equation 17) and observed wavelet coefficient at scale i and position k. Four scales are used for the χ^2. Note that this fit can also be used to discard point source candidates: if a source candidate at pixel k does not have an acceptably low value of χ^2, it could be rejected as a point source.

The MHW technique has been implemented to deal with simulated Planck observations in flat patches of the sky [87] and also on the whole sphere [89]. The procedure to detect point sources on spherical data is very similar to the one outlined before, but, in this case, the spherical Mexican Hat wavelet is used to convolve the data, which is given by [64]

$$\Psi_S(y,R) = \frac{1}{\sqrt{2\pi}N(R)}\left[1+\left(\frac{y}{2}\right)^2\right]^2\left[2-\left(\frac{y}{R}\right)^2\right]e^{-y^2/2R^2},$$
$$N(R) \equiv R\left(1+\frac{R^2}{2}+\frac{R^4}{4}\right)^{1/2}. \tag{20}$$

y is the distance to the tangent plane which is given by $y \equiv 2\tan\frac{\theta}{2}$, where θ is the latitude angle. Another important point when dealing with spherical data is the estimation of the optimal scale. In CMB observations, the contaminants can be very anisotropic in the sky. This means that the properties of the background change significantly for different areas of the sky and therefore the optimal scale for filtering with the SMHW has to be obtained locally. This is simply done by projecting small patches of the sky on a plane and obtaining the optimal scale for each of them. The maps are then convolved with SMHW of different scales and the detection is performed on the sphere but with the corresponding R_0 for each patch.

Table 1 shows the catalogue predicted by [89] using simulated Planck observations of the whole sky. In addition to CMB and point sources, galactic foregrounds, thermal SZ and instrumental noise were included in the simulations. Using the recovered catalogue, mean spectral indices can also be estimated with good accuracy. The SMHW has also been adapted to deal with realistic asymmetric beams as those expected for the Planck Mission.

It is also interesting to note that the MHW technique has been combined with the maximum-entropy method [88]. Using Planck simulated data, it was shown that the joint method improved the quality of both the reconstruction of the diffuse components and the point source catalogue.

Table 1. Predicted point source catalogue using the SMHW from simulated Planck observations. The point sources have been detected outside a galactic cut that varies from channel to channel (from no cut at the lowest frequency channels up to a galactic cut with $b = 25°$ for 857 GHz). The different columns correspond to Planck frequency channel, number of detections (above the minimum flux), minimum flux, mean error, mean bias, number of optimal scales needed for the algorithm and completeness of the catalogue above the minimum flux (see [89] for details)

Frequency (GHz)	Number	Min. flux (Jy)	$\bar{E}$ (%)	$\bar{b}$ (%)	N_{R_o}	Completeness (%)
857	27257	0.48	17.7	−4.4	17	70
545	5201	0.49	18.7	4.0	15	75
353	4195	0.18	17.7	1.4	10	70
217	2935	0.12	17.0	−2.5	4	80
143	3444	0.13	17.5	−4.3	2	90
100	3342	0.16	16.3	−7.0	4	85
70	2172	0.24	17.1	−6.7	6	80
44	1987	0.25	16.4	−6.4	9	85
30	2907	0.21	18.7	1.2	7	85

Finally, we would like to point out that the MHW has been used for the detection of objects in X-ray images [22, 23] and, more recently, in SCUBA [6, 53] and Boomerang data [21].

3.3 Neyman–Pearson Detector and the Biparametric Scale-Adaptive Filter

As mentioned in Sect. 1, filtering helps the detection process because it amplifies the sought signal over the background. However, whether we filter or not, we still need a detection criterion – the detector – to decide if a given signal belongs to the background or to a true source. In addition, the final performance of the filter will clearly depend on the choice of the detector. A criterion that has been extensively used in astronomy is thresholding, i.e. those pixels of the data above a given value (e.g. 5σ) are identified as the signal. Thresholding has a number of advantages, including its simplicity and the fact that it has a precise meaning in the case of Gaussian backgrounds in the sense of controlling the probability of spurious detections. However, it only uses a limited part of the information contained in the data, the intensity, to perform decisions.

An example of detector – based on the Neyman–Pearson rule – that takes into account additional information has been recently proposed in [7, 57, 58] for one-dimensional signals and [59] for the two-dimensional case. The first step of the procedure is to identify maxima as point source candidates. To decide then whether the maxima are due to the presence of background on its own or to a combination of background plus source, a Neyman–Pearson detector is applied, which is given by (for two-dimensional signals)

$$L(\nu, \kappa, \epsilon) \equiv \frac{n(\nu, \kappa, \epsilon)}{n_b(\nu, \kappa, \epsilon)} \geq L_* \,, \tag{21}$$

where $n_b(\nu, \kappa, \epsilon)\mathrm{d}\boldsymbol{x}\mathrm{d}\nu\mathrm{d}\kappa\mathrm{d}\epsilon$ is the expected number of maxima of the background in the intervals $(\boldsymbol{x}, \boldsymbol{x}+\mathrm{d}\boldsymbol{x})$, $(\nu, \nu+\mathrm{d}\nu)$, $(\kappa, \kappa+\mathrm{d}\kappa)$ and $(\epsilon, \epsilon+\mathrm{d}\epsilon)$, whereas $n(\nu, \kappa, \epsilon)\mathrm{d}\boldsymbol{x}\mathrm{d}\nu\mathrm{d}\kappa\mathrm{d}\epsilon$ corresponds to the same number in the presence of background plus source. ν, κ and ϵ are the normalized intensity, normalized curvature and normalized shear of the field, respectively, and $\boldsymbol{x}$ is the spatial variable. For a homogeneous and isotropic Gaussian background and point sources with spherical symmetry, it can be shown that the previous detector is equivalent to [59]

$$\varphi(\nu, \kappa) = a\nu + b\kappa \geq \varphi_* \,, \tag{22}$$

where a and b are constants that depend on the properties of the background and the profile of the source and φ_* is a constant that needs to be fixed. Therefore, if the considered maximum satisfies $\varphi \geq \varphi_*$, we decide that the signal is present, otherwise we consider that the maximum is due to the presence of only background.

Using this detector, the paper [59] compares the performance of different filters. In order to do this, φ_* is fixed to produce the same number of spurious sources for all the filters and then the number of true detections is compared. In their study they consider the matched filter, the Mexican Hat wavelet, the scale-adaptive filter and the biparametric scale-adaptive filter (BSAF). In addition, the scale of the filter is allowed to vary (similarly to what is done in the Mexican Hat wavelet technique) through the introduction of a parameter α. For the case of a background with a scale-free power spectrum ($P(q) \propto q^{-\gamma}$) and a source with Gaussian profile, the BSAF is given by

$$\psi_{BSAF}(q) = N(\alpha) z^{\gamma} \exp\left(z^2/2\right)\left(1 + cz^2\right), \qquad z \equiv q\alpha R \,, \tag{23}$$

where c is a free parameter. α and c are optimized in order to produce the maximum number of true detections given a fixed number of spurious detections. For $c = 0$ and $\alpha = 1$ the MF is recovered. Using this approach the performance of the considered filters has been studied for the case of a Gaussian white noise background. The results predict that, in certain cases, the BSAF can obtain up to around 40% more detections than the other filters. Although in CMB observations the background is not usually dominated by white noise, this case can also be of interest to detect point sources on CMB maps that have been previously processed using a component separation technique such as the maximum-entropy method [88]. In this case, the expected contribution of foregrounds and CMB is subtracted from each of the frequency maps, leaving basically the emission of extragalactic point sources and white noise (as well as some residuals). In this type of maps, the application of this technique could be useful. In any case, a test of the BSAF on realistic CMB simulations would be necessary to establish how well this approach would perform on real data.

3.4 Bayesian Approach to Discrete Object Detection

Many of the standard techniques for the detection of point sources are based on the design of linear filters. However, other methods – usually more complicated – are also possible. For instance, [51] has recently proposed a Bayesian approach for the detection of compact objects. The method is based on the evaluation of the (unnormalized) posterior distribution $\overline{Pr}(\theta|D)$ for the parameters θ that characterize the unknown objects (such as position, amplitude or size), given the observed data D. The unnormalized posterior probability is given in terms of the likelihood $\Pr(D|\theta)$ and the prior $\Pr(\theta)$ as

$$\overline{Pr}(\theta|D) \equiv Pr(D|\theta)Pr(\theta) \,. \tag{24}$$

Two different strategies are proposed for the detection of compact sources: an exact approach that tries to detect all the objects present in the data simultaneously and an iterative – much faster – approach (called McClean algorithm). In both cases an estimation of the parameters of the sources as well as their errors is provided. For both methods, a Markov chain Monte Carlo technique is used to explore the parameter space characterizing the objects.

As an illustration of the performance of the method, [51] studies the performance of both algorithms for a simple example that contains eight discrete objects with a Gaussian profile embedded on a Gaussian white noise background. The test image has 200×200 pixels and the signal-to-noise ratio of the objects ranges from 0.25 to 0.5. Using the exact method, where the number of objects is an additional parameter to be determined by the algorithm, all the objects are detected with no spurious detections. However, two of the objects (which overlapped in the noiseless data) are identified as a single detection. The results show that the parameters have been estimated with reasonably good accuracy. Unfortunately, although this method seems to perform very well, it is also very computationally demanding, which can make it unfeasible in many realistic applications.

The iterative approach tries to detect the objects one by one. This significantly reduces the CPU time necessary for the algorithm but provides a convenient approximation to the exact method. For the simple example previously considered, the McClean algorithm provides quite similar results to the exact approach, with only one less object detected than the exact approach.

This approach is certainly very promising and can be very useful for the detection of compact sources in future CMB data. However, it assumes the knowledge of the functional form of the likelihood, the prior of the parameters and the profile of the objects, which will not be known in many realistic situations. In addition, the presence of anisotropic contaminants (such as instrumental noise or galactic foregrounds) would introduce additional complexity that would make the algorithm more computationally demanding. The paper [51] gives some hints on how to deal with some of these problems. In any case, it would be very useful to test the performance of the method under realistic conditions in order to establish the real potentiality of the technique.

4 Techniques for Extraction of the Thermal SZ Effect

Another important application of the compact source extraction techniques is the detection of the thermal SZ effect due to galaxy clusters in microwave observations. The resolution of most CMB experiments (e.g. 5 arcminutes for the best Planck channels) is usually not enough to resolve the structure of the clusters of galaxies. Thus, the SZ emission appears in the CMB maps as compact sources whose shape is given by the convolution of the beam of the experiment with the profile of cluster. This means that most of the techniques used for the detection of extragalactic point sources could be easily adapted to detect the SZ effect (in one single map) just by including the correct profile of the sought source in the algorithm. This type of studies have been done for instance for the SAF [45], the MF [75] or the Bayesian approach [51] discussed in the previous section.

However, the thermal SZ effect has a very characteristic frequency signature that can be used to extract this emission, provided multifrequency observations are available. One alternative to recover the SZ emission is to apply a component separation technique – such as the maximum-entropy method, Wiener filter or blind source separation – that tries to recover simultaneously all the different emissions of the microwave sky. The second possibility is to design specific methods to extract the SZ signal that make use of the multifrequency information [30, 43, 71]. Regarding this second strategy, we will describe two of the methods that have been devised for the extraction of the thermal SZ effect from multifrequency CMB observations. Both methods have been tested using Planck simulated data.

4.1 Filtering Techniques

Reference [43] presents different filtering techniques for the detection of SZ clusters in multifrequency maps. Two alternative strategies are proposed: a combination technique and the design of a multifrequency filter (or multifilter). In both cases, the spatial profile of the clusters is assumed to be known. In the combination method, the individual frequency maps are linearly combined in an optimal way. The weights of the linear combination can be determined from the data, and they are optimal in the sense of giving the maximum amplification of objects that have the required spatial profile and the correct frequency dependence. This combined map is then filtered either with the MF or with the SAF constructed taking into account the characteristics of this new map. In the second approach, each frequency map is filtered separately but the filters are constructed taking into account the cross-correlations between frequency channels as well as the spectral dependence of the SZ effect. Then, the filtered maps are added together. This second method can also be implemented for two different kinds of filters: the matched multifilter and the scale-adaptive multifilter. The paper [43] performs a comparison of all these techniques finding that the matched multifilter provides the best results. Another interesting point is that the combination method is appreciably faster than the multifilter technique, whereas it still detects a large fraction of the clusters found by the matched multifilter.

Taking these results into account, we will discuss in more detail the matched multifilter (MMF) approach.[2] Let us consider a set of N observed maps given by

$$y_\nu(\mathbf{x}) = f_\nu A \tau_\nu(\mathbf{x}) + n_\nu(\mathbf{x}), \quad \nu = 1, \ldots, N \ , \tag{25}$$

[2] A similar technique has also been independently developed in [67], but in the context of the detection of extragalactic point sources in multifrequency observations.

where, for illustration, it is assumed that the SZ signal is due to the presence of a single cluster located in the origin of the image. The first term of the right hand side describes the contribution of the sought signal (the thermal SZ effect in this case) whereas n_ν is a generalized noise term that includes the sum of all the other components present in the map. f_ν is the frequency dependence of the thermal SZ effect (normalized to be 1 at a reference frequency), τ_ν is the shape of the cluster at each frequency (i.e. the profile of the cluster convolved with the corresponding antenna beam) and A is the amplitude of the SZ effect at the reference frequency. For simplicity, the profile of the cluster is assumed to be spherically symmetric and is parametrized by a characteristic scale – the core radius r_c – but a generalization to more complex profiles can be easily done. The background is assumed to be a homogeneous and isotropic random field with zero mean value and cross-power spectrum $P_{\nu_1\nu_2}(q)$ defined as

$$\langle n_{\nu_1}(\mathbf{q}) n^*_{\nu_2}(\mathbf{q}')\rangle = P_{\nu_1\nu_2}(q)\delta^2_D(\mathbf{q}-\mathbf{q}'), \quad q \equiv |\mathbf{q}| , \tag{26}$$

where $n_\nu(\mathbf{q})$ is the Fourier transform of $n_\nu(\mathbf{x})$ and δ^2_D is the two-dimensional Dirac distribution.

The MMF is given (in matrix notation) by

$$\mathbf{\Upsilon}(q) = \alpha \mathbf{P}^{-1} F, \quad \alpha^{-1} = \int d\mathbf{q} F^t \mathbf{P}^{-1} F \tag{27}$$

where F is the column vector $F = [f_\nu \tau_\nu]$ and $\mathbf{P}^{-1}$ is the inverse of the cross-power spectrum matrix $P \equiv [P_{\nu_1\nu_2}]$.

The output map, where the detection of the SZ effect is finally performed, is obtained by filtering each frequency map with its corresponding filter ψ_ν and then adding together all the filtered maps. The detection is then performed by looking for regions (five or more pixels) of connected pixels above a 3σ threshold. The maximum of the region determines the position of the cluster. In addition, the MMF is constructed so that the value of the intensity of the output map in the position of the source is an unbiased estimator of the amplitude. Therefore, the estimated amplitude of the SZ is simply given by the value of the output field in the considered maximum. Another interesting point is that the scale of the clusters r_c will not be known a priori. To overcome this problem, the data are multifiltered using different values of r_c. When the scale of the cluster coincides with that of the filter, the amplification of the signal will be maximum and that gives an estimation of the core radius r_c.

The MMF has been tested on Planck simulated data of small patches ($12.8° \times 12.8°$) of the sky containing CMB, thermal and kinetic SZ effects, galactic foregrounds (synchrotron, free–free, thermal dust and spinning dust), extragalactic point sources and instrumental noise. The simulated Planck data are shown in Fig. 7. Note that the SZ emission of clusters is completely masked by the rest of the components present in the data.

Figure 8 shows the input thermal SZ emission included in the simulations and the reconstructed SZ map after filtering the data with a MMF of r_c=1

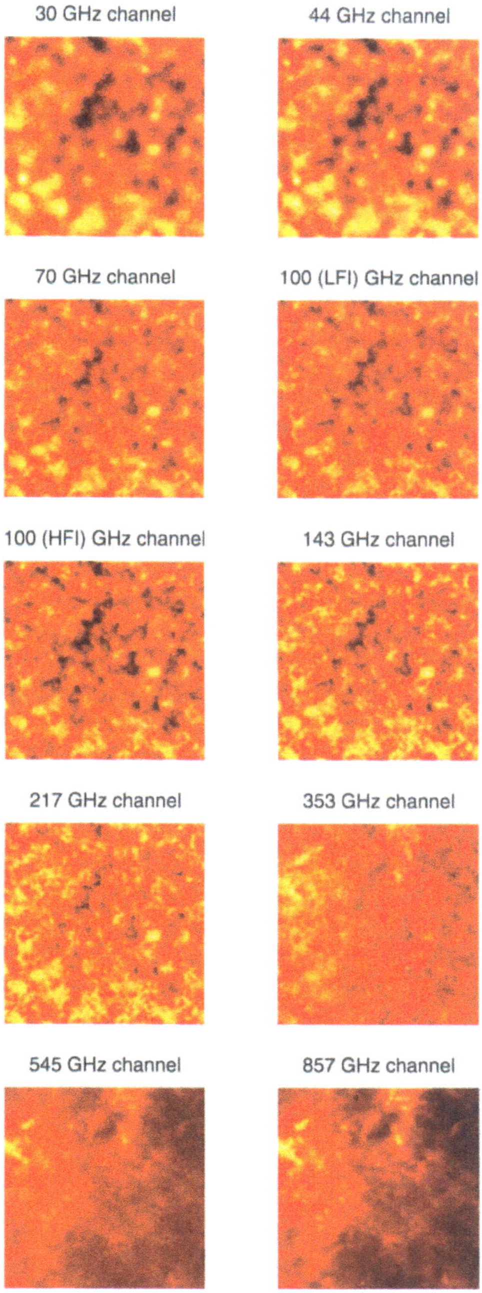

Fig. 7. Simulated Planck channels used to test the performance of the MMF

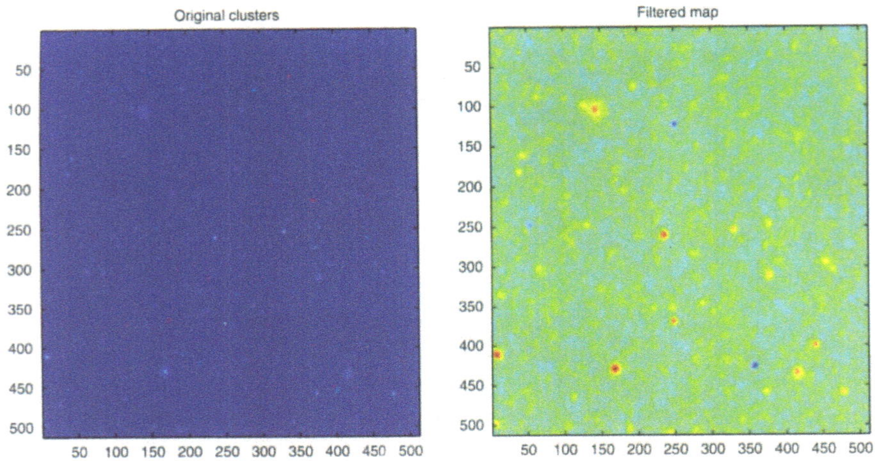

Fig. 8. The input (*left panel*) and the reconstructed (*right panel*) SZ effect after filtering the data with a MMF of $r_c = 1$ pixel

pixel. Clusters with scales similar to the chosen one are clearly visible in the output map. Reference [43] finds that the mean error in the determination of the position of the clusters is around 1 pixel whereas the core radii are determined with an error of 0.30 pixels. Regarding the determination of the cluster amplitudes, the mean error is around 30% for the brightest clusters, whereas there is a bias in the estimation of the weakest clusters. This bias can be understood since most weak clusters will only reach the detection threshold if they are on top of a positive fluctuation of the background, which will lead to an overestimate of the amplitude. The bias could be reduced by improving the method of amplitude estimation (for instance, by performing a fit to the profile of the cluster) that was simply given by the value of the maximum of the detection. Using this technique it is expected that Planck detects around 10,000 clusters in 2/3 of the sky.

An extension of the multifilter technique to spherical data has been carried out by [74]. They test the method on realistic simulations of Planck in the whole sky that, in addition to the main microwave components, also include the effect of non-uniform noise, sub-millimetric emission from celestial bodies of the Solar system and galactic CO-line radiation. It is again found that the multifilter approach can significantly reduce the background, allowing the cluster signal to be detected.

4.2 Bayesian Non-parametric Technique

The paper [30] proposes an alternative method to detect SZ clusters in Planck data. The method has been also tested on the simulated data set of Fig. 7. The procedure is as follows. First of all, the frequency maps are significantly cleaned from the most damaging contaminants. In particular, extragalactic point sources are subtracted with the MHW and subsequently the emission from dust and CMB is removed using the information of the 857 and 217 GHz channels, respectively. The next step consists in obtaining a map of the Compton parameter y_c in Fourier space by maximizing, mode by mode, the posterior probability $P(y_c|d)$. Taking into account Bayes' theorem, this probability is given by

$$P(y_c|d) \propto P(d|y_c)P(y_c) \; . \tag{28}$$

In order to perform this maximization we need to know the likelihood function $P(d|y_c)$ and the prior $P(y_c)$. Since the residuals left in the frequency maps are mainly dominated by the instrumental noise, the likelihood can be well approximated by a multivariate Gaussian distribution. In addition, one needs to assume a form for the prior $P(y_c)$. Using SZ simulations, [30] finds that the prior follows approximately the form $P(y_c) \propto \exp(-|y_c|^2/P_{y_c})$ at each k-mode, where Py_c is the power spectrum of the SZ map. Taking these results into account one gets, after maximizing the posterior probability, the following solution for the y_c map at each mode:

$$y_c = \frac{\mathbf{d}\mathbf{C}^{-1}\mathbf{R}^\dagger}{\mathbf{R}\mathbf{C}^{-1}\mathbf{R}^\dagger + P_{y_c}^{-1}} \; , \tag{29}$$

where d is the data, R is the response vector (that includes the information from the beam at each frequency and the frequency dependence of thermal SZ effect) and C is the cross-correlation matrix of the residuals. This result coincides with the multifrequency Wiener filter solution for the Compton y_c parameter. Note that the recovered y_c map will depend on the assumed power spectrum P_{y_c}. However, [30] shows that the final results do not depend significantly on the particular choice of $P(y_c)$, provided its form satisfies some general conditions. We would like to remark that this method does not need to make any assumption about the profile of the SZ clusters.

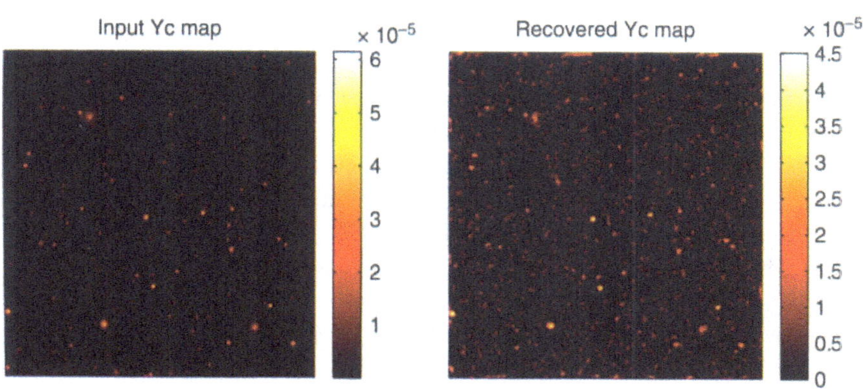

Fig. 9. The input (*left panel*) and the reconstructed (*right panel*) Compton parameter map after applying the Bayesian non-parametric method

The recovered y_c map after applying this approach to the Planck simulated data of Fig. 7 is shown in Fig. 9. The detection and the estimation of the flux of clusters are performed in the recovered map using the package SExtractor [12], which selects connected pixels above a given threshold. Using a 3σ threshold, this method predicts that Planck will detect around 9000 SZ clusters over four-fifths of the sky. In addition, it is shown that the flux of the clusters is estimated with no significant bias, which is important to carry out cosmological studies with the recovered catalogue.

5 Techniques for Extraction of the Kinetic SZ Effect

The determination of peculiar velocities of individual clusters through the kinetic SZ effect is a very challenging task. This is mainly due to the fact that the kinetic SZ emission is very weak – around 1 order of magnitude weaker than the thermal SZ effect – and that it has the same frequency dependence as the CMB, meaning that both signals cannot be separated using only multifrequency information. In addition, the presence of all the other components of the microwave sky and the instrumental noise makes even more difficult to detect this tiny signal. On the other side, we should take advantage of some characteristics of the kinetic SZ that could help to extract this emission. An important point is to use the available information about the thermal SZ. Since both SZ effects are produced by clusters of galaxies, there is a strong spatial correlation between them. Multifrequency observations are also important to separate the kinetic SZ signal from the components of the microwave sky (except from the CMB). In particular, it is useful to consider observations at the frequency of 217 GHz, where the contribution of the thermal SZ is expected to be negligible. Finally, the probability distribution of the kinetic SZ signal (expected to be highly non-Gaussian) and its power spectrum are very different from the ones of the cosmological signal, which could also be used to separate it from the CMB fluctuations.

Due to the complexity of the problem, only a few methods have been proposed and tested with the aim of extracting this emission from microwave observations. For instance, [42] studied the performance of an optimal filter (which is actually a matched filter for the cluster profile) to detect the kinetic SZ effect, concluding that peculiar velocities could be measured only for a few fast-moving clusters at intermediate redshift. The authors of [51] also applied their Bayesian algorithm to detect the kinetic SZ effect on CMB simulations at 217 GHz, but including only CMB and instrumental noise in the background. They claim that their technique is around twice as sensitive as the optimal linear filter. An alternative approach has been proposed by [38], which makes use of spatial correlation between the thermal and the kinetic SZ effects. The method is tested in ideal conditions, using as starting point a map of the Compton parameter and a second map containing only CMB and kinetic SZ emission. In these ideal conditions the method provides very promising results. However, a detailed study of the performance of the method under realistic conditions should be done before establishing the true potentiality of this approach. Recently, [48] tested a modification of the matched multifilter on Planck simulated data for the detection of the kinetic SZ effect that we discuss here in more detail.

5.1 Unbiased Matched Multifilter

If multifrequency information is available, we can also construct a MMF adapted to the kinetic SZ emission. As for the thermal SZ, the shape of the sought source will be the convolution of the antenna beam with the cluster profile but the frequency dependence will now follow that of the kinetic SZ emission. However, [48] found that the estimation of the kinetic SZ effect (and in fact that of the thermal SZ effect) using the MMF is intrinsically biased. It can be shown that this is due to the presence of two signals (the thermal and kinematic SZ effects in this case) that have basically the same spatial profile. Given the difference in amplitude of both effects, this bias is negligible for the case of the thermal SZ but it can be very important for the kinetic one. In order to correct this bias, a new family of filters, the unbiased matched multifilter (UMMF), has been constructed. For the case of the kinetic SZ effect, the UMMF is given by [48]

$$\mathbf{\Phi} = \frac{1}{\alpha\gamma - \beta^2}(-\beta\mathbf{F} + \alpha\tau) ,$$
$$\beta = \int d\mathbf{q}\tau^t\mathbf{P}^{-1}\mathbf{F} , \quad \gamma = \int d\mathbf{q}\tau^t\mathbf{P}^{-1}\tau . \tag{30}$$

This new multifilter leads to a slightly lower amplification of the sources than the MMF, but is intrinsically unbiased. The UMMF has been tested using Planck simulated data of small patches of the sky including CMB, galactic foregrounds (synchrotron, free–free, thermal and spinning dust) and point sources. In order to test just the effect of the intrinsic bias on the estimation of the amplitude, simplistic simulations of clusters have been used, and the knowledge of the profile and position of the clusters have been assumed.

Figure 10 shows the normalized histogram of the recovered parameter $V = (v_r m_e c)/(k_B T_e)$ using the MMF and the UMMF obtained from simulations that contained clusters with r_c=1.5 arcminutes, $y_c = 10^{-4}$ and $V = -0.1$. For a temperature of the electrons of $T_e \simeq 5\,\mathrm{keV}$, this value of V corresponds to a radial velocity along the line of sight of $v_r \simeq 300\,\mathrm{km\,s^{-1}}$. As predicted, the estimation of the amplitude of the kinetic SZ effect is strongly biased when using the MMF. However, this bias is corrected when the data are filtered with the UMMF. The work [48] also shows that this result remains valid for smaller values of y_c or V. Unfortunately, the error in the determination of peculiar velocities remains very large even for bright clusters. For instance, for $y_c = 10^{-4}$ and $T_e \simeq 5\,\mathrm{keV}$, the statistical error in the determination of

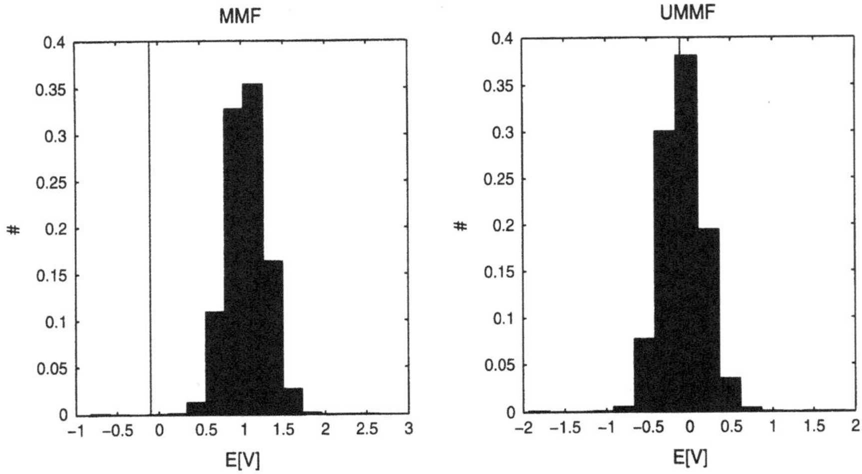

Fig. 10. Normalized histogram of the V parameter using the MMF (*left panel*) and the UMMF (*right panel*). The vertical line indicates the input value of V

v_r is $\sim$ 800 km s^{-1}. This means that Planck will not be able to measure, in general, the peculiar velocities of individual clusters, at least using just UMMF. Nonetheless, since the UMMF provides an unbiased estimation of v_r, it could be possible to measure mean peculiar velocities on large scales by averaging over many clusters.

6 Extraction of Statistical Information from Undetected Sources

In some cases the sought signal may be too weak to be individually detected or we may have found the brightest sources but we are unable to go down in flux. However, in this case, it may still be possible to extract some valuable statistical information from the background of sources. In this section we briefly discuss some works that have pursued this objective.

The differential number counts of extragalactic point sources are usually parametrized as

$$n(S) = kS^{-\eta}, \quad S > 0, \tag{31}$$

where S is the flux of the source and k and η are the normalization and slope parameters, respectively. The paper [70] uses the information of the high order moments of simulated data containing CMB, noise and residual point sources to estimate these two parameters. The paper [47] determines k and η by fitting the characteristic function of the point sources distribution to an α-stable model. The method is tested in the presence of a Gaussian background. These approaches offer interesting possibilities for the extraction of statistical information from an unresolved background of point sources.

However, the study of their performance on more realistic conditions – that take into account, for instance, the presence of anisotropic noise and non-Gaussian foregrounds – remains to be done.

Another interesting possibility is the study of the bispectrum of undetected point sources since this quantity will depend on the characteristics of the underlying population of extragalactic point sources [2, 54]. It also provides an estimation of the level of the contamination introduced by the residual point sources. In particular, [55] estimated the power spectrum of residual point sources in the WMAP data through the measurement of the bispectrum.

One may also use statistical information of the unresolved sources to identify which type of emission is present. In particular, [72] presents a detailed study of the contribution of the thermal SZ emission and of the extragalactic point sources to the probability distribution of the brightness map. The different imprints left by these two emissions would allow one to discriminate whether the excess of power found at small scales in some CMB data is due to an unresolved background of point sources or to the presence of unresolved SZ clusters. An alternative study based on the analysis of the Gaussianity of the wavelet coefficients has also been carried out to explain this excess of power [31].

Regarding the kinetic SZ effect, we have already mentioned that the detection of peculiar velocities of individual clusters will be a very challenging task. However, one could infer bulk flows on large scales that would provide valuable cosmological information. This point has been addressed by different authors, including [1, 3, 52].

7 Conclusions

The development of techniques for the extraction of compact sources in CMB observations has become a very relevant and active topic. This is not only due to the necessity of cleaning the CMB maps from astrophysical contaminants that would impair our ability to extract all the valuable information encoded in the cosmological signal but also because the recovered catalogues of point sources and/or SZ clusters would contain themselves extremely relevant astrophysical and cosmological information.

An important effort has been done in the last years towards the development of more powerful and sophisticated tools to extract compact sources. Many of them have been tested on simulated Planck observations, showing their potentiality. However, some important work still remains to be done. First of all, in some cases, quite ideal conditions have been assumed. For instance, it is commonly assumed that the cluster profile is known but, in general, this will not be the case for real data. Other methods have been applied to simulations that do not include foreground emissions. Therefore, these and other problematics – beam asymmetry, extension to the sphere, relativistic effects, anisotropic noise, etc. – present on real data should be taken into account to establish the true performance of the developed methods. Also, the methods do not always use all the available information present in the data. For instance, if multifrequency observations are available, it would be useful to include this multifrequency information in the detection of extragalactic point sources even if they do not follow a simple well-known spectral law. The final and most important step would be to apply these techniques to real CMB data (e.g. WMAP) as they become available.

We would also like to point out that many polarization experiments are currently planned (or already operating) which will provide a wealth of information about our universe [19]. Given the weakness of the cosmological signal in polarization and the current lack of knowledge regarding the foreground emissions, a careful process of cleaning of the CMB polarization maps is even more critical than for the intensity case. However, no techniques have been yet specifically developed to extract compact sources from polarization CMB observations. Therefore, it is crucial to extend some of the current methods – or to develop new ones – to deal with this type of maps.

Finally, a very critical issue is to assess which is the impact of possible residuals left in the CMB data after applying these techniques [9]. In particular, it is very important to control the effect of undetected sources, or even possible artefacts introduced in the image after subtracting the signals, on the estimation of the power spectrum of the CMB. In addition, this process should not modify the underlying CMB temperature distribution, since it would impair our ability to perform Gaussianity analysis of the CMB (or even lead us to wrong conclusions), which is of great importance to learn about the structure formation of our universe.

Acknowledgements

I acknowledge the Universidad de Cantabria and the Ministerio de Educación y Ciencia for a Ramón y Cajal contract and J.M. Diego, D. Herranz, M. López-Caniego, E. Martínez-González, J.L. Sanz and P. Vielva for their help in preparation of these lecture notes. I would also like to thank the organizers for inviting me to participate in a very fruitful and interesting school.

References

1. Aghanim, N., Górski, K.M., Puget, J.-L.: A&A, **374**, 1 (2001)
2. Argüeso, F., González-Nuevo, J., Toffolatti, L.: ApJ, **598**, 86 (2003)
3. Atrio-Barandela, F., Kashlinsky, A., Mücket, J.P.: ApJ, **601**, 111 (2004)
4. Baccigalupi, C., Bedini, L., Burigana, C., De Zotti, G., Farusi, A., Maino, D., Maris, M., Perrota, F., Salerno, E., Toffolatti, L., Tonazzini, A.: MNRAS, **318**, 769 (2000)
5. Banday, A.J., Dickinson, C., Davies, R.D., Davis, R.J., Górski, K.M.: MNRAS, **345**, 897 (2003)

6. Barnard, V.E., Vielva, P., Pierce-Price, D.P.I., Blain, A.W., Barreiro, R.B., Richer, J.S., Qualtrough, C.: MNRAS, **352**, 961 (2004)
7. Barreiro, R.B., Sanz, J.L., Herranz, D., Martínez-González, E.: MNRAS, **342**, 119 (2003)
8. Barreiro, R.B., Hobson, M.P., Banday, A.J., Lasenby, A.N., Stolyarov, V., Vielva, P., Górski, K.M.: MNRAS, **351**, 515 (2004)
9. Barreiro, R.B., Martínez-González, E., Vielva, P., Hobson, M.P.: MNRAS **368**, 226, (2006)
10. Bedini, L., Herranz, D., Salerno, E., Baccigalupi, C., Kuruoglu, E.E., Tonazzini, A.: EURASIP J. Appl. Signal Process. (Special Issue on Applications of Signal Processing in Astrophysics and Cosmology), **15**, 2400 (2005)
11. Bennett, C.L., et al.: ApJS, **148**, 97 (2003)
12. Bertin, E., Arnouts, S.: A&AS, **117**, 393 (1996)
13. Birkinshaw, M.: PhR, **310**, 97 (1999)
14. Bouchet, F.R., Gispert, R., Puget, J.L.: The MM/SUB-MM foregrounds and future CMB space missions. In: Dwek, E. (ed.) The COBE Workshop: Unveiling the Cosmic Infrared Background, AIP Conference Proceedings, vol. 348, (1996), pp. 255–270
15. Bouchet, F.R., Gispert, R.: New Astr., **4**, 443 (1999)
16. Burrus, C.S., Gopinath, R.A., Guo, H.: Introduction to Wavelets and Wavelet Transforms. A Primer. Prentince-Hall, Upper Saddle River, New Jersey, (1998)
17. Carlstrom, J.E., Holder, G.P., Reese, E.D.: AR&A&A, **40**, 643 (2002)
18. Cayón, L., Sanz, J.L., Barreiro, R.B., Martínez-González, E., Vielva, P., Toffolatti, L., Silk, J., Diego, J.M., Argüeso, F.: MNRAS, **315**, 757 (2000)
19. Challinor, A.: Cosmic microwave background polarization analysis. In: Martinez, V.J. (ed.) *Data Analysis in Cosmology.* Lect. Notes Phys. **665**, Springer, Heidelberg (2008)
20. Chiang, L.-Y., Jorgensen, H.E., Naselsky, I.P., Naselsky, P.D., Novikov, I.D., Christensen, P.R.: MNRAS, **335**, 1054 (2002)
21. Coble, K., et al., ApJS, submitted, astro-ph/0301599 (2003)
22. Damiani, F., Maggio, A., Micela, G., Sciortino, S.: ApJ, **483**, 350 (1997)
23. Damiani, F., Maggio, A., Micela, G., Sciortino, S.: ApJ, **483**, 370 (1997)
24. Daubechies, I.: Ten Lectures on Wavelets. S.I.A.M., Philadelphia (1992)
25. Dellabrouille, J., Cardoso, J.F., Patanchon, G.: MNRAS, **346**, 1089 (2003)
26. de Oliveira-Costa, A., et al.: ApJ, **567**, 363 (2002)
27. de Oliveira-Costa, A., Tegmark, M., Davies, R.D., Gutiérrez, C.M., Lasenby, A.N., Rebolo, R., Watson, R.A.: ApJ, **606**, L89 (2004)
28. de Zotti, G., Toffolatti, L., Argüeso, F., Davies, R.D., Mazzotta, P., Partridge, R.B., Smoot, G.F., Vittorio, N.: The Planck surveyor mission: Astrophysical prospects. In: Maiani, L., Melchiorri, F., Vittorio, N. (eds.) 3K Cosmology, Proceedings of the EC-TMR Conference held in Rome, Italy, October, 1998, vol. 476, pp. 204–223. American Institute of Physics, Woodbury, N.Y. (1999)
29. de Zotti, G., Ricci, R., Mesa, D., Silva, L., Mazzotta, P., Toffolatti, L., González-Nuevo, J.: A&A, **431**, 893 (2005)
30. Diego, J.M., Vielva, P., Martínez-González, E., Silk, J., Sanz, J.L.: MNRAS, **336**, 1351 (2002)
31. Diego, J.M., Vielva, P., Marínez-González, E., Silk, J.: astro-ph/0403561 (2004)
32. Dole, H., Lagache, G., Puget, J.-L.: ApJ, **585**, 617 (2003)
33. Draine, B.T., Lazarian, A.: ApJ, **494**, L19 (1998)

34. Draine, B.T., Lazarian, A.: ApJ, **508**, 157 (1998)
35. Eriksen, H.K., Banday, A.J., Górski, K.M., Lilje, P.B.: ApJ, **612**, 633 (2004)
36. Eriksen, H.K., et al.: ApJ, **641**, 665, (2005)
37. Finkbeiner, D.P., Langston, G.I., Minter, A.H.: ApJ, **617**, 350 (2004)
38. Forni, O., Aghanim, N.: A&A, **420**, 49 (2004)
39. González-Nuevo, J., Toffolatti, L., Argüeso, F.: ApJ, **621**, 1 (2005)
40. Granato, G.L., de Zotti, G., Silva, L., Bressan, A., Danese, L.: ApJ, **600**, 580 (2004)
41. Guiderdoni, B.: Far-infrared point sources. In: de Oliveira-Costa, A., Tegmark, M. (eds) Microwave Foregrounds, ASP Conference Series, vol. 181, pp. 173–198. ASP, San Francisco, CA (1999)
42. Haehnelt, M.G., Tegmark, M.: MNRAS, **279**, 545 (1996)
43. Herranz, D., Sanz, J.L., Hobson, M.P., Barreiro, R.B., Diego, J.M., Martínez-González, E., Lasenby, A.N.: MNRAS, **336**, 1057 (2002)
44. Herranz, D., Gallegos, J., Sanz, J.L., Martínez-González, E.: MNRAS, **334**, 353 (2002)
45. Herranz, D., Sanz, J.L., Barreiro, R.B., Martínez-González, E.: ApJ, **580**, 610 (2002)
46. Herranz, D.: Analysis of the anisotropies in the cosmic microwave background radiation using adaptive filters. PhD Thesis, Universidad de Cantabria, Spain (2002)
47. Herranz, D., Kuruoglu, E.E., Toffolatti, L.: A&A, **424**, 1081 (2004)
48. Herranz, D., Sanz, J.L., Barreiro, R.B., López-Caniego, M.: MNRAS, **356**, 944 (2005)
49. Hobson, M.P., Jones, A.W., Lasenby, A.N., Bouchet, F.R.: MNRAS, **300**, 1 (1998)
50. Hobson, M.P., Barreiro, R.B., Toffolatti, L., Lasenby, A.N., Sanz, J.L., Jones, A.W., Bouchet, F.R.: MNRAS, **306**, 232 (1999)
51. Hobson, M.P., McLachlan, C.: MNRAS, **338**, 765 (2003)
52. Kashlinsky, A., Atrio-Barandela, F.: ApJ, **536**, 67 (2000)
53. Knudsen, K.K., Barnard, V.E., van der Werf, R.P., Vielva, P., Kneib, J.-P., Blain, A.W., Barreiro, R.B., Ivison, R.J., Smail, I.R., Peacock, J.A.: MNRAS, **368**, 487, (2005)
54. Komatsu, E., Spergel, D.N.: PhRvD, **63**, 063002 (2001)
55. Komatsu, E., et al.: ApJS, **148**, 119 (2003)
56. Lagache, G.: A&A, **405**, 813 (2003)
57. López-Caniego, M., Herranz, D., Barreiro, R.B., Sanz, J.L.: SPIE, **5299**, 145 (2004)
58. López-Caniego, M., Herranz, D., Barreiro, R.B., Sanz, J.L.: MNRAS, **359**, 993 (2005)
59. López-Caniego, M., Herranz, D., Sanz, J.L., Barreiro, R.B.: EURASIP J. Appl. Signal Process. (Special Issue on Applications of Signal Processing in Astrophysics and Cosmology), **15**, 1 (2005)
60. López-Caniego, M., Sanz, J.L., Herranz, D., González-Nuevo, J., Barreiro, R.B., Kuruoglu, E.E.: Non-linear fusion of images and the detection of point sources. In: IEEE proceedings of the International Workshop on Nonlinear Signal and Image Processing, in press (2005)
61. López-Caniego, M., Sanz, J.L., Herranz, D., Barreiro, R.B., González-Nuevo, J.: Linear and quadratic fusion of images: detection of point sources. In: Proceedings of the 13th European Signal Processing Conference (EUSIPCO 2005), in press (2005)

62. Maino, D., Farusi, A., Baccigalupi, C., Perrotta, F., Banday, A.J., Bedini, L., Burigana, C., G. de Zotti, G., Górski, K.M., Salerno, E.: MNRAS, **334**, 53 (2002)
63. Maino, D., Banday, A.J., Baccigalupi, C., Perrotta, F., Górski, K.M.: MNRAS, **344**, 544 (2003)
64. Martínez-González, E., Gallegos, J.E., F. Argüeso, L. Cayón, Sanz, J.L.: MNRAS, **336**, 22 (2002)
65. Martínez-González, E., Diego, J.M., Vielva, P., Silk, J.: MNRAS, **345**, 1101 (2003)
66. Mukherjee, P., Dennison, B., Ratra, B., Simonetti, J.H., Ganga, K., Hamilton, J.-C.: ApJ, **579**, 83 (2002)
67. Naselsky, P., Novikov, D., Silk, J.: MNRAS, **335**, 550 (2002)
68. Patanchon, G., Cardoso, J.F., Delabrouille, J., Vielva, P.: MNRAS, **364**, 1185, (2005)
69. Planck Mission: http://www.rssd.esa.int/Planck
70. Pierpaoli, E.: ApJ, **589**, 58 (2003)
71. Pierpaoli, E., Anthoine, S., Huffenberger, K., Daubechies, I.: MNRAS, **359**, 261 (2005)
72. Rubiño-Martín, J.A., Sunyaev, R.A.: MNRAS, **344**, 1155 (2003)
73. Sanz, J.L., Herranz, D., Martínez-González, E.: ApJ, **552**, 484 (2001)
74. Schaefer, B.M., Pfrommer, C., Hell, R., Bartelmann, M.: MNRAS, **370**, 1713, (2006)
75. Schulz, A.E., White, M.: ApJ, **586**, 723 (2003)
76. Stolyarov, V., Hobson, M.P., Ashdown, M.A.J., Lasenby, A.N.: MNRAS, **336**, 97 (2002)
77. Stolyarov, V., Hobson, M.P., Lasenby, A.N., Barreiro, R.B.: MNRAS, **357**, 145 (2005)
78. Sunyaev, R.A., Zel'dovich, Ya.B.: Ap&SS, **7**, 3 (1970)
79. Sunyaev, R.A., Zel'dovich, Ya.B.: CoASP, **4**, 173 (1972)
80. Sunyaev, R.A., Zel'dovich, Ya.B.: MNRAS, **190**, 413 (1980)
81. Tegmark, M., Efstathiou, G.: MNRAS, **281**, 1297 (1996)
82. Tegmark, M., de Oliveira-Costa, A.: ApJ, **500**, L83 (1998)
83. Tegmark, M., Eisenstein, D.J., Hu, W., de Oliveira-Costa, A.: Overview of the foregrounds and their impact. In: de Oliveira-Costa, A., Tegmark, M. (eds.) Microwave Foregrounds, ASP Conference Series, vol 181, pp. 3–58. ASP, San Francisco, CA (1999)
84. Tegmark, M., de Oliveira-Costa, A., Hamilton, A.J.S.: Phys. Rev. D., **68**, 123523 (2003)
85. Toffolatti, L., Argüeso-Gómez, F., De Zotti, G., Mazzei, P., Francheschini, A., Danese, L., Burigana, C.: MNRAS, **297**, 117 (1998)
86. Tucci, M., Martínez-González, E., Toffolatti, L., González-Nuevo, J., De Zotti, J.G.: MNRAS, **349**, 1267 (2004)
87. Vielva, P., Martínez-González, E., Cayón, L., Sanz, J.L., Toffolatti, L.: MNRAS, **326**, 181 (2001)
88. Vielva, P., Barreiro, R.B., Hobson, M.P., Martínez-González, E., Lasenby, A.N., Sanz, J.L., Toffolatti, L.: MNRAS, **328**, 1 (2001)
89. Vielva, P., Martínez-González, E., Gallegos, J.E., Toffolatti, L., Sanz, J.L.: MNRAS, **344**, 89 (2003)

Determination of Cosmological Parameters from Cosmic Microwave Background Anisotropies

A. Melchiorri, C. Odman, and P. Serra

Dipartimento di Fisica, and Sezione INFN Roma, Universita' di Roma
"La Sapienza", Ple Aldo Moro 5, 00185, Rome, Italy
alessandro.melchiorri@roma1.infn.it
carolina.odman@roma1.infn.it
paolo.serra@roma1.infn.it

Abstract. New cosmic microwave background anisotropies data are increasingly improving our understanding of the evolution and formation of structures in the universe. Here we briefly discuss the cosmological parameters extraction methods, review the latest results and comment about possible modifications to the standard model.

1 Introduction

The cosmic microwave background (hereafter CMB) provides an unexcelled probe of the early universe. Its spectrum, a close approximation to that of a blackbody, constrains the thermal history of the universe since an epoch of approximately 1 year after the Big Bang.

Gravitational instability due to primordial irregularities in density explains the origin of galaxies and large-scale structure (LSS) and is our standard model of cosmic evolution. These early density perturbations must have imprinted trace fluctuations in the CMB, visible as angular anisotropies.

Coherent oscillations in the CMB anisotropies angular power spectrum were predicted a long time ago from simple assumptions about scale invariance and linear perturbation theory (see, e.g. [1, 2, 3, 4, 5]). The physics of these oscillations and their dependence on the various cosmological parameters have been described in great detail in many reviews ([6, 7, 8, 9, 10]). Prior to recombination, photons and baryons formed a tightly coupled fluid that performed acoustic oscillations on sub-horizon scales, driven by gravitational potential. These acoustic oscillations defined a structure that translates into peaks in the CMB anisotropy angular power spectrum and can be measured today.

Melchiorri, A. et al.: *Determination of Cosmological Parameters from Cosmic Microwave Background Anisotropies*. Lect. Notes Phys. **665**, 237–259 (2009)
DOI 10.1007/978-3-540-44767-2_8

Recent years have been an exciting period for the field of the CMB research. Let us here briefly review some main results.

In 1999 with the TOCO–97/98 [11, 12] and BOOMERanG-97 [13] experiments a firm detection of a first peak on about degree scales was obtained. In the framework of adiabatic cold dark matter (CDM) models, the position, amplitude and width of this peak provide strong supporting evidence for the inflationary predictions of a low curvature (flat) universe and a scale-invariant primordial spectrum [14, 15, 16].

About a year later, the experimental data from BOOMERanG LDB [20, 17], DASI [18] and MAXIMA ([19]) provided further evidence for the presence of the first peak and refined the data at higher multipoles. The combined data suggest the presence of a second and third peaks in the spectrum, confirming the standard model prediction of acoustic oscillations in the primeval plasma and shedding new light on various cosmological and inflationary parameters [20, 21, 22].

Just few months later, the CBI and ACBAR experiments, pushing the measurements on smaller angular scales (1–10 arcmin) ([23, 24]), provided evidence for the expected "damping tail" in the CMB anisotropy power spectrum and hints for further acoustic oscillations beyond the first three acoustic peaks. The acoustic oscillations and the presence of a first acoustic peak were also confirmed by the VSA [25, 26] and Archeops [27, 28] experiments, respectively.

Around the end of 2002, the DASI experiment reported the first detection of a polarization of the CMB anisotropies (see [29] and, more recently, [30] and CBI [31]) in agreement with the standard theoretical picture (see the contribution by Anthony Challinor in these Proceedings).

At the beginning of 2003, the measurements of the CMB anisotropies provided by the Wilkinson Microwave Anisotropy Probe (WMAP) satellite [32] truly marked the beginning of the era of precision cosmology. In particular, the shape of the measured temperature and temperature–polarization angular power spectra are in spectacular agreement with the expectations of the standard model of structure formation, based on primordial adiabatic and nearly scale-invariant initial perturbations. Assuming this model of structure formation, accurate albeit *indirect* constraints on several cosmological parameters have been reported [33].

The goal of this review was to give an introduction on the methods used in the derivation of constraints on cosmological parameters from current CMB anisotropy and polarization power spectra. Then we report the latest results on the several parameters of the model discussing the presence of physics beyond the standard model. Finally we draw our conclusions.

2 Bayesian Paradigm

In this section, let us briefly recall the method adopted to obtain constraints on the cosmological parameters from CMB observations. Each measurement is a realization of an experiment which suffers from both random and systematic errors. Because of the errors, the step of inverting the measured data into constraints on the free parameters of the underlying theory has no unique solution.

Uncertainties can be accounted for by considering each result as a probability and by giving information on the errors. Bayesian inference techniques are used to obtain results from data while trying to maximize the understanding of the errors and their impact on the inferred results.

The results inferred from data in the form of a likelihood function are subject to assumptions, the prior. They also depend on how well the theory describes the problem and that can be assessed through an evidence function which is a normalization of the likelihood function [34].

The errors of the measurement are encoded as $P(\mathbf{x}|\theta)$, the probability of measuring $\mathbf{x}$ given a model θ. Bayes' theorem tells us how to invert the conditional probability:

$$P(\theta|\mathbf{x}) = \frac{P(\mathbf{x}|\theta)P(\theta)}{P(\mathbf{x})} , \tag{1}$$

where θ represents the values of the free parameters in the underlying theory and $\mathbf{x}$ represents the data. $P(\theta)$ is the prior. $P(\theta|\mathbf{x})$ is the posterior probability of the model given the data and $P(\mathbf{x})$ is the evidence.

In the case of near-Gaussian errors, which is often a reasonable assumption for well-controlled systematic errors, the fit of a model to data is best described by a chi-squared distribution

$$\begin{aligned} P(\theta|\mathbf{x}) \propto \mathcal{L}(\theta) &\equiv \exp\left(-\frac{\chi^2(\theta)}{2}\right) \\ \chi^2(\theta) &= \left(\mathbf{x} - \mathbf{x}_\theta^{\text{pred}}\right)^T C^{-1} \left(\mathbf{x} - \mathbf{x}_\theta^{\text{pred}}\right) , \end{aligned} \tag{2}$$

where $\mathbf{x}_\theta^{\text{pred}}$ is the data expected for the model θ, $\mathcal{L}(\theta)$ is the likelihood function and C^{-1} is the inverse correlation matrix between the data points.

Suppose an N-dimensional likelihood function is known from some parameter estimation technique. In order to get one- or two-dimensional probability distributions, the joint probability distribution is integrated over the other parameters. This is called *marginalization*:

$$\mathcal{L}(\alpha, \beta) = \int_{\mathbf{V}_\Omega} \mathcal{L}(\alpha, \beta, \gamma, \dots) d\Omega , \tag{3}$$

where $\theta = (\alpha, \beta, \gamma, \dots)$ and $\mathcal{L}(\alpha, \beta, \gamma, \dots)$ is the joint likelihood function, proportional to the probability distribution. The integral is performed over the $N-2$ dimensional volume $\mathbf{V}_\Omega$ in parameter space defined by all remaining parameters $\Omega = \{\gamma, \dots)$ and allowed by the prior.

Alternatively a parameter space can be cut after *maximization* of the likelihood function by fixing the values of some parameters at the overall maximum of the likelihood. The resulting likelihood function on $n < N$ dimensions is the projection of the joint likelihood function onto an n-dimensional slice of parameter space. This is equivalent to an additional prior on the values of the remaining parameters

$$\mathcal{L}(\alpha,\beta) = \mathcal{L}(\alpha,\beta,\tilde{\Omega}) \, , \tag{4}$$

where $\tilde{\Omega} = \{\tilde{\gamma},\dots\}$ are the coordinates of the maximum of the likelihood function.

Note that the likelihood function obtained by maximization is equivalent to the marginalized likelihood function in the case of Gaussian probability distributions.

If a likelihood function is not presented graphically, results are often given in terms of confidence intervals. The one-dimensional $X\%$ confidence limit on a parameter α is given by $\alpha_{\rm min} \le \alpha \le \alpha_{\rm max}$ where

$$\frac{X}{100} = \int_{\alpha_{\rm min}}^{\alpha_{\rm max}} P(\alpha')d\alpha' \, ,$$
$$P(\alpha_{\rm min}) = P(\alpha_{\rm max}) \, . \tag{5}$$

If the likelihood function is not smooth and/or has multiple extrema, the confidence levels can be calculated with

$$\mathcal{L}\left(\alpha^{n\sigma}_{\rm min}\right) = \mathcal{L}\left(\alpha^{n\sigma}_{\rm max}\right) = \exp\left(-\frac{n^2}{2}\right)\mathcal{L}^{\rm max} \, , \tag{6}$$

where $n\sigma$ is the level, $\mathcal{L}^{\rm max} = \mathcal{L}(\tilde{\alpha})$ is the overall maximum likelihood, and $\alpha^{n\sigma}_{\rm min} \le \alpha \le \alpha^{n\sigma}_{\rm max}$ is the allowed range of the parameter α around $\tilde{\alpha}$ at $n\sigma$. This is exact only in the case of a Gaussian probability distribution but represents a useful convention.

2.1 Parameter Estimation Techniques

Several methods are used to estimate the values of parameters from cosmological data. Let us here present the two most popular methods.

Likelihood Function Estimation on a Grid of Models

First a grid is defined on the parameter space. The likelihood of each model according to the data, the likelihood function, is evaluated on those grid points. When marginalizing the likelihood function, i.e. when summing over all other grid points, the computing time scales typically like N^n, where N is the number of grid points in each dimension and n is the number of dimensions.

The likelihood function outside grid points is estimated using a suitable interpolation.

Markov Chain Monte Carlo Simulations

The Markov chain Monte Carlo (MCMC) approach is very different. The idea is to generate a random walk through parameter space that converges towards the most likely value of the parameters and samples the parameter space following the posterior probability distribution.

A Markov chain is characterized by a range of states with a transition probability from each state to each other state:

$$P_{i,j} = P(\theta_i \to \theta_j)\ . \tag{7}$$

Mathematically speaking the MCMC algorithm generates a Markov chain whose equilibrium transition probabilities are the joint posterior probability distribution of the parameters. Note that a condition of stationarity of the Markov chain is that $P_{i,j} = P_{j,i}$. For a more detailed description of Markov chains see [35].

As with the grid approach the number of parameters and their priors have to be defined. MCMC is an iterative process. At each step a sample θ_{n+1} is randomly selected (Monte Carlo) from a proposal probability density. The proposal can be uniform to investigate a range of values, or uniform in logarithm (Jeffrey's prior) to pin down an order of magnitude.

The likelihood of the sample is evaluated and whether that sample is accepted into the chain or not depends on an acceptance criterion that depends only on the value of the likelihood function at the previous point (Markov chain).

The criterion is the following: A random number u is generated between 0 and 1, and θ_{n+1} is accepted into the walk if it satisfies

$$u \leq \min\left\{1, \frac{p(\theta_{n+1})\mathcal{L}(\theta_{n+1})}{p(\theta_n)\mathcal{L}(\theta_n)}\right\}\ , \tag{8}$$

where $p(\theta)$ is the prior and $\mathcal{L}(\theta)$ is the likelihood, containing the information from the data. This is called the Metropolis–Hastings algorithm.

During a first period of time called the burn-in the random walk simply makes its way through parameter space and converges towards the maximum likelihood. Eventually the Markov chain reaches equilibrium and samples the parameter space according to the posterior probability distribution of the parameters. A major drawback of MCMC methods is that there is no formal proof of the convergence of the chain so the burn-in time cannot be determined and must be guessed. For an interesting discussion about convergence criteria see [36]. At the end of the MCMC routine, the number density of the samples taken after the burn-in period represents the already marginalized (but unnormalized) posterior distribution of each parameter.

Computing time required for use of an MCMC algorithm depends essentially on the number of samples. It scales like $\sim m$, where m is the number of dimensions of parameter space. It also depends on the overall shape of the likelihood function. A high number of local extrema may increase the time it takes for the Markov chain to reach equilibrium. Nevertheless for large problems involving a high number of parameters to evaluate, MCMC algorithms allow for simultaneous Bayesian parameter estimation and likelihood analysis to be performed at minimal computational cost.

The grid-based approach and MCMC are asymptotically identical in the sense that both an infinite sampling of the parameter space and an infinitely fine grid tend towards a continuous coverage of the likelihood surface.

3 Constraints on the Standard Λ-CDM Model

In principle, the CDM scenario of structure formation based on adiabatic primordial fluctuations can depend on at least 11 parameters.

For a first analysis, it possible to restrict ourselves to a flat universe with just four parameters: the tilt of primordial spectrum of perturbations n_S, the optical depth of the universe τ_c, the density in baryons and dark matter $\omega_b = \Omega_b a h^2$ and $\omega_{dm} = \Omega_{dm} h^2$. The assumption of flatness is motivated by observations. The position and width of the CMB anisotropy peaks can yield strong constraints on the so-called "shift parameter" $\mathcal{R}$ which is related to the geometry of the Universe through (see [37], [38])

$$\mathcal{R} = 2\sqrt{|\Omega_k|/\Omega_m}/\chi(y) , \tag{9}$$

where $\Omega_m = \Omega_b + \Omega_{dm}$, $\Omega_k = 1 - \Omega_m - \Omega_\Lambda$, the function $\chi(y)$ is y, $\sin(y)$ or $\sinh(y)$ for flat, closed and open universes, respectively, and

$$y = \sqrt{|\Omega_k|} \int_0^{z_{dec}} [\Omega_m(1+z)^3 + \Omega_k(1+z)^2 + \Omega_\Lambda]^{-1/2} dz . \tag{10}$$

Unfortunately, constraining $\mathcal{R}$ does not constrain the overall curvature Ω_k due to the degeneracy between Ω_m and Ω_Λ. However, combination of CMB observations with complementary data sets like type Ia Supernovae measurements (SNe-Ia), LSS and measurements of the Hubble parameter from the HST telescope can break the angular diameter distance degeneracy. A combination of these data sets yields $\Omega_k = -0.02 \pm 0.02$ ([33]), strongly suggesting a flat universe but preferring slightly closed models. It may be possible to have a closed universe model from inflation (see for example [39]) and, if the geometry is positive, one could speculate about possible non-standard topologies (see cf. [40]) or dark energy scenarios. Given the lack of strong physical motivations for considering those models and the relatively small evidence for a positive curvature it remains reasonable to assume flatness. Moreover the preferred positive curvature may be an artefact due to the angular diameter distance degeneracy which makes flat, Λ-dominated models degenerate with closed, matter-dominated models [38].

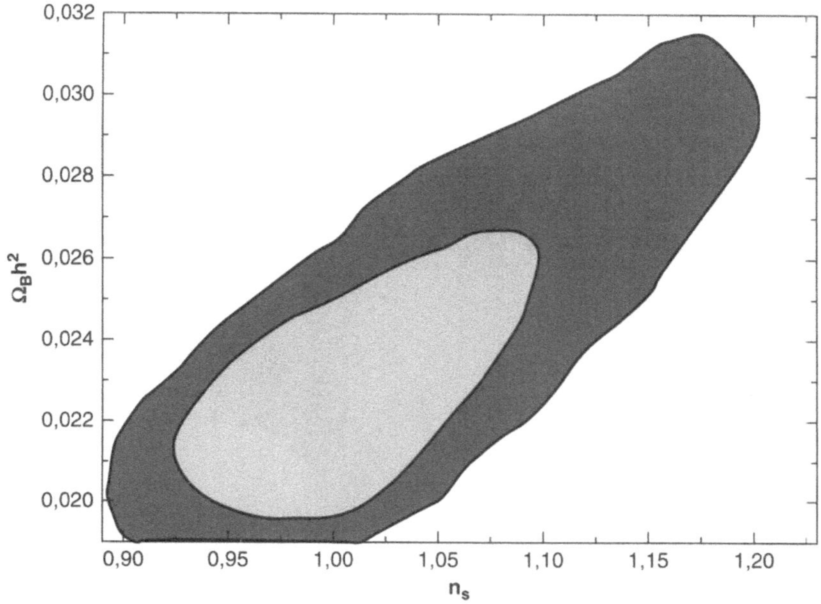

Fig. 1. Likelihood contours in the $(\Omega_b h^2, n_S)$ plane for the joint CMB+LSS analysis. The contours correspond to 0.32 and 0.05 of the peak value of the likelihood, which are the 68 and 95% confidence levels, respectively

Under the assumption of flatness one can derive constraints on the remaining four parameters but degeneracies are still present. It is useful to project those constraints on a two-dimensional parameters space. In Fig. 1 we plot the likelihood contours in the $\Omega_b h^2 - n_S$ plane from the recent CMB data. The likelihood contours are in beautiful agreement with *both* a scale-invariant spectrum of primordial fluctuations, as predicted by inflation, and the value for the baryon density $\omega_b = 0.020 \pm 0.002$ predicted by standard Big Bang nucleosynthesis (see, e.g. [41]). However, a degeneracy between the two parameters is clear: varying the spectral index and the baryon density changes the relative amplitude of the first two peaks in an opposite way and therefore "blue" ($n_S > 1$) and high baryon density models still fit the data well.

An increase in the optical depth τ_c after recombination by reionization (see, e.g. [42] for a review) damps the amplitude of the CMB peaks and is therefore degenerate with the spectral index n_S if one analyses temperature data alone. However, it is also well known that reionization produces a "bump" on the polarization and cross temperature–polarization power spectra on large angular scales due to the new last-scattering surface at late redshifts. This "bump" was detected in the first year WMAP $< TE >$ spectra, yielding a $1 - \sigma$ constraint $\tau = 0.17 \pm 0.04$. This value is too high to be consistent with standard reionization models and a large number of works have investigated this point.

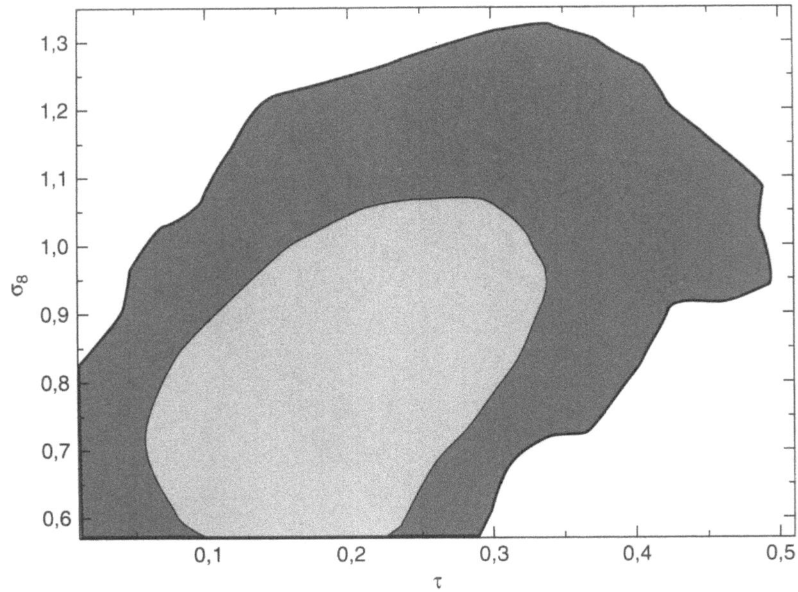

Fig. 2. Likelihood contours in the (τ_c, σ_8) plane for the joint CMB+LSS analysis. The contours correspond to 0.32 and 0.05 of the peak value of the likelihood, which are the 68 and 95% confidence levels, respectively

In Fig. 2 we plot the constraints from a combination of CMB and galaxy clustering data from the 2dF survey [43]. This analysis constrains τ and σ_8, a number which defines the amount of dark matter clustering on spheres of $\sim 8\,\mathrm{Mpc}/h$. Once the four parameters of the model are fixed one can indeed derive the corresponding expected value of σ_8. As shown in the contour plot a degeneracy is present between τ and σ_8. An increase in τ makes models with high n_s (and therefore higher clustering on small scales) more compatible with the CMB data. Values of $\tau \sim 0.2$ are compatible with values of $\sigma_8 \sim 1$ and, in principle, an accurate measurement of σ_8 could bring more information to the "high" τ problem. However, recent analyses of the local cluster number counts give different results for σ_8 mainly due to systematics in the calibration between cluster virial mass and temperature: a *high* value $\sim \Omega_m^{0.6}\sigma_8 = 0.55 \pm 0.05$ is in agreement with the results of [44, 45] and a *lower* one, $\sim \Omega_m^{0.6}\sigma_8 = 0.40 \pm 0.05$ follows from the analyses of [46] and [47]. If future cluster temperature or cosmic shear analyses (see, e.g. [48, 49, 50, 51]) were to converge towards a high σ_8 value, this could lead to a possible discrepancy with the standard reionization model [52, 53]. Future experiments will analyse this interesting result.

The amount of non-baryonic dark matter is also constrained by the CMB data with $\Omega_{dm}h^2 = 0.134 \pm 0.006$ at 68% c.l. [33]. The presence of power around the third peak is crucial in this sense, since it cannot be easily accommodated in models based on just baryonic matter (see, e.g. [53, 54, 55]).

Furthermore, under the assumption of flatness, we can derive important constraints on the age of the universe t_0 given by (see, e.g. [56, 57])

$$t_0 = 9.8Gy \int_0^1 \frac{ada}{[\omega_m a + \omega_\Lambda a^4]^{1/2}} \,. \tag{11}$$

Again, the WMAP team reported an age of $t_0 = 13.7 \pm 0.2$ Gyrs from a combined CMB+LSS analysis consistent with other independent constraints derived for example from stellar populations in bright ellipticals [57], ^{238}U age measurement of old halo star in our galaxy [58] and age of the oldest halo globular cluster [59].

4 Open Questions

Even if the present CMB observations can be explained with just four parameters it is interesting to extend the analysis to other parameters allowed by the theory. Here we will just summarize a few of them and discuss how well we can constrain them and what the effects on the results obtained in the previous section would be.

4.1 Gravity Waves

The metric perturbations created during inflation belong to two types: *scalar* perturbations, which couple to the stress-energy of matter in the universe and form the "seeds" for structure formation and *tensor* perturbations, also known as gravitational wave perturbations. Both scalar and tensor perturbations contribute to CMB anisotropy. A sizable background of gravity waves is expected in most inflationary scenarios. Furthermore, in the simplest models, a detection of the GW background can provide information on the second derivative of the inflaton potential and shed light on the physics at $\sim 10^{16}$ Gev (see, e.g. [60]).

The shape of the C_ℓ^T spectrum from tensor modes is drastically different from the one expected from scalar fluctuations, affecting only large angular scales (see, e.g. [61]). The effect of including tensor modes is similar to just a rescaling of the degree-scale COBE normalization and/or a removal of the corresponding data points from the analysis.

This further increases the degeneracies among cosmological parameters, affecting mainly the estimates of the baryon and cold dark matter densities and the scalar spectral index n_S (see, e.g. [62, 63]).

The amplitude of the GW background is therefore weakly constrained by the CMB data alone, however, when information from BBN, local cluster abundance and galaxy clustering are included, an upper limit of about $r = A^T/A^S < 0.4$ is obtained (see, e.g. [33, 64]).

4.2 Scale-Dependence of the Spectral Index

The theoretical possibility of a scale dependence of the scalar spectral index, $n_S(k)$, has been considered in various works (see, e.g. [65, 66, 67, 68]). Even though this dependence is considered to have small effects on CMB scales in most of the slow-roll inflationary models, it is worthwhile to see if any useful constraint can be obtained. Allowing the power spectrum to bend nearly erases the ability of the CMB data to measure the tensor-to-scalar perturbation ratio and enlarges the uncertainties on many cosmological parameters.

In several recent works (see, e.g. [33, 64, 69, 70]) the case of a running spectral index has been studied, expanding the power spectrum $P(k)$ to second order in $ln(k)$. These analyses indicate that a *negative* running is consistent with CMB and galaxy clustering data but with no significant evidence for it.

On the other hand, a *positive* running, predicted by running-mass inflation models, is strongly constrained from the present data ([71, 72])

Furthermore, phase transitions associated with spontaneous symmetry breaking during the inflationary era could result in the breaking of the scale invariance of the primordial density perturbation. In [73, 74] and [75] the possibility of having step or bump-like features in the spectrum has also been considered.

4.3 Dark Energy

We saw in the previous section that current data strongly suggest the presence of a dark energy component. This discovery may turn out to be one of the most important contribution to physics in our generation. Hence it is particularly important to consider all possible schemes.

A true cosmological constant Λ may be at work here. However, as it is well known it is difficult to associate the small observed value of the cosmological constant $\rho_\Lambda \sim (10^{-3}\,\mathrm{eV})^4$ with vacuum fluctuations in scalar field theories, which, for example, for bosonic and fermionic fields would lead to an effective cosmological constant of $\rho_\Lambda \sim 10^{76}\,\mathrm{GeV}^4$, i.e. 123 orders of magnitude larger. Moreover, the cosmological constant immediately introduces a "why now" problem, since an extreme fine-tuning of initial conditions is required in order to obtain $\rho_\Lambda \sim \rho_m$ *today*: already at redshift $z \sim 2$ the cosmological constant is subdominant, while at the time of the electroweak phase transition $\rho_\Lambda/\rho_m \sim 10^{-55}$.

Systematics in the data are most probably under control: combined analyses of CMB, LSS and SNe-Ia data yield $\Omega_\Lambda = 0.74 \pm 0.04$, better than a 14σ detection. The SNe-Ia alone is highly inconsistent with $\Omega_\Lambda = 0$ if one considers flat universes or open with $\Omega_M > 0.1$. The CMB data alone is also inconsistent with $\Omega_\Lambda = 0$ unless one considers closed models with $\Omega_M \sim 1.3$ and a very low Hubble parameter $h \sim 0.4$ which, again, are incompatible with several complementary data sets.

Assuming modifications to the model of structure formation which are not connected with a new form of energy, like, for example, a contribution from isocurvature perturbations, does not seems able to mimic Λ or a dark energy component (see, e.g. [76]).

Alternatives to Λ

A complete treatment of the possible dark energy candidates can be found in several and excellent recent reviews (see, e.g. [77, 78, 79, 80]). The important point is that dark energy candidates have an equation of state parameter which can be different from -1 and varies with time compared to that of a cosmological constant which remains fixed at $w_\Lambda = -1$. Thus, observationally distinguishing a time variation in the equation of state or finding w_X different from -1 will rule out a pure cosmological constant as an explanation for the data, but be consistent with a dynamical solution.
Here we mention few models according to the expected values of their equation of state.

Topological Defects, $-1/3 \geq w_X \geq -2/3$

Dark energy can receive contributions from topological defects produced at phase transitions in the early universe (see, e.g. [81, 82]). However, despite a well-established theoretical framework, topological defects have not been thoroughly explored due to technical difficulties in the numerical simulations. More recently, a plausible version of dark energy made of a frustrated network of domain walls was proposed by [83] (see also [84]). These models have several appealing features: First, topological defects are ubiquitous in field theory and unavoidable in models with spontaneously broken symmetries. Second, the scale of spontaneous symmetry breaking responsible for the walls is expected to lie in the 10–100 KeV range and can arise naturally in supersymmetric theories [85]. Finally, a firm phenomenological prediction can be made for domain wall models: an equation of state strictly $-1/3 \geq w_X \geq -2/3$ (see, e.g. [85]). These models are therefore predictive in the value of the equation of state parameter and distinguishable from a cosmological constant even at zero order on w_X (while, for example, scalar field models can also produce $w_X \sim -1$ although they differ from a cosmological constant which in the first-order variation has $\dot{w}_X = 0$).

Scalar Fields - Quintessence $-2/3 \geq w_X \geq -1$

It is entirely possible that a dynamic mechanism is giving rise to the observed acceleration of the present universe. Some of the popular proposed candidates to explain the observations are a slowly-rolling scalar field, "quintessence" [86, 87], or a "k-essence" scalar field with non-canonical kinetic terms [88, 89]. An important property of these models is that the fine-tuning ("why now") problem can be in principle alleviated as the equation of state is time dependent. Several models have been proposed and a complete study of all the related potentials goes well beyond the present work. As mentioned, the most general prediction is a value for the equation of state $w(z)$ that differs from unity and varies with redshift z. A second way to distinguish between scalar field candidates is to measure the sound speed of the dark energy component that affects the perturbations in its energy distribution. The sound speed in many models of quintessence is equal to the speed of light, however, it can be different from c, for example, in k-essence models, where it varies, triggered by the transformations in the background equation of state.

Phantom or Super-Quintessence, $-1 \geq w_X$

As we will see in the section, the present data sets do not rule out but seem to prefer $w_X < -1$. Scalar field models with such equation of state (known as "phantom" or super-quintessence models) deserve a separate discussion since they cannot be achieved by scalar fields with positive kinetic energy term. The limitation to $w_X > -1$ is indeed a theoretical consideration motivated, for example, by imposing on matter (for positive energy densities) the null energy condition, which states that $T_{\mu\nu}N^{\mu}N^{\nu} > 0$ for all null 4-vectors N^{μ}. Such energy conditions are often demanded in order to ensure stability of the theory. However, theoretical attempts to obtain $w_X < -1$ have been considered [90, 91, 92, 93, 94]. Unstable at quantum level, a careful analysis of their potential instabilities has been performed in [95]. Moreover, the expansion factor of a universe dominated by phantom energy diverges in a finite amount of cosmic time, culminating in a future curvature singularity (Big Rip [96] or Big Smash [97] phase).

Chaplygin Gases $w_X = -1$ Today, $w_X = 0$ Yesterday

The Chaplygin Gas (CG) (see, e.g. [98, 99, 100]) provides an interesting possibility for a unified picture of dark energy and dark matter. Such a component interpolates in time between dust ($w_X = 0$) and a cosmological constant ($w_X = -1$), with an intermediate behaviour as $p = \alpha\rho$. Perturbations of this fluid are stable on small scales, but behave in a very different way with respect to standard quintessence. Analysis of the effect of those perturbations on CMB and LSS data, in particular, has strongly constrained CG, disfavouring it as an unified dark matter candidate (see, e.g. [101]).

Current Constraint on w Constant with Redshift

In Fig. 3 we plot the likelihood contours in the (Ω_M, w_X) plane from our joint analyses of CMB+SNe-Ia+HST+LSS data. As we can see, there is strong supporting evidence for dark energy. A cosmological constant with $w_X = -1$ is in good agreement with all the data. However, the $2-\sigma$ confidence levels are $-1.32 < w_X < -0.82$ with a best-fit value of $w_X \sim 1.04$, slightly preferring "phantom" models.

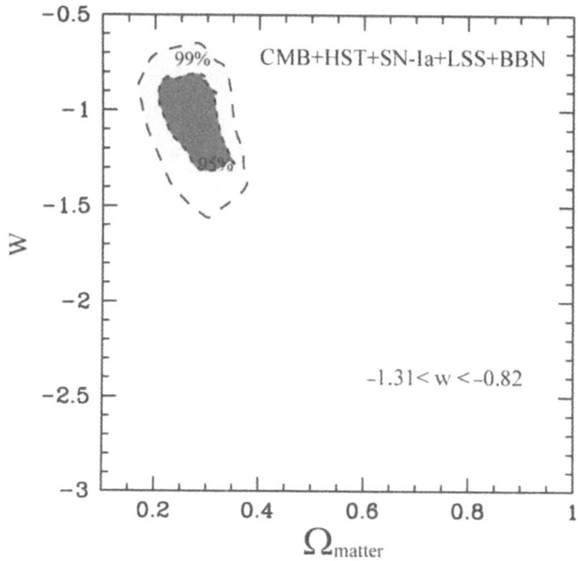

Fig. 3. Likelihood contours in the (Ω_M, w_X) plane for the joint CMB+HST+SNe-Ia+LSS analysis described in the text. We take the best-fit values for the remaining parameters. The contours correspond to 0.05 and 0.01 of the peak value of the likelihood, which are the 95 and 99% confidence levels, respectively

While the analysis rules out topological defects as dark energy, it is important to note that this result is almost completely due to the inclusion of the Supernovae Type-Ia dataset. Topological defects can provide a good fit to the WMAP data for a different choice of priors with "lower" values of the Hubble parameter ($h < 0.65$) (as indicated by Sunyaev–Zeldovich and time delays for gravitational lensing observations) and "higher" values of the matter density ($\Omega_m > 0.35$) (in agreement with recent measurements of the temperature–luminosity relation of distant clusters observed with the XMM-Newton satellite) (see [102]).

Fit with a Varying-with-Redshift Equation of State

A cosmological constant is compatible with our analysis but this result may be biased by the assumption that the value of the equation of state remains constant with redshift. Such a behaviour is indeed not expected in most plausible models of dark energy. For example in models with power-law potentials like $V = V_0/\phi^\alpha$, the equation of state generally remains closer to the background value $w_X = \alpha w_B/(\alpha + 2)$ while the ratio of the energy density of the scalar field to that of the dominant component gradually increases. In models based on exponential potentials $V = V_0 e^{-\lambda\phi}$ $w_X(z)$ mimics exactly the scaling of the dominant background in the attractor regime ($w_X = w_B$) and if the background component scales as $\rho_B = \rho_0(\frac{a_0}{a})^n$, then the scaling field approaches an attractor solution, and its fractional energy density is given by $\Omega_X = \frac{n}{\lambda^2}$, constant with redshift. In "k-essence" models, the dark energy undergoes two transitions: one at the onset of matter domination and the second when the k-essence density begins to dominate over the matter density. During the radiation-dominated era, the k-essence energy tracks the radiation, falling as $1/a^4$ where a is the scale factor. At the onset of the matter-dominated era it begins to act as an energy component with $w_X(z) \leq 0$. When k-essence overtakes the matter density, $w_X(z)$ changes to another value around -1, the precise value of which depends on the detailed model.

Let us now analyse the data using a phenomenological model with a value of the equation of state that varies with redshift. We consider a model where the energy density evolves with redshift as

$$\rho_X(z) = A\left(1+\frac{1+z}{1+z_s}\right)\left[1+\left(\frac{1+z}{1+z_b}\right)^3\right] , \tag{12}$$

where A is a normalization constant, related to the present dark energy density Ω_X, and z_s and z_b are two free parameters that identify the three epochs of scaling. The corresponding dark energy equation of state becomes

$$w_X(z) = \frac{\left[\left(\frac{1+z}{1+z_b}\right)^3 - 2\right]\frac{1+z}{1+z_s} - 3}{3\left(1+\frac{1+z}{1+z_s}\right)\left[1+\left(\frac{1+z}{1+z_b}\right)^3\right]} , \tag{13}$$

which depends on the parameters z_s and z_b such that

$$w_X \sim 1/3 \text{ for } z >> z_s ,$$

$$w_X \sim 0 \text{ for } z_b << z << z_s ,$$

$$w_X \sim -1 \text{ for } z << z_b .$$

The model we obtain is able to mimic a fluid following first a radiation equation of state, then a matter phase and then approaches a cosmological constant phase with constant energy. This model is useful to identify and/or constrain a cosmological imprint of scaling dark energy in the data. The same model can be further extended to a fluid with transitions between more generic equations of state.

In Fig. 4, likelihood contours are plotted in the (z_s, z_b) plane. Those were obtained from CMB data only. The current data does not show any evidence for scaling dark energy and we can only derive weak lower limits on the two parameters of the model. We find that $z_s > 60$ and $z_b > 3.8$ at $1-\sigma$. An interesting correlation exists between the two parameters. Namely, the current CMB observations do not favour or provide evidence for an extra matter or radiation component: if z_s is too small then it is necessary to consider larger

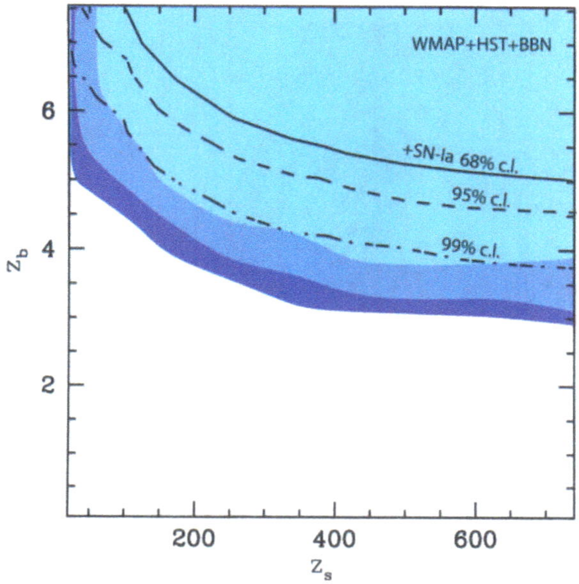

Fig. 4. Likelihood probability contours at 68% (*light blue*), 95% (*blue*) and 99% (*dark blue*) in the $z_s - z_b$ plane from WMAP. The solid, long-dashed and short-dashed lines are the 68, 95 and 99% from a WMAP+SNe-Ia analysis

values of z_b in order to have the extra dark energy component not dominant in the past. At high z_s, we have an asymptotic value 95% c.l. of $z_b > 3.0$. Likelihood contours derived from a combined CMB+SNe-Ia analysis are also plotted. The inclusion of the SNe-Ia data improves the constraints to $z_b > 5$ and $z_s > 100$ at $1-\sigma$. SNe-Ia data are insensitive to variations in z_s since they are probing only redshifts $z < 1.5$, but they provide anyway complementary constraints on z_b and the matter density Ω_m.

This is better explained in Fig. 5, where we superimpose the likelihood contours in the (Ω_m, z_b) plane from CMB and SNe-Ia analysis. The current SNe-Ia data does not provide evidence for dark energy evolution. Low values of z_b are compatible with the SNe-Ia data if one decreases the amount of the matter component. The SNe-Ia data is consistent with $\Omega_m = 0$ but in this case the dark energy model behaves like an unified dark energy model with transition redshift $z_b \sim 0.5$. On the other hand, lower values of z_b are compatible with CMB data if one increases the matter density. This is easily explained from the fact that a lower z_b results in higher values for the effective dark energy equation of state. The direction of degeneracy in the plot in the case of the CMB data is therefore only a consequence of the geometrical degeneracy present in angular diameter distance data at high redshift. The lower limit on z_b from the CMB data comes mainly from our assumptions on the possible values of the Hubble parameter. Combining the CMB and SNe-Ia

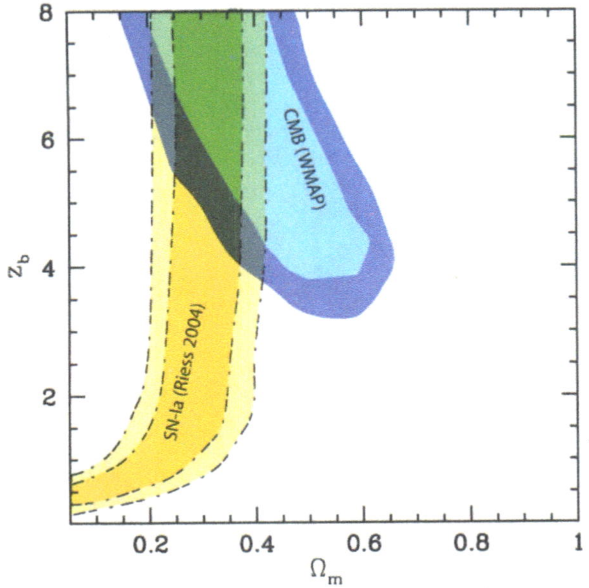

Fig. 5. 68 and 95% likelihood probability contours in the $\Omega_m - z_b$ plane from WMAP and SNe-Ia

further breaks this degeneracy, improving the lower limit on z_b and excluding at high significance an unified dark energy model with $\Omega_m = 0$.

4.4 Big Bang Nucleosynthesis and Neutrinos

In the previous section we saw that the SBBN 95% C.L. region, corresponding to $\Omega_b h^2 = 0.020 \pm 0.002$ (95% c.l.), has a large overlap with the analogous CMBR contour. This fact, if confirmed by future experiments on CMB anisotropies, can be seen as one of the greatest successes of the standard hot Big Bang model so far.

SBBN is well known to provide strong bounds on the number of relativistic species N_ν. On the other hand, degenerate BBN (DBBN), first analysed in [103, 104, 105, 106], gives a very weak constraint on the effective number of massless neutrinos, since an increase in N_ν can be compensated by a change in both the chemical potential of the electron neutrino, $\mu_{\nu_e} = \xi_e T$, and $\Omega_b h^2$. Practically, SBBN relies on the theoretical assumption that background neutrinos have negligible chemical potential, just like their charged lepton partners. Although this hypothesis is justified by Occam's razor, models have been proposed [107, 108, 109, 110, 111, 112, 113] where large neutrino chemical potentials can be generated. It is therefore an interesting issue for our understanding of fundamental interactions, to try to constrain the neutrino–antineutrino asymmetry with the cosmological observables. It is well known that degenerate BBN yields severe constraints on the electron neutrino chemical potential, $-0.06 \leq \xi_e \leq 1.1$, and weaker bounds on the chemical potentials of both the μ and τ neutrino, $|\xi_{\mu,\tau}| \leq 5.6$–$6.9$ [106], since electron neutrinos are directly involved in neutron to proton conversion processes which eventually fix the total amount of ^{4}He produced in nucleosynthesis, while $\xi_{\mu,\tau}$ only enters via their contribution to the expansion rate of the universe. We recall that the neutrino chemical potentials contribute to the total neutrino effective degrees of freedom N_ν as

$$N_\nu = 3 + \Sigma_\alpha \left[\frac{30}{7} \left(\frac{\xi_\alpha}{\pi} \right)^2 + \frac{15}{7} \left(\frac{\xi_\alpha}{\pi} \right)^4 \right] . \tag{14}$$

Note that in order to get a bound on ξ_α we assume that all relativistic degrees of freedom, other than photons, are given by three (possibly) degenerate active neutrinos.

A comparison of BBN and CMB constraints in the N_ν, ω_b plane is useful in order to test the overall BBN scenario. Such an analysis was performed in [114, 115, 116, 117] using the first data release of BOOMERanG and MAXIMA ([118, 119]) and more recently in [120, 121] and [122]. The current data does not require modifications to the standard Big Bang nucleosynthesis with $N_\nu <$ 7.1.

Some caution is naturally necessary when comparing the effective number of neutrino degrees of freedom from BBN and CMB, since they may be related to different physics. In fact the energy density in relativistic species may change from the time of BBN ($T \sim$ MeV) to the time of last rescattering ($T \sim$ eV). In this respect the bound on N_ν can be more generally considered as an upper limit on a possible background of extra relativistic particles.

4.5 Varying α

There are several experimental constraints on the value of the fine-structure constant α. These measurements cover a wide range of timescales (see [123] for a review of this subject): present-day laboratories ($z \sim 0$), geophysical tests ($z << 1$), quasars ($z \sim 1$–3), the CMB ($z \sim 10^3$) and BBN ($z \sim 10^{10}$) bounds.

The recent analysis of [124] of fine splitting of quasar doublet absorption lines gives a 4σ evidence for a time variation of α, $\Delta\alpha/\alpha = (-0.72 \pm 0.18) 10^{-5}$, for the redshift range $z \sim 0.5$–3.5. This positive result was obtained using a many-multiplet method, which achieves an order of magnitude greater precision than the alkali doublet method. Some of the initial ambiguities of the method have been tackled by the authors with an improved technique, in which a range of ions is considered, with varying dependence on α. This helps reduce possible problems such as varying isotope ratios, calibration errors and possible Doppler shifts between different populations of ions [125, 126, 127, 128].

The results of present analyses of the dependence of α on relevant cosmological observables like the anisotropy of CMB, large-scale structure and the light element primordial abundances does not support variations of the fine-structure constant (see [129] and [130] and references therein).

4.6 Isocurvature Modes

Another assumption in the analyses above is that the primordial fluctuations were adiabatic. Adiabaticity is not a necessary consequence of inflation and many inflationary models have been constructed where isocurvature perturbations are a natural consequence (see, e.g. [131, 132, 133]).

In a phenomenological approach one should consider the most general primordial perturbation, introduced by [134] and described by a $5X5$ symmetric matrix-valued generalization of the power spectrum. As showed by [134], the inclusion of isocurvature perturbations with auto and cross-correlations modes has dramatic effects on standard parameter estimation with relative uncertainties becoming of order one.

Even assuming priors such as flatness, the inclusion of isocurvature modes significantly enlarges our constraints on the baryon density [135, 136]. Pure isocurvature perturbations are highly excluded by present CMB data [137].

It is also possible to have *active* and *decoherent* perturbations such as those produced by an inhomogeneously distributed form of matter like topological defects. Models based on global defects like cosmic strings and textures are excluded at high significance by the present data (see, e.g. [138]). However a mixture of adiabatic+defects is still compatible with the observations [139, 138]. In principle, toy models based on *active* perturbations can be constructed [140] that can mimic inflation and retain a good agreement with observations [141].

5 Conclusions

Recent CMB data represent a successful confirmation of the standard cosmological model. The acoustic oscillations in the CMB angular power spectrum, a major prediction of the model, have now been detected with high statistical significance. Furthermore, when constraints on cosmological parameters are derived under the assumption of adiabatic primordial perturbations their values are in agreement with the predictions of the theory and/or with independent observations.

Modifications to the standard model such as isocurvature modes or topological defects are still compatible with current CMB observations but are not required and can be reasonably constrained when complementary data sets are included.

Since the inflationary scenario is in agreement with the data and all the most relevant parameters are starting to be constrained within a few per cent accuracy, the CMB is becoming a powerful laboratory for the investigation of new physics. With the promise of large data sets from WMAP, Planck and SNAP satellites and from the SLOAN digital sky survey, opportunities may be open, for example, to constrain dark energy models, variations in fundamental constants and neutrino physics.

Acknowledgements

We like to thank the director of the school "Data Analysis in Cosmology", Vicent J. Martinez and the co-directors Enrique Martinez-Gonzalez and Enn Saar. CJO is supported by a Marie Curie Intra-European Fellowship, grant 501007. Our research is supported by the MIUR "GEMINI" program. We acknowledge the use of the Legacy Archive for Microwave Background Data Analysis (LAMBDA) from where some figures for this work have been taken. Support for LAMBDA is provided by the NASA Office of Space Science.

References

1. Peebles, P.J.E., Yu, J.T.: Astrophys. J. **162**, 815 (1970)
2. Sunyaev, R.A., Zeldovich, Ya.B.: Astrophys. Space Sci. **7**, 3 (1970)
3. Wilson, M.L., Silk, J.: Astrophys. J. **243**, 14 (1981)
4. Vittorio, N., Silk, J., Astrophys. J. **285L**, 39 (1984)
5. Bond, J.R., Efstathiou, G.: Astrophys. J. **285**, L45 (1984)
6. Hu, W., Sugiyama, N., Silk, J., Nature **386**, 37 (1997), astro-ph/9604166
7. Hu, W., Scott, D., Sugiyama, N., White, M.J.: Phys. Rev. **D 52**, 5498 (1995), astro-ph/9505043
8. White, M.J., Scott, D., Silk, J.: Ann. Rev. Astron. Astrophys. **32**, 319 (1994)
9. Bond, J.R.: Class. Quant. Grav. **15**, 2573 (1998)
10. Durrer, R.: J. Phys. Stud. **5**, 177 (2001), astro-ph/0109522
11. Torbet, E., et al.: Astrophys. J. **521**, L79 (1999), astro-ph/9905100
12. Miller, A.D., et al.: Astrophys. J. **524**, L1 (1999), astro-ph/9906421
13. Mauskopf, P.D., et al. (Boomerang Collaboration): Astrophys. J. **536**, L59 (2000), astro-ph/9911444
14. Dodelson, S., Knox, L.: Phys. Rev. Lett. **84**, 3523 (2000), astro-ph/9909454
15. Melchiorri, A., et al. (Boomerang Collaboration): Astrophys. J. **536**, L63 (2000), astro-ph/9911445
16. Tegmark, M.: Astrophys. J. **514**, L69 (1999), astro-ph/9809201
17. Netterfield, C.B., et al. (Boomerang Collaboration): Astrophys. J. **571**, 604 (2002), astroph/0104460
18. Halverson, N.W., et al.: Astrophys. J. **568**, 38 (2001), astro-ph/0104489
19. Lee, A.T., et al.: Astrophys. J. **561**, L1 (2001), astro-ph/0104459
20. de Bernardis, P., et al. (Boomerang Collaboration): Astrophys. J. **564**, 559 (2002), astroph/0105296
21. Wang, X., Tegmark, M., Zaldarriaga, M.: Phys. Rev. **D 65**, 123001 (2002), astroph/0105091

22. Pryke, C., Halverson, N.W., Leitch, E.M., Kovac, J., Carlstrom, J.E., Holzapfel, W.L., Dragovan, M.: Astrophys. J. **568**, 46 (2002), astro-ph/0104490
23. Pearson, T.J., et al.: Astrophys. J. **591**, 556 (2003), astro-ph/0205388
24. Kuo, C.l.: et al. (ACBAR collaboration): Astrophys. J. **600**, 32 (2004), astroph/0212289
25. Grainge, K., et al.: Mon. Not. Roy. Astron. Soc. **341**, L23 (2003), astroph/0212495
26. Dickinson, C., et al.: MNRAS **353**, 732 (2004), astro-ph/0402498
27. Benoit, A., et al.: (Archeops Collaboration), Astron. Astrophys. **399**, L19 (2003), astro-ph/0210305
28. Tristram, M., et al.: Astron. Astrophys. **436**, 785 (2005), astro-ph/0411633
29. Kovac, J., Leitch, E.M., Pryke, C., Carlstrom, J.E., Halverson, N.W., Holzapfel, W.L., Nature **420**, 772 (2002), astro-ph/0209478
30. Leitch, E.M., Kovac, J.M., Halverson, N.W., Carlstrom, J.E., Pryke, C., Smith, M.W.E.: Astrophys. J. **624**, 10 (2005), astro-ph/0409357
31. Readhead, A.C.S., et al.: Science **306**, 836 (2004), astro-ph/0409569
32. Bennett, C.L., et al.: Astrophys. J. Suppl. **148**, 1 (2003), astro-ph/0302207
33. Spergel, D.N., et al.: Astrophys. J. Suppl. **148**, 175 (2003), astro-ph/0302209
34. Marshall, P., Rajguru, N., Slosar, A.: Phys. Rev. D. **73**, 067302 (2006), astroph/0412535
35. Häggström, O.: Finite Markov Chains and Algorithmic Applications London Math. Soc. Student Texts 52, Cambridge University Press, Cambridge (2002)
36. Gelman, A., Rubin, D.: Stat. Sci. **7**, 457 (1992)
37. Efstathiou, G., Bond, J.R.: MNRAS **304**, 75 (1999), astro-ph/9807103
38. Melchiorri, A., Griffiths, L.M., New Astron. Rev. **45**, 321 (2001), astroph/0011147
39. Uzan, J.P., Kirchner, U., Ellis, G.F.R.: Mon. Not. Roy. Astron. Soc. **344**, L65 (2003), astro-ph/0302597
40. Luminet, J.P., Weeks, J., Riazuelo, A., Lehoucq, R., Uzan, J.P., Nature **425**, 593 (2003), astro-ph/0310253
41. Burles, S., Nollett, K.M., Turner, M.S.: Astrophys. J. **552**, L1 (2001), astro-ph/0010171
42. Haiman, Z., Knox, L.: In: de Oliveira, A., Tegmark, M. (eds.) Microwave Foregrounds. ASP, San Francisco, p. 227 (1999), astro-ph/9902311
43. Tegmark, M., Hamilton, A.J.S., Xu, Y.: MNRAS **335**, 887 (2002), astroph/0111575 (2001)
44. Pierpaoli, E., Scott, D., White, M.: MNRAS **325**, 77 (2001), astro-ph/0010039
45. Eke, V.R., Cole, S., Frenk, C.S.: MNRAS, **282**, 263 (1996)
46. Borgani, S., et al.: Astrophys. J. **561**, 13, astro-ph/0106428 (2001)
47. Viana, P.T.P., Nichol, R.C., Liddle, A.R.: Astrophys. J. **569**, L75 (2002), astroph/0111394
48. Van Waerbeke, L., et al.: Astron. Astrophys. **374**, 757 (2001)
49. Maoli, R., et al.: Astron. Astrophys. **368**, 766, 2001
50. Refregier, A., et al.: Astrophys. J. **572**, L131 (2002), astro-ph/0203131
51. Bacon, D., et al.: MNRAS **344**, 673 (2003), astro-ph/0203134
52. Lahav, O., et al.: MNRAS **333**, 961 (2002), astro-ph/0112162
53. Melchiorri, A., Silk, J.: Phys. Rev. **D 66**, 041301 (2002), astro-ph/0203200
54. Griffiths, L.M., Melchiorri, A., Silk, J.: Astrophys. J. **553**, L5 (2001), astroph/0101413

55. McGaugh, S.S.: Astrophys. J. **541**, L33 (2000), astro-ph/0008188
56. Knox, L., Christensen, N., Skordis, C.: Astrophys. J. **563**, L95 (2001), astro-ph/0109232
57. Ferreras, I., Melchiorri, A., Silk, J.: MNRAS, **327**, L47 (2001), astroph/0105384
58. Cayrel, R., et al.: Nature **409**, 691–692 (2001)
59. Salaris, M., Weiss, A.: Astron. Astrophys. **335**, 943–953 (1998)
60. Hoffman, M.B., Turner, M.S.: Phys. Rev. **D64**, 023506 (2001), astro-ph/0006312
61. Crittenden, R., Bond, J.R., Davis, R.L., Efstathiou, G., Steinhardt, P.J.: Phys. Rev. Lett. **71**, 324 (1993), astro-ph/9303014
62. Melchiorri, A., Sazhin, M.V., Shulga, V.V., Vittorio, N.: Astrophys. J. **518**, 562 (1999), astro-ph/9901220
63. Efstathiou, G.: MNRAS **332**, 193 (2002), astro-ph/0109151
64. Kinney, W.H., Kolb, E.W., Melchiorri, A., Riotto, A.: Phys. Rev. **D 69**, 103516 (2004), hep-ph/0305130
65. Kosowsky, A., Turner, M.S.: Phys. Rev. **D 52**, 1739 (1995), astroph/9504071
66. Copeland, E.J., Grivell, I.J., Liddle, A.R.: MNRAS **298**, 1233 (1998), astroph/9712028
67. Lyth, D.H., Covi, L.: Phys. Rev. **D 62**, 103504 (2000), astro-ph/0002397
68. Dodelson, S., Stewart, E.: Phys. Rev. **D 65**, 101301 (2002), astro-ph/0109354
69. Hannestad, S., Hansen, S.H., Villante, F.L.: Astropart. Phys. **16**, 137 (2001), astro-ph/0012009
70. Hannestad, S., Hansen, S.H., Villante, F.L., Hamilton, A.J.: Astropart. Phys. **17**, 375 (2002), astro-ph/0103047
71. Covi, L., Lyth, D.H.: MNRAS **326**, 885 (2001), astro-ph/0008165
72. Covi, L., Lyth, D.H., Melchiorri, A., Odman, C.J.: Phys. Rev. **D 70**, 123521 (2004), astro-ph/0408129
73. Barriga, J., Gaztanaga, E., Santos, M.G., Sarkar, S.: Mon. Not. Roy. Astron. Soc. **324**, 977 (2001), astro-ph/0011398
74. Griffiths, L.M., et al.: MNRAS **324**, 712 (2001), astro-ph/0010571
75. Wang, Y., Mathews, G.: Astrophys. J. **573**, 1 (2002), astro-ph/0011351
76. Trotta, R., Riazuelo, A., Durrer, R.: Phys. Rev. **D 67**, 063520 (2003), astroph/0211600
77. Peebles, P.J.E., Ratra, B.: Rev. Mod. Phys. **75**, 559 (2003), astroph/0207347
78. Sahni, V.: Lect. Notes Phys. **653**, 141 (2004), astro-ph/0403324
79. Padmanabhan, T.: Phys. Rept. **380**, 235 (2003), hep-th/0212290
80. Trodden, M., Carroll, S.M., SU-GP-04/1-1 (2004), astro-ph/0401547
81. Vilenkin, A.: Phys. Rev. Lett. **53**, 1016 (1984)
82. Durrer, R., Kunz, M., Melchiorri, A.: Phys. Rev. **D 59**, 123005 (1999)
83. Bucher, M., Spergel, D.N.: Phys. Rev. **D 60**, 043505 (1999), astroph/9812022
84. Eichler, D., Astrophys. J. **468**, 75 (1996)
85. Friedland, A., Murayama, H., Perelstein, M.: Phys. Rev. **D 67**, 043519 (2003), astro-ph/0205520
86. Wetterich, C.: Nucl. Phys. **B 302**, 668 (1988)
87. Caldwell, R.R., Dave, R., Steinhardt, P.J.: Phys. Rev. Lett. **80**, 1582 (1998), astro-ph/9708069
88. Armendariz-Picon, C., Damour, T., Mukhanov, V.: Phys. Lett. **B 458**, 209 (1999), hep-th/9904075
89. Chiba, T., Okabe, T., Yamaguchi, M.: Phys. Rev. **D 62**, 023511 (2000), astro-ph/9912463

90. Caldwell, R.R.: Phys. Lett. **B 545**, 23 (2002), astro-ph/9908168
91. Schulz, A.E., White, M.J.: Phys. Rev. **D 64**, 043514 (2001), astroph/0104112
92. Parker, L., Raval, A.: Phys. Rev. **D60**, 063512 (1999)
93. Frampton, P.H.: Phys. Lett. **B 555**, 139 (2003), astro-ph/0209037
94. Ahmed, M., Dodelson, S., Greene, P.B., Sorkin, R.: Phys. Rev. **D 69**, 103523 (2004), astro-ph/0209274
95. Carroll, S.M., Hoffman, M., Trodden, M.: Phys. Rev. **D 68**, 023509 (2003), astro-ph/0301273
96. Caldwell, R.R., Kamionkowski, M., Weinberg, N.N.: Phys. Rev. Lett. **91**, 071301 (2003)
97. McInnes, B., JHEP **0208**, 029 (2002)
98. Kamenshchik, A., Moschella, U., V Pasquier, Phys. Lett. **B511**, 265 (2001)
99. Bilic, N., Tupper, G.B., Viollier, R.D.: Phys. Lett. **B535**, 17 (2002)
100. Bento, M.C., Bertolami, O., Sen, A.A.: Phys. Rev. **D66**, 043507 (2002)
101. Bean, R., Dore, O.: Phys. Rev. **D 68**, 023515 (2003), astro-ph/0301308
102. Conversi, L., Melchiorri, A., Mersini, L., Silk, J.: Astropart. Phys. **21**, 443 (2004), astro-ph/0402529
103. Doroshkevich, A.G., Novikov, I.D., Sunaiev, R.A., Zeldovich, Y.B.: In: de Jager (ed.) Highlights of Astronomy, p. 318 (1971)
104. Fowler, W.A., Accademia Nazionale dei Lincei, Roma **157**, 115 (1971)
105. Beaudet, G., Goret, P.: Astron. Astrophys. **49**, 415 (1976)
106. Kang, H., Steigman, G.: Nucl. Phys. **B372**, 494 (1992)
107. Di Bari, P., Foot, R.: Phys. Rev. **D63**, 043008 (2001)
108. Affleck, I., Dine, M.: Nucl. Phys **B249**, 361 (1985)
109. Dolgov, A.D., Kirilova, D.P., Moscow, J. Phys. Soc. **1**, 217 (1991)
110. Casas, A., Cheng, W.Y., Gelmini, G.: Nucl. Phys. **B538**, 297 (1999)
111. March-Russell, J., Murayama, H., Riotto, A., JHEP **11**, 015 (1999)
112. McDonald, J.: Phys. Rev. Lett. **84**, 4798 (2000)
113. Foot, R., Thomson, M.J., Volkas, R.R.: Phys. Rev. **D53**, 5349 (1996)
114. Esposito, S., Mangano, G., Melchiorri, A., Miele, G., Pisanti, O.: Phys. Rev. **D 63**, 043004 (2001), astro-ph/0007419
115. Lesgourgues, J., Peloso, M.: Phys. Rev. **D 62**, 081301 (2000), astroph/0004412
116. Hannestad, S.: Phys. Rev. Lett. **85**, 4203 (2000), astro-ph/0005018
117. Orito, M., Kajino, T., Mathews, G.J., Boyd, R.N., astro-ph/0005446
118. de Bernardis, P., et al. (Boomerang Collaboration): Nature **404**, 955 (2000), astro-ph/0004404
119. Hanany, S., et al.: Astrophys. J. **545**, L5 (2000), astro-ph/0005123
120. Hannestad, S., JCAP **0305**, 004 (2003), astro-ph/0303076
121. Pierpaoli, E.: Mon. Not. Roy. Astron. Soc. **342**, L63 (2003), astro-ph/0302465
122. Crotty, P., Lesgourgues, J., Pastor, S.: Phys. Rev. **D 67**, 123005 (2003), astro-ph/0302337
123. Varshalovich, D.A., Potekhin, A.Y., Ivanchik, A.V.: In: Dunford, R.W., Gemmet, D.S., Kanter, E.P., Kraessig, B., Southworth, S.H., Young, L. (eds.) X-Eay and Inner-Shell Processes. AIP, Melville, p. 503 (2000), physics/0004062
124. Murphy, M.T., Webb, J.K., Flambaum, V.V., Dzuba, V.A., Churchill, C.W., Prochaska, J.X., Barrow, J.D., Wolfe, A.M.: MNRAS **327**, 1208 (2001), astroph/0012419
125. Murphy, M.T., Webb, J.K., Flambaum, V.V., Prochaska, J.X., Wolfe, A.M.: MNRAS **327**, 1237 (2001), astro-ph/0012421

126. Carilli, C.L., et al.: Phys. Rev. Lett. **85**, 5511 (2001)
127. Webb, J.K., Murphy, M.T., Flambaum, V.V., Dzuba, V.A., Churchill, C.W., Prochaska, J.X., Barrow, J.D., Wolfe, A.M., Phys, Rev. Lett. **87**, 091301 (2001), astro-ph/0012539
128. Murphy, M.T., Webb, J.K., Flambaum, V.V., Drinkwater, M.J., Combes, F., Wiklind, T.: MNRAS **327**, 1244 (2001), astro-ph/0101519
129. Avelino, P.P., et al.: Phys. Rev. **D 64**, 103505 (2001), astro-ph/0102144
130. Martins, C.J.A., Melchiorri, A., Rocha, G., Trotta, R., Avelino, P.P., Viana, P.: Phys. Lett. **B 585**, 29 (2004), astro-ph/0302295
131. Langlois, D., Riazuelo, A.: Phys. Rev. **D 62**, 043504 (2000)
132. Gordon, C., Wands, D., Bassett, B.A., Maartens, R.: Phys. Rev. **D 63** (2001) 023506, astro-ph/0009131
133. Bartolo, N., Matarrese, S., Riotto, A.: Phys. Rev. **D 64**, 123504 (2001), astro-ph/0107502
134. Bucher, M., Moodley, K., Turok, N.: Phys. Rev. **D 62**, 083508 (2000), astroph/9904231
135. Trotta, R., Riazuelo, A., Durrer, R.: Phys. Rev. Lett. **87**, 231301 (2001)
136. Dunkley, J., Bucher, M., Ferreira, P.G., Moodley, K., Skordis, C.: MNRAS **356**, 925 (2006), astro-ph/0405462
137. Enqvist, K., Kurki-Suonio, H., Valiviita, J.: Phys. Rev. **D 65**, 043002 (2001), astro-ph/0108422
138. Durrer, R., Kunz, M., Melchiorri, A.: Phys. Rept. **364**, 1 (2002), astroph/0110348
139. Bouchet, F.R., Peter, P., Riazuelo, A., Sakellariadou, M.: Phys. Rev. **D 65**, 021301 (2002), astro-ph/0005022
140. Turok, N.: Phys. Rev. Lett. **77**, 4138 (1996), astro-ph/9607109
141. Durrer, R., Kunz, M., Melchiorri, A.: Phys. Rev. **D 63**, 081301 (2001), astroph/0010633

Cosmic Microwave Background Data Analysis: From Time-Ordered Data to Angular Power Spectra

J. Borrill

Computational Research Division, Lawrence Berkeley Laboratory & Space Sciences Laboratory, UC Berkeley, Berkeley, CA, USA
jdborrill@lbl.gov

Abstract. I describe the process of generating sky maps and estimating the power spectra from the raw CMB time stream data.

1 Introduction

The cosmic microwave background (CMB) is made up of primordial photons last scattered when the previously free electrons were first captured by protons some 400,000 years after the Big Bang. The tiny fluctuations in the temperature and polarization of the CMB encode a wealth of cosmological information, testing the viability of competing classes of cosmologies and constraining the fundamental parameters of those that survive. These anisotropies are not only extraordinarily tiny – $O(10^{-5})$ in temperature and $O(10^{-6})$ to $O(10^{-8})$ or less in E- and B-mode polarization, respectively – but also imprinted on an extremely cold photon field, with the expansion of the Universe reducing its characteristic (black-body) temperature from around 3000 K at last scattering to 3 K at detection today. Moreover to realize the full scientific potential of the CMB we need to know the power of the anisotropies in each of its components over the entire range of angular scales, from all sky to arcminute. Increases in the sensitivity and resolution of CMB observations in pursuit of this goal have been accompanied by corresponding increases in data volumes, and this trend will only accelerate as we push towards the elusive B-mode polarization. Table 1 shows this for CMB satellite missions, but the same evolution is also seen in ground-based and balloon-borne experiments. As CMB data sets grow their analysis becomes increasingly challenging, requiring new algorithms and efficient implementations on the most massively parallel supercomputers.

Borrill, J.: *Cosmic Microwave Background Data Analysis: From Time-Ordered Data to Angular Power Spectra*. Lect. Notes Phys. **665**, 261–265 (2009)
DOI 10.1007/978-3-540-44767-2_9

Table 1. The evolution of the order of magnitude numbers of time samples, sky pixel and spectral multipoles for past, present and future CMB satellite missions

Year	Experiment	Time samples $\mathcal{N}_t$	Sky pixels $\mathcal{N}_p$	Spectral multipoles $\mathcal{N}_l$
1989	COBE	10^9	10^3	10^1
2001	WMAP (4 yr)	10^{11}	10^7	10^3
2007	PLANCK	10^{12}	10^8	10^4
2020	CMBpol	10^{14}	10^9	10^4

2 Overview

CMB data analysis is performed typically by iterating over four steps. First the data are pre-processed, including determining where on the sky each detector was pointing for each observation, removing glitches such as cosmic ray hits and characterizing each detector's time stream noise correlations. Second, the time stream data is made into one or more pixelized sky maps. Third, some or all of the six angular power spectra are calculated from the maps. And finally the parameters of a chosen class of cosmologies are constrained from the power spectra. Here we focus on the basic algorithms and some implementation issues for the second and third stages – map making and power spectrum estimation. In what follows upper case variables are matrices and lower case are vectors, except for the vector of spectral multipoles which is represented by $\mathcal{C}$ for consistency with CMB convention. $\mathcal{N}$ is used for the number of a particular data type in descriptions of an algorithm/implementation pair's scaling behaviour; recall that general matrix–matrix operations scale as $\mathcal{N}^3$, matrix–vector as $\mathcal{N}^2$ and vector–vector as $\mathcal{N}$.

3 Map Making

The time-ordered data d_t comprises detector noise and CMB signal

$$d_t = n_t + s_t = n_t + A_{tp}\, s_p \ , \tag{1}$$

where the pointing matrix A_{tp} gives the weight of each pixel p in time sample t. Typically A is a very sparse matrix, with, for example, a single non-zero entry for a total power temperature observation, two non-zeros for a differencing temperature observation and three non-zeros for a total power polarization observation. This simple representation turns out to be very flexible, and can encompass both sky-asynchronous systematics and sky-synchronous but frequency-dependent foregrounds, with the only restriction being that the final pointing matrix must be full rank for the system to be separable.

The Gaussian noise likelihood function can be maximized over all possible sky signals to yield the map-making equations

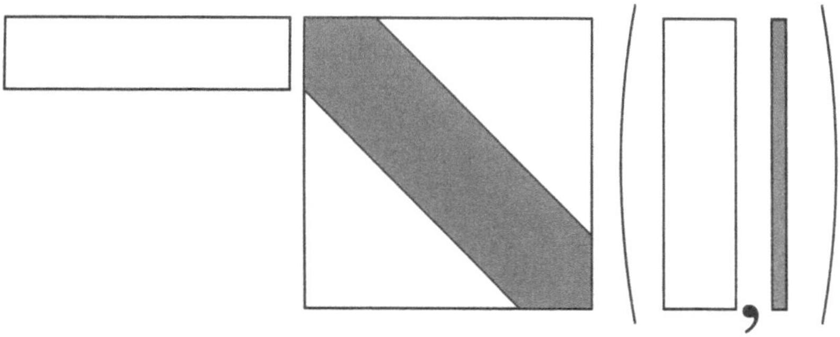

Fig. 1. The structure of the CMB map-making equations

$$\begin{aligned} N^{-1}_{pp'} &= A^T_{tp}\, N^{-1}_{tt'}\, A_{t'p'} \;, \\ z_p &= A^T_{tp}\, N^{-1}_{tt'}\, d_{t'} \;, \\ d_p &= N_{pp'}\, z_{p'} \;. \end{aligned} \tag{2}$$

Naively we would expect a solver for this system to scale as $O(\mathcal{N}_t^2\,\mathcal{N}_p + \mathcal{N}_t^2 + \mathcal{N}_p^3)$ from the matrix–matrix multiplication, matrix–vector multiplication and matrix inversion, respectively. However, as Fig. 1 shows, the matrices are very structured – specifically A_{tp} is sparse and $N_{tt'}$ is Toeplitz (i.e. $N_{tt'} = f(|t-t'|)$ and $f(\Delta) = 0\ \forall\ \Delta > \tau$). The sparsity of A, with $\mathcal{N}_z$ non-zeros, allows the first map-making equation to be solved in $O(\mathcal{N}_t\,\mathcal{N}_z^2)$ operations; the Toeplitz nature of $N_{tt'}$ allows the second map-making equation to be solved in $O(\mathcal{N}_t\,\log\tau)$ operations in Fourier space; and the positive-definiteness of $N^{-1}_{pp'}$ allows the third map-making equation to be solved in $O(\mathcal{N}_i\,\mathcal{N}_t\,\log\tau)$ operations using $\mathcal{N}_i$ iterations of a preconditioned conjugate gradient solver. In practice the white-noise approximation to $N^{-1}_{pp'}$ makes a good preconditioner, giving $\mathcal{N}_i \sim 10^2$ for residuals at the 10^{-6} level, and with a typical noise bandwidth $\tau \sim 10^4$ the floating-point operation (flop) count for Planck-scale map making is reduced from $O(10^{32})$ to $O(10^{15})$ flops, bringing it back into the realm of computational tractability.

4 Power Spectrum Estimation

Given a map of the CMB d_p and its pixel–pixel noise correlations $N_{pp'}$ the log-likelihood of the data given the underlying angular power spectra – under the assumption of a Gaussian, azimuthally symmetric CMB and a uniform Bayesian prior – is

$$\mathcal{L}(d|\mathcal{C}) = -\frac{1}{2}\left(d^T\, D^{-1}\, d + \mathrm{Tr}\left[\ln D\right]\right) \;, \tag{3}$$

where the data correlation matrix $D = N+S$ is the sum of the noise and signal correlations. Since there is no closed-form solution for the spectral coefficients maximizing this function we must use iterative search techniques such as Newton–Raphson. This requires the evaluation of first two derivatives to the log-likelihood functions in order to calculate a quadratic correction to the current estimate:

$$\frac{\partial \mathcal{L}}{\partial \mathcal{C}_l} = \frac{1}{2}\left(d^T D^{-1} \frac{\partial S}{\partial \mathcal{C}_l} D^{-1} d - \mathrm{Tr}\left[D^{-1} \frac{\partial S}{\partial \mathcal{C}_l} \right] \right) ,$$
$$\mathcal{F}_{ll'} \equiv - \left\langle \frac{\partial^2 \mathcal{L}}{\partial \mathcal{C}_l \, \partial \mathcal{C}_{l'}} \right\rangle = \mathrm{Tr}\left[D^{-1} \frac{\partial S}{\partial \mathcal{C}_l} D^{-1} \frac{\partial S}{\partial \mathcal{C}_{l'}} \right] ,$$
$$\delta \mathcal{C}_l = \mathcal{F}_{ll'}^{-1} \frac{\partial \mathcal{L}}{\partial \mathcal{C}_{l'}} . \tag{4}$$

Solving this system scales as $O(\mathcal{N}_l \mathcal{N}_p^3)$ from the $D^{-1}\,\partial S/\partial \mathcal{C}_l$ matrix–matrix multiplications – requiring an impractical $O(10^{28})$ flops for the full Planck data. Although some advantage can be taken of the block structure of the S-derivative matrices (Fig. 2), this only saves an order of magnitude or so in flop count. Given its scaling, maximum-likelihood power spectrum estimation can only practically be used for up to $O(10^5)$ pixels. For data sets larger than this it is restricted to the analysis of reduced resolution maps (critical for low l spectral analysis), full resolution sub-maps (for example chosen for their low foreground contamination) or a hierarchical hybrid or these. In this case an extra level of parallelism can be employed by performing many of the independent matrix–matrix multiplications simultaneously, which allows very large numbers of processors to be used very efficiently.

For situations where maximum-likelihood analysis is intractable Monte Carlo pseudo-spectral methods may be used. In this case we assume that the pseudo-spectrum of the signal and noise are independent and that the pseudo-spectra are related to the real spectra by an invertible linear transform

$$\hat{\mathcal{C}}_l^d = \hat{\mathcal{C}}_l^n + \hat{\mathcal{C}}_l^s \qquad \& \qquad \hat{\mathcal{C}}_l = T_{ll'} \, \mathcal{C}_{l'} , \tag{5}$$

at which point Monte Carlo realizations of simulated signal+noise and noise only observations with the same properties as the actual observation can be used to reconstruct the transfer matrix and hence obtain the real spectra. This approach scales as $O(\mathcal{N}_r \mathcal{N}_i \mathcal{N}_t \log \tau)$ for $\mathcal{N}_r$ Monte Carlo realizations and $\mathcal{N}_i$ map-making iterations or $O(10^{18})$ flops for 1000 realizations of the

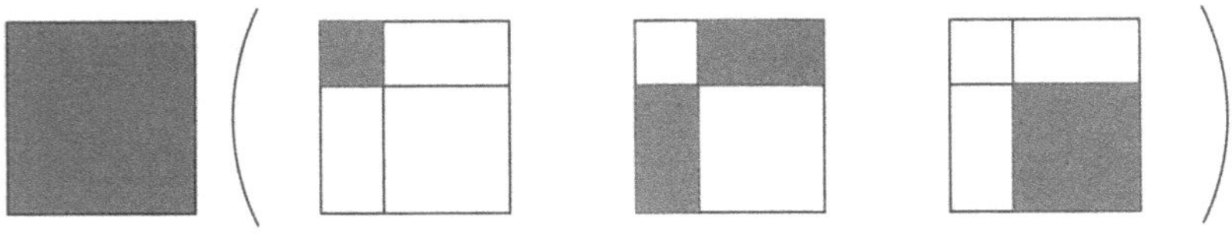

Fig. 2. The structure of the $D^{-1}\,\partial S/\partial \mathcal{C}_l$ matrix–matrix multiplications for the $\mathcal{C}^{TT}$, $\mathcal{C}^{TE,TB}$ and $\mathcal{C}^{EE,BB,EB}$ spectra, respectively

Planck data (although the limiting factor may now be simply generating this number of simulations with sufficient accuracy). It is discussed in greater detail elsewhere in this volume.

Further Reading

1. Bond, J. R., et al.: Radical Compression of Cosmic Microwave Background Data, astro-ph/9808264
2. Borrill, J.: MADCAP – The Microwave Anisotropy Computational Analysis Package, astro-ph/9911389
3. Stompor, R., et al.: Making Maps Of The Cosmic Microwave Background: The MAXIMA Example, astro-ph/0106451
4. Poutanen, T., et al.: Comparison of map-making algorithms for CMB experiments, astro-ph/0501504
5. Hivon, E., et al.: MASTER of the CMB Anisotropy Power Spectrum: A Fast Method for Statistical Analysis of Large and Complex CMB Data Sets, astro-ph/0105302

Part III

Statistics of Large-Scale Structure

The Large-Scale Structure in the Universe: From Power Laws to Acoustic Peaks

V.J. Martínez

Observatori Astronòmic, Universitat de València, Edifici d'Instituts d'Investigació, Polígon La Coma, 46980 Paterna, València, Spain
martinez@uv.es

Abstract. The most popular tools for analysing the large-scale distribution of galaxies are second-order spatial statistics such as the two-point correlation function or its Fourier transform, the power spectrum. In this review, we explain how our knowledge of cosmic structures, encapsulated by these statistical descriptors, has evolved since their first use when applied on the early galaxy catalogues to the present generation of wide and deep redshift surveys.[1]

1 Introduction

As the reader can learn from this volume, there are mainly two astronomical observations that provide the most relevant cosmological data needed to probe any cosmological model: the cosmic microwave background radiation and the large-scale structure of the universe. This review deals with the second of these cosmological fossils. The statistical analysis of galaxy clustering has been progressing in parallel with the development of the observations of the galaxy distribution (for a review see, e.g. Jones et al. [23]). Since the pioneering works by Hubble, measuring the distribution of the number counts of galaxies in telescope fields and finding a log-Gaussian distribution [20], many authors have described the best available data at each moment making use of the then well-established statistical tools. For example, Zwicky [58] used the ratio of clumpiness, the quotient between the variance of the number counts and the expected quantity for a Poisson distribution.

The first map of the sky revealing convincing clustering of galaxies was the Lick survey undertaken by Shane and Wirtanen [52]. While the catalogue was in progress, two different approaches to its statistical description were

[1] Being the first editor of this volume gives me the opportunity of updating this review taking into account the most recent developments in the field. I have used this opportunity trying to incorporate the most challenging discovery in the study of the galaxy distribution: the detection of baryon acoustic oscillations.

Martínez, V.J.: *The Large-Scale Structure in the Universe: From Power Laws to Acoustic Peaks*. Lect. Notes Phys. **665**, 269–289 (2009)
DOI 10.1007/978-3-540-44767-2_10

developed: The Neyman–Scott approach and the Correlation Function school named in this way by Bernard Jones [22].

Jerzy Neyman and Elisabeth Scott were the first to consider the galaxy distribution as a realization of a homogeneous random point process [33]. They formulated a priori statistical models to describe the clustering of galaxies and later they tried to fit the parameters of the model by comparing it with observations. In this way, they modelled the distribution of galaxy clusters as a random superposition of groups following what now is known in spatial statistics as a Neyman–Scott process, i.e. a Poisson cluster process constructed in two steps: first, a homogeneous Poisson process is generated by randomly distributing a set of centres (or parent points); second, a cluster of daughter points are scattered around each of the parent points, according to a given density function. This idea [34, 40] is the basis of the recent halo model [53] that successfully describes the statistics of the matter distribution in structures of different sizes at different scales: at small scales the halo model assumes that the distribution is dominated by the density profiles of the dark matter halos, and therefore correlations come mainly from intra-halo pairs. The most popular density profile is that of Navarro, Frenk and White [32].

The second approach based on the correlation function was envisaged first by Vera Rubin [48] and by Nelson Limber [28]. They thought that the galaxy distribution was in fact a set of points sampled from an underlying continuous density distribution that later was called the Poisson model by Peebles [43]. In spatial statistics this is known as a Cox process [31]. They derived the auto-correlation function from the variance of the number counts of the ongoing Lick survey. Moreover, Limber provided an integral equation relating the angular and the spatial correlation functions valid for small angle separation (a special version of this equation appears also in the paper by Rubin). The correlation function measures the clustering in excess $[\xi(r) > 0]$ or in defect $[\xi(r) < 0]$ compared with a Poisson distribution. It can be defined in terms of the probability dP of finding a galaxy in a small volume dV lying at a distance r of a given galaxy:

$$dP = n[1 + \xi(r)]dV, \tag{1}$$

where n is the mean number density over the whole sample volume (see Sect. 3 for a more formal definition). Totsuji and Kihara [55] were the first to obtain a power-law behaviour for the spatial correlation function $\xi(r) = (r/r_0)^{-1.8}$ on the basis of angular data taken from the Lick survey and making use of the Limber equation. Moreover, as we can see in Fig. 1 reproduced from their paper, the observed correlation function of the Lick survey is fitted to an early halo model – the Neyman–Scott process. The evidence of the power-law behaviour was extended by Pebbles [41, 42] and demonstrated in Groth and Peebles [16]. The contributions by Jim Peebles and co-workers and, in particular, his influential book published in 1980 "The Large-Scale Structure of the Universe" [43] were crucial to make theses statistics widely used in cosmology.

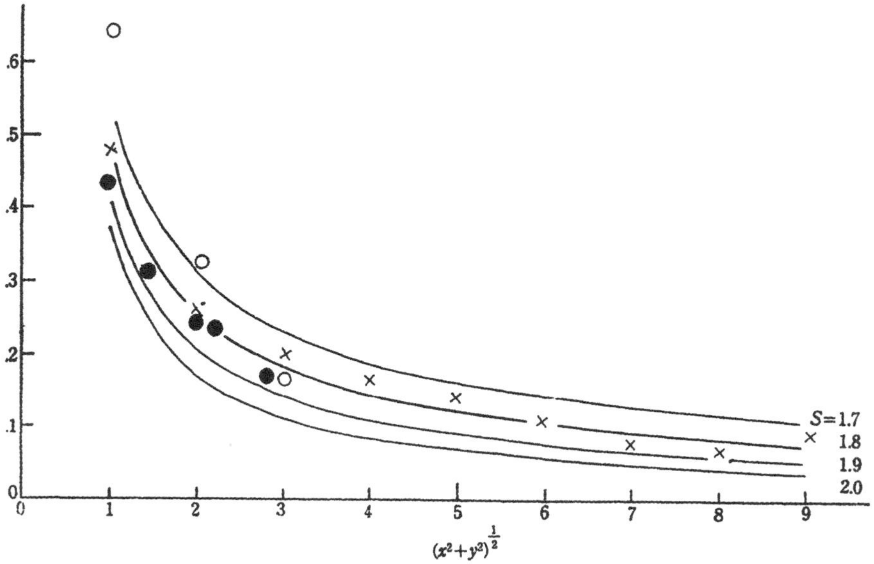

Fig. 1. The first power law fitting the spatial correlation function of the distribution of galaxies after deprojecting from an angular catalogue (reproduced from [55]). The filled circles were obtained by Totsuji and Kihara, while the open circles and crosses were derived by Neyman, Scott and Shane under the assumption of their clustering model. The solid lines correspond to power-law correlation functions $xi(r) = (r_0/r)^s$ with the value of the exponent s indicated in the legend

This remarkable power law for the two-point correlation function has dominated many of the analyses of the large-scale structure for the past three decades and more.

Complementary to the Lick catalogue, other surveys mapped the large-scale distribution of clusters of galaxies, for example, the Palomar Observatory Sky Survey was used by George Abell to publish a catalogue of 2,712 clusters of galaxies [1]. Some of them turned out not to be real clusters, but the majority were genuine. Analyses of this and other samples of galaxy clusters has yielded also power-law fits to the cluster–cluster correlation function $\xi_{cc}(r)$ but with exponents and amplitudes varying in a wider range, depending on selection effects, richness class, etc. [4, 7, 35, 45, 46].

2 Redshift Surveys

Listing extragalactic objects and magnitudes as they appear projected onto the celestial sphere was just the first step towards obtaining a cartography of the universe. The second step was to obtain distances by measuring redshifts using spectroscopy for a large number of galaxies mapping large areas of the sky. This task provided information about how the universe is structured now and in the recent past. In the 1980s, the Center for Astrophysics surveys played a leading role in the discovery of very large cosmic structures in the distribution of the galaxies. The first "slice of the universe" compiled by de Lapparent et al. [10] extended up to $150\,h^{-1}$ Mpc, a deep distance at that time. The calculation of the correlation function – now in redshift space – of the CfA catalogue confirmed the power-law behaviour discovered by Totsuji and Kihara 14 years before [9]. It is worth to mention however that redshift distortions affect severely the correlation function at small separations, and a distinction between redshift and real space became necessary.

The present wide field surveys are much deeper as it can be appreciated in Figs. 2 and 3. Figure 2 illustrates our local neighbourhood (up to $400\,h^{-1}$ Mpc) from the two-degree Field Galaxy Redshift Survey (2dF-GRS) in a three-dimensional view, where large superclusters surround more empty regions, delineated by long filaments. Figure 3 shows the first CfA slice with cone diagrams from the 2dFGRS and the Sloan Digital Sky Survey (SDSS). The first one contains redshifts of about 250,000 galaxies in wide regions around the north and south galactic poles with a median redshift $z = 0.11$. It extends up to $z \simeq 0.3$. Galaxies in this survey go down to apparent blue magnitude $b_{\rm lim} = 19.45$; therefore, this is a magnitude-limited survey that misses faint galaxies at large distances, as it can be seen in Fig. 3. The SDSS survey is also magnitude-limited, but the limit has been selected to be red, $r_{\rm lim} = 17.77$. The present release of the SDSS (DR6) covers an area almost five times as big as the area covered by the 2dFGRS.

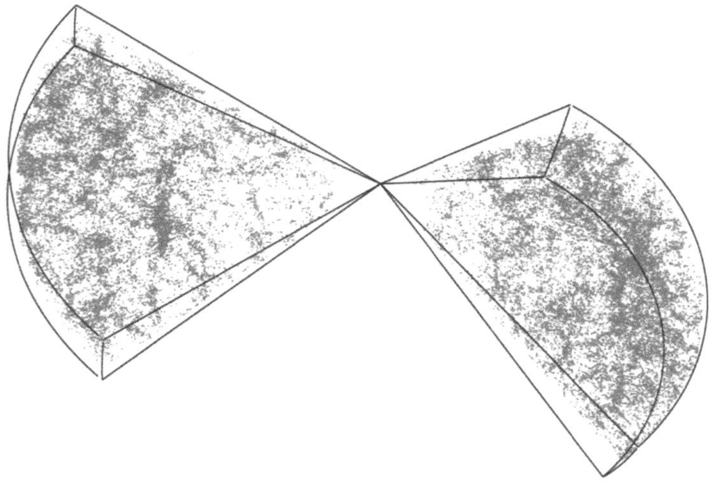

Fig. 2. The two slices that conform the 2dFGRS showing the galaxy distribution up to a distance of $400\,h^{-1}$ Mpc. The *left slice* lies in the north galactic hemisphere, while the *right* one lies in the south galactic hemisphere

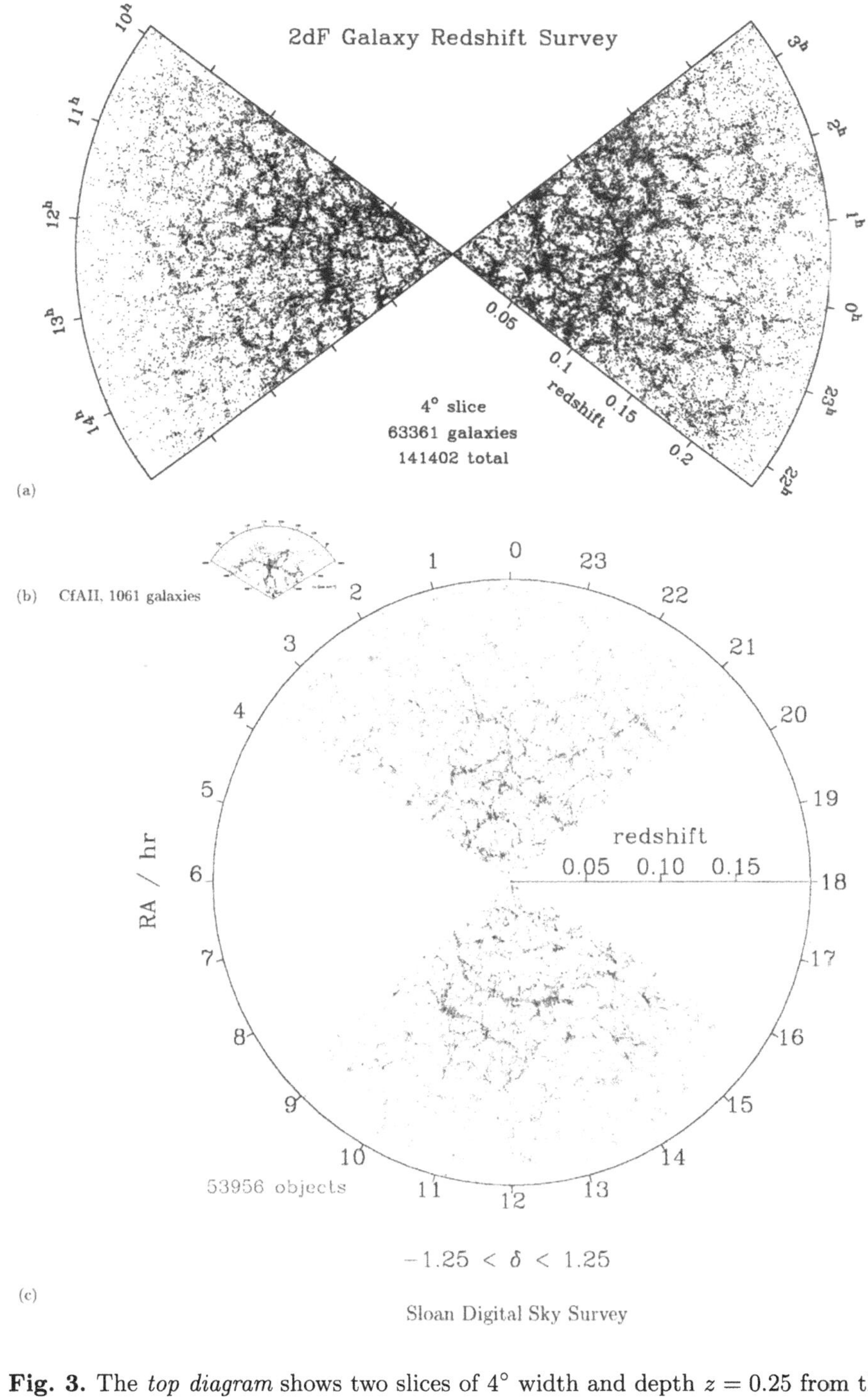

Fig. 3. The *top diagram* shows two slices of 4° width and depth $z = 0.25$ from the 2dF Galaxy Redshift Survey (from [39]). The *circular diagram* at the bottom has a radius Corresponding to redshift $z = 0.2$ and shows 55,958 galaxies from the SDSS survey (from [29]). As an inset, the first CfA slice from [10] is shown to scale

More information about these surveys can be found in their web pages: `http://www.mso.anu.edu.au/2dFGRS/` for the 2dF survey and `http://www.sdss.org/` for the SDSS survey.

3 The Two-Point Correlation Function

After measuring the two-point correlation function over projected galaxy samples, the great challenge was to do it directly for redshift surveys where the distance inferred from the recession velocities was used, providing a three-dimensional space. As it has been already mentioned, we have to bear in mind that measured redshifts are contaminated by the peculiar velocities. This 3D space, the so-called redshift space, is a distorted view of the real space. Figure 4 shows a simulation with the effect of the peculiar velocities distorting the real space (left panel), squeezing the structures to produce the radial stretched structures pointing to the observer, known as fingers of God (right panel). For details see Praton et al. [47] and the web page `http://kusmos.phsx.ku.edu/~melott/redshift-distortions.html`. These fingers of God appear strongest where the galaxy density is largest, and are attributable to the extra "peculiar" (i.e. non-Hubble) component of the velocity of individual galaxies in the galaxy clusters [18, 21, 24, 50].

Considering two infinitesimal volume elements dV_1 and dV_2 separated by a vector distance $\mathbf{r}_{12}$, the joint probability of there being a galaxy lying in each of these volumes is

$$dP_{12} = n^2[1 + \xi(\mathbf{r}_{12})]dV_1 dV_2. \quad (2)$$

Assuming homogeneity (the point process is invariant under translation) and isotropy (the point process is invariant under rotation) for the galaxy

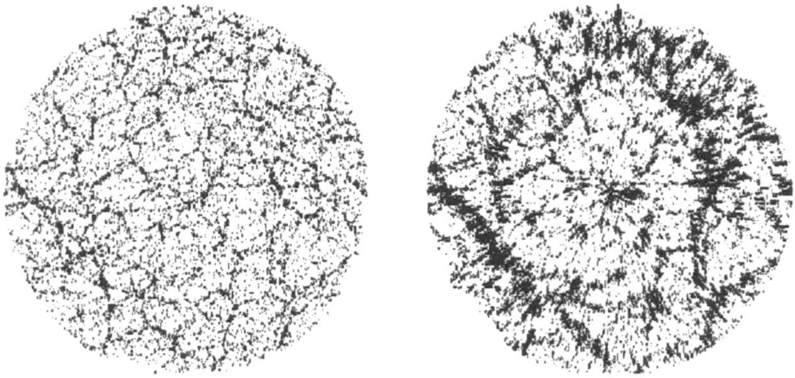

Fig. 4. Illustration by a two-dimensional simulation of the effect of the peculiar velocities distorting the real space (*left panel*) to produce the redshift space (*right panel*). Figures courtesy of Adrian Melott (see also [47])

distribution, the quantity depends only on the distance $r_{12} = |\mathbf{r}_{12}|$ and (2) becomes (1).

Apart from the formal definitions given in the previous equations, to estimate the correlation function for a particular complete galaxy sample with N objects, several formulae providing appropriate estimators have been introduced. The most widely used are the Hamilton estimator [17] and the Landy and Szalay estimator [27]. For both, a Poisson catalogue, a binomial process with N_{rd} points, has to be generated within the same boundaries of the real data set. The estimators can be written as

$$\widehat{\xi}_{\text{HAM}}(r) = \frac{DD(r) \cdot RR(r)}{[DR(r)]^2} - 1, \tag{3}$$

$$\widehat{\xi}_{\text{LS}}(r) = 1 + \left(\frac{N_{\text{rd}}}{N}\right)^2 \frac{DD(r)}{RR(r)} - 2\frac{N_{\text{rd}}}{N}\frac{DR(r)}{RR(r)}, \tag{4}$$

where $DD(r)$ is the number of pairs of galaxies of the data sample with separation within the interval $[r - dr/2, r + dr/2]$, $DR(r)$ is the number of pairs between a galaxy and a point of the Poisson catalogue, and $RR(r)$ is the number of pairs between points from the Poisson catalogue [25, 44].

As explained in the contributions by Hamilton and Szapudi [20], the above formulae have to be corrected due to the selection effects. These effects could be radial due to the fact that redshift surveys are built as apparent magnitude catalogues, and therefore fainter galaxies are lost at larger distances, and could be angular due to the galactic absorption that makes the sky not equally transparent in all directions or to the fact that different areas of the sky within the sample boundaries are not equally covered by the observations, therefore providing varying apparent magnitude limits depending on the direction. Moreover, some areas could not be covered at all because of the presence of nearby stars, or because of fiber collisions in the spectrograph. In order to account for this complexity the best solution is to use the freely available MANGLE software (`http://space.mit.edu/home/tegmark/mangle/`), a generic tool for managing angular masks on a sphere [54].

3.1 The Projected Correlation Function

Since at small scales peculiar velocities strongly distort the correlation function, it has become customary to calculate and display the so-called projected correlation function

$$w_p(r_p) = 2\int_0^\infty \xi(\pi, r_p)\mathrm{d}\pi, \tag{5}$$

where the two-dimensional correlation function $\xi(\pi, r_p)$ is computed on a grid of pair separations parallel (π) and perpendicular (r_p) to the line of sight. Figure 5 shows this function calculated by Peacock et al. [39] for the 2dFGRS.

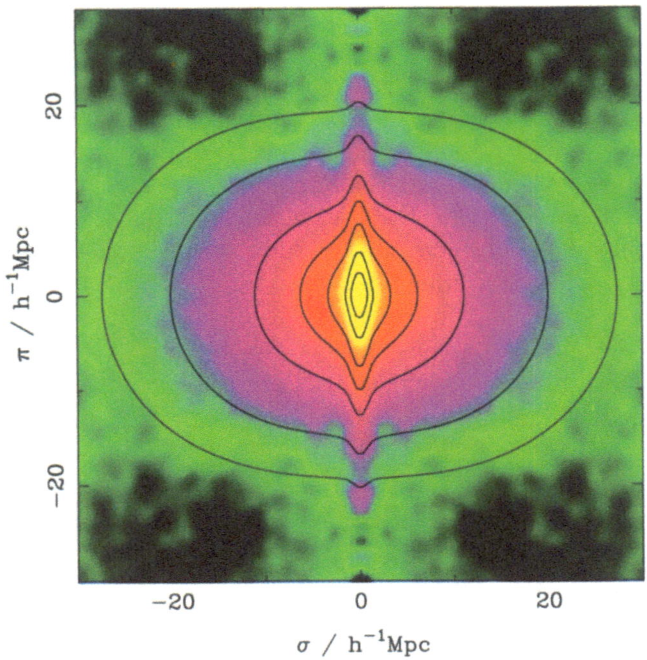

Fig. 5. The galaxy correlation function $\xi(\pi, r_p)$ for the 2dFGRS (transverse distance r_p is represented here by σ). This diagram shows the two sources of anisotropy in the correlation function: the radial smearing due to random velocities within groups and clusters at small distances and the large-scale flattening produced by coherent infall velocities. In this diagram the calculation has been performed by counting pairs in boxes and then smoothing with a Gaussian. The results obtained for the first quadrant are repeated with reflection in both axes to show deviations from circular symmetry. Overplotted lines correspond to the function calculated for a given theoretical model. Figure from [39]

If the separation vector between two positions in redshift space is $\boldsymbol{s} = \boldsymbol{s_2} - \boldsymbol{s_1}$, and the line-of-sight vector is $\boldsymbol{l} = \boldsymbol{s_1} + \boldsymbol{s_2}$, the parallel and perpendicular distances of the pair are (see Fig. 6)

$$\pi = \frac{|\boldsymbol{s} \cdot \boldsymbol{l}|}{|\boldsymbol{l}|}, \quad r_p = \sqrt{\boldsymbol{s} \cdot \boldsymbol{s} - \pi^2}.$$

Figure 7 shows the projected correlation function calculated for the Sloan Digital Sky Survey by Zehavi et al. [57]. The relation between the projected correlation function and the three-dimensional real correlation function (not affected by redshift distortions) is, for an isotropic distribution [9],

$$w_p(r_p) = 2 \int_{r_p}^{\infty} \xi(r) \frac{r \mathrm{d}r}{\left(r^2 - r_p^2\right)^{1/2}}. \tag{6}$$

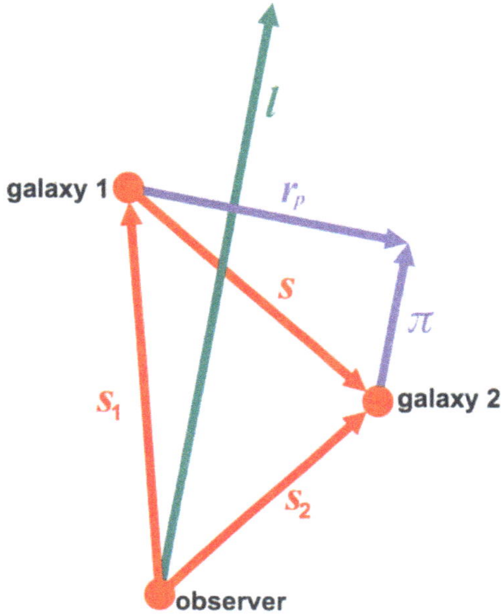

Fig. 6. Illustration of the parallel and perpendicular separations between two objects

From the previous equation it is seen that if $\xi(r)$ fits well a power law, i.e. $\xi(r) = (r/r_0)^{-\gamma}$, $w_p(r_p)$ also does, $w_p(r_p) = Ar_p^{-\alpha}$, with

$$\alpha = \gamma - 1, \quad \text{and} \quad A = \frac{r_0^\gamma \Gamma(0.5)\Gamma[0.5(\gamma-1)]}{\Gamma(0.5\gamma)}.$$

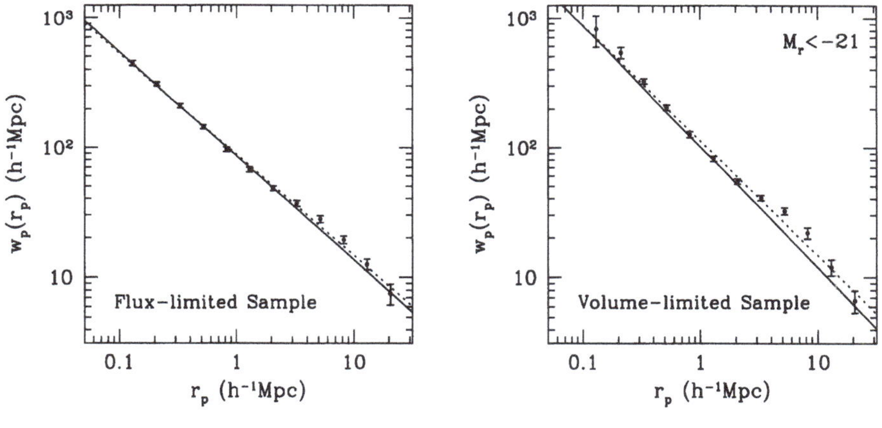

Fig. 7. The projected correlation function $w_p(r_p)$ for the SDSS data. *Left panel* shows the result for the flux-limited sample and *right panel* for the volume-limited sample. Two different power-law fits to the data have been performed. Solid lines make use of the full covariance matrix while dashed lines only use the diagonal elements. Figure from Zehavi et al. [57]

In practice, the integration in (5) is performed up to a fixed value $\pi_{\max}$ which depends on the survey. For the SDSS, Zehavi et al. [57] used $\pi_{\max} = 40\,\mathrm{h}^{-1}$ Mpc, a value considered large enough by the authors to include the relevant information to measure $w_p(r_p)$ in the range $0.1\,\mathrm{h}^{-1}\,\mathrm{Mpc} < r_p < 20\,\mathrm{h}^{-1}$ Mpc. The assumed cosmological model for the calculation of distances is the concordance model for which $\Omega_m = 0.3$ and $\Omega_\Lambda = 0.7$.

The function shown in the left panel of Fig. 7 has been calculated making use of a subset containing 118,149 galaxies drawn from the flux-limited sample selected by Blanton et al. [2]. The estimator of the correlation function makes use of the radial selection function that incorporates the luminosity evolution model of Blanton et al. [2]. On the right panel, the calculation has been performed over a volume-limited sample containing only galaxies bright enough to be seen within the whole volume (up to $462\,\mathrm{h}^{-1}$ Mpc, the limit of the sample). This subsample contains 21,659 galaxies with absolute red magnitude $M_r < -21$ (for $h = 1$). The solid line on the left panel of Fig. 7 shows the fit to $w_p(r_p)$ which corresponds to a real-space correlation function $\xi(r) = (r/5.77\,\mathrm{h}^{-1}\,\mathrm{Mpc})^{-1.80}$. For the volume-limited sample the fit shows a slightly steeper power law $\xi(r) = (r/5.91\,\mathrm{h}^{-1}\,\mathrm{Mpc})^{-1.93}$. This is an expected consequence of the segregation of luminosity as we will show later, since galaxies in this subsample are 0.56 magnitudes brighter than the characteristic value of the Schechter [51] luminosity function [2].

Although it is remarkable from the power-law fits shown in Fig. 7 how the scaling holds for about three orders of magnitude in scale, the main point stressed in this analysis was precisely the unambiguous detection of a systematic departure from the simple power-law behaviour. A similar result was also obtained by Hawkins et al. [19] for the 2dFGRS, although the best-fit power law for the correlation function of 2dF galaxies is $\xi(r) = (r/5.05\,\mathrm{h}^{-1}\,\mathrm{Mpc})^{-1.67}$ with a less steep slope than the one found for SDSS galaxies and with a value of the correlation length $r_0 = 5.05 \pm 0.26\,\mathrm{h}^{-1}$ Mpc, substantially smaller than the SDSS result. Again, this can be explained as a consequence of the different galaxy content of both surveys; SDSS are red-magnitude selected while 2dF are blue-magnitude selected.

Error bars for the correlation function in Fig. 7 have been calculated in two different ways which illustrate the two main methods currently used. For the flux-limited sample, jackknife resampling of the data has been used. The sample is divided into N disjoint subsamples covering each approximately the same area of the sky; then the calculation of $\xi(r)$ is performed on each of the jackknife samples created by summing up the N subsamples except one, which is omitted in turn. The ij element of the covariance matrix is computed by [56]

$$C_{ij} = \frac{N-1}{N} \sum_{k=1}^{N} ({\xi_i}^k - \bar{\xi}_i)({\xi_j}^k - \bar{\xi}_j), \tag{7}$$

where $\bar{\xi}_i$ is the average value of ξ_i measured on the jackknife samples. Statistical errors can be calculated using the whole covariance matrix, or just making use of the elements in the diagonal, and thus ignoring the correlation amongst the errors. The other possibility consists in using mock catalogues from N-body simulations or semi-analytical models of structure formation with a recipe for allocating galaxies. These mock catalogues can be used as the subsamples in which (7) can be applied to obtain the covariance matrix.

The variation of the slope in the two-point correlation function of galaxies with the scale might be ascribed to the existence of two different clustering regimes: the small-scale regime dominated by pairs of galaxies within the same dark matter halo and a second regime where pairs of galaxies belonging to different halos contribute to the downturn of the power law in $w_p(r_p)$.

3.2 Galaxy Properties and Clustering

The photometric and spectral information provided by surveys like SDSS and 2dFGRS allows to study how the clustering of galaxies depends on different factors such as luminosity, morphology, colour and spectral type, although these factors are certainly not independent. For example, it is well known [8, 11] that early-type galaxies show more pronounced clustering at small separations than late-type galaxies, the first kind displaying steeper power-law fits to their correlation than the latter. This segregation plays an interesting role in the understanding of the galaxy formation process, since galaxies are biased tracers of the total matter distribution in the universe (mainly dark) and the bias also depends on the scale [26]. Madgwick et al. [30] have recently divided the 2dFGRS into two subsets: passive galaxies with a relatively low star formation rate and active galaxies with higher current star formation rate. This division correlates well with colour and morphology, being passive galaxies mainly red old ellipticals.

Figure 8 (left panel) shows the projected correlation function for these two subsets. As it can be appreciated, passive galaxies present a two-point correlation function with steeper slope and larger amplitude than active galaxies, being the best fit for each subsample $\xi(r) = (r/6.10 \pm 0.34\,\mathrm{h}^{-1}\,\mathrm{Mpc})^{-1.95\pm0.03}$ for passive galaxies and $\xi(r) = (r/3.67 \pm 0.30\,\mathrm{h}^{-1}\,\mathrm{Mpc})^{-1.60\pm0.04}$ for active galaxies. A similar analysis was also performed by Zehavi et al. [56] dividing an early release of the SDSS galaxies into two subgroups by colours, red and blue, using the value of the colour $u^* - r^* = 1.8$ for the division. The blue subset contains mainly late morphological types while the red group is formed mainly by bulge galaxies, as it should be expected. Again, as it can be appreciated in Fig. 8 (right panel), red galaxies cluster stronger than blue galaxies, being their best fit to a power law in the range $[0.1\,\mathrm{h}^{-1}\,\mathrm{Mpc} < r_p < 16\,\mathrm{h}^{-1}\,\mathrm{Mpc}]$, $\xi(r) = (r/6.78 \pm 0.23\,\mathrm{h}^{-1}\,\mathrm{Mpc})^{-1.86\pm0.03}$, while for blue galaxies the best fit is $\xi(r) = (r/4.02 \pm 0.25\,\mathrm{h}^{-1}\,\mathrm{Mpc})^{-1.41\pm0.04}$. Blanton et al. [3] have shown that large amplitudes in the correlation function corresponding to subsets selected by luminosity or colour are typically accompanied with steeper slopes.

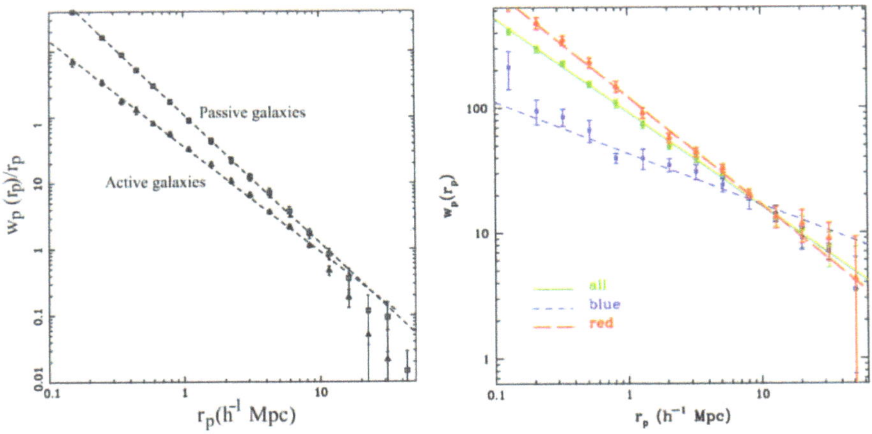

Fig. 8. In the *left panel*, the projected correlation function $w_p(r_p)$ for two subsamples of the 2dFGRS data where the division has been performed in terms of current star formation rate is shown. Passive galaxies cluster stronger than their active counterparts. Figure adapted from Madgwick et al. [30]. In the *right panel*, the projected correlation function of subsamples divided by colour drawn from the SDSS is shown. Different lines show the best-fit power laws for $w_p(r_p)$. The short-dashed, long-dashed and solid lines correspond to the blue, red and full samples, respectively. Figure from Zehavi et al. [56]

4 The Power Spectrum

The power spectrum $P(k)$ is a clustering descriptor depending on the wavenumber k that measures the amount of clustering at different scales. It is the Fourier transform of the correlation function, and therefore both functions contain equivalent information, although it can be said that they describe different sides of the same process. For a Gaussian random field, the Fourier modes are independent, and the field gets completely characterized by its power spectrum. As the initial fluctuations from the inflationary epoch in the universe are described as a Gaussian field, the model predictions in cosmology are typically made in terms of power spectra.

The power spectrum and the correlation function are related through a Fourier transform:

$$P(\mathbf{k}) = \int \xi(\mathbf{r}) e^{i\mathbf{k}\cdot\mathbf{r}}\, d^3r,$$

$$\xi(\mathbf{r}) = \int P(\mathbf{k}) e^{-i\mathbf{k}\cdot\mathbf{r}} \frac{d^3k}{(2\pi)^3}.$$

Assuming isotropy, the last equation can be rewritten as

$$\xi(r) = 4\pi \int_0^\infty P(k) \frac{\sin(kr)}{kr} \frac{k^2\, dk}{(2\pi)^3}.$$

Some authors [38] prefer to use the following normalization for the power spectrum

$$\Delta^2(k) = \frac{1}{2\pi^2} P(k) k^3$$

in such a way that the total variance of the density field is just

$$\sigma_\delta^2 = \int_0^\infty \Delta^2(k)\, d(\ln k).$$

One of the advantages of the power spectrum over the correlation function is that amplitudes for different wavenumbers are statistically orthogonal (for a more detailed discussion see the contributions by Andrew Hamilton [19]):

$$E\left\{\widetilde{\delta}(\mathbf{k})\widetilde{\delta}^\star(\mathbf{k}')\right\} = (2\pi)^3 \delta_D(\mathbf{k} - \mathbf{k}') P(\mathbf{k}). \tag{8}$$

Here $\widetilde{\delta}(\mathbf{k})$ is the Fourier amplitude of the overdensity field $\delta = (\rho - \bar{\rho})/\bar{\rho}$ at a wavenumber $\mathbf{k}$, ρ is the matter density, a star denotes complex conjugation, $E\{\}$ denotes expectation values over realizations of the random field and $\delta_D(\mathbf{x})$ is the three-dimensional Dirac delta function.

If we have a sample (catalogue) of galaxies with the coordinates $\mathbf{x}_j$, we can write the estimator for a Fourier amplitude of the overdensity distribution [14] (for a finite set of frequencies $\mathbf{k}_i$) as

$$F(\mathbf{k}_i) = \sum_j \frac{\psi(\mathbf{x}_j)}{\bar{n}(\mathbf{x}_j)} e^{i\mathbf{k}_i \cdot \mathbf{x}} - \widetilde{\psi}(\mathbf{k}_i),$$

where $\bar{n}(\mathbf{x})$ is the position-dependent selection function (the observed mean number density) of the sample and $\psi(\mathbf{x})$ is a weight function that can be selected at will.

The raw estimator for the spectrum is

$$P_R(\mathbf{k}_i) = F(\mathbf{k}_i) F^\star(\mathbf{k}_i),$$

and its expectation value

$$E\left\{\langle |F(\mathbf{k}_i)|^2 \rangle\right\} = \int G(\mathbf{k}_i - \mathbf{k}') P(\mathbf{k}') \frac{d^3k'}{(2\pi)^3} + \int_V \frac{\psi^2(\mathbf{x})}{\bar{n}(\mathbf{x})} d^3x,$$

where $G(\mathbf{k}) = |\widetilde{\psi}(\mathbf{k})|^2$ is the window function that also depends on the geometry of the sample volume. The reader can learn more about the estimation of the power spectrum in the contributions by Andrew Hamilton [19].

4.1 Acoustic Peak in ξ and Acoustic Oscillations in $P(k)$

Prior to the epoch of recombination, the universe is filled by a plasma where photons and baryons are coupled. Due to the pressure of photons, sound speed is relativistic at this time and the sound horizon has a comoving radius of 150 Mpc. Cosmological fluctuations produce sound waves in this plasma.

At about 380,000 years after the Big Bang, when the temperature has fallen down to 3000 K, and recombination takes place, the universe loses its ionized stage and neutral gas dominates. At this state, sound speed drops off abruptly and acoustic oscillations in the fluid become frozen. Their signature can be detected in both the cosmic microwave background (CMB) radiation and the large-scale distribution of galaxies. Figure 9 shows a representation of the temperature at the last scattering surface from WMAP. These fluctuations have been analysed in detail to obtain a precise estimation of the anisotropy power spectrum of the CMB. The acoustic peaks in this observed

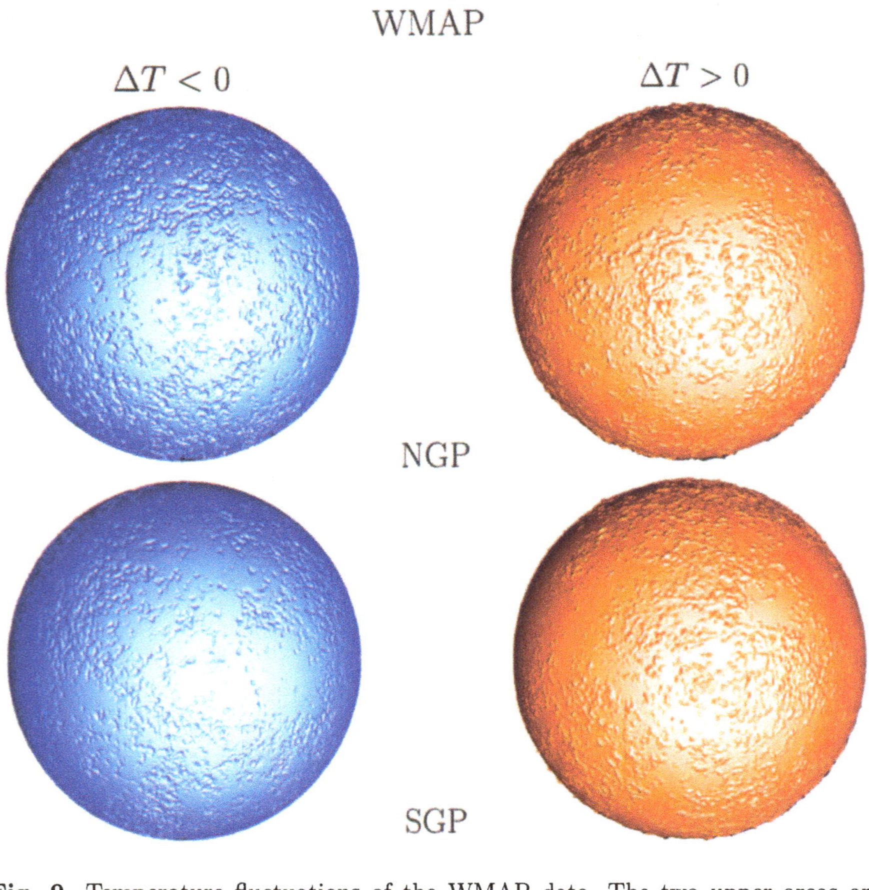

Fig. 9. Temperature fluctuations of the WMAP data. The two upper areas are centred in the north galactic pole (NGP), while the bottom two are in the south galactic pole (SGP). On the left hand side, in *blue*, pixels where $\Delta T < 0$ are depicted as depths, while on the right hand side, *red* pixels with $\Delta T > 0$ are displayed as elevations. The "sea level" in blue spheres corresponds to the pixels where $\Delta T \geq 0$ and in the red sphere, where $\Delta T \leq 0$

angular power spectrum (see contribution by Enrique Martínez-Gónzalez [35]) have become a powerful cosmological probe. In particular, the CMB provides an accurate way to measure the characteristic length scale of the acoustic oscillations, which depends on the speed of sound, c_s, in the photon-baryon fluid and the cosmic time when this takes place. The distance that a sound wave has travelled at the age of the universe at that time is

$$\int_0^{t_{rec}} c_s(t)(1+z)dt \simeq 147 \pm 2 \text{ Mpc}, \tag{9}$$

for the standard flat Λ-CDM model. This fixed scale imprinted in the matter distribution at recombination can be used as a "standard ruler" for cosmological purposes.

The imprint in the matter distribution of this acoustic feature should be detected in both the correlation function and the matter power spectrum. However, the amplitude of the acoustic peaks in the CMB angular power spectrum is much larger than the expected amplitude of the oscillations in the matter power spectrum, which are called for obvious reasons baryonic acoustic oscillations (BAOs). Moreover, the feature should manifest as a single peak in the correlation function at about $100\,\mathrm{h}^{-1}$ Mpc, while in the power spectrum it should be detected as a series of small-amplitude oscillations as shown in Fig. 10. Baryons represent only a small fraction of the matter in the universe, and therefore, as it can be appreciated in the figure, the amplitude of the oscillations in the power spectrum is rather tiny for the concordance model (green dashed line in the top panel of Fig. 10). We can see how increasing the baryon fraction increases the amplitude of the oscillations, while wiggles disappear for a pure Λ-CDM model (with no baryonic content). At small scales the oscillations are erased by Silk damping; therefore one needs to accurately measure the power spectrum or the correlation function on scales between 50 and $150\,\mathrm{h}^{-1}$ Mpc to detect theses features.

Eisenstein et al. (2005) [12] announced the detection of the acoustic peak in the two-point redshift-space correlation function of the SDSS LRG survey (see Fig. 11). More or less simultaneously, Cole et al. (2005) [5] discovered the corresponding feature in the form of wiggles of about 10% amplitude in the power spectrum of the 2dF Galaxy Redshift Survey. We have also calculated the redshift correlation function for a nearly volume-limited sample of the 2dFGRS extracted by Croton et al. [6]. There are about 32,000 galaxies in this sample with absolute magnitude within the range $-20 > M_{B_J} - 5\log_{10} h > -21$. The correlation function displayed in the right panel of Fig. 11 shows a prominent peak around $100\,\mathrm{h}^{-1}$ Mpc which expands for a wider scale range than the bump observed in the SDSS-LRG sample (left panel). This could be due to scale-dependent differences between the clustering of the two samples. A similar effect has been recently observed in the power spectrum [49] of the two surveys (see also the figure caption of Fig. 12). Of course, the statistical significance of this feature is still to be tested. Interestingly enough is the fact that the mock catalogues generated by Norberg et al. [37] to mimic the

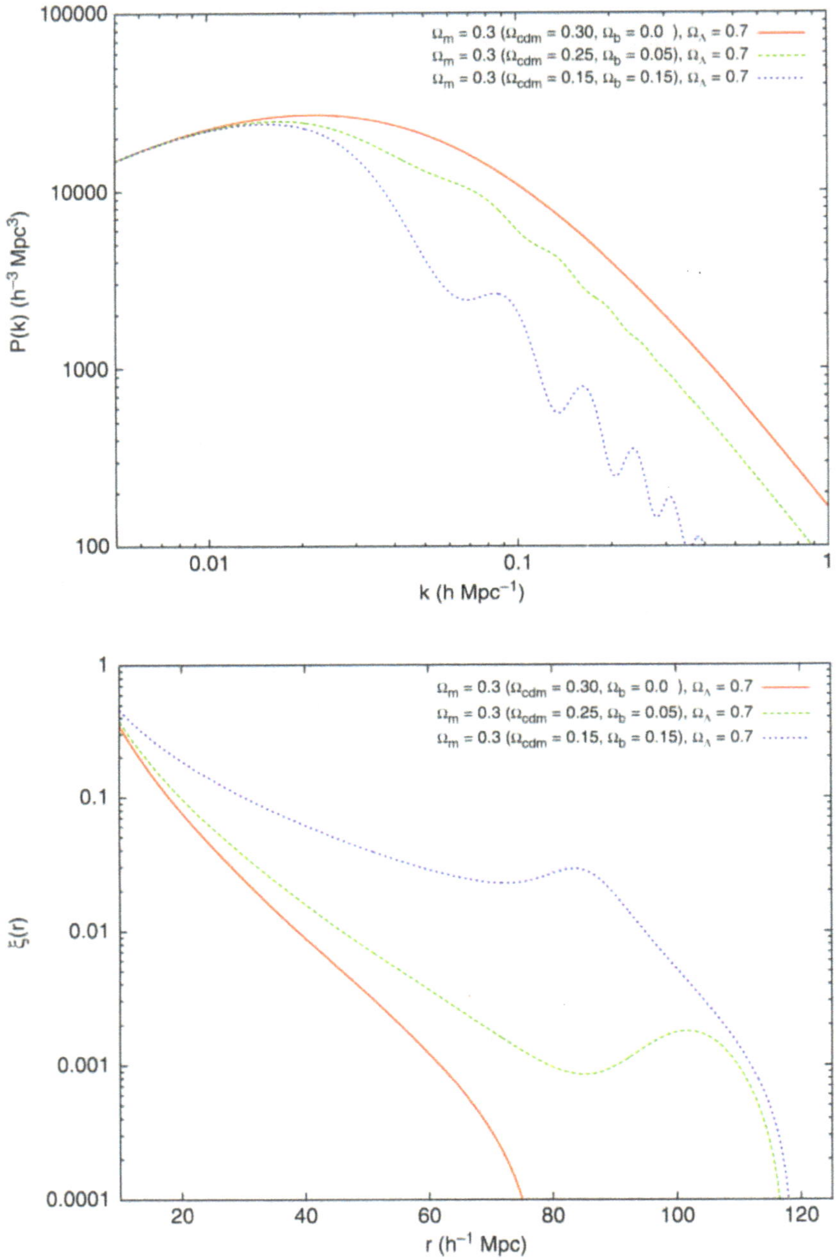

Fig. 10. (*Continued*)

properties of the 2dFGRS at small scales do not show the acoustic peak. Moreover, we can see a large scatter in the correlation function of the mocks, with average values that do not follow the data (mocks show larger correlations at intermediate scales and smaller at large scales).

Figure 12 shows the power spectrum calculated recently by Sánchez and Cole [49] for the 2dFGRS and the SDSS-DR5 survey. The expected acoustic oscillations are clearly detected within the error bands. These errors have been calculated using mock catalogues generated from lognormal density fields with a given theoretical power spectrum.

4.2 Concluding Remarks and Challenges

The expected value of the sound horizon at recombination (9) determined from the CMB observations can be compared with the observed BAO scale in the radial direction at a given redshift to estimate the variation of the Hubble parameter with redshift $H(z)$. High accurate redshifts are needed to carry on this test. Likewise, the BAO scale observed in redshift surveys compared with its expected value provides us with a way to measure the angular diameter distance, as a function of redshift $D_A(z)$. As Nichol [36] points out, this is similar, in a sense, to the measurement of the correlation function in the parallel and perpendicular directions to the line of sight, $\xi(\pi, r_p)$, explained in Sect. 3.1.

There are several ongoing observational projects that will map a volume large enough to accurately measure BAOs in the galaxy distribution, some of them making use of spectroscopic redshifts (e.g. AAT WiggleZ, SDSS BOSS, HETDEX and WFMOS) and others making use of photometric redshifts (e.g. DES, LSST, Pan-STARRS and PAU), all of them surveying large areas of the sky and encompassing volumes of several Gpc^3. For an updated review see [15]. To deal with the uncertainties of the BAO measurement due to different effects (non-linear gravitational evolution, biasing of galaxies with respect to dark matter, redshift distortions, etc.) is not easy, and accurate cosmological simulations are required for this purpose.

Fig. 10. *Top panel*: The linear-regime power spectrum of the matter in the universe for different flat models with $\Omega_{\text{total}} = 1$, $h = 0.7$, $\Omega_m = 0.3$ and $\Omega_\Lambda = 0.7$. The three curves correspond to different proportions of baryonic and cold dark matter: from top to bottom $\Omega_b = 0, 0.05, 0.15$. As we see, increasing the baryon content (at fixed Ω_m) increases the amplitude of the acoustic oscillations, while suppressing power on small scales (large wavenumber). In the *bottom panel*, the corresponding correlation function to each model displayed with the same line style is shown. For no baryons (pure cold dark matter), the acoustic peak is missing, while the peak amplitude is larger with a larger proportion of baryons. Data for the figure courtesy of Gert Hütsi. A similar diagram can be found in [36] and [13]

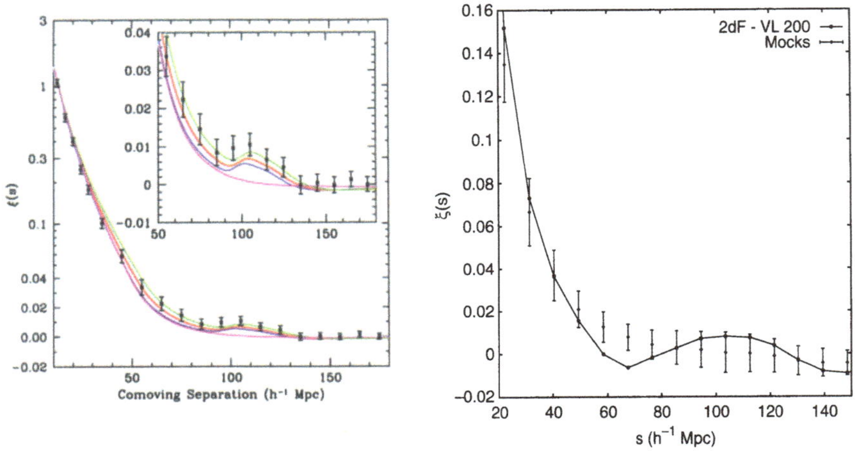

Fig. 11. *Left.* The redshift-space galaxy correlation function measured for the LRG SDSS sample containing 46,748 luminous read galaxies in redshift space. The vertical axis mixes logarithmic and linear scales. The inset shows an expanded view around the peak ($\sim 100\,\mathrm{h}^{-1}\,\mathrm{Mpc}$) with the vertical axis in linear scale. The different solid curves correspond to a Λ-CDM model with $\Omega_m h^2 = 0.12$ (*green*), 0.13 (*red*), 0.14 (*blue*); in all cases the baryon content is fixed to $\Omega_b h^2 = 0.024$. The magenta line corresponds to a pure Λ-CDM model with no baryons. Figure from Eisenstein et al. [12]. *Right.* The redshift-space galaxy correlation function measured for a volume-limited sample extracted from the 2dFGRS (*solid discs* joined by a *solid line*). The same function has been calculated on the 22 mocks models explained in the text. The average correlation function together with 1-σ deviations is shown in the diagram. Mocks do not show the peak detected in the real galaxy survey

The correlation function can be generalized to higher order (see the contribution by Istvan Szapudi [60]): the N-point correlation functions. This allows to statistically characterize the galaxy distribution with a hierarchy of quantities which progressively provide us with more and more information about the clustering of the point process. These measures, however, had been difficult to derive with reliability from the scarcely populated galaxy catalogues. The new generation of surveys will surely overcome this problem.

There are, nevertheless, other clustering measures which provide complementary information to the second-order quantities described above. For example, the topology of the galaxy distribution measured by the genus statistic provides information about the connectivity of the large-scale structure. The topological genus of a surface is the number of holes minus the number of isolated regions plus 1. This quantity is calculated for the isodensity surfaces of the smoothed data corresponding to a given density threshold (excursion sets). The genus can be considered as one of the four Minkowski functionals used commonly in stochastic geometry to study the shape and connectivity of union of convex three-dimensional bodies. In three dimension there are four functionals: volume, surface area, integral mean curvature, and Euler–Poincaré

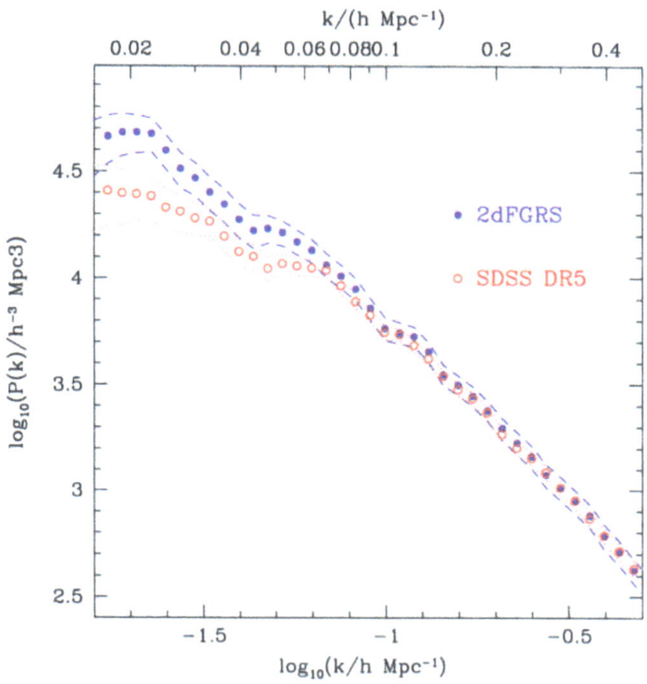

Fig. 12. The matter power spectrum $P(k)$ for the 2dFGRS and the SDSS-DR5. The agreement is good at small scales, while at larger scales there is a clear evidence of SDSS having more small-scale power than 2dFGRS. Sánchez and Cole [49] interpret this result as a consequence of the stronger scale-dependent bias shown by the red galaxies that dominate the SDSS catalogue. Figure adapted from Sánchez and Cole [49]

characteristic, related with the genus of the boundary (see the contribution by Enn Saar [53]).

The use of wavelets and related integral transforms is an extremely promising tool in the clustering analysis of three-dimensional catalogues. Some of these techniques are introduced in the contributions by Bernard Jones, Enn Saar and Belen Barreiro [25].

Acknowledgements

I thank Enn Saar, Carlos Peña and Pablo Arnalte for useful comments and suggestions on the manuscript, Gert Hütsi for the data for Fig. 10, Fernando Ballesteros and Silvestre Paredes for help with the figures and Pablo de la Cruz and María Jesús Pons-Bordería for allowing me to use unpublished common work on the analysis of the large-scale correlation function of the 2dFGRS in this review. I acknowledge financial support from the Spanish Ministerio de Educación y Ciencia project AYA2006-14056 (including FEDER) and the Acción Complementaria AYA2004-20067-E.

References

1. Abell, G.O.: ApJS, **3**, 211 (1958)
2. Blanton, M.R., et al.: ApJ, **592**, 819 (2003)
3. Blanton, M.R., et al.: ApJ, **629**, 143 (2005)
4. Borgani, S., Guzzo, L.: Nature, **409**, 39 (2001)
5. Cole, S., et al.: MNRAS, **362**, 505 (2005)
6. Croton, D.J., et al.: MNRAS, **352**, 828 (2004)
7. Dalton, G.B., Croft, R.A.C., Efstathiou, G., Sutherland, W.J., Maddox, S.J., Davis, M.: MNRAS, **271**, L47 (1994)
8. Davis, M., Geller, M.J.: ApJ, **208**, 13 (1976)
9. Davis, M., Peebles, P.J.E.: ApJ, **267**, 465 (1983)
10. de Lapparent, V., Geller, M.J., Huchra, J.P.: ApJ, **302**, L1 (1986)
11. Dressler, A.: ApJ, **236**, 351 (1980)
12. Eisenstein, D.J. et al.: ApJ, **633**, 560 (2005)
13. Eisenstein, D.J., Hu, W.: ApJ, **496**, 605 (1998)
14. Feldman, H.A., Kaiser, N., Peacock, J.A.: ApJ, **426**, 23 (1994)
15. Frieman, J.A., Turner, M.S., Huterer, D.: ArXiv e-prints:0803.0982 (2008)
16. Groth, E.J., Peebles, P.J.E.: ApJ, **217**, 385 (1977)
17. Hamilton, A.J.S.: ApJ, **417**, 19 (1993)
18. Hamilton, A.J.S.: In: Hamilton, D. (ed.) The Evolving Universe, Astrophysics and Space Science Library, vol. 231, p. 185. Kluwer, Dordrecht (1998)
19. Hamilton, A.J.S.: *Data Analysis in Cosmology*, Lect. Notes Phys. **665**. Springer, Heidelberg (2008)
20. Hamilton, A.J.S., Szapudi, I.: *Data Analysis in Cosmology*, Lect. Notes Phys. **665**. Springer, Heidelberg (2008)
21. Hawkins, E., et al.: MNRAS, **346**, 78 (2003)
22. Hubble, E.: ApJ, **79**, 8 (1934)
23. Jackson, J.C.: MNRAS, **156**, 1P (1972)
24. Jones, B.J.T.: In: Martínez, V.J. (eds.) Historical Development of Modern Cosmology, vol. 252, p. 245. ASP conference series (2001)
25. Jones, B.T.J., Saar, E., Barreiro, R.B.: *Data Analysis in Cosmology*, Lect. Notes Phys. **665**. Springer, Heidelberg (2008)
26. Jones, B.J.T., Martínez, V.J., Saar, E., Trimble, V.: Rev. Mod. Phys., **76**, 1211 (2004)
27. Kaiser, N.: MNRAS, **227**, 1 (1987)
28. Kerscher, M., Szapudi, I., Szalay, A.S.: ApJ, **535**, L13 (2000)
29. Lahav, O., et al.: MNRAS, **333**, 961 (2002)
30. Landy, S.D., Szalay, A.S.: ApJ, **412**, 64 (1993)
31. Limber, D.N.: ApJ, **119**, 655 (1954)
32. Loveday, J.: Contemp. Phys., **43**, 437 (2002)
33. Madgwick, D.S., et al.: MNRAS, **344**, 847 (2003)
34. Martínez, V.J., Saar, E.: Statistics of the Galaxy Distribution, Chapman & Hall/CRC Press, Boca Raton (2002)

35. Martínez-González, E.: *Data Analysis in Cosmology*, Lect. Notes Phys. **665**. Springer, Heidelberg (2008)
36. Navarro, J.F., Frenk, C.S., White, S.D.M.: ApJ, **490**, 493 (1997)
37. Neyman, J., Scott, E.L.: ApJ, **116**, 144 (1952)
38. Neyman, J., Scott, E.L., Shane, C.D.: ApJ, **117**, 92 (1953)
39. Nichol, R.C., Collins, C.A., Guzzo, L., Lumsden, S.L.: MNRAS, **255**, 21P (1992)
40. Nichol, R.C.: ArXiv e-prints:0708.2824 (2007)
41. Norberg, P., et al.: MNRAS, **336**, 907 (2002)
42. Peacock, J.A.: Cosmological Physics, Cambridge University Press, Cambridge (1999)
43. Peacock, J.A., et al.: Nature, **410**, 169 (2001)
44. Peebles, P.J.E.: A& A, **32**, 197 (1974)
45. Peebles, P.J.E.: ApJ, **189**, L51 (1974)
46. Peebles, P.J.E.: ApJ, **196**, 647 (1974)
47. Peebles, P.J.E.: The Large-Scale Structure of the Universe, Princeton University Press, Princeton (1980)
48. Pons-Bordería, M.-J., Martínez, V.J., Stoyan, D., Stoyan, H., Saar, E.: ApJ, **523**, 480 (1999)
49. Postman, M., Huchra, J.P., Geller, M.J.: ApJ, **384**, 404 (1992)
50. Postman, M.: In: Banday, A.J., Sheth, R.K., da Costa, L.N. (eds.) Evolution of Large Scale Structure: From Recombination to Garching, European Southern Observatory, Garching p. 270 (1999)
51. Praton, E.A., Melott, A.L., McKee, M.Q.: ApJ, **479**, L15 (1997)
52. Rubin, V.C.: Proc. Natl. Acad. Sci., **40**, 541 (1954)
53. Saar, E.: *Data Analysis in Cosmology*, Lect. Notes Phys. **665**. Springer, Heidelberg (2008)
54. Sánchez, A.G., Cole, S.: MNRAS, **385**, 830 (2008)
55. Sargent, W.L.W., Turner, E.L.: ApJ, **212**, L3 (1977)
56. Schechter, P.: ApJ, **203**, 297 (1976)
57. Shane, C.D., Wirtanen, C.A.: Publ. Lick Obs. **22**, 1 (1967)
58. Seljak, U.: MNRAS, **318**, 203 (2000)
59. Swanson, M.E.C., Tegmark, M., Hamilton, A.J.S., Hill, J.C.: ArXiv e-prints: 0711.4352 (2007)
60. Szapudi, I.: *Data Analysis in Cosmology*, Lect. Notes Phys. **665**. Springer, Heidelberg (2008)
61. Totsuji, H., Kihara, T.: Pub. Astron. Soc. Japan, **21**, 221 (1969)
62. Zehavi, I., et al.: ApJ, **571**, 172 (2002)
63. Zehavi, I., et al.: ApJ, **608**, 16 (2004)
64. Zwicky, F.: Helv. Phys. Acta, **26**, 241 (1953)

The Cosmic Web: Geometric Analysis

R. van de Weygaert and W. Schaap

Kapteyn Astronomical Institute, University of Groningen, P.O. Box 800, 9700 AV Groningen, the Netherlands
weygaert@astro.rug.nl

1 Outline: Patterns in the Cosmic Web

The spatial cosmic matter distribution on scales of a few up to more than a hundred megaparsec displays a salient and pervasive foam-like pattern. Revealed through the painstaking efforts of redshift survey campaigns, it has completely revised our view of the matter distribution on these cosmological scales. The web-like spatial arrangement of galaxies and mass into elongated filaments, sheet-like walls and dense compact clusters, the existence of large near-empty void regions and the hierarchical nature of this mass distribution – marked by substructure over a wide range of scales and densities – are three major characteristics we have come to know as the *cosmic web*.

The vast megaparsec cosmic web is undoubtedly one of the most striking examples of complex geometric patterns found in nature and the largest in terms of sheer size. In a great many physical systems, the spatial organization of matter is one of the most readily observable manifestations of the forces and processes forming and moulding them. Richly structured morphologies are usually the consequence of the complex and nonlinear collective action of basic physical processes. Their rich morphology is therefore a rich source of information on the combination of physical forces at work and the conditions from which the systems evolved. In many branches of science, the study of geometric patterns has therefore developed into a major industry for exploring and uncovering the underlying physics (see, e.g., Balbus & Hawley [16]).

Computer simulations suggest that the observed cellular patterns are a prominent and natural aspect of cosmic structure formation through gravitational instability (Peebles [174]), the standard paradigm for the emergence of structure in our Universe. Structure in the Universe is the result of the gravitational growth of tiny density perturbations and the accompanying tiny velocity perturbations in the primordial Universe. Supported by an impressive body of evidence, primarily those of temperature fluctuations in the cosmic microwave background (Smoot et al. [223], Bennett et al. [20], Spergel et al.

van de Weygaert, R., Schaap, W.: *The Cosmic Web: Geometric Analysis*. Lect. Notes Phys. **665**, 291–413 (2009)
DOI 10.1007/978-3-540-44767-2_11

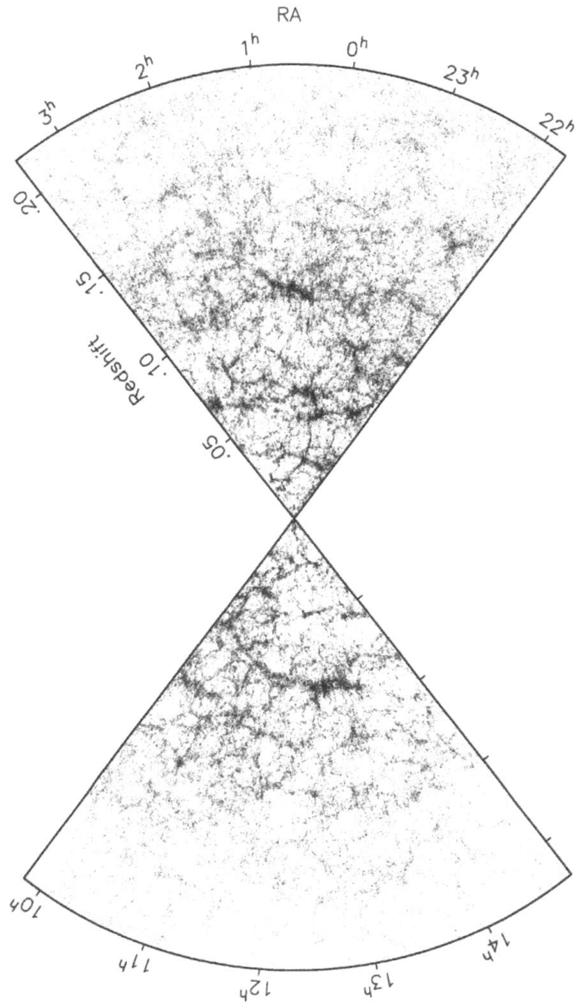

Fig. 1. The galaxy distribution uncovered by the 2dF galaxy redshift survey. Depicted are the positions of 221414 galaxies in the final 2dFGRS catalogue. Clearly visible is the foam-like geometry of walls, filaments and massive compact clusters surrounding near-empty void regions (from Colless et al. [57])

[226]), the character of the primordial random density and velocity perturbation field is that of a *homogeneous and isotropic spatial Gaussian process.* Such fields of primordial Gaussian perturbations in the gravitational potential are a natural product of an early inflationary phase of our Universe.

The early linear phase of pure Gaussian density and velocity perturbations has been understood in great depth. This knowledge has been exploited extensively in extracting a truly impressive score of global cosmological parameters. Notwithstanding these successes, the more advanced phases of cosmic structure formation are still in need of substantially better understanding. Mildly nonlinear structures do contain a wealth of information on the emergence of cosmic structure at a stage features start to emerge as individually recognizable objects. The anisotropic filamentary and planar structures, the characteristic large underdense void regions and the hierarchical clustering of matter marking the web-like spatial geometry of the megaparsec matter distribution are typical manifestations of mildly advanced gravitational structure formation. The existence of the intriguing foam-like patterns representative of this early nonlinear phase of evolution got revealed by major campaigns to map the galaxy distribution on megaparsec scales revealed while ever larger computer N-body simulations demonstrated that such matter distributions are indeed typical manifestations of the gravitational clustering process. Nonetheless, despite the enormous progress, true insight and physical understanding have remained limited. The lack of readily accessible symmetries and the strong non-local influences are a major impediment towards the development-relevant analytical descriptions. The hierarchical nature of the gravitational clustering process forms an additional complication. While small structures materialize before they merge into large entities, each cosmic structure consists of various levels of substructure so that instead of readily recognizing one characteristic spatial scale we need to take into account a range of scales. Insight into the complex interplay of emerging structures throughout the Universe and at a range of spatial scales has been provided through a variety of analytical approximations. Computer simulations have provided us with a good impression of the complexities of the emerging matter distribution, but for the analysis of the resulting patterns and hierarchical substructure the toolbox of descriptive measures is still largely heuristic, ad hoc and often biased in character.

While cosmological theories are describing the development of structure in terms of continuous density and velocity fields, our knowledge stems mainly from discrete samplings of these fields. In the observational reality, galaxies are the main tracers of the cosmic web and it is through measuring the redshift distribution of galaxies that we have been able to map its structure. Likewise, simulations of the evolving cosmic matter distribution are almost exclusively based on N-body particle computer calculation, involving a discrete representation of the features we seek to study.

Both the galaxy distribution and the particles in an N-body simulation are examples of spatial point processes in that they are (1) *discretely sampled* and (2) have an *irregular spatial distribution.* The translation of *discretely sampled* and *spatially irregularly distributed* sampled objects into the related continuous fields is not necessarily a trivial procedure. The standard procedure is to use a filter to process the discrete samples into a representative reconstruction of the underlying continuous field. It is the design of the filter which determines the character of the reconstruction.

Astronomical applications are usually based on a set of user-defined filter functions. Nearly without exception, the definition of these includes

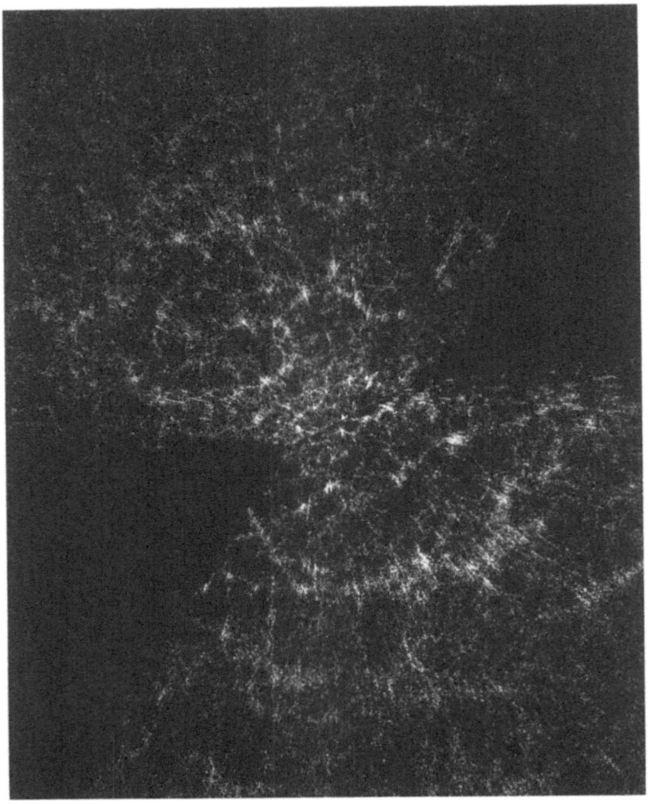

Fig. 2. SDSS is the largest and most systematic sky survey in the history of astronomy. It is a combination of a sky survey in five optical bands of 25% of the celestial (northern) sphere. Each image is recorded on CCDs in these five bands. On the basis of the images/colours and their brightness, a million galaxies are subsequently selected for spectroscopic follow-up. The total sky area covered by SDSS is 8452 square degrees. Objects will be recorded to $m_{lim} = 23.1$. In total the resulting atlas will contain 10^8 stars, 10^8 galaxies and 10^5 quasars. Spectra are taken of around 10^6 galaxies, 10^5 quasars and 10^5 unusual stars (in our galaxy). Of the five public data releases four have been accomplished, i.e. 6670 square degrees of images is publicly available, along with 806,400 spectra. In total, the sky survey is now completely done, 107%), the spectroscopic survey for 68%. This image is taken from a movie made by Subbarao, Surendran & Landsberg (see website: http://astro.uchicago.edu/cosmus/projects/sloangalaxies/). It depicts the resulting redshift distribution after the third public data release. It concerns 5282 square degrees and contained 528,640 spectra, of which 374,767 galaxies

pre-conceived knowledge about the features one is looking for. A telling example is the use of a Gaussian filter. This filter will suppress the presence of any structures on a scale smaller than the characteristic filter scale. Moreover, nearly always it is a spherically defined filter which tends to smooth out any existing anisotropies. Such procedures may be justified in situations in which we are particularly interested in objects of that size or in which physical understanding suggests the smoothing scale to be of particular significance. On the other hand, they may be crucially inept in situations of which we do not know in advance the properties of the matter distribution. The gravitational clustering process in the case of hierarchical cosmic structure formation scenarios is a particularly notorious case. As it includes structures over a vast range of scales and displays a rich palette of geometries and patterns, any filter design tends to involve a discrimination against one or more – and possibly interesting – characteristics of the cosmic matter distribution it would be preferable to define filter and reconstruction procedures that tend to be defined by the discrete point process itself.

A variety of procedures that seek to define and employ more **natural** filters have been put forward in recent years. The scale of smoothing kernels can be adapted to the particle number density, yielding a density field that retains to a large extent the spatial information of the sampled density field. While such procedures may still have the disadvantage of a rigid user-defined filter function and filter geometry, a more sophisticated, versatile and particularly promising class of functions is that of wavelet-defined filters (see contribution of B.J.T. Jones). These can be used to locate contributions on a particular scale, and even to trace features of a given geometry. While its successes have been quite remarkable, the success of the application is still dependent on the particular class of employed wavelets.

In this contribution we will describe in extension the technique of the Delaunay Tessellation Field Estimator. DTFE is based on the use of the Voronoi and Delaunay tessellations of a given spatial point distribution to form the basis of a natural, fully self-adaptive filter for discretely sampled fields in which the Delaunay tessellations are used as multidimensional interpolation intervals. The method has been defined, introduced and developed by [202] and forms an elaboration of the velocity interpolation scheme introduced by [27].

Our focus on DTFE will go along with a concentration on the potential of spatial **tessellations** as a means of estimating and interpolating discrete point samples into continuous field reconstructions. In particular we will concentrate on the potential of *Voronoi* and *Delaunay tessellations*. Both tessellations – each other's *dual* – are fundamental concepts in the field of stochastic geometry. Exploiting the characteristics of these tessellations, we will show that the DTFE technique is capable of delineating the hierarchical and anisotropic nature of spatial point distributions and in outlining the presence and shape of void-like regions. The spatial structure of the cosmic matter distribution is marked by precisely these, and precisely this potential has been the incentive for analysing cosmic large-scale structure. DTFE exploits three particular properties of Voronoi and Delaunay tessellations. The tessellations are very sensitive to the local point density, in that the volume of the tessellation cells is a strong function of the local (physical) density. The DTFE method uses this fact to define a local estimate of the density. Equally important is their sensitivity to the local geometry of the point distribution. This allows them to trace anisotropic features such as encountered in the cosmic web. Finally it uses the adaptive and minimum triangulation properties of Delaunay tessellations to use them as adaptive spatial interpolation intervals for irregular point distributions. In this it is the first-order version of the *natural neighbour method* (NN method). The theoretical basis for the NN method, a generic smooth and local higher order spatial interpolation technique developed by experts in the field of computational geometry, has been worked out in great detail by Sibson [220, 221] and Watson [253]. As has been demonstrated by telling examples in geophysics (Braun & Sambridge [41]) and solid mechanics (Sukumar [233]), NN methods hold tremendous potential for grid-independent analysis and computations.

Following the definition of the DTFE technique, we will present a systematic treatment of various virtues of relevance to the cosmic matter distribution. The related performance of DTFE will be illustrated by means of its success in analysing computer simulations of cosmic structure formation as well as that of the galaxy distribution in large-scale redshift surveys such as the 2dFGRS and SDSS surveys. Following the definition and properties of DTFE, we will pay attention to extensions of the project. Higher order natural neighbour renditions of density and velocity fields involve improvements in terms of smoothness and discarding artefacts of the underlying tessellations.

Following the determination of the "raw" DTFE produces density and velocity fields – and/or other sampled fields, e.g. temperature values in SPH simulations – the true potential of DTFE is realized in the subsequent stage in which the resulting DTFE fields get processed and analysed. We will discuss an array of techniques which have been developed to extract information on aspects of the megaparsec matter distribution. Straightforward processing involves simple filtering and the production of images from the reconstructed field. Also, we will shortly discuss the use of the tessellation fields towards defining new measures for the topology and geometry of the underlying matter distribution. The determination of the Minkowski functionals of isodensity surfaces, following the SURFGEN formalism (Sahni et al. [198], Sheth et al. [217], Shandarin et al. [213]), can be greatly facilitated on the basis of the Delaunay triangulation itself and the estimated DTFE density field. Extending the topological information contained in the Minkowski functionals leads us to the concept of *Betti* numbers and *α-shapes*, a filtration of the Delaunay complex of a data set (Edelsbrunner et al. [68], Edelsbrunner & Mücke [69], Edelsbrunner et al. [72]). With the associated persistence diagrams the α-shapes encode the evolution of the Betti numbers. We have started to add to the arsenal of tools to quantify the patterns in the megaparsec matter distribution (Vegter et al. [251], Eldering [76]). Particularly interesting are the recently developed elaborate and advanced techniques of *watershed void identification* (Platen, van de Weygaert & Jones [177]) and *multiscale morphology filter* (Aragón-Calvo [9], Aragón-Calvo et al. [10]). These methods enable the unbiased identification and measurement of voids, walls, filaments and clusters in the galaxy distribution.

Preceding the discussion on the DTFE and related tessellation techniques, we will first have to describe the structure, dynamics and formation of the *cosmic web*. The complex structure of the cosmic web, and the potential information it does contain, has been the ultimate reason behind the development of the DTFE.

2 Introduction: The Cosmic Web

Macroscopic patterns in nature are often due the collective action of basic, often even simple, physical processes. These may yield a surprising array of complex and genuinely unique physical manifestations. The macroscopic organization into complex spatial patterns is one of the most striking. The rich

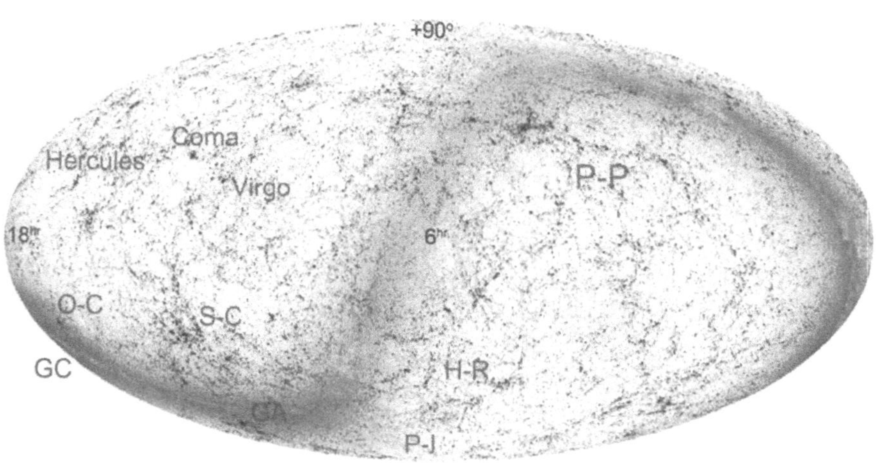

Fig. 3. Equatorial view of the 2MASS galaxy catalogue (6h RA at centre). The grey scale represents the total integrated flux along the line of sight – the nearest (and therefore brightest) galaxies produce a vivid contrast between the Local Supercluster (centre left) and the more distant cosmic web. The dark band of the Milky Way clearly demonstrates where the galaxy catalogue becomes incomplete due to source confusion. Some well-known large-scale structures are indicated: P-P=Perseus-Pisces supercluster; H-R=Horologium-Reticulum supercluster; P-I=Pavo-Indus supercluster; GA=Great Attractor; GC=Galactic Centre; S-C=Shapley Concentration; O-C=Ophiuchus Cluster; Virgo, Coma and Hercules=Virgo, Coma and Hercules superclusters. The Galactic "anti-centre" is front and centre, with the Orion and Taurus giant molecular clouds forming the dark circular band near the centre. From Jarrett [109]

morphology of such systems and patterns represents a major source of information on the underlying physics. This has made them the subject of a major and promising area of inquiry.

2.1 Galaxies and the Cosmic Web

One of the most striking examples of a physical system displaying a salient geometrical morphology, and the largest in terms of sheer size, is the Universe as a whole. The past few decades have revealed that on scales of a few up to more than a hundred megaparsec, the galaxies conglomerate into intriguing

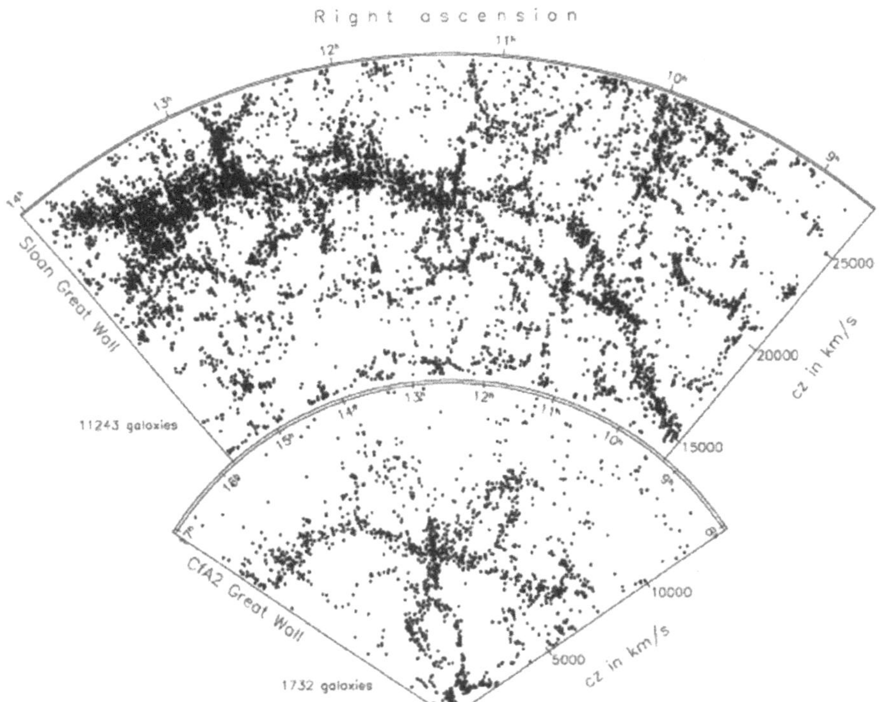

Fig. 4. The CfA Great Wall (*bottom slice*, Geller & Huchra [90]) compared with the Sloan Great Wall (*top slice*). Both structures represent the largest coherent structural in the galaxy redshift surveys in which they were detected, the CfA redshift survey and the SDSS redshift survey. The (CfA) Great Wall is a huge planar concentration of galaxies with dimensions that are estimated to be of the order of $60\,h^{-1} \times 170\,h^{-1} \times 5\,h^{-1}$ Mpc. Truly mind-boggling is the Sloan Great Wall, a huge conglomerate of clusters and galaxies. With a size in the order of $400\,h^{-1}$Mpc it is at least three times larger than the CfA Great Wall. It remains to be seen whether it is a genuine physical structure or mainly a stochastic arrangement and enhancement, at a distance coinciding with the survey's maximum in the radial selection function. Courtesy: J. Richard Gott and M. Juric, Princeton University

cellular or foam-like patterns that pervade throughout the observable cosmos. An initial hint of this cosmic web was seen in the view of the local Universe offered by the first CfA redshift slice (de Lapparent, Geller & Huchra [128]). In recent years, this view has been expanded dramatically to the present grand vistas offered by the hundred thousands of galaxies in the 2dF – two-degree Field – Galaxy Redshift Survey, the 2dFGRS (Colless et al. [57]), and SDSS (e.g. Tegmark et al. [235]) galaxy redshift surveys.[1] Galaxies are found in dense, compact clusters, in less dense filaments and in sheet-like walls which surround vast, almost empty regions called *voids*. This is most dramatically illustrated by the 2dFGRS and SDSS maps. The published maps of the distribution of nearly 250,000 galaxies in two narrow "slice" regions on the sky yielded by the 2dFGRS surveys reveal a far from homogeneous distribution (Fig. 1). Instead, we recognize a sponge-like arrangement, with galaxies aggregating in striking geometric patterns such as prominent filaments, vaguely detectable walls and dense compact clusters on the periphery of giant voids.[2] The three-dimensional view emerging from the SDSS redshift survey provides an even more convincing image of the intricate patterns defined by the cosmic web (Fig. 1). A careful assessment of the galaxy distribution in our immediate vicinity reveals us how we ourselves are embedded and surrounded by beautifully delineated and surprisingly sharply defined web-like structures. In particular the all-sky nearby infrared 2MASS survey (see Fig. 3) provides us with a meticulously clear view of the web surrounding us. The cosmic web is outlined by galaxies populating huge *filamentary* and *wall-like* structures, the sizes of the most conspicuous one frequently exceeding $100\,h^{-1}$ Mpc. The closest and best studied of these massive anisotropic matter concentrations can be identified with known supercluster complexes, enormous structures comprising one or more rich clusters of galaxies and a plethora of more modestly sized clumps of galaxies. A prominent and representative nearby specimen is the Perseus-Pisces supercluster, a $5\,h^{-1}$ wide ridge of at least $50\,h^{-1}$ Mpc length, possibly extending out to a total length of $140\,h^{-1}$ Mpc. While such giant elongated structures are amongst the most conspicuous features of the megaparsec matter distribution, filamentary features are encountered over a range of scales and seem to represent a ubiquitous and universal state of concentration of matter. In addition to the presence of such filaments, the galaxy distribution also contains vast planar assemblies. A striking local example is the *Great Wall*, a huge planar concentration of galaxies with dimensions that are estimated to be of the order of $60\,h^{-1} \times 170\,h^{-1} \times 5\,h^{-1}$ Mpc (Geller & Huchra [90]). In both the SDSS and 2dF surveys even more impressive planar

[1] See `http://www.mso.anu.edu.au/2dFGRS/` and `http://www.sdss.org/`

[2] It is important to realize that the interpretation of the megaparsec galaxy distribution is based on the tacit yet common assumption that it forms a fair reflection of the underlying matter distribution. While there are various indications that this is indeed a reasonable approximation, as long as the intricate and complex process of the formation of galaxies has not been properly understood this should be considered as a plausible yet heuristic working hypothesis.

complexes were recognized, with dimensions substantially in excess of those of the local Great Wall. At the moment, the so-called *SDSS Great Wall* appears to be the largest known structure in the Universe (see Fig. 4). Gradually galaxy surveys are opening the view onto the large-scale distribution of galaxies at high redshifts. The Subaru survey has even managed to map out a huge filamentary feature at a redshift of $z \sim 3.1$, perhaps the strongest evidence for the existence of pronounced cosmic structure at early cosmic epochs (Fig. 5).

Of utmost significance for our inquiry into the issue of cosmic structure formation is the fact that the prominent structural components of the galaxy distribution – clusters, filaments, walls and voids – are not merely randomly and independently scattered features. On the contrary, they have arranged themselves in a seemingly highly organized and structured fashion, the *cosmic foam* or *cosmic web*. They are woven into an intriguing *foam-like* tapestry that permeates the whole of the explored Universe. The vast underpopulated *void* regions in the galaxy distribution represent both contrasting and complementary spatial components to the surrounding *planar* and *filamentary* density enhancements. At the intersections of the latter we often find the most prominent density enhancements in our Universe, the *clusters* of galaxies.

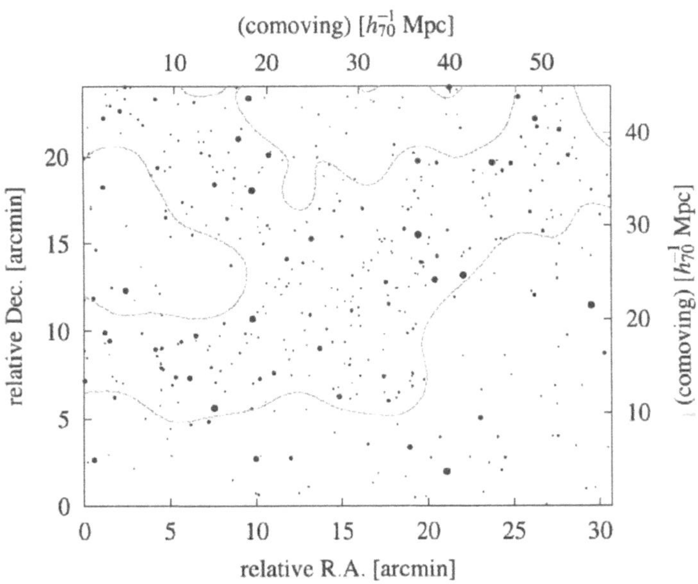

Fig. 5. The cosmic web at high redshifts: a prominent web-like feature at a redshift $z \sim 3.1$ found in a deep view obtained by the Subaru telescope. Large-sky distribution of 283 strong Lyα emitters (*black filled circles*), the Lyα absorbers (*gray filled circles*), and the extended Lyα emitters (*gray open squares*).The dashed lines indicate the high-density region of the strong Lyα emitters. From Hayashino et al. [101]

2.2 Cosmic Nodes: Clusters

Within and around these anisotropic features we find a variety of density condensations, ranging from modest groups of a few galaxies up to massive compact *galaxy clusters*. The latter stand out as the most massive, and most recently, fully collapsed and virialized objects in the Universe. Approximately 4% of the mass in the Universe is assembled in rich clusters. They may be regarded as a particular population of cosmic structure beacons as they typically concentrate near the interstices of the cosmic web, *nodes* forming a recognizable tracer of the cosmic matter distribution (Borgani & Guzzo [38]). Clusters function not only as wonderful tracers of structure over scales of dozens up to hundred of megaparsec (Fig. 7) but also as useful probes for precision cosmology on the basis of their unique physical properties.

The richest clusters contain many thousands of galaxies within a relatively small volume of only a few megaparsec size. For instance, in the nearby Virgo and Coma clusters, more than a thousand galaxies have been identified within a radius of a mere $1.5\,h^{-1}$ Mpc around their core (see Fig. 6). Clusters are first and foremost dense concentrations of dark matter, representing overdensities $\Delta \sim 1000$. In a sense galaxies and stars only form a minor constituent of clusters. The cluster galaxies are trapped and embedded in the deep gravitational wells of the dark matter. These are identified as a major source of X-ray emission, emerging from the diffuse extremely hot gas trapped in them. While it fell into the potential well, the gas got shock-heated to temperatures in excess of $T > 10^7$ K, which results in intense X-ray emission due to the bremsstrahlung radiated by the electrons in the highly ionized intracluster gas. In a sense clusters may be seen as hot balls of X-ray radiating gas. The amount of intracluster gas in the cluster is comparable to that locked into stars and stands for $\Omega_{ICM} \sim 0.0018$ (Fukugita & Peebles [88]). The X-ray emission

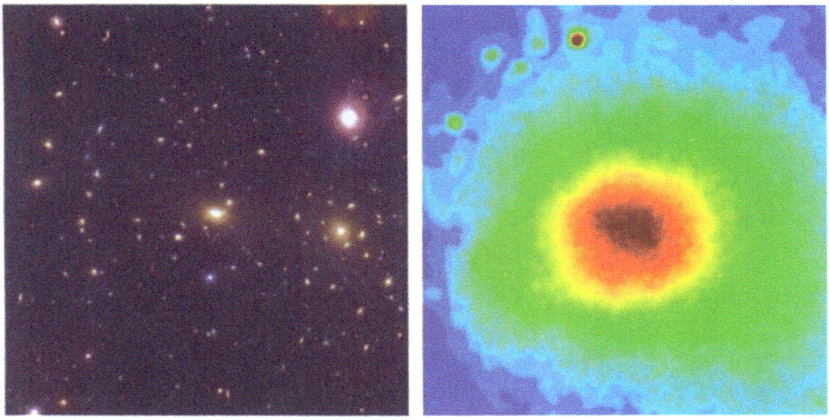

Fig. 6. Comparison of optical and X-ray images of Coma cluster. *Top*: optical image (courtesy: O. Lopez-Cruz; *lower*: X-ray image (ROSAT)

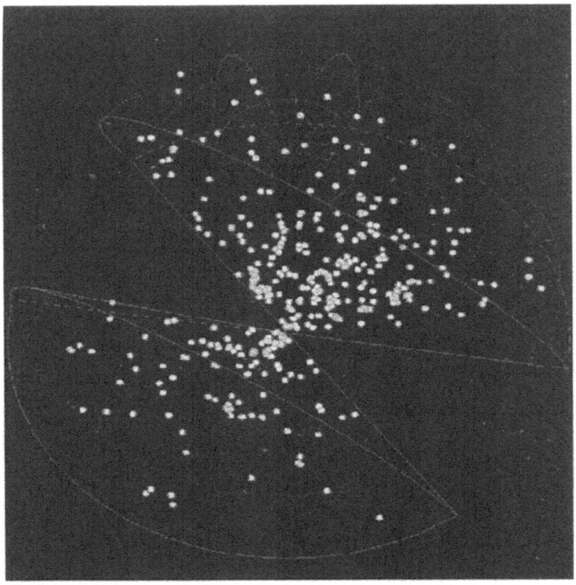

Fig. 7. The spatial cluster distribution. The full volume of the X-ray REFLEX cluster survey within a distance of $600\,h^{-1}$Mpc. The REFLEX galaxy cluster catalogue (Böhringer et al. [34]) contains all clusters brighter than an X-ray flux of $3 \times 10^{-12}\,\mathrm{ergs}^{-1}\,\mathrm{cm}^{-2}$ over a large part of the southern sky. The missing part of the hemisphere delineates the region highly obscured by the galaxy. Courtesy: Borgani & Guzzo [38]). Reproduced with permission from Nature

represents a particularly useful signature, an objective and clean measure of the potential well depth, directly related to the total mass of the cluster (see, e.g., Reiprich & Böhringer [187]). Through their X-ray brightness they can be seen out to large cosmic depths. The deep gravitational dark matter wells also strongly affect the path of passing photons. While the resulting strong lensing arcs form a spectacular manifestation, it has been the more moderate distortion of background galaxy images in the weak lensing regime (Kaiser [112], Kaiser & Squires [113]) which has opened up a new window onto the Universe. The latter has provided a direct probe of the dark matter content of clusters and the large-scale Universe (for a review see, e.g., Mellier [152], Refregier [184]).

2.3 Cosmic Depressions: Voids

Complementing this cosmic inventory leads to the existence of large *voids*, enormous regions with sizes in the range of 20–50 h^{-1} Mpc that are practically devoid of any galaxy, usually roundish in shape and occupying the major share of space in the Universe. Forming an essential ingredient of the *cosmic*

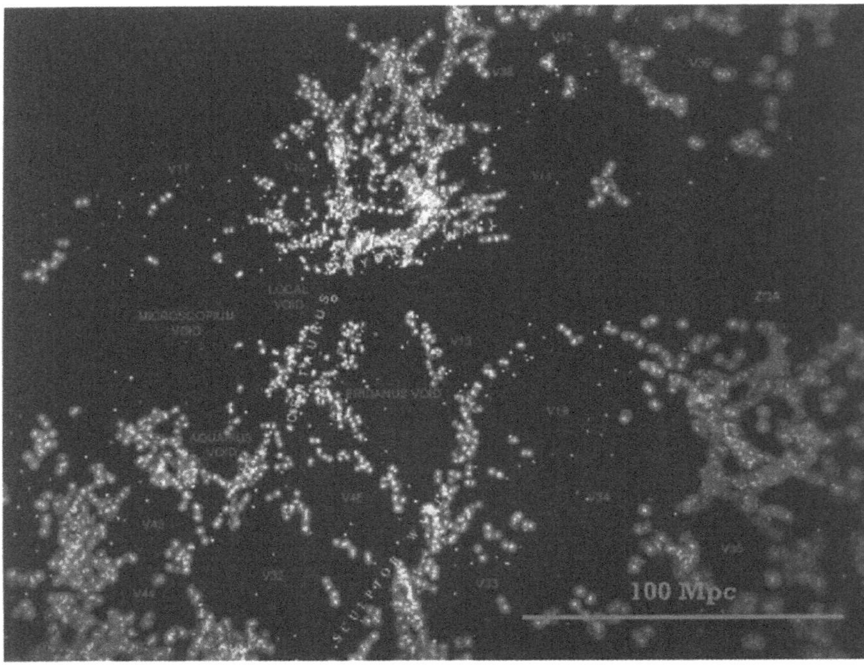

Fig. 8. A region of the 6dF redshift survey marked by the presence of various major voids. The image concerns a three-dimensional rendering of the galaxy distribution in a 1000 km/s thick slice along the supergalactic SGX direction, at $\mathrm{SGX} = -2500\,\mathrm{km\ s}^{-1}$. Courtesy: A. Fairall

web, they are surrounded by elongated filaments, sheet-like walls and dense compact clusters.

Voids have been known as a feature of galaxy surveys since the first surveys were compiled (Chincarini & Rood [47], Gregory & Thompson [99], Einasto et al. [73]). Following the discovery by Kirshner et al. [122, 123] of the most outstanding specimen, the Boötes void, a hint of their central position within a web-like arrangement came with the first CfA redshift slice (de Lapparent et al. [128]). This view has been expanded dramatically as maps of the spatial distribution of hundreds of thousands of galaxies in the 2dFGRS (Colless et al. [57]) and SDSS redshift survey (Abazajian et al. [1]) became available, recently supplemented with a high-resolution study of voids in the nearby Universe based on the 6dF survey (Heath Jones et al. [102], Fairall [80]). The 2dFGRS and SDSS maps of Fig. 1 and the void map of the 6dF survey in Fig. 8 form telling illustrations.

Voids in the galaxy distribution account for about 95% of the total volume (see Kauffmann & Fairall [116], El Adi et al. [74], El Ad & Piran [75], Hoyle & Vogeley [104], Plionis & Basilakos [180], Rojas et al. [192]). The typical sizes of voids in the galaxy distribution depend on the galaxy population used to

define the voids. Voids defined by galaxies brighter than a typical L_* galaxy tend to have diameters of order 10–20 h^{-1} Mpc, but voids associated with rare luminous galaxies can be considerably larger; diameters in the range of 20 h^{-1}–50 h^{-1} Mpc are not uncommon (e.g. Hoyle & Vogeley [104], Plionis & Basilakos [180]). These large sizes mean that only now we are beginning to probe a sufficiently large cosmological volume to allow meaningful statistics with voids to be done. As a result, the observations are presently ahead of the theory.

3 Cosmic Structure Formation: From Primordial Noise to the Cosmic Web

The fundamental cosmological importance of the *cosmic web* is that it comprises features on a typical scale of tens of megaparsec, scales at which the Universe still resides in a state of moderate dynamical evolution. Structures have only freshly emerged from the almost homogeneous pristine Universe and have not yet evolved beyond recognition. Therefore, they still retain a direct link to the matter distribution in the primordial Universe and thus still contain a wealth of direct information on the cosmic structure formation process.

In our exploration of the cosmic web and the development of appropriate tools towards the analysis of its structure, morphology and dynamics, we start from the assumption that the cosmic web is traced by a population of discrete objects, either galaxies in the real observational world or particles in that of computer simulations. The key issue will be to reconstruct the underlying continuous density and velocity field, retaining the geometry and morphology of the web-like structures in all its detail.

In this we will pursue the view that filaments are the basic elements of the cosmic web, the key features around which most matter will gradually assemble and the channels along which matter is transported towards the highest density knots within the network, the clusters of galaxies. Likewise we will emphasize the crucial role of the voids – the large underdense and expanding regions occupying most of space – in the spatial organization of the various structural elements in the cosmic web. One might even argue that it is the voids which should be seen as the key ingredients of the cosmic matter distribution. This tends to form the basis for geometrical models of the megaparsec scale matter distribution, with the Voronoi model as its main representative (see, e.g., van de Weygaert [242, 244, 245, 246]).[3]

[3] These Voronoi models are spatial models for cellular/web-like galaxy distributions, not to be confused with the application of Voronoi tessellations in DTFE and the tessellation-based methods towards spatial interpolation and reconstruction.

3.1 Gravitational Instability

The standard paradigm of cosmic structure formation is that of gravitational instability scenarios (Peebles [174]), Zeldovich [259]). Structure in the Universe is the result of the gravitational growth of tiny density perturbations and the accompanying tiny velocity perturbations in the primordial Universe. Supported by an impressive body of evidence, primarily that of temperature fluctuations in the cosmic microwave background (Smoot et al. [223], Bennet et al. [20], Spergel et al. [226]), the character of the primordial random density and velocity perturbation field is that of a *homogeneous and isotropic spatial Gaussian process.* Such fields of primordial Gaussian perturbations in the gravitational potential are a natural product of an early inflationary phase of our Universe.

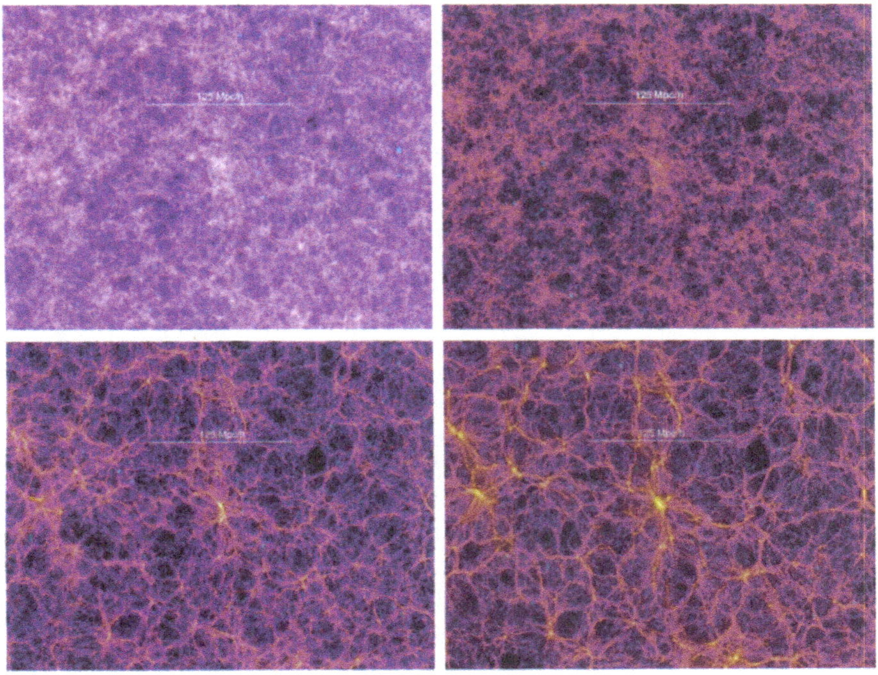

Fig. 9. The cosmic web in a box: a set of four time slices from the Millennium simulation of the ΛCDM model. The frames show the projected (*dark*) matter distribution in slices of thickness $15\,h^{-1}$ Mpc, extracted at $z = 8.55, z = 5.72, z = 1.39$ and $z = 0$. These redshifts correspond to cosmic times of 600 Myr, 1, 4.7 and 13.6 Gyr after the Big Bang. The four frames have a size of $125\,h^{-1}$ Mpc. The evolving mass distribution reveals the major characteristics of gravitational clustering: the formation of an intricate filamentary web, the hierarchical build-up of ever more massive mass concentrations and the evacuation of large underdense voids. Courtesy: V. Springel & Virgo consortium

The formation and moulding of structure is a result of the gravitational growth of the primordial density- and velocity perturbations. Gravity in slightly overdense regions will be somewhat stronger than the global average gravitational deceleration, as will be the influence they exert over their immediate surroundings. In these regions the slowdown of the initial cosmic expansion is correspondingly stronger, and when the region is sufficiently overdense it may even come to a halt, turn around and start to contract. If or as long as pressure forces are not sufficient to counteract the infall, the overdensity will grow without bound, assemble more and more matter by accretion of matter from its surroundings and ultimately fully collapse to form a gravitationally bound and virialized object. In this way the primordial overdensity finally emerges as an individual recognizable denizen of our Universe, their precise nature (galaxy, cluster, etc.) and physical conditions determined by the scale, mass and surroundings of the initial fluctuation.

3.2 Nonlinear Clustering

Once the gravitational clustering process has progressed beyond the initial linear growth phase we see the emergence of complex patterns and structures in the density field. Highly illustrative of the intricacies of the structure formation process is that of the state-of-the-art N-body computer simulation, the Millennium simulation by Springel et al. [229]. Figure 9 shows four time frames out of this massive 10^{10} particle simulation of a ΛCDM matter distribution in a $500\,h^{-1}$ Mpc box. The time frames correspond to redshifts $z = 8.55$, $z = 5.72$, $z = 1.39$ and $z = 0$ (i.e. at epochs 600 Myr, 1, 4.7 and 13.6 Gyr after the Big Bang). The earliest time frame is close to that of the condensation of the first stars and galaxies at the end of the *Dark Ages* and the reionization of the gaseous IGM by their radiation. The frames contain the dark matter particle distribution in a $15\,h^{-1}$ Mpc thick slice of a $125\,h^{-1}$ Mpc region centred on the central massive cluster of the simulation.

The four frames provide a beautiful picture of the unfolding cosmic web, starting from a field of mildly undulating density fluctuations towards that of a pronounced and intricate filigree of filamentary features, dented by dense compact clumps at the nodes of the network. Clearly visible is the hierarchical nature in which the filamentary network builds up. At first consisting of a multitude of small-scale edges, they quickly merge into a few massive elongated channels.

Large N-body simulations like the Millennium simulation and the many others currently available all reveal a few "universal" characteristics of the (mildly) nonlinear cosmic matter distribution. Three key characteristics of the megaparsec Universe stand out:

- Hierarchical clustering
- Web-like spatial geometry
- Voids

Hierarchical clustering implies that the first objects to condense first are small and that ever larger structures form through the gradual merging of smaller structures. Usually an object forms through the accretion of all matter and the fusion of all substructures within its realm, including that of the small-scale objects which had condensed out at an earlier stage. The *second* fundamental aspect is that of *anisotropic gravitational collapse.* Aspherical overdensities, on any scale and in any scenario, will contract such that they become increasingly anisotropic. At first they turn into a flattened *pancake*, rapidly followed by contraction into an elongated filament and possibly, dependent on scale, total collapse into a galaxy or a cluster may follow. This tendency to collapse anisotropically finds its origin in the intrinsic primordial flattening of the overdensity, augmented by the anisotropy of the gravitational force field induced by the external matter distribution (i.e. by tidal forces). It is evidently the major agent in shaping the web-like cosmic geometry. The *third* manifest feature of the megaparsec Universe is the marked and dominant presence of large roundish underdense regions, the *voids*. They form in and around density troughs in the primordial density field. Because of their lower interior gravity they will expand faster than the rest of the Universe, while their internal matter density rapidly decreases as matter evacuates their interior. They evolve in the nearly empty void regions with sharply defined boundaries marked by filaments and walls. Their essential role in the organization of the cosmic matter distribution got recognized early after their discovery. Recently, their emergence and evolution has been explained within the context of hierarchical gravitational scenarios [218].

The challenge for any viable analysis tool is to trace, highlight and measure each of these aspects of the cosmic web. Ideally it should be able to do so without resorting to user-defined parameters or functions and without affecting any of the other essential characteristics. We will argue in this contribution that the *DTFE* method, a linear version of *natural neighbour* interpolation, is indeed able to deal with all three aspects (see Fig. 11).

4 Spatial Structure and Pattern Analysis

Many attempts to describe, let alone identify, the features and components of the cosmic web have been of a mainly heuristic nature. There is a variety of statistical measures characterizing specific aspects of the large-scale matter distribution (for an extensive review see Martínez & Saar [147]). For completeness and comparison, we list briefly a selection of methods for structure characterization and finding. It is perhaps interesting to note two things about this list:

(a) each of the methods tends to be specific to one particular structural entity
(b) there are no explicit wall-finders.

This emphasizes an important aspect of our scale space approach: it provides a uniform approach to finding blobs, filaments and walls as individual objects that can be catalogued and studied.

4.1 Structure from Higher Moments

The clustering of galaxies and matter is most commonly described in terms of a hierarchy of correlation functions (Peebles [174]). The two-point correlation function – and the equivalent power spectrum, its Fourier transform (Peacock & Dodds [173], Tegmark et al. [235]) – remains the mainstay of cosmological clustering analysis and has a solid physical basis. However, the nontrivial and nonlinear patterns of the cosmic web are mostly a result of the phase correlations in the cosmic matter distribution (Ryden & Gramann [196], Chiang & Coles [46], Coles & Chiang [56]). While this information is contained in the moments of cell counts (Peebles [174], de Lapparent et al. [129], Gaztañaga [89]) and, more formally so, in the full hierarchy of M-point correlation functions ξ_M, except for the lowest orders their measurement has proven to be practically unfeasible (Peebles [174], Szapudi [234], Jones et al. [110]). Problem remains that these higher order correlation functions do not readily translate into a characterization of identifiable features in the cosmic web.

The void probability function (White [225], Lachieze-Rey et al. [125]) provides a characterization of the "voidness" of the Universe in terms of a function that combined information from many higher moments of the point distribution. But, again, this did not provide any identification of individual voids.

4.2 Topological Methods

The shape of the local matter distribution may be traced on the basis of an analysis of the statistical properties of its inertial moments (iBabul & Starkman [14], Luo & Vishniac [142], Basilakos et al. [19]). These concepts are closely related to the full characterization of the topology of the matter distribution in terms of four Minkowski functionals (Mecke et al. [150], Schmalzing et al. [207]). They are solidly based on the theory of spatial statistics and also have the great advantage of being known analytically in the case of Gaussian random fields. In particular, the *genus* of the density field has received substantial attention as a strongly discriminating factor between intrinsically different spatial patterns (Gott et al. [98], Hoyle & Vogeley [105]).

The Minkowski functionals provide global characterizations of structure. An attempt to extend its scope towards providing locally defined topological measures of the density field has been developed in the SURFGEN project defined by Sahni and Shandarin and their co-workers (Sahni et al. [198], Sheth et al. [217], Shandarin et al. [213]). The main problem remains the user-defined, and thus potentially biased, nature of the continuous density field inferred from the sample of discrete objects. The usual filtering techniques suppress substructure on a scale smaller than the filter radius, introduce artificial topological features in sparsely sampled regions and diminish the flattened or elongated morphology of the spatial patterns. Quite possibly the introduction of more advanced geometry-based methods to trace the density field may prove a major advance towards solving this problem. Martínez et al. [148] have generalized the use of Minkowski functionals by calculating their values in variously smoothed volume-limited subsamples of the 2dF catalogue.

4.3 Cluster and Filament Finding

In the context of analysing distributions of galaxies we can think of cluster finding algorithms. There we might define a cluster as an aggregate of neighbouring galaxies sharing some localized part of velocity space. Algorithms like HOP attempt to do this. However, there are always issues arising such as how to deal with substructure: that perhaps comes down to the definition of what a cluster is. Nearly always coherent structures are identified on the basis of particle positions alone. Velocity-space data are often not used since there is no prior prejudice as to what the velocity space should look like.

The connectedness of elongated supercluster structures in the cosmic matter distribution was first probed by means of percolation analysis, introduced and emphasized by Zel'dovich and co-workers (Zeldovich et al. [260]), while a related graph-theoretical construct, the minimum spanning tree of the galaxy distribution, was extensively probed and analysed by Bhavsar and collaborators (Barrow et al. [18], Colberg [54]) in an attempt to develop an objective measure of filamentarity.

Finding filaments joining neighbouring clusters has been tackled, using quite different techniques, by Kolberg et al. [52] and Pimbblet [177]. More general filament finders have been put forward by a number of authors. Skeleton analysis of the density field (Novikov et al. [167]) describes continuous density fields by relating density field gradients to density maxima and saddle points. This is computationally intensive but quite effective, though it does depend on the artefacts in the reconstruction of the continuous density field.

Stoica et al. [230] use a generalization of the classical Candy model to locate and catalogue filaments in galaxy surveys. This approach has the advantage that it works directly with the original point process and does not require the creation of a continuous density field. However, it is very computationally intensive.

A recently developed method, the multiscale morphology filter (Aragón-Calvo [9], Aragón-Calvo et al. [10], see Sect. 15.6), seeks to identify different morphological features over a range of scales. Scale space analysis looks for structures of a mathematically specified type in a hierarchical, scale-independent, manner. It is presumed that the specific structural characteristic is quantified by some appropriate parameter (e.g. density, eccentricity, direction, curvature components). The data are filtered to produce a hierarchy of maps having different resolutions, and at each point, the dominant parameter value is selected from the hierarchy to construct the scale independent map. While this sounds relatively straightforward, in practice a number of things are required to execute the process. There must be an unambiguous definition of the structure-defining characteristic. The implementation of [10] uses the principal components of the local curvature of the density field at each point as a morphology type indicator. This requires that the density be defined at all points of a grid, and so there must be a method for going from a discrete point set to a grid sampled continuous density field. This is done using the DTFE methodology since that does minimal damage to the structural morphology of the density field (see Sect. 8).

4.4 Void Finding

Voids are distinctive and striking features of the cosmic web, yet finding them systematically in surveys and simulations has proved rather difficult. There have been extensive searches for voids in galaxy catalogues and in numerical simulations (see Sect. 2.3). Identifying voids and tracing their outline within the complex spatial geometry of the cosmic web appear to be a nontrivial issue. The fact that voids are almost empty of galaxies means that the sampling density plays a key role in determining what is or is not a void (Schmidt et al. [208]). There is not an unequivocal definition of what a void is and as a result there is considerable disagreement on the precise outline of such a region (see, e.g., Shandarin et al. [214]).

Moreover, void finders are often predicated on building void structures out of cubic cells (Kaufmann & Fairall [116]) or out of spheres (e.g. Patiri et al. [171]). Such methods attempt to synthesize voids from the intersection of cubic or spherical elements and do so with varying degrees of success. Because of the vague and different definitions, and the range of different interests in voids, there is a plethora of void identification procedures (Kauffmann & Fairall [116], El-Ad & Piran [75], Aikio & Mähönen [3], Hoyle & Vogeley [104], Arbabi-Bidgoli & Müller [11], Plionis & Basilakos [180], Patiri et al. [171], Colberg et al. [53], Shandarin et al. [214]). The "voidfinder" algorithm of El-Ad & Piran [75] has been at the basis of most void-finding methods. However, this successful approach will not be able to analyse complex spatial configurations in which voids may have arbitrary shapes and contain a range and variety of substructures. The *Void Finder Comparison Project* of Colberg et al. [55] will clarify many of these issues.

The watershed-based WVF algorithm of Platen, van de Weygaert & Jones [177] (see Sect. 15.7) aims to avoid issues of both sampling density and shape. This new and objective void-finding formalism has been specifically designed to dissect in a self-consistent manner the multiscale character of the void network and the web-like features marking its boundaries. The *Watershed Void Finder* (WVF) is based on the watershed algorithm (Beucher & Lantuejoul [30], Meyer & Beucher [154], Beucher & Meyer [31]). This is a concept from the field of *mathematical morphology* and *image analysis*. The WVF is defined with respect to the DTFE density field of a discrete point distribution (Schaap & van de Weygaert [202]), assuring optimal sensitivity to the morphology of spatial structures and an unbiased probe over the full range of substructure in the mass distribution. Because the WVF void finder does not impose a priori constraints on the size, morphology and shape of voids it has the potential to analyse the intricacies of an evolving void hierarchy.

5 Structural Reconstruction

In the real world it is impossible to get exhaustive values of data at every desired point of space. Also astronomical observations, physical experiments and computer simulations often produce discretely sampled data sets in two, three or more dimensions. This may involve the value of some physical quantity measured at an irregularly distributed set of reference points. Also cosmological theories describe the development of structure in terms of continuous (dark matter) density and velocity fields while to a large extent our knowledge stems from a discrete sampling of these fields.

In the observational reality galaxies are the main tracers of the cosmic web, and it is mainly through measuring of the redshift distribution of galaxies that we have been able to map its structure. Another example is that of the related study of cosmic flows in the nearby Universe, based on the measured peculiar velocities of a sample of galaxies located within this cosmic volume. Likewise, simulations of the evolving cosmic matter distribution are almost exclusively based on N-body particle computer calculation involving a discrete representation of the features we seek to study. Both the galaxy distribution and the particles in an N-body simulation are examples of *spatial point processes* in that they are

- *discretely sampled*
- have an *irregular spatial distribution.*

The principal task for any formalism seeking to process the discretely sampled field is to optimally retain or extract the required information. Dependent on the purpose of a study, various different strategies may be followed. One strategy is to distil various statistical measures, or other sufficiently descriptive cosmological measures, characterizing specific aspects of the large-scale matter distribution (Martínez & Saar [147]; see also see Sect. 4). In essence this involves the compression of the available information into a restricted set of parameters or functions, with the intention to compare or relate these to theoretical predictions. The alternative is to translate the *discretely sampled* and *spatially irregularly distributed* sampled objects into related continuous fields. While demanding in itself, it is complicated by the highly inhomogeneous nature of the sample point distribution. The translation is far from trivial procedure. If forms the subject of an extensive literature in computer science, visualization and applied sciences. An interesting comparison and application of a few different techniques is shown in Fig. 12. It shows how some methods to be discussed in the following sections fare when applied to the reconstruction of a Martian landscape from measurements by the MOLA instrument on the Mars Global Surveyor (Abramov & McEven [2]).

5.1 Spatial Data: Filtering and Interpolation

Instead of direct statistical inference from data sets one can seek to reconstruct the underlying continuous field(s). For a meaningful analysis and interpretation of spatial data, it is indeed often imperative and/or preferable to use methods of parameterization and interpolation to obtain estimates of the related field values throughout the sample volume. The *reconstructed* continuous field may subsequently be processed in order to yield a variety of interesting parameters.

It involves issues of *smoothing* and *spatial interpolation* of the measured data over the sample volume, of considerable importance and interest in many different branches of science. Interpolation is fundamental to graphing, analysing and understanding of spatial data. Key references on the involved problems and solutions include those by Ripley [188], Watson [253] and Cressie [59]. While of considerable importance for astronomical purposes, many available methods escaped attention. A systematic treatment and discussion within the astronomical context is the study by Rybicki & Press [195], who focused on linear systems as they developed various statistical procedures related to linear prediction and optimal filtering, commonly known as Wiener filtering. An extensive, systematic and more general survey of available mathematical methods can be found in a set of publications by Lombardi & Schneider [137, 138, 139].

A particular class of spatial point distributions is the one in which the point process forms a representative reflection of an underlying smooth and continuous density/intensity field. The spatial distribution of the points itself may then be used to infer the density field. This forms the basis for the interpretation and analysis of the large-scale distribution of galaxies in galaxy redshift surveys. The number density of galaxies in redshift survey maps and N-body particles in computer simulations is supposed to be proportional to the underlying matter density.

5.2 Local Interpolation: Natural Neighbour Methods

The complex spatial geometry and large density variations marking the cosmic web ideally should be analysed by a technique which would (1) not lose information against the backdrop of a highly inhomogeneous spatial resolution and (2) which is capable of tracing hierarchically structured and anisotropic spatial patterns in an entirely objective fashion. Nearly all existing techniques for analysing galaxy redshift surveys or numerical simulations of cosmic structure formation have important shortcomings with respect to how they treat

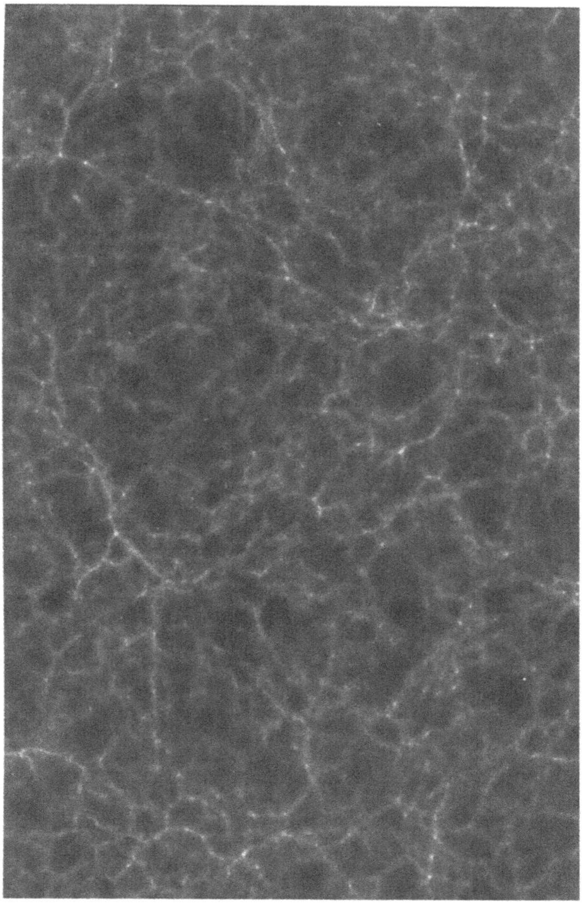

Fig. 10. DTFE processed image of the cosmic web: GIF N-body simulation of structure formation in a ΛCDM cosmology. Part of the central X-slice. Simulation courtesy: J. Colberg

the web-like geometry of the large-scale matter distribution and trace the cosmic matter distribution over a wide range of densities. The limited available mathematical machinery has often been a major obstacle in exploiting the potentially large information content of the cosmic web.

The various aspects characterizing the complex and nontrivial spatial structure of the cosmic web have proven to be notoriously difficult to quantify and describe. For the analysis of web-like patterns the toolbox of descriptive measures is still largely ad hoc and is usually biased towards preconceived notions of their morphology and scale. None of the conventional, nor even specifically designed, measures of the spatial matter distribution have succeeded in describing all relevant features of the cosmic web. Even while they

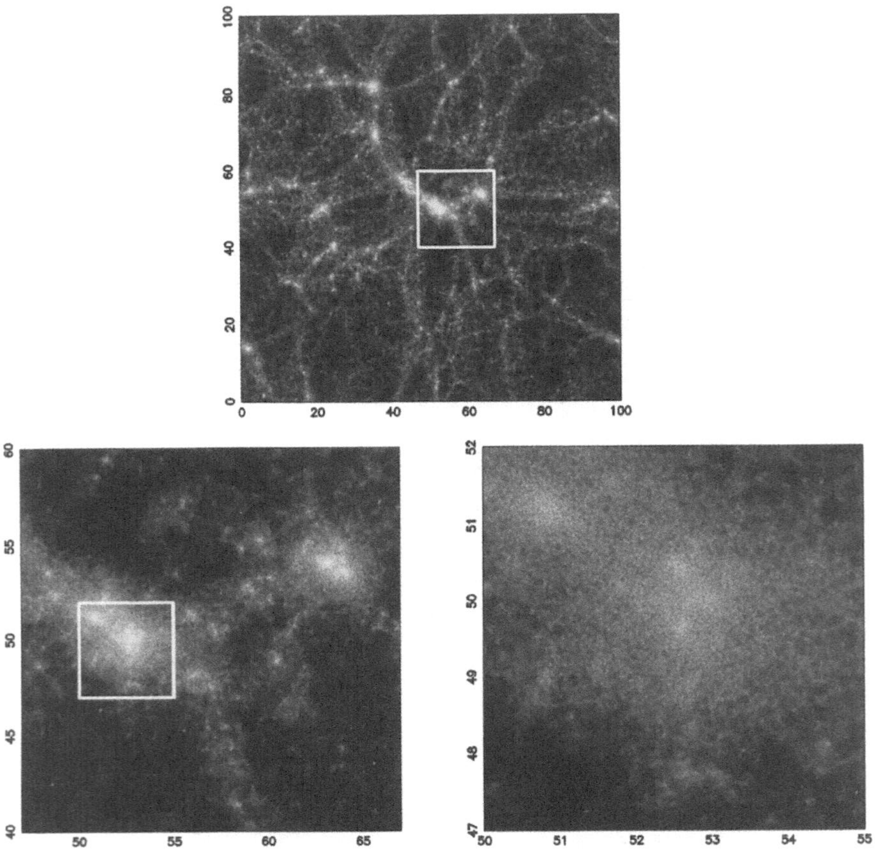

Fig. 11. Cosmic density field illustrating the large dynamic range which is present in the large-scale matter distribution. In the left-hand frame the density field in a $10\,h^{-1}$ Mpc wide slice through a cosmological simulation is depicted. In the subsequent frames zoom-ins focusing on a particular structure are shown. On all depicted scales structures are present

may succeed in quantifying one particular key aspect it usually excludes the ability to do so with other characteristics.

For many applications a "local" interpolation and reconstruction method appears to provide the preferred path. In this case the values of the variable at any point depend only on the data in its neighbourhood (see, e.g., Sambridge, Braun & McQueen [197]). Local schemes usually involve a discretization of the region into an adaptive mesh. In data interpolation this usually represents a more realistic approach, and generically it also tends to be independent of specific model assumptions while they are very suited for numerical modelling applications. When the points have a regular distribution many local methods are available for smooth interpolation in multidimensional space. Smooth, local methods also exist for some specific irregular point distributions. A telling example are the "locally constrained" point distributions employed in applications of the finite element method.

In this review we specifically concentrate on a wide class of tessellation-based *multidimensional* and entirely *local* interpolation procedures, commonly known as *natural neighbour interpolation* (Watson [253] Chap. 6, Braun & Sambridge [41], Sukumar [233], Okabe et al. [169]). The local *natural neighbour* methods are based on the *Voronoi and Delaunay tessellations* of the point sample, basic concepts from the field of stochastic and computational geometry (see Okabe et al. [169] and references therein). These spatial volume-covering divisions of space into mutually disjunct triangular (two-dimensional) or tetrahedral (three-dimensional) cells adapt to the local *density* and the local *geometry* of the point distribution (see Figs. 10 and 11). The natural neighbour interpolation schemes exploit these virtues and thus adapt automatically and in an entirely natural fashion to changes in the density or the geometry of the distribution of sampling points. For the particular requirements posed by astronomical and cosmological data sets, for which it is not uncommon to involve millions of points, we have developed a linear first-order version of *natural neighbour interpolation*, the *Delaunay Tessellation Field Estimator* (DTFE, Bernardeau & van de Weygaert [27], Schaap & van de Weygaert [202], Schaap [201]). Instead of involving user-defined filters which are based on artificial smoothing kernels, the main virtue of natural neighbour methods is that they are intrinsically *self-adaptive* and involve filtering kernels which are defined by the *local density and geometry* of the point process or object distribution.

5.3 Meshless Methods

The natural neighbour schemes, including DTFE, are mesh based. With current technology this is computationally feasible as long as the domain of the matter density field has at most three dimensions. However, also interesting would be to extend attention to six-dimensional phase space, incorporating not only the location of galaxies/particles but also their velocities. This will double the number of dimensions and makes mesh-based methods a real challenge. While the application of DTFE to the analysis of the phase space of dark haloes has been shown to lead to very good results (Arad et al. [8]), studies by Ascasibar & Binney [13] and Sharma & Steinmetz [215] argued it would be far more efficient and reasonably accurate to resort to the simpler construct of a *k-d tree* (for an early astronomical implementation see Bentley [21], Friedmann et al. [87], Bentley & Friedmann [22], van de Weygaert [241]). While Ascasibar & Binney [13] and Sharma & Steinmetz [215] do not take into account that phase space is not a simple metric space but a symplectic one, it may indeed be a real challenge to develop mesh-based methods for the analysis of structure in phase space. Although algorithms for the computation of higher dimensional Voronoi and Delaunay tessellations have been implemented (e.g. in `CGAL`), the high running time and memory use make further processing computationally unfeasible.

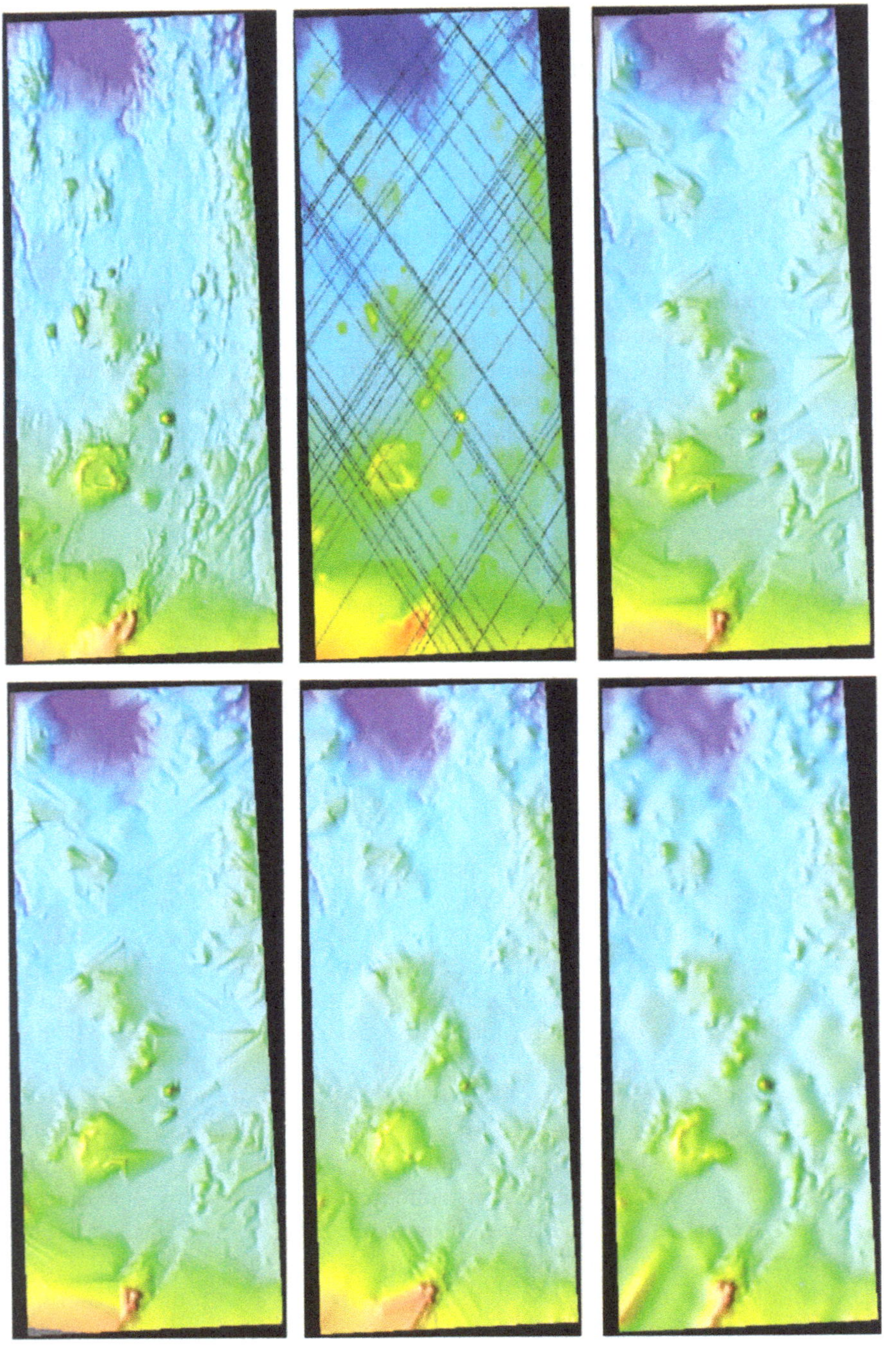

Fig. 12. A comparison of landscape reconstructions of a terrain near the Korolev crater in the north polar region of planet Mars. Altitude data were acquired by the Mars Orbiter Laser Altimeter (MOLA), on board the Mars Global Surveyor.

Meshless spatial interpolation and approximation methods for data sets in spaces of dimensions greater than three may therefore provide the alternative of choice. There are a variety of meshless multivariate data interpolation schemes. Examples are Shepard's interpolant (iShepard [216]), moving least squares approximants (Lancaster & Salkauskas [127]) or Hardy's multiquadrics (Hardy [100]).

Spline Interpolation

Spline interpolation (Schoenberg [209, 210]) is based on interpolating between sampling points by means of higher order polynomials. The coefficients of the polynomial are determined "slightly" non-locally, such that a global smoothness in the interpolated function is guaranteed up to some order of derivative. The order of the interpolating polynomials is arbitrary, but in practice cubic splines are most widely used. Cubic splines produce an interpolated function that is continuous through the second derivative. To obtain a cubic spline interpolation for a data set of $N + 1$ points, N separate cubics are needed. Each of these cubics should have each end point match up exactly with the end points to either side. At the location of these points the two adjacent cubics should also have equal first and second derivatives. A full mathematical derivation can be found in, e.g., Gerald & Wheatley [91, 182].

Spline interpolation is a widely used procedure. Equalizing the derivatives has the effect of making the resulting interpolation appear smooth and visually pleasing. For this reason splines are, for example, frequently used in graphical applications. Splines can provide extremely accurate results when the original sample rate is notably greater than the frequency of fluctuation in the data. Splines however cannot deal very well with large gaps in the data set. Because the gap between two points is represented by a cubic, these result in peaks or troughs in the interpolation. Also, splines are rather artificially defined constructs.

Radial Basis Functions

One of the most promising schemes may be that of radial basis functions, (RBFs, see, e.g., Powell [181], Arnold [12], Wendland [254]). RBFs may be

Fig. 12. (continued) The "original" highly resolved image of the terrain is shown in the top left frame. The comparison concerns data that were measured at the track points indicated in the top centre image. The medium resolution (209 × 492) interpolations are: natural neighbour (*top right*), linear (DTFE) interpolation (*bottom left*), nearest neighbour interpolation (*bottom centre*) and spline interpolation (*bottom right*). Evidently, the nn-neighbour map is the most natural-looking reconstruction. From Abramov & McEwen [2] used to determine a smooth density field interpolating three-dimensional spatial and four-dimensional spatio-temporal data sets, or even data sets in six-dimensional phase space. In the first step of this approach the implicit function is computed as a linear combination of translates of a single radial basis function. This function is determined by the geometric constraint that the input sample points belong to its zero set. If the input is a density map, the geometric constraint boils down to the implicit function interpolating the densities at the input points (and some additional constraints preventing the construction of the zero function). The construction of radial basis functions with suitable interpolation properties is discussed in [204], while an early review of the mathematical problems related to RBF interpolation may be found in [181]. A nice overview of the state of the art in scattered data modelling using radial basis functions may be obtained from the surveys [43, 108, 136].

In practice variational implicit surfaces, based on radial basis functions which minimize some global energy or curvature functional, turn out to be very flexible (Dink et al. [64], Turk & O'Brien [238]): they are adaptive to curvature variations, can be used for enhancement of fine detail and sharp features that are missed or smoothed out by other implicit techniques and can overcome noise in the input data since they are approximating rather than interpolating; especially the use of parameter dependent or anisotropic radial basis functions allows for graceful treatment of sharp features and provides multiple orders of smoothness (Turk & O'Brien [239]).

6 Spatial Tessellations

6.1 Stochastic and Computational Geometry

Random spatial tessellations are a fundamental concept in the fields of *stochastic geometry* and *computational geometry.*

Stochastic geometry, or geometric probability theory, is the subject in mathematics concerned with the problems that arise when we ascribe probability distributions to geometric objects such as points, lines and planes (usually in Euclidian spaces) or to geometric operations such as rotations or projections (see, e.g., Stoyan et al. [231, 232]). A formal and restricted definition of stochastic geometry was given by Stoyan, Kendall & Mecke [231], who defined the field as the branch of mathematics *devoted to the study of geometrical structures which can be described by random sets, in particular by point processes, in suitable spaces.* Since many problems in stochastic geometry are either not solvable analytically, or only in a few non-generic cases, we have to resort to the help of the computer to find a (partial) answer to the problems under consideration. This makes it necessary to find efficient algorithms for constructing the geometrical objects involved.

Computational geometry is the branch of computer science that is concerned with finding the computational procedures for solving geometric problems, not just the geometric problems arising from stochastic geometry but

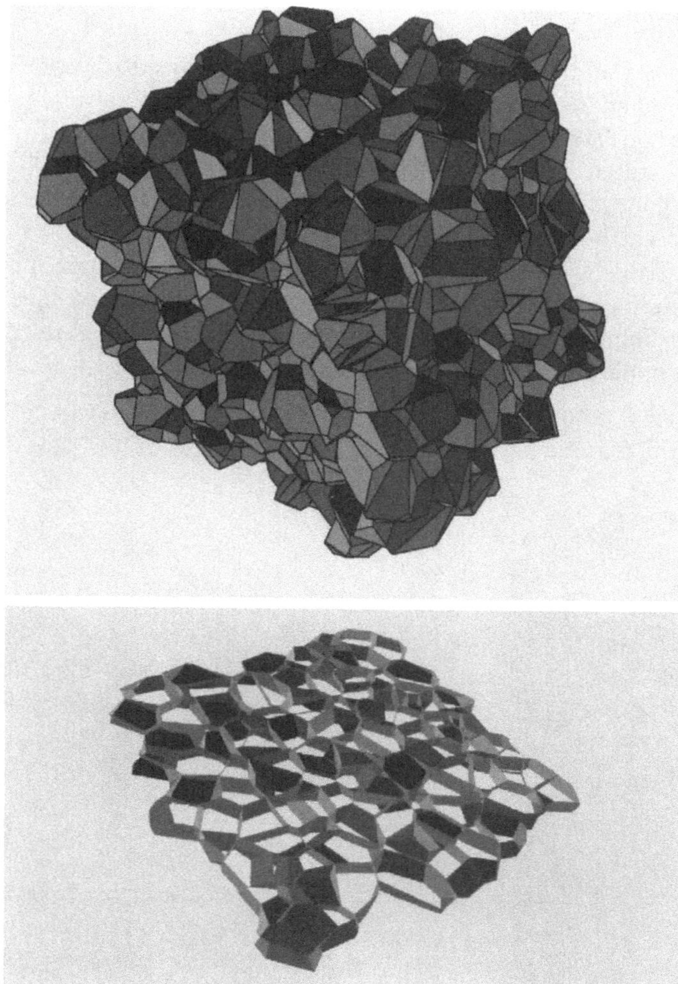

Fig. 13. A full three-dimensional tessellation comprising 1000 Voronoi cells/ polyhedra generated by 1000 Poissonian distributed nuclei. Courtesy: Jacco Dankers

geometric problems in general (Boissonnat et al. [35], de Berg et al. [60], Goodman & O'Rourke [96]). It is concerned with the design and analysis of algorithms and software for processing geometric objects and data. Typical problems are the construction of spatial tessellations, like Voronoi diagrams and Delaunay meshes, the reconstruction of objects from finite point samples or finding nearest neighbours in point sets. Methods from this field have many applications in applied areas like computer graphics, computer vision, robotics, computer aided design and manufacturing.

6.2 Random Tessellations

Random tessellations or mosaics occur as primary objects of various processes of division of space $\Re^d$ into convex cells or as secondary or auxiliary objects in, e.g., various statistical problems.

Simply stated, a *tessellation* is an arrangement of polytopes (two dimensions polygons, three dimensions: polyhedra) fitting together without overlapping so as to cover d-dimensional space $\Re^d$ $(d = 1, 2, \ldots)$, or a subset $X \subset \Re^d$. Usually one requires that the cells are convex and compact with disjoint interiors, although tessellations may also involve non-convex cells (an example is Johnson-Mehl tessellations). Posing the additional requirement of convexity implies that all interfaces separating pairs of cells are planar (for

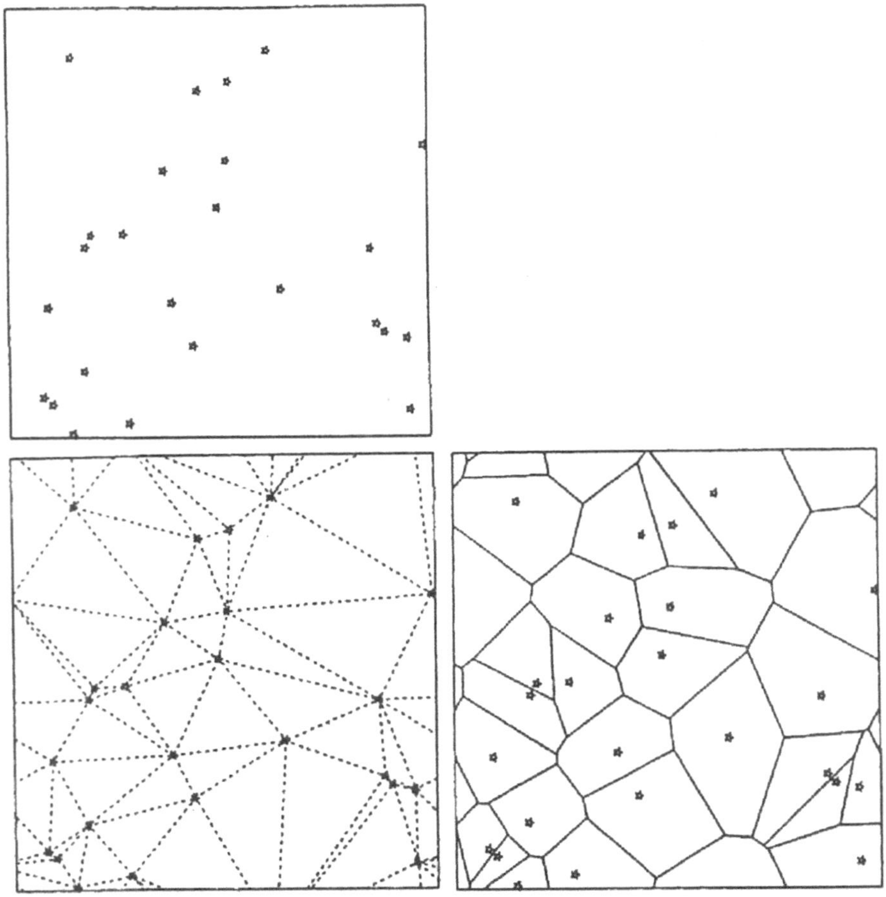

Fig. 14. An illustration of a distribution of nuclei (*stars*) in a square (*top*) and its corresponding Delaunay triangulation (*bottom left*) and Voronoi tessellation (*bottom right*), assuming periodic boundary conditions

three dimensions), and all edges common to the boundaries of three or more cells are linear, implying each cell to be a (convex) polyhedron. Formally, a *tessellation* in $\Re^d$ is a set $\mathcal{T} = \{X_i\}$ of d-dimensional sets $X_i \subset \Re^d$ called *cells* such that

$$\begin{aligned} &\tilde{X}_i \cap \tilde{X}_j = \emptyset \qquad \text{for } i \neq j, \\ &\bigcup_i X_i = \Re^d, \\ &\#\{X_i \in \mathcal{T} : X_i \cap B \neq \emptyset\} < \infty \qquad \forall \text{ bounded } B \subset \Re^d, \end{aligned} \tag{1}$$

with $\tilde{X}_i$ the interior of cell X_i (Møeller [162, 163]). The first property implies the interiors of the cells to be disjoint, the second one that the cell aggregate $\{X_i\}$ is space filling and the third one that $\mathcal{T}$ is a countable set of cells.

6.3 Voronoi Tessellation

The Voronoi tessellation $\mathcal{V}$ of a point set $\mathcal{P}$ is the division of space into mutually disjunct polyhedra, each *Voronoi polyhedron consisting of the part of space closer to the defining point than any of the other points* (Voronoi [252], Okabe et al. [169]). Assume that we have a distribution of a countable set $\mathcal{P}$ of nuclei $\{\mathbf{x}_i\}$ in $\Re^d$. Let $\mathbf{x}_1, \mathbf{x}_2, \mathbf{x}_3, \ldots$ be the coordinates of the nuclei. Then the *Voronoi region* $\mathcal{V}_i$ of nucleus i is defined by the points $\mathbf{x}$ (Fig. 14),

$$\mathcal{V}_i = \{\mathbf{x} | d(\mathbf{x}, \mathbf{x}_i) < d(\mathbf{x}, \mathbf{x}_j \quad \forall j \neq i\}, \tag{2}$$

where $d(\mathbf{x}, \mathbf{y})$ is the Euclidian distance between $\mathbf{x}$ and $\mathbf{y}$. In other words, $\mathcal{V}_i$ is the set of points which is nearer to $\mathbf{x}_i$ than to $\mathbf{x}_j$, $\quad j \neq i$. From this basic definition, we can directly infer that each Voronoi region $\mathcal{V}_i$ is the intersection of the open half-spaces bounded by the perpendicular bisectors (bisecting planes in three dimensions) of the line segments joining the nucleus i and any of the other nuclei. This implies a Voronoi region $\mathcal{V}_i$ to be a convex polyhedron (a polygon when in two dimensions), a *Voronoi polyhedron*. Evidently, the concept can be extended to any arbitrary distance measure (Icke, priv. comm.). The relation between the point distribution $\mathcal{P}$ and its Voronoi tessellation can be clearly appreciated from the two-dimensional illustration in Fig. 14. The complete set of Voronoi polyhedra constitutes a space-filling tessellation of mutually disjunct cells, the *Voronoi tessellation* $\mathcal{V}(\mathcal{P})$ relative to $\mathcal{P}$. A good impression of the morphology of a complete Voronoi tessellation can be seen in Fig. 13, a tessellation of 1000 cells generated by a Poisson distribution of 1000 nuclei in a cubic box. The Voronoi foam forms a packing of Voronoi cells, each cell being a convex polyhedron enclosed by the bisecting planes between the nuclei and their *natural neighbours*.

Voronoi Elements

Taking the three-dimensional tessellation as the archetypical representation of structures in the physical world, the Voronoi tessellation $\mathcal{V}(\mathcal{P})$ consists of four constituent *elements*: *Voronoi cells, Voronoi walls, Voronoi edges* and *Voronoi vertices.* Table 1 provides a listing of these elements together with a description of their relation with respect to the nuclei of the generating point set $\mathcal{P}$, augmented by the accompanying illustration in Fig. 15.

Generalized Voronoi Tessellations

The Voronoi tessellation can be generalized. Reference [155] defined the *generalized Voronoi tessellation* $\mathcal{V}_n$. The original Voronoi tessellation is $\mathcal{V}_1 = \mathcal{V}$.

This involves the extension of the definition of the Voronoi cell $\mathcal{V}_i$ generated by one nucleus i to that of a *higher order* Voronoi cell $\mathcal{V}^k(i_1, \ldots, i_k)$ generated by a set of k nuclei $\{i_1, \ldots, i_k\} \in \mathcal{P}$. Each *k-order* Voronoi cell $\mathcal{V}^k(i_1, \ldots, i_k)$ consists of that part of space in which the points $\mathbf{x}$ have the k nuclei $\{i_1, \ldots, i_k\} \in \mathcal{P}$ as their k nearest neighbours. In addition to Miles [155, 156, 158] see Okabe et al. [169], Chap. 3, for more details and references.

Table 1. Three-dimensional Voronoi tessellation elements (see also Fig. 15)

$\mathcal{V}_i$	*Voronoi cell*	- Polyhedron - Defined by nucleus $i \in \mathcal{P}$ - Volume of space closer to i than any other nucleus $m \in \mathcal{P}$
Σ_{ij}	*Voronoi wall (facet)*	- Polygon - Defined by nuclei $(i,j) \in \mathcal{P}$ - All points $\mathbf{x}$ with equal distance to (i,j) and larger distance to any other nucleus $m \in \mathcal{P}$ - Constitutes part surface cells: $\mathcal{V}_i$, $\mathcal{V}_j$
Λ_{ijk}	*Voronoi edge*	- Line segment - Defined by nuclei $(i,j,k) \in \mathcal{P}$ - All points $\mathbf{x}$ with equal distance to (i,j,k) and larger distance to any other nucleus $m \in \mathcal{P}$ - Constitutes part rim Voronoi cells: $\mathcal{V}_i$, $\mathcal{V}_j$, $\mathcal{V}_k$ - Constitutes part rim Voronoi walls: Σ_{ij}, Σ_{ik} and Σ_{jk}
$\mathcal{D}_{ijkl}$	*Voronoi vertex*	- Point - Defined by nuclei $(i,j,k,l) \in \mathcal{P}$ - Equidistant to nuclei (i,j,k,l) - Closer to (i,j,k,l) than to any other nucleus $m \in \mathcal{P}$ - Circumcentre of (Delaunay) tetrahedron (i,j,k,l)

Voronoi Uniqueness

An inverse look at tessellations reveals the special and degenerate nature of Voronoi tessellations. Given a particular tessellation one might wonder whether there is a point process which would have the tessellation as its Voronoi tessellation. One may demonstrate that in general this is not true. By defining a geometric procedure to reconstruct the generating nucleus distribution, one may straightforwardly infer that there is no unique solution for an arbitrary tessellation.

This may be inferred from the study by Chiu et al. [48], who defined and developed a nucleus reconstruction procedure to demonstrate that a two-dimensional section through a three- or higher dimensional Voronoi tessellation will itself not be a Voronoi tessellation. By doing so their work clarified the special and degenerate nature of these tessellations.

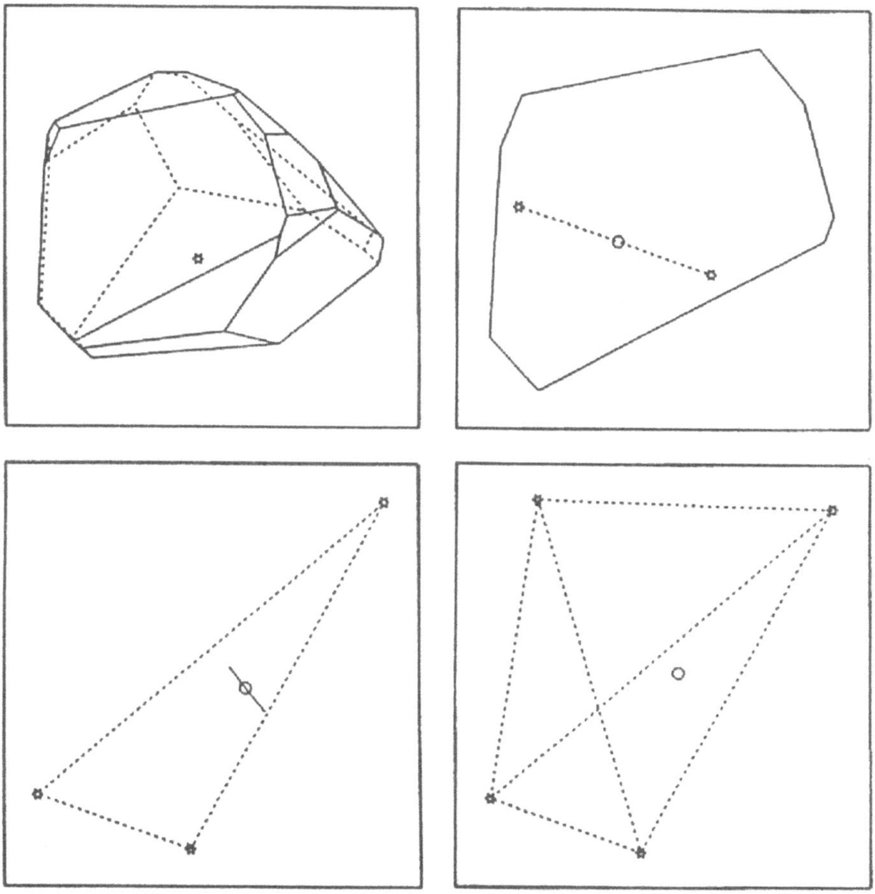

Fig. 15. The four Voronoi elements of a Voronoi tessellation generated by a nucleus set P. See Table 1

6.4 Delaunay Tessellation

Pursuing our census of Voronoi tessellation elements (Table 1), we found that each set of nuclei i, j, k, l corresponding to a Voronoi vertex $\mathcal{D}(i, j, k, l)$ defines a unique tetrahedron. This is known as *Delaunay tetrahedron* (Delaunay [61]).

Each *Delaunay tetrahedron* is defined by the *set of four points whose circumscribing sphere does not contain any of the other points* in the generating set (triangles in two dimensions: Delaunay [61], see Fig. 14). For the countable set $\mathcal{P}$ of points $\{\mathbf{x}_i\}$ in $\Re^d$, a Delaunay tetrahedron $\mathcal{D}_m$ is the simplex T defined by $(1+d)$ points $\{\mathbf{x}_{i1}, \ldots, \mathbf{x}_{i(d+1)}\} \in \mathcal{P}$ (the vertices of this d-dimensional tetrahedron) such that the corresponding circumscribing sphere $\mathcal{S}_m(\mathbf{y}_m)$ with circumcentre $\mathbf{C}_m$ and radius R_m does not contain any other point of $\mathcal{P}$,

$$\mathcal{D}_M = T(\mathbf{x}_{i1}, \ldots, \mathbf{x}_{i(d+1)}) \qquad \text{with } d(\mathbf{C}_m, \mathbf{x}_j) > R_m \quad \forall \ j \neq i1, \ldots, i(d+1)\,. \tag{3}$$

Following this definition, the *Delaunay tessellation* of a point set $\mathcal{P}$ is the uniquely defined and volume-covering tessellation of mutually disjunct *Delaunay tetrahedra.*

Figure 14 depicts the Delaunay tessellation resulting from a given (two-dimensional) distribution of nuclei. On the basis of the figure we can immediately observe the intimate relation between a *Voronoi tessellation* and its *Delaunay tessellation.* The Delaunay and Voronoi tessellations are like the opposite sides of the same coin; they are each other's *dual*: one may directly infer one from the other and vice versa. The combinatorial structure of either tessellation is completely determined from its dual.

The duality between Delaunay and Voronoi tessellations may be best appreciated on behalf of the following properties:

- *Circumcentre and Voronoi vertex*:
  The centre of the circumsphere of a Delaunay tetrahedron is a vertex of the Voronoi tessellation. This follows from the definition of the Voronoi tessellation, wherein the four nuclei which form the Delaunay tetrahedron are equidistant from the vertex.
- *Contiguity condition*:
  The circumsphere of a Delaunay tetrahedron is empty and cannot contain any nucleus in the set $\mathcal{P}$. If there would be such a nucleus it would be closer to the centre than the four tetrahedron-defining nuclei. This would render it impossible for it being the vertex of all corresponding Voronoi cells.

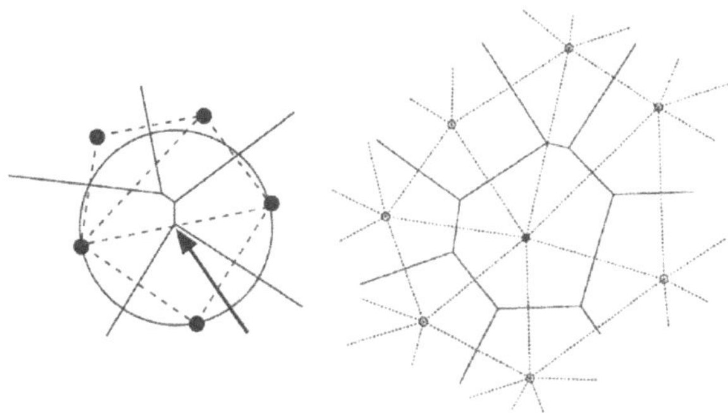

Fig. 16. The *dual* relationship between Voronoi (*solid*) and Delaunay (*dashed*) tessellations of a set of nuclei (*circles*). Left: Zoom-in on the three Delaunay triangles corresponding to a set of five nuclei (*black dots*) and the corresponding Voronoi edges. The circle is the circumcircle of the lower Delaunay triangle; its centre (*arrow*) is a vertex of the Voronoi cell. Note that the circle does not contain any other nucleus in its interior! Right: a zoom-in on the Voronoi cell $\mathcal{V}_i$ of nucleus i (*black dot*). The Voronoi cell is surrounded by its related Delaunay triangles and clearly delineates its *natural neighbours* (*open circles*)

Natural Neighbours

A pair of nuclei i and j whose Voronoi polyhedra $\mathcal{V}_i$ and $\mathcal{V}_j$ have a face in common is called a *contiguous pair* and a member of the pair is said to be *contiguous* to the other member. Contiguous pairs of nuclei are each other's *natural neighbour.*

Natural neighbours of a point i are the points j with whose Voronoi $\mathcal{V}_j$ its Voronoi cell $\mathcal{V}_i$ shares a face or, in other words, the points with which it is connected via a Delaunay tetrahedron. This unique set of neighbouring points defines the neighbourhood of the point and represents the cleanest definition of the surroundings of a point (see Fig. 16), an aspect which turns out to be of *seminal* importance for the local interpolation method(s) discussed in this contribution.

6.5 Voronoi and Delaunay Statistics

In particular, for practical applications, the knowledge of statistical properties of Voronoi and Delaunay tessellations as a function of the generating stochastic point processes is of considerable interest. However, despite their seemingly simple definition, it has proven remarkably difficult to obtain solid analytical expressions for statistical properties of Voronoi tessellations. Moreover, with the exception of some general tessellation properties, nearly all analytical work on Voronoi and Delaunay tessellations has concentrated on those generated by homogeneous Poissonian point distributions.

Statistical knowledge of Delaunay tessellations generated by Poissonian nuclei in d-dimensional space is relatively complete. Some important distribution functions are known, mainly due to the work of Miles [155, 156, 157], Kendall [118] and Moeller [162, 163]. For Voronoi tessellations, even for Poissonian nuclei analytical results are quite rare. Only very few distribution functions are known; most results are limited to a few statistical moments: expectation values, variances and correlation coefficients. Most of these results stem from the pioneering works of Meijering [151], Gilbert [92], Miles [155] and Moeller [162]. The Moeller's paper [162] provides analytical formulae for a large number of first-order moments of d-dimensional Poisson–Voronoi tessellations, as well as of s-dimensional section through them, following the work of Miles [156, 157, 158] (see also van de Weygaert [242]). Numerical Monte Carlo evaluations have proven to be the main source for a large range of additional statistical results. For an extensive listing and discussion we refer the reader to [169].

Of prime importance for the tessellation interpolation techniques discussed in this review are two fundamental characteristics: (1) the number of natural neighbours of the generating nuclei and (2) the volume of Voronoi and Delaunay cells.

In two dimensions each point has on average exactly six natural neighbours (see, e.g., Icke & van de Weygaert [106]) irrespective of the character of the spatial point distribution. This also implies that each point belongs to on average six Delaunay triangles. Going from two to three dimensions, the character of the tessellation changes fundamentally:

- *Dependence point process*:
  The average number of natural neighbours is no longer independent of the underlying point distribution.
- *Integer*:
  The average number of natural neighbours is not an integer: for a Poisson distribution it is ~ 13.4!
- *Delaunay tetrahedra/Voronoi vertices*:
  For a Poisson distribution, the number of vertices per Voronoi cell is ~ 27.07. While in the two-dimensional case the number of vertices per Voronoi cell has same value as the number of *natural neighbours*, in three dimensions it is entirely different!

As yet it has not been possible to derive, from first principle, a closed analytical expression for the distribution functions of the volumes of Voronoi polyhedra and Delaunay cells in a Poisson Voronoi tessellation. However, the fitting formula suggested by Kiang [119] has proven to represent a reasonably good approximation. Accordingly, the probability distribution of the volume of a Voronoi polyhedron in d-dimensional space $\Re^d$ follows a gamma distribution

$$f_{\mathcal{V}}(V_{\mathcal{V}})\,\mathrm{d}V_{\mathcal{V}} = \frac{q}{\Gamma(q)}\left(q\frac{V_{\mathcal{V}}}{\langle V_{\mathcal{V}}\rangle}\right)^{(q-1)} \exp\left(-q\frac{V_{\mathcal{V}}}{\langle V_{\mathcal{V}}\rangle}\right)\mathrm{d}\left(\frac{V_{\mathcal{V}}}{\langle V_{\mathcal{V}}\rangle}\right), \tag{4}$$

with $V_{\mathcal{V}}$ the size of the Voronoi cell and $\langle V_{\mathcal{V}}\rangle$ the average cell size. The conjecture of Kiang [119] is that the index has a value $q = 2d$ for a tessellation in d-dimensional space (i.e. $q = 4$ for two-dimensional space and $q = 6$ for three-dimensional space). Even though other studies indicated slightly different values, we have found the Kiang suggestion to be quite accurate (see Sect. 11.2, see also Schaap [201]).

While the distribution for Voronoi cell volumes involves a conjecture, for Delaunay cells $\mathcal{D}$ it is possible to derive the ergodic distribution from first principle. Its $d+1$ vertices completely specify the Delaunay tetrahedron $\mathcal{D}$. If $\mathbf{c}$, R its circumradius, its vertices are the points $\{\mathbf{c} + R\mathbf{u}_i\}$. In this the unit vectors $\{\mathbf{u}_i\}$ $(i = 0, \ldots, d+1)$, directed towards the vertices, determine the shape of the Delaunay tetrahedron. Given that Δ_d is the volume of the unity simplex $\{\boldsymbol{u}_0, \ldots, \boldsymbol{u}_d\}$, the volume $V_{\mathcal{D}}$ of the Delaunay tetrahedron is given by

$$V = \Delta_d \, R^d \, . \tag{5}$$

Miles [157] and Moeller [162, 163] found that for a Poisson point process of intensity n in d-dimensional space $\Re^d$ the distribution is specified by

$$\begin{aligned} f_{\mathcal{D}}(D) &= f_{\mathcal{D}}(\{\boldsymbol{u}_0, \ldots, \boldsymbol{u}_d\}, R) \\ &= a(n,d) \, \Delta_d \, R^{d^2-1} \, \exp\left(-n\omega_d \, R^d\right) , \end{aligned} \tag{6}$$

with ω_d the volume of the unity sphere in d-dimensional space ($\omega_2 = 2\pi$ and $\omega_3 = 4\pi$) and $a(n,d)$ is a constant dependent on number density n and dimension d. In other words, the circumradius R of the Delaunay tetrahedron is ergodically independent of its shape, encapsulated in Δ_d. From this, one finds that the distribution law of $n\omega_d R^d$ is the $\chi^2_{2d}/2$ distribution (Kendall [118]):

$$f(R)\,\mathrm{d}R = \frac{(n\omega_d R^d)^{d-1}}{(d-1)!} \, \exp\left(-n\omega_d R^d\right) \, \mathrm{d}(n\omega_d R^d) \, . \tag{7}$$

It is of considerable importance to realize that even for a uniform density field, represented by a discrete point distribution of intensity n, neither the Voronoi nor the Delaunay tessellation will involve a volume distribution marked by a considerable spread and skewness.

6.6 Spatial Adaptivity

The distribution of the Delaunay and Voronoi cells adjusts itself to the characteristics of the point distribution: in sparsely sampled regions the distance between the *natural neighbours* is large, and also if the spatial distribution is anisotropic this will be reflected in their distribution. This may be readily appreciated from Fig. 17. At three successive spatial scales, the Delaunay tessellations have traced the density and geometry of the local point distribution to a remarkable degree. On the basis of these observations it is straightforward to appreciate that the corresponding Delaunay tessellation forms an ideal adaptive multidimensional interpolation grid.

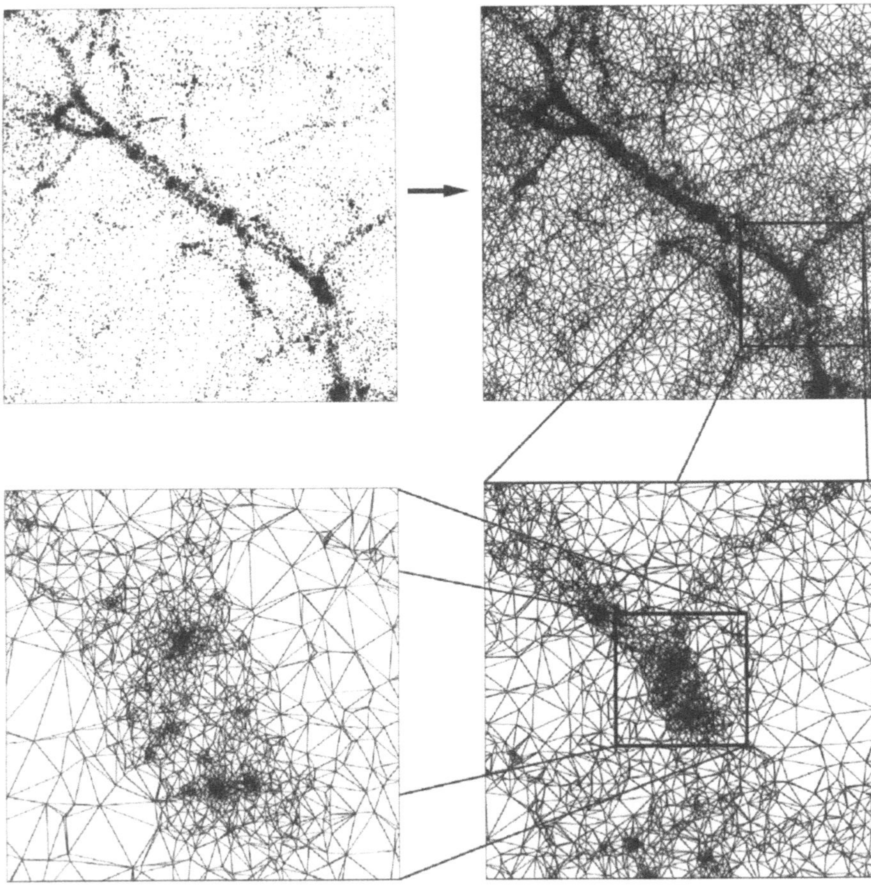

Fig. 17. The Delaunay tessellation of a point distribution in and around a filamentary feature. The generated tessellations are shown at three successive zoom-ins. The frames form a testimony of the strong adaptivity of the Delaunay tessellations to local density and geometry of the spatial point distribution

Note that not only the size but also the shape of the Delaunay simplices is fully determined by the spatial point distribution. The density and geometry of the local point distribution will therefore dictate the resolution of spatial interpolation and reconstruction procedures exploiting the Delaunay tessellation. The prime representatives of such methods are the natural neighbour techniques (see Sect. 7) and the Delaunay Tessellation Field Estimator.

These techniques exploit the fact that a Delaunay tetrahedron may be regarded as the optimal multidimensional interpolation interval. The corresponding minimal coverage characteristics of the Delaunay tessellation thus imply it to be optimal for defining a network of multidimensional interpolation intervals. The resulting interpolation kernels of natural neighbour and

DTFE interpolation not only embody an optimal spatial resolution but also involve a high level of adaptivity to the local geometry of the point distribution (see Sects. 7 and 8). Within this context it is no coincidence that in many computer visualization applications Delaunay tessellations have acquired the status of optimal triangulation. Moreover, the superb spatial adaptivity of the volumes of Delaunay and Voronoi polyhedra to the local point density may be readily translated into measures for the value of the local density. This forms a crucial ingredient of the DTFE formalism (see Sect. 8).

6.7 Voronoi and Delaunay Tessellations: Context

The earliest significant use of Voronoi regions seems to have occurred in the work of Dirichlet [65] and Voronoi [252] in their investigations on the reducibility of positive definite quadratic forms. However, Dirichlet and Voronoi tessellations as applied to random point patterns appear to have arisen independently in various fields of science and technology (Okabe et al. [169]). For example, in crystallography, one simple model of crystal growth starts with a fixed collection of sites in two- and three-dimensional space, and allows crystals to begin growing from each site, spreading out at a uniform rate in all directions, until all space is filled. The "crystals" then consist of all points nearest to a particular site, and consequently are just the Voronoi regions for the original set of points. Similarly, the statistical analysis of meteorological data led to the formulation of Voronoi regions under the name *Thiessen polygons* (Thiessen [236]).

Applications of Voronoi tessellations can therefore be found in fields as diverse as agriculture and forestry (Fischer & Miles [81]), astrophysics (e.g. Kiang [119], Icke & van de Weygaert [106], Ebeling & Wiedenmann [67], Bernardeau & van de Weygaert [27], Molchanov et al. [161], Schaap & van de Weygaert [202], Cappellari & Copin [44], Ritzerveld & Icke [190]), ecology, zoology and botany, cell biology (Gor et al. [97]), protein research (Liang et al. [133, 134], Liang et al. [135]), cancer research (iKansal et al. [114], Schaller & Meyer-Hermann [206]), chemistry, crystal growth and structure, materials science (see, e.g., Torquato [237], incl. many references), geophysics (Sambridge et al. [197]), geography and geographic information systems (Boots [37], Gold et al. [95]), communication theory (Baccelli & Zuyev [15]) and art and archaeology (Kimia & Leymarie [121], Leymarie & Kimia [131]). Due to the diversity of these applications it has acquired a set of alternative names, such as Dirichlet regions, Wigner-Seitz cells, and Thiessen figures.

7 Natural Neighbour Interpolation

The *natural neighbour interpolation* formalism is a generic higher order multidimensional interpolation, smoothing and modelling procedure utilizing the concept of natural neighbours to obtain locally optimized measures of system

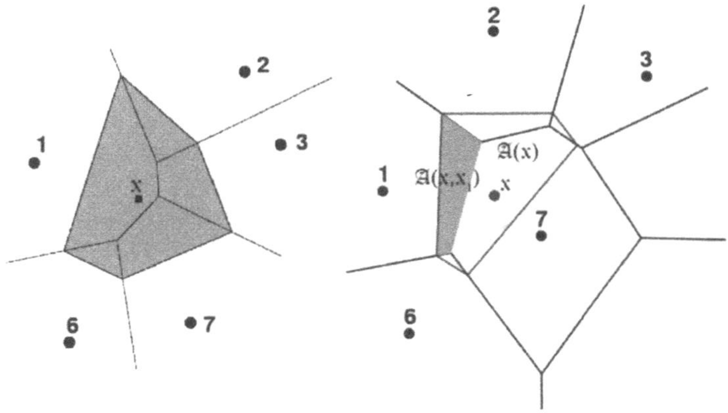

Fig. 18. Natural neighbour interpolants. Example of nn-interpolation in two dimensions. *Left*: the Voronoi cell $\mathcal{V}(\mathbf{x})$ generated by a point $\mathbf{x}$. *Right*: the second-order Voronoi cell $\mathcal{V}_2(\mathbf{x}, \mathbf{x}_1)$, the region of space for which the points $\mathbf{x}$ and $\mathbf{x}_j$ are the closest points

characteristics. Its theoretical basis was developed and introduced by Sibson [221], while extensive treatments and elaborations of nn-interpolation may be found in [253, 233]. Natural neighbour interpolation produces a conservative, artifice-free, result by finding weighted averages, at each interpolation point, of the functional values associated with that subset of data which are natural neighbours of each interpolation point. Unlike other schemes, like Shepard's interpolant (Shepard [216]), where distance-based weights are used, the Sibson natural neighbour interpolation uses area-based weights. According to the nn-interpolation scheme the interpolated value $\widehat{f}(\mathbf{x})$ at a position $\mathbf{x}$ is given by

$$\widehat{f}(\mathbf{x}) = \sum_i \phi_{nn,i}(\mathbf{x})\, f_i\,, \tag{8}$$

in which the summation is over the natural neighbours of the point $\mathbf{x}$, i.e. the sample points j with whom the order-2 Voronoi cells $\mathcal{V}_2(\mathbf{x}, \mathbf{x}_j)$ are not empty (Figs. 18 and 19). Sibson interpolation is based on the interpolation kernel $\phi(\mathbf{x}, \mathbf{x}_j)$ to be equal to the normalized order-2 Voronoi cell,

$$\phi_{nn,i}(\mathbf{x}) = frac\mathcal{A}_2(\mathbf{x}, \mathbf{x}_i)\mathcal{A}(\mathbf{x})\,, \tag{9}$$

in which $\mathcal{A}(\mathbf{x}) = \sum_j \mathcal{A}(\mathbf{x}, \mathbf{x}_j)$ is the volume of the potential Voronoi cell of point $\mathbf{x}$ if it had been added to the point sample $\mathcal{P}$ and the volume $\mathcal{A}_2(\mathbf{x}, \mathbf{x}_i)$ concerns the order-2 Voronoi cell $\mathcal{V}_2(\mathbf{x}, \mathbf{x}_i)$, the region of space for which the points $\mathbf{x}$ and $\mathbf{x}_i$ are the closest points. Notice that the interpolation kernels ϕ are always positive and sum to one.

The resulting function is continuous everywhere within the convex hull of the data and has a continuous slope everywhere except at the data themselves

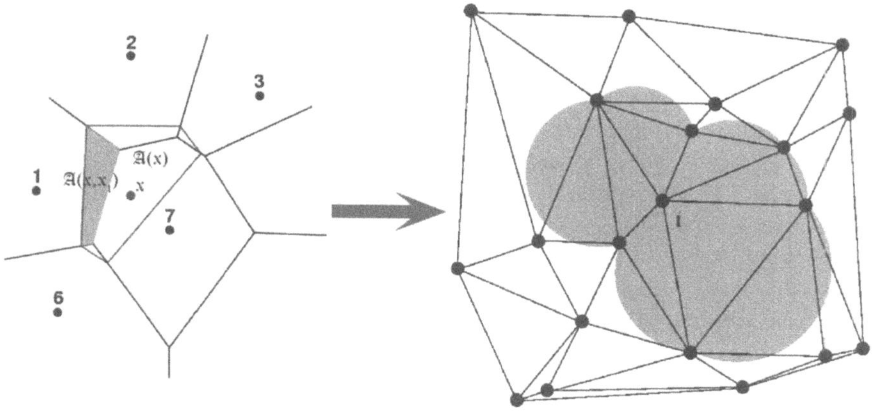

Fig. 19. Natural neighbour interpolants. Example of nn-interpolation in two dimensions. The basis of the nn-interpolation kernel $\phi_{nn,1}$, the contribution of nucleus $\mathbf{x}_1$ to the interpolated field values

(Figs. 20 and 21). Beautiful two-dimensional examples of nn-interpolation applications testify of its virtues (see Fig. 12).

An interesting study of its performance, in comparison with other interpolation/approximation methods, concerned the test involving data acquired by the Mars Orbiter Laser Altimeter (MOLA), one of the instruments on the Mars Global Surveyor (deceased November 2006). Applying various schemes towards the reconstruction of a terrain near the Korolev crater, a large crater in the north polar region of Mars, demonstrated that nn-neighbour interpolation produced the most impressive reproduction of the original terrain

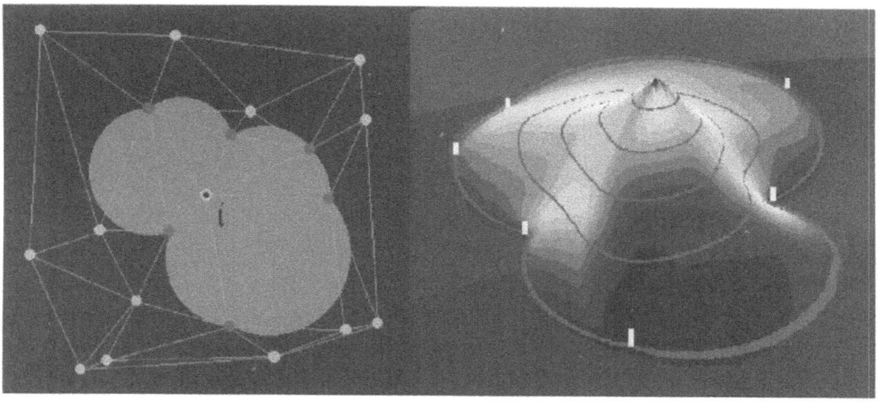

Fig. 20. Natural neighbour interpolants. Example of nn-interpolation in two dimensions. Three-dimensional illustration of the nn-interpolation kernel $\phi_{nn,1}$, the contribution of nucleus $\mathbf{x}_1$ to the interpolated field values. From Braun & Sambridge [41]

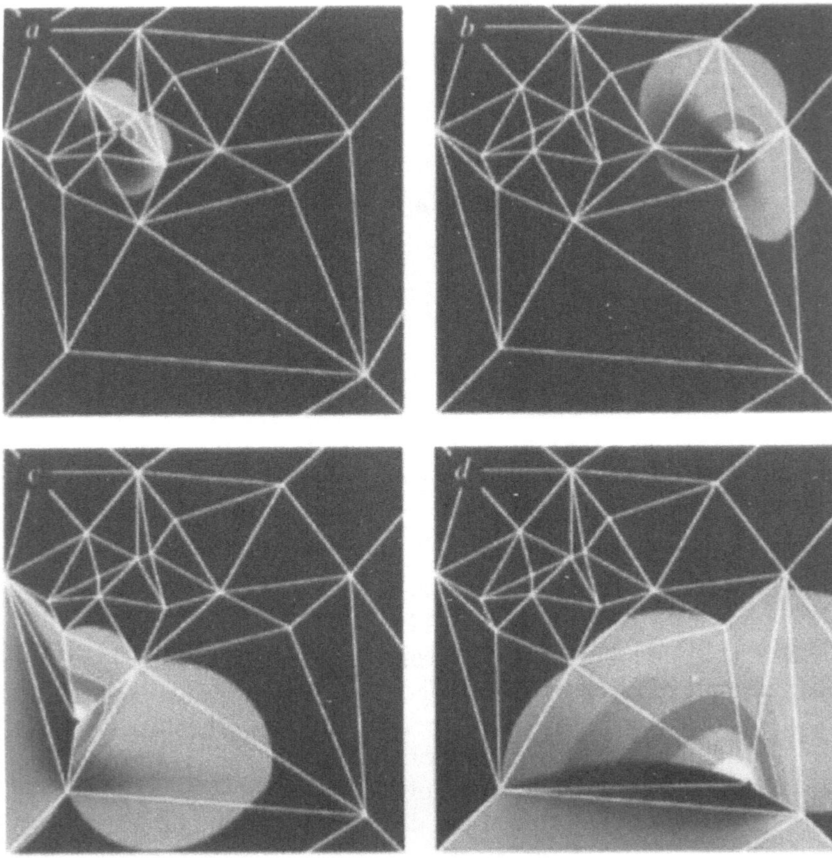

Fig. 21. Natural neighbour kernels: illustration the locally defined natural neighbour kernels $\phi_{nn,j}$ for four different nuclei j. From Braun & Sambridge [41]

(Abramov & McEven [2], Fig. 12). In comparison to the other methods – including nearest neighbour interpolation, spline interpolation and linear DTFE type interpolation – the nn-neighbour map not only looks most natural but also proves to contain fewer artefacts, in both number and severity.

Natural neighbour interpolation may rightfully be regarded as the most general and robust method for multidimensional interpolation available to date. This smooth and local spatial interpolation technique has indeed gradually acquired recognition as a dependable, optimal, local method. For the two-dimensional applications it has seen highly interesting applications in geophysical fluid dynamics calculations (Braun & Sambridge [41], Sambridge et al. [197]), and in equivalent schemes for solid mechanics problems (Sukumar [233]). Both applications used nn-interpolation to solve partial differential equations, showing its great potential in the field of computational fluid mechanics.

While almost optimal in the quality of its reconstructions, it still involves a heavy computational effort. This is certainly true for three- or higher dimensional spaces. This has prodded us to define a related local, adaptive method which can be applied to large astronomical data sets.

8 DTFE: Delaunay Tessellation Field Estimator

For three-dimensional samples with large number of points, akin to those found in large cosmological computer simulations, the more complex geometric operations involved in the pure nn-neighbour interpolation still represent a computationally challenging task. To deal with the large point samples consisting of hundreds of thousands to several millions of points we chose to follow a related nn-neighbour-based technique that restricts itself to pure linear interpolation. DTFE uses to obtain optimal *local* estimates of the spatial density (see Okabe et al. [169], Sect 8.5), while the tetrahedra of its *dual* Delaunay tessellation are used as multidimensional intervals for *linear* interpolation of the field values sampled or estimated at the location of the sample points (Okabe et al. [169], Chap. 6). The DTFE technique allows us to follow the same geometrical and structural adaptive properties of the higher order nn-neighbour methods while allowing the analysis of truely large data sets. The presented tests in this review will demonstrate that DTFE indeed is able to highlight and analyse essential elements of the cosmic matter distribution.

8.1 DTFE: Point Sample

The DTFE procedure, outlined in the flow diagram in Fig. 22, involves a sequence of steps. It starts with outlining the discrete point sample $\mathcal{P}$ in d-dimensional space Re^d,

$$\mathcal{P} = \{\mathbf{x}_1, \ldots, \mathbf{x}_N\}\,. \tag{10}$$

In the applications described in this study we restrict ourselves to Euclidian spaces, in particular two- or three-dimensional Euclidian spaces. At the locations of the points in the countable set $\mathcal{P}$ the field values $\{f(\mathbf{x}_i), i = 1, \ldots, N\}$ are sampled or can be estimated on the basis of the spatial point distribution itself. The prime example of the latter is when the field f involves the density field itself.

On the basis of the character of the sampled field $f(\mathbf{x})$ we need to distinguish two options. The first option is the one defined in [27], the straight forward multidimensional linear interpolation of measured field values on the basis of the Delaunay tessellation. The papers [202] and [201] extended this to the recovery of the density or intensity field. DTFE is therefore characterized by a second option, the ability to reconstruct the underlying density field from the discrete point process itself.

The essential extension of DTFE with respect to [27] is that it allows the option of using the *point sample process* $\mathcal{P}$ itself as a measure for the value of the *density* at its position. The latter poses some stringent requirements on the sampling process. It is crucial for the discrete spatial point distribution to constitute a fair sample of the underlying continuous density field. In other words, the discrete point sample $\mathcal{P}$ needs to constitute a general Poisson point process of the density field.

Such stringent requirements on the spatial point process $\mathcal{P}$ are not necessary when the sampled field has a more generic character. As long as the sample points are spread throughout most of the sample volume, the interpolation procedure will yield a properly representative field reconstruction. It was this situation with respect to cosmic velocity fields which leads to the original definition of the Delaunay spatial interpolation procedure forming the basis of DTFE (Bernardeau & van de Weygaert [27, 28]).

8.2 DTFE: Linear Interpolation

At the core of the DTFE procedure is the use of the Delaunay tessellation of the discrete point sample process (see Sect. 6.4) as an adaptive spatial linear interpolation grid. Once the Delaunay tessellation of $\mathcal{P}$ is determined, it is used as a *multidimensional linear interpolation grid* for a field $f(\mathbf{r})$. That is, each Delaunay tetrahedron is supposed to be a region with a constant field gradient, ∇f.

The linear interpolation scheme of DTFE exploits the same spatially adaptive characteristics of the Delaunay tessellation generated by the point sample $\mathcal{P}$ as that of regular natural neighbour schemes (see Sect. 7, (8)). For DTFE the interpolation kernel $\phi_{dt,i}(\mathbf{x})$ is that of regular linear interpolation within the Delaunay tetrahedron in which $\mathbf{x}$ is located,

$$\widehat{f}_{dt}(\mathbf{x}) = \sum_i \phi_{dt,i}(\mathbf{x})\, f_i, \tag{11}$$

in which the sum is over the four sample points defining the Delaunay tetrahedron. Note that not only the size but also the shape of the Delaunay simplices is fully determined by the spatial point distribution. As a result the resolution of the DTFE procedure depends on both the density and geometry of the local point distribution. Not only does the DTFE kernel embody an optimal spatial resolution, but it also involves a high level of adaptivity to the local geometry of the point distribution (see Sect. 11.1).

Also note that for both the nn-interpolation and the linear DTFE interpolation, the interpolation kernels ϕ_i are unity at sample point location $\mathbf{x}_i$ and equal to zero at the location of the other sample points j (e.g. see Fig. 20),

$$\phi_i(\mathbf{x}_j) = \begin{cases} 1 & \text{if } i = j\,, \\ 0 & \text{if } i \neq j\,, \end{cases} \tag{12}$$

where $\mathbf{x}_j$ is the location of sample point j.

In practice, it is convenient to replace (11) with its equivalent expression in terms of the (linear) gradient $\widehat{\nabla f}\big|_m$ inside the Delaunay simplex m,

$$\widehat{f}(\mathbf{x}) = \widehat{f}(\mathbf{x}_i) + \widehat{\nabla f}\big|_m \cdot (\mathbf{x} - \mathbf{x}_i) . \tag{13}$$

The value of $\widehat{\nabla f}\big|_m$ can be easily and uniquely determined from the $(1 + D)$ field values f_j at the sample points constituting the vertices of a Delaunay simplex. Given the location $\mathbf{r} = (x, y, z)$ of the four points forming the Delaunay tetrahedron's vertices, $\mathbf{r}_0$, $\mathbf{r}_1$, $\mathbf{r}_2$ and $\mathbf{r}_3$, and the value of the sampled field at each of these locations, f_0, f_1, f_2 and f_3, and defining the quantities

$$\begin{aligned} \Delta x_n &= x_n - x_0 ; \\ \Delta y_n &= y_n - y_0 ; \qquad \text{for } n = 1, 2, 3 \\ \Delta z_n &= z_n - z_0 \end{aligned} \tag{14}$$

as well as $\Delta f_n \equiv f_n - f_0 \, (n = 1, 2, 3)$ the gradient ∇f follows from the inversion

$$\nabla f = \begin{pmatrix} \frac{\partial f}{\partial x} \\ \\ \frac{\partial f}{\partial y} \\ \\ \frac{\partial f}{\partial z} \end{pmatrix} = \mathbf{A}^{-1} \begin{pmatrix} \Delta f_1 \\ \\ \Delta f_2 \\ \\ \Delta f_3 \end{pmatrix} ; \qquad \mathbf{A} = \begin{pmatrix} \Delta x_1 & \Delta y_1 & \Delta z_1 \\ \\ \Delta x_2 & \Delta y_2 & \Delta z_2 \\ \\ \Delta x_3 & \Delta y_3 & \Delta z_3 \end{pmatrix} . \tag{15}$$

Once the value of ∇_f has been determined for each Delaunay tetrahedron in the tessellation, it is straightforward to determine the DTFE field value $\widehat{f}(\mathbf{x})$ for any location $\mathbf{x}$ by means of straightforward linear interpolation within the Delaunay tetrahedron in which $\mathbf{x}$ is located (13).

The one remaining complication is to locate the Delaunay tetrahedron $\mathcal{D}_m$ in which a particular point $\mathbf{x}$ is located. This is not as trivial as one might naively think. It does not necessarily concern a tetrahedron of which the nearest nucleus is a vertex. Fortunately, a very efficient method, the *walking triangle algorithm* (Lawson [130], Sloan [222]), has been developed. Details of the method may be found in [197, 201].

8.3 DTFE: Extra Features

While DTFE in essence is a first-order version of *natural neighbour interpolation* procedure, following the same adaptive multidimensional interpolation characteristics of the Delaunay grid as the higher order nn-neighbour techniques, it also incorporates significant extensions and additional aspects. In particular, DTFE involves two extra and unique features which are of crucial importance for the intended cosmological context and application:

- *Volume weighted*:
  The interpolated quantities are *volume-weighted*, instead of the implicit *mass-weighted* averages yielded by conventional grid interpolations.
- *Density estimates*:
  The spatial adaptivity of the Delaunay/Voronoi tessellation to the underlying point distribution is used to estimate the local density.

9 DTFE: Field Reconstruction Procedure

The complete DTFE reconstruction procedure, essential steps of which are illustrated in Fig. 23, can be summarized in terms of the flow diagram in Fig. 22 and consists of the following sequence of steps.

- **Point sample:**
  Defining the spatial distribution of the point sample:
  + *Density field:*
    Point sample needs to be a general Poisson process of the (supposed) underlying density field, i.e. it needs to be an unbiased sample of the underlying density field.
  + *General (non-density) field*:
    No stringent requirements upon the stochastic representativeness of the sampling process will be necessary except that the sample volume is adequately covered.
- **Boundary conditions:**
  An important issue, regarding the subsequent Delaunay tessellation computation and the self-consistency of the DTFE density and velocity field reconstructions, is that of the assumed boundary conditions. These will determine the Delaunay and Voronoi cells that overlap the boundary of the sample volume. Depending on the sample at hand, a variety of options exist:
  + *Vacuum boundary conditions:*
    Outside the sample volume there are no points. This will lead to infinitely extended (contiguous) Voronoi cells surrounding sample points near the boundary. Evidently, these cells cannot be used for DTFE density field estimates and field interpolations: the volume of the DTFE reconstruction is smaller than that of the sample volume.
  + *Periodic boundary conditions:*
    The point sample is supposed to be repeated periodically in boundary boxes, defining a toroidal topology for the sample volume. The resulting Delaunay and Voronoi tessellations are also periodic, their total volume exactly equal to the sample volume. While specific periodic tessellation algorithms do exist (van de Weygaert [243]), this is not yet true for most available routines in standard libraries. For the analysis of N-body simulations this is the most straightforward and preferable choice.

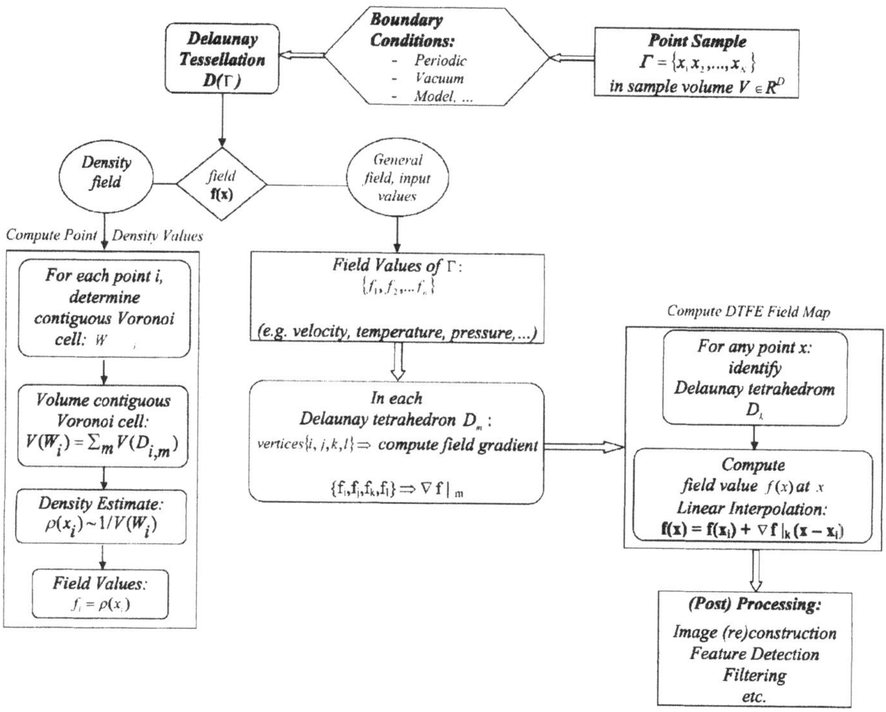

Fig. 22. Flow diagram illustrating the essential ingredients of the DTFE procedure

+ *Buffer conditions:*
  The sample volume box is surrounded by a buffer zone filled with a synthetic point sample. The density of the synthetic buffer point sample should be related to the density in the nearby sample volume. The depth of the buffer zone depends on the density of the synthetic point sample; it should be sufficiently wide for any Delaunay or Voronoi cell related to a sample point not to exceed the buffer zone. A clear condition for a sufficiently deep buffer zone has been specified by Neyrinck, Gnedin & Hamilton [166].
  When involving velocity field analysis, the velocities of the buffer points should also follow the velocity distribution of the sample and be in accordance with the continuity equation. Relevant examples of possible choices are
  - *internal:* the analysed sample is a subsample embedded within a large sample volume; a sufficient number of these sample points outside the analysis volume is taken along in the DTFE reconstruction.
  - *random cloning technique:* akin to the technique described by [257].
  - *constrained random field:* realizations employing the existing correlations in the field (Bertschinger [17], Hoffman & Ribak [103], van de Weygaert & Bertschinger [248], Zaroubi et al. [258]).

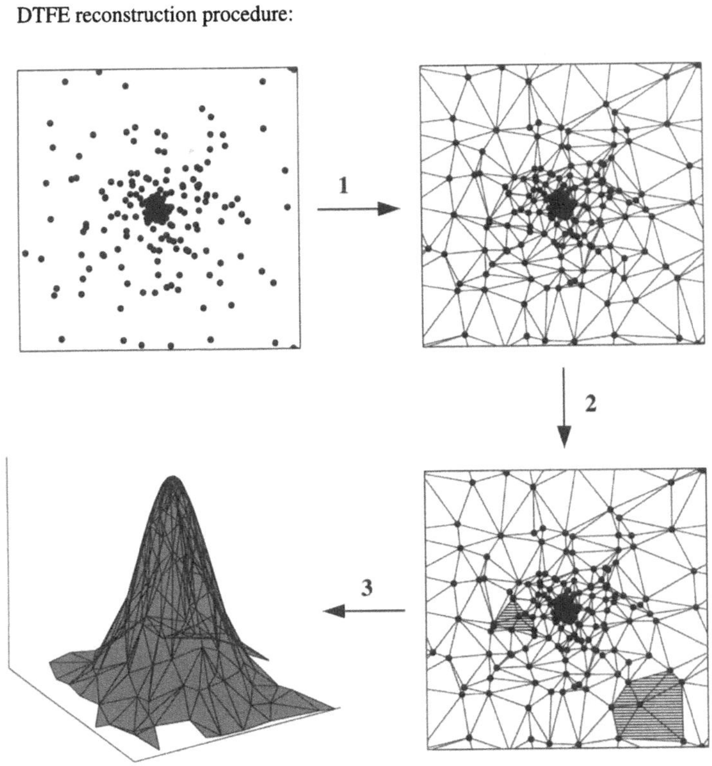

Fig. 23. Summary: overview of the essential steps of the DTFE reconstruction procedure. Given a point distribution (*top left*), one has to construct its corresponding Delaunay tessellation (*top right*), estimate the density at the position of the sampling points by taking the inverse of the area of their corresponding contiguous Voronoi cells (*bottom right*) and finally assume that the density varies linearly within each Delaunay triangle, resulting in a volume-covering continuous density field (*bottom left*)

- **Delaunay tessellation:**
  Construction of the Delaunay tessellation from the point sample (see Fig. 17). While we still use our own Voronoi–Delaunay code (van de Weygaert [243]), at present there is a score of efficient library routines available. Particularly noteworthy is the `CGAL` initiative, a large library of computational geometry routines.[4]
- **Field values for point samples:**
  Depending on whether it concerns the densities at the sample points or a measured field value, there are two options:

[4] `CGAL` is a `C++` library of algorithms and data structures for computational geometry; see **`www.cgal.org`**

+ *General (non-density) field:*
  (Sampled) value of field at sample point.
+ *Density field:*
  The density values at the sampled points are determined from the corresponding Voronoi tessellations. The estimate of the density at each sample point is the normalized inverse of the volume of its *contiguous* Voronoi cell $\mathcal{W}_i$ of each point i. The *contiguous Voronoi cell* of a point i is the union of all Delaunay tetrahedra of which point i forms one of the four vertices (see Fig. 27 for an illustration). We recognize two applicable situations:
  - *uniform sampling process*: the point sample is an unbiased sample of the underlying density field. Typical example is that of N-body simulation particles. For D-dimensional space the density estimate is
  $$\widehat{\rho}(\mathbf{x}_i) = (1+D)\,\frac{w_i}{V(\mathcal{W}_i)}\,, \tag{16}$$
  with w_i the weight of sample point i; usually we assume the same "mass" for each point.
  - *systematic non-uniform sampling process*: sampling density according to specified selection process quantified by an a priori known selection function $\psi(\mathbf{x})$, varying as function of sky position (α, δ) as well as depth/redshift. For D-dimensional space the density estimate is
  $$\widehat{\rho}(\mathbf{x}_i) = (1+D)\,\frac{w_i}{\psi(\mathbf{x}_i)\,V(\mathcal{W}_i)}\,. \tag{17}$$

- **Field gradient**:
  Calculation of the field gradient estimate $\widehat{\nabla f}|_m$ in each D-dimensional Delaunay simplex m ($D = 3$: tetrahedron; $D = 2$: triangle) by solving the set of linear equations for the field values at the positions of the $(D+1)$ tetrahedron vertices,
  $$\widehat{\nabla f}|_m \iff \begin{cases} f_0 & f_1 & f_2 & f_3 \\ \\ \mathbf{r}_0 & \mathbf{r}_1 & \mathbf{r}_2 & \mathbf{r}_3 \end{cases}\,. \tag{18}$$

Evidently, linear interpolation for a field f is only meaningful when the field does not fluctuate strongly. Particularly relevant for velocity field reconstructions is that there should be no orbit crossing flows within the volume of the Delaunay cell which would involve multiple velocity values at any one location. In other words, DTFE velocity field analysis is only significant for *laminar* flows.

Note that in the case of the sampled field being the velocity field $\mathbf{v}$ we may not only infer the velocity gradient in a Delaunay tetrahedron but also the directly related quantities such as the *velocity divergence*, *shear* and *vorticity*.

- **Interpolation**:
The final basic step of the DTFE procedure is the field interpolation. The processing and post-processing steps involve numerous interpolation calculations for each of the involved locations $\mathbf{x}$. Given a location $\mathbf{x}$, the Delaunay tetrahedron m in which it is embedded is determined. On the basis of the field gradient $\widehat{\nabla f}|_m$ the field value is computed by (linear) interpolation (see (13)):

$$\widehat{f}(\mathbf{x}) = \widehat{f}(\mathbf{x}_i) + \widehat{\nabla f}\Big|_m \cdot (\mathbf{x} - \mathbf{x}_i)\,. \tag{19}$$

In principle, higher order interpolation procedures are also possible. Two relevant high-order procedures are
  - Spline interpolation.
  - Natural neighbour interpolation (see (8) and (9)).
  Implementation of natural neighbour interpolation, on the basis of CGAL routines, is presently in progress.
- **Processing**:
Though basically of the same character for practical purposes we make a distinction between straightforward processing steps concerning the production of images and simple smoothing filtering operations on the one hand and more complex post-processing on the other hand. The latter are treated in the next item. Basic to the processing steps is the determination of field values following the interpolation procedure(s) outlined above.
Straightforward "first line" field operations are *image reconstruction* and, subsequently, *smoothing/filtering*.
  + *Image reconstruction.*
  For a set of *image points*, usually grid points, determine the *image value*: formally the average field value within the corresponding grid cell. In practice a few different strategies may be followed, dictated by accuracy requirements. These are

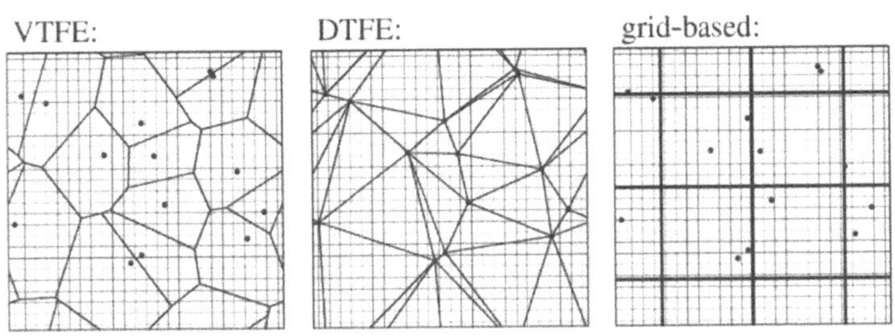

Fig. 24. Two-dimensional display grids in the VTFE, DTFE and grid-based reconstruction methods. The grid is overlaid on top of the basic underlying structure used to reconstruct the density field. SPH-like methods are not shown, because of the inherent difficulty in visualizing their underlying structure, which does not consist of a subdivision of space in distinct non-overlapping structural elements, but of circles of different radius at each position in space

- *Formal geometric approach*: integrate over the field values within each grid cell. This implies the calculation of the intersection of the relevant Delaunay tetrahedra and integration of the (linearly) running field values within the intersection. Subsequently the integrands of each Delaunay intersection are added and averaged over the grid-cell volume.
- *Monte Carlo approach*: approximate the integral by taking the average over a number of (interpolated) field values probed at randomly distributed locations within the grid cell around an *image point*. Finally average over the obtained field values within a grid cell.
- *Singular interpolation approach*: a reasonable and usually satisfactory alternative to the formal geometric or Monte Carlo approach is the shortcut to limit the field value calculation to that at the (grid) location of the *image point*. This offers a reasonable approximation for grid cells which are smaller or comparable to that of intersecting Delaunay cells, on the condition the field gradient within the cell(s) is not too large.

+ *Smoothing* and *filtering*:
  - Linear filtering of the field $\widehat{f}$: convolution of the field $\widehat{f}$ with a filter function $W_s(\mathbf{x}, \mathbf{y})$, usually user specified,

$$f_s(\mathbf{x}) = \int \widehat{f}(\mathbf{x}')\, W_s(\mathbf{x}', \mathbf{y})\, d\mathbf{x}' \qquad (20)$$

  - Median (natural neighbour) filtering: the DTFE density field is adaptively smoothed on the basis of the median value of densities within the *contiguous Voronoi cell*, the region defined by the

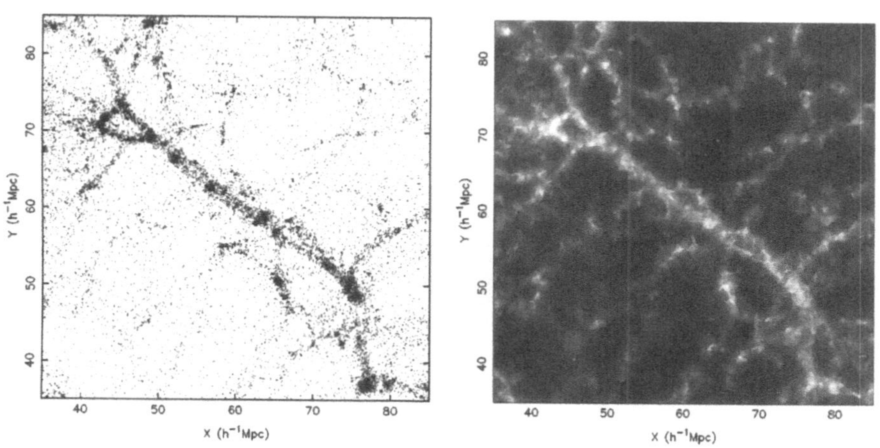

Fig. 25. Image of a characteristic filamentary region in the outcome of a cosmological N-body simulation. *Left-hand frame*: particle distribution in a *thin slice* through the simulation box. *Right-hand frame*: two-dimensional slice through the three-dimensional DTFE density field reconstruction. From Schaap [201]

point and its *natural neighbours* (see Platen, van de Weygaert & Jones [179]).
  - (Nonlinear) diffusion filtering: filtering of (sharply defined) features in images by solving appropriately defined diffusion equations (see, e.g., Mitra & Sicuranza [160]).

- **Post-processing**:
  The real potential of DTFE fields may be found in sophisticated applications tuned towards uncovering characteristics of the reconstructed fields. An important aspect of this involves the analysis of structures in the density field. Some notable examples are
  + Advanced filtering operations. Potentially interesting applications are those based on the use of wavelets (Martínez et al. [148]).
  + Cluster, filament and wall detection by means of the *multiscale morphology filter* (Aragón-Calvo [9], Aragón-Calvo et al. [10]).
  + Void identification on the basis of the *cosmic watershed* algorithm (Platen et al. [179]).
  + Halo detection in N-body simulations (Neyrinck et al. [166]).
  + The computation of two-dimensional surface densities for the study of gravitational lensing (Bradac et al. [39]).

  In addition, DTFE enables the simultaneous and combined analysis of density fields and other relevant physical fields. As it allows the simultaneous determination of *density* and *velocity* fields, it can serve as the basis for studies of the dynamics of structure formation in the cosmos. Its ability to detect substructure as well as reproduce the morphology of cosmic features and objects implies DTFE to be suited for assessing their dynamics without having to invoke artificial filters.
  + DTFE as basis for the study of the full *phase-space* structure of structures and objects. The phase-space structure dark haloes in cosmological structure formation scenarios has been studied by Arad et al. [8].

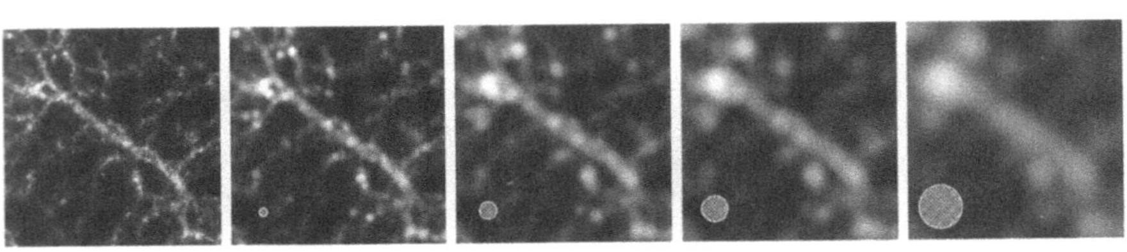

Fig. 26. Filtering of density fields. The left-hand frame depicts the original DTFE density field. The subsequent frames show filtered DTFE density fields. The FWHM of the Gaussian filter is indicated by the shaded circle in the lower left-hand corner of these frames. From Schaap [201]

10 DTFE: Densities and Velocities

10.1 DTFE Density Estimates

The DTFE procedure extends the concept of interpolation of field values sampled at the point sample $\mathcal{P}$ to the estimate of the density $\widehat{\rho}(\mathbf{x})$ from the spatial point distribution itself. This is only feasible if the spatial distribution of the discrete point sample forms a fair and unbiased reflection of the underlying density field.

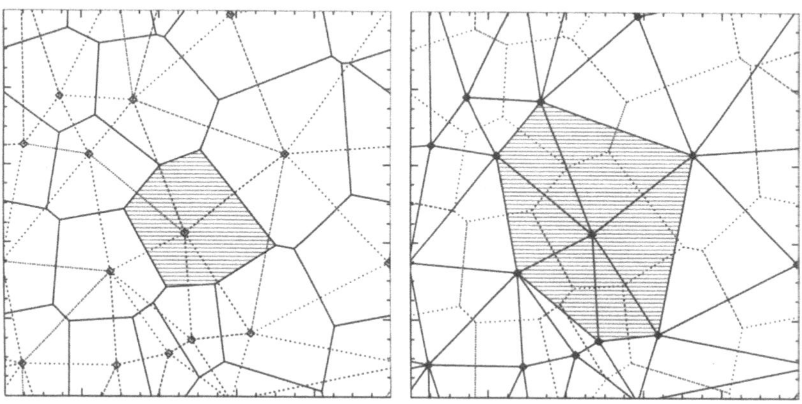

Fig. 27. Illustration of a contiguous Voronoi cell. A contiguous Voronoi cell is the union of all Delaunay tetrahedra of which a point i is one of the vertices

It is commonly known that an optimal estimate for the spatial density at the location of a point $\mathbf{x}_i$ in a discrete point sample $\mathcal{P}$ is given by the inverse of the volume of the corresponding Voronoi cell (see for references Okabe et al. [169], Sect. 8.5). Tessellation-based methods for estimating the density have been introduced by Brown [42] and Ord [170]. In astronomy, Ebeling & Wiedenmann [67] were the first to use tessellation-based density estimators for the specific purpose of devising source detection algorithms. This work has recently been applied to cluster detection algorithms by Ramella et al. [183], Kim et al. [120], Marinoni et al. [146] and Lopes et al. [140]. Along the same lines, Ascalbar & Binney [13] suggested that the use of a multidimensional binary tree might offer a computationally more efficient alternative. However, these studies have been restricted to raw estimates of the local sampling density at the position of the sampling points and have not yet included the more elaborate interpolation machinery of the DTFE and natural neighbour interpolation methods.

Density Definition

The density field reconstruction of the DTFE procedure consists of two steps, the zeroth-order estimate $\widehat{\rho}_0$ of the density values at the location of the points in $\mathcal{P}$ and the subsequent linear interpolation of those zeroth-order density estimates over the corresponding Delaunay grid throughout the sample volume. This yields the DTFE density field estimate $\widehat{\rho}(\mathbf{x})$.

It is straightforward to infer (see next Sect. 10.1) that if the zeroth-order estimate of the density values would be the inverse of the regular Voronoi volume the condition of mass conservation would not be met. Instead, the DTFE

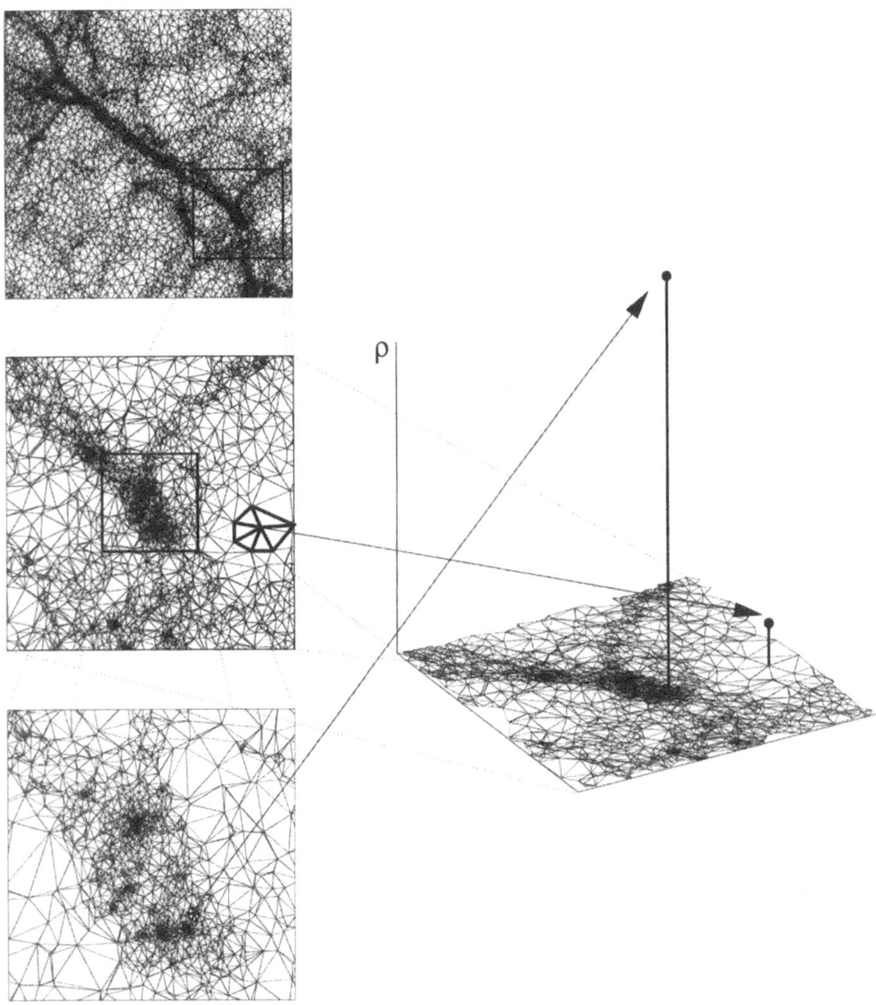

Fig. 28. Relation between density and volume contiguous Voronoi cells. Two example points embedded within a filamentary structure (see Fig. 17)

procedure employs a slightly modified yet related zeroth-order density estimate, the normalized inverse of the volume $V(\mathcal{W}_i)$ of the *contiguous Voronoi cell* $\mathcal{W}_i$ of each point i. For D-dimensional space this is

$$\widehat{\rho}(\mathbf{x}_i) = (1 + D) \frac{m_i}{V(\mathcal{W}_i)} . \tag{21}$$

The *contiguous Voronoi cell* of a point i is the union of all Delaunay tetrahedra of which point i forms one of the four vertices (see Fig. 27).

Mass Conservation

An *essential* requirement for the cosmological purposes of our interpolation scheme is that the estimated DTFE density field $\widehat{\rho}(\mathbf{x})$ should guarantee *mass conservation*: the total mass corresponding to the density field should be equal to the mass represented by the sample points. Indeed, this is an absolutely crucial condition for many applications of a physical nature. Since the mass M is given by the integral of the density field $\rho(\mathbf{x})$ over space, this translates into the integral requirement

$$\begin{aligned} \widehat{M} &= \int \widehat{\rho}(\mathbf{x})\, d\mathbf{x} \\ &= \sum_{i=1}^{N} m_i = M = \text{cst.}, \end{aligned} \tag{22}$$

with $m_i = m$ the mass per sample point: the interpolation procedure should conserve the mass M.

The integral (22) is equal to the volume below the linearly varying $\widehat{\rho}$-surface in $(\mathbf{x}, \widehat{\rho})$-space. In this space each Delaunay tetrahedron m is the base "hyper-plane" of a polyhedron $\mathcal{D}^+_m$ (here we use the term "tetrahedron" for any multidimensional Delaunay simplex). The total mass corresponding to the density field may therefore be written as the sum of the volumes of these polyhedra,

$$M = \sum_{m=1}^{N_T} V(\mathcal{D}^*_m) , \tag{23}$$

with N_T being the total number of Delaunay tetrahedra in the tessellation and $V(\mathcal{D}^*_m)$ the volume of polyhedron $\mathcal{D}^+_m$. This volume may be written as the average density at the vertices of Delaunay tetrahedron $\mathcal{D}_m$ times its volume $V(\mathcal{D}_m)$:

$$V(\mathcal{D}^*_m) = \frac{1}{D+1} \left(\widehat{\rho}_{m1} + \widehat{\rho}_{m2} + \ldots + \widehat{\rho}_{m(D+1)}\right) V(\mathcal{D}_m) . \tag{24}$$

The points $\{m1, m2, \ldots, m(D+1)\}$ are the nuclei which are vertices of the Delaunay tetrahedron $\mathcal{D}^*_m$. The total mass M contained in the density field is the sum over all Delaunay tetrahedra within the sample volume:

$$M = \frac{1}{D+1} \sum_{m=1}^{N_T} \left(\widehat{\rho}_{m1} + \widehat{\rho}_{m2} + \ldots + \widehat{\rho}_{m(D+1)}\right) V(\mathcal{D}_m). \tag{25}$$

A simple reordering of this sum yields

$$M = \frac{1}{D+1} \sum_{i=1}^{N} \widehat{\rho}_i \sum_{m=1}^{N_{D,i}} V(\mathcal{D}_{m,i}), \tag{26}$$

in which $\mathcal{D}_{m,i}$ is one of the $N_{D,i}$ Delaunay tetrahedra of which nucleus i is a vertex. The complete set $\mathcal{D}_{m,i}$ constitutes the *contiguous Voronoi cell* $\mathcal{W}_i$ of nucleus i. Mass M may therefore be written as

$$M = \frac{1}{D+1} \sum_{i=1}^{N} \widehat{\rho}_i \, V(\mathcal{W}_i). \tag{27}$$

This equation immediately implies that M is equal to the total mass of the sampling points (22) on the condition that the density at the location of the sampling points is

$$\widehat{\rho}(\mathbf{x}_i) = (D+1) \frac{m_i}{V(\mathcal{W}_i)}. \tag{28}$$

This shows that the DTFE density estimate (21), proportional to the inverse of contiguous Voronoi cell volume, indeed guarantees mass conservation. The corresponding normalization constant $(1+D)$ stands for the number of times each Delaunay tetrahedron is used in defining the DTFE density field, equal to the number of (1+D) sample points constituting its vertices.

Non-uniform Sampling

Astronomical situations often involve a non-uniform sampling process. Often the non-uniform selection may be quantified by an a priori known selection function $\psi(\mathbf{x})$. This situation is typical for galaxy surveys: $\psi(\mathbf{x})$ may encapsulate differences in sampling density as function of

- Sky position (α, δ)
  In practice, galaxy (redshift) surveys are hardly ever perfectly uniform. Various kinds of factors – survey definition, observing conditions, instrumental aspects, incidental facts – will result in a non-uniform coverage of the sky. These may be encapsulated in a *survey mask* $\psi(\alpha, \delta)$.
- Distance/redshift: $\psi_z(r)$
  Magnitude- or flux-limited surveys induce a systematic redshift selection. At higher redshifts, magnitude-limited surveys can only probe galaxies whose luminosity exceeds an ever increasing value $L(z)$. The corresponding radial selection function ψ_z is given by

$$\psi_z(z) = \frac{\int_{L(z)}^{\infty} \Phi(L) \mathrm{d}L}{\int_{L_0}^{\infty} \Phi(L) \mathrm{d}L}, \tag{29}$$

where $\Phi(L)$ is the galaxy luminosity function.

Both selection effects will lead to an underestimate of density value. To correct for these variations in sky completeness and/or redshift/depth completeness the estimated density at a sample point/galaxy i is weighted by the inverse of the selection function,

$$\psi(\mathbf{x}_i) = \psi_s(\alpha_i, \delta_i)\, \psi_z(r_i)\,, \tag{30}$$

so that we obtain

$$\widehat{\rho}(\mathbf{x}_i) = (1+D)\, \frac{m_i}{\psi(\mathbf{x}_i)\, V(\mathcal{W}_i)}\,. \tag{31}$$

While it is straightforward to correct the density estimates accordingly, a complication occurs for locations where the completeness is very low or even equal to zero. In general, regions with redshift completeness zero should be excluded from the correction procedure, even though for specific cosmological contexts it is feasible to incorporate field reconstruction procedures utilizing field correlations. A *constrained random field* approach (Bertschinger [17], Hoffman & Ribak [103], Zaroubi et al. [258]) uses the autocorrelation function of the presumed density field to infer a realization of the corresponding Gaussian field. Recently, Aragón-Calvo [9] developed a DTFE-based technique which manages to reconstruct a structured density field pattern within selection gaps whose size does not exceed the density field's coherence scale.

10.2 DTFE Velocity Fields

DTFE Density and Velocity Field Gradients

The value of the density and velocity field gradient in each Delaunay tetrahedron is directly and uniquely determined from the location $\mathbf{r} = (x, y, z)$ of the four points forming the Delaunay tetrahedron's vertices, $\mathbf{r}_0$, $\mathbf{r}_1$, $\mathbf{r}_2$ and $\mathbf{r}_3$, and the value of the estimated density and sampled velocities at each of these locations, $(\widehat{\rho}_0, \mathbf{v}_0)$, $(\widehat{\rho}_1, \mathbf{v}_1)$, $(\widehat{\rho}_2, \mathbf{v}_2)$ and $(\widehat{\rho}_3, \mathbf{v}_3)$:

$$\begin{array}{c} \widehat{\nabla \rho}\big|_m \\ \\ \widehat{\nabla \mathbf{v}}\big|_m \end{array} \quad \Longleftarrow \quad \left\{ \begin{array}{cccc} \widehat{\rho}_0 & \widehat{\rho}_1 & \widehat{\rho}_2 & \widehat{\rho}_3 \\ \mathbf{v}_0 & \mathbf{v}_1 & \mathbf{v}_2 & \mathbf{v}_3 \\ & & & \\ \mathbf{r}_0 & \mathbf{r}_1 & \mathbf{r}_2 & \mathbf{r}_3 \end{array} \right. \tag{32}$$

The four vertices of the Delaunay tetrahedron are both necessary and sufficient for computing the entire 3×3 velocity gradient tensor $\partial v_i / \partial x_j$. Evidently, the same holds for the density gradient $\partial \rho / \partial x_j$. We define the matrix $\mathbf{A}$ is defined on the basis of the vertex distances $(\Delta x_n, \Delta y_n, \Delta z_n)$ (n=1,2,3),

$$\begin{aligned} \Delta x_n &= x_n - x_0 \\ \Delta y_n &= y_n - y_0 \\ \Delta z_n &= z_n - z_0 \end{aligned} \quad \Longrightarrow \quad \mathbf{A} = \begin{pmatrix} \Delta x_1 & \Delta y_1 & \Delta z_1 \\ & & \\ \Delta x_2 & \Delta y_2 & \Delta z_2 \\ & & \\ \Delta x_3 & \Delta y_3 & \Delta z_3 \end{pmatrix} . \tag{33}$$

Similarly defining $\Delta \mathbf{v}_n \equiv \mathbf{v}_n - \mathbf{v}_0$ $(n = 1, 2, 3)$ and $\Delta \rho_n \equiv \rho_n - \rho_0$ $(n = 1, 2, 3)$ it is straightforward to compute directly and simultaneously the density field gradient $\nabla \rho|_m$ and the velocity field gradient $\nabla \mathbf{v}|_m = \partial v_i / \partial x_j$ in Delaunay tetrahedron m via the inversion

$$\begin{aligned} \begin{pmatrix} \frac{\partial \rho}{\partial x} \\ \\ \frac{\partial \rho}{\partial y} \\ \\ \frac{\partial \rho}{\partial z} \end{pmatrix} &= \mathbf{A}^{-1} \begin{pmatrix} \Delta \rho_1 \\ \\ \Delta \rho_2 \\ \\ \Delta \rho_3 \end{pmatrix} ; \\ \begin{pmatrix} \frac{\partial v_x}{\partial x} & \frac{\partial v_y}{\partial x} & \frac{\partial v_z}{\partial x} \\ \\ \frac{\partial v_x}{\partial y} & \frac{\partial v_y}{\partial y} & \frac{\partial v_z}{\partial y} \\ \\ \frac{\partial v_x}{\partial z} & \frac{\partial v_y}{\partial z} & \frac{\partial v_z}{\partial z} \end{pmatrix} &= \mathbf{A}^{-1} \begin{pmatrix} \Delta v_{1x} & \Delta v_{1y} & \Delta v_{1z} \\ \\ \Delta v_{2x} & \Delta v_{2y} & \Delta v_{2z} \\ \\ \Delta v_{3x} & \Delta v_{3y} & \Delta v_{3z} \end{pmatrix} . \end{aligned} \tag{34}$$

Velocity Divergence, Shear and Vorticity

From the nine velocity gradient components $\partial v_i / \partial x_j$ we can directly determine the three velocity deformation modes, the velocity divergence $\nabla \cdot \mathbf{v}$, the shear σ_{ij} and the vorticity ω:

$$\begin{aligned} \nabla \cdot \mathbf{v} &= \left(\frac{\partial v_x}{\partial x} + \frac{\partial v_y}{\partial y} + \frac{\partial v_z}{\partial z} \right) , \\ \sigma_{ij} &= \frac{1}{2} \left\{ \frac{\partial v_i}{\partial x_j} + \frac{\partial v_j}{\partial x_i} \right\} - \frac{1}{3} (\nabla \cdot \mathbf{v}) \, \delta_{ij} , \\ \omega_{ij} &= \frac{1}{2} \left\{ \frac{\partial v_i}{\partial x_j} - \frac{\partial v_j}{\partial x_i} \right\} . \end{aligned} \tag{35}$$

where $\omega = \nabla \times \mathbf{v} = \epsilon^{kij} \omega_{ij}$ (and ϵ^{kij} is the completely antisymmetric tensor). In the theory of gravitational instability, there will be no vorticity contribution as long as there has not been shell crossing (i.e. in the linear and mildly nonlinear regime).

10.3 DTFE: Velocity Field Limitations

DTFE interpolation of velocity field values is only feasible in regions devoid of multi-stream flows. As soon as there are multiple flows – notably in high-density cluster concentrations or in the highest density realms of the filamentary and planar caustics in the cosmic web – the method breaks down and cannot be applied.

In the study presented here this is particularly so in high-density clusters. The complication can be circumvented by filtering the velocities over a sufficiently large region, imposing an additional resolution constraint on the DTFE velocity field. Implicitly this has actually already been accomplished in the linearization procedure of the velocity fields preceding the DTFE processing. The linearization of the input velocities involves a kernel size of $\sqrt{5}\,\mathrm{h}^{-1}$ Mpc so that the resolution of the velocity field is set to a lower limit of $\sqrt{5}\,\mathrm{h}^{-1}$ Mpc. This is sufficient to assure the viability of the DTFE velocity field reconstructions.

10.4 DTFE: Mass Weighted Versus Volume Weighted

A major and essential aspect of DTFE field estimates is that it concerns *volume-weighted* field averages

$$\widehat{f}_{vol}(\mathbf{x}) \equiv \frac{\displaystyle\int d\mathbf{y}\, f(\mathbf{y})\, W(\mathbf{x}-\mathbf{y})}{\displaystyle\int d\mathbf{y}\; W(\mathbf{x}-\mathbf{y})} \tag{36}$$

instead of the more common *mass-weighted* field averages

$$\widehat{f}_{mass}(\mathbf{x}) \equiv \frac{\displaystyle\int d\mathbf{y}\, f(\mathbf{y})\, \rho(\mathbf{y})\, W(\mathbf{x}-\mathbf{y})}{\displaystyle\int d\mathbf{y}\, \rho(\mathbf{y})\, W(\mathbf{x}-\mathbf{y})} \tag{37}$$

where $W(\mathbf{x}, \mathbf{y})$ is the adopted filter function defining the weight of a mass element in a way that is dependent on its position y with respect to the position $\mathbf{x}$.

Analytical calculations of physical systems in an advanced stage of evolution do quite often concern a perturbation analysis. In a cosmological context we may think of the nonlinear evolution of density and velocity field perturbations. In order to avoid unnecessary mathematical complications most results concern *volume-weighted* quantities. However, when seeking to relate these to the results of observational results or numerical calculations, involving a discrete sample of measurement points, nearly all conventional techniques implicitly involve *mass-weighted* averages. This may lead to considerable confusion, even to wrongly motivated conclusions.

Conventional schemes for translating a discrete sample of field values f_i into a continuous field $\hat{f}(\mathbf{x})$ are usually based on a suitably weighted sum over the discretely sampled field values involving the kernel weight functions $W(\mathbf{x},\mathbf{y})$. It is straightforward to convert this discrete sum into an integral over Dirac delta functions:

$$\begin{aligned}\hat{f}(\mathbf{x}) &= \frac{\sum_{i=1}^{N} \tilde{f}_i\, W(\mathbf{x}-\mathbf{x}_i)}{\sum_{i=1}^{N} W(\mathbf{x}-\mathbf{x}_i)} \\ &= \frac{\displaystyle\int d\mathbf{y}\, f(\mathbf{y})\, W(\mathbf{x}-\mathbf{y})\, \sum_{i=1}^{N} \delta_D(\mathbf{y}-\mathbf{x}_i)}{\displaystyle\int d\mathbf{y}\, W(\mathbf{x}-\mathbf{y})\, \sum_{i=1}^{N} \delta_D(\mathbf{y}-\mathbf{x}_i)} \\ &= \frac{\displaystyle\int d\mathbf{y}\, f(\mathbf{y})\, \rho(\mathbf{y})\, W(\mathbf{x}-\mathbf{y})}{\displaystyle\int d\mathbf{y}\, \rho(\mathbf{y})\, W(\mathbf{x}-\mathbf{y})}\,.\end{aligned} \tag{38}$$

Evidently, a weighted sum implicitly yields a mass-weighted field average. Notice that this is true not only for rigid grid-based averages but also for spatially adaptive SPH-like evaluations.

A reasonable approximation of volume-averaged quantities may be obtained by volume averaging over quantities that were mass-filtered with, in comparison, a very small scale for the mass-weighting filter function. This prodded Juszkiewicz et al. [111] to suggest a (partial) remedy in the form of a two-step scheme for velocity field analysis. First, the field values are interpolated onto a grid according to (39). Subsequently, the resulting grid of field values is used to determine volume averages according to (36). We can then make the interesting observation that the asymptotic limit of this, using a filter with an infinitely small filter length, yields

$$f_{mass}(\mathbf{x}_0) = \frac{\sum_i w_i f(\mathbf{x}_i)}{\sum_i w_i} = \frac{f(\mathbf{x}_1) + \sum_{i=2}^{N} \frac{w_i}{w_1} f(\mathbf{x}_i)}{1 + \sum_{i=2}^{N} \frac{w_i}{w_1}} \longrightarrow f(\mathbf{x}_1)\,, \tag{39}$$

where we have ordered the locations i by increasing distance to $\mathbf{x}_0$ and thus by decreasing value of w_i.

The interesting conclusion is that the resulting estimate of the volume-weighted average is in fact the field value at the location of the closest sample point $\mathbf{x}_1$, $f(\mathbf{x}_1)$. This means we should divide up space into regions consisting of that part of space closer to a particular sample point than to any of the other sample points and take the field value of that sample point as the value of the field in that region. This is nothing but dividing up space according to the Voronoi tessellation of the set of sample points $\mathcal{P}$. This observation formed the rationale behind the introduction and definition of Voronoi and Delaunay tessellation interpolation methods for velocity fields by Bernardeau & van de Weygaert [27].

While the resulting *Voronoi Tessellation Field Estimator* would involve a discontinuous field, the step towards a multidimensional linear interpolation procedure would guarantee the continuity of field values. The subsequent logical step, invoking the dual Delaunay tessellation as equally adaptive spatial linear interpolation grid, leads to the definition of the DTFE interpolation scheme.

10.5 DTFE Density and Velocity Maps: Non-uniform Resolution

For a proper quantitative analysis and assessment of DTFE density and velocity field reconstructions it is of utmost importance to take into account the inhomogeneous spatial resolution of raw DTFE maps.

Cosmic density and velocity fields, as well as possible other fields of relevance, are composed of contributions from a range of scales. By implicitly filtering out small-scale contributions in regions with a lower sampling density, DTFE will include a smaller range of spatial scales contributing to a field reconstruction. Even while selection function corrections are taken into account, the DTFE density and velocity fields will therefore involve lower amplitudes. DTFE velocity fields are less affected than DTFE density fields (van de Weygaert & Romano-Díaz [249]), a manifestation of the fact that the cosmic velocity field is dominated by larger scale modes than the density field.

In addition, it will also lead to a diminished "morphological resolution". The sampling density may decrease to such a level lower than required to resolve the geometry or morphology of web-like features. Once this level has been reached, DTFE will no longer be suited for an analysis of the cosmic web.

11 DTFE: Technical Issues

11.1 DTFE Kernel and Effective Resolution

DTFE distributes the mass m_i corresponding to each sampling point i over its corresponding contiguous Voronoi cell. A particular sampling point i will therefore contribute solely to the mass of those Delaunay simplices of which it is a vertex. This restricted spatial smoothing is the essence of the strict locality of the DTFE reconstruction scheme.

The expression for the interpolation kernel of DTFE provides substantial insight into its local spatial resolution. Here we concentrate on the density field kernel; the shape and scale of the interpolation kernels of other fields are comparable.

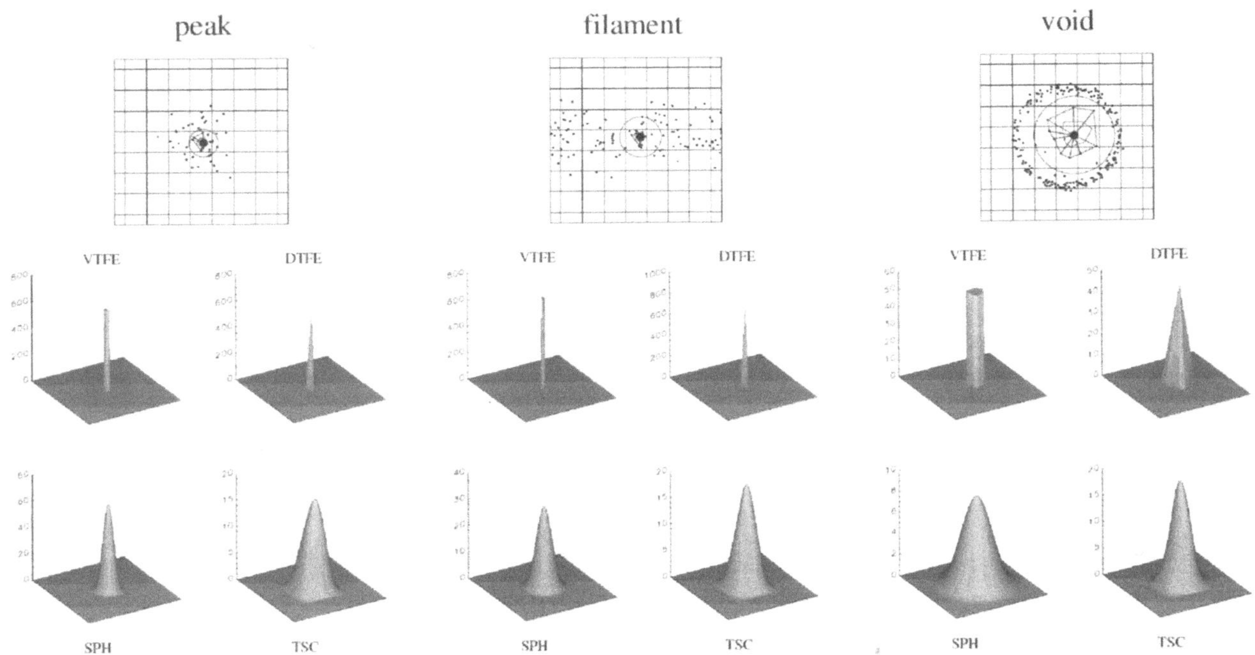

Fig. 29. Typical interpolation kernels for points embedded within three different morphological environments: a point in a high-density peak (*bottom*), a point in a filament (*centre*) and a point in a low-density field environment (*top*). The three kernels depicted concern three different environments: a point embedded within a Gaussian density peak (*left hand*), a point within an elongated filamentary concentration of points (*centre*) and a point in a low-density environment (*right hand*)

In essence, a density reconstruction scheme distributes the mass m_i of a sampling particle i over a region of space according to a distributing function $\mathcal{F}_i(\mathbf{x})$,

$$\rho(\mathbf{x}) = \sum_{i=1}^{N} m_i \mathcal{F}_i(\mathbf{x}), \tag{40}$$

with $\int d\mathbf{x}\,\mathcal{F}_i(\mathbf{x}) = 1 \;\forall i$. Generically, the shape and scale of the *interpolation kernel* $\mathcal{F}_i(\mathbf{x})$ may be different for each sample point. The four examples of nn-neighbour kernels in Fig. 21 are a clear illustration of this.

For the linear interpolation of DTFE we find that (see (11)) for a position $\mathbf{x}$ within a Delaunay tetrahedron m defined by $(1+D)$ sample points $\{\mathbf{x}_{m0}, \mathbf{x}_{m1}, \ldots, \mathbf{x}_{mD}\}$ the interpolation indices $\phi_{dt,i}$ are given by

$$\phi_{dt,i}(\mathbf{x}) = \begin{cases} 1 + t_1 + t_2 + \ldots + t_D & i \in \{m0, m1, \ldots, mD\} \\ \\ 0 & i \notin \{m0, m1, \ldots, mD\} \end{cases} . \tag{41}$$

In this, for $i \in \{m0, m1, \ldots, mD\}$, the D parameters $\{t_1, \ldots, t_D\}$ are to be solved from

$$\mathbf{x} = \mathbf{x}_i + t_1(\mathbf{x}_{m1} - \mathbf{x}_i) + t_2(\mathbf{x}_{m2} - \mathbf{x}_i) + \ldots + t_D(\mathbf{x}_{mD} - \mathbf{x}_i), \tag{42}$$

On the basis of eqns. (40) and (11) the expression for the DTFE kernel is easily derived:

$$\mathcal{F}_{\mathrm{DTFE},i}(\mathbf{x}) = \begin{cases} \dfrac{(D+1)}{V(\mathcal{W}_i)}\,\phi_{dt,i}(\mathbf{x}) & \text{if} \quad \mathbf{x} \in \mathcal{W}_i \\ \\ 0 & \text{if} \quad \mathbf{x} \notin \mathcal{W}_i \end{cases} , \tag{43}$$

in which $\mathcal{W}_i$ is the contiguous Voronoi cell of sampling point i. In this respect we should realize that in two dimensions the contiguous Voronoi cell is defined by on average seven sample points: the point itself and its *natural neighbours*. In three dimensions this number is, on average, 13.04. Even with respect to spatially adaptive smoothing such as embodied by the kernels used in smooth particle hydrodynamics, defined by a certain number of nearest neighbours (usually in the order of 40), the DTFE kernel is indeed highly localized.

A set of DTFE smoothing kernels are depicted in Fig. 29. Their local and adaptive nature may be best appreciated from the comparison with the corresponding kernels for regular (rigid) grid-based TSC interpolation, a scale-adaptive SPH smoothing kernel (based on 40 nearest neighbours) and a zeroth-order Voronoi (VTFE) kernel (where the density at a point is set equal to the inverse of the volume of the corresponding Voronoi cell). The three kernels depicted concern three different environments: a point embedded within a Gaussian density peak (left hand), a point within an elongated filamentary concentration of points (centre) and a point in a low-density environment (right hand). The figure clearly illustrates the adaptivity in scale and geometry of the DTFE kernel.

11.2 Noise and Sampling Characteristics

In order to appreciate the workings of DTFE one needs to take into account the noise characteristics of the method. Tessellation-based reconstructions do start from a discrete random sampling of a field. This will induce noise and errors. Discrete sampling noise will propagate even more strongly into DTFE density reconstructions as the discrete point sample itself will be the source for the density value estimates.

In order to understand how sampling noise affects the reconstructed DTFE fields it is imperative to see how intrinsically uniform fields are affected by the discrete sampling. Even though a uniform Poisson process represents a discrete reflection of a uniform density field, the stochastic nature of the Poisson sample will induce a non-uniform distribution of Voronoi and Delaunay volumes (see Sect. 6.5). Because the volumes of Voronoi and/or Delaunay cells are at the basis of the DTFE density estimates, their non-uniform distribution propagates immediately into fluctuations in the reconstructed density field.

This is illustrated in Fig. 29, in which three uniform Poisson point samples, each of a different sampling density n, are shown together with the corresponding Voronoi and Delaunay tessellations. In addition, we have shown the first-order DTFE reconstructed density fields, along with the zeroth-order VTFE density field (where the density at a point is set equal to the inverse of the volume of the corresponding Voronoi cell). The variation in both the Delaunay and Voronoi cells, as well as in the implied VTFE and DTFE density field reconstructions, provides a reasonable impression of the fluctuations going along with these interpolation schemes.

Following the suggestion by Kiang [119] for the Gamma-type volume distribution for the volumes of Voronoi cells (4), we may find an analytical expression for the density distribution for the zeroth-order VTFE field:

$$f(\tilde{\rho}) = \begin{cases} \frac{128}{3}\,\tilde{\rho}^{-6}\,\exp\left(-\frac{4}{\tilde{\rho}}\right) & 2D \\ \\ \frac{1944}{5}\,\tilde{\rho}^{-8}\,\exp\left(-\frac{6}{\tilde{\rho}}\right) & 3D \end{cases} \quad . \tag{44}$$

The normalized density value $\tilde{\rho}$ is the density estimate ρ in units of the average density $\langle\rho\rangle$. While in principle this only concerns the zeroth-order density estimate, it turns out that these expressions also provide an excellent description of the one-point distribution function of the first-order DTFE density field, both in two and three dimensions. This may be appreciated from Fig. 31, showing the pdfs for DTFE density field realizations for a Poisson random field of $10,000$ points (two dimensions) and $100,000$ point (three dimensions). The superimposed analytical expressions (44) represent excellent fits.

The two-dimensional and three-dimensional distributions are clearly non-Gaussian, involving a tail extending towards high-density values. The positive value of the skewness ($2\sqrt{3}$ for two dimensions and $\sqrt{5}$ for three dimensions)

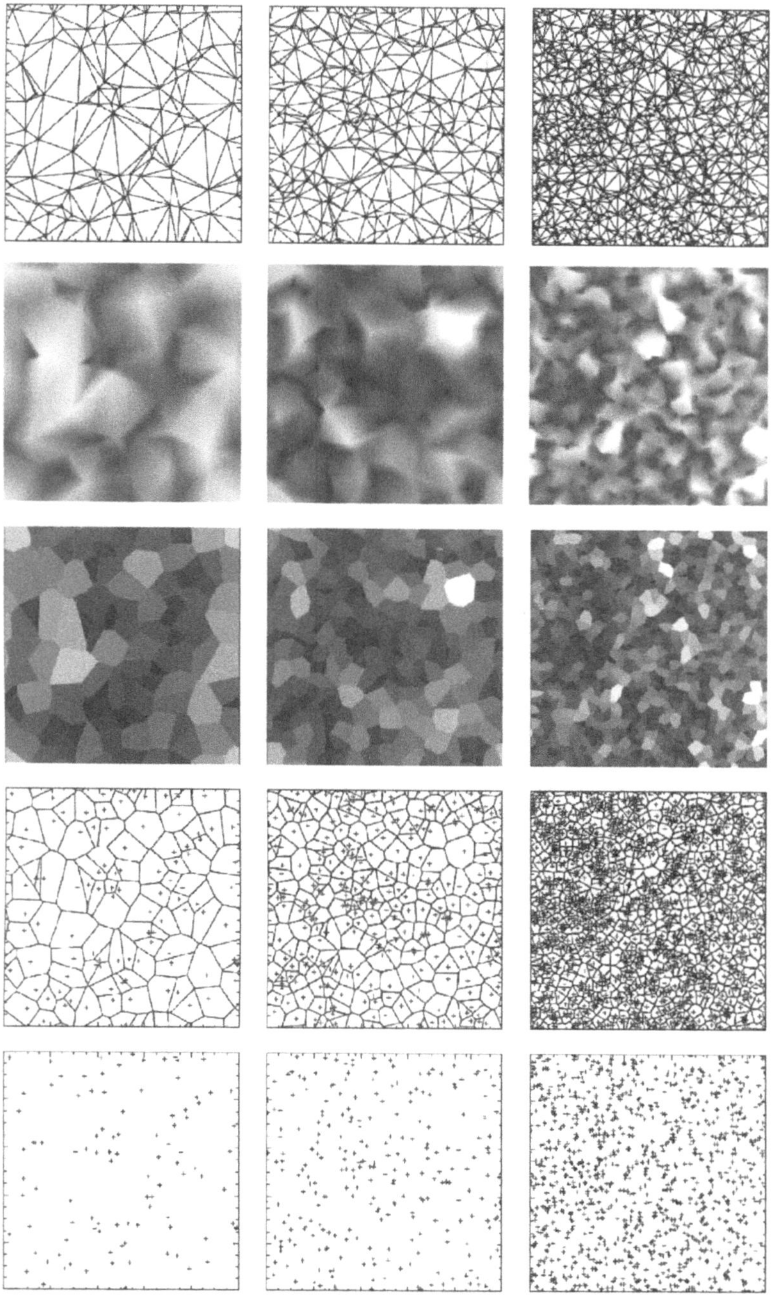

Fig. 30. Poisson sampling noise in uniform fields. The rows illustrate three Poisson point samplings of a uniform field with increasing sampling density (from top to bottom consisting of 100, 250 and 1000 points). From left to right, the point distribution, the corresponding Voronoi tessellation, the zeroth-order VTFE reconstructed density field, the first-order DTFE reconstructed density field and the corresponding Delaunay tessellation are shown. From Schaap [201]

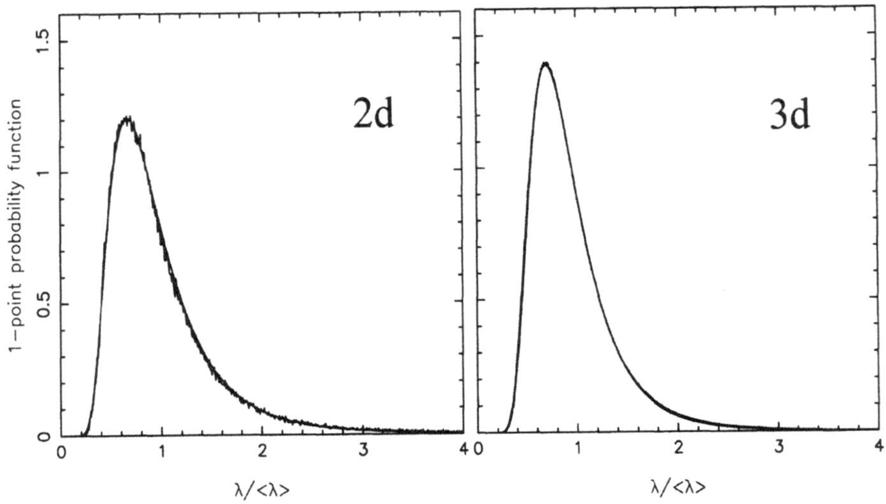

Fig. 31. One-point distribution functions of the DTFE density field for a Poisson point process of 10 000 points (for two-dimensional space, *left hand*) and 100 000 points (for three-dimensional space, *right hand*). Superimposed are the analytical approximations given by (44)

confirms the presence of this tail. Interestingly, the distribution falls off considerably more rapid for $d = 3$ than for $d = 2$. Also we see that the distributions are more narrow than in the case of a regular Gaussian: the variance is 1/3 for $d = 2$ and 1/5 for d=3, confirmed by the strongly positive value of the kurtosis. The larger value for $d = 2$ indicates that it is more strongly peaked than the distribution for $d = 3$.

On the basis of the above one may also evaluate the significance of DTFE density field reconstructions, even including that for intrinsically inhomogeneous fields. For details we refer to [201].

11.3 Computational Cost

The computational cost of DTFE is not overriding. Most computational effort is directed towards the computation of the Delaunay tessellation of a point set of N particles. The required CPU time is in the order of $\mathcal{O}(N \log N)$, comparable to the cost of generating the neighbour list in SPH. The different computational steps of the DTFE procedure, along with their scaling as a function of number of sample points N, are:

1. Construction of the Delaunay tessellation: $\mathcal{O}(N \log N)$;
2. Construction of the adjacency matrix: $\mathcal{O}(N)$;
3. Calculation of the density values at each location: $\mathcal{O}(N)$;
4. Calculation of the density gradient inside each Delaunay triangle: $\mathcal{O}(N)$;
5. Interpolation of the density to an image grid: $\mathcal{O}(ngrid^2 \cdot N^{1/2})$.

Step 2, involving the calculation of the adjacency matrix necessary for the *walking triangle* algorithm used for Delaunay tetrahedron identification, may be incorporated in the construction of the Delaunay tessellation itself and therefore omitted. The last step, the interpolation of the density to an image grid, is part of the post-processing operation and could be replaced by any other desired operation. It mainly depends on the number of grid cells per dimension.

12 DTFE: Hierarchical and Anisotropic Patterns

To demonstrate the ability of the Delaunay Tessellation Field Estimator to quantitatively trace key characteristics of the cosmic web we investigate in some detail two aspects, its ability to trace the hierarchical matter distribution and its ability to reproduce the shape of anisotropic – filamentary and planar – web-like patterns.

12.1 Hierarchy and Dynamic Range

The fractal Soneira–Peebles model (Soneira & Peebles [224]) has been used to assess the level to which DTFE is able to trace a density field over the full range of scales represented in a point sample. The Soneira–Peebles model is an analytic self-similar spatial point distribution which was defined for the purpose of modelling the galaxy distribution, such that its statistical properties would be tuned to reality. An important property of the Soneira–Peebles model is that it is one of the few nonlinear models of the galaxy distribution whose statistical properties can be fully and analytically evaluated. This concerns its power-law two-point correlation function, correlation dimension and its Hausdorff dimension. The Soneira–Peebles model is specified by three parameters. The starting point of the model is a level-0 sphere of radius R. At each level m, a number of η subspheres are placed randomly within their parent level-m sphere: the level-$(m+1)$ spheres have a radius R/λ where $\lambda > 1$, the size ratio between parent sphere and subsphere. This process is repeated for L successive levels, yielding η^L level-L spheres of radius R/λ^L. At the centre of each of these spheres a point is placed, producing a point sample of η^L points. While this produces a pure *singular* Soneira–Peebles model, usually a set of these is superimposed to produce a somewhat more realistically looking model of the galaxy distribution, an *extended* Soneira–Peebles model.

An impression of the self-similar point sets produced by the Soneira–Peebles process may be obtained from Fig. 32. The top row contains a series of three-point distributions, zoomed in at three consecutive levels on a singular Soneira–Peebles model realization with $(\eta = 4, \lambda = 1.90, L = 8)$. The next three columns depict the density field renderings produced by three different interpolation schemes, a regular rigid grid-based TSC scheme, a spatially adaptive SPH scheme and finally the DTFE reconstruction. The figure

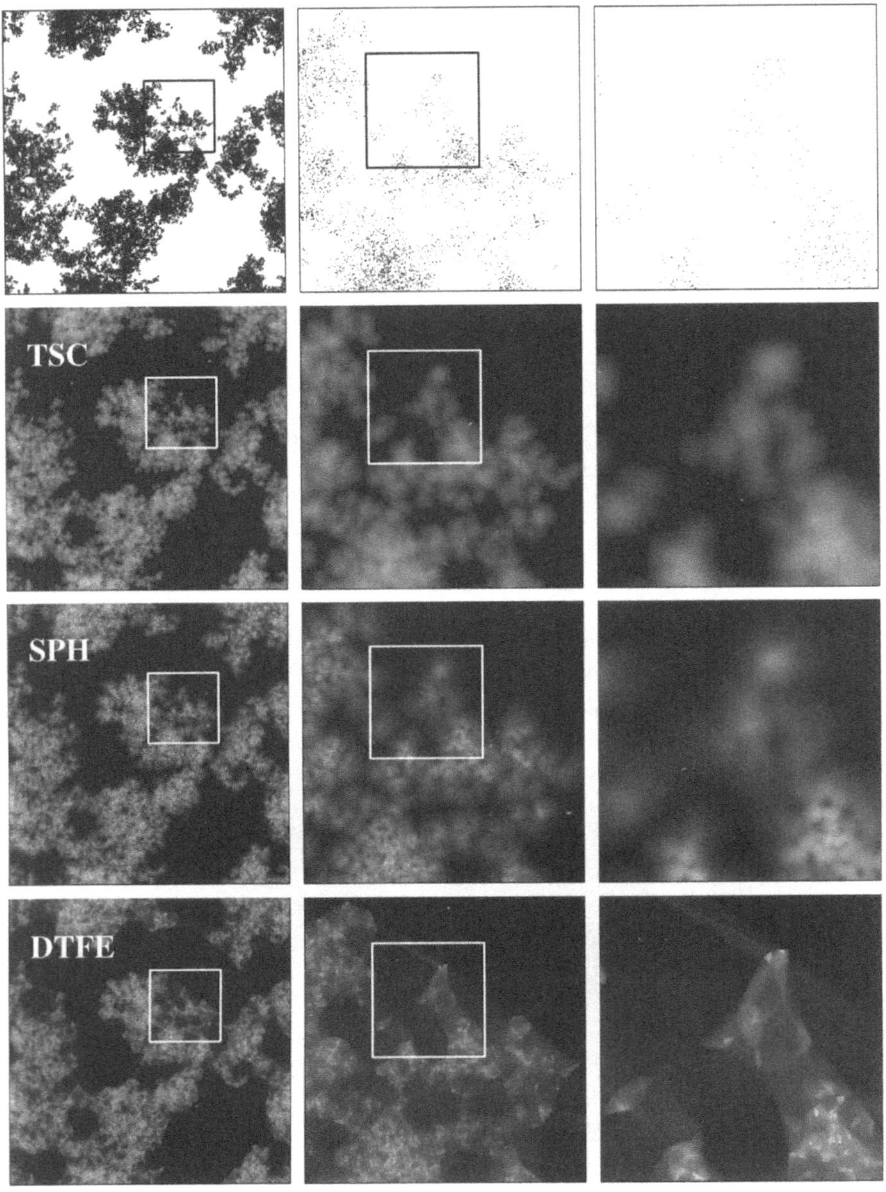

Fig. 32. Singular Soneira–Peebles model with $\eta = 4$, $\lambda = 1.9$ and $L = 8$. *Top row*: full point distribution (*left-hand frame*) and zoom-ins focusing on a particular structure (*central* and *right-hand frames*). Rows 2–4: corresponding density field reconstructions produced using the TSC, SPH and DTFE methods

clearly shows the supreme resolution of the latter. By comparison, even the SPH reconstructions appear to fade for the highest resolution frame.

Self-Similarity

One important manifestation of the self-similarity of the defined Soneira–Peebles distribution is reflected in the power-law two-point correlation function. For M dimensions it is given by

$$\xi(r) \sim r^{-\gamma}, \qquad \gamma = M - \left(\frac{\log \eta}{\log \lambda}\right) \quad \text{for} \quad \frac{R}{\lambda^{L-1}} < r < R. \tag{45}$$

The parameters η and λ may be adjusted such that they yield the desired value for the correlation slope γ.

To probe the self-similarity we look at the one-point distribution of the density field, for both the point distribution and the TSC, SPH and DTFE density field reconstructions. Mathematically, the condition of self-similarity implies that the PDF corresponding to a density field $\rho(\mathbf{x})$ inside an n-level circle of radius R/λ^n should be equal to the PDF inside the reference circle of radius R, after the density field in the n-level circle has been scaled according to

$$\rho(\mathbf{x}) \rightarrow \rho_n(\mathbf{x}) = \rho(\mathbf{x})/\lambda^{Mn}, \tag{46}$$

in which M is the dimension of space. Yet another multiplication factor of λ^{Mn} has to be included to properly normalize the PDF (per density unit). In total this results in an increase by a factor λ^{2Mn}. Self-similar distributions would look exactly the same at different levels of resolution once scaled accordingly. This self-similarity finds its expression in a *universal* power-law behaviour of the PDFs at different levels.

We have tested the self-similar scaling of the pdfs for a range of Soneira–Peebles models, each with a different fractal dimension (Schaap & van de Weygaert [203]). For a Soneira–Peebles model with parameters ($\eta = 2, \lambda = 1.75, L = 14$) we show the resulting scaled PDFs for the TSC, SPH and DTFE density field reconstructions in the top row of Fig. 33. The self-similar scaling of the TSC density field reconstructions is hardly convincing. On the other hand, while the SPH reconstruction performs considerably better, it is only the DTFE rendering which within the density range probed by each level displays an almost perfect power-law scaling! Also notice that the DTFE field manages to probe the density field over a considerably larger density range than, e.g., the SPH density field. Subsequently we determined the slope α of the *universal* power-law PDF and compared it with the theoretical predictions. The set of three frames in the bottom row of Fig. 33 shows the resulting distributions with the fitted power laws. Going from left to right, the frames in this figure

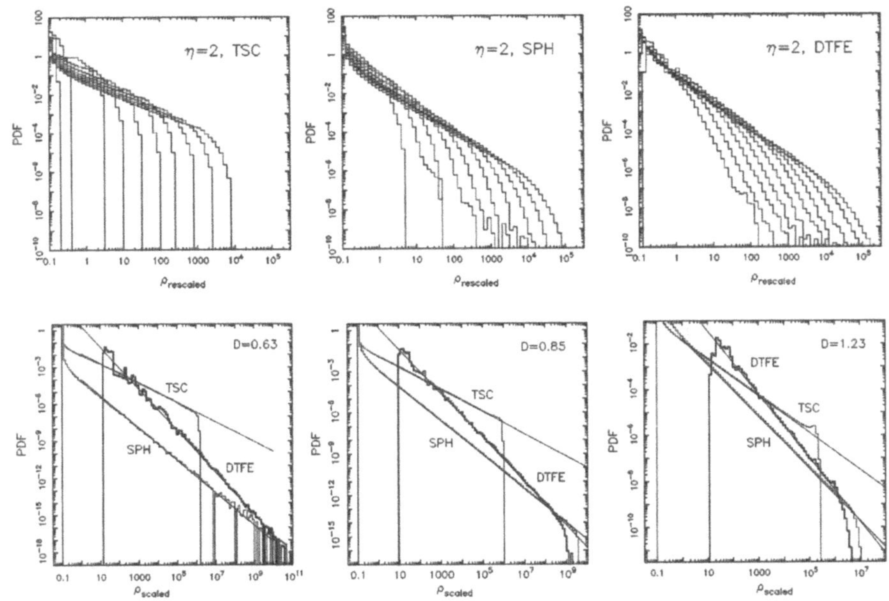

Fig. 33. *Top row*: Average PDFs of the density field in circles of different level (see text for a description) for the TSC, SPH and DTFE density field reconstructions. Model with $\eta = 2$, $\lambda = 1.75$ and $L = 14$. *Bottom row*: scaled PDFs of Soneira–Peebles density field reconstructions. Each frame corresponds to a Soneira–Peebles realization of a different fractal dimension, denoted in the upper right-hand corner. In each frame the TSC, SPH and DTFE reconstructed PDFs are shown

correspond to Soneira–Peebles realizations with fractal dimensions of $D = 0.63$, $D = 0.85$ and $D = 1.23$. The slope α of the PDF can be found by comparing the PDF at two different levels,

$$\begin{aligned} \alpha &= \frac{\log \mathrm{PDF}(\rho_1) - \log \mathrm{PDF}(\rho_0)}{\log \rho_1 - \log \rho_0} \\ &= \frac{\log(\lambda^{2Mn}/\eta^n)}{\log\,(1/\lambda^{Mn})} = \frac{D}{M} - 2\,, \end{aligned} \tag{47}$$

in which D is the fractal dimension of the singular Soneira–Peebles model. When turning to Table 2 we find not only that the values of α derived from the TSC, SPH and DTFE fields do differ significantly, a fact which has been clearly borne out by Fig. 33, but also that the DTFE density field PDFs do reproduce to an impressively accurate level the analytically expected power-law slope of the model itself (Schaap & van de Weygaert [203]). It is hardly possible to find a more convincing argument for the ability of DTFE to deal with hierarchical density distributions!

Table 2. Slopes of the power-law region of the PDF of a Soneira–Peebles density field as reconstructed by the TSC, SPH and DTFE procedures. The theoretical value (48) is also listed. Values are listed for three different Soneira–Peebles realizations, each with a different fractal dimension D

D	α(theory)	α(TSC)	α(SPH)	α(DTFE)
0.63	−1.69	−0.81	−1.32	−1.70
0.86	−1.57	−0.82	−1.24	−1.60
1.23	−1.39	−0.79	−1.13	−1.38

12.2 Shapes and Patterns

DTFE's ability to trace anisotropic web-like patterns is tested on the basis of a class of heuristic models of cellular matter distributions, *Voronoi clustering models* (Schaap & van de Weygaert [203], van de Weygaert [245]). Voronoi models use the frame of a Voronoi tessellation as the skeleton of the galaxy distribution, identifying the structural frame around which matter will gradually assemble during the emergence of cosmic structure. The interior of Voronoi *cells* correspond to voids and the Voronoi *planes* with sheets of galaxies. The *edges* delineating the rim of each wall are identified with the filaments in the galaxy distribution. What is usually denoted as a flattened "supercluster" will comprise an assembly of various connecting walls in the Voronoi foam, as elongated "superclusters" of "filaments" will usually consist of a few coupled edges. The most outstanding structural elements are the *vertices*, corresponding to the very dense compact nodes within the cosmic web, the rich clusters of galaxies.

The Voronoi clustering models offer flexible templates for cellular patterns, and they are easy to tune towards a particular spatial cellular morphology. To investigate the shape performance of DTFE we use pure *Voronoi element models*, tailor-made heuristic "galaxy" distributions either and exclusively in and around (1) the Voronoi walls, (2) the Voronoi edges, (3) the Voronoi vertices. Starting from a random initial distribution, all model galaxies are projected onto the relevant wall, edge or vertex of the Voronoi cell in which they are located.

The three boxes in the top row of Fig. 34 depict a realization for a three-dimensional *Voronoi wall model*, a *Voronoi filament model* and a *Voronoi cluster model.* Planar sections through the TSC, SPH and DTFE density field reconstructions of each of these models are shown in three consecutive rows by means of greyscale maps. The distinctly planar geometry of the *Voronoi wall model* and the one-dimensional geometry of the *Voronoi filament model* are clearly recognizable from the sharply defined features in the DTFE density field reconstruction. While the SPH reconstructions outline the same distinct patterns, in general the structural features have a more puffy appearance. The grid-based TSC method is highly insensitive to the intricate web-like Voronoi

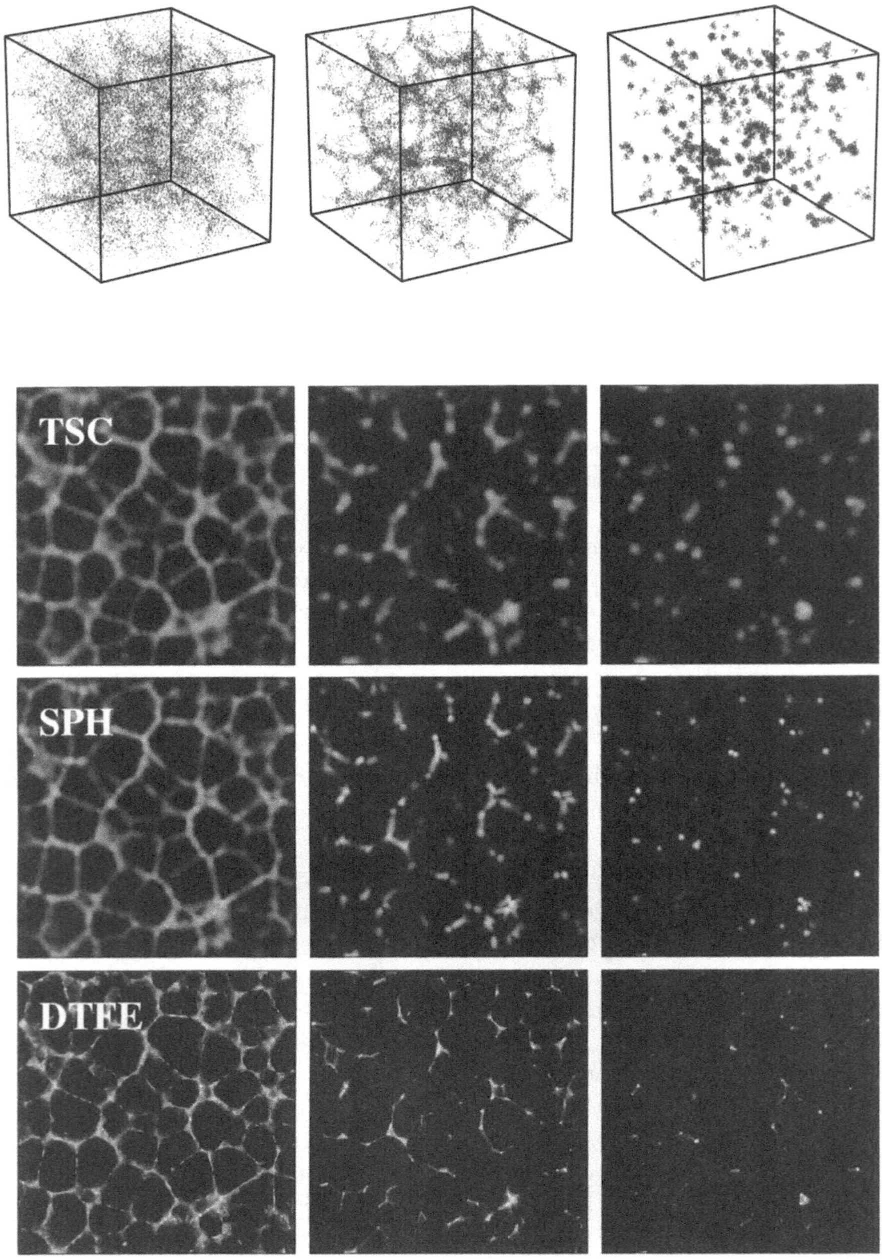

Fig. 34. Three Voronoi element clustering models. *Top row*: box with periodic boundary conditions, box size $100\,h^{-1}$Mpc. Left: Wall Model; *Centre*: Filament model; *Right*: Cluster model. Second to fourth row: Corresponding density reconstructions of the three Voronoi element clustering models. Second: TSC; third: SPH; fourth: DTFE

features; often the effective rigid grid-scale TSC filter is not able to render and detect them.

The DTFE reconstruction of the *Voronoi cluster models* (Fig. 34, lower right hand) does depict some of the artefacts induced by the DTFE technique. DTFE is successful in identifying nearly every cluster, even the poor clusters which SPH cannot find. The compact dense cluster also presents a challenge in that they reveal low-density triangular wings in the regions between the clusters. Even though these wings include only a minor fraction of the matter distribution, they are an artefact which should be accounted for. Evidently, the SPH reconstruction of individual clusters as spherical blobs is visually more appealing.

Voronoi Filament Model

The best testbench for the ability of the different methods to recover the patterns and morphology of the fully three-dimensional density field is that of the contrast-rich *Voronoi filament models.* In Fig. 35 the central part of a sample box of the *Voronoi filament model* realization is shown. Isodensity contour levels are chosen such that 65% of the mass is enclosed within regions of density equal to or higher than the corresponding density value. The resulting TSC, SPH and DTFE density fields are depicted in the lower left hand (TSC), lower right hand (SPH) and top frame (DTFE). The galaxy distribution in the upper left-hand frame does contain all galaxies within the region. Evidently, the galaxies have distributed themselves over a large range of densities and thus occupy a larger fraction of space than that outlined by the density contours. The appearances of the TSC, SPH and DTFE patterns do differ substantially. Part of this is due to a different effective scale of the filter kernel. The 65% mass contour corresponds to a density contour $\rho = 0.55$ in the TSC field, $\rho = 1.4$ in the SPH reconstruction and $\rho = 2.0$ in the DTFE reconstruction (ρ in units of the average density). The fine filamentary maze seen in the galaxy distribution is hardly reflected in the TSC grid-based reconstruction even though the global structure, the almost ring-like arrangement of filaments, is clearly recognizable. The SPH density field fares considerably better, as it outlines the basic configuration of the filamentary web. Nonetheless, the SPH rendered filaments have a width determined by the SPH kernel scale, resulting in a pattern of tubes. Bridging substantial density gradients is problematic for SPH reconstructions, partly due to the lack of directional information.

It is the DTFE reconstruction (top frame Fig. 35) which yields the most outstanding reproduction of the filamentary web-like character of the galaxy distribution. A detailed comparison between the galaxy distribution and the density surfaces shows that it manages to trace the most minute details in the cosmic web. Note that the density contours do enclose only 65% of the mass, and thus relates to a smaller volume than suggested by the features in the galaxy distribution itself. The success of the DTFE method is underlined by

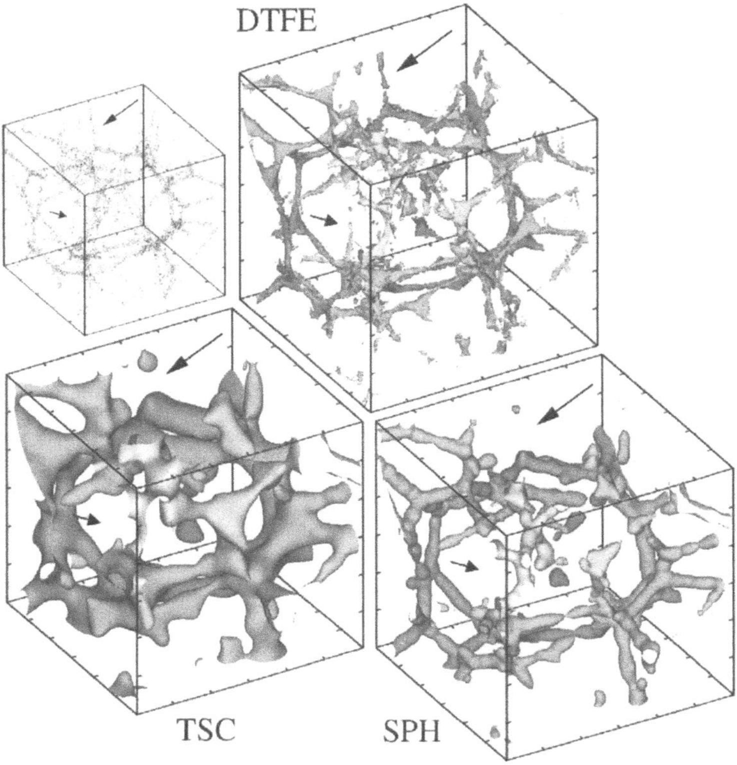

Fig. 35. Three-dimensional visualization of the Voronoi filament model and the corresponding TSC, SPH and DTFE density field reconstructions. The density contours have been chosen such that 65% of the mass is enclosed. The arrows indicate two structures which are visible in both the galaxy distribution and the DTFE reconstruction, but not in the TSC and SPH reconstructions

identifying a few features in the galaxy distribution which were identified by DTFE but not by SPH and TSC. The arrows in Fig. 35 point at two tenuous filamentary features visible in the galaxy distribution as well as in the DTFE field, yet entirely absent from the TSC and SPH fields. In comparison to the inflated contours of the SPH and TSC reconstructions, the structure outlined by the DTFE density field has a more intricate, even somewhat tenuous, appearance marked by a substantial richness in structural detail and contrast. Some artefacts of the DTFE method are also visible: in particular near intersections of filaments we tend to find triangular features which cannot be identified with similar structures in the galaxy distribution. Nearby filaments are connected by relatively small tetrahedra, translating into high-density features of such shape.

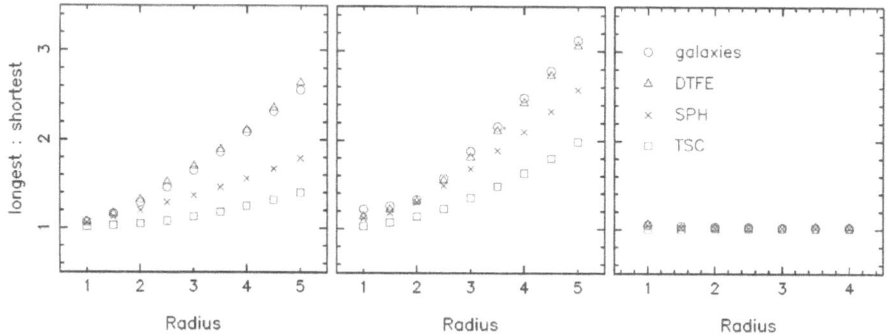

Fig. 36. Anisotropy measurements for the Voronoi models. Plotted is the longest-to-shortest axis ratio of the inertia tensor inside a series of concentric spheres centred on a characteristic structure as a function of the radius of the sphere. The radius is given in units of the standard deviation (σ) of the corresponding Gaussian density profiles. The *left-hand frame* corresponds to the Voronoi wall model, the *central frame* to the Voronoi filament model and the *right-hand frame* to the Voronoi cluster model. In each frame the results are shown for the TSC, SPH and DTFE reconstructions, as well as for the galaxy distribution. The meaning of the symbols is depicted in the *right-hand frame*

Shape and Morphology Analysis

An important measure of the local density distribution concerns the shape of the density contour levels. Various representative features in the three Voronoi element models were identified, followed by a study of their shape over a range of spatial scales. The axis ratios of the local density distribution, within a radius R, was computed from the eigenvalues of the mass inertia tensor. The results of the shape analysis are shown in Fig. 36. From left to right, the three frames present the axis ratio of the longest over the smallest axis, a_1/a_3, for walls, filaments and clusters, as a function of the scale R over which the shape was measured. The open circles represent the shape of the particle distribution, the triangles the shape found in the equivalent DTFE density field, while crosses and squares stand for the findings of SPH and TSC.

In the case of the Voronoi cluster models, all three density field reconstructions agree with the sphericity of the particle distributions. In the central and right-hand frame of Fig. 36 we see that the intrinsic shapes of the walls and filaments become more pronounced as the radius R increases. The uniform increase of the axis ratio a_1/a_3 with R is a reflection of the influence of the intrinsic width of the walls and filaments on the measured shape. For small radii the mass distribution around the centre of one of these features is largely confined to the interior of the wall or filament and thus near-isotropic. As the radius R increases in value, the intrinsic shape of these features comes to the fore, resulting in a revealing function of shape as function of R.

The findings of our analysis are remarkably strong and unequivocal: over the complete range of radii we find a striking agreement between DTFE and the corresponding particle distribution. SPH reveals systematic and substantial differences, in that they tend to be more spherical than the particle distribution, in particular for the strongly anisotropic distributions of the walls and filaments. In turn, the SPH shapes are substantially better than those obtained from the TSC reconstructions. The rigidity of the grid-based TSC density field reconstructions renders them the worst descriptions of the anisotropy of the local matter distribution. These results show that DTFE is indeed capable of an impressively accurate description of the shape of walls, filaments and clusters. It provides a striking confirmation of the earlier discussed visual impressions.

13 DTFE: Velocity Field Analysis

De facto the DTFE method has been first defined in the context of a description and analysis of cosmic flow fields which have been sampled by a set of discretely and sparsely sampled galaxy peculiar velocities. Bernardeau & van de Weigaert [27] demonstrated the method's superior performance with respect to conventional interpolation procedures. In particular, they also proved that the obtained field estimates involved those of the proper *volume-weighted* quantities, instead of the usually implicit *mass-weighted* quantities (see Sect. 10.4). This corrected a few fundamental biases in estimates of higher order velocity field moments.

The potential of the DTFE formalism may be fully exploited within the context of the analysis of N-body simulations and the galaxy distribution in galaxy redshift surveys. It allows not only a study of the patterns in the nonlinear matter distribution but also a study of the related velocity flows. Because DTFE manages to follow both the density distribution and the corresponding velocity distribution into nonlinear features, it opens up the window towards a study of the dynamics of the formation of the cosmic web and its corresponding elements. Evidently, such an analysis of the dynamics is limited to regions and scales without multi-stream flows (see Sect. 10.3).

A study of a set of GIF ΛCDM simulations has opened up a detailed view of the dynamics in and around filaments and other components of the cosmic web by allowing an assessment of the detailed density and velocity field in and around them (see Fig. 37). DTFE density and velocity fields may be depicted at any arbitrary resolution without involving any extra calculation: zoom-ins represent themselves a real magnification of the reconstructed fields. This is in stark contrast to conventional reconstructions in which the resolution is arbitrarily set by the users and whose properties are dependent on the adopted resolution.

13.1 Case Study: Void Outflow

In Fig. 38 a typical void-like region is shown, together with the resulting DTFE density and velocity field reconstructions. It concerns a void-like region

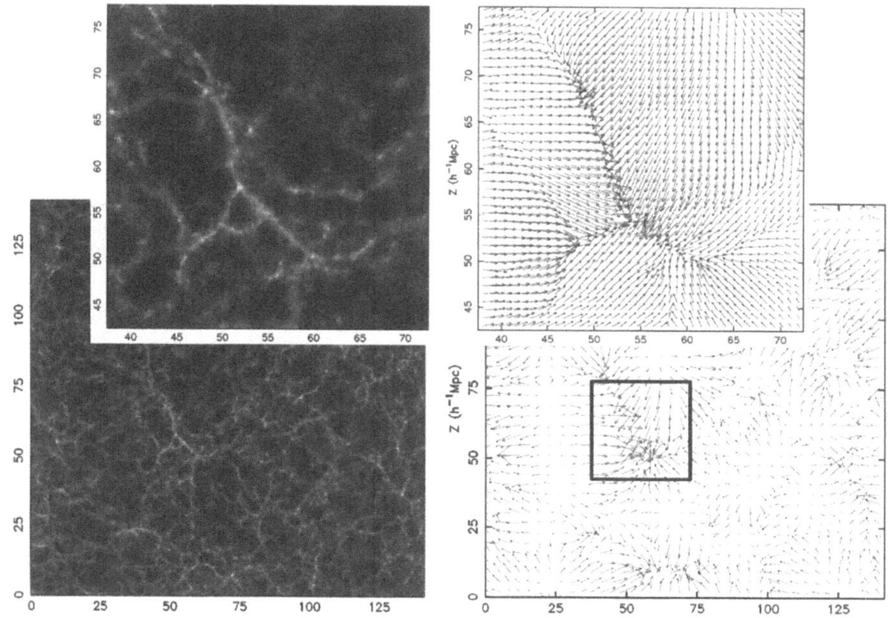

Fig. 37. The density and velocity field of the LCDM GIF N-body simulation, computed by DTFE. The bottom two frames show the density (*bottom left*) and velocity field (*bottom right*) in a central slice through the complete simulation box. The density map is a greyscale map. The velocity field is shown by means of velocity vectors: the vectors are the velocities component within the slice, their size proportional to the velocity's amplitude. By means of DTFE the images in both top frames zoom in on the density structure (*left*) and flow field (*bottom*) around a filamentary structure (whose location in the box is marked by the *solid square* in the *bottom right-hand panel*). The shear flow along the filaments is meticulously resolved. Note that DTFE does not need extra operations to zoom in; one DTFE calculation suffices to extract all necessary information. From Schaap [201]

selected from a ΛCDM GIF simulation (Kauffmann et al. [117]). Figure 10 shows a major part of the (DTFE) density field of the entire GIF simulation which contains the void. It concerns a 256^3 particles GIF N-body simulation, encompassing a ΛCDM ($\Omega_m = 0.3, \Omega_\Lambda = 0.7, H_0 = 70\,\mathrm{km\ s^{-1}/Mpc}$) density field within a (periodic) cubic box with length $141\,h^{-1}$ Mpc and produced by means of an adaptive P^3M N-body code.

The top left frame shows the particle distribution in and around the void within this $42.5\,h^{-1}$ Mpc wide and $1\,h^{-1}$ Mpc thick slice through the simulation box. The corresponding density field (top right-hand frame) and velocity vector field (bottom left-hand frame) are a nice illustration of the ability of DTFE to translate the inhomogeneous particle distribution into highly resolved continuous and volume-filling fields.

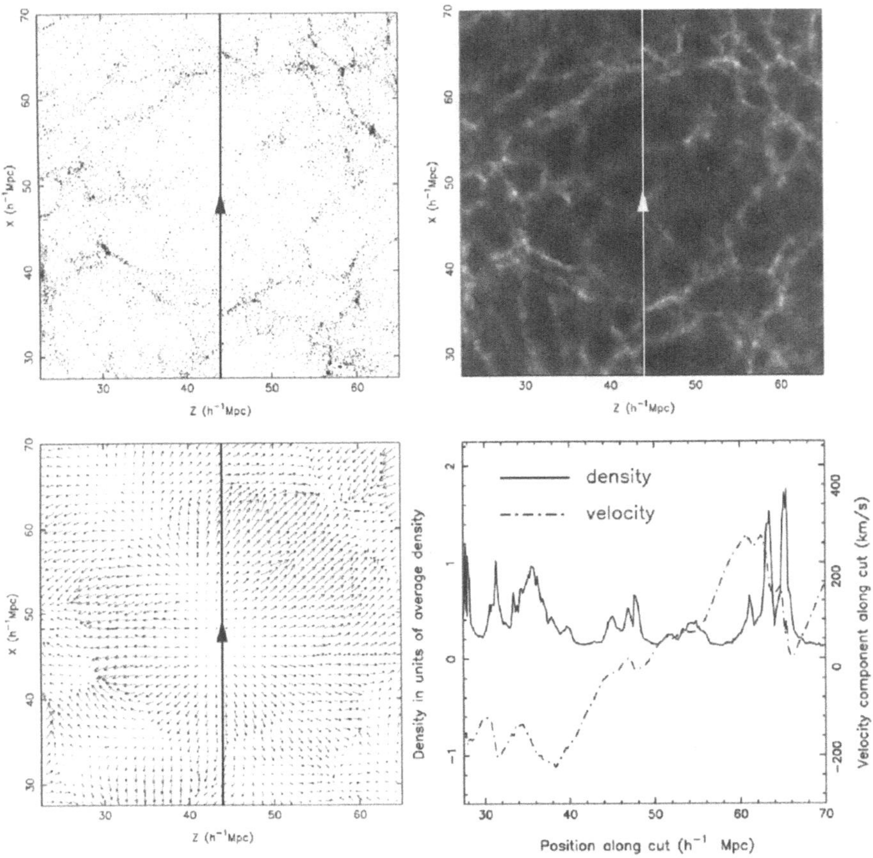

Fig. 38. The density and velocity fields around a void in the GIF LCDM simulation. The *top right-hand panel* shows the N-body simulation particle distribution within a slice through the simulation box, centred on the void. The top right-hand panel shows the greyscale map of the DTFE density field reconstruction in and around the void; the corresponding velocity vector plot is shown in the *bottom left-hand panel*. Notice the detailed view of the velocity field: within the almost spherical global outflow of the void, features can be recognized that can be identified with the diluted substructure within the void. Along the solid line in these panels we determined the linear DTFE density and velocity profiles (*bottom right-hand frame*). We can recognize the global "bucket"-shaped density profile of the void, be it marked by substantial density enhancements. The velocity field reflects the density profile in detail, dominated by a global super-Hubble outflow. From Schaap [201]

The DTFE procedure clearly manages to render the void as a slowly varying region of low density. Notice the clear distinction between the empty (dark) interior regions of the void and its edges. In the interior of the void several smaller *subvoids* can be distinguished, with boundaries consisting of low-density filamentary or planar structures. Such a hierarchy of voids, with large voids containing the traces of the smaller ones from which it formed earlier through merging, has been described by theories of void evolution (Regős & Geller [186], Dubinski et al. [66], van de Weygaert & van Kampen [247], Sheth & van de Weygaert [218]).

The velocity vector field in and around the void represents a nice illustration of the qualities of DTFE to recover the general velocity flow. It also demonstrates its ability identify detailed features within the velocity field. The flow in and around the void is dominated by the outflow of matter from the void, culminating into the void's own expansion near the outer edge. The comparison with the top two frames demonstrates the strong relation with features in the particle distribution and the DTFE density field. Not only is it slightly elongated along the direction of the void's shape, but it is also sensitive to some prominent internal features of the void. Towards the "SE" direction the flow appears to slow down near a ridge, while near the centre the DTFE reconstruction identifies two expansion centres.

The general characteristics of the void expansion are most evident by following the density and velocity profiles along a one-dimensional section through the void. The bottom-left frame of Fig. 38 shows these profiles for the linear section along the solid line indicated in the other three frames. The first impression is that of the *bucket-like* shape of the void, be it interspersed by a rather pronounced density enhancement near its centre. This profile shape does confirm to the general trend of low-density regions to develop a near uniform interior density surrounded by sharply defined boundaries. Because initially emptier inner regions expand faster than the denser outer layers the matter distribution gets evened out while the inner layers catch up with the outer ones. The figure forms a telling confirmation of DTFE being able to recover this theoretically expected profile by interpolating its density estimates across the particle diluted void (see Sect. 13.1).

The velocity profile strongly follows the density structure of the void. The linear velocity increase is a manifestation of its general expansion. The near constant velocity divergence within the void conforms to the *super-Hubble flow* expected for the near uniform interior density distribution (see Sect. 13.1). Because voids are emptier than the rest of the Universe they will expand faster than the rest of the Universe with a net velocity divergence equal to

$$\theta = \frac{\nabla \cdot \mathbf{v}}{H} = 3(\alpha - 1), \qquad \alpha = H_{\rm void}/H, \tag{48}$$

where α is defined to be the ratio of the super-Hubble expansion rate of the void and the Hubble expansion of the Universe.

Evidently, the highest expansion ratio is that for voids which are completely empty, i.e. $\Delta_{\rm void} = -1$. The expansion ratio α for such voids may be inferred from Birkhoff's theorem, treating these voids as empty FRW universes whose expansion time is equal to the cosmic time. For a matter-dominated Universe with zero cosmological constant, the maximum expansion rate that a void may achieve is given by

$$\theta_{\max} = 1.5\Omega_m^{0.6}, \tag{49}$$

with Ω_m the cosmological mass density parameter. For empty voids in a Universe with a cosmological constant a similar expression holds, be it that the value of α will have to be numerically calculated from the corresponding equation. In general the dependence on Λ is only weak. Generic voids will not be entirely empty, their density deficit $|\Delta_{\rm void}| \approx 0.8 - 0.9$ (cf. e.g. the linear density profile in Fig. 38). The expansion rate $\theta_{\rm void}$ for such a void follows from numerical evaluation of the expression

$$\theta_{\rm void} = \frac{3}{2}\,\frac{\Omega_m^{0.6} - \Omega_{m,{\rm void}}^{0.6}}{1 + \frac{1}{2}\Omega_{m,{\rm void}}^{0.6}}\,; \qquad \Omega_{m,{\rm void}} = \frac{\Omega_m(\Delta_{\rm void}+1)}{(1+\frac{1}{3}\theta)^2} \tag{50}$$

in which $\Omega_{m,{\rm void}}$ is the effective cosmic density parameter inside the void.

13.2 Velocity Field Evolution

On a mildly nonlinear scale the velocity field contains important information on the cosmic structure formation process. Density and velocity perturbations are supposed to grow gravitationally out of an initial field of Gaussian density and velocity perturbations. Once the fluctuations start to acquire values in the order of unity or higher, the growth rapidly becomes more complex. The larger gravitational acceleration exerted by the more massive structures in the density field induces the infall of proportionally large amounts of matter and hence to an increase of the growth rate, while the opposite occurs in and around the underdense regions.

The overdensities collapse into compact massive objects whose density excess may achieve values exceeding unity by many orders of magnitude. Meanwhile the underdensities expand and occupy a growing fraction of space while evolving into empty troughs with a density deficit tending towards a limiting value of -1.0. The migration flows which are induced by the evolving matter distribution are evidently expected to strongly reflect the structure formation process.

In practice a sufficiently accurate analysis of the nonlinear cosmic velocities is far from trivial. The existing estimates of the peculiar velocities of galaxies are still ridden by major uncertainties and errors. This is aggravated by the fact that the sampling of the velocity field is discrete, rather poor and diluted and highly inhomogeneous. While the conventional approach involves a smoothing of the velocity field over scales larger than $10\,h^{-1}$ Mpc in order to attain a map of the linear velocity field, we have argued in Sect. 10.4 that implicitly this usually yields a *mass-weighted* velocity field. DTFE represents a major improvement on this. Not only does it offer an interpolation which guarantees an optimal resolution, thereby opening the window onto the nonlinear regime, it also guarantees a *volume-weighted* flow field (see (36)).

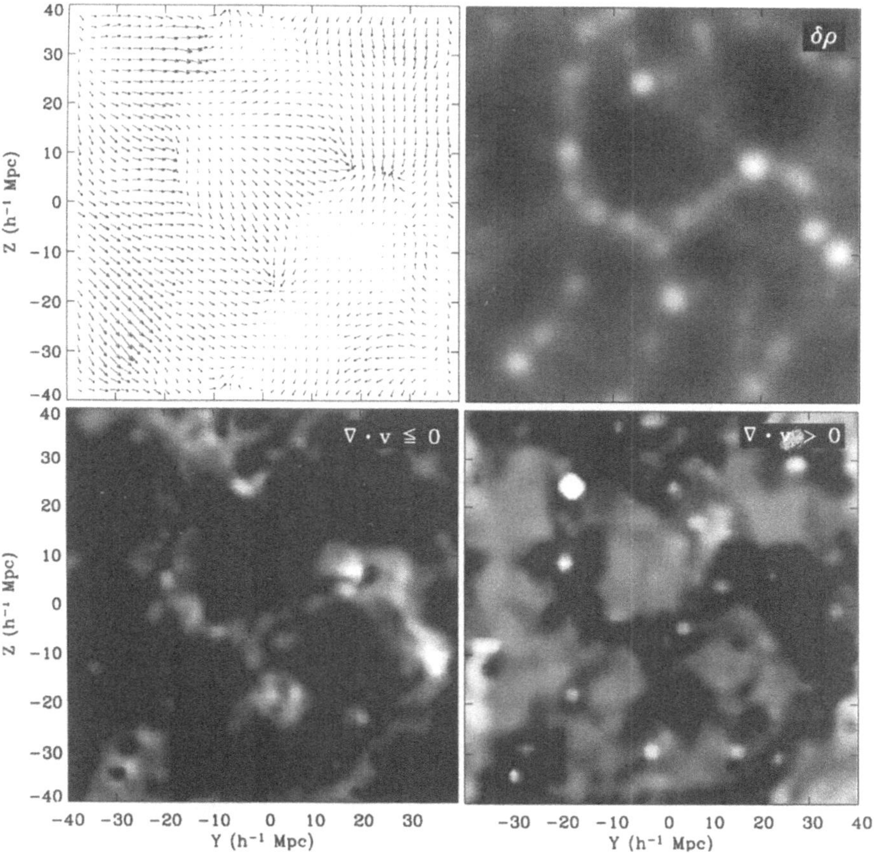

Fig. 39. The DTFE reconstructed velocity and velocity divergence field of a (low-resolution) SCDM N-body simulation, compared with the corresponding DTFE density field (*top right-hand frame*). The velocity divergence field is split into two parts. The negative velocity divergence regions, those marked by inflow, are shown in the *bottom left-hand frame*. They mark both the high-density cluster regions and the more moderate filamentary regions. The positive divergence outflow regions in the *bottom right-hand panel* not only assume a larger region of space but also have a more roundish morphology. They centre on the large (expanding) void regions in the matter distribution. The figure represents a nice illustration of DTFE's ability to successfully render the non-Gaussian nonlinear velocity field. From Romano-Díaz [194]

Density and Velocity Divergence

The implied link between the matter distribution and the induced migration flows is most strongly expressed through the relation between the density field $\delta(\mathbf{x})$ and the velocity divergence field. The bottom frames of Fig. 39 contain greyscale maps of the DTFE normalized velocity divergence estimate $\widehat{\theta}$ (with H_0 the Hubble constant),

$$\widehat{\theta} \equiv \frac{\widehat{\nabla \cdot \mathbf{v}}}{H_0} = \frac{1}{H_0}\left(\widehat{\frac{\partial v_x}{\partial x}} + \widehat{\frac{\partial v_y}{\partial y}} + \widehat{\frac{\partial v_z}{\partial z}}\right), \tag{51}$$

for an N-body simulation. For illustrative purposes we have depicted the regions of negative and positive velocity divergence separately. The comparison between these maps and the corresponding density field (upper right-hand frame) and velocity field (upper left hand) provides a direct impression of their intimate relationship. The two bottom frames clearly delineate the expanding and contracting modes of the velocity field. The regions with a negative divergence are contracting, matter is falling in along one or more directions. The inflow region delineates elongated regions coinciding with filaments and peaks in the density field. The highest inflow rates are seen around the most massive matter concentrations. Meanwhile the expanding regions may be seen to coincide with the surrounding large and deep underdense voids, clearly occupying a larger fraction of the cosmic volume.

The strong relation between the density contrast and velocity divergence is a manifestation of the continuity equation. In the linear regime this is a strict linear one-to-one relation,

$$\frac{\nabla \cdot \mathbf{v}(\mathbf{x}, \mathbf{t})}{H} = -a(t) f(\Omega_m, \Omega_\Lambda)\, \delta(\mathbf{x}, t)\,, \tag{52}$$

linking the density perturbation field δ to the peculiar velocity field $\mathbf{v}$ via the factor $f(\Omega_m)$ (see Peebles [174]). There remains a one-to-one relation between velocity divergence and density into the mildly nonlinear regime (see (53)). This explains why the map of the velocity divergence in Fig. 39 is an almost near perfect negative image of the density map.

Even in the quasi-linear and mildly nonlinear regime the one-to-one correspondence between velocity divergence and density remains intact, be it that it involves higher order terms (see Bernardeau [29] for an extensive review). Within the context of Eulerian perturbation theory, Bernardeau [23] derived an accurate second-order approximation from the relation between the divergence and the density perturbation $\delta(\mathbf{x})$. Nusser et al. [168] derived a similar quasi-linear approximation within the context of the Lagrangian Zel'dovich approximation. According to these approximate nonlinear relations,

$$\frac{1}{H}\nabla \cdot \mathbf{v}(\mathbf{x}) = \begin{cases} \frac{3}{2} f(\Omega_m)\left[1 - (1+\delta(\mathbf{x}))^{2/3}\right] & \text{(B)} \\ \\ -f(\Omega_m)\,\dfrac{\delta(\mathbf{x})}{1 + 0.18\,\delta(\mathbf{x})} & \text{(N)} \end{cases} \tag{53}$$

for a Universe with Hubble parameter $H(t)$ and matter density parameter Ω_m. The studies by Bernardeau & van de Weygaert [27], van de Weygaert & Romano-Díaz [249] and Schaap [201] have shown that the DTFE velocity field reconstructions are indeed able to reproduce these quasi-linear relations.

Velocity Field Statistics

The asymmetric nonlinear evolution of the cosmic velocity and density field manifests itself in an increasing shift of the statistical distribution of density and velocity perturbations away from the initial Gaussian probability distribution function. The analytical framework of Eulerian perturbation theory provides a reasonably accurate description for the early nonlinear evolution (see Bernardeau [29] for a review). As for the velocity field, robust results on the evolution and distribution of the velocity divergence, $\nabla \cdot \mathbf{v}$, were derived in a series of papers by Bernardeau (e.g. Bernardeau et al. [26]). The complete probability distribution function (pdf) of the velocity divergence may be evaluated via the summation of the series of cumulants of the non-Gaussian distribution function. The velocity divergence pdf is strongly sensitive to the underlying cosmological parameters, in particular the cosmological density parameter Ω_m. It represents a considerable advantage to the more regular analysis of the cosmic velocity flows on large linear scales. While the latter yields constraints on a combined function of Ω_m and the bias b between the galaxy and matter distribution, access to nonlinear scales would break this degeneracy.

An impression of the generic behaviour and changing global shape of the resulting non-Gaussian pdf as a function of Ω_m and the amplitude σ_θ of the velocity divergence perturbations may be obtained from the top frame of Fig. 40. The curves correspond to the theoretically predicted velocity divergence pdf for three different (matter-dominated) FRW cosmologies. Not only does Ω_m influence the overall shape of the pdf, but it also changes the location of the peak – indicated by $\theta_{\max}$ – as well as that of the cut-off at the high positive values of θ. By means of the Edgeworth expansion one may show that the velocity divergence pdf reaches its maximum for a peak value $\theta_{\rm peak}$ (Juszkiewicz et al. [111], Bernardeau & Kofman [25]),

$$\theta_{\rm peak} = -\frac{T_3(\theta)}{2}\,\sigma_\theta\,; \qquad \langle\theta^3\rangle = T_3\,\langle\theta^2\rangle^2, \tag{54}$$

where the coefficient T_3 relates the third-order moment of the pdf to the second moment (see, e.g., Bernardeau [24]). While the exact location of the peak is set by the bias-independent coefficient T_3, it does include a dependence not only on Ω_m but also on the shape of the power spectrum, the geometry of the window function that has been used to filter the velocity field and the value of the cosmological constant Λ. To infer the value of Ω_m extra assumptions need to be invoked. Most direct therefore is the determination of Ω_m via the sharp cut-off of the pdf related to the expansion velocity of the deepest voids in a particular cosmology (see (49)).

While conventional techniques may reproduce the pdf for moderate values of the velocity divergence θ, they tend to have severe difficulties in tracing the distribution for the more extreme positive values and the highest inflow velocities. An illustration of the tendency to deviate from the analytically

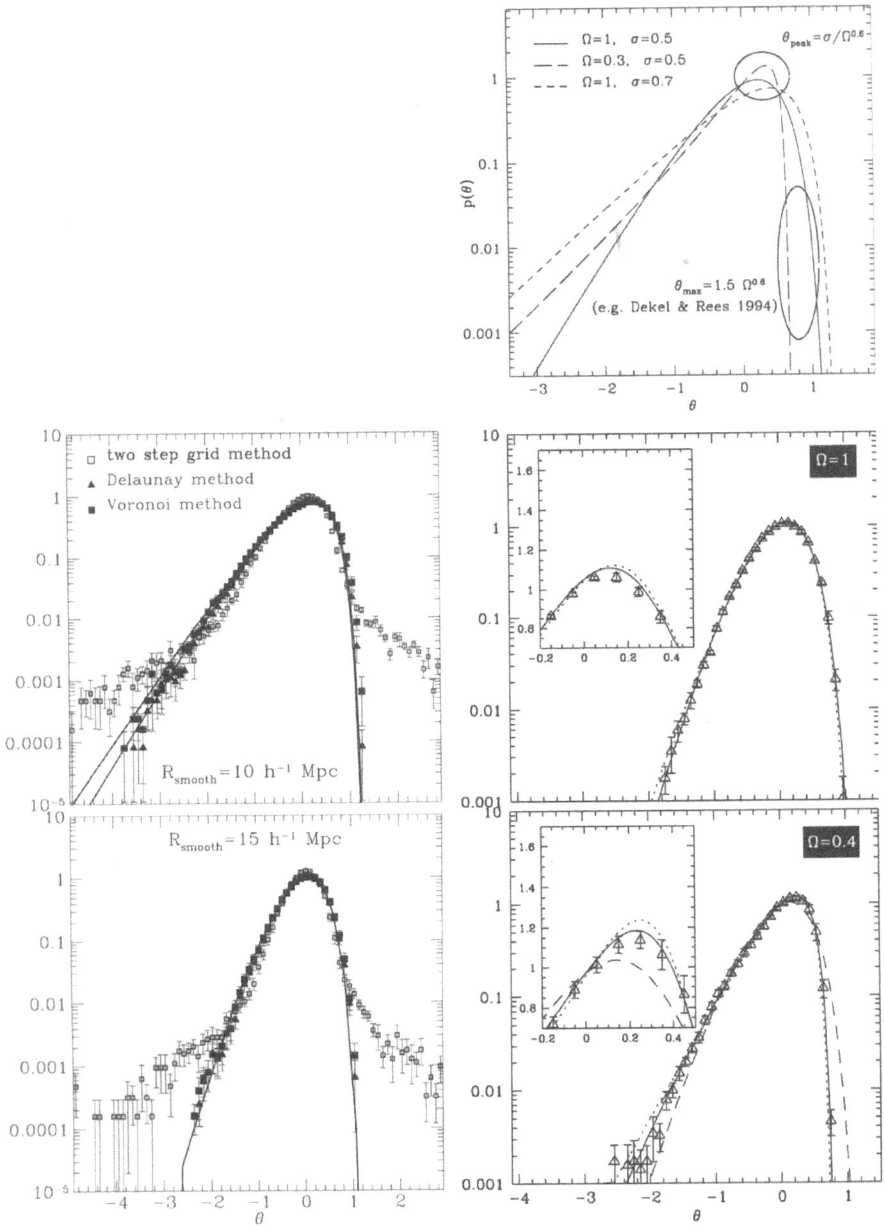

Fig. 40. DTFE determination of the probability distribution function (pdf) of the velocity divergence θ. *Top frame* illustrates the sensitivity of the *theta* pdf to the underlying cosmology. The superposition of the theoretical pdf curves of three cosmologies immediately shows that the location of the maximum of the pdf and its sharp cut-off at the positive end are highly sensitive to Ω. The other four frames demonstrate DTFE's ability to successfully trace these marks. *Left-hand frames*: confrontation of the DTFE velocity divergence pdf and that determined by more

predicted distribution can be seen in the two left-hand frames of Fig. 40, showing the velocity divergence pdf determined from a SCDM N-body simulation for two scales of tophat-filtered velocity fields ($R = 10\,\mathrm{h}^{-1}\,\mathrm{Mpc}$ and $R = 15\,\mathrm{h}^{-1}\,\mathrm{Mpc}$). The open squares depict the velocity divergence pdf determined from an N-body simulation through a two-step grid procedure (see Sect. 10.4). Conventional grid interpolation clearly fails by huge margins to recover inflow onto the high-density filamentary structures as well as the outflow from voids.

What came as the first evidence for the success of tessellation-based interpolation is the rather tight correspondence of the Delaunay tessellation interpolation results with the analytically predicted pdf. This finding by Bernardeau & van de Weygaert [27] suggested that it would indeed be feasible to probe the nonlinear velocity field and infer directly accurate estimates of Ω_m. In a follow-up study Bernardeau et al. [28] successfully tested this on a range of N-body simulations of structure formation, showing Delaunay in terpolation indeed recovered the right values for Ω_m. The centre and bottom right-hand frames depict two examples: the pdf's for a $\Omega = 1$ and a $\Omega = 0.4$ Universe accurately traced by the Delaunay interpolated velocity field.

13.3 PSCz Velocity Field

In a recent study, van de Weygaert & Romano-Díaz [249] applied DTFE towards reconstructing the velocity flow map throughout the nearby Universe volume sampled by the PSCz galaxy redshift survey (see also Romano-Díaz [193]).

PSCz Sample of the Local Universe

The IRAS-PSC*z* catalogue (Saunders et al. [200]) is an extension of the 1.2 Jy catalogue (Fisher et al. [82]). It contains $\sim 15,500$ galaxies with a flux at $60\,\mu\mathrm{m}$

Fig. 40. (continued) conventional two-step fixed grid interpolation method. Both curves concern the same 128^3 SCDM N-body particle simulation (i.e. $\Omega = 1$). *Top left-hand frame*: tophat filter radius $R_{\mathrm{TH}} = 10\,\mathrm{h}^{-1}\,\mathrm{Mpc}$. *Bottom left-hand panel*: $R_{\mathrm{TH}} = 15\,\mathrm{h}^{-1}\,\mathrm{Mpc}$. The solid lines represent theoretical predictions of the PDF for the measured values of σ_θ (Bernardeau [24]). Black triangles are the pdf values measured by the DTFE method, the black squares by the equivalent VTFE Voronoi method. The open squares show the results obtained by a more conventional two-step fixed grid interpolation method. For especially low and high θ values the tessellation methods prove to produce a much more genuine velocity divergence pdf. From Bernardeau & van de Weygaert [27]. *Right-hand frames*: on the basis of DTFE's ability to trace the velocity divergence pdf we have tested its operation in a $\Omega = 1$ and a $\Omega = 0.4$ CDM N-body simulation. For both configurations DTFE manages to trace the pdf both at its high-value edge and near its peak. From Bernardeau et al. [28]

larger than 0.6 Jy. For a full description of the catalogue, selection criteria and the procedures used to avoid stellar contamination and galactic cirrus, we refer the reader to [200]. For our purposes the most important characteristics of the catalogue are the large area sampled ($\sim$ 84% of the sky), its depth with a median redshift of $8,500\,\mathrm{km\ s^{-1}}$ and the dense sampling (the mean galaxy separation at $10,000\,\mathrm{km\ s^{-1}}$ is $\langle l \rangle = 1,000\,\mathrm{km\ s^{-1}}$). It implies that PSCz contains most of the gravitationally relevant mass concentrations in our local cosmic neighbourhood, certainly sufficient to explain and study the cosmic flows within a local volume of radius $\sim 120\,h^{-1}\,\mathrm{Mpc}$.

To translate the redshift space distribution of galaxies in the PSCz catalogue into galaxy positions and velocities in real space, the study based itself on an iterative processing of the galaxy sample by Branchini et al. [40] based on the linear theory of gravitational instability (Peebles [174]). The method involved a specific realization of an iterative technique to minimize redshift-space distortions (Yahil et al. [257]). While the resulting galaxy velocities relate by construction to the linear clustering regime, the reconstruction of the velocity field throughout the sample volume does appeal to the capability of DTFE to process a discretely sampled field into a continuous volume-filling field and its ability to resolve flows in both high-density regions and the outflows from underdense voids.

PSCz Velocity and Density Field

The central circular field of Fig. 41 presents the DTFE velocity field in the supergalactic plane. For comparison, the bottom left-hand insert shows the discrete sample of galaxy velocities which formed the input for the reconstruction. The velocity field is shown by means of the projected velocity vectors within the Z-supergalactic plane, superposed upon the corresponding DTFE density contour maps inferred from the same PSCz galaxy sample. The length of the velocity arrows can be inferred from the arrow in the lower left-hand corner, which corresponds to a velocity of $650\,\mathrm{km\ s^{-1}}$. The Local Group is located at the origin of the map.

The map of Fig. 41 shows the success of DTFE in converting a sample of discretely sampled galaxy velocities and locations into a sensible volume-covering flow and density field. The processed DTFE velocity field reveals intricate details along the whole volume. The first impression is that of the meticulously detailed DTFE flow field, marked by sharply defined flow regions over the whole supergalactic plane. Large-scale bulk flows and distorted flow patterns such as shear, expansion and contraction modes of the velocity field are clear features revealed through by the DTFE technique. DTFE recovers clearly outlined patches marked by strong bulk flows, regions with characteristic shear flow patterns around anisotropically shaped supercluster complexes, radial inflow towards a few massive clusters and, perhaps most outstanding, strong radial outflows from the underdense void regions.

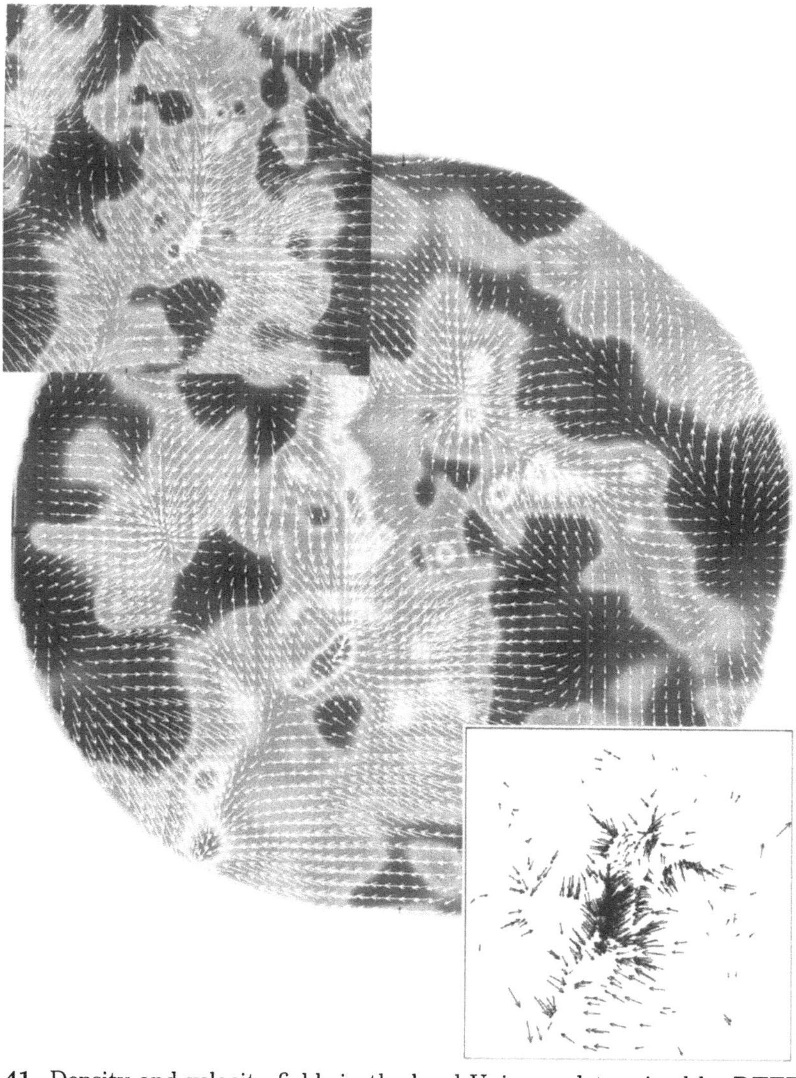

Fig. 41. Density and velocity fields in the local Universe determined by DTFE on the basis of the PSCz galaxy redshift survey. Our Local Group is at the centre of the map. To the left we see the Great Attractor region extending out towards the Shapley supercluster. To the left-hand side we can find the Pisces-Perseus supercluster. The peculiar velocities of the galaxies in the PSCz galaxy redshift catalogue were determined by means of a linearization procedure (Branchini et al. [40]). The resulting galaxy positions and velocities (vectors) of the input sample for the DTFE reconstruction are shown in the bottom left-hand frame. The density values range from ~ 4.9 down to ~ -0.75 (*black*), with light gray regions having a density near the global cosmic average ($\delta \sim 0$]. The velocity vectors are scaled such that a vector with a length of $\approx 1/33$rd of the region's diameter corresponds to $650\,\mathrm{km\ s^{-1}}$. The density and velocity fields have an effective Gaussian smoothing radius of $R_G \sim \sqrt{5}\,\mathrm{h^{-1}\,Mpc}$. The top right-hand insert zooms in on the Local Supercluster and Great Attractor complex. From Romano-Díaz [193]

Overall, there is a tight correspondence with the large-scale structures in the underlying density distribution. While the density field shows features down to a scale of $\sqrt{5}\,h^{-1}\,$Mpc, the patterns in the flow field clearly have a substantially larger coherence scale, nearly all in excess of $10\,h^{-1}\,$Mpc. The DTFE velocity flow sharply follows the elongated ridge of the Pisces-Perseus supercluster. In addition we find the DTFE velocity field to contain markedly sharp transition regions between void expansion and the flows along the body of a supercluster.

The local nature of the DTFE interpolation guarantees a highly resolved flow field in high-density regions. Massive bulk motions are concentrated near and around the massive structure extending from the Local Supercluster (centre map) towards the Great Attractor region and the Shapley concentration. The DTFE map nicely renders this pronounced bulk flow towards the Hydra-Centaurus region and shows that it dominates the general motions at our Local Group and Local Supercluster. The top right-hand insert zooms in on the flow in and around this region. The most massive and coherent bulk flows in the supergalactic plane appear to be connected to the Sculptor void and the connected void regions (towards the left-hand side of the figure). They are the manifestation of the combination of gravitational attraction by the heavy matter concentration of the Pavo-Indus-Telescopium complex (at $[SGX, SGY] \approx [-40, -10]\,h^{-1}\,$Mpc), the more distant "Hydra-Centaurus-Shapley ridge", and the effective push by the Sculptor void region. Conspicuous shear flows can be recognized along the ridge defined by the Cetus wall towards the Pisces-Perseus supercluster ($[SGX, SGY] \approx [20, -40]\,h^{-1}\,$Mpc). A similar strong shear flow is seen along the extension of the Hydra-Centaurus supercluster towards the Shapley concentration.

The influence of the Coma cluster is beautifully outlined by the strong and near perfect radial infall of the surrounding matter, visible at the top-centre of Fig. 41. Also the velocity field near the Perseus cluster, in the Pisces-Perseus supercluster region, does contain a strong radial inflow component.

Perhaps most outstanding are the radial outflow patterns in and around voids. In particular its ability to interpolate over the low-density and thus sparsely sampled regions is striking: the voids show up as regions marked by a near-spherical outflow. The intrinsic suppression of shot noise effects through the adaptive spatial interpolation procedure of DTFE highlights these important components of the megaparsec flow field and emphasizes the dynamical role of voids in organizing the matter distribution of the large-scale cosmic web. By contrast, more conventional schemes, such as TSC or SPH (see Schaap & van de Weygaert [203]), meet substantial problems in defining a sensible field reconstruction in low-density regions without excessive smoothing and thus loss of resolution.

14 DTFE Meets Reality: 2dFGRS and the Cosmic Web

To finish the exposé on the potential of the Delaunay Tessellation Field Estimator, we present a reconstruction of the foamy morphology of the galaxy distribution in the 2dF Galaxy Redshift Survey (2dFGRS). DTFE was used to reconstruct the projected galaxy surface density field as well as the full three-dimensional galaxy density field.

14.1 2dF Galaxy Redshift Survey

The 2dFGRS is one of the major spectroscopic surveys in which the spectra of 245, 591 objects have been obtained, with the scope of obtaining a representative picture of the large-scale distribution of galaxies in the nearby universe (Colless et al. [57]). It is a magnitude-limited survey, with galaxies selected down to a limiting magnitude of $b_J \sim 19.45$ from the extended APM Galaxy Survey (Maddox et al. [143, 144, 145]). The galaxy redshifts were measured with the 2dF multi-fibre spectrograph on the Anglo-Australian Telescope capable of simultaneously observing 400 objects in a 2° diameter field. The galaxies were confined to three regions, together covering an area of approximately 1500 squared degrees. These regions include two declination strips, each consisting of overlapping 2° fields, as well as a number of "randomly" distributed 2° control fields. One strip (the SGP strip) is located in the southern galactic hemisphere and covers about $80° \times 15°$ close to the south galactic pole ($21^h40^m < \alpha < 03^h40^m$, $-37.5° < \delta < -22.5°$). The other strip (the NGP strip) is located in the northern galactic hemisphere and covers about $75° \times 10°$ ($09^h50^m < \alpha < 14^h50^m$, $-7.5° < \delta < +2.5°$). Reliable redshifts were obtained for 221 414 galaxies.

These data have been made publicly available in the form of the 2dFGRS final data release (available at `http://msowww.anu.edu.au/2dFGRS/`).

14.2 Galaxy Surface Density Field Reconstructions

The galaxy distribution in the 2dFGRS is mainly confined to the two large strips, NGP and SGP. Since their width is reasonably thin, a good impression of the spatial patterns in the galaxy distribution may be obtained from the two-dimensional projection shown in Fig. 1.

We have reconstructed the galaxy surface density field in redshift space in the 2dFGRS NGP and SGP regions. All density field reconstructions are DTFE reconstructions on the basis of the measured galaxy redshift space positions. As no corrections were attempted to translate these into genuine positions in physical space, the density reconstructions in this section concern redshift space. In order to warrant a direct comparison with the galaxy distribution in Fig. 1, the results shown in this section were not corrected for any observational selection effect, also not for the survey radial selection function. For selection-corrected density field reconstructions, we refer to the analysis in [201, 203].

Figure 42 shows the resulting DTFE reconstructed density field. DTFE manages to reveal the strong density contrasts in the large-scale density distribution. The resolution is optimal, in that the smallest interpolation units

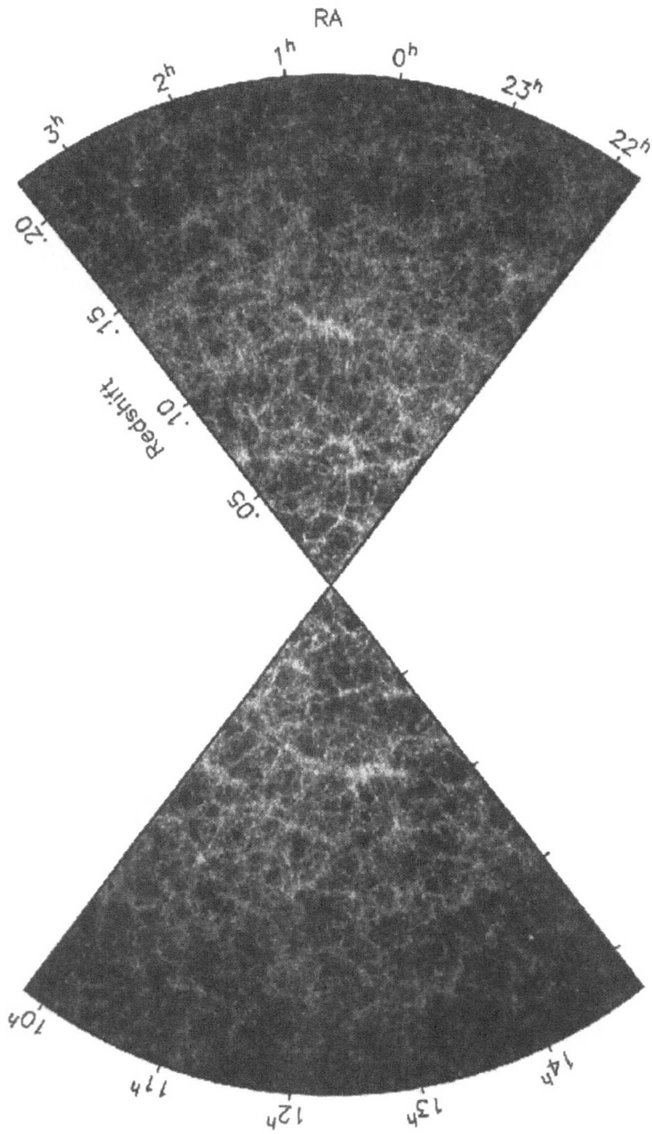

Fig. 42. DTFE galaxy surface density reconstructions of the projected galaxy distribution in the two 2dF slices (north and south). For comparison see the galaxy distribution in Fig. 1. A description may be found in the text (Sect. 14.2). From Schaap [201] and Schaap & van de Weygaert [203]

are also the smallest units set by the data. At the same time the DTFE manages to bring out the fine structural detail of the intricate and often tenuous filamentary structures. Particularly noteworthy are the thin sharp edges surrounding void-like regions.

Individual Structures

The impressive resolution and shape sensitivity of the DTFE reconstruction becomes particularly visible when zooming in on structural details in the cosmic web. Figure 43 zooms in on a few interesting regions in the map of Fig. 1.

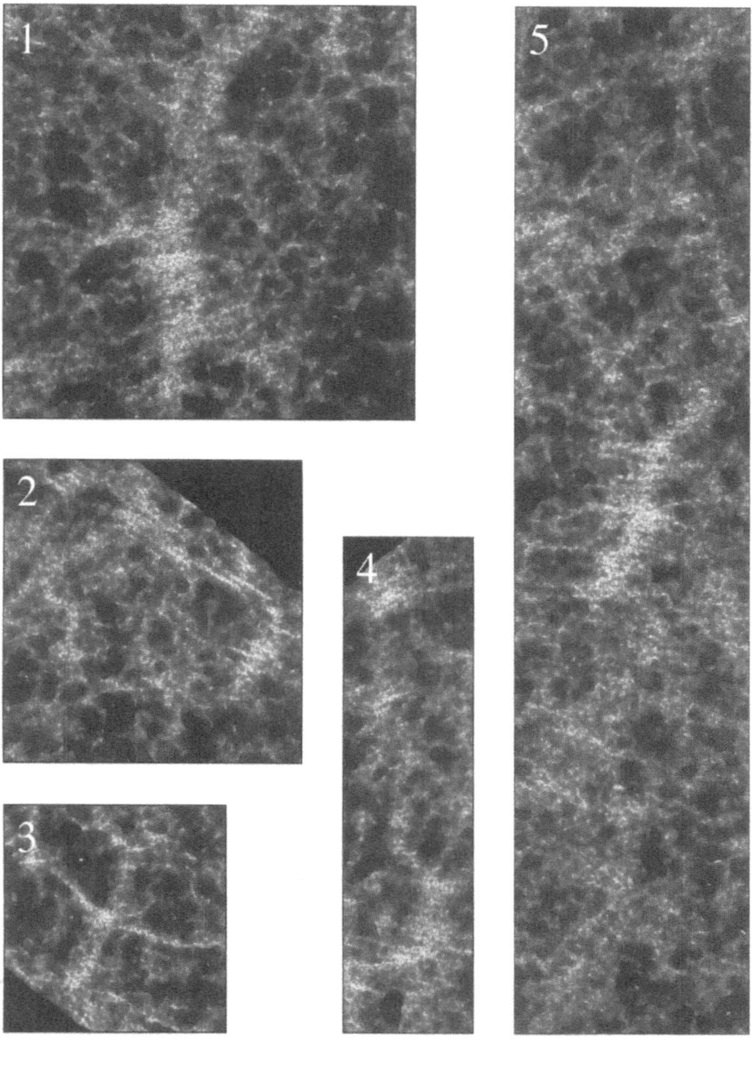

Fig. 43. DTFE galaxy surface density in selected regions in the 2dFGRS galaxy surface density field. For the density field in the total 2dFGRS region, see Fig. 42. For a discussion see Sect. 14.2. Frame 1 zooms in on the Great Wall in the southern (SGP) 2dF slice, frame 5 on the one in the northern (NGP) slice. Note the sharply rendered "fingers of God" marking the sites of clusters of galaxies. Perhaps the most salient feature is the one seen in frame 3, the cross-like configuration in the lower half of the NGP region. From Schaap [201] and Schaap & van de Weygaert [203]

Region 1 focuses on the major mass concentration in the NGP region, the Sloan Great Wall. Various filamentary regions emanate from the high-density core. In region 2 a void-like region is depicted. The DTFE renders the void as a low-density area surrounded by various filamentary and wall-like features. Two fingers of God are visible in the upper right-hand corner of region 2, which show up as very sharply defined linear features. Many such features can be recognized in high-density environments. Note that the void is not a totally empty or featureless region. The void is marked by substructure and contains a number of smaller subvoids, reminding of the hierarchical evolution of voids (Dubinski et al. [66], Sheth & van de Weygaert [218]). Region 3 is amongst the most conspicuous structures encountered in the 2dF survey. The cross-shaped feature consists of four tenuous filamentary structures emanating from a high-density core located at the centre of the region. Region 4 zooms in on some of the larger structures in the SGP region. Part of the Pisces-Cetus supercluster is visible near the bottom of the frame, while the concentration at the top of this region is the upper part of the Horologium-Reticulum supercluster. Finally, region 5 zooms in on the largest structure in the SGP region, the Sculptor supercluster.

DTFE Artefacts

Even though the DTFE offers a sharp image of the cosmic web, the reconstructions also highlight some artefacts. At the highest resolution we can directly discern the triangular imprint of the DTFE kernel. Also a considerable amount of noise is visible in the reconstructions. This is a direct consequence of the high resolution of the DTFE reconstruction. Part of this noise is natural, a result of the statistical nature of the galaxy formation process. An additional source of noise is due to the fact that the galaxy positions have been projected onto a two-dimensional plane. Because DTFE connects galaxies which lie closely together in the projection, it may involve objects which in reality are quite distant. A full three-dimensional reconstruction followed by projection or smoothing on a sufficiently large scale would alleviate the problem.

14.3 Three-Dimensional Structure of 2dFGRS

We have also reconstructed the full three-dimensional galaxy density field in the NGP and SGP regions of the 2dFGRS. The result is shown in Fig. 44. It shows the three-dimensional rendering of the NGP (left) and SGP slices (right) out to a redshift $z = 0.1$. The maps demonstrate that DTFE is able to recover the three-dimensional structure of the cosmic web as well as its individual elements. Although less obvious than for the two-dimensional reconstructions, the effective resolution of the three-dimensional reconstructions is also varying across the map.

The NGP region is dominated by the large supercluster at a redshift of about 0.08, the Sloan Great Wall. The structure near the upper edge at a

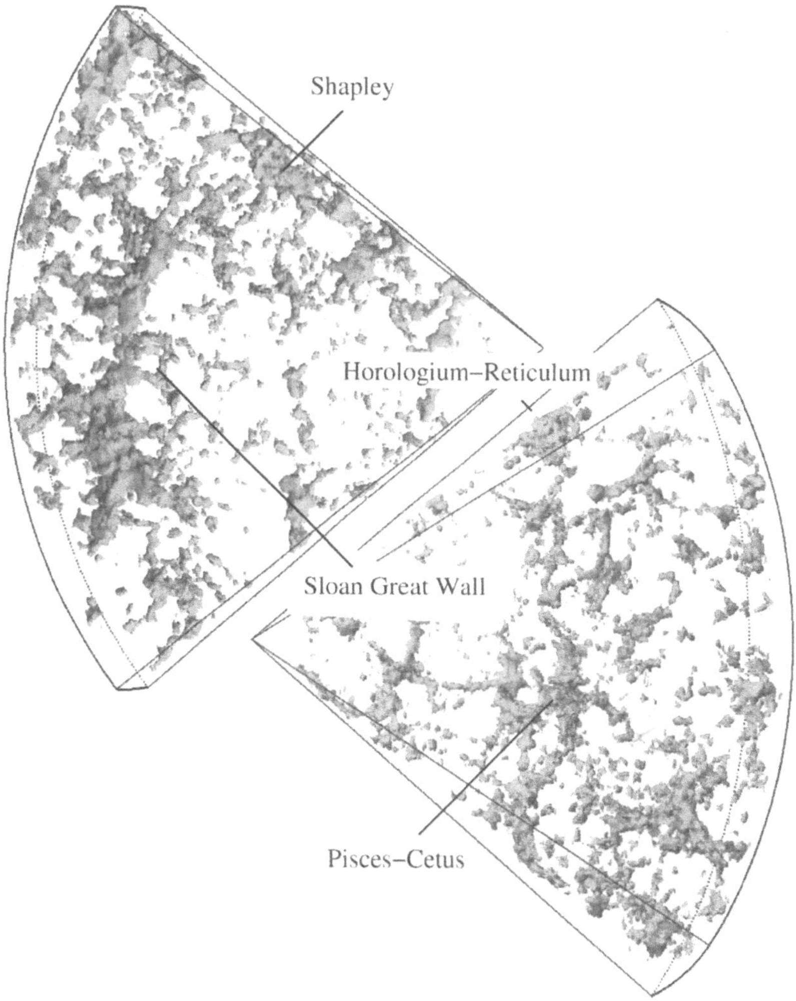

Fig. 44. Isodensity surface of the galaxy distribution in the north (*top*) and south regions (*bottom*) of the 2dFGRS. The densities are determined by means of the DTFE technology, subsequently Gaussian smoothed on a scale of $2\,h^{-1}$ Mpc. Several well-known structures are indicated. From Schaap [201] and Schaap & van de Weygaert [203]

redshift of 0.05–0.06 is part of the upper edge of the Shapley supercluster. In the SGP region several known superclusters can be recognized as well. The supercluster in the centre of this region is part of the Pisces-Cetus supercluster. The huge concentration at a redshift of about 0.07 is the upper part of the enormous Horologium-Reticulum supercluster.

15 Extensions, Applications and Prospects

In this study we have demonstrated that DTFE density and velocity fields are optimal in the sense of defining a continuous and unbiased representation of the data while retaining all information available in the point sample.

In the present review we have emphasized the prospects for the analysis of web-like structures or, more general, any discretely sampled complex spatial pattern. In the meantime, DTFE has been applied in a number of studies of cosmic structure formation. These studies do indeed suggest a major improvement over the more conventional analysis tools. Evidently, even though density/intensity field analysis is one of the primary objectives of the DTFE formalism, one of its important features is its ability to extend its Delaunay tessellation-based spatial interpolation to any corresponding spatially sampled physical quantity. The true potential of DTFE and related adaptive random tessellation-based techniques will become more apparent as further applications and extensions will come to the fore. The prospects for major developments based on the use of tessellations are tremendous. As yet we may identify a diversity of astrophysical and cosmological applications which will benefit substantially from the availability of adaptive random tessellation-related methods. A variety of recent techniques have recognized the high dynamic range and adaptivity of tessellations to the spatial and morphological resolutions of the systems they seek to analyse.

The first major application of DTFE concerns its potential towards uncovering morphological and statistical characteristics of spatial patterns. A second major class of applications is that of issues concerning the dynamics of many particle systems. They may prove to be highly useful for the analysis of phase-space structure and evolution of gravitational systems in their ability to efficiently trace density concentrations and singularities in higher dimensional space. Along similar lines a highly promising avenue is that of the application of similar formalisms within the context of computational hydrodynamics. The application of Delaunay tessellations as a fully self-adaptive grid may finally open a Lagrangian grid treatment of the hydrodynamical equations describing complex multiscale systems such as encountered in cosmology and galaxy formation. Various attempts along these lines have already been followed and with the arrival of efficient adaptive Delaunay tessellation calculations they may finally enable a practically feasible implementation. A third and highly innovative application, also using both the *adaptive* and *random* nature of Delaunay tessellations, is their use as Monte Carlo solvers for systems of complex equations describing complex physical systems. Particularly interesting has been the recent work by Ritzerveld & Icke [190]. They managed to exploit the random and adaptive nature of Delaunay tessellations as a stochastic grid for Monte Carlo lattice treatment of radiative transfer in the case of multiple sources. In the following we will describe a few of these applications in some detail.

15.1 Gravitational Dynamics

Extrapolating the observation that DTFE is able to simultaneously handle spatial density and velocity field (e.g. Bernardeau & Weygaert [27], Romano-Díaz *RWemiliophd2004*, van de Weygaert & Romano-Díaz [249]), and encouraged by the success of Voronoi-based methods in identifying dark halos in N-body simulations (Neyrinck, Gnedin & Hamilton [166]), Arad et al. [8] used DTFE to assess the six-dimensional phase-space density distribution of dark halos in cosmological N-body simulations. While a fully six-dimensional analysis may be computationally cumbersome (Ascalbar & Binney [13]), and not warranted because of the symplectic character of phase space, the splitting of the problem into a separate spatial and velocity-space three-dimensional tessellation may indeed hold promise for an innovative analysis of the dynamical evolution of dark halos.

Gravitational Lensing and Singularity Detection

A related promising avenue seeks to use the ability of DTFE to trace sharp density contrasts. This impelled Bradac et al. [39] to apply DTFE to gravitational lensing. They computed the surface density map for a galaxy from the projection of the DTFE volume density field and used the obtained surface density map to compute the gravitational lensing pattern around the object. Recently, Li et al. [132] evaluated the method and demonstrated that it is indeed a promising tool for tracing higher order singularities.

15.2 Computational Hydrodynamics

Ultimately, the ideal fluid dynamics code would combine the advantages of the Eulerian as well as the Lagrangian approach. In their simplest formulation, Eulerian algorithms cover the volume of study with a fixed grid and compute the fluid transfer through the faces of the (fixed) grid cell volumes to follow the evolution of the system. Lagrangian formulations, on the other hand, compute the system by following the ever-changing volume and shape of a particular individual element of gas.[5] Their emphasis on mass resolution makes Lagrangian codes usually better equipped to follow the system into the highest density regions at the price of a decreased resolution in regions of a lower density.

As we may have appreciated from the adaptive resolution and morphological properties of Delaunay tessellations in DTFE and the more general class of *natural neighbour* interpolations, the recruitment of Delaunay tessellations may define an appropriate combination of Eulerian and Lagrangian description of fluid dynamical systems.

---

[5] Interestingly, the *Lagrangian* formulation is also due to Euler [78] who employed this formalism in a letter to Lagrange, who later proposed these ideas in a publication by himself (Lagrange [126]).

Particle Hydrodynamics

A well-known strategy in computational hydrodynamics is to follow the Lagrangian description by discretizing the fluid system in terms of a (large) number of fluid particles which encapsulate, trace and sample the relevant dynamical and thermodynamical characteristics of the entire fluid (see Koumoutsakos [124] for a recent review). Smooth particle hydrodynamics (SPH) codes (Lucy [141], Gingold & Monaghan [93], Monaghan [164]) have found widespread use in many areas of science and have arguably become the most prominent computational tool in cosmology (e.g. Katz et al. [115], Springel et al. [227], Springel [228]). SPH particles should be seen as discrete balls of fluid, whose shape, extent and properties are specified according to user-defined criteria deemed appropriate for the physical system at hand. Notwithstanding substantial virtues of SPH – amongst which one should include its ability to deal with systems of high dynamic range, its adaptive spatial resolution and its flexibility and conceptual simplicity – it also involves various disadvantages.

One straightforward downside concerns its comparatively bad performance in low-density regions. In a highly topical cosmological issue such as that of the *reionization of the Universe* upon the formation of the first galaxies and stars it may therefore not be able to appropriately resolve the void patches which may be relatively important for the transport of radiation. Even in high-density regions it may encounter problems in tracing singularities. As a result of its user-defined finite size it may not succeed in a sufficiently accurate outlining of shocks. The role of the user-dependent and artificial particle kernel has also become apparent when comparing the performance of SPH versus the more natural DTFE kernels (see Fig. 32, Fig. 33 and Sect. 12.1).

In summary, SPH involves at least four user-defined aspects which affect its performance:

1. SPH needs an arbitrary user-specified kernel function W.
2. SPH needs a smoothing length h;
   Even though the standard choice is a length adapting to the particle density, it does imply a finite extent.
3. SPH kernel needs a shape; a spherical shape is usually presumed.
4. SPH needs an (unphysical) artificial viscosity to stabilize solutions and to be able to capture shocks.

Given the evidently superior density and morphology-tracing abilities of DTFE and related methods based on adaptive grids, Pelupessy, Schaap & van de Weygaert [175] investigated the question what the effect would be of replacing a regular SPH kernel by an equivalent DTFE-based kernel. In an application to a TreeSPH simulation of the (neutral) ISM, they managed to identify various density and morphological environments where the natural adaptivity of DTFE proves to yield substantially better results. They concluded with various suggestions for the formulation of an alternative particle-based adaptive hydrodynamics code in which the kernel would be defined by DTFE.

A closely related and considerably advanced development concerns the formulation of a complete and self-consistent particle hydrodynamics code by Español, Coveney and collaborators (Español [77], Flekkøy et al. [83], Serrano & Español [211], Serrano et al. [212], iDe Fabritiis & Coveney [79]). Their (*multiscale*) *dissipative particle hydrodynamics* code is entirely formulated on the basis of *Voronoi fluid particles*. Working out the Lagrangian equations for a fluid system, they demonstrate that the subsequent compartmentalization of the system into discrete thermodynamic systems – *fluid particles* – leads to a set of equations whose solution would benefit best if they are taken to be defined by the Voronoi cells of the corresponding Voronoi tessellation. In other words, the geometrical features of the Voronoi model are directly connected to the physics of the system by interpretation of the Voronoi volumes as coarse-grained "fluid clusters". Not only does their formalism capture the basic physical symmetries and conservation laws and reproduce the continuum for (sampled) smooth fields, but it also suggests a striking similarity with turbulence. Their formalism has been invoked in the modelling *mesoscale* systems, simulating the molecular dynamics of complex fluid and soft condense matter systems which are usually marked fluctuations and Brownian motion. While the absence of long-range forces such as gravity simplifies the description to some extent, the *Voronoi particle* descriptions do provide enough incentive for looking into possible implementations within an astrophysical context.

Adaptive Grid Hydrodynamics

For a substantial part the success of the DTFE may be ascribed to the use of Delaunay tessellations as optimally covering grid. This implies them to also be ideal for the use in *moving and adaptive grid* implementations of computational hydrodynamics.

Unstructured grid methods originally emerged as a viable alternative to structured or block-structured grid techniques for discretization of the Euler and Navier–Stokes equations for structures with complex geometries (see Mavripilis [149] for an extensive review). The use of unstructured grids provides greater flexibility for discretizing systems with a complex shape and allows the adaptive update of mesh points and connectivities in order to increase the resolution. A notable and early example of the use of unstructured grids may be found in general relativity in the formulation of Regge calculus (Regge [185]). Its applications include quantum gravity formulations on the basis of Voronoi and Delaunay grids (e.g. Miller [159]).

One class of unstructured grids is based on the use of specific geometrical elements (triangles in two dimensions, tetrahedra in three dimensions) to cover in a non-regular fashion a structure whose evolution one seeks to compute. It has become a prominent technique in a large variety of technological applications, with those used in the design of aerodynamically optimally shaped cars

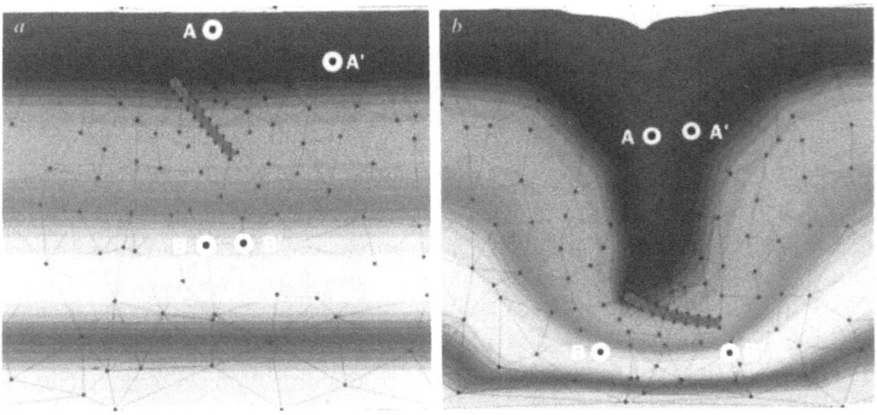

Fig. 45. Application of the natural neighbour scheme solution of partial differential equations on highly irregular evolving Delaunay grids, described by Braun & Sambridge [41]. It involves the natural-element method (NEM) solution of the Stokes flow problem in which motion is driven by the fall of an elasto-plastic plate denser than the viscous fluid. The problem is solved by means of a natural neighbour scheme; the Delaunay grid is used as unstructured computation mesh. The Stokes equation is solved at the integration points in the linear fluid; the equations of force balance at the integration points located within the plate. The solution is shown at two timesteps, 1 and 10,000 of the integration). From Braun & Sambridge [41]

and the design of aeroplanes are perhaps the most visible ones. The definition and design of optimally covering and suitable meshes is a major industry in computational science, including issues involved with the definition of optimal Delaunay meshes (see, e.g., Amenta et al. [6], Amenta & Bern [7], Shewchuk [219], Cheng et al. [45], Dey [63], Alliez et al. [4, 5], Boissonnat et al. [36]). A second class of unstructured grids is based on the use of structural elements of a mixed type with irregular connectivity. The latter class allows the definition of a self-adaptive grid which follows the evolution of a physical system. It is in this context that one may propose the use of Voronoi and Delaunay tessellations as adaptive hydrodynamics lattices.

Hydrocodes with Delaunay tessellations at their core warrant a close connection to the underlying matter distribution and would represent an ideal compromise between an Eulerian and a Lagrangian description, guaranteeing an optimal resolution and dynamic range while taking care of an appropriate coverage of low-density regions. Within astrophysical context there have been a few notable attempts to develop *moving grid* codes. Although these have shown their potential (Gnedin [94], Pen [176]), their complexity and the implicit complications raised by the dominant presence of the long-range force of gravity have as yet prevented their wide-range use.

It is here that Delaunay tessellations may prove to offer a highly promising alternative. The advantages of a moving grid fluid dynamics code based on

Delaunay tessellations have been explicitly addressed in a detailed and enticing study by Whitehurst [256]. His two-dimensional FLAME Lagrangian hydrocode used a first-order solver and proven to be far superior to all tested first-order and many second-order Eulerian codes. The adaptive nature of the Langrangian method and the absence of preferred directions in the grid proved to be key factors for the performance of the method. Whitehurst [256] illustrated this with impressive examples such as the capturing of shock and collision of perfectly spherical shells. The related higher order natural neighbour hydrocodes used by Braun & Sambridge [41], for a range of geophysical problems, and Sukumar [233] are perhaps the most convincing examples and applications of Delaunay grid-based hydrocodes. The advantages of Delaunay grids in principle apply to any algorithm invoking them, in particular also for three-dimensional implementations (of which we are currently unaware).

Kinetic and Dynamic Delaunay Grids

If anything, the major impediment towards the use of Delaunay grids in evolving (hydro)dynamical systems is the high computational expense for computing a Delaunay grid. In the conventional situation one needs to completely upgrade a Delaunay lattice at each timestep. It renders any Delaunay-based code prohibitively expensive.

In recent years, considerable attention has been devoted towards developing kinetic and dynamic Delaunay codes which manage to limit the update of a grid to the few sample points that have moved so far that the identity of their natural neighbours has changed. The work by Meyer-Hermann and collaborators (Schaller & Meyer-Hermann [205], Beyer et al. [32], Beyer & Meyer-Hermann [33]) may prove to represent a watershed in having managed to define a parallel code for the kinetic and dynamic calculation of Delaunay grids. Once the Delaunay grid of the initial point configuration has been computed, subsequent timesteps involve a continuous dynamical upgrade via local Delaunay simplex upgrades as points move around and switch the identity of their natural neighbours.

The code is an elaboration on the more basic step of inserting one point into a Delaunay grid and evaluating which Delaunay simplices need to be upgraded. Construction of Delaunay triangulations via incremental insertion was introduced by Fortune [85] and Edelsbrunner & Shah [70]. Only sample points which experience a change in identity of their natural neighbours need an update of their Delaunay simplices, based on *vertex flips*. The update of sample points that have retained the same natural neighbours is restricted to a mere recalculation of the position and shape of their Delaunay cells.

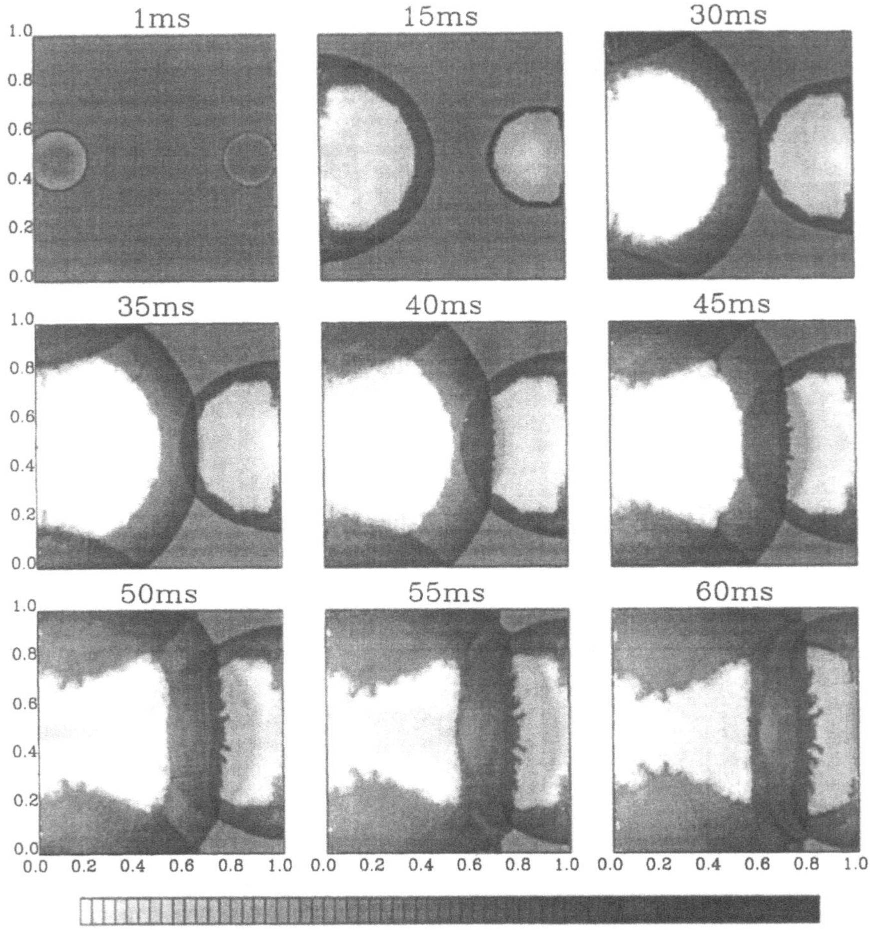

Fig. 46. Nine nonlinear greyscale images of the density evolution of the two-dimensional interacting blast waves test of the FLAME Delaunay grid hydrocode of Whitehurst [256]. The circular shape of the shock fronts is well represented; the contact fronts are clearly visible. Of interest is the symmetry of the results, which is not enforced and so is genuine, and the instability of the decelerated contact fronts. From Whitehurst [256]

15.3 Random Lattice Solvers and SimpleX

The use of Delaunay tessellations as adaptive and dynamic meshes for evaluating the fluid dynamical equations of an evolving physical system emphasizes their adaptive nature. Perhaps even more enticing is the use of the random nature of the these meshes. Taking into account their spatial adaptivity as well as their intrinsic stochastic nature, they provide an excellent template for following physical systems along the lines of Monte Carlo methods.

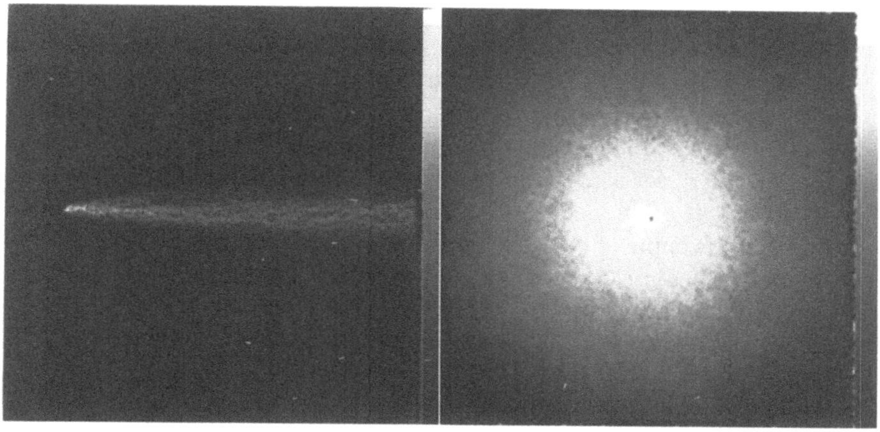

Fig. 47. The result of two simple `SimpleX` radiative transfer tests on a two-dimensional Poisson–Delaunay random lattice with $N = 5\times10^4$ points. Both are logarithmic plots of the number of particles at each site. *Left*: Illustration of the conservation of momentum by means of the transport of particles with constant momentum vectors. *Right*: Illustration of a scattering transport process. From Ritzerveld [189]

Generally speaking, Monte Carlo methods determine the characteristics of many-body systems as statistical averages over a large number of particle histories, which are computed with the use of random sampling techniques. They are particularly useful for transport problems since the transport of particles through a medium can be described stochastically as a random walk from one interaction event to the next. The first to develop a computer-oriented Monte Carlo method for transport problems was Metropolis & Ulam [153]. Such transport problems may in general be formulated in terms of a stochastic master equation which may evaluated by means of Monte Carlo methods by simulating large numbers of different particle trajectories or histories (see Ritzerveld [189] for a nice discussion).

While the asymmetry of regular meshes and lattices for Monte Carlo calculations introduces various undesirable effects, random lattices may alleviate or solve the problematic lack of symmetry. Lattice Boltzmann studies were amongst the first to recognize this (see, e.g., Ubertini & Succi [240]). In a set of innovative publications Christ et al. [49, 50, 51] applied random lattices, including Voronoi and Delaunay tessellations, to solve (QCD) field theoretical equations.

Recently, a highly innovative and interesting application to (astrophysical) radiative transfer problems followed the same philosophy as the random lattice gauge theories. Ritzerveld & Icke [190] and Ritzerveld [189] translated the problem of radiative transfer into one which would be solved on an appropriately defined (Poisson)-Delaunay grid. It leads to the formulation of the `SimpleX` radiation transfer technique which translates the transport of radiation through a medium into the *deterministic* displacement of photons from one vertex to another vertex of the *stochastic* Delaunay mesh. The perfect random and isotropic nature of the Delaunay grid assures an unbiased Monte Carlo sampling of the photon paths. The specification of appropriate absorption and scattering coefficients at the nodes completes the proper Monte Carlo solution of the master equation for radiative transfer.

One of the crucial aspects of **SimpleX** is the sampling of the underlying density field $\rho(\mathbf{x})$ such that the Delaunay grid edge lengths L_{D} represent a proper random sample of the (free) paths λ_γ of photons through a scattering and absorbing medium. In other words, **SimpleX** is built on the assumption that the (local) mean Delaunay edge length $\langle L_{\mathrm{D}} \rangle$ should be proportional to the (local) mean free path of the photon. For a medium of density $\rho(\mathbf{x})$ in d-dimensional space ($d=2$ or $d=3$), with absorption/scattering coefficient σ, sampled by a Delaunay grid generated by a point sample with local number density $n_{\mathrm{X}}(\mathbf{x})$,

$$\langle L_{\mathrm{D}} \rangle \propto \lambda_\gamma \quad \Longleftarrow \quad \begin{cases} \langle L_{\mathrm{D}} \rangle = \zeta(d)\, n_{\mathrm{X}}^{-1/d} \\ \\ \lambda_\gamma = \dfrac{1}{\rho(\mathbf{x})\,\sigma} \end{cases} , \tag{55}$$

where $\zeta(d)$ is a dimension-dependent constant. By sampling of the underlying density field $\rho(\mathbf{x})$ by a point density,

$$n_{\mathrm{X}}(\mathbf{x}) \propto \rho(\mathbf{x})^d . \tag{56}$$

SimplexX produces a Delaunay grid whose edges are guaranteed to form a representative stochastic sample of the photon paths through the medium.

To illustrate the operation of **SimpleX**, Fig. 47 shows the outcome of two-dimensional **SimpleX** test calculations. One involves a test of the ability of **Simplex** to follow a beam of radiation, the second one its ability to follow the spherical spread of radiation emitted by an isotropically emitting source. The figure nicely illustrates the success of **SimpleX**, meanwhile providing an impression of the effect and its erratic action at the scale of Delaunay grid cells.

The **SimpleX** method has been applied to the challenging problem of the reionization of the intergalactic medium by the first galaxies and stars. A major problem for most radiative transfer techniques is to deal with multiple sources of radiation. While astrophysical problems may often be approximated by a description in terms of a single source, a proper evaluation of reionization should take into account the role of a multitude of sources. **SimpleX** proved its ability to yield sensible answers in a series of test comparisons between different radiative transfer codes applied to aspects typical for reionization (Illiev et al. [107]). The first results of the application of **SimpleX** to genuine cosmological circumstances, by coupling it to a cosmological SPH code, do yield highly encouraging results (Ritzerveld et al. [191]). Figure 48 is a beautiful illustration of the outcome of one of the reionization simulations.

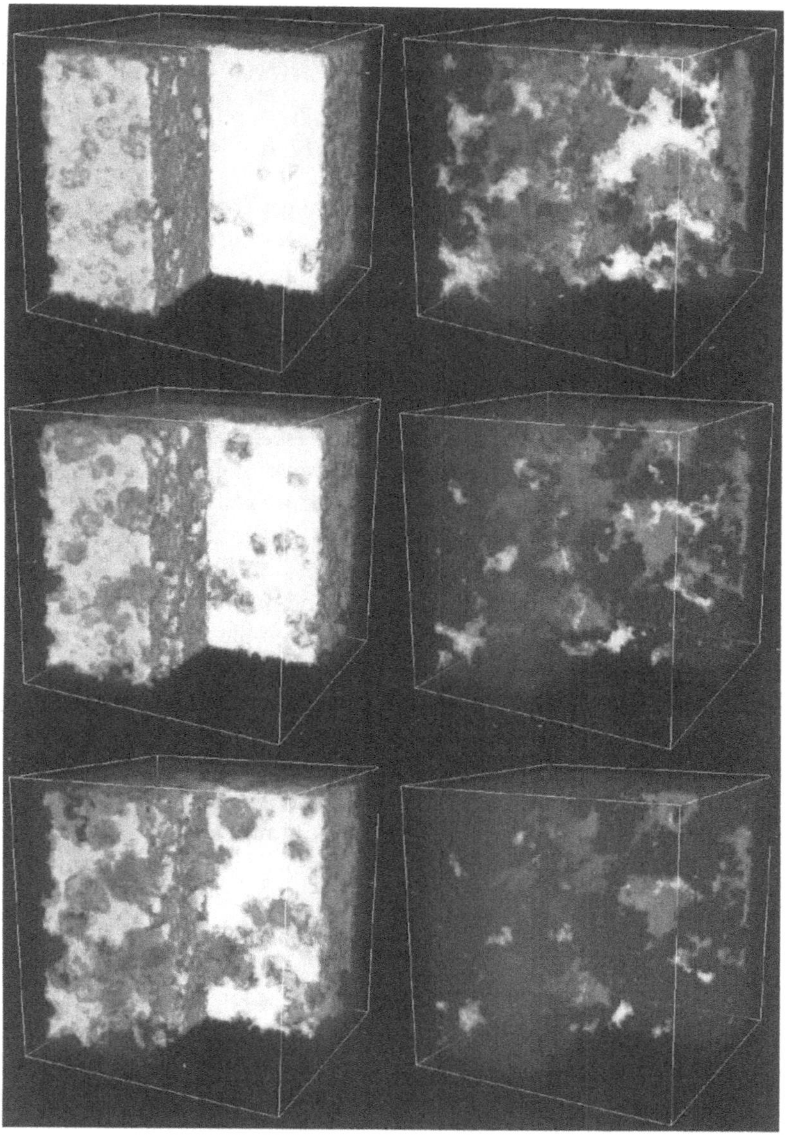

Fig. 48. A volume rendering of the result of using the `SimpleX` method to transport ionizing photons through the intergalactic medium at and around the epoch of reionization. The `SimpleX` method was applied to a PMFAST simulation of the large-scale structure in a ΛCDM universe. The results of six different timesteps are plotted, in which white corresponds to hydrogen that is still neutral (*opaque*), the blue to the already ionized hydrogen (*transparent*). From Ritzerveld [189]

If anything, `SimpleX` proves that the use of random Delaunay grids has the potential of representing a genuine breakthrough for addressing a number of highly complex astrophysical problems.

15.4 Spatial Statistics and Pattern Analysis

Within its cosmological context DTFE will meet its real potential in more sophisticated applications tuned towards uncovering morphological characteristics of the reconstructed spatial patterns.

A potentially interesting application would be its implementation in the SURFGEN machinery. SURFGEN seeks to provide locally defined topological measures, e.g. local Minkowski functionals, of the density field (Sahni et al. [198], Sheth et al. [217], Shandarin et al. [213]). A recent new sophisticated technique for tracing the cosmic web is the *skeleton* formalism developed by Novikov et al. [167] based on the framework of Morse theory (Colombi et al. [58]). The *skeleton* formalism seeks to identify filaments in the web by identifying ridges along which the gradient of the density field is extremal along its isocontours (see also Sousbie et al. [225]). The use of unbiased web-like density fields traced by DTFE would undoubtedly sharpen the ability of the skeleton formalism to trace intricate web-like features over a wider dynamical range.

Pursuing an alternative yet related track concerns the direct use of tessellations themselves in outlining topological properties of a complex spatial distribution of points. *Alpha shapes* of a discrete point distribution are subsets of Delaunay triangulations which are sensitive tracers of its topology and may be exploited towards inferring Minkowski functionals and Betti numbers (Edelsbrunner et al. [68], Edelsbrunner & Mücke [69], Edelsbrunner et al. [72]). A recent and ongoing study seeks their application to the description of the cosmic web (Vegter et al. [251]) (see also Sect. 15.5).

Two major extensions of DTFE already set out to the identification of key aspects of the cosmic web. Specifically focused on the hierarchical nature of the megaparsec matter distribution is the detection of web-like anisotropic features over a range of spatial scales by the multiscale morphology filter (MMF) introduced and defined by Aragón-Calvo et al. [10]. The *Watershed Void-Finding* (WVF) algorithm of Platen, van de Weygaert & Jones [179] is a void detection algorithm meant to outline the hierarchical nature of the cosmic void population. We will shortly touch upon these extensions in order to illustrate the potential of DTFE.

15.5 Alpha Shapes

Alpha shapes are a description of the (intuitive) notion of the shape of a discrete point set. *Alpha shapes* of a discrete point distribution are subsets of a Delaunay triangulation and were introduced by Edelsbrunner and collaborators (Edelsbrunner et al. [68], Mücke [165], Edelsbrunner & Mücke [69],

Edelsbrunner et al. [72]). Alpha shapes are generalizations of the convex hull of a point set and are concrete geometric objects which are uniquely defined for a particular point set. Reflecting the topological structure of a point distribution, it is one of the most essential concepts in the field of computational topology (Dey [62], Vegter [250], Zomorodian [261]).

If we have a point set S and its corresponding Delaunay triangulation, we may identify all *Delaunay simplices* – tetrahedra, triangles, edges, vertices – of the triangulation. For a given non-negative value of α, the *alpha complex* of a point set consists of all simplices in the Delaunay triangulation which have an empty circumsphere with squared radius less than or equal to α, $R^2 \leq \alpha$. Here "empty" means that the open sphere does not include any points of S. For an extreme value $\alpha = 0$ the alpha complex merely consists of the vertices of the point set. The set also defines a maximum value $\alpha_{\max}$ such that for $\alpha \geq \alpha_{\max}$ the alpha shape is the convex hull of the point set.

The *alpha shape* is the union of all simplices of the alpha complex. Note that it implies that although the alpha shape is defined for all $0 \leq \alpha < \infty$ there are only a finite number of different alpha shapes for any one point set. The alpha shape is a polytope in a fairly general sense; it can be concave and even disconnected. Its components can be three-dimensional patches of tetrahedra, two-dimensional ones of triangles, one-dimensional strings of edges and even single points. The set of all real numbers α leads to a family of shapes capturing the intuitive notion of the overall versus fine shape of a point set. Starting from the convex hull of a point set and gradually decreasing α, the shape of the point set gradually shrinks and starts to develop cavities. These cavities may join to form tunnels and voids. For sufficiently small α the alpha shape is empty.

Following this description, one may find that alpha shapes are intimately related to the topology of a point set. As a result they form a direct and unique way of characterizing the topology of a point distribution. A complete quantitative description of the topology is that in terms of Betti numbers β_{k} and these may indeed be directly inferred from the alpha shape. The first Betti number β_0 specifies the number of independent components of an object. In $\mathbb{R}$, β_1 may be interpreted as the number of independent tunnels and β_2 as the number of independent enclose voids. The kth Betti number effectively counts the number of independent k-dimensional holes in the simplicial complex.

Applications of alpha shapes have as yet focused on biological systems. Their use is in characterizing the topology and structure of macromolecules. The work by Liang and collaborators (Edelsbrunner et al. [71], Liang et al. [133, 134], Liang et al. [135]) uses alpha shapes and Betti numbers to assess the voids and pockets in an effort to classify complex protein structures, a highly challenging task given the 10,000–30,000 protein families involving 1000–4000 complicated folds. Given the interest in the topology of the cosmic mass distribution (e.g. Gott et al. [98], Mecke et al. [150], Schmalzing et al. [207]), it is evident that *alpha shapes* also provide a highly interesting tool for studying the topology of the galaxy distribution and N-body simulations of cosmic

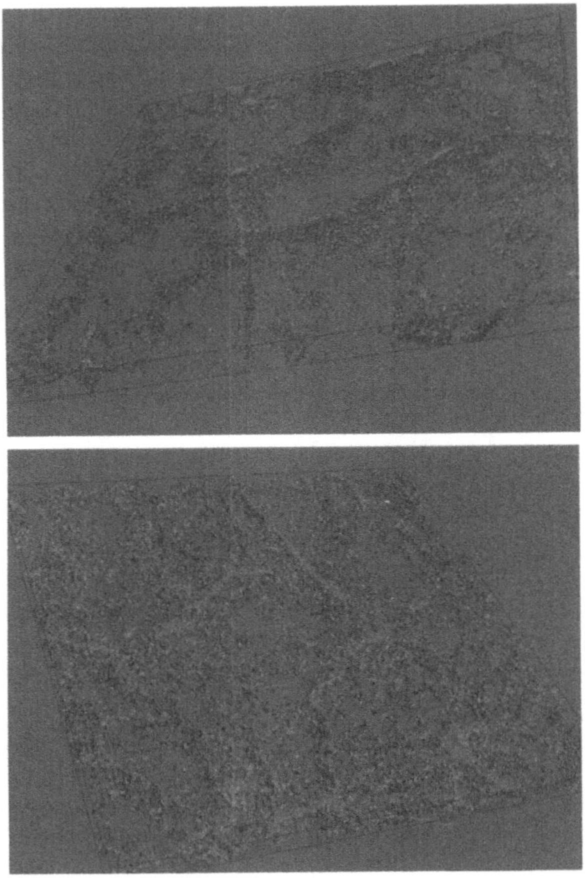

Fig. 49. Examples of *alpha shapes* of the LCDM GIF simulation. Shown are central slices through two alpha shapes (*top*: low alpha; *bottom*: high alpha). The image shows the sensitivity of alpha shapes to the topology of the matter distribution. From: Vegter et al. [251]. Courtesy: Bob Eldering

structure formation. Directly connected to the topology of the point distribution itself, it would discard the need of user-defined filter kernels.

In a recent study Vegter et al. [251] computed the alpha shapes for a set of GIF simulations of cosmic structure formation (see Fig. 49). On the basis of a calibration of the inferred Minkowski functionals and Betti numbers from a range of Voronoi clustering models, their study intends to refine the knowledge of the topological structure of the cosmic web.

15.6 Multiscale Morphology Filter

The multiscale detection technique – MMF – is used for characterizing the different morphological elements of the large-scale matter distribution in the

Universe (Aragón-Calvo et al. [10]). The method is ideally suited for extracting catalogues of clusters, walls, filaments and voids from samples of galaxies in redshift surveys or particles in cosmological N-body simulations.

The multiscale filter technology is particularly oriented towards recognizing and identifying features in a density or intensity field on the basis of an assessment of their coherence along a range of spatial scales and with the virtue of providing a generic framework for characterizing the local morphology of the density field and enabling the selection of those morphological features which the analysis at hand seeks to study. The multiscale morphology filter (MMF) method has been developed on the basis of visualization and feature extraction techniques in computer vision and medical research (Florack et al. [84]). The technology, finding its origin in computer vision research, has been optimized within the context of feature detections in medical imaging. Frangi et al. [86] and Sato et al. [199] presented its operation for the specific situation of detecting the web of blood vessels in a medical image. This defines a notoriously complex pattern of elongated tenuous features whose branching makes it closely resemble a fractal network.

The density or intensity field of the sample is smoothed over a range of multiple scales. Subsequently, this signal is processed through a morphology response filter. Its form is dictated by the particular morphological feature it seeks to extract and depends on the local shape and spatial coherence of the intensity field. The morphology signal at each location is then defined to be the one with the maximum response across the full range of smoothing scales. The MMF translates, extends and optimizes this technology towards the recognition of the major characteristic structural elements in the megaparsec matter distribution. It yields a unique framework for the combined identification of dense, compact blob-like clusters of the salient and moderately dense elongated filaments and of tenuous planar walls. Figure 50 includes a few of the stages involved in the MMF procedure.

Crucial for the ability of the method to identify anisotropic features such as filaments and walls is the use of a morphologically unbiased and optimized continuous density field retaining all features visible in a discrete galaxy or particle distribution. Accordingly, DTFE is used to process the particle

Table 3. Behaviour of the eigenvalues for the characteristic morphologies. The lambda conditions describe objects with intensity higher than their background (as clusters, filaments and walls). For voids we must reverse the sign of the eigenvalues. From the constraints imposed by the λ conditions we can describe the blob morphology as a subset of the line which is itself a subset of the wall

Structure	λ ratios	λ constraints
Blob	$\lambda_1 \simeq \lambda_2 \simeq \lambda_3$	$\lambda_3 < 0$; $\lambda_2 < 0$; $\lambda_1 < 0$
Line	$\lambda_1 \simeq \lambda_2 \gg \lambda_3$	$\lambda_3 < 0$; $\lambda_2 < 0$
Sheet	$\lambda_1 \gg \lambda_2 \simeq \lambda_3$	$\lambda_3 < 0$

distribution into a continuous density field (top centre frame, Fig. 50). The morphological intentions of the MMF method renders DTFE a key element for translating the particle or galaxy distribution into a representative continuous density field $f_{\rm DTFE}$.

In the implementation of Aragón-Calvo et al. [10], the scaled representations of the data are obtained by repeatedly smoothing the DTFE reconstructed density field $f_{\rm DTFE}$ with a hierarchy of spherically symmetric Gaussian filters $W_{\rm G}$ having different widths R:

$$f_{\rm S}(\boldsymbol{x}) = \int {\rm d}\boldsymbol{y}\, f_{\rm DTFE}(\boldsymbol{y})\, W_{\rm G}(\boldsymbol{y},\boldsymbol{x}),$$

where $W_{\rm G}$ denotes a Gaussian filter of width R. A pass of the Gaussian smoothing filter attenuates structure on scales smaller than the filter width. The scale space itself is constructed by stacking these variously smoothed data sets, yielding the family Φ of smoothed density maps f_n:

$$\Phi = \bigcup_{levels\ n} f_n \tag{57}$$

In essence the *scale space* structure of the field is the $(D+1)$-dimensional space defined by the D-dimensional density or intensity field smoothed over a continuum of filter scales R_G. As a result a data point can be viewed at any of the scales where scaled data has been generated.

The crux of the concept is that the neighbourhood of a given point will look different at each scale. While there are potentially many ways of making a comparison of the scale dependence of local environment, Aragón-Calvo et al. [10] chose to calculate the Hessian matrix in each of the smoothed replicas of the original data to describe the local "shape" of the density field in the neighbourhood of that point. In terms of the Hessian, the local variations around a point $\boldsymbol{x}_0$ of the density field $f(\boldsymbol{x})$ may be written as the Taylor expansion

$$f(\boldsymbol{x}_0+\boldsymbol{s}) = f(\boldsymbol{x}_0) + \boldsymbol{s}^T\nabla f(\boldsymbol{x}_0) + \frac{1}{2}\boldsymbol{s}^T\mathcal{H}(\boldsymbol{x}_0)\boldsymbol{s} + ... \tag{58}$$

where

$$\mathcal{H} = \begin{pmatrix} f_{xx} & f_{yx} & f_{zx} \\ f_{xy} & f_{yy} & f_{zy} \\ f_{xz} & f_{yz} & f_{zz} \end{pmatrix} \tag{59}$$

is the Hessian matrix. Subscripts denote the partial derivatives of f with respect to the named variable. There are many possible algorithms for evaluating these derivatives. In practice, the scale space procedure evaluates the Hessian directly for a discrete set of filter scales by smoothing the DTFE density field by means of Mexican Hat filter,

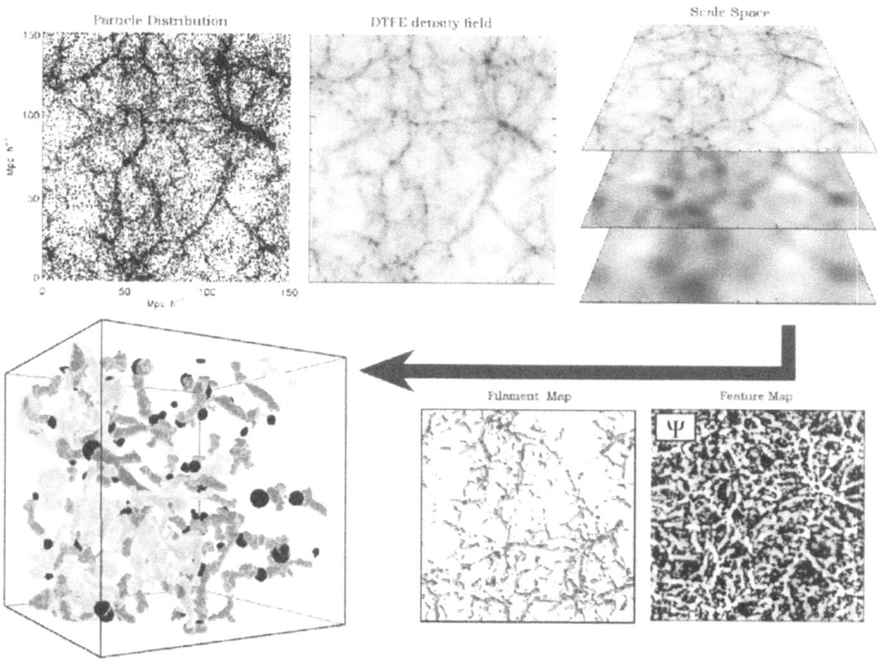

Fig. 50. Schematic overview of the multiscale morphology filter (MMF) to isolate and extract elongated filaments (*dark grey*), sheet-like walls (*light grey*) and clusters (*black dots*) in the web-like pattern of a cosmological N-body simulation. The first stage is the translation of a discrete particle distribution (*top left-hand frame*) into a DTFE density field (*top centre*). The DTFE field is filtered over a range of scales (*top right-hand* stack of filtered fields). By means of morphology filter operations defined on the basis of the Hessian of the filtered density fields, the MMF successively selects the regions which have a blob-like (*cluster*) morphology, a filamentary morphology and a planar morphology at the scale at which the morphological signal is optimal. This produces a feature map (*bottom left hand*). By means of a percolation criterion the physically significant filaments are selected (*bottom centre*). Following a sequence of blob, filament and wall filtering finally produces a map of the different morphological features in the particle distribution (*bottom left hand*). The three-dimensional isodensity contours in the bottom left-hand frame depict the most pronounced features. From Aragón-Calvo et al. [10]

$$\begin{aligned}\frac{\partial^2}{\partial x_i \partial x_j} f_{\mathrm{S}}(\boldsymbol{x}) &= f_{\mathrm{DTFE}} \otimes \frac{\partial^2}{\partial x_i \partial x_j} W_{\mathrm{G}}(R_{\mathrm{S}}) \\ &= \int \mathrm{d}\boldsymbol{y}\, f(\boldsymbol{y})\, \frac{(x_i - y_i)(x_j - y_j) - \delta_{ij} R_{\mathrm{S}}^2}{R_{\mathrm{S}}^4}\, W_{\mathrm{G}}(\boldsymbol{y}, \boldsymbol{x}), \quad (60)\end{aligned}$$

with $x_1, x_2, x_3 = x, y, z$ and δ_{ij} the Kronecker delta. In other words, the scale space representation of the Hessian matrix for each level n is evaluated by means of a convolution with the second derivatives of the Gaussian filter, also known as the *Marr or Mexican Hat* wavelet.

The eigenvalues of the Hessian matrix at a point encapsulate the information on the local shape of the field. Eigenvalues are denoted as $\lambda_a(\boldsymbol{x})$ and arranged so that $\lambda_1 \geq \lambda_2 \geq \lambda_3$:

$$\left| \frac{\partial^2 f_n(\boldsymbol{x})}{\partial x_i \partial x_j} - \lambda_a(\boldsymbol{x})\, \delta_{ij} \right| = 0, \quad a = 1,2,3, \qquad \text{with} \quad \lambda_1 > \lambda_2 > \lambda_3. \tag{61}$$

The $\lambda_i(\boldsymbol{x})$ are coordinate-independent descriptors of the behaviour of the density field in the locality of the point $\boldsymbol{x}$ and can be combined to create a variety of morphological indicators. They quantify the rate of change of the field gradient in various directions about each point. A small eigenvalue indicates a low rate of change of the field values in the corresponding eigendirection and vice versa. The corresponding eigenvectors show the local orientation of the morphology characteristics.

Evaluating the eigenvalues and eigenvectors for the renormalized Hessian $\tilde{\mathcal{H}}$ of each data set in a scale space shows how the local morphology changes with scale. First regions in scale space are selected according to the appropriate morphology filter, identifying blob-like, filamentary and wall-like features at a range of scales. The selections are made according to the eigenvalue criteria listed in Table 3. Subsequently, a sophisticated machinery of filters and masks is applied to assure the suppression of noise and the identification of the locally relevant scales for the various morphologies.

Finally, for the cosmological or astrophysical purpose at hand the identified spatial patches are tested by means of an *erosion* threshold criterion. Identified blobs should have a critical overdensity corresponding to virialization, while identified filaments should fulfil a percolation requirement (bottom central frame). By successive repetition for the identification of blobs, filaments and sheets – each with their own morphology filter – the MMF has dissected the cosmological density field into the corresponding features. The box in the bottom left-hand frame shows a segment from a large cosmological N-body simulation: filaments are coloured dark grey, the walls light grey and the clusters indicated by the black blobs.

Once these features have been marked and identified by MMF, a large variety of issues may be addressed. An important issue is that of environmental influences on the formation of galaxies. The MMF identification of filaments allowed Aragón-Calvo et al. [10] to show that galaxies in filaments and walls do indeed have a mass-dependent alignment.

15.7 Watershed Void Finder

The Watershed Void Finder (WVF) is an implementation of the *Watershed transform* for segmentation of images of the galaxy and matter distribution

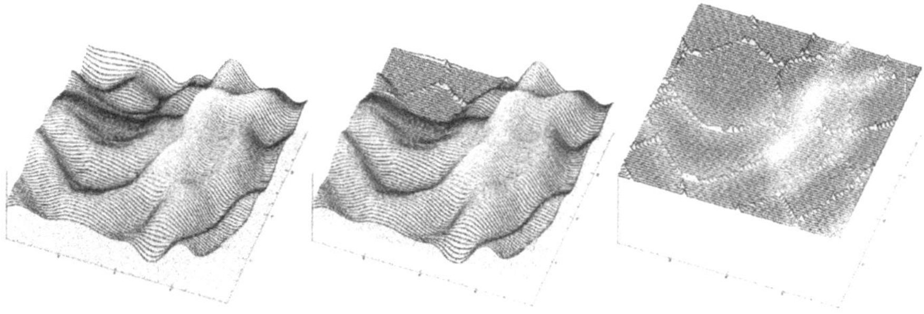

Fig. 51. Three frames illustrating the principle of the watershed transform. The left-hand frame shows the surface to be segmented. Starting from the local minima the surrounding basins of the surface start to flood as the water level continues to rise (*dotted plane* initially below the surface). Where two basins meet up near a ridge of the density surface, a "dam" is erected (*central frame*). Ultimately, the entire surface is flooded; leaving a network of dams defines a segmented volume and delineates the corresponding cosmic web (*right-hand frame*). From Platen et al. [179]

into distinct regions and objects and the subsequent identification of voids (Platen [178], Platen et al. [179]). The watershed transform is a concept defined within the context of mathematical morphology. The basic idea behind the watershed transform finds its origin in geophysics. It delineates the boundaries of the separate domains, the *basins*, into which yields of, e.g., rainfall will collect. The analogy with the cosmological context is straightforward: *voids* are to be identified with the *basins*, while the *filaments* and *walls* of the cosmic web are the ridges separating the voids from each other.

The identification of voids in the cosmic matter distribution is hampered by the absence of a clearly defined criterion of what a void is. Unlike overdense and virialized clumps of matter voids are not genuinely defined physical objects. The boundary and identity of voids are therefore mostly a matter of definition. As a consequence there is a variety of void-finding algorithms (Kauffmann & Fairall [116], El Ad et al., [74], Plionis & Basilakos [180], Hoyle & Vogeley [104], Shandarin et al. [214], Colberg et al. [53], Patiri et al. [171]). A recent study (Colberg et al. [55]) contains a balanced comparison of the performance of the various void finders with respect to a small region taken from the Millennium simulation (Springel et al. [229]).

Watersheds

With respect to the other void finders, the watershed algorithm seeks to define a natural formalism for probing the hierarchical nature of the void distribution in maps of the galaxy distribution and in N-body simulations of cosmic structure formation. The WVF has several advantages (see, e.g., Meyer & Beucher [154]). Because it identifies a void segment on the basis of the crests in a density field surrounding a density minimum, it is able to trace the void

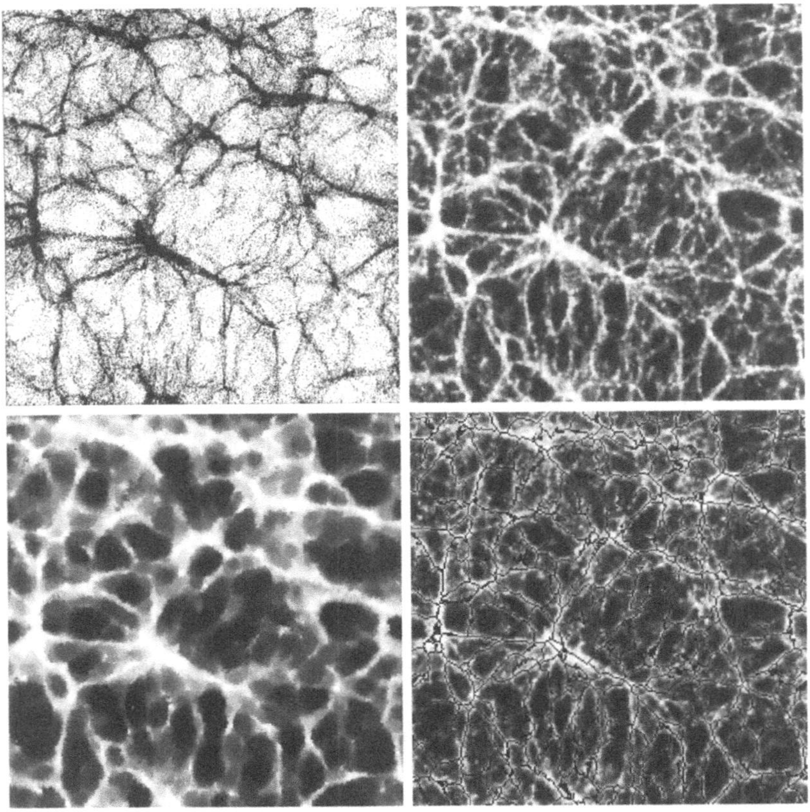

Fig. 52. A visualization of several intermediate steps of the Watershed Void-Finding (WVF) method. The *top left-hand frame* shows the particles of a slice in the LCDM GIF simulation. The corresponding DTFE density field is shown in the *top right-hand frame.* The next, *bottom left-hand frame* shows the resulting n-th order median-filtered image. Finally, the *bottom right-hand frame* shows the resulting WVF segmentation computed on the basis of the median-filtered image. From Platen et al. [179].

boundary even though it has a distorted and twisted shape. Also, because the contours around well-chosen minima are by definition closed the transform is not sensitive to local protrusions between two adjacent voids. The main advantage of the WVF is that for an ideally smoothed density field it is able to find voids in an entirely parameter-free fashion.

The word *watershed* finds its origin in the analogy of the procedure with that of a landscape being flooded by a rising level of water. Figure 51 illustrates the concept. Suppose we have a surface in the shape of a landscape. The surface is pierced at the location of each of the minima. As the water level rises a growing fraction of the landscape will be flooded by the water in the expanding basins. Ultimately basins will meet at the ridges corresponding to saddle points in the density field. These define the boundaries of the basins, enforced by means of a sufficiently high *dam*. The final result of the completely immersed landscape is a division of the landscape into individual cells, separated by the *ridge dams*.

Formalism

The WVF consists of eightfold crucial steps. The two essential first steps relate directly to DTFE. The use of DTFE is essential to infer a continuous density field from a given N-body particle distribution of galaxy redshift survey. For the success of the WVF it is of utmost importance that the density field retains its morphological character, i.e. the hierarchical nature, the web-like morphology dominated by filaments and walls and the presence of voids. In particular, in and around low-density void regions, the raw density field is characterized by a considerable level of noise. In an essential second step, noise gets suppressed by an adaptive smoothing algorithm which in a consecutive sequence of repetitive steps determines the median of densities within the *contiguous Voronoi cell* surrounding a point. The determination of the median density of the natural neighbours turns out to define a stable and asymptotically converging smooth density field fit for a proper watershed segmentation.

Figure 52 is an illustration of four essential stages in the WVF procedure. Starting from a discrete point distribution (top left), the continuous density field is determined by the DTFE procedure (top right). Following the natural smoothing by *natural neighbour median* filtering (bottom left), the watershed formalism is able to identify the void segments in the density field (bottom right).

16 Concluding Remarks

This extensive presentation of tessellation-based machinery for the analysis of web-like patterns in the spatial matter and galaxy distribution intends to provide the interested reader with a framework to exploit and extend the large potential of Voronoi and Delaunay tessellations. Even though conceptually not too complex, they do represent an intriguing world by themselves. Their cellular geometry paves the way towards a natural analysis of the intricate fabric known as cosmic web.

Acknowledgements

RvdW wishes to thank Vicent Martinez and Maria Pons for their invitation and hospitality during this fine and inspiring September week in Valencia, and their almost infinite patience regarding my ever-shifting deadlines. The work described in this lecture is the result of many years of work, involving numerous collaborations. In particular we wish to thank Vincent Icke, Francis Bernardeau, Dietrich Stoyan, Sungnok Chiu, Jacco Dankers, Inti Pelupessy, Emilio Romano-Díaz, Jelle Ritzerveld, Miguel Aragón-Calvo, Erwin Platen, Sergei Shandarin, Gert Vegter, Niko Kruithof and Bob Eldering for invaluable contributions and discussions, covering nearly all issues touched upon in this study. Particularly we are grateful to Bernard Jones for his enthusiastic and crucial support and inspiration and the many original ideas, already over two decades, for all that involved tessellations, DTFE, the Universe, and much more. RvdW is especially grateful to Manolis and Menia for their overwhelming hospitality during the 2005 Greek Easter weeks in which a substantial fraction of this manuscript was conceived. What better setting to wish for than the view from the Pentelic mountain overlooking the cradle of western civilization, the city of Pallas Athena. Finally, this work could not have been finished without the patience and support of Marlies.

References

1. Abazajian, K., et al. (The SDSS collaboration): Astron. J. **126**, 2081 (2003)
2. Abramov, O., McEwen, A.S.: Int. J. Rem. Sens. **25**, 669 (2004)
3. Aikio, J., Mähönen, P.: Astrophys. J. **497**, 534 (1998)
4. Alliez, P., Cohen-Steiner, D., Yvinec, M., Desbrun, M.: ACM Trans. Graph. 24, 517 (2005)
5. Alliez, P., Ucelli, G., Gotsman, M., Attene, M.: Recent Advances in Remeshing of Surfaces. Report of the AIM@SHAPE EU network (2005)
6. Amenta, N., Bern, M., Kamvysselis, M.: Siggraph '98, 415 (1998)
7. Amenta, N., Bern, M.: Discrete and Comp. Geometry 22, 481 (1999)
8. Arad, I., Dekel, A., Klypin, A.: Mon. Not. R. Astron. Soc. **353**, 15 (2004)
9. Aragón-Calvo, M.A.: Morphology and Dynamics of the Cosmic Web, Ph.D. thesis, Groningen University (2007)
10. Aragón-Calvo, M.A., Jones, B.J.T.: van de Weygaert, R.: van der Hulst, J.M.: Mon. Not. R. Astron. Soc., subm. Astron. Astrophys., **474**, 315 (2007)
11. Arbabi-Bidgoli, S., Müller, V.: Mon. Not. R. Astron. Soc. **332**, 20 (2002)
12. Arnol'd, V.I.: The Theory of Singularities and Its Applications. Cambridge University Press, Cambridge (1991)
13. Ascasibar, Y., Binney, J.: Mon. Not. R. Astron. Soc. **356**, 872 (2005)
14. Babul, A., Starkman, G.D.: Astrophys. J. **401**, 28 (1992)
15. Baccelli, F., Zuyev, S.: Oper. Res. **47**, 619 (1999)
16. Balbus, S.A., Hawley, J.F.: Revs. Mod. Phys. **70**, 1 (1998)
17. Bertschinger, E.: Astrophys. J. **323**, L103 (1987)
18. Barrow, J.D., Bhavsar, S.P., Sonoda, D.H.: Mon. Not. R. Astron. Soc. **216**, 17 (1985)
19. Basilakos, S., Plionis, M., Rowan-Robinson, M.: Mon. Not. R. Astron. Soc. **323**, 47 (2001)
20. Bennett, C.L., et al.: Astrophys. J. Suppl. **148**, 1 (2003)

21. Bentley, J.L.: Comm. ACM **18**, 509 (1975)
22. Bentley, J.L., Friedmann, J.H.: IEEE Trans. Comput. **C-27**, 97 (1978)
23. Bernardeau, F.: Astrophys. J. **390**, L61 (1992)
24. Bernardeau, F.: Astrophys. J. **433**, 1 (1994)
25. Bernardeau, F., Kofman, L.: Astrophys. J. **443**, 479 (1995)
26. Bernardeau, F., Juszkiewicz, R., Dekel, A., Bouchet, F. R.: Mon. Not. R. Astron. Soc. **274**, 20 (1995)
27. Bernardeau, F., van de Weygaert, R.: Mon. Not. R. Astron. Soc. **279**, 693 (1996)
28. Bernardeau, F., van de Weygaert, R., Hivon, E., PBouchet, F.: Mon. Not. R. Astron. Soc. **290**, 566 (1997)
29. Bernardeau, F., Colombi, S., Gaztanaga, E., Scoccimarro, R.: Phys. Rep. **367**, 1 (2002)
30. Beucher, S., Lantuejoul, C. In: Proc. Intern. Workshop on Image Processing, CCETT/IRISA, Rennes (1979)
31. Beucher, S., Meyer, F.: The morphological approach to segmentation, the watershed transformation. In: Dougherty, E. (ed.) Mathematical Morphology in Image Processing. M. Dekker Inc., New York (1993)
32. Beyer, T., Schaller, G., Deutsch, A., Meyer-Hermann, M.: Comput. Phys. Commun. **172**, 86 (2005)
33. Beyer, T., Meyer-Hermann, M.: WSEAS Trans. Syst. **5**, 1100 (2006)
34. Böhringer, H., Schuecker, P., Guzzo, L., Collins, C.A., Voges, W., Schindler, S., Neumann, D.M., Cruddace, R.G., De Grandi, S., Chincarini, G., Edge, A.C., MacGillivray, H.T., Shaver, P.: Astron. Astrophys., **369**, 826 (2001)
35. Boissonnat, J.-D., Yvinec, M., Bronniman, H.: Algorithmic Geometry. Cambridge University Press, Cambridge (1998)
36. Boissonnat, J.-D., Cohen-Steiner, D., Mourain, B., Rote, G., Vegter, G.: Meshing of surfaces. In: Boissonnat, J.-D., Teillaud, M. (eds.) Effective Computational Geometry for Curves and Surfaces. Springer-Verlag, Berlin (2006)
37. Boots, B.N.: Metallography **17**, 411 (1984)
38. Borgani, S., Guzzo, L.: Nature **409**, 39 (2001)
39. Bradač M., Schneider, P., Lombardi, M., Steinmetz, M., Koopmans, L.V.E., Navarro, J.F.: Astron. Astrophys. **423**, 797 (2004)
40. Branchini, E., Teodoro, L., Frenk, C.S., Schmoldt, I., Efstathiou, G., White, S.D.M., Saunders, W., Sutherland, W., Rowan-Robinson, M., Keeble, O., Tadros, H., Maddox, S., Oliver, S.: Mon. Not. R. Astron. Soc. **308**, 1 (1999)
41. Braun, J., Sambridge, M.: Nature **376**, 660 (1995)
42. Brown, G.S.: New Zealand Forestry Service Research Notes **38**, 1 (1965)
43. Buhmann, M.D.: Approximation and interpolation with radial functions. In: Dyn, N., Leviatan, D., Levin, D., Pinkus, A. (eds.) Multivariate Approximation and Applications, vol. 25, Cambridge University Press, Cambridge (2001)
44. Cappellari, M., Copin, Y.: Mon. Not. R. Astron. Soc. **342**, 345 (2003)
45. Cheng, S.-W., Dey, T.K., Ramos, E., Ray, T.: Sampling and meshing a surface with guaranteed topology and geometry. In: Proc. 20th Sympos. Comput. Geom. (2004)
46. Chiang, L.-Y., Coles, P.: Mon. Not. R. Astron. Soc. **311**, 809 (2000)
47. Chincarini, G., Rood, H.J.: Nature **257**, 294 (1975)
48. Chiu, S.N., van de Weygaert, R., Stoyan, D.: Journ. Appl. Prob. **28**, 356 (1996)
49. Christ, N.H., Friedberg, R., Lee, T.D.: Nucl. Phys. **B202**, 89 (1982)

50. Christ, N.H., Friedberg, R., Lee, T.D.: Nucl. Phys. **B210**, 310 (1982)
51. Christ, N.H., Friedberg, R., Lee, T.D.: Nucl. Phys. **B210**, 337 (1982)
52. Colberg, J.M., Krughoff, K.S., Connolly, A.J.: Mon. Not. R. Astron. Soc. **359**, 272 (2005)
53. Colberg, J.M., Sheth, R.K., Diaferio, A., Gao, L., Yoshida, N.: Mon. Not. R. Astron. Soc. **360**, 216 (2005)
54. Colberg, J.M.: Mon. Not. R. Astron. Soc. **375**, 337 (2007)
55. Colberg, J.M., Pearce, F., Brunino, R., Foster, C., Platen, E., Basilakos, S., Fairall, A., Feldman, H., Gottlöber, S., Hahn, O., Hoyle, F., Müller, V., Nelson, L., Neyrinck, M., Plionis, M., Porciani, C., Shandarin, S., Vogeley, M., van de Weygaert, R.: Mon. Not. R. Astron. Soc., **387**, 933 (2008)
56. Coles, P., Chiang, L.-Y.: Nature **406**, 376 (2000)
57. Colless, M., et al.: astro-ph/0306581 (2003)
58. Colombi, S., Pogosyan, D., Souradeep, T.: Phys. Rev. Let. **85**, 5515 (2000)
59. Cressie, N.: Statistics for Spatial Data, rev. ed., John Wiley & Sons, Chichester (1993)
60. de Berg, M., van Kreveld, M., Overmars, M., Schwarzkopf, O.: Computational Geometry: Algorithms and Applications, 2nd ed., Springer-Verlag, Heidelberg (2000)
61. Delaunay, B. N.: Bull. Acad. Sci. USSR: Clase Sci. Mat. **7**, 793 (1934)
62. Dey, T., Edelsbrunner, H., Guha, S.: Computational topology. In: Chazelle, B., Goodman, J.E., Pollack, R. (eds.) Advances in Discrete and Computational Geometry. Contemporary Mathematics, AMS, Providence (1998)
63. Dey, T.K.: Curve and surface reconstruction. In: Goodman, J.E., O'Rourke, J. (eds.) Handbook of Discrete and Computational Geometry. 2nd ed., CRC Press LLC, Boca Raton (2004)
64. Dinh, H.Q., Turk, G., Slabaugh, G.: IEEE Trans. Pattern Anal. Mach. Intell. **24**, 1358 (2002)
65. Dirichlet, G.L.: Journal für die Reine und Angwandte Mathematik **40**, 209 (1850)
66. Dubinski, J., da Costa, L.N., Goldwirth, D.S., Lecar, M., Piran, T.: Astrophys. J. **410**, 458 (1993)
67. Ebeling, H., Wiedenmann, G.: Phys. Rev. **E 47**, 704 (1993)
68. Edelsbrunner, H., Kirkpatrick, D., Seidel, R.: IEEE Trans. Inform. Theory IT-**29**, 551, 193
69. Edelsbrunner, H., Mücke, E.P.: ACM Trans. Graph. **13**, 43
70. Edelsbrunner, H., Shah, N.R.: Algorithmica **15**, 223 (1996)
71. Edelsbrunner, H., Facello, M., Liang, J.: Discrete Appl. Math. **88**, 83 (1998)
72. Edelsbrunner, H., Letscher, D., Zomorodian, A.: Discrete Comput. Geom. **28**, 511
73. Einasto, J., Joeveer, M., Saar, E., Nature **283**, 47 (1980)
74. El-Ad, H., Piran, T., da Costa, L.N.: Astrophys. J. **462**, L13 (1996)
75. El-Ad, H., Piran, T.: Astrophys. J. **491**, 421 (1997)
76. Eldering, B.: Topology of galaxy models, M.Sc. thesis, Groningen University (2006)
77. Español, P.: Phys. Rev. E. **57**, 2930 (1998)
78. Euler, L.: Letter dated 1759 October 27 from Euler to Lagrange, Opera Postuma **2**, 559 (1862)

79. De Fabritiis, G., Coveney, P.V.: arXiv:cond-mat/0301378 (2003)
80. Fairall, A.P.: Voids in the 6dF Galaxy Survey. In: van de Weygaert, R., Vogeley, M., Sheth, R., Peebles, P.J.E., Hoyle, F. (eds.) Cosmic Voids: Much Ado About Nothing, Edita (2008)
81. Fischer, R.A., Miles, R.E.: Math. Biosci. **18**, 335 (1973)
82. Fisher, K. B., Huchra, J. P., Strauss, M. A., Davis, M., Yahil, A., Schlegel, D.: Astrophys. J. Suppl. **100**, 69 (1995)
83. Flekkøy, E.G., Coveney, P.V.: De Fabritiis, G.: Phys. Rev. E. **62**, 2140 (2000)
84. Florack, L.M.J., ter Haar Romeny, B.M., Koenderink, J.J., Viergever, M.A.: Image Vision Comput. **10**, 376 (1992)
85. Fortune, S.: Voronoi diagrams and delaunay trinagulations. In: Du, D.-Z., Hwang, F. (eds.) Computing in Euclidian Geometry. Lecture Notes Series on Computing vol. 1, World Scientific, Singapore (1992)
86. Frangi, A.F., Niessen, W.J., Vincken, K.L., Viergever, M.A.: Multiscale vessel enhancement filtering. In: Medical Image Computing and Computer-Assisted Interventation – MICCAI'98. Lecture Notes in Computer Science, vol. 1496, Springer-Verlag, Berlin, Heidelberg (1998)
87. Friedman, J.H., Bentley, J.L., Finkel, R.A.: ACM Trans. Math. Softw. **3**, 209 (1977)
88. Fukugita, M., Peebles, P.J.E.: Astrophys. J. **616**, 643 (2004)
89. Gaztañaga, E.: Astrophys. J. **398**, L17 (1992)
90. Geller, M., Huchra, J. P.: Nature **246**, 897 (1989)
91. Gerald, C. F., Wheatley, P.O.: Applied Numerical Analysis. Addison Wesley, Reading (1999)
92. Gilbert, E.N.: Ann. Math. Stat. **33**, 958 (1962)
93. Gingold, R.A., Monaghan, J.J.: Mon. Not. R. Astron. Soc. **181**, 375 (1977)
94. Gnedin, N.Y.: Astrophys. J. Suppl. **97**, 231 (1995)
95. Gold, C.M., Remmele, P.R., Roos, T.: Voronoi methods in GIS. In: Lecture Notes in Computer Science, vol. 1340, p. 21. Springer-Verlag, Berlin (1997)
96. Goodman, J. E., O'Rourke, J. (eds.): Handbook of Discrete and Computational Geometry, 2nd ed., Chapman & Hall/CRC Press, Boca Raton (1980)
97. Gor, V., Shapiro, B.E., Jönsson, H., Heisler, M., Reddy, G.V., Meyerowitz, sE.M., Mjolsness, E.: A Software Architecture for Developmental Modelling in Plants: The Computable Plant Project. In: Bioinformatics of Genome Regulation and Structure II, Springer, U.S. (2006)
98. Gott, J.R. III, Dickinson, M., Melott, A.L.: Astrophys. J. **306**, 341 (1986)
99. Gregory, S.A., Thompson, L.A.: Astrophys. J. **222**, 784 (1978)
100. Hardy, R.L.: J. Geophys. Res. **76** (1905)
101. Hayashino, T. et al.: Astron. J. **128**, 2073 (2004)
102. Heath Jones, D. et al.: Mon. Not. R. Astron. Soc. **355**, 747 (2004)
103. Hoffman, Y., Ribak, E.: Astrophys. J. **380**, L5 (1991)
104. Hoyle, F., Vogeley, M.: Astrophys. J. **566**, 641 (2002)
105. Hoyle, F., Vogeley, M.: Astrophys. J. **580**, 663 (2002)
106. Icke, V., van de Weygaert, R.: Astron. Astrophys. **184**, 16 (1987)
107. Illiev, I.T., Ciardi, B., Alvarez, M.A., Maselli, A., Ferrara, A., Gnedin, N.Y., Mellema, G., Nakamoto, M.L., Norman, M.L., Razoumov, A.O., Rijkhorst, E.-J., Ritzerveld, J., Shapiro, P.R., Susa, H., Umemura, M., Whalen, D.J.: Mon. Not. R. Astron. Soc. **371**, 1057 (2006)

108. Iske, A.: Scattered data modelling using radial basis functions. In: Iske, A., Quak, E., Floater, M.S. (eds.) Tutorials on Multiresolution in Geometric Modelling, Mathematics and Visualization, p. 287, Springer-Verlag, Heidelberg (2002)
109. Jarrett, T.: Publ. Astr. Soc. Australia, **21**, 396 (2004)
110. Jones, B.J.T., Martínez, V.J., Saar, E., Trimble, V.: Rev. Mod. Phys. **76**, 1211 (2005)
111. Juszkiewicz, R., Weinberg, D.H., Amsterdamski, P., Chodorowski, M., Bouchet, F.: Astrophys. J. **442**, 39 (1995)
112. Kaiser, N.: Statistics of gravitational lensing 2: Weak lenses. In: Martinez, V.J., Portilla, M., Saez, D. (eds.) *New Insights into the Universe*. Lect. Notes Phys. **408**, 279. Springer-Verlag, Berlin, Heidelberg, New York (1992)
113. Kaiser, N., Squires, G.: Astrophys. J. **404**, 441 (1993)
114. Kansal, A.R., Torquato, S., Harsh, G.R., Chiocca, E.A., Deisboeck, T.S.: J. Theor. Biol. **203**, 367
115. Katz, N., Weinberg, D.H., Hernquist, L.: Astrophys. J. Suppl. **105**, 19 (1996)
116. Kauffmann, G., Fairall, A.P.: Mon. Not. R. Astron. Soc. **248**, 313 (1991)
117. Kauffmann, G., Colberg, J.M., Diaferio, A., White, S.D.M.: Mon. Not. R. Astron. Soc. **303**, 188 (1999)
118. Kendall, D.G.: Stat. Sci. **4**, 87 (1989)
119. Kiang, T.: Z. Astrophys. **64**, 433 (1966)
120. Kim, R. S. J., Kepner, J.V., Postman, M., Strauss, M.A., Bahcall, N.A., Gunn, J.E., Lupton, R.H., Annis, J., Nichol, R.C., Castander, F.J., Brinkmann, J., Brunner, R.J., Connolly, A., Csabai, I., Hindsley, R.B., Izević Ž., Vogeley, M.S., York, D.G.: Astron. J. **123**, 20 (2002)
121. Kimia, B.B., Leymarie, F.F., Symmetry-based representations of 3D data. In: Proc. 2001 Intern. Conf. on Image Processing 2, 581 (2001)
122. Kirshner, R.P., Oemler, A., Schechter, P.L., Shectman, S.A.: Astrophys. J. **248**, L57 (1981)
123. Kirshner, R.P., Oemler, A., Schechter, P.L., Shectman, S.A.: Astrophys. J. **314**, 493 (1987)
124. Koumoutsakos, P.: Ann. Rev. Fluid Mech. **37**, 457 (2005)
125. Lachieze-Rey, M., da Costa, L.N., Maurogordata, S.: Astrophys. J. **399**, 10 (1992)
126. Lagrange, J.L.: Oeuvres de Lagrange **1**, 151 (1762)
127. Lancaster, P., Salkauskas, K.: Math. Comput. **37**, 141 (1981)
128. de Lapparent, V., Geller, M.J., Huchra, J.P.: Astrophys. J. **302**, L1 (1986)
129. de Lapparent, V., Geller, M.J., Huchra, J.P.: Astrophys. J. **369**, 273 (1991)
130. Lawson, C.L.: Software for C1 surface interpolation. In: Rice, J.R. (eds.) Mathematical Software III, vol. 3, Academic Press, New York (1977)
131. Leymarie, F.F., Kimia, B.B.: Computation of the shock scaffold for unorganized point clouds in 3D. In: Proc. 2003 Comp. Soc. Conf. on Computer Vision and Pattern Recognition 1, I-821 (2003)
132. Li, G.-L., Mao, S., Jing, Y.P., Kang, X., Bartelmann, M.: Astrophys. J. **652**, 43 (2006)
133. Liang, J., Edelsbrunner, H., Fu, P., Sudhakar, P.V., Subramaniam, S.: Proteins: Struct. Funct. Genet. **33**, 1 (1998)
134. Liang, J., Edelsbrunner, H., Fu, P., Sudhakar, P.V., Subramaniam, S.: Proteins: Struct. Funct. Genet. **33**, 18 (1998)

135. Liang, J., Edelsbrunner, H., Woodward, C.: Protein Sci. **7**, 1884 (1998)
136. Lodha, S.K., Franke, R.: Scattered Data Techniques for Surfaces. In: Proc. Dagstuhl Conf. Scientific Visualization, IEEE Computer Society Press, 182 (1999)
137. Lombardi, M., Schneider, P.: Astron. Astrophys. **373**, 359 (2001)
138. Lombardi, M., Schneider, P.: Astron. Astrophys. **392**, 1153 (2002)
139. Lombardi, M., Schneider, P.: Astron. Astrophys. **407**, 385 (2003)
140. Lopes, P.A.A., de Carvalho, R.R., Gal, R.R., Djorgovski, S.G., Odewahn, S.C., Mahabal, A.A., Brunner, R.J.: Astron. J. **128**, 1017 (2004)
141. Lucy, L.B.: Astron. J. **82**, 1013 (1977)
142. Luo, R., Vishniac, E.: Astrophys. J. Suppl. **96**, 429 (1995)
143. Maddox, S.J., Efstathiou, G., Sutherland, W.J., Loveday, J.: Mon. Not. R. Astron. Soc. **242**, 43 (1990)
144. Maddox, S.J., Efstathiou, G., Sutherland, W.J., Loveday, J.: Mon. Not. R. Astron. Soc. **243**, 692 (1990)
145. Maddox, S.J., Efstathiou, G., Sutherland, W.J.: Mon. Not. R. Astron. Soc. **246**, 433 (1990)
146. Marinoni, C., Davis, M., Newman, J.A., Coil, A.L.: Astrophys. J. **580**, 122 (2002)
147. Martínez, V., Saar, E.: Statistics of the Galaxy Distribution. Chapman & Hall/CRC Press, Boca Raton (2002)
148. Martínez, V., Starck, J.-L., Saar, E., Donoho, D.L., Reynolds, S.C., de la Cruz, P., Paredes, S.: Astrophys. J. **634**, 744 (2005)
149. Mavripilis, D.J.: Ann. Rev. Fluid Mech. **29**, 473 (1997)
150. Mecke, K.R., Buchert, T., Wagner, H.: Astron. Astrophys. **288**, 697 (1994)
151. Meijering, J.L.: Philips Res. Rep. **8**, 270 (1953)
152. Mellier, Y.: Ann. Rev. Astron. Astrophys. **37**, 127 (1999)
153. Metropolis, N., Ulam, S.: J. Am. Stat. Assoc. **44**, 247 (1949)
154. Meyer, F., Beucher, S.: J. Visual Comm. Image Rep. **1**, 21 (1990)
155. Miles, R.E.: Math. Biosci. **6**, 85 (1970)
156. Miles, R.E.: Adv. Appl. Prob. (Suppl.) **4**, 243 (1972)
157. Miles, R.E.: A synopsis of 'Poisson Flats in Euclidian Spaces'. In: Harding, E.F., Kendall, D.G. (eds.) Stochastic Geometry, p. 202. John Wiley, New York (1974)
158. Miles, R.E., Maillardet, R.J.: J. Appl. Prob. **19A**, 97 (1982)
159. Miller, W.: Class. Quant. Grav. **14**, L199 (1997)
160. Mitra, S.K., Sicuranza, G.L.: Nonlinear Image Processing. Academic Press Inc., U.S. (2000)
161. Molchanov, S.A., Surfailis, D., Woyczynski, W.A.: Ann. Appl. Prob. **7**, 200 (1997)
162. Møller, J.: Adv. Appl. Prob. **21**, 37 (1989)
163. Møller, J.: Lectures on Random Voronoi Tessellations. In: Lecture Notes in Statistics, p. 87. Springer-Verlag, New York (1994)
164. Monaghan, J.J.: Ann. Rev. Astron. Astrophys. **30**, 543 (1992)
165. Mücke, E.P.: Shapes and Implementations in three-dimensional geometry, Ph.D. thesis, University of Illinois Urbana-Champaign (1993)
166. Neyrinck, M.C., Gnedin, N.Y., Hamilton, A.J.S.: Astrophys. J. **356**, 1222 (2005)
167. Novikov, D., Colombi, S., Doré, O.: Mon. Not. R. Astron. Soc. **366**, 1201 (2006)

168. Nusser, A., Dekel, A., Bertschinger, E., Blumenthal, G. R.: Astrophys. J. **379**, 6 (1991)
169. Okabe, A., Boots, B., Sugihara, K., Chiu, S. N.: Spatial tessellations: concepts and applications of Voronoi diagrams, 2nd ed., John Wiley & Sons, Chichester, Toronto (2000)
170. Ord, J.K.: Math. Scientist **3**, 23 (1978)
171. Patiri, S.G., Betancort-Rijo, J.E., Prada, F., Klypin, A., Gottlöber, S.: Mon. Not. R. Astron. Soc. **369**, 335 (2006)
172. Patiri, S.G., Prada, F., Holtzman, J., Klypin, A., Betancort-Rijo, J.E.: Mon. Not. R. Astron. Soc. **372**, 1710 (2006)
173. Peacock, J.A., Dodds, S.J.: Mon. Not. R. Astron. Soc. **267**, 1020 (1994)
174. Peebles, P.J.E.: The Large-Scale Structure of the Universe. Princeton University Press, Princeton, NJ (1980)
175. Pelupessy, F.I., Schaap, W.E., van de Weygaert, R.: Astron. Astrophys. **403**, 389 (2003)
176. Pen, Ue.-Li.: Astrophys. J. Suppl. **115**, 19 (1998)
177. Pimbblet, K.A.: Mon. Not. R. Astron. Soc. **358**, 256 (2005)
178. Platen, E.: Segmenting the Universe, M.Sc. thesis, Groningen University (2005)
179. Platen, E., van de Weygaert, R., Jones, B.J.T.: Mon. Not. R. astr. Soc., **387**, 128 (2008)
180. Plionis, M., Basilakos, S.: Mon. Not. R. Astron. Soc. **330**, 399 (2002)
181. Powell, M.J.D.: The theory of radial basis function approximation in 1990. In: Light, W.A. (ed.) Advances in Numerical Analysis II: Wavelets, Subdivision, and Radial Functions, p. 105. Oxford University Press, Oxford (1992)
182. Press, W.H., Teukolsky, S.A., Vetterling, W.T., Flannery, B.P.: Numerical Recipes. Cambridge University Press, Cambridge, U.K. (1992)
183. Ramella, M., Boschin, W., Fadda, D., Nonino, M.: Astron. Astrophys. **368**, 776 (2001)
184. Refregier, Y.: Ann. Rev. Astron. Astrophys. **41**, 645 (2003)
185. Regge, T.: Nuovo Cimentto **A 19**, 558
186. Regös, E., Geller, M.J.: Astrophys. J. **373**, 14 (1991)
187. Reiprich, T.H., Böhringer, H.: Astron. Nachr. **320**, 296 (1999)
188. Ripley, B.D.: Spatial Statistics, Wiley & Sons, Chichester (1981)
189. Ritzerveld, J.: The simplicity of transport, Ph.D. thesis, Leiden University (2007)
190. Ritzerveld, J., Icke, V.: Phys. Rev. E. **74**, 026704 (2006)
191. Ritzerveld, J., Pawlik, A., Schaye, J.: in prep. (2007)
192. Rojas, R.R., Vogeley, M.S., Hoyle, F., Brinkmann, J.: Astrophys. J. **624**, 571 (2005)
193. Romano-Díaz, E.: Probing Cosmic Velocity Flows in the Local Universe, Ph.D. thesis, Groningen University (2007)
194. Romano-Díaz (2004)
195. Rybicki, G.B., Press, W.H.: Astrophys. J. **398**, 169 (1992)
196. Ryden, B.S., Gramann, M.: Astrophys. J. **383**, 33 (1991)
197. Sambridge, M., Braun, J., McQueen, H.: Geophys. J. Int. **122**, 837 (1995)
198. Sahni, V., Sathyaprakash, B.S., Shandarin, S.F.: Astrophys. J. **495**, 5 (1998)
199. Sato, Y., Nakajima, D., Atsumi, H., Koller, T., Gerig, G., Yoshida, S., Kikinis, R.: IEEE Medical Image Analysis, vol. 2, p. 143 (1998)

200. Saunders, W., Sutherland, W.J., Maddox, S.J., Keeble, O., Oliver, S.J., Rowan-Robinson, M., McMahon, R.G., Efstathiou, G.P., Tadros, H., White, S.D.M., Frenk, C.S., Carramiñana, A., Hawkins, M.R.S.: Mon. Not. R. Astron. Soc. **317**, 55 (2000)
201. Schaap, W.E.: The Delaunay Tessellation Field Estimator, Ph.D. thesis, Groningen University (2007)
202. Schaap, W.E., van de Weygaert, R.: Astron. Astrophys. **363**, L29 (2000)
203. Schaap, W.E., van de Weygaert, R.: Astron. Astrophys., subm. (2007)
204. Schaback, R., Wendland, H.: Characterization and construction of radial basis functions. In: Dyn, N., Leviatan, D., Levin, D., Pinkus, A. (eds.) Multivariate Approximation and Applications, p. 1. Cambridge University Press, Cambridge (2001)
205. Schaller, G., Meyer-Hermann, M.: Comput. Phys. Commun. **162**, 9 (2004)
206. Schaller, G., Meyer-Hermann, M.: Phys. Rev. E. **71**, 051910.1 (2005)
207. Schmalzing, J., Buchert, T., Melott, A.L., Sahni, V., Sathyaprakash, B.S., Shandarin, S.F.: Astrophys. J. **526**, 568
208. Schmidt, J.D., Ryden, B.S., Melott, A.L.: Astrophys. J. **546**, 609
209. Schoenberg, I.J.: Quart. Appl. Math. **4**, 45 (1946)
210. Schoenberg, I.J.: Quart. Appl. Math. **4**, 112 (1946)
211. Serrano, M., Español, P.: Phys. Rev. E. **64**, 046115 (2001)
212. Serrano, M., De Fabritiis, G., Español, P., Flekkøy, E.G., Coveney, P.V.: J. Phys. A. Math. Gen. **35**, 1605 (2002)
213. Shandarin, S.F., Sheth, J.V., Sahni, V.: Mon. Not. R. Astron. Soc. **353**, 162 (2004)
214. Shandarin, S., Feldman, H.A., Heitmann, K., Habib, S.: Mon. Not. R. Astron. Soc. **376**, 1629 (2006)
215. Sharma, S., Steinmetz, M.: Mon. Not. R. Astron. Soc. **373**, 1293 (2006)
216. Shepard, D.: A two-dimensional interpolation function for irregularly spaced points. In: ACM National Conference, 517 (1968)
217. Sheth, J.V., Sahni, V., Sathyaprakash, B.S., Shandarin, S.F.: Mon. Not. R. Astron. Soc. **343**, 22 (2003)
218. Sheth, R. K., van de Weygaert, R.: Mon. Not. R. Astron. Soc. **350**, 517 (2004)
219. Shewchuk, J.R.: Comp. Geom. Theor. Appl. **22**, 21 (2002)
220. Sibson, R.: Math. Proc. Cambridge Phil. Soc. **87**, 151 (1980)
221. Sibson, R.: A brief description of natural neighbor interpolation. In: Barnett, V. (ed.) Interpreting Multi-Variate Data, p. 21. Wiley, Chichester (1981)
222. Sloan, S.W.: Adv. Eng. Software **9**, 34 (1987)
223. Smoot, G.F., et al.: Astrophys. J. **396**, L1 (1992)
224. Soneira, R.M., Peebles, P.J.E.: Astrophys. J. **211**, 1 (1977)
225. Sousbie, T., Pichon, C., Courtois, H., Colombi, S., Novikov, D.: astroph/0602628 (2006)
226. Spergel, D.N., et al.: Astrophys. J. Suppl. **170**, 377 (2007)
227. Springel, V., Yoshida, N., White, S.D.M.: New Astron. **6**, 79 (2001)
228. Springel, V.: Mon. Not. R. Astron. Soc. **364**, 1105 (2005)
229. Springel, V., White, S.D.M., Jenkins, A., Frenk, C.S., Yoshida, N., Gao, L., Navarro, J., Thacker, R., Croton, D., Helly, J., Peacock, J.A., Cole, S., Thomas, P., Couchman, H., Evrard, A., Colberg, J.M., Pearce, F.: Nature **435**, 629 (2005)

230. Stoica, R.S., Martínez, V.J., Mateu, J., Saar, E.: Astron. Astrophys. **434**, 423 (2005)
231. Stoyan, D., Kendall, W.S., Mecke, J.: Stochastic Geometry and its Applications, 1st ed., John Wiley, New York (1987)
232. Stoyan, D., Kendall, W.S., Mecke, J.: Stochastic Geometry and its Applications, 2nd ed., John Wiley, New York (1995)
233. Sukumar, N.: The Natural Element Method in Solid Mechanics, Ph.D. thesis, Northwestern University (1998)
234. Szapudi, I.: Astrophys. J. **497**, 16 (1998)
235. Tegmark, M., et al.: Astrophys. J. **606**, 702 (2004)
236. Thiessen, A.H.: Mon. Wea. Rev. **39**, 1082 (1911)
237. Torquato, S.: Random Heterogeneous Materials: Microstructure and Macroscopic Properties. Springer, New York (2002)
238. Turk, G., O'Brien, J.F.: Shape transformation using variational implicit functions. In: Computer Graphics. Annual Conference Series (SIGGRAPH 1999), 335 (1999)
239. Turk, G., O'Brien, J.F.: Reconstructing Surfaces Using Anisotropic Basis Functions. In: Proc. Intern. Conference Computer Vision (ICCV), Vancouver, 606 (2001)
240. Ubertini, S., Succi, S.: Prog. Comp. Fluid Dyn. **5**, 85 (2005)
241. van de Weygaert, R.: The Multidimensional binary tree and its use in astronomy, (minor) M.Sc. Thesis, Leiden University (1987)
242. van de Weygaert, R.: Voids and the Geometry of Large Scale Structure, Ph.D. thesis, Leiden University (1991)
243. van de Weygaert, R.: Astron. Astrophys. **283**, 361 (1994)
244. van de Weygaert, R.: Froth across the universe, dynamics and stochastic geometry of the cosmic foam. In: Plionis, M., Cotsakis, S. (eds.) Modern Theoretical and Observational Cosmology, Proc. 2nd Hellenic Cosmology Meeting, ASSL 276, p. 119. Kluwer, Dordrecht (2002)
245. van de Weygaert, R.: Astron. Astrophys., subm. (2007)
246. van de Weygaert, R., Astron. Astrophys., subm. (2007)
247. van de Weygaert, R., van Kampen, E.: Mon. Not. R. Astron. Soc. **263**, 481 (1993)
248. van de Weygaert, R., Bertschinger, E.: Mon. Not. R. Astron. Soc. **281**, 84 (1996)
249. Romano-Díaz, E., van de Weygaert, R.: Mon. Not. R. Astr. Soc., **382**, 2 (2007)
250. Vegter, G.: Computational topology. In: Goodman, J.E., O'Rourke, J. (eds.) Handbook of Discrete and Computational Geometry, 2nd. ed., p. 719. CRC Press LLC, Boca Raton (2004)
251. Vegter, G., van de Weygaert, R., Platen, E., Kruithof, N., Eldering, B.: Mon. Not. R. Astron. Soc. in prep. (2007)
252. Voronoi, G.: J. Reine Angew. Math. **134**, 167 (1908)
253. Watson, D.F., Contouring: A Guide to the Analysis and Display of Spatial Data. Pergamon Press, Oxford (1992)
254. Wendland, H.: Scattered Data Approximation. Cambridge Mon. Appl. Comput. Math., Cambridge University Press, Cambridge (2005)
255. White, S.D.M.: Mon. Not. R. Astron. Soc. **186**, 145 (1979)
256. Whitehurst, R.: Mon. Not. R. Astron. Soc. **277**, 655 (1995)
257. Yahil, A., Strauss, M.A., Davis, M., Huchra, J.P.: Astrophys. J. **372**, 380 (1991)

258. Zaroubi, S., Hoffman, Y., Fisher, K.B., Lahav, O.: Astrophys. J. **449**, 446 (1995)
259. Zel'dovich Ya. B.: Astron. Astrophys. **5**, 84 (1970)
260. Zel'dovich Ya.B., Einasto, J., Shandarin, S.F.: Nature **300**, 407 (1982)
261. Zomorodian, A.J.: Topology for Computing, Cambr. Mon. Appl. Comp. Math. (2005)

Power Spectrum Estimation. I. Basics

A.J.S. Hamilton

JILA and Dept. Astrophysical & Planetary Sciences, Box 440, U. Colorado, Boulder, CO 80309, USA
Andrew.Hamilton@colorado.edu, http://casa.colorado.edu/~ajsh/

Abstract. This chapter and its companion form an extended version of notes provided to participants in the Valencia September 2004 summer school on Data Analysis in Cosmology. The lectures offer a pedagogical introduction to the problem of estimating the power spectrum from galaxy surveys. The intention is to focus on concepts rather than on technical detail, but enough mathematics is provided to point the student in the right direction.

This first lecture presents background material. It collects some essential definitions, discusses traditional methods for measuring power, notably the Feldman–Kaiser–Peacock [2] method, and introduces Bayesian analysis, Fisher matrices, and maximum likelihood. For pedagogy and brevity, several derivations are set as exercises for the reader. At the summer school, multiple choice questions, included herein, were used to convey some didactic ideas, and provoked a little lively debate.

1 Introduction

It was a flawlessly organized September summer school in the historic Mediterranean city of Valencia, whose narrow, marble-paved streets are so randomly variable that you got lost in them as easily as in one of the lectures on "Data Analysis in Cosmology" going on at the Palacio Pineda.

The lecture on power estimation was one of the first lectures at the summer school, and it seemed sensible to make available to the students in advance a reference set of notes containing essential definitions and background material that would prove useful throughout the summer school. The present chapter is a somewhat extended version of those notes. The background material in the notes was not presented at the lecture, but rather was left as homework for the student during the long hours of siesta. To facilitate self-study, several of the derivations are posed as exercises for the reader. Solutions are not included, but the derivations contain enough guidance that the persistent student should be able to solve them.

The power spectrum is the most important statistic that can be measured from large-scale structure (LSS). During the lecture, the reasons for this being

Hamilton, A.J.S.: *Power Spectrum Estimation. I. Basics*. Lect. Notes Phys. **665**, 415–431 (2009)
DOI 10.1007/978-3-540-44767-2_12

so were conveyed through the device of multiple choice questions, which are included in this chapter, along with answers at the end of the chapter.

This chapter is arranged as follows. Section 2 collects some essential definitions of correlation functions, power spectra, and shot noise. Section 3 discusses traditional methods for measuring power, notably the Feldman–Kaiser–Peacock [2] method. Section 4 introduces Bayesian analysis, Fisher matrices, and maximum likelihood.

A separate companion chapter focuses on the actual designated topic of the lecture. It covers the practical issues of measuring power spectra from observations, with an emphasis on using maximum likelihood techniques to measure power at large, linear scales.

2 Definitions

This section collects definitions of some of the jargon that you will encounter not only in this lecture but repeatedly throughout this summer school. It is a good idea to assimilate the jargon.[1]

2.1 Correlation Function

Let $n(\boldsymbol{r})$ denote the observed number **density** of particles (galaxies) at position $\boldsymbol{r}$ in a survey.

Let $\bar{n}(\boldsymbol{r})$ denote the **selection function**, the expected mean number of particles (galaxies) at position $\boldsymbol{r}$ given the selection criteria of the survey. Often but not always, the selection function is separable into a product of an **angular selection function** $\bar{n}(\hat{\boldsymbol{r}})$ and a **radial selection function** $\bar{n}(r)$. The determination or measurement of the angular and radial selection functions of a survey is a non-trivial enterprise which is an essential prerequisite for measuring correlation functions or power spectra.

The **overdensity** $\delta(\boldsymbol{r})$ of particles (galaxies) at position $\boldsymbol{r}$ is defined by

$$\delta(\boldsymbol{r}) \equiv \frac{n(\boldsymbol{r}) - \bar{n}(\boldsymbol{r})}{\bar{n}(\boldsymbol{r})} . \tag{1}$$

The **correlation function** $\xi(r_{ij})$ (or 2-point correlation function) is the covariance of overdensities at separation $r_{ij} \equiv |\boldsymbol{r}_i - \boldsymbol{r}_j|$

$$\xi(r_{ij}) \equiv \langle \delta(\boldsymbol{r}_i) \delta(\boldsymbol{r}_j) \rangle . \tag{2}$$

[1] I have added some optional footnotes, like this, on Hilbert space. A Hilbert space is an infinite-dimensional vector space equipped with an inner product. Hilbert space provides a compact, powerful, and unifying mathematical formalism, just as ordinary vectors do in finite-dimensional geometry. A density field is a vector in a Hilbert space; a covariance function is a matrix in Hilbert space; an n-point correlation function is a rank-n tensor in Hilbert space.

In large-scale structure (LSS), the correlation function is often, though not always, conventionally taken to refer to the covariance function in real space (as opposed to Fourier space or some other space). The assumption that the Universe is **statistically homogeneous** (= statistically translation invariant) means that the correlation function is a function only of the vector separation $\boldsymbol{r}_{ij} \equiv \boldsymbol{r}_i - \boldsymbol{r}_j$ of two points. The assumption that the Universe is **statistically isotropic** (= statistically rotation invariant) means that the correlation function is a function only of the magnitude of the separation $r_{ij} \equiv |\boldsymbol{r}_{ij}|$ of two points.

2.2 Power Spectrum

A **Fourier mode** $\delta(\boldsymbol{k})$ is the Fourier transform of the overdensity:[2]

$$\delta(\boldsymbol{k}) \equiv \int \delta(\boldsymbol{r}) \mathrm{e}^{\mathrm{i}\boldsymbol{k}.\boldsymbol{r}} \, \mathrm{d}^3 r \, , \quad \delta(\boldsymbol{r}) = \int \delta(\boldsymbol{k}) \mathrm{e}^{-\mathrm{i}\boldsymbol{k}.\boldsymbol{r}} \, \frac{\mathrm{d}^3 k}{(2\pi)^3} \, . \tag{5}$$

[2] You may not be familiar with the practice of using the same symbol δ for both real and Fourier space; but δ is the same vector in Hilbert space, with components $\delta_{\boldsymbol{r}}$ in real space or $\delta_{\boldsymbol{k}}$ in Fourier space. The essential property of Hilbert space is the existence of an inner product (= scalar product). In the present case, the inner product of two real-valued vectors $a_{\boldsymbol{r}}$ and $b_{\boldsymbol{r}}$ is defined to be

$$a \cdot b = a^{\boldsymbol{r}} b_{\boldsymbol{r}} = \int a(\boldsymbol{r}) b(\boldsymbol{r}) \, \mathrm{d}^3 r \, . \tag{3}$$

In the index notation $a^{\boldsymbol{r}} b_{\boldsymbol{r}}$, repeated indices imply implicit summation (which becomes integration over the infinite-dimensional space of positions), just as in relativity and quantum mechanics. In repeated pairs of indices, one index is always raised, while the other is always lowered (though it is also common, for notational simplicity, to keep all indices lowered, which causes no ambiguity as long as it is implicitly understood that in contracting over paired indices, one index is always raised and the other lowered). In real space, the raised components of a real-valued vector are numerically equal to the lowered components, $a^{\boldsymbol{r}} = a_{\boldsymbol{r}}$, but this is a special feature of real space, and is not true in Fourier space, spherical harmonic space, or other spaces.

Exercise 1. Show from (3) and the definition $a(\boldsymbol{k}) \equiv \int a(\boldsymbol{r}) \mathrm{e}^{\mathrm{i}\boldsymbol{k}.\boldsymbol{r}} \, \mathrm{d}^3 r$ of the Fourier transform that the inner product of vectors $a_{\boldsymbol{k}}$ and $b_{\boldsymbol{k}}$ in Fourier space is

$$a \cdot b = a^{\boldsymbol{k}} b_{\boldsymbol{k}} = \int a(\boldsymbol{k})^* b(\boldsymbol{k}) \, \frac{\mathrm{d}^3 k}{(2\pi)^3} \, , \tag{4}$$

which is called **Parseval's theorem**. Once you have set up the formalism, you can deduce by inspection that $a^{\boldsymbol{k}} b_{\boldsymbol{k}} = a^{\boldsymbol{r}} b_{\boldsymbol{r}}$, since the inner product is by construction a scalar, independent of the representation of the vectors. Notice that in Fourier space, the raised components of a vector are equal to the complex conjugate of its lowered components, $a^{\boldsymbol{k}} = (a_{\boldsymbol{k}})^* = a_{-\boldsymbol{k}}$. □

The allocation of factors of 2π here follows the standard convention in cosmology, which you would be wise to stick to even if you do not like it. Other disciplines have their own conventions.

The **power spectrum** $P(k)$ is the Fourier transform of the correlation function

$$P(k) \equiv \int \xi(r) e^{i\boldsymbol{k}.\boldsymbol{r}} \, d^3r \, , \quad \xi(r) = \int P(k) e^{-i\boldsymbol{k}.\boldsymbol{r}} \frac{d^3k}{(2\pi)^3} \, . \tag{6}$$

Exercise 2. From the definitions (2) of the correlation function and (5) of Fourier modes, and the relation (6) between the power spectrum and the correlation function, show that the covariance of Fourier modes is[3]

$$\langle \delta(\boldsymbol{k}_i) \delta(\boldsymbol{k}_j) \rangle = (2\pi)^3 \delta_D(\boldsymbol{k}_i + \boldsymbol{k}_j) P(k_i) \, , \tag{7}$$

where δ_D denotes the (here three-dimensional) Dirac delta-function. Show that the delta-function arises from the assumption of statistical translation invariance. Show that the fact that the power spectrum $P(k_i)$ is a function of the magnitude $k_i \equiv |\boldsymbol{k}_i|$ of its argument follows from the assumption of statistical rotation invariance. □

2.3 2-Point Function

The correlation function or power spectrum are both representations, expressed respectively in real space and Fourier space, of the **covariance function**, also known as the **2-point function**.

The 2-point function is the second member of an infinite sequence of n-**point functions**, which are proportional to the n**th-order irreducible moments**. The first irreducible moment is the **mean**. The key property of the irreducible moments is that they are **additive over sums of independent density fields**.

2.4 Shot Noise

Typically, a galaxy survey samples only some fraction of the galaxies present in any volume element of the survey. To proceed, one makes the assumption that the galaxies surveyed are selected randomly from some continuous underlying population.

Exercise 4. Convince yourself of the theorem that **The correlation function $\xi(r)$ of a discrete random sampling of a density field is equal to the correlation function of the original field.** □

[3] **Exercise 3.** Show that the quantity $(2\pi)^3\delta_D(\boldsymbol{k}_i+\boldsymbol{k}_j)$ in (7) is just the unit matrix $\mathbf{1}_{\boldsymbol{k}_i}^{\boldsymbol{k}_j}$ in Hilbert space. In other words, show that $\mathbf{1}_{\boldsymbol{k}_i}^{\boldsymbol{k}_j} a_{\boldsymbol{k}_j} = a_{\boldsymbol{k}_i}$ for any vector $a_{\boldsymbol{k}}$. □

Actually, there is a catch to the above theorem, which is that the correlation function of the randomly subsampled field is equal to that of the parent field at all separations **except at zero separation**. If it is allowed that a particle is considered to be a neighbour of itself, then the correlation function of the randomly subsampled field acquires an extra contribution, a delta-function at zero separation, which is the shot noise.

As a general definition, the **shot noise is the self-particle contribution to any statistic**. In the case of the correlation function or power spectrum, the shot noise is the self-pair contribution, that is, the contribution from pairs consisting of a particle (galaxy) and itself.

Exercise 5. Argue that the shot noise contribution to the correlation function at a point where the selection function is $\bar{n}(\boldsymbol{r})$ is

$$\langle \delta(\boldsymbol{r}_i)\delta(\boldsymbol{r}_j)\rangle_{\text{shot}} = \frac{\delta_D(\boldsymbol{r}_{ij})}{\bar{n}(\boldsymbol{r}_i)} \ . \tag{8}$$

□

You might think that this is trickery. Can't you just exclude self-pairs and disregard this shot noise nonsense? The answer is that if you go to another space, such as Fourier space, then the shot noise shows up in a way that is not so trivial to remove.

Exercise 6. Show from (8) that the shot noise contribution to the covariance of Fourier modes is

$$\langle \delta(\boldsymbol{k}_i)\delta(\boldsymbol{k}_j)\rangle_{\text{shot}} = (1/\bar{n})(\boldsymbol{k}_i + \boldsymbol{k}_j) \ , \tag{9}$$

where $(1/\bar{n})(\boldsymbol{k})$ is the Fourier transform of $1/\bar{n}(\boldsymbol{r})$

$$(1/\bar{n})(\boldsymbol{k}) \equiv \int [1/\bar{n}(\boldsymbol{r})]\, \mathrm{e}^{\mathrm{i}\boldsymbol{k}.\boldsymbol{r}}\, \mathrm{d}^3 r \ . \tag{10}$$

□

According to (9) and (10), the shot noise contribution to the variance of Fourier modes is $\langle \delta(\boldsymbol{k})\delta(\boldsymbol{k})^*\rangle_{\text{shot}} = (1/\bar{n})(0)$. For any finite survey, this shot noise contribution is infinite. This simply reflects the fact that Fourier modes are waves extending to infinity, and that it would require an infinite survey to measure the amplitude of a wave whose wavenumber is specified with infinite precision. Fourier modes of real finite surveys are subject to an uncertainty principle: the wavenumbers of their Fourier modes are not precise, but rather are smeared over some finite width $\Delta k \sim 1/R$, where R is a measure of the linear size of the survey. You will discover more about what happens in real surveys in the exercise in Sect. 3.1.

Question 1. **Why is the 2-point function (either the correlation function or the power spectrum) the statistic of choice in characterizing LSS?** All of the following are true, but which is the most important?

A. Because it has a simple physical meaning: the correlation function $\xi(r)$ is the average excess over random of the probability of finding a particle (galaxy) at given separation r from another particle (galaxy).
B. Because the 2-point function can be measured relatively easily from observations, essentially by counting pairs.
C. Because the correlation function or power spectrum is the (co)variance of density (the second irreducible moment), which is the lowest order irreducible moment after the mean (the first irreducible moment).
D. Because the central limit theorem implies that a density distribution is asymptotically Gaussian in the limit where the density results from the average of many independent random processes; and a Gaussian is completely characterized by its mean and variance (the first and second irreducible moments).
E. Because the 2-point function satisfies a dynamical equation, a low-order member of the BBGKY hierarchy of equations.

Answer at end of chapter. □

Question 2. **What is the advantage of the power spectrum $P(k)$ over the correlation function $\xi(r)$?** Which of the following is the most important?

A. During the linear growth of fluctuations, the evolution of the Fourier mode $\delta(\boldsymbol{k})$ at each wavevector $\boldsymbol{k}$ is independent of every other.
B. The covariance matrix of Fourier modes $\delta(\boldsymbol{k})$ is a diagonal matrix, (7), whereas the covariance matrix of real space modes $\delta(\boldsymbol{r})$ is *not* a diagonal matrix, (2).
C. The power spectrum is the covariance of Fourier modes; Fourier modes are eigenmodes of the translation operator; the density distribution is statistically translation invariant; hence the cosmic covariance matrix must commute with the translation operator.
D. Estimates of the power $P(k)$ at different wavenumbers k are uncorrelated, for Gaussian fluctuations, whereas estimates of the correlation function $\xi(r)$ at different separations r are correlated.
E. The power spectrum is easier to measure than the correlation function.

Answer at end of chapter. □

3 Traditional Methods for Measuring Power

Yu & Peebles [12] and Peebles [9] were the first to characterize LSS with the power spectrum. Their methodology was complicated by the fact that they had only positions on the sky, not full three-dimensional positions.

Baumgart & Fry [1] first pointed out that you could measure the galaxy power spectrum $P(k)$ in a redshift survey by the simple method of enclosing the survey in a box and Fourier transforming **without having to bother about the detailed boundaries of the survey**. This astonishing result appeared to be in stark contrast to measurements of the correlation function $\xi(r)$, where it was essential to worry about the survey boundaries.

In an influential paper, Feldman, Kaiser & Peacock [2] proposed a variant of the Baumgart & Fry [1] method, in which each galaxy i is first weighted by

$$w_i = \frac{1}{1 + \bar{n}(\boldsymbol{r}_i)P(k)} , \tag{11}$$

where $\bar{n}(\boldsymbol{r}_i)$ is the selection function at the position $\boldsymbol{r}_i$ of galaxy i, before Fourier transforming. FKP showed that this procedure provided an optimal estimate of power $P(k)$ in the case that

(1) The wavelength $2\pi/k$ is small compared to the scale of the survey, and
(2) fluctuations are Gaussian.

Physically, the FKP weighting (11) is an approximation to the inverse variance weighting. It weights volumes by $1/\left[\bar{n}^{-1}(\boldsymbol{r}) + P(k)\right]$, which one recognizes as the reciprocal of the sum of the shot noise $\bar{n}^{-1}(\boldsymbol{r})$ and the cosmic power $P(\boldsymbol{k})$. This approximation is valid only in the "classical" limit where position $\boldsymbol{r}$ and wavenumber $\boldsymbol{k}$ are simultaneously measurable, which is condition (1) above. The condition (2) of Gaussianity comes from the fact that the thing being measured is power, a 2-point statistic, and the uncertainty in power involves, in addition to a product of 2-point terms, a 4-point term which vanishes only for Gaussian fluctuations.

Notice that the FKP weighting (11) depends on the power spectrum $P(k)$, which is the same thing that the FKP method aims to measure. In Bayesian analysis the $P(k)$ in the FKP weight would be recognized as being part of the model (part of the prior). However, the FKP approach is not Bayesian, but rather follows traditional statistical methods.

The FKP method is excellent for the intuition, and for quick, approximate estimates. However, it is not adequate for precision cosmology and the estimation of cosmological parameters. The FKP method is inadequate both at large scales where assumption (1) fails, and at small scales, where assumption (2) fails. The problem is not so much that the FKP weighting is suboptimal (though that is true), but rather that the FKP method does not yield a precise estimate of the variance and covariance of measured power. Moreover, it is not powerful enough to deal with all of the real world issues of actual galaxy surveys.

3.1 The Baumgart & Fry [1] Miracle

Exercise 7. Let $\tilde{\delta}(\boldsymbol{r})$ denote a weighted overdensity of galaxies at position $\boldsymbol{r}$ in a survey:

$$\tilde{\delta}(\boldsymbol{r}) \equiv w(\boldsymbol{r})\left[n(\boldsymbol{r}) - \bar{n}(\boldsymbol{r})\right] = w(\boldsymbol{r})\bar{n}(\boldsymbol{r})\delta(\boldsymbol{r}) = W(\boldsymbol{r})\delta(\boldsymbol{r}) \,. \tag{12}$$

Here $w(\boldsymbol{r})$ is some arbitrary weighting (such as the FKP weighting) that you choose [with the proviso that the weighting must be chosen a priori, independent of the observed galaxy density $n(\boldsymbol{r})$]; and $W(\boldsymbol{r}) \equiv w(\boldsymbol{r})\bar{n}(\boldsymbol{r})$ is the product of the weighting function and the selection function.

(a) **Fourier modes of the weighted overdensity**
The Fourier transform of the weighted overdensity $\tilde{\delta}(\boldsymbol{r})$ is, by definition,

$$\tilde{\delta}(\boldsymbol{k}) = \int \tilde{\delta}(\boldsymbol{r}) \mathrm{e}^{\mathrm{i}\boldsymbol{k}.\boldsymbol{r}} \, \mathrm{d}^3 r = \int W(\boldsymbol{r})\delta(\boldsymbol{r}) \mathrm{e}^{\mathrm{i}\boldsymbol{k}.\boldsymbol{r}} \, \mathrm{d}^3 r \,. \tag{13}$$

Show that $\tilde{\delta}(\boldsymbol{k})$ equals the convolution of the Fourier transform $W(\boldsymbol{k})$ of the window with the Fourier transform $\delta(\boldsymbol{k})$ of the overdensity:

$$\tilde{\delta}(\boldsymbol{k}) = \int W(\boldsymbol{k} - \boldsymbol{k}')\delta(\boldsymbol{k}') \frac{\mathrm{d}^3 k}{(2\pi)^3} \,. \tag{14}$$

This is just the standard result that multiplication in real space becomes convolution in Fourier space.

(b) **Covariance of Fourier modes of the weighted overdensity**
Assume that the covariance $\langle \delta(\boldsymbol{k}_i)\delta(\boldsymbol{k}_j) \rangle$ of (unweighted) overdensities in Fourier space is a sum of a cosmic term $(2\pi)^3\delta_D(\boldsymbol{k}_i{+}\boldsymbol{k}_j)P(k_i)$, (7), and a shot noise term $(1/\bar{n})(\boldsymbol{k}_i{+}\boldsymbol{k}_j)$, (9),

$$\langle \delta(\boldsymbol{k}_i)\delta(\boldsymbol{k}_j) \rangle = (2\pi)^3\delta_D(\boldsymbol{k}_i + \boldsymbol{k}_j)P(k_i) + (1/\bar{n})(\boldsymbol{k}_i + \boldsymbol{k}_j) \,. \tag{15}$$

Show that the covariance of Fourier modes of the weighted overdensity is

$$\langle \tilde{\delta}(\boldsymbol{k}_i)\tilde{\delta}(\boldsymbol{k}_j) \rangle = \int W(\boldsymbol{k}_i - \boldsymbol{k})W(\boldsymbol{k}_j + \boldsymbol{k})P(\boldsymbol{k}) \frac{\mathrm{d}^3 k}{(2\pi)^3} \; + \; N(\boldsymbol{k}_i + \boldsymbol{k}_j) \,, \tag{16}$$

where the shot noise $N(\boldsymbol{k})$ is[4]

[4] Actually it is more accurate to use the **actual** value of the shot noise, which is

$$N(\boldsymbol{k}) = \sum_{\text{galaxies } i} w(\boldsymbol{r}_i)^2 \, \mathrm{e}^{\mathrm{i}\boldsymbol{k}.\boldsymbol{r}_i} \,, \tag{17}$$

as opposed to (18) which merely gives the **expectation value** of the shot noise. Shot noise is, by definition, the contribution to the covariance from self-pairs (pairs consisting of a particle and itself). Equation (15), from which follows (18), is a statement about the average excess of neighbours of a particle. But in fact we know more about the shot noise than just its average: we know that each particle always has exactly one of itself as a neighbour, not merely on average. In statistics, an estimate that uses more prior information is always better than an estimate that uses less.

$$N(\boldsymbol{k}) = \int w(\boldsymbol{r})^2 \bar{n}(\boldsymbol{r}) e^{i\boldsymbol{k}.\boldsymbol{r}} \, d^3r \, . \tag{18}$$

Hence conclude that the variance of Fourier modes of the weighted overdensity is

$$\langle \tilde{\delta}(\boldsymbol{k}) \tilde{\delta}(\boldsymbol{k})^* \rangle = \int \left| W(\boldsymbol{k} - \boldsymbol{k}') \right|^2 P(\boldsymbol{k}') \frac{d^3 k'}{(2\pi)^3} + N(0) \, . \tag{19}$$

Equation (19) says that the variance of Fourier modes of weighted overdensity is, after subtraction of the shot noise $N(0)$, equal to the power spectrum $P(\boldsymbol{k})$ smoothed over a smoothing function given by the magnitude square of $|W(\boldsymbol{k})|^2$ of the Fourier transform $W(\boldsymbol{k})$ of the window. Let W^2 denote the integral over the window (a notation suggested by the fact that W^2 is the scalar product of the Hilbert space vector $W_{\boldsymbol{k}}$ with itself)

$$W^2 = \int |W(\boldsymbol{k})|^2 \frac{d^3 k}{(2\pi)^3} = \int W(\boldsymbol{r})^2 \, d^3r \, . \tag{20}$$

Then a smoothed estimate $\widetilde{P}(\boldsymbol{k})$ of power at wavevector $\boldsymbol{k}$ is

$$\widetilde{P}(\boldsymbol{k}) \equiv \frac{\int \left| W(\boldsymbol{k} - \boldsymbol{k}') \right|^2 P(\boldsymbol{k}') \, d^3k'/(2\pi)^3}{W^2} = \frac{\langle \tilde{\delta}(\boldsymbol{k}) \tilde{\delta}(\boldsymbol{k})^* \rangle - N(0)}{W^2} \, . \tag{21}$$

Equation (21) is essentially Baumgart & Fry's [1] remarkable result.

(c) **What does it mean?**

Suppose that the survey window $W(\boldsymbol{r})$ has a characteristic size R. Approximately what is the width of the smoothing window $|W(\boldsymbol{k})|^2$ in the smoothed power spectrum, (21)? At what wavenumber k would you cease to trust the smoothed power spectrum $\widetilde{P}(k)$ as a reasonable estimate of the true power $P(k)$? What happens as the size R of the survey gets larger? [The important thing here is the concept rather than the mathematics. But if you want to see how this plays out mathematically, you might consider a survey window $W(\boldsymbol{r})$ which happens to be a Gaussian $W(\boldsymbol{r}) = \exp\left[-r^2/(2R^2)\right]$, centred at the origin, with a $1-\sigma$ width of R.]

□

3.2 The Feldman–Kaiser–Peacock [2] Method

In the previous exercise you obtained an estimate, (21), of (smoothed) power $\widetilde{P}(\boldsymbol{k})$ at wavevector $\boldsymbol{k}$. The estimate involved an arbitrarily adjustable weighting $W(\boldsymbol{r})$, (12), of volume elements in the survey. It is natural to try to choose this weighting $W(\boldsymbol{r})$ to try to minimize the variance of the resulting estimate (21) of power. The FKP weighting, already given as (11), is an approximation to the desired minimum variance weighting, valid under the two conditions stated immediately after (11).

It proves surprisingly tricky to derive the FKP weighting in a rigorous way with a minimum of unnecessary assumptions. It would be nice to take you through the derivation in an exercise, but I could not devise an approach that was satisfactorily clean, insightful, and brief. You might like to consult the original Feldman, Kaiser, & Peacock [2] paper to see how they did it. A more general derivation can be found in Hamilton [5]. Perhaps the most elegant approach is to use the quadratic method of Tegmark [10], discussed in Sect. 12 of Paper 2.

A core part of the FKP argument is the following. If the survey has characteristic linear size $\sim R$, then the Fourier transform $W(\boldsymbol{k})$ of the survey window will be a ball of width $\Delta k \sim 1/R$ around the origin $\boldsymbol{k} = 0$. It follows that at wavenumbers much larger than the reciprocal size of the survey, $k \gg 1/R$, the smoothing window $\left|W(\boldsymbol{k} - \boldsymbol{k}')\right|^2$ in the Baumgart & Fry estimate (21) will be narrowly concentrated around the target wavenumber $\boldsymbol{k}$. To the extent that the power spectrum $P(\boldsymbol{k}')$ is slowly varying over the narrow window, it can be approximated by a constant,

$$P(\boldsymbol{k}') \approx P(k) = \text{constant} . \tag{22}$$

If the power spectrum is interpreted as literally constant, then the covariance matrix of overdensities is diagonal in real space

$$\langle \delta(\boldsymbol{r}_i)\delta(\boldsymbol{r}_j)\rangle = \delta_D\left(\boldsymbol{r}_i - \boldsymbol{r}_j\right)\left[P(k) + \frac{1}{\bar{n}(\boldsymbol{r}_i)}\right] . \tag{23}$$

This indicates that each volume element of the survey can be approximated as being statistically uncorrelated with all other volume elements. If you buy into the notion that minimum variance weighting is inverse variance weighting (Sect. 4.2 explains where that notion comes from), then (23) suggests that each volume element should be weighted by

$$W(\boldsymbol{r}) = \frac{1}{P(k) + 1/\bar{n}(\boldsymbol{r})} . \tag{24}$$

In the present case, the thing of interest is not single volume elements, but rather pairs of volume elements. For the specific case of Gaussian fluctuations, the covariance of pairs is a product of covariances of singles (e.g. Hamilton [5] Sect. 2.1)

$$\langle (\delta_i\delta_j - \langle\delta_i\delta_j\rangle)(\delta_k\delta_l - \langle\delta_k\delta_l\rangle)\rangle = \langle\delta_i\delta_k\rangle\langle\delta_j\delta_l\rangle + \langle\delta_i\delta_l\rangle\langle\delta_j\delta_k\rangle , \tag{25}$$

which is true in both real space and Fourier space (it is a covariant expression in Hilbert space). It follows that, for Gaussian fluctuations, the inverse variance weighting of pairs of volume elements is

$$W(\boldsymbol{r}_i)W(\boldsymbol{r}_j) = \frac{1}{\left[P(k) + 1/\bar{n}(\boldsymbol{r}_i)\right]\left[P(k) + 1/\bar{n}(\boldsymbol{r}_j)\right]} . \tag{26}$$

Equation (26) is the FKP pair weighting.

4 Bayes, Fisher, and Maximum Likelihood

Bayesian statistics provides the modern mathematical framework for rigorous statistics. As explained below, Sect. 4.3, it gives special status to maximum likelihood as yielding the best estimate of a parameter or set of parameters.

Fisher, Scharf & Lahav [4] were the first to apply a likelihood approach to large-scale structure. Heavens & Taylor [6] may be credited with accomplishing the first likelihood analysis designed to retain as much information as possible at linear scales. With the work of Heavens & Taylor, maximum likelihood methods appeared on the LSS scene essentially fully fledged.

4.1 Bayesian Statistics

Traditional statistics. Measure the mean by measuring the mean; measure the variance by measuring the variance; and such-like naïvety.

Bayesian statistics. Measure the mean (or variance) by asking what is the probability that the mean (or variance) takes such and such a value, given this set of observations and this set of assumptions (the prior). Commonly, the prior is subdivided into (a) assumptions that you assert are true and (b) a model equipped with parameters whose values you wish to estimate.

The foundation of Bayesian statistics is Bayes' theorem.

Bayes' Theorem: It states that the **posterior probability** $P(p|x, X)$ that the parameters p take on certain values, given the observational data x and prior assumptions X, is proportional to the **likelihood function**, the probability $P(x|p, X)$ of the observations x given parameters p and prior assumptions X, multiplied by the **prior probability** $P(p|X)$ of the parameters p given the prior assumptions X

$$P(p|x, X) = \frac{P(x|p, X)\, P(p|X)}{P(x|X)} \ . \tag{27}$$

□

The probability $P(x|X)$ of the observations given the prior assumptions is an overall normalization constant which plays no role except to ensure that the integral over the posterior probability is 1.

The likelihood function $P(x|p, X)$, which converts by multiplication a prior probability into a posterior probability, encapsulates all the information provided by a set of observations. In recognition of this fundamental role, the likelihood is given its own letter, $\mathcal{L}$. R. A. Fisher, who developed much of the formalism of likelihoods in the first half of the twentieth century, never subscribed to Bayesian statistics – after all, as just remarked, the likelihood encapsulates all the information provided by a set of observations. Nevertheless, if you want to convert a likelihood into a (posterior) probability that the parameters take on a certain range of values, then you have to assume a prior probability distribution of parameters.

"If the prior matters, then you are not learning much from the data" – from an Aspen Center for Physics workshop in summer 1997.

Question 3. **Rank each of the following prior assumptions in order of probability of being true:**

A. The Universe is statistically homogeneous and isotropic.
B. The growth of fluctuations is driven primarily by gravity.
C. Fluctuations at linear scales are Gaussian.
D. The Universe is spatially flat.
E. The ΛCDM model, with $\Omega_\Lambda \approx 0.7$, $\Omega_c \approx 0.26$, and $\Omega_b \approx 0.04$, is correct (Λ is the cosmological constant, c is cold dark matter, and b is baryons).
F. Galaxy bias $b(k)$, defined to be the square root of the ratio of galaxy–galaxy power $P_{gg}(k)$ to matter–matter power $P_{mm}(k)$,

$$b(k)^2 = \frac{P_{gg}(k)}{P_{mm}(k)} , \tag{28}$$

is constant at linear scales.

Answer at end of chapter. □

4.2 Fisher Information Matrix

The **Fisher information matrix** (Fisher [3]) plays a fundamental role in Bayesian statistical analysis. Many of us in the field of large-scale structure learnt about Fisher matrices from the superb paper by Tegmark, Taylor & Heavens [11], and I can offer little better advice than to go read that paper!

The term "optimal", applied to some statistical estimate of a quantity, has acquired a bad reputation thanks to misuse. The Fisher matrix puts what is meant by "optimal" on a sound mathematical basis. It is well worth getting your brain around the Fisher matrix, because it will raise your understanding of statistics to a new level.

The Fisher information matrix $F_{\alpha\beta}$ of a set of parameters p_α is formally defined to be minus the expectation value of the second derivative of the log-likelihood function with respect to the parameters:

$$F_{\alpha\beta} \equiv - \left\langle \frac{\partial^2 \ln \mathcal{L}}{\partial p_\alpha \partial p_\beta} \right\rangle . \tag{29}$$

Expectation value here means averaged over an ensemble of observational data x predicted by the likelihood function

$$\langle t \rangle \equiv \int t \, \mathcal{L}(x|p) \, \mathrm{d}x . \tag{30}$$

Since the likelihood $\mathcal{L}$ is multiplicative over statistically independent sets of observations, it follows that the Fisher matrix, (29), is additive over statistically independent observations, a sensible property for information to have.

The power of the Fisher matrix derives from the Cramér–Rao inequality (Kendall & Stuart [7] Sect. 17.15), which states that the variance $\langle \Delta\hat{p}_\alpha^2 \rangle$ of any unbiased (see (32) below) estimate $\hat{p}_\alpha$ of a parameter p_α must exceed the reciprocal of the diagonal element of the Fisher matrix:

$$\langle \Delta\hat{p}_\alpha^2 \rangle \geq \frac{1}{F_{\alpha\alpha}} \quad \text{(no summation over } \alpha) \,. \tag{31}$$

You will derive the Cramér–Rao inequality in Exercise 9 below.

To the extent that the likelihood function $\mathcal{L}$ is a Gaussian about its maximum (this is distinct from the proposition that the likelihood function is Gaussian in the data), often a good approximation thanks to the central limit theorem, the Fisher matrix is approximately equal to the inverse of the covariance matrix of the parameters. You are probably familiar from your earliest statistical training with the notion that the "best" way to weight a set of data is by their inverse variance (an idea already encountered in Sect. 3.2 on the FKP weighting). Inverse variance weighting effectively weights data in proportion to the amount of information in each part.

4.3 Maximum Likelihood

An **estimator** $\hat{p}$ of a parameter p (the hat on the estimator $\hat{p}$ distinguishes it from the true value p) is some function $\hat{p}(x)$ of the observational data x. An estimate $\hat{p}$ of a parameter p is **unbiased** if the expectation value, (30), of the estimate equals the true value

$$\langle \hat{p} \rangle = p \,. \tag{32}$$

A theorem of fundamental importance (Kendall & Stuart [7] Sect. 18.5) states that if an unbiased estimator attaining the Cramér–Rao bound exists, then it is the **maximum likelihood** estimator, the values $\hat{p}_\alpha$ of the parameters for which the likelihood attains its maximum value given the observed data:

$$\left. \frac{\partial \ln \mathcal{L}}{\partial p_\alpha} \right|_{p_\alpha = \hat{p}_\alpha} = 0 \,. \tag{33}$$

It is this theorem that gives the maximum likelihood method its special status.

Yet more theorems give conditions (see Kendall & Stuart [7], and Exercise 9(d) below) under which an unbiased estimator attaining the Cramér–Rao bound exists. For example, such an estimator exists if the likelihood function $\mathcal{L}$ is a Gaussian about its maximum. The central limit theorem ensures that $\mathcal{L}$ is asymptotically Gaussian in the limit of a large amount of data. Thus an unbiased estimator attaining the Cramér–Rao bound exists in the asymptotic limit of a large amount of data.

Exercise 8. The Schwarz inequality.

The basis of the Cramér–Rao inequality is the Schwarz inequality, which states, (35) below, that the correlation coefficient between two estimates must be less than or equal to one in absolute value. The Schwarz inequality is a powerful general result in statistics, and it is well worth knowing how to derive it:

(a) Consider two estimators $\hat{p}$ and $\hat{q}$, with (co)variances $\langle \Delta\hat{p}^2 \rangle$, $\langle \Delta\hat{p}\Delta\hat{q} \rangle$, and $\langle \Delta\hat{q}^2 \rangle$. It is evidently true that

$$\langle (\Delta\hat{p} + \lambda\Delta\hat{q})^2 \rangle \geq 0 \tag{34}$$

for any real number λ. For what value of λ, in terms of the (co)variances, is the left-hand side of (34) a minimum?

(b) Hence derive the **Schwarz inequality**

$$\left| \frac{\langle \Delta\hat{p}\Delta\hat{q} \rangle}{\langle \Delta\hat{p}^2 \rangle^{1/2} \langle \Delta\hat{q}^2 \rangle^{1/2}} \right| \leq 1 \, . \tag{35}$$

The quantity inside the vertical bars on the left-hand side of (35) is called the **correlation coefficient** of the estimators $\hat{p}$ and $\hat{q}$. The Schwarz inequality states that correlation coefficient must lie in the interval $[-1, 1]$. A correlation coefficient of 1 means that the estimators are perfectly correlated; a correlation coefficient of -1 means that the estimators are perfectly anti-correlated.

(c) What relation between $\hat{p}$ and $\hat{q}$ must be satisfied for the Schwarz inequality to become an equality? [Answer: $\Delta\hat{p}$ must be proportional to $\Delta\hat{q}$.]

□

Question 4. **Which of the following is true according to the Schwarz inequality?** In the following, the subscripts g and m denote galaxies and matter, so that, for example, $\xi_{\rm gg}(r)$ and $\xi_{\rm mm}(r)$ are the galaxy–galaxy and matter–matter correlation functions, while $\xi_{\rm gm}(r)$ is the galaxy–matter cross-correlation.

A. $\left| \xi_{\rm gm}(r) / \left[\xi_{\rm gg}(r)\xi_{\rm mm}(r)\right]^{1/2} \right| \leq 1$.

B. $\left| P_{\rm gm}(k) / \left[P_{\rm gg}(k)P_{\rm mm}(k)\right]^{1/2} \right| \leq 1$, where P is the shot-noise-subtracted power spectrum.

C. $\left| P_{\rm gm}(k) / \left[P_{\rm gg}(k)P_{\rm mm}(k)\right]^{1/2} \right| \leq 1$, where P is the power spectrum before shot noise is subtracted.

D. All of the above.

E. None of the above.

Answer at end of chapter.

□

Exercise 9. The Cramér–Rao inequality.

In this exercise you will derive the Cramér–Rao inequality, (31). The derivation follows Kendall & Stuart [7] Sect. 17.15, which you should consult for more detail and generality. For simplicity, the exercise considers just a single parameter p.

(a) The likelihood function $\mathcal{L}(x|p)$ is the probability of the data x given the parameter p, and so satisfies the normalization condition

$$\int \mathcal{L}(x|p)\,\mathrm{d}x = 1 \tag{36}$$

for any value of the parameter p. Differentiate this with respect to the parameter p to obtain $\int (\partial\mathcal{L}/\partial p)\,\mathrm{d}x = 0$, or equivalently

$$\left\langle \frac{\partial \ln \mathcal{L}}{\partial p} \right\rangle = 0\,. \tag{37}$$

Differentiate again with respect to p to obtain

$$-\left\langle \frac{\partial^2 \ln \mathcal{L}}{\partial p^2} \right\rangle = \left\langle \left(\frac{\partial \ln \mathcal{L}}{\partial p} \right)^2 \right\rangle\,. \tag{38}$$

You recognize the left-hand side of (38) as the Fisher information in the parameter p, (29), and you see from (38) that this information must be positive.

(b) Consider an unbiased estimator $\hat{p}$. Being unbiased, (32), the estimator must satisfy

$$\langle \hat{p} - p \rangle = \int (\hat{p} - p)\mathcal{L}\,\mathrm{d}x = 0\,. \tag{39}$$

Differentiate this with respect to p to show that

$$\left\langle (\hat{p} - p)\frac{\partial \ln \mathcal{L}}{\partial p} \right\rangle = 1\,. \tag{40}$$

(c) Apply the Schwarz inequality to deduce from (40), coupled with (38), the Cramér–Rao inequality

$$\langle (\hat{p} - p)^2 \rangle \geq \frac{1}{1 - \langle \partial^2 \ln \mathcal{L}/\partial p^2 \rangle}\,. \tag{41}$$

(d) In question 8(d) you obtained a condition for the Schwarz inequality to become an equality. What is the condition on the likelihood function $\mathcal{L}$ for the Cramér–Rao bound to be attained? [Before you rush to the conclusion that $\mathcal{L}$ must be Gaussian, consider (and prove) the fact that the Cramér–Rao bound is also attained by a Poisson distribution, for which the likelihood of observing x counts over an interval during which the expected number of counts is p is $(p)^x \mathrm{e}^{-p}/x!$.]

□

Acknowledgements

This work was supported in part by NASA ATP award NAG5-10763 and by NSF grant AST-0205981, and of course by this wonderful summer school.

Appendix: Answers to Multiple Choice Questions

Question 1. The correct answer is D. Whereas the other answers have to do with humanistics (A & B) or mathematics (C & E), D has to do with physics. If density fluctuations in the Universe were originally generated as a superposition of many independent random processes, then the resulting primordial density field will be Gaussian. This is a generic, albeit not universal, prediction of inflation, where density fluctuations are seeded by quantum fluctuations of the inflaton field. The prediction of Gaussianity remains consistent with observations (Komatsu et al. [8]).

Question 2. All of these are true except perhaps for E, the truth of which depends on methodology. For example, the Baumgart & Fry [1] method described in Sect. 3 is about as easy as could be, but it is far from the best method. Answers A–D are all important. However, answer C is the most insightful, because it gives the fundamental reason – statistical homogeneity – for the power spectrum's superiority over the correlation function. Answers A, B, and D can all be construed as consequences of the fundamental assumption of statistical homogeneity. It is for essentially the same reason that CMB folk use the spherical harmonic power spectrum C_l rather than the angular correlation function to characterize fluctuations in the CMB. The power spectrum C_l is the covariance of spherical harmonics; spherical harmonics are eigenmodes of the rotation operator; CMB fluctuations are statistically rotation invariant; hence the covariance matrix of CMB fluctuations must commute with the rotation operator. All else (the spherical harmonic analogue of answers A, B, & D) follow.

Question 3. This question generated some debate at the school. My own ordering was A–F in the same order as written, but many respectable people opined that B should come before A. I might even agree with them.

Question 4. The correct answer is C. Answer A is *not* true, and B is true only in the limit of vanishing shot noise. Only the non-shot-noise-subtracted power spectra can be expressed in the form of the Schwarz inequality (35), as

$$\left| \frac{\langle \delta_g(\boldsymbol{k}) \delta_m(-\boldsymbol{k}) \rangle}{\langle |\delta_g(\boldsymbol{k})|^2 \rangle^{1/2} \langle |\delta_m(\boldsymbol{k})|^2 \rangle^{1/2}} \right| \leq 1 \ . \tag{42}$$

If you are concerned about the mixture of $\boldsymbol{k}$ and $-\boldsymbol{k}$ in (42), then split $\delta(\boldsymbol{k})$ into its real and imaginary parts (which are uncorrelated, with equal variances), and consider an estimator which is a sum of the real and imaginary parts. If you are concerned with the appearance of the wavevector $\boldsymbol{k}$ rather than its absolute value k, then consider an estimator that is an arbitrarily weighted sum of modes $\delta(\boldsymbol{k})$ having the same wavenumber k.

References

1. Baumgart, D.J., Fry, J.N.: Fourier spectra of three-dimensional data. ApJ **375**, 25 (1991)
2. Feldman, H.A., Kaiser, N., Peacock, J.A.: Power spectrum analysis of three-dimensional redshift surveys. ApJ **426**, 23–37 (1994)
3. Fisher, R.A.: The logic of inductive inference. J. Roy. Stat. Soc. **98**, 39–54 (1935)
4. Fisher, K.B., Scharf, C.A., Lahav, O.: A spherical harmonic approach to redshift distortion and a measurement of Ω_0 from the 1.2 Jy *IRAS* redshift survey. MNRAS, **266**, 219 (1994)
5. Hamilton, A.J.S.: Towards optimal measurement of power spectra - I. Minimum variance pair weighting and the Fisher matrix. MNRAS **289**, 285 (1997)
6. Heavens, A.F., Taylor, A.N.: A spherical harmonic analysis of redshift space. MNRAS **275**, 483–497 (1995)
7. Kendall, M.G., Stuart, A.: The Advanced Theory of Statistics. Hafner Publishing, New York (1967)
8. Komatsu, E., et al. (15 authors; WMAP collaboration): First-Year Wilkinson Microwave Anisotropy Probe (WMAP) observations: tests of Gaussianity, ApJS **148**, 119 (2003)
9. Peebles, P.J.E.: Statistical analysis of catalogues of extragalactic objects. I. Theory, ApJ **185**, 413–440 (1973)
10. Tegmark, M.: How to measure CMB power spectra without losing information. Phys. Rev. D **55**, 5895 (1997)
11. Tegmark, M., Taylor, A., Heavens, A.: Karhunen-Loeve eigenvalue problems in cosmology: how should we tackle large data sets?. ApJ **480**, 22 (1997)
12. Yu, J.T., Peebles, P.J.E.: Superclusters of galaxies?. ApJ **158**, 103–113 (1969)

Power Spectrum Estimation II. Linear Maximum Likelihood

A.J.S. Hamilton

JILA and Dept. Astrophysical & Planetary Sciences, Box 440, U. Colorado, Boulder, CO 80309, USA
Andrew.Hamilton@colorado.edu, http://casa.colorado.edu/~ajsh/

Abstract. This second chapter of two companion chapters on the estimation of power spectra specializes in the topic of estimating galaxy power spectra at large, linear scales using maximum likelihood methods. As in the first chapter, the aims are pedagogical, and the emphasis is on concepts rather than technical detail. The chapter covers most of the salient issues, including selection functions, likelihood functions, Karhunen–Loève compression, pair-integral bias, Local Group flow, angular or radial systematics (arising for example from extinction), redshift distortions, quadratic compression, decorrelation, and disentanglement. The procedures are illustrated with results from the *IRAS* PSCz survey. Most of the PSCz graphics included in this chapter have not been published elsewhere.

1 Introduction

This chapter addresses the problem of estimating galaxy power spectra at linear scales using maximum likelihood methods. As discussed in Paper 1, the power spectrum is the most important statistic that can be measured from large-scale structure (LSS), and, again as elaborated in Paper 1, maximum likelihood has a special status in statistics: if a best method exists, then it is the maximum likelihood method (Tegmark et al. [43]).

Fisher et al. [14] were the first to apply a likelihood approach to large-scale structure. Heavens & Taylor [22] may be credited with accomplishing the first likelihood analysis designed to retain as much information as possible at linear scales. The primary goal of both Fisher et al. [14] and Heavens & Taylor [22] was to measure the linear redshift distortion parameter $\beta \approx \Omega_m^{4/7}/b$ from the *IRAS* 1.2 Jy survey. Shortly thereafter, Ballinger, Heavens & Taylor [1] extended the analysis to include a measurement of the real space (as opposed to redshift space) linear power spectrum of the 1.2 Jy survey in several bins of wavenumber.

Several authors have built on and applied the Heavens & Taylor's [22] method. Tadros et al. [35] and Taylor et al. [36] applied improved versions

Hamilton, A.J.S.: *Power Spectrum Estimation II. Linear Maximum Likelihood.* Lect. Notes Phys. **665**, 433–456 (2009)
DOI 10.1007/978-3-540-44767-2_13

of the method to measure the power spectrum and redshift distortions in the *IRAS* Point Source Catalogue redshift survey (PSCz; Saunders et al. [33]). More recently, Percival et al. [31] applied the method to the final version of the two-degree Field Galaxy Redshift Survey (2dFGRS; Colless et al. [8]).

The present chapter aims at a pedagogical presentation of the various issues involved in carrying out a maximum likelihood analysis of the galaxy power spectrum and its redshift distortions. The chapter is based largely on our own work over the last several years (Hamilton et al. [21]; Padmanabhan et al. [29]; Tegmark et al. [41]; Tegmark et al. [39]). The procedures are illustrated here mainly with measurements from the *IRAS* PSCz survey (Saunders et al. [33]). The measurements are based on the work reported by Hamilton et al. [21], but most of the graphics in the present chapter appear here for the first time.

The chapter starts, Sect. 2, by showing the final real space linear power spectra measured by the methods described herein from the PSCz, 2dF, and SDSS surveys. The remainder of the chapter is organized into sections each dealing with a specific aspect of measuring the linear power spectrum: Sect. 3 selection function; Sect. 4 linear vs. nonlinear regimes; Sect. 5 Gaussian likelihood function; Sect. 6 numerical obstacle; Sect. 7 Karhunen–Loève compression; Sect. 8 removing pair-integral bias; Sect. 9 Local Group flow; Sect. 10 isolating angular and radial systematics; Sect. 11 redshift distortions and logarithmic spherical waves; Sect. 12 quadratic compression; Sect. 13 decorrelation; and Sect. 14 disentanglement. Finally, Sect. 15 summarizes the conclusions.

2 Results

Figure 1, taken from Tegmark et al. [39], shows the linear galaxy power spectra measured, by the methods described in this chapter, from the PSCz (Saunders et al. [33]), 2dF (Colless et al. [8]), and SDSS (York et al. [48]) surveys.

Two important points to note about these power spectra are, first, that the power spectra have redshift distortions removed (Sect. 11), and are therefore real space power spectra and, second, that the power spectra have been decorrelated (Sect. 13) so that each point with its error bar represents a statistically independent object. Each point represents not the power at a single wavenumber, but rather the power in a certain well-defined band (Sect. 14).

3 Selection Function

A prerequisite for measuring galaxy power spectra, whether at linear or nonlinear scales, is to measure the angular and radial selection functions of a survey. All I really want to say here is that selection functions do not grow on trees, but require a lot of sometimes unappreciated hard work to measure.

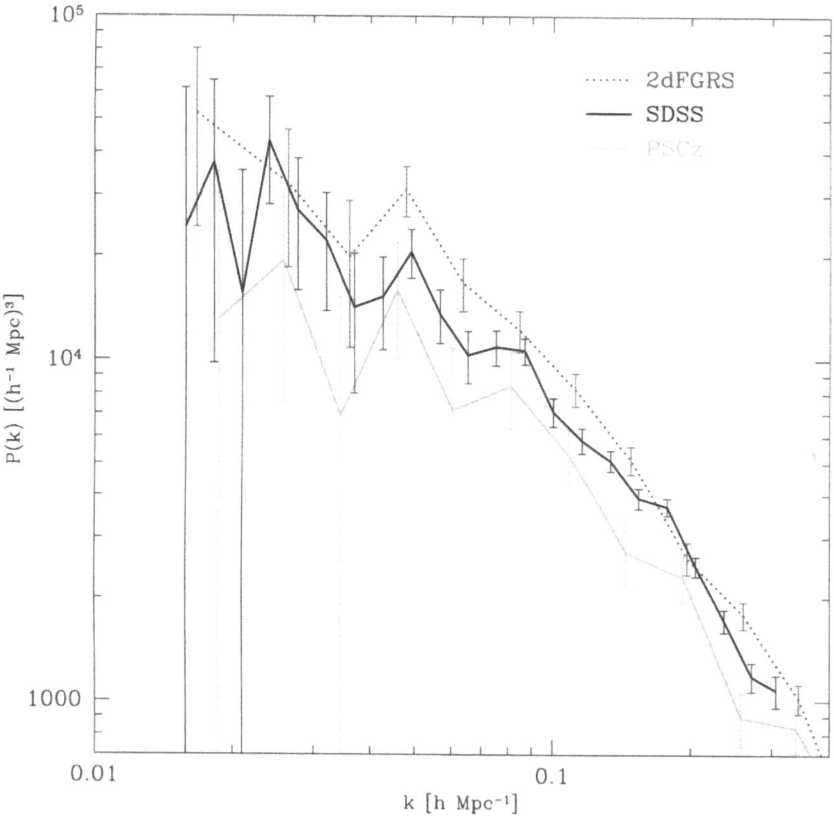

Fig. 1. From Tegmark et al. [39]. Decorrelated, real space (not redshift space) linear power spectra measured from the PSCz survey (Hamilton, Tegmark & Padmanabhan [21]), the 2dF 100k survey (Tegmark et al. [41]), and the SDSS survey (Tegmark et al. [39])

As regards the angular selection function, one thing that can help cut down the work and improve accuracy is a software package MANGLE that we recently published (Hamilton & Tegmark [20]). Max tells me that MANGLE has become the de facto software with which the SDSS team are characterizing the angular selection function of the SDSS.

Measuring the radial selection function is similarly onerous. The principal goal of all the methods is to separate out the smooth radial variation of the selection function from the large variations in galaxy density caused by galaxy clustering. Invariably, the essential assumption made to effect this separation is that the galaxy luminosity function is a universal function, independent of position. One of the best papers (in terms of content, as opposed to comprehensibility) on the subject remains the seminal paper by Lynden-Bell [28], who applied a maximum likelihood method. Other papers include Sandage,

Tammann & Yahil [32], Chołoniewski [6, 7], Binggeli et al. [4], Efstathiou et al. [11], SubbaRao et al. [34], Heyl et al. [23], Willmer [47], Tresse [45].

4 Linear vs. Nonlinear Regimes

There is a big difference between the linear and nonlinear regimes of galaxy clustering, and it makes sense to measure the power spectrum using different techniques in the two regimes.

At linear scales, it is legitimate to assume a much tighter prior than at nonlinear scales. On the other hand, at linear scales there are far fewer modes than at nonlinear scales.

At linear scales, you can reasonably assume (with various degrees of confidence):

- Gaussian fluctuations;
- redshift distortions conform to the linear Kaiser [24] model;
- linear bias between galaxies and matter.

At nonlinear scales, all of the above assumptions are false, and it would be an error to assume that they are true. There is however one useful assumption that is a better approximation at nonlinear scales than at linear scales:

- redshift distortions are plane-parallel (the distant observer approximation).

The remainder of this chapter devotes itself to the problem of measuring power at linear scales.

5 Gaussian Likelihood Function

Measuring the power spectrum of a galaxy survey at linear scales starts with the fundamental prior assumption that the density field is Gaussian. With this assumption, one has the luxury of being able to write down an explicit Gaussian likelihood function

$$\mathcal{L} \propto \frac{1}{|C|^{1/2}} \exp\left(-\frac{1}{2} C_{ij}^{-1} \delta_i \delta_j\right) , \tag{1}$$

where δ_i is a vector of measured overdensities, C_{ij} is their covariance matrix (part of the prior), and $|C|$ is the determinant of the covariance matrix. Normally, the covariance matrix is assumed to be a sum of two parts, a cosmic part and a shot noise part – see Sect. 7 below.

It should be emphasized that the assumption of Gaussian fluctuations is *not* valid at nonlinear scales, and it would be wrong to assume that the likelihood (1) holds at nonlinear scales. If you use the likelihood (1) at nonlinear scales, then you will substantially underestimate the true error bars on the power spectrum.

6 Numerical Obstacle

Perhaps the biggest single obstacle to overcome in carrying out a Gaussian likelihood analysis is a numerical one, the limit on the size of the covariance matrix C that the numerics can deal with. Suppose that you have N modes in the likelihood function. Manipulating the $N \times N$ covariance matrix C is typically an N^3 process. Thus doubling the number of modes takes $2^3 = 8$ times as much computer time.

To make matters worse, consider the fact that the number of modes in a survey increases as roughly the cube of the maximum wavenumber (smallest scale) to which you choose to probe, $N \sim k_{\max}^3$. The numerical problem thus scales as $k_{\max}^9$. In other words, if you want to push to half the scale, by doubling $k_{\max}$, then you need $2^3 = 8$ times as many modes, and $2^9 = 512$ times as much computer time.

Fortunately, the linear regime is covered with a tractable number of modes. The boundary between linear and nonlinear regimes is at $k \approx 0.3\,\mathrm{h\ Mpc}^{-1}$. If you prefer to remain safely in the linear regime, you may prefer to stick to $k \lesssim 0.1\,\mathrm{h\ Mpc}^{-1}$, as did Heavens & Taylor [22]. Max Tegmark and I typically have used about 4000 modes (a week of computer time on a workstation), which took us to $k \approx 0.25\,\mathrm{h\ Mpc}^{-1}$ in the PSCz survey, and $k \approx 0.15\,\mathrm{h\ Mpc}^{-1}$ in the 2dF 100k and SDSS surveys.

The previous paragraph would seem to suggest that 4000 modes is fine, but I have to confess that although our coverage of information is close to 100% at the largest scales, we lose progressively more information at smaller scales. To catch all the information available at say $k \approx 0.1\,\mathrm{h\ Mpc}^{-1}$, we should really push to 10^4 or 10^5 modes.

How does one deal with the numerical limitation of being able to use only a finite number of modes? One thing is to make sure that your code is as fast as can be. Several years ago Max Tegmark taught me some tricks that can speed up matrix manipulations by 1 or 2 orders of magnitude (without which our code would take a year to run instead of a week). It helps to time the various steps in the code and to oil the bottlenecks.

The other big thing to get around the numerical limitations is to use the Karhunen–Loève technique to compress the information of interest – power at linear scales – into a modest number of modes.

7 Karhunen–Loève Compression

The idea of Karhunen–Loève (KL) compression is to keep only the highest signal-to-noise modes in the likelihood function. This idea was first proposed by Vogeley & Szalay [46] for application to LSS.

Suppose that the covariance matrix C is a sum of a signal S (the cosmic variance) and noise N (the shot noise)

Fig. 2. (*Continued*)

$$C = S + N \ . \tag{2}$$

Prewhiten the covariance matrix, that is, transform the covariance matrix so that the noise matrix is the unit matrix:

$$N^{-1/2} C N^{-1/2} = N^{-1/2} S N^{-1/2} + 1 \ , \tag{3}$$

where the 1 on the RHS is to be interpreted as the unit matrix. Diagonalize the prewhitened signal:

$$N^{-1/2} S N^{-1/2} = O \Lambda O^{\top} \ , \tag{4}$$

where O is an orthogonal matrix and Λ is diagonal. Since the unit matrix remains the unit matrix under any diagonalization, the covariance matrix is

$$N^{-1/2} C N^{-1/2} = O(\Lambda + 1) O^{\top} \ . \tag{5}$$

The resulting eigenmodes, the columns of the orthogonal matrix O, are **Karhunen–Loève**, or **signal-to-noise**, eigenmodes, with eigenvalues λ equal to the signal-to-noise ratio of each mode.

The KL procedure thus parcels the information in a signal into a discrete set of modes ordered by their signal to noise. This is definitely a very neat trick.

However, there is a big drawback to KL compression, which is that you typically want to extract N modes from a much larger pool of $\mathcal{N} \gg N$ modes – that is why it is called compression. And that requires diagonalizing a large $\mathcal{N} \times \mathcal{N}$ matrix. But the whole point of KL compression is to avoid having to mess with a large matrix.

Fortunately, there is a way out of this loop. The first point to note is that the thing of interest is the ensemble of KL modes, not each mode individually (though they are cute to look at – see Figs. 2 and 3). Thus the KL modes do not have to be perfect. The second point is that what you call "signal" is up to you. In the case under consideration the signal is "the power spectrum at linear scales", which is actually not a single signal but a whole suite of signals.

Fig. 2. Selection of angular Karhunen–Loève modes in the PSCz survey with the high-latitude mask. The projection is Hammer-Aitoff in galactic coordinates, with the Galactic Centre at centre. The top four modes are special, while the rest are constructed from the KL procedure. The angular modes from top left to bottom right are given: mode 1, the (cut) monopole mode; modes 2–4, the three (cut) dipole modes (with small admixtures of cut monopole to make them orthogonal to the monopole); then modes 5, 10, 20, 40, 80, and 160. All modes are mutually orthogonal over the unmasked part of the sky. The modes are finite sums of harmonics up to $l = 39$. Mode 5 (*left middle*) is the (nonspecial) KL mode containing the most information about large-scale angular power. The mode evidently "knows about" the PSCz angular mask: the mode has low amplitude in masked regions of the survey and high amplitude in unmasked regions

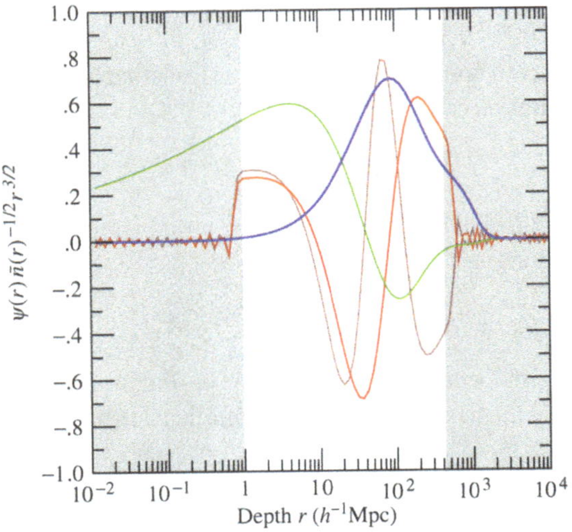

Fig. 3. The first five radial Karhunen–Loève modes associated with the (cut) monopole angular mode in the PSCz survey. The first two radial modes are special: mode 1 (thick blue) is the mean mode, and mode 2 (*medium green*) is the Local Group (LG) motion mode (with an admixture of the mean mode to make it orthogonal to the mean). The remaining radial modes are constructed from the KL procedure. The modes are all mutually orthogonal over the (*unshaded*) interval from $1\,\mathrm{h}^{-1}$ Mpc to $420\,\mathrm{h}^{-1}$ Mpc. The modes are finite sums of logarithmic radial waves (Sect. 11) defined, to avoid aliasing, over the extended logarithmic interval $10^{-2}\,\mathrm{h}^{-1}$ Mpc to $10^4\,\mathrm{h}^{-1}$ Mpc.

Radial KL modes associated with other angular KL modes are similar but not identical. For nonspecial angular modes (i.e. angular modes other than the cut monopole and dipole) it is optional whether or not to force the first mode(s) to be special. Nowadays we tend to keep the mean but not the LG radial mode in nonspecial angular modes. Keeping the mean radial mode makes it possible to test for possible purely angular systematics, such as might be associated with extinction

Our strategy to solve the KL crunch is to compress first into a set of angular KL modes (Fig. 2); and then within each angular KL mode to compress into a set of radial KL modes (Fig. 3). The result is a set of "pseudo-Karhunen–Loève" (PKL) modes each of which is the product of an angular and a radial profile. The PKL modes are not perfect, but they cover the relevant subspace of Hilbert space without gaps, and that is all we need.

Since the goal is to measure power at linear scales, we choose the "signal" to be not a realistic power spectrum, but rather an artificial power spectrum $P(k) \propto k^{-2.5}$ which increases steeply to large scales. Thus our procedure favours modes that are sensitive to power at large scales; but a low-noise small-scale mode can beat out a noisy large-scale mode.

Typically, we start with a few thousand angular modes in spherical harmonic space, and apply KL diagonalization to these angular modes. We then

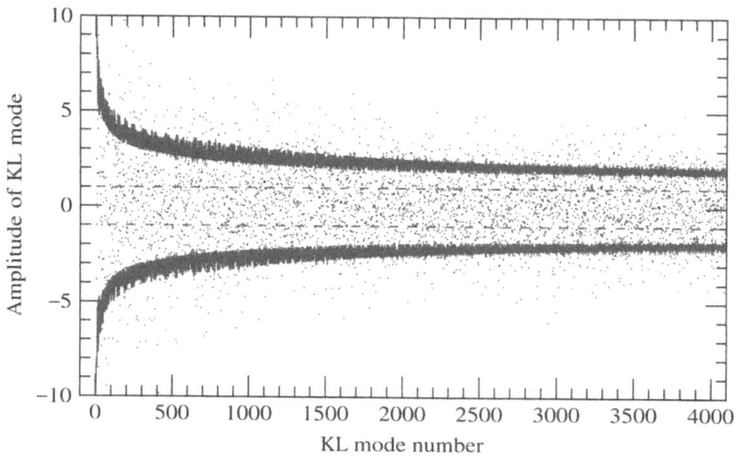

Fig. 4. Amplitudes of 4095 of the 4096 pseudo-Karhunen–Loève modes in the PSCz survey (the missing mode is the mean mode, whose amplitude on this scale is huge). The dots are the measured amplitudes, which according to the prior are expected to be Gaussianly distributed about zero, with expected standard deviation as given by the solid line. The dashed line is the expected standard deviation from shot noise alone

march through each angular KL mode one by one. Within each angular KL mode, we resolve the radial direction into several hundred logarithmic spherical waves and apply KL diagonalization to those. We keep a running pool of the best 4000 modes so far. There is no need to go through all the angular KL modes. The later angular KL modes contain little information, and when we have gone through 10 successive angular KL modes and found no new mode good enough to make it into the pool of best modes, we stop. The procedure effectively compresses 10^5–10^6 modes into 4000, but remains well within the capabilities of a modern workstation.

Figure 4 shows the amplitudes x_i of 4095 of 4096 PKL modes (the omitted mode is the mean (7), whose amplitude (8), is huge compared to all other modes) measured in the PSCz survey. According to the prior, the amplitudes should be Gaussianly distributed about zero (excepting the mean mode), with variances given by the diagonal elements $C_{ii} = \langle \Delta x_i^2 \rangle$ of the (prior) covariance matrix of PKL modes. Indeed, the measured amplitudes are consistent with this prior.

8 Removing Pair-Integral Bias

You are undoubtedly familiar with the notion that if you measure both the mean and the variance from a data set, then the measurement of the variance will be biased low. The usual fix-up is, if you have N independent data, to divide the sum of the squared deviations by $N - 1$ rather than N.

Applied to LSS, this bias is known in the literature as the "pair integral constraint" (the observed number of neighbours of a galaxy in a survey equals the number of galaxies in the survey minus one).

The simplistic procedure of dividing by $N-1$ instead of N does not work in LSS, but Fisher et al. [13] pointed out a delightfully simple trick that does completely solve the problem. The Fisher et al. trick is to isolate the mean, the selection function $\bar{n}(\boldsymbol{r})$, into a single "mean mode" and to make all other modes orthogonal to the mean.

In the present context, let $\psi_i(\boldsymbol{r})$ denote a PKL mode. The observed amplitude x_i of a PKL mode is defined as

$$x_i \equiv \int \psi_i(\boldsymbol{r}) \frac{n(\boldsymbol{r})}{\bar{n}(\boldsymbol{r})} \, \mathrm{d}^3 r = \sum_{\text{galaxies } g} \frac{\psi_i(\boldsymbol{r}_g)}{\bar{n}(\boldsymbol{r}_g)} . \tag{6}$$

The mean mode $\psi_1(\boldsymbol{r})$ is defined to be the mode whose shape is the mean

$$\psi_1(\boldsymbol{r}) \equiv \bar{n}(\boldsymbol{r}) . \tag{7}$$

The amplitude of the mean mode is

$$x_1 = \int n(\boldsymbol{r}) \, \mathrm{d}^3 r = N_{\text{gal}} , \tag{8}$$

the number of galaxies in the survey. Fisher et al.'s [13] trick is to arrange all modes other than the mean to be orthogonal to the mean

$$\langle x_i \rangle = \int \psi_i(\boldsymbol{r}) \, \mathrm{d}^3 r = 0 . \tag{9}$$

This trick ensures that the amplitudes of all modes except the mean mode are unaffected (to linear order, anyway) by uncertainty in the mean. The resulting power spectrum is unbiased by the pair integral constraint. Neat!

The observed amplitude of the mean mode is used in computing the maximum likelihood normalization of the selection function, but is then discarded from the analysis, because it is impossible to measure the fluctuation of the mean mode just as it is impossible to measure the fluctuation of the monopole mode in the CMB.

9 Local Group Flow

The motion of the Local Group (LG) induces a dipole in the density distribution around it (Hamilton [16], (4.42)) (or rather, since the LG is going with the flow, the motion of the LG removes the dipole present in the CMB frame). Although the LG mode is a single mode, we choose to project the effect, along with the mean mode, into a set of eight modes, whose angular parts are (cut) monopole and dipole ($1+3=4$ modes), illustrated in the top four panels of Fig. 2, and whose radial parts are the mean mode $\bar{n}(r)$ and the LG radial mode $\partial\left[r^2\bar{n}(r)\right]/r^2\partial r$ $(1+1=2$ modes), illustrated in Fig. 3.

Since the motion of the LG through the CMB is known (Bennett et al. [3]; Courteau & van den Bergh [9]; Lineweaver et al. [27]), the amplitudes of the LG modes can be corrected for this motion and included in the analysis. This is unlike the CMB, where the fluctuations of the dipole modes cannot be measured separately from the motion of the Sun, and must be discarded.

10 Isolating Angular and Radial Systematics

A similar trick can be used to isolate other potential problems into specific modes or sets of modes.

For example, possible angular systematics, associated for example with uncertainties in angular extinction across the sky, can be projected into a set of purely angular modes (whose radial part is the mean radial mode). Other modes should be unaffected (to linear order) by such systematics, because they are orthogonal to purely angular variations. If a systematic effect arising from extinction were present, then it would show up as a systematic enhancement of power in the purely angular modes.

Similarly, possible radial systematics, associated perhaps with uncertainties in the radial selection function, or with evolution as a function of redshift, can be projected into a set of purely radial modes (whose angular part is the cut monopole).

We did not project out purely angular or radial modes in our PSCz analysis, but we have been doing this in our more recent work (Tegmark, Hamilton & Xu [41]; Tegmark et al. [39]).

11 Redshift Distortions and Logarithmic Spherical Waves

Large-scale coherent motions towards overdense regions induce a **linear squashing** effect on the correlation function of galaxies observed in redshift space, as illustrated in Fig. 5 for the PSCz survey. At smaller scales, collapse and virialization gives rise to the so-called **fingers-of-god**, visible in Fig. 5 as a mild extension of the correlation function along the line-of-sight axis.

Kaiser [24] first pointed out the celebrated result that, at linear scales, and in the plane-parallel (distant observer) approximation, the Fourier amplitude $\delta^{(s)}(\boldsymbol{k})$ of galaxies in redshift space is amplified over the Fourier amplitude $\delta(\boldsymbol{k})$ of mass in real space by a factor $b+f\mu_{\boldsymbol{k}}^2$:

$$\delta^{(s)}(\boldsymbol{k})=(b+f\mu_{\boldsymbol{k}}^2)\delta(\boldsymbol{k})\ , \tag{10}$$

where $\mu_{\boldsymbol{k}}=\hat{\boldsymbol{z}}.\hat{\boldsymbol{k}}$ is the cosine of the angle between the wavevector $\boldsymbol{k}$ and the line of sight $\hat{\boldsymbol{z}}$, the quantity b is the linear galaxy-to-mass bias, and $f\approx\Omega_m^{4/7}$

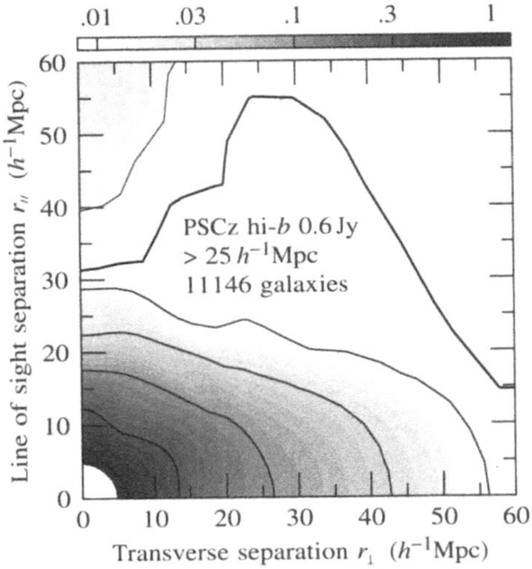

Fig. 5. Contour plot of the redshift space two-point correlation function in the PSCz survey with the high galactic latitude angular mask. Unlike the analysis discussed in the rest of this chapter, this plot assumes that redshift distortions are plane-parallel (to which end only galaxies beyond $25\,\mathrm{h}^{-1}$ Mpc are included). The expected linear squashing effect is plainly visible, while nonlinear fingers-of-god show up as a mild extension along the line-of-sight axis. Thin, medium, and thick contours represent negative, positive, and zero values, respectively. The correlation function has been smoothed over pair separation $r = (r_\perp^2 + r_\parallel^2)^{1/2}$ with a tophat window of width 0.2 dex, and over angles $\theta = \tan^{-1}(r_\perp/r_\parallel)$ to the line of sight with a Gaussian window with a 1σ width of $10°$. From Hamilton et al. [21]

is the dimensionless linear growth rate of fluctuations. It follows immediately from Kaiser's formula that, again at linear scales, and in the plane-parallel approximation, the redshift space galaxy power spectrum $P^s(\boldsymbol{k})$ is amplified over the real space matter power spectrum $P(\boldsymbol{k})$ by

$$P^{(s)}(\boldsymbol{k}) = (b + f\mu_{\boldsymbol{k}}^2)^2 P(\boldsymbol{k}) \ . \tag{11}$$

Translated from Fourier space into real space, Kaiser's formula predicts the large-scale squashing effect visible in Fig. 5.

For the linear likelihood analysis being considered here, the assumption of linear redshift distortions is fine, but the plane-parallel approximation is not adequate. Kaiser [24], already in his original paper, presented formulae for radial redshift distortions. A pedagogical derivation can be found in the review by Hamilton [16]. Unfortunately, when the radial character of the redshift distortions is taken into account, the formula for the amplification of modes ceases to be anything like as simple as Kaiser's formula (10). Indeed, the full, correct formula is more complicated in Fourier space than in real space.

In our pipeline, we use logarithmic spherical waves as the fundamental basis with respect to which we express PKL modes, in part because radial redshift distortions take a simple form in that basis, as first pointed out by Hamilton & Culhane [18]. Logarithmic spherical waves are products of logarithmic radial waves $e^{i\omega \ln r}$ and spherical harmonics $Y_{lm}(\hat{\boldsymbol{r}})$

$$Z_{\omega lm}(\boldsymbol{r}) = e^{i\omega \ln r} Y_{lm}(\hat{\boldsymbol{r}}) \tag{12}$$

and are eigenmodes of the complete set of commuting Hermitian operators

$$-i\left(\frac{\partial}{\partial \ln r} + \frac{3}{2}\right) , \quad L^2 , \quad L_z . \tag{13}$$

If you wonder why no one ever told you about logarithmic spherical waves in quantum mechanics, so do I! They are beautiful things. For example, the logarithmic radial frequency ω is the radial analogue of the angular harmonic number l (see Fig. 6)

$$\omega \leftrightarrow \text{radial} \quad \text{as} \quad l \leftrightarrow \text{angular} , \tag{14}$$

something that everyone ought to know.

In logarithmic spherical wave space, the overdensity $\delta^{(s)}_{\omega lm}$ of galaxies in redshift space is related to the overdensity $\delta_{\omega lm}$ of mass in real space by

$$\delta^{(s)}_{\omega lm} = \left[b + f \frac{(i\omega + 1/2)(i\omega - 1/2) - \alpha(r)(i\omega - 1/2)}{(i\omega + l - 1/2)(i\omega - l - 3/2)} \right] \delta_{\omega lm} . \tag{15}$$

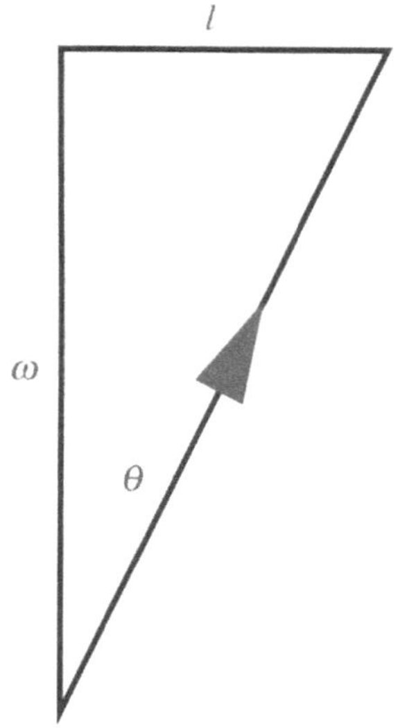

Fig. 6. The dimensionless log frequency ω is the radial analogue of the dimensionless spherical harmonic number l. A logarithmic spherical wave $Z_{\omega lm}$ with log frequency ω and harmonic number l has wavevector angled effectively at $\theta = \tan^{-1}(l/\omega)$ to the line of sight

Equation (15) may look complicated, but the amplification factor in square brackets on the RHS is just a number, as in the pretty Kaiser formula (10). OK, so I lied; if you look closely, you will see that (15) contains, in the expression in the square brackets, a function $\alpha(r)$ of radial depth r, defined to be the logarithmic derivative of the radial selection function

$$\alpha(r) \equiv \frac{\partial \ln \left[r^2 \bar{n}(r)\right]}{\partial \ln r} . \tag{16}$$

So the expression (15) is not as pretty as Kaiser's. But in practice, it turns out that the $\alpha(r)$ factor gets absorbed into another factor of $\bar{n}(r)$ at a previous step of the pipeline and so proves not to pose any special difficulty. [If you are wondering whether (15) has a rigorous mathematical meaning, the answer is yes it does: in real space, $\alpha(r)$ is a diagonal matrix with eigenvalues $\alpha(r)$; the symbol $\alpha(r)$ in (15) is the same matrix, but expressed in ωlm space.]

12 Quadratic Compression

Quadratic compression (Tegmark [37]; Bond et al. [5]) is a beautiful idea in which the information in a set of modes is losslessly (or almost losslessly) compressed not all the way to cosmological parameters, but rather to a set of power spectra. For galaxy power spectra the result is not one power spectrum but (at least) three: the galaxy–galaxy, galaxy–velocity, and velocity–velocity power spectra. I say "at least" because I expect that in the future the drive to reduce systematics from luminosity bias (more luminous galaxies are more clustered than faint galaxies) will warrant resolution of power spectra into "luminous" and "faint" components.

The point of reducing to power spectra rather than cosmological parameters is that the covariance matrix C in the Gaussian likelihood function depends, by definition, *linearly* on the prior cosmic power spectrum p_α:

$$C = \sum_\alpha C_{,\alpha}\, p_\alpha + N . \tag{17}$$

Here p_α denotes a set of cosmic galaxy–galaxy, galaxy–velocity, and velocity–velocity powers at various wavenumbers, $C_{,\alpha}$ is shorthand for the derivative $\partial C / \partial p_\alpha$, and N is the shot noise. For example, we have typically estimated the power in 49 bins of logarithmically spaced wavenumbers, so there are $49 \times 3 = 147$ power spectrum parameters p_α to estimate. Normally it would be intractable to find the maximum likelihood solution for 147 cosmological parameters, but because the covariance C depends linearly on the powers p_α the solution is analytic.

Although the solution is analytic, it involves 147 matrices $C_{,\alpha}$ each 4000×4000 in size (for 4000 modes), and it still takes a bit of cunning to accomplish that solution numerically. A good trick is to decompose the (symmetric) covariance matrix C as the product of a lower triangular matrix L and its transpose $L^\top$

$$C = LL^\top \,. \tag{18}$$

The jargon name for this is Cholesky decomposition, and there are fast ways to do it. The Fisher matrix of the parameters p_α is

$$F_{\alpha\beta} = \frac{1}{2}\left(L^{-1}C_{,\alpha}L^{-\top}\right)\cdot\left(L^{-1}C_{,\beta}L^{-\top}\right) \,. \tag{19}$$

Then, from the measured amplitudes δ of the PKL modes, form the shot noise-subtracted **quadratic estimator** $\hat{q}_\alpha$

$$\hat{q}_\alpha = \frac{1}{2}\left(L^{-1}\delta\right)^\top\left(L^{-1}C_{,\alpha}L^{-\top}\right)\left(L^{-1}\delta\right) - \hat{N}_\alpha \,, \tag{20}$$

in which $\hat{N}_\alpha$ denotes the shot noise, the self-pair contribution to the main term on the right hand side. The expected mean and variance of the quadratic estimators $\hat{q}_\alpha$ are

$$\langle \hat{q}_\alpha \rangle = F_{\alpha\beta}\, p_\beta \,, \tag{21}$$

$$\langle \Delta\hat{q}_\alpha \Delta\hat{q}_\beta \rangle = F_{\alpha\beta} \,. \tag{22}$$

Suitably scaled, the quadratic estimates $\hat{q}_\alpha$ can be regarded as smoothed-but-correlated estimates of the parameters p_α.

Given (21), an estimator of power $\hat{p}_\alpha$, which we call the "raw" estimator, can be defined by

$$\hat{p}_\alpha = F^{-1}_{\alpha\beta}\, \hat{q}_\beta \,, \tag{23}$$

which is an unbiased estimator because $\langle \hat{p}_\alpha \rangle = p_\alpha$. The raw estimates are anti-correlated, with covariance

$$\langle \Delta\hat{p}_\alpha \Delta\hat{p}_\beta \rangle = F^{-1}_{\alpha\beta} \,. \tag{24}$$

This raw estimator $\hat{p}_\alpha$ exhausts the Cramér–Rao inequality (Lecture 1, (31)), and therefore no better estimator of power exists.

The raw estimates $\hat{p}_\alpha$, along with their full covariance matrix, contain all the information available from the observational data and can be used in a maximum likelihood analysis of cosmological parameters. However, if the raw estimates $\hat{p}_\alpha$ are plotted on a graph, with error bars given by the square root of the diagonal element of the covariance matrix, $(F^{-1}_{\alpha\alpha})^{1/2}$, then the result gives a misleadingly pessimistic impression of the true uncertainties. This is because in plotting errors only from the diagonal elements of the covariance matrix, one is effectively discarding useful information in the cross-correlation between bins. Thus the raw estimates of power, plotted on a graph, appear unnecessarily pessimistic and noisy.

Whereas the quadratic estimates $\hat{q}_\alpha$ are correlated, and the raw estimates $\hat{p}_\alpha$ are anti-correlated, there are compromise estimators that are, like

Goldilocks' porridge, just right. These are the decorrelated estimators discussed in the next section.

It was stated above that the raw estimates $\hat{p}_\alpha$ and their full covariance matrix contain all the information available from the observational data. Actually this is not quite true, if one uses a Gaussian approximation to the likelihood function as a function of the parameters p_α, as opposed to the full likelihood function. A principal idea behind **radical compression** (Bond et al. [5]) is to take functions of the parameters arranged so as to make the likelihood function as Gaussian in the remapped parameters as possible, and hence to extract the last (well, almost the last) ounce of information from the data.

13 Decorrelation

Decorrelation, introduced by Hamilton [15], and first applied (to the COBE power spectrum) by Tegmark & Hamilton [40], is another delightful concept, yielding estimates of power at each wavenumber that are uncorrelated with all others. A detailed exposition is given by Hamilton & Tegmark [19]. We assume throughout this section that the likelihood function is well approximated as a Gaussian in the parameters p_α so that the Fisher matrix equals the inverse covariance matrix.

The idea of decorrelation applies quite generally to any set of correlated estimates of parameters, not just to power spectra. The left panel of Fig. 7 illustrates an example of two correlated parameter estimates $\hat{p}_1$ and $\hat{p}_2$. The fact that the error ellipse is tilted from horizontal indicates that the parameter estimates are correlated.

There are infinitely many linear combinations of the parameter estimates $\hat{p}_1$ and $\hat{p}_2$ of Fig. 7 that are uncorrelated. Most obviously, the eigenvectors of the covariance matrix – the major and minor axes of the error ellipse – are uncorrelated. The decomposition of a set of parameter estimates into eigenvectors is called **principal component decomposition**.

However, there are infinitely many other ways to form uncorrelated linear combinations of correlated parameters. The right panel of Fig. 7 is the same as the left panel, but stretched out along the minor axis so that the error ellipse becomes an error circle. Any two orthogonal vectors on this error circle are uncorrelated. For example, one possibility, shown as thick solid lines in Fig. 7, is to choose the two vectors on the error circle to be parallel to the original parameter axes. This choice of uncorrelated parameters has the merit that it is in a sense "closest" to the original parameters. When the error circle is squashed back to the original error ellipse in the left panel of Fig. 7, the decorrelated parameters (the thick solid lines) are no longer perpendicular, but they are nonetheless uncorrelated.

Exercise 1. Show that these decorrelated parameter estimates (the ones corresponding to the thick solid lines in Fig. 7) are given by $F^{1/2}\hat{\boldsymbol{p}}$. □

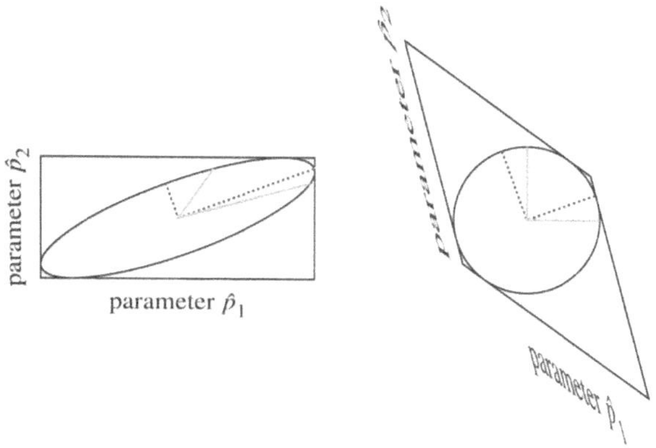

Fig. 7. A vector $\hat{\boldsymbol{p}} = (\hat{p}_1, \hat{p}_2)$ of correlated parameter estimators. The thick dashed lines represent the principal component decomposition of $\hat{p}_\alpha$, the eigenvectors of their covariance matrix. They are uncorrelated. The thick solid lines represent $F^{1/2}\hat{\boldsymbol{p}}$, the parameter estimates decorrelated with the square root of the Fisher matrix. They are also uncorrelated. The diagram on the right is the same as that on the left, but stretched out along the minor axis of the error ellipse, so that the error ellipse becomes an error circle. Any two vectors that are orthogonal on the error circle are uncorrelated

Mathematically, decorrelating a set of correlated parameter estimates is equivalent to decomposing their Fisher matrix as

$$F = M^\top \Lambda M \ , \tag{25}$$

where Λ is diagonal. The matrix M, which need not be orthogonal, is called a **decorrelation matrix**. The parameter combinations $M\hat{\boldsymbol{p}}$ are uncorrelated because their covariance matrix is diagonal:

$$\langle \Delta(M\hat{\boldsymbol{p}})\Delta(M\hat{\boldsymbol{p}})^\top \rangle = M \left\langle \Delta\hat{\boldsymbol{p}}\Delta\hat{\boldsymbol{p}}^\top \right\rangle M^\top = M F^{-1} M^\top = \Lambda^{-1} \ . \tag{26}$$

Thus each row of the decorrelation matrix M represents a parameter combination that is uncorrelated with all other rows. It is possible to rescale each row of M so the diagonal matrix Λ is the unit matrix, so that the parameter combinations $M\hat{\boldsymbol{p}}$ have unit covariance matrix. Usually however one prefers a more physically motivated scaling.

In the case of the power spectrum, the rows of the decorrelation matrix M represent **band-power windows**, and it is sensible to normalize them to unit area (sum of each row is one) so that a measured power $M\hat{\boldsymbol{p}}$ can be interpreted as the power averaged over the band-power window.

Figure 7 illustrates the case where the decorrelation matrix is taken to be the **square root of the Fisher matrix**

$$M = F^{1/2} \ . \tag{27}$$

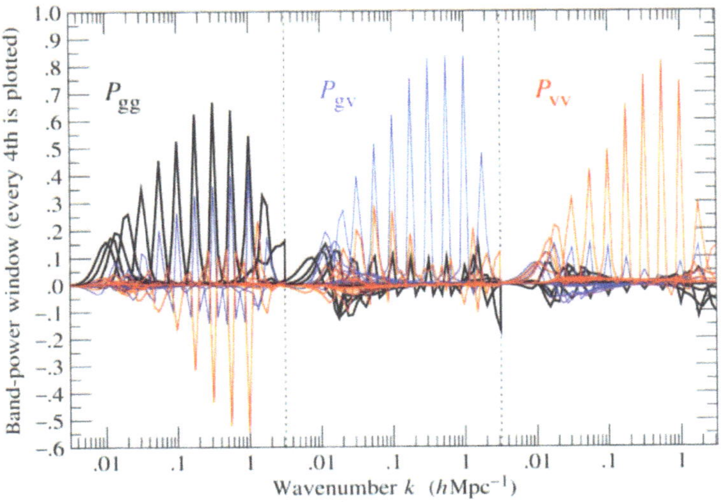

Fig. 8. Band-power windows for the disentangled, decorrelated power spectra measured from the PSCz survey for (*left*) the galaxy–galaxy power spectrum P_{gg}, (*middle*) the galaxy–velocity power spectrum P_{gv}, and (*right*) the velocity–velocity power spectrum P_{vv}. Each band power includes contributions from all three power spectra, the (*thick black*) gg, (*medium blue*) gv, and (*thin red*) gg powers, but the off-type contributions cancel, according to the prior. For example, in the left panel, the contributions to the gg band power from each of the gv and vv powers should sum to zero, if the prior is correct

Applied to the power spectrum, this choice (or rather, a version thereof scaled with the prior power – see Hamilton & Tegmark [19] for details) provides nicely behaved band-power windows that are (at least at linear scales) everywhere positive, and concentrated narrowly about each target wavenumber, as illustrated in Fig. 8.

By contrast, principal component decomposition yields broad, wiggly, non-positive band-power windows that mix power at small and large scales in a physically empty fashion.

14 Disentanglement

As described in Sect. 12, quadratic compression yields estimates of not one but three power spectra, the galaxy–galaxy (gg), galaxy–velocity (gv), and velocity–velocity (vv) power spectra. The three power spectra are related to the true underlying matter power spectrum $P(k)$ by Kolatt & Dekel [25]; Tegmark [38]; Pen [30]; Tegmark & Peebles [42]; Dekel & Lahav [10]:

$$\begin{aligned} \text{galaxy–galaxy power} : P_{gg}(k) &= b(k)^2 P(k) \ , \\ \text{galaxy–velocity power} : P_{gv}(k) &= r(k) b(k) f P(k) \ , \\ \text{velocity–velocity power} : P_{vv}(k) &= f^2 P(k) \ , \end{aligned} \tag{28}$$

where $b(k)$ is the (possibly scale-dependent) galaxy-to-mass bias factor, $r(k) \in [-1, 1]$ is a (possibly scale-dependent) galaxy–velocity correlation coefficient, and f is the dimensionless linear growth rate, which is well approximated by (Lahav et al. [26]; Hamilton [17])

$$f \approx \Omega_{\rm m}^{4/7} + (1 + \Omega_{\rm m}/2)\,\Omega_\Lambda/70 \,. \tag{29}$$

More correctly, the "velocity" here refers to minus the velocity divergence, which in linear theory is related to the mass (not galaxy) overdensity δ by

$$f\delta + \nabla \cdot \boldsymbol{v} = 0 \,, \tag{30}$$

where ∇ denotes the comoving gradient in velocity units.

The linear velocity–velocity power spectrum $P_{\rm vv}(k)$ is of particular interest because, to the extent that galaxy flows trace dark matter flows on large scales (which should be true), it offers an unbiased measure of the shape of the matter power spectrum $P(k)$. Indeed if the dimensionless linear growth rate f is taken as a known quantity, then $P_{\rm vv}(k)$ provides a direct measurement of the shape and amplitude of the matter power spectrum $P(k)$. It should be cautioned that $P_{\rm vv}(k)$ is a direct measure of matter power only at linear scales, where redshift distortions conform to the linear Kaiser [24] model. At nonlinear scales, fingers-of-god are expected to enhance the velocity–velocity power above the predicted linear value.

Although each quadratic estimate $\hat{q}_\alpha$, (20), is targeted to measure a single power type and a single wavenumber (a single parameter p_α), inevitably the nature of real surveys causes each estimate $\hat{q}_\alpha$ to contain a mixture of all three power spectra at many wavenumbers, in accordance with (21).

What one would like to do is to disentangle the three power spectra, projecting out an unmixed version of each. The problem is somewhat similar to the CMB problem of forming disentangled TT, TE, EE, and BB power spectra from the amplitudes of fluctuations in temperature (T), and in electric (E) and magnetic (B) polarization modes (TB and EB cross-power spectra are expected to vanish because B modes have opposite parity to T and E modes).

Now the raw estimates of power $\hat{p}_\alpha$, (23), are already disentangled, and as remarked in Sect. 12, they and their covariance matrix contain (almost) all the information in the observational data. Thus if the aim is merely to do a cosmological parameter analysis, then it is fine to stop at the raw powers $\hat{p}_\alpha$.

However, before leaping to cosmological parameters, it is wise to plot the gg, gv, and vv power spectra explicitly.

One possibility would be to plot each power spectrum marginalized over the other two. The drawback with this is that marginalization effectively means discarding information contained in the correlations between the power spectra, and discarding this information leads to unnecessarily noisy estimates of power.

Instead, we have adopted the following procedure. First, we decorrelate the entire set of estimates of power (three power spectra at each of many wavenumbers) with the square root of the Fisher matrix (prescaled with the prior power – see Hamilton & Tegmark [19]). The result is three power spectra every point of which is uncorrelated with every other (with respect to both type and wavenumber). The three power spectra are uncorrelated, but not disentangled: each uncorrelated power spectrum contains contributions from all three types: gg, gv, and vv. To disentangle them, we multiply the three uncorrelated power spectra at each wavenumber with a matrix that, if the prior is correct, yields pure gg, gv, and vv power spectra. The resulting band-power spectra and their error bars represent the information in the survey about as fairly as can be. The three power spectra at each wavenumber are correlated, but uncorrelated with the power spectra at all other wavenumbers.

Figure 8 illustrates the resulting band-power windows for each of the gg, gv, and vv power spectra in the PSCz survey. The gg band-power window, for example, contains contributions from gv and vv powers, but those contributions should cancel out if the prior power is correct. Note that cancellation

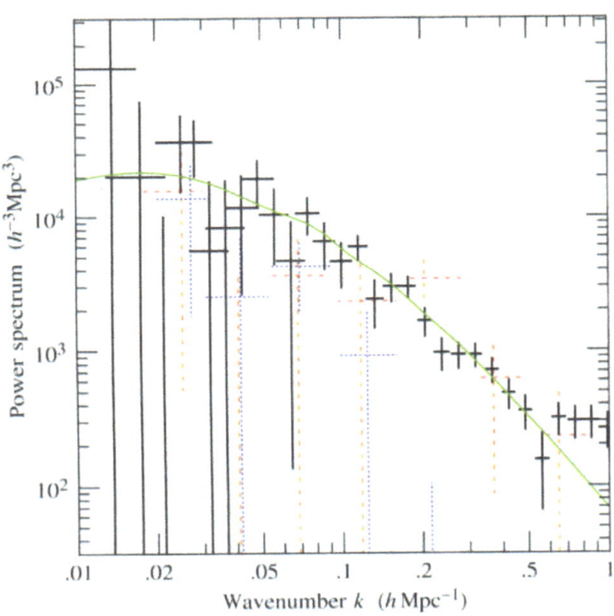

Fig. 9. Points with error bars show the decorrelated power spectra measured from the PSCz survey: *thick black*, the galaxy–galaxy power spectrum P_{gg}; *medium blue dotted*, the galaxy–velocity power spectrum P_{gv}; and *thin red dashed*, the velocity–velocity power spectrum P_{gv}. The gv and vv powers are plotted with coarser binning because they are noisier than the gg power. Each point is plotted at the median wavenumber of its band-power window; the horizontal error bar gives the FWHM of the band-power window. The smooth solid (*green*) line running through the galaxy–galaxy power spectrum is the linear concordance ΛCDM model of Tegmark et al. [44]

invokes the prior only weakly: the off-type contributions will cancel provided only that the true power has the same shape (not the same normalization) as the prior power over a narrow range (because the band-power windows are narrow).

Figure 9 shows the galaxy–galaxy, galaxy–velocity, and velocity–velocity power spectra P_{gg}, P_{gv}, and P_{vv} measured in this fashion from the PSCz survey (the galaxy–galaxy power spectrum P_{gg} shown here appears in Hamilton et al. [21], but P_{gv} and P_{vv} have not appeared elsewhere). As can be seen, the galaxy–velocity and velocity–velocity power spectra are noisy, but well detected. Nonlinear fingers-of-god cause the galaxy–velocity power P_{gv} to go negative at $k \gtrsim 0.2\,\mathrm{h\ Mpc}^{-1}$ and at the same time enhance the velocity–velocity power P_{vv} somewhat.

15 Conclusion

In its simplest form, measuring the galaxy power spectrum is almost trivial: bung the survey in a box, Fourier transform it, and measure the resulting power spectrum (Baumgart & Fry) [2]. A variant of this method, the Feldman et al. [12] method, in which the galaxy density is weighted by a weighting which is near-minimum-variance subject to some not-necessarily-true assumptions before Fourier transforming it, offers a fast and not-so-bad way to measure power spectra, good for when maximum likelihood would be overkill.

However, the "right" way to measure power spectra, at least at large, linear scales, is to use Bayesian maximum likelihood analysis. Maximum likelihood methods for measuring galaxy power spectra were pioneered by Fisher et al. [14] and Heavens & Taylor [22] and have seen considerable refinement in recent years. Maximum likelihood methods require a lot more work than traditional methods. The pay-off is not necessarily more precision (the error bars are all too often larger than claimed error bars from more primitive methods), but more precision in the precision. That is, you can have some confidence that the derived uncertainties reliably reflect the information in the data, no more and no less. This is essential when measurements are to be used in estimating cosmological parameters. Moreover maximum likelihood methods provide a general way to deal with all the complications of real galaxy surveys, as spherical redshift distortions, Local Group flow, variable extinction, and suchlike.

Acknowledgements

This work was supported in part by NASA ATP award NAG5-10763 and by NSF grant AST-0205981, and of course by this wonderful summer school.

References

1. Ballinger, W.E., Heavens, A.F., Taylor, A.N.: The Real-Space Power Spectrum of *IRAS* Galaxies on Large Scales and the Redshift Distortion, MNRAS **276**, L59–L63 (1995)
2. Baumgart, D.J., Fry, J.N.: Fourier spectra of three-dimensional data, ApJ **375**, 25 (1991)
3. Bennett, C.L. et al. (21 authors; WMAP collaboration): First year *Wilkinson Microwave Anisotropy Probe* (*WMAP*) observations: preliminary maps and basic results, ApJS **148**, 1
4. Binggeli, B., Sandage, A., Tammann, G.A.: The luminosity function of galaxies, Ann. Rev. Astr. Astrophys. **26**, 509 (1988)
5. Bond, J.R., Jaffe, A.H., Knox, L.E.: Radical compression of cosmic microwave background data, ApJ **533**, 19 (2000)
6. Chołoniewski, J.: New method for the determination of the luminosity function of galaxies, MNRAS **223**, 1 (1986)
7. Chołoniewski, J.: On Lynden-Bell's method for the determination of the luminosity function, MNRAS **226**, 273 (1987)
8. Colless, M. et al. (28 authors; the 2dFGRS team) The 2dF Galaxy Redshift Survey: Final Data Release astro-ph/0306581 (2003)
9. Courteau, S., van den Bergh, S., S.: The Solar Motion Relative to the Local Group, AJ **118**, 337 (1999)
10. Dekel, A., Lahav, O.: Stochastic nonlinear galaxy biasing, ApJ **520**, 24 (1999)
11. Efstathiou, G., Ellis, R.S., Peterson, B.A.: Analysis of a complete galaxy redshift survey – II. The field-galaxy luminosity function, MNRAS **232**, 431 (1988)
12. Feldman, H.A., Kaiser, N., Peacock, J.A.: Power spectrum analysis of three-dimensional redshift surveys, ApJ **426**, 23–37 (1994)
13. Fisher, K.B., Davis, M., Strauss, M.A., Yahil, A., Huchra, J.P.: The power spectrum of *IRAS* galaxies, ApJ **402**, 42–57 (1993)
14. Fisher, K.B., Scharf, C.A., Lahav, O.: A spherical harmonic approach to redshift distortion and a measurement of Ω_0 from the 1.2 Jy *IRAS* redshift survey, MNRAS, 266, 219 (1994)
15. Hamilton, A.J.S.: Towards optimal measurement of power spectra – II. A basis of positive, compact, statistically orthogonal kernels, MNRAS **289**, 295 (1997)
16. Hamilton, A.J.S.: Linear redshift distortions: a review, in *The Evolving Universe*, ed by D. Hamilton (Kluwer, Dordrecht), 185–275 (1998); astro-ph/9708102
17. Hamilton, A.J.S.: Formulae for growth factors in expanding universes containing matter and a cosmological constant, MNRAS **322**, 419 (2001)
18. Hamilton, A.J.S., Culhane, M.: Spherical redshift distortions, MNRAS **278**, 73 (1996)
19. Hamilton, A.J.S., Tegmark, M.: Decorrelating the power spectrum of galaxies, MNRAS **312**, 285–294 (2000)
20. Hamilton, A.J.S., Tegmark, M.: A scheme to deal accurately and efficiently with complex angular masks in galaxy surveys, MNRAS **349**, 115–128 (2004); software available at `http://casa.colorado.edu/~ajsh/mangle/`
21. Hamilton, A.J.S., Tegmark, M., Padmanabhan, N.: Linear redshift distortions and power in the PSCz survey, MNRAS 317, L23 (2000)
22. Heavens, A.F., Taylor, A.N.: A spherical harmonic analysis of redshift space, MNRAS **275**, 483–497 (1995)

23. Heyl, J., Colless, M., Ellis, R.S., Broadhurst, T.: Autofib Redshift Survey – II. Evolution of the galaxy luminosity function by spectral type, MNRAS **285**, 613 (1997)
24. Kaiser, N.: Clustering in real space and in redshift space, MNRAS **227**, 1–21 (1987)
25. Kolatt, T., Dekel, A.: Large-scale power spectrum from peculiar velocities, ApJ **479**, 592 (1997)
26. Lahav, O., Lilje, P.B., Primack, J.R., Rees, M.J.: Dynamical effects of the cosmological constant, MNRAS **251**, 128 (1991)
27. Lineweaver, C.H., Tenorio, L., Smoot, G.F., Keegstra, P., Banday, A.J., Lubin, P.: The dipole observed in the COBE DMR 4 year data, ApJ **470**, 38 (1996)
28. Lynden-Bell, D.: A method of allowing for known observational selection in small samples applied to 3CR quasars, MNRAS **155**, 95 (1971)
29. Padmanabhan, N., Tegmark, M., Hamilton, A.J.S.: The power spectrum of the CFA/SSRS UZC galaxy redshift survey, ApJ **550**, 52 (2001)
30. Pen, U.-L.: Reconstructing nonlinear stochastic bias from velocity space distortions, ApJ **504**, 601 (1998)
31. Percival, W.J. et al. (29 authors; the 2dFGRS team): The 2dF Galaxy Redshift Survey: spherical harmonics analysis of fluctuations in the final catalogue MNRAS **353**, 1201 (2004)
32. Sandage, A., Tammann, G.A., Yahil, A.: The velocity field of bright nearby galaxies. I – The variation of mean absolute magnitude with redshift for galaxies in a magnitude-limited sample, ApJ **232**, 352 (1979)
33. Saunders, W., Sutherland, W.J., Maddox, S.J., Keeble, O., Oliver, S.J., Rowan-Robinson, M., McMahon, R.G., Efstathiou, G.P., Tadros, H., White, S.D.M., Frenk, C.S., Carramiñana, A., Hawkins, M.R.S.: The PSCz catalogue, MNRAS **317**, 55 (2000)
34. SubbaRao, M.U., Connolly, A.J., Szalay, A.S., Koo, D.C.: Luminosity functions from photometric redshifts. I. Techniques, AJ **112**, 929 (1996)
35. Tadros, H., Ballinger, W.E., Taylor, A.N., Heavens, A.F., Efstathiou, G., Saunders, W., Frenk, C.S., Keeble, O., McMahon, R., Maddox, S.J., Oliver, S., Rowan-Robinson, M., Sutherland, W.J., White, S.D.M.: Spherical harmonic analysis of the PSCz galaxy catalogue: redshift distortions and the real-space power spectrum, MNRAS **305**, 527 (1999)
36. Taylor, A.N., Ballinger, W.E., Heavens, A.F., Tadros, H.: Application of data compression methods to the redshift-space distortions of the PSCz galaxy catalogue, MNRAS **327**, 689 (2001)
37. Tegmark, M.: How to measure CMB power spectra without losing information, Phys. Rev. D **55**, 5895 (1997)
38. Tegmark, M.: Bias and beyond, in Wide Field Surveys in Cosmology, ed by S. Colombi, Y. Mellier (Editions Frontières, 1998) p. 43
39. Tegmark, M. et al. (62 authors; SDSS collaboration) The 3D power spectrum of galaxies from the SDSS, ApJ **606**, 702–740 (2004)
40. Tegmark, M., Hamilton, A.J.S.: in Relativistic Astrophysics and Cosmology, 18th Texas Symposium on Relativistic Astrophysics, ed by A. V. Olinto, J. A. Frieman, D. Schramm (World Scientific), 270 (1998);
41. Tegmark, M., Hamilton, A.J.S., Xu, Y.: The power spectrum of galaxies in the 2dF 100k redshift survey, MNRAS **335**, 887–908 (2002)
42. Tegmark, M., Peebles, P.J.E.: The time evolution of bias, ApJ **500**, L79 (1998)

43. Tegmark, M., Taylor, A., Heavens, A.: Karhunen-Loeve eigenvalue problems in cosmology: how should we tackle large data sets?, ApJ **480**, 22 (1997)
44. Tegmark, M., Zaldarriaga, M., Hamilton, A.J.S.: Towards a refined cosmic concordance model: joint 11-parameter constraints from the cosmic microwave background and large-scale structure, PhysRev. D. **63**, 3007 (2001)
45. Tresse, L.: Luminosity Functions and Field Galaxy Population Evolution, in Formation and Evolution of Galaxies, Les Houches School Series, ed by O. Le Fèvre, S. Charlot, astro-ph/9902209 (1999)
46. Vogeley, M.S., Szalay, A.S.: Eigenmode Analysis of Galaxy Redshift Surveys. I. Theory and Methods, ApJ **465**, 34–53 (1996)
47. Willmer, C.N.A.: Estimating galaxy luminosity functions, AJ **114**, 898 (1997)
48. York, D.G., et al. (144 authors; SDSS collaboration): The Sloan Digital Sky Survey: technical summary, AJ **120**, 1579 (2000)

Introduction to Higher Order Spatial Statistics in Cosmology

I. Szapudi

Institute for Astronomy, University of Hawaii, 2680 Woodlawn Dr, Honolulu, HI 96822, USA
szapudi@ifa.hawaii.edu

1 Introduction

Higher order spatial statistics characterize non-Gaussian aspects of random fields, which are ubiquitous in cosmology: from the cosmic microwave background (CMB) to the large-scale structure (LSS) of the universe. These random fields are rich in their properties; they can be continuous or discrete; can have one through three, or even more dimensions; their degree of non-Gaussianity ranges from tiny to significant. Yet, there are several techniques and ideas, which are applicable to virtually all cosmological random fields, be it Lyman-α forests, LSS, or CMB.

In this lecture notes, I concentrate on the classic and widely applicable characterization of higher order statistics by joint moments, a.k.a. higher order correlation functions, and directly related statistics. These statistics are very powerful, although they have a perturbative nature to them in that they constitute an infinite (formal) expansion. Clearly, when only the first N terms of this expansion are extracted from data, interesting information might remain in even higher order terms. This is why a host of alternative statistics (void probabilities, wavelets, Minkowski functionals, minimal spanning trees, phase correlations, etc. just to name a few) is introduced and used extensively in the literature, in complementary fashion to N-point functions. More on these alternatives appear in other lecture notes of this volume, and if your appetite is whet you should read, e.g. [1] for more information. The present lecture notes serve as an informal introduction to the subject, a starting point for further studies, rather than a full-blown review.

Higher order statistics are complicated due to several factors. Most of them arise from the large configuration space of parameters which results in a "combinatorial explosion" of terms. For instance, the first non-trivial three-point correlation function can in principle depend on nine coordinates. Taking into account applicable symmetries still leaves three parameters in real space and six parameters in redshift space (five if one uses the distant observer

Szapudi, I.: *Introduction to Higher Order Spatial Statistics in Cosmology*. Lect. Notes Phys. **665**, 457–492 (2009)
DOI 10.1007/978-3-540-44767-2_14

approximation). This dependence on a large number of parameters renders higher order statistics quite cumbersome, their measurement, interpretation, and visualization surprisingly complex and CPU intensive. For the same reason, theoretical prediction of higher order statistics correspondingly involved non-linearities, redshift distortions, or projection effects, and bias affects them in subtle and non-trivial ways. In addition, higher order statistics are more sensitive to geometry and systematics than their low-order cousin, the two-point correlation function. This means that a comparatively larger and cleaner sample is needed for effective studies. Even with the presently available large surveys, such as SDSS and 2dF, the overlap between well-understood theory and reliable measurement is in fact disquietingly small.

Despite the above difficulties, the study of higher order statistics is a rewarding one, both intellectually and scientifically. The simple idea that small initial fluctuations grew by gravitational amplification is enough to predict higher order correlation functions, at least on large scales, to impressive accuracy. Analytical predictions and simulations that are in contrast with data have already provided a strong support for Gaussian initial conditions and show that our basic picture of structure formation is correct.

On the other hand, the two-point correlation function alone gives a quite simplistic view of LSS, which has been visually demonstrated first by Alex Szalay, through randomizing the phases of a highly non-Gaussian simulation. This procedure erases a large portion of (although not all) higher order correlations, while keeping the two-point statistics (cf. Fig. 1. in Peter Coles' lecture notes in the same volume). It is striking how most structure we see in the simulation disappears when higher order information is erased, despite that the two-point function (or power spectrum) is the same.

In addition, the two-point correlation function is degenerate with respect to the bias (the difference between the statistics of the observed "light" from

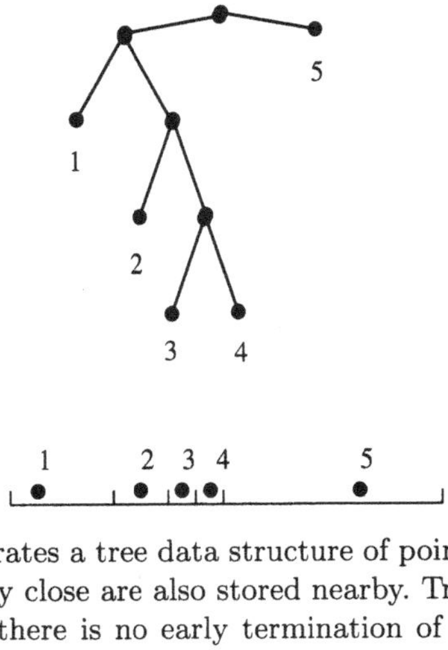

Fig. 1. This figure illustrates a tree data structure of points in 1 dimension. In this construct, points spatially close are also stored nearby. Tree structure is similar for continuous field, except there is no early termination of the tree (such as that of number 5 on the figure)

galaxies and the underlying dark matter) and the amplitude of primordial fluctuations. Most methods which extract information from the two-point correlation function or power spectrum alone draw on assumptions about the bias to avoid this degeneracy problem and/or combine with CMB measurements. Higher order statistics not only yields a more accurate picture of the distribution, but resolves the degeneracy of the two-point function by constraining the bias.

These notes focus on phenomenological aspects of higher order correlation functions and related statistics. In particular, I emphasize the relationship of these statistics with symmetries. Some aspects of this are quite new, and it is the only way I know how to ease the pain of the above-mentioned "combinatorial explosion", at least partially. In the next sections I review the most important theoretical and statistical information, in particular I present the definitions of the statistics, estimators, how errors are calculated, algorithms to realize the estimators, and some points on bias, redshift distortions. I develop symmetry considerations in slightly more detail throughout. For completeness, I include the foundations of perturbation theory; a detailed recent review is available by [2] which is highly recommended. Finally, I illustrate most of the above through an example by working out the theory of conditional cumulants. Finally, while LSS is emphasized in examples, most of the theory is directly applicable to CMB or any other random field as well.

2 Basic Definitions

The learning curve of higher order statistics is initially steep, partly due to the large number of new concepts and definitions which need to be absorbed. Here I attempt to collect some of the most basic definitions which will be used later.

A Spatial Random Field A is a field which has random values at all spatial points. Its corresponding measure is a functional, $\mathcal{P}(A)\mathcal{D}A$, the "probability" of a realization. This is analogous to a probability density. Averages can be calculated via functional integrals, which although not always well defined mathematically are very pictorial.

Ensemble Average We denote with $\langle A \rangle$ the ensemble average of some quantity A. The averaging corresponds to a functional integral over the density measure. Physically, this means simply the average of the quantity over infinite (or many) independent realizations.

Ergodicity In the study of random fields, it is often assumed that ensemble averages can be replaced with spatial averages. Physically this means that distant parts of the random field are uncorrelated, and thus a large realization of the random field (a "representative volume", hopefully the chunk of our universe which is observable) can be used for measurements of statistical quantities.

Joint Moments The most basic set of statistics we can consider to characterize a random field T are the joint moments of the field taken at N distinct points $x_1, \ldots, x_N$,

$$F^{(N)}(x_1, \ldots, x_N) = \langle T(x_1), \ldots, Tx_N \rangle \,. \tag{1}$$

We often work with fluctuation fields which have been normalized with the average: $\delta = \frac{T}{\langle T \rangle} - 1$. Note that the spatial dependence on coordinate x_i is often abbreviated as δ_i. No assumption is made on the dimensionality of the random field. One (Lyman alpha forest), two (CMB, LSS-projected catalogues), or three (LSS redshift surveys) dimensional fields are typical in cosmology.

Connected Moments Connected moments are denoted with $\langle\rangle_c$ and are defined recursively with the following formula:

$$\langle \delta_1, \ldots, \delta_N \rangle_c = \langle \delta_1, \ldots, \delta_N \rangle - \sum_P \langle \delta_1 \ldots \delta_i \rangle_c \ldots \langle \delta_j \ldots \delta_k \rangle_c \ldots \,, \tag{2}$$

where P denotes symbolically a summation of all possible partitions. In other words, connected moment of all possible partitions have to be subtracted from the full (or disconnected) joint moment. What is left is the connected moment.

N-point correlation functions are defined as the connected joint moments of the normalized fluctuation field δ

$$\xi^{(N)}(1, \ldots, N) = \langle \delta_1, \ldots, \delta_N \rangle_c \,, \tag{3}$$

where we introduce yet another short hand for denoting $\xi^N(x_1, \ldots, x_N)$. The three-point correlation function is often denoted with ζ instead of $\xi^{(3)}$, and the two-point correlation function with $\xi \equiv \xi^{(2)}$. Most statistics we are dealing with in this review are special cases of N-point correlation functions.

Gaussian A random field is called Gaussian if its first and second moments fully determine it statistically. We will often use the field δ, which denotes a random field whose average is 0. Then a Gaussian field is fully determined by its two-point correlation function $\xi = \langle \delta_1 \delta_2 \rangle$. Gaussian random fields have trivial higher order moments in that all their connected moments vanish for $N \geq 3$ (Wick theorem).

Non-Gaussian A non-Gaussian field has at least one non-vanishing higher-than-second connected moment. Note that the definition of non-Gaussian is highly general (like "non-Elephant"); everything that is not Gaussian is non-Gaussian. Therein lies one of the highest difficulty of the subject.

Cumulants These correspond to the simplest special case of N-point functions: the joint moment of a degenerate configuration, where all the field values are taken at the same place. The usual definition in cosmology is

$$S_N = \frac{\langle \delta^N \rangle_c}{\langle \delta^2 \rangle^{N-1}} \,, \tag{4}$$

where the normalization with the average correlation function $\bar{\xi} = \langle \delta^2 \rangle$ is motivated by the dynamics of perturbation theory. Other normalizations might be more suitable, e.g. for CMB. Cumulants depend only on one scale: the smoothing kernel of δ which implicit in the above equation. This fact accounts for most of their popularity, and indeed they are the only statistics (along with cumulant correlators), which have been measured as high order as $N = 10$ for galaxies. An alternative definition is $Q_N = S_N/N^{N-2}$, which corresponds to the division with the number of possible tree graphs. This definition typically ensures that for galaxies and dark matter all Q_Ns are of order unity.

Cumulant Correlators These correspond to the next simplest special case of $N+M$-point functions: the joint moment of a doubly degenerate configuration, where the N and M field values are taken at two places, respectively. The usual definition in cosmology is

$$Q_{N,M} = \frac{\langle \delta_1^N \delta_2^M \rangle_c}{\langle \delta^2 \rangle^{N+N-1} \langle \delta_1 \delta_2 \rangle N^{N-2} M^{M-2}} . \tag{5}$$

Other generalizations, e.g. with triple degeneracy, are obvious but not clear whether fruitful. In the above, it is assumed that the smoothing kernel is the same for both points, which is usual, although not necessary. When the normalization of the possible number of trees is not done, the quantity is often denoted with $C_{N,M}$.

Conditional Cumulants These are defined as the joint connected moment of one unsmoothed and $N-1$ smoothed density fluctuation fields. They can be realized by integrals of the N-point correlation function through $N-1$ spherical top hat windows,

$$U_N(R_1, \ldots, R_{N-1}) = \int \xi_N(x_1, \ldots, x_{N-1}, 0) \prod_{i=1}^{N-1} d^3 x_i \frac{W_{R_i}(x_i)}{V_i} , \tag{6}$$

where $V_i = 4\pi/3R_i^3$ is the volume of the window function W_{R_i}. In the most general case, each top hat window might have a different radius. The conditional cumulants are analogous to ordinary cumulants, but subtly different: they have an advantageous smoothing kernel which enables their measurement with an edge-corrected estimator (more on this later).

Power Spectrum Since space is assumed to be *statistically* invariant under translations, it is convenient to introduce the Fourier transform of the D-dimensional field

$$\delta(x) = \int \frac{d^D k}{(2\pi)^D} \delta_k \, e^{i\,kx} . \tag{7}$$

The translation invariance causes the independent k-modes to be uncorrelated, which renders the correlation function of the Fourier modes "diagonal"

$$\langle \delta_{k_1} \delta_{k_2} \rangle \equiv \delta_D(k_1 + k_2)(2\pi)^D P(k) , \tag{8}$$

which is the definition of the power spectrum. (δ_D is the Dirac-δ function.) It can be shown that the power spectrum is the Fourier transform of the two-point correlation function ξ. Note that the $(2\pi)^D$ normalization is customary, but it is not followed by everyone.

Poly-spectra Similar to the power spectrum, the joint connected moments of N-Fourier modes are called a poly-spectra ($N-1$ spectrum) modulo of a Dirac-δ function. The most important special case is the bispectrum, the Fourier transform of the three-point correlation function $\zeta \equiv \xi^{(3)}$. The next order is called trispectrum.

A Discrete Random Field This corresponds to all possible arrangements of discrete objects (whose number can also vary) with the corresponding measure. It is a generalization over the continuous random field which is needed to describe the statistics of the distribution objects in space, such as galaxies.

Poisson Sampling Often it is a good approximation to derive a discrete random field from a continuous one via a simple assumption: take an infinitesimally small volume, such that the probability of having more than one object in the volume is negligible, and simply assume that the probability of having one object is proportional to the value of the field within the small volume. This is called an infinitesimal Poisson process.

Discreteness Effects These correspond to the difference between the statistics of an observed, discrete random field and an assumed underlying, continuous random field. It is also known as shot noise or Poisson noise. It can be especially simply calculated under the assumption of infinitesimal Poisson sampling.

3 Estimators

The main idea underlying the construction of estimators for spatial statistics is the ergodic theorem. If we replace the ensemble averages with spatial averages, we obtain estimators for the above-defined quantities.

To put the above simple idea into practice, one has to deal with some subtleties as well. I list a few interesting ones: discreteness, counts in cells, edge correction, and optimality.

3.1 Discreteness

Imagine a galaxy catalogue with average count $\bar{N}$ in a given cell of size R, our chosen smoothing length. We can estimate approximately the density at each cell as $\tilde{\delta}_i \simeq N_i/\bar{N} - 1$. The above simple idea suggests to estimate the Kth-order cumulant with (for a quantity A we denote its estimator $\tilde{A}$ throughout)

$$\tilde{S}_K \propto \frac{1}{N_{tot}} \sum \tilde{\delta}_i^K|_c \, . \tag{9}$$

A little more thought, however, reveals that while our estimator for density field is unbiased, our estimator for the cumulant is not. While $\langle N \rangle = \bar{N}\langle 1+\delta \rangle = \bar{N}$, $\langle N^K \rangle \neq \bar{N}^K \langle (1+\delta)^K \rangle$. The reason for this is the "self-contribution" to correlations, a typical discreteness effect.

To see this, let us follow [3] and imagine for $K = 2$ that our cell of size R is divided into T infinitesimally small cells, each of them small enough that the probability of having more than one galaxy in it would be negligible. Let us call the number of galaxies in each tiny cells μ_i. Since $\mu_i = 0, 1$, $\mu_i^K = \mu_i$ for any K. This means that all moments $\langle \mu_i^K \rangle = \langle \mu_i \rangle = \bar{N}/T$. The total number of objects in our original cell is $\sum \mu_i$. Now it is easy to see that

$$\langle N^2 \rangle = (\sum \mu_i)^2 = \sum_i \langle \mu_i \rangle + \sum_{i \neq j} \langle \mu_i \rangle \langle \mu_j \rangle (1 + \xi_{ij}) \, . \tag{10}$$

With $T \to \infty$ the above expression yields the final results

$$\begin{aligned} \langle N^2 \rangle &= \bar{N} + \bar{N}^2 (1 + \bar{\xi}) \\ &= \bar{N} + \bar{N}^2 \langle (1 + \delta)^2 \rangle \, . \end{aligned} \tag{11}$$

The first term corresponds to the Poisson noise bias of our estimator. Note that we already simplified the case assuming that $\bar{N}$ and all the normalization required to obtain S_K from the above is known a priori.

Higher orders can be calculated either with the above suggestive, but tedious method, or with generating functions. One can prove the following simple rule [4]

$$\langle (N)_K \rangle = \bar{N}^K \langle (1 + \delta)^K \rangle \, , \tag{12}$$

where $\langle (N)_K \rangle = \langle N(N-1) \cdots (N-K+1) \rangle$ are the factorial moments of N (the quantity inside the average is called falling factorial). In other words, if you replace regular moments with factorial moments, discreteness effects magically disappear. With this simple rule, and taking into account that discreteness can only affect overlap, you can construct estimators free from discreteness biases for any of our defined estimators, e.g.

$$\begin{aligned} \langle (N)_2 \rangle &= \bar{N}^2 (1 + \bar{\xi}) \, , \\ \langle (N)_3 \rangle &= \bar{N}^3 \langle (1 + 3\delta + 3\delta^2 + \delta^3 \rangle \\ &= \bar{N}^3 (1 + 3\bar{\xi} + 3 S_3 \bar{\xi}^2) \, , \end{aligned} \tag{13}$$

and all you have to do is express algebraically S_3 as a function of the factorial moments. It is a good exercise, for solutions see [5].

Note that both intuition and halo models [6] suggest that Poisson sampling might not be a good approximation on very small scales for galaxy surveys. When in doubt (i.e. no well defined and trusted model for the discreteness exist), try to use estimators without self-overlap, since discreteness manifests itself through self-correlations only.

3.2 Counts in Cells

We have shown in the previous subsection how to construct estimators for cumulants, cumulant correlators, etc. even in the discrete case. For cumulants, however, a more efficient method exists based on counts in cells.

According to the previous subsection, if we can estimate factorial moments $(N)_K$, it is a simple matter of calculation to obtain cumulants. While one could directly estimate the factorial moments from data, the usual procedure is to estimate first the probability P_N that a cell of a given volume (or area) contains N objects. Then the factorial moments can be calculated as

$$\langle (N)_K \rangle = \sum_{N \geq 0} P_N (N)_K \ . \tag{14}$$

The advantage of this technique is that there exist many fast and efficient algorithms to estimate P_N from data, and once the counts-in-cells probabilities are estimated, arbitrary high-order cumulants can be obtained in a straightforward manner.

3.3 Edge Correction and Heuristic Weights

If signal and noise were uniform, uniform weighting of objects would be optimal as required by symmetry. On scales approaching the characteristic size of a survey, uneven (or more generally suboptimal) weighting of objects near the edges of the survey will often dominate the errors. The optimal weight has contributions from survey geometry and correlations (signal and noise), and it is not even diagonal. These two effects are only separate conceptually, but they are often coupled in non-trivial ways.

Edge correction with heuristic weights fully accounts for the effect of survey geometry (including uneven sampling, etc.); however, it does not include a prior on the correlations. For instance, the heuristic weights used in the SpICE method [7] are exact in the noise-dominated regime, however, they are only approximate when the signal is important.

A class of edge-corrected estimators [8, 9] can be written symbolically as

$$\tilde{\xi}^N = \sum_{i_1,\ldots,i_N} w_{i_1,\ldots,i_N} (\delta_{i_1} \ldots \delta_{i_N})_c \ , \tag{15}$$

which is formally the same as the quantity we want to measure, except we replace $\langle\rangle$ with an average over the sample itself according to the ergodic principle. The heuristic weights are such that their total sum is 1, and usually correspond to an inverse variance. This would be exact if the bins were uncorrelated, which is usually not the case. It can be a good approximation power spectra as shown by [10] (cf. lecture notes of Andrew Hamilton in this volume).

Note that the above estimator introduces additional subtleties due to taking of connected moments and due to the non-linear nature of the estimator. The non-linearity is introduced by calculating the fluctuations via a division with the average density. If we estimate the average density from the same sample, the bias due to this effect is termed the "integral constraint".

In practice, the above estimator is often replaced with a Monte Carlo version ([11] for the two-point correlation function and [8, 9] for the N-point functions). Let D and R denote the data set and a random set, respectively. The latter is a random distribution over the geometry and sampling variations in the survey. If we define symbolically an estimator $D^p R^q$, with $p + q = N$ for a function Φ symmetric in its arguments

$$D^p R^q = \sum \Phi(x_1, \ldots, x_p, y_1, \ldots, y_q) , \tag{16}$$

with $x_i \neq x_j \in D$, $y_i \neq y_j \in R$. As an example, the two-point correlation function corresponds to $\Phi(x,y) = [x, y \in D, r \leq d(x,y) \leq r + dr]$, where $d(x,y)$ is the distance between the two points, and [*condition*] equals 1 when *condition* holds, 0 otherwise.

The estimator for the N-point correlation functions is written with the above symbolic notation as $(\hat{D} - \hat{R})^N$, or more precisely,

$$\tilde{\xi}_N = \frac{1}{S} \sum_i \binom{N}{i} (-)^{N-i} \left(\frac{D}{\lambda}\right)^i \left(\frac{R}{\rho}\right)^{N-i} , \tag{17}$$

where $S = \int \Phi \mu_N$, a simple phase space integral, and λ, ρ are the densities of data and random sets (for details see [9]). This is the Monte Carlo version of the edge-corrected estimator.

It is worthwhile to note that counts in cells cannot be properly edge corrected, although approximate procedures exist in the regime where the probabilities have a weak dependence on the cell shape, or one could integrate the above estimators numerically.

3.4 Optimality

For the measurement of the two-point function for a Gaussian process we know that maximum likelihood methods are optimal in the sense that no other method can achieve smaller variance. This can be shown with the use of the Fisher matrix and the Cramer–Rao inequality [12, 13] (see also the lecture notes of Andrew Hamilton in this volume). Since our previous prescription for edge effects is different (in particular it does not know about correlations), it is clearly suboptimal. Given that we know the optimal method, why do not we use it for all measurements? The reason is that there are caveats, which, especially for galaxy surveys, severely limit the applicability (and optimality) of the optimal method.

- Computational issues: The quadratic incarnation of the maximum likelihood method for two-point correlation function (or power spectrum) amounts to sandwiching the projection matrix $P_l = \partial C / \partial l$, the derivative of the full (signal plus noise) correlation matrix according to the lth parameter C_l between the data vectors weighted by the inverse correlation matrix C^{-1}, yielding

$$\tilde{C}_l = < C^{-1}x|P_l|C^{-1}x > , \quad (18)$$

where $\tilde{C}_l$ is the estimator for C_l. We can see from this that the inverse of the correlation matrix needs to be calculated, which typically scales as $O(N^3)$ with the number of data items.[1] For $N \simeq 1000$ this can be done on a laptop computer (about 17 s on mine), but for $N \gtrsim 10^6$ it becomes prohibitive even for the fastest supercomputers (not to mention storage capacity, which might also become unrealistic for a large data set). Iterative, and brute force maximum likelihood methods, although seemingly more complicated, scale the same way.

State-of-the-art data sets ("mega-pixel" CMB maps, such as WMAP and Planck, as well as decent LSS surveys) are way beyond present-day computational capabilities (unless some special tricks and approximations are involved, e.g. [14]).

- Caveats: The "lossless and optimal" property of the maximum likelihood estimator is subject to practical limitations. Data reduction and analysis involve many subtle choices, such as systematic corrections, binning, pixelization, etc. In the past these turned into differences between analyses which all used the same optimal method. In addition, non-iterative quadratic maximum likelihood estimators are posteriors based on a prior, thus they are only optimal in the limit when we have full prior knowledge of what we want to measure.
- The Gaussianity condition: While it is an excellent approximation for the CMB, it breaks down for the LSS even on very large scales. In the Bayesian sense, the Gaussian prior is not justified, thus the estimators might become suboptimal and might even be biased. Those measurements, where variants of the maximum likelihood method are used for LSS, take great pain at controlling "non-linear contamination", e.g. with filtering. Besides the computational aspects, this is the main reason why many LSS analyses still use heuristic weights. However, some aspects of the maximum likelihood method have been adapted to deal with non-Gaussianities [15] and even for estimating three-point correlation functions under the assumption of Gaussianity [16, 17] (this is not a contradiction: when the non-Gaussianity is tiny, as in CMB, correlation matrices can be well approximated with a Gaussian assumption).

4 Errors

The formal calculation of errors (or more generally covariance matrices) is straightforward from (15): one simply takes the square of the equation and performs ensemble averages. In general, for two bins represented by the weights w^a and w^b, the errors will scale as

[1] There are methods, such as preconditioned conjugate gradients, with which $C^{-1}x$ can be calculated in $O(N^2)$. However, the Fisher matrix still needs $O(N^3)$ computations.

$$\Delta\tilde{\xi}^N = \sum_{i_1,\dots,i_N} w^a_{i_1,\dots,i_N} w^b_{j_1,\dots,j_N} \langle \delta_{i_1} \dots \delta_{i_N} \delta_{j_1} \dots \delta_{j_N} \rangle \,. \tag{19}$$

In practice, complications will arise from (i) the large number of terms, (ii) the complicated summation (or integral in the continuum limit), (iii) the overlaps between indices and the corresponding discreteness terms, and (iv) the theoretical expression of the ensemble averages. However, we can draw the general idea from the expression that the covariance matrix of N-point estimators depends on $2N$-point correlation functions. Next we demonstrate this formula with the calculation of the errors on the two-point correlation function.

To specify more our symbolic estimator, let us assume that the survey is divided into K pixels, each of them with fluctuations δ_i, with i running from $1, \dots, K$. For this configuration our estimator can be written as

$$\tilde{\xi} = w_{12}\delta_1\delta_2. \tag{20}$$

The above equation uses a "shorthand" Einstein convention: $1, 2$ substituting for i_1, i_2, and repeated indices summed, and it is assumed that the two indices cannot overlap.

The ensemble average of the above estimator is clearly $w_{12}\xi_{12}$. The continuum limit (co)variance between bins a and b can be calculated by taking the square of the above, and taking the ensemble average:

$$\langle \delta\tilde{\xi}^a \delta\tilde{\xi}^b \rangle = w^a_{12} w^b_{34} \left(\langle \delta_1\delta_2\delta_3\delta_4 \rangle - \langle \delta_1\delta_2 \rangle \langle \delta_3\delta_4 \rangle \right) \,. \tag{21}$$

Note that the averages in this formula are not connected moments, which are distinguished by $\langle\rangle_c$.

The above equation yields only the continuum limit terms. To add Poisson noise contribution to the error, note that it arises from the possible overlaps between the indices (indices between two pair weights *can* still overlap!). In the spirit of infinitesimal Poisson models, we replace each overlap with a $1/\lambda$ term and express the results in terms of connected moments. There are three possibilities:

(i) no overlap (continuum limit)

$$w^a_{12} w^b_{34} \left(\xi_{1234} + \xi_{13}\xi_{24} + \xi_{14}\xi_{23} \right) \,, \tag{22}$$

(ii) one overlap (four possibilities)

$$\frac{4}{\lambda} w^a_{12} w^b_{13} \left(\xi_{123} + \xi_{23} \right) \,, \tag{23}$$

(iii) two overlaps (two possibilities)

$$\frac{2}{\lambda^2} w^a_{12} w^b_{12} \left(1 + \xi_{12} \right) \,. \tag{24}$$

In these equations, for the sake of the Einstein convention we used $\xi(i,j,k,l) \rightarrow \xi_{ijkl}$. The above substitutions (rigorously true only in the infinitesimal Poisson sampling limit) become increasingly accurate with decreasing cell size. For the Monte Carlo estimator of (17) a slightly more general formula is valid, where the summation is replaced with integrals over a bin function $\Phi(1,2)$

$$\begin{aligned}\langle\delta\tilde{\xi}_2^a\delta\tilde{\xi}_2^b\rangle = \frac{1}{S^2}\Big\{ &\int \Phi_a(1,2)\Phi_b(3,4)\,[\xi_4(1,2,3,4) + 2\xi(1,3)\xi(2,4)] \\ &+\frac{4}{\lambda}\int \Phi_a(1,2)\Phi_b(1,3)\,[\xi(2,3) + \xi_3(1,2,3)] \\ &+\frac{2}{\lambda^2}\int \Phi_a(1,2)\Phi_b(1,2)\,[1+\xi(1,2)]\Big\}\,. \end{aligned} \tag{25}$$

For completeness, we present the result for the three-point correlation function as well:

$$\begin{aligned}\langle\delta\tilde{\xi}_3^a\delta\tilde{\xi}_3^b\rangle = \frac{1}{S^2}\Big\{ &\int \Phi_a(1,2,3)\Phi_b(4,5,6)\big[\xi(1,2,3,4,5,6) \\ &+3\xi(1,2)\xi(3,4,5,6) + 9\xi(1,4)\xi(2,3,5,6) \\ &+3\xi(4,5)\xi(1,2,3,6) + 9\xi(1,5,6)\xi(2,3,4) \\ &+9\xi(1,4)\xi(2,3)\xi(5,6) + 6\xi(1,4)\xi(2,5)\xi(3,6)\big] \\ =& \frac{9}{\lambda}\int \Phi_a(1,2,3)\Phi_b(1,4,5)\big[\xi(1,2,3,4,5) \\ &+\xi(2,3,4,5) + 2\xi(1,2)\xi(3,4,5) \\ &+2\xi(1,4)\xi(2,3,5) + \xi(2,3)\xi(1,4,5) \\ &+4\xi(2,5)\xi(1,3,4) + \xi(4,5)\xi(1,2,3) \\ &+\xi(2,3)\xi(4,5) + 2\xi(2,4)\xi(3,5)\big] \\ &+\frac{18}{\lambda^2}\int \Phi_a(1,2,3)\Phi_b(1,2,4)\big[\xi(1,2,3,4) \\ &+2\xi(1,3,4) + \xi(1,2)\xi(3,4) + 2\xi(1,3)\xi(2,4) + \xi(3,4)\big] \\ &\frac{6}{\lambda^3}\int \Phi_a(1,2,3)\Phi_b(1,2,3)\big[\xi(1,2,3) + 3\xi(1,2) + 1\big]\Big\}\,. \end{aligned} \tag{26}$$

In the above formula, ξ with N variables is an N-point correlation function, Φ_a is a bin function corresponding to a triangle (its value is 1 when the triplet is in the bin, 0 otherwise).

The above integral is fairly complicated to calculate in practice, usually some approximations are necessary. In series of papers [5, 7, 18] has worked out practical approximations for up to $N = 4$ for the moments of CIC, obtained (lengthy) semi-analytical formulae, and checked validity of those approximations against simulations. The resulting code (FORtran for Cosmic Errors, FORCE is available publicly.[2] The errors of CIC have the special property

[2] `http://www.ifa.hawaii.edu/users/szapudi/force.html`

that they contain terms $\propto 1/C$, where C is the number of cells used for the CIC estimation (measurement errors). This is the motivation for algorithms that have $C = \infty$.

From the above error calculations an intuitive physical picture has emerged. We can distinguish three classes of errors, often approximately separated in the final expressions: finite volume errors (the term cosmic variance is used in CMB literature), corresponding to the fact that the universe might have fluctuations on scales larger than the survey size (the smallness of such fluctuations often termed qualitatively as "representative volume"); discreteness errors, arising from the fact that we are sampling the dark matter distribution (presumably continuous) at finite points with galaxies; and edge effect errors, arising from the uneven weights given to galaxies during the estimation. For typical galaxy surveys, edge effects dominate on large scales and discreteness on small scales. Finite volume effects change slower, and they might dominate in the transition region on intermediate scales. In the above example for the two-point function, the first term contains mainly finite volume effects (the integral of the four-point function), and terms with powers of $1/\lambda$ are discreteness effects. Edge effects are due to the complicated weight summation blended with the other effects.

These qualitative observations are valid for CMB as well with some caveats: (i) the CMB is Gaussian to good approximation, therefore typically only two-point functions need to be taken into account, (ii) the CMB is continuous, i.e. there are no discreteness effects, and (iii) instead, there are terms arising from the noise of the detection with similar role to discreteness (noise) in LSS. Note that for high-precision cosmology applications constraining cosmological parameters, "error on the error" (or uncertainty of the uncertainty as sometimes called) is as important as the size of the error.

5 Symmetry Considerations

A class of functions subject to (Lie-group) symmetries is best represented as an expansion over irreducible representations. In Euclidean (flat) space translation invariance is the most obvious symmetry. The appropriate transform is the Fourier transform, and homogeneity can be taken into account as a Dirac-delta function in transform space. The customary definition of the power spectrum and bispectrum are with the Fourier transform of the random fluctuation field $\delta(k)$ as

$$\begin{aligned} \langle \delta(k_1)\delta(k_2) \rangle &= (2\pi)^D \delta_D(k_1 + k_2) P(k_1) \\ \langle \delta(k_1)\delta(k_2)\delta(k_3) \rangle &= (2\pi)^D \delta_D(k_1 + k_2 + k_3) B(k_1, k_2, k_3) \,, \end{aligned} \tag{27}$$

where δ_D is a Dirac-delta function and D is the spatial dimension. Thus, because of homogeneity, the two-point function becomes the Fourier transform of the power spectrum

$$\xi(x_1, x_2) = \int \frac{d^D k}{(2\pi)^D} P(k) \exp\left[\mathrm{i}\, k(x_1 - x_2)\right] , \qquad (28)$$

and a similar equation is true for the three-point correlation function

$$\xi(x_1, x_2, x_3) = \int \Pi_{i=1}^{3} \frac{d^D k_i}{(2\pi)^D} B(k_1, k_2, k_3) \times \exp\left[\mathrm{i}\,(k_1 x_1 + k_2 x_2 + k_3 x_3)\right] \delta_D(k_1 + k_2 + k_3) . \qquad (29)$$

From these equations, the two-point correlation function is only a function of the difference of its two arguments. If the statistical properties of the underlying field are isotropic as well, these equations can be further simplified. We quote the results for two and three spatial dimensions:

$$\xi(r) = \int \frac{kdk}{2\pi} P(k) J_0(kr) , \qquad 2\mathrm{D} , \qquad (30)$$

$$\xi(r) = \int \frac{k^2 dk}{2\pi^2} P(k) j_0(kr) , \qquad 3\mathrm{D} , \qquad (31)$$

where J_0 and j_0 are ordinary and spherical Bessel functions, respectively.

As a consequence of spherical symmetry the three-point correlation function and the bispectrum depend only on the shape of a triangle. Szapudi [19] has observed that the three-point statistics can be expressed with two unit vectors, thus zero angular momentum bipolar expansion is suitable in three spatial dimensions under SO(3) symmetry. Zero angular momentum bipolar functions are proportional to the Legendre polynomials, thus in turn this becomes multi-pole expansion of the bispectrum. If we parametrize the bispectrum as depending on two vectors, and the angle between, this can be written as

$$B(k_1, k_2, \theta) = \sum_l B_l(k_1, k_2) P_l(\cos\theta) \frac{2l+1}{4\pi} , \qquad (32)$$

with an entirely similar expansion for the three-point correlation function. It is then simple matter to show [19] that the multi-poles of the bispectrum are related to the multi-poles of the three-point correlation function via a double Hankel transform

$$\xi_l^3(r_1, r_2) = \int \frac{k_1^2}{2\pi^2} dk_1 \frac{k_2^2}{2\pi^2} dk_2 (-1)^l B_l(k_1, k_2) j_l(k_1 r_1) j_l(k_2 r_2) . \qquad (33)$$

B_ls and ξ_l are analogous to the C_ls of the angular power spectrum, they correspond to an angular power spectrum of a shell of three-point statistics.

In two dimensions, the situation is entirely analogous: the symmetry is just U(1) rotations on a ring, thus the correct expansion of Fourier (actually cosine) transform is

$$B(k_1, k_2, \theta) = \frac{B_0(k_1, k_2)}{2} + \sum_{n<0} B_n(k_1, k_2) \cos(n\theta) , \qquad (34)$$

with analogous expansion for the three-point correlation functions. Note that the sine modes are killed by parity. This expansion is generally applicable for any three-point statistics in isotropic two-dimensional spaces. This in turn renders the connection of the Fourier coefficients of the three-point correlation function and bispectrum as

$$\xi_n^3(r_1, r_2) = \int \frac{k_1}{2\pi} dk_1 \frac{k_2}{2\pi} dk_2 (-1)^n B_n(k_1, k_2) J_n(k_1 r_1) J_n(k_2 r_2) . \qquad (35)$$

This latter equation, derived for two-dimensional Euclidean space, is also applicable to the spherical bispectrum in the flat sky approximation. This provides surprisingly accurate results, even for relatively low multi-poles. A form equivalent to the expansion of (35) was proposed independently by [20] as well for the (redshift space) projected bispectrum; this is yet another application of the flat sky approximation.

6 Algorithms

Estimation of higher order statistics is a daunting computational problem. Naive solutions typically range from the inefficient (for two-point correlations and counts in cells) to the impossible, hence the application of advanced computer science is a must.

For counts in cells, the naive algorithms scale as MN_C which is preferably $N_C \gtrsim 10^{9-12}$, the number of cells used for the estimation.

Naive estimation of N-point quantities typically scales M^N where M is the number of data points. For modern data sets $M \gtrsim 10^{6-10}$, this becomes prohibitive for higher than second-order correlation functions. This is a fast developing field. Next we present a set of algorithms to illustrate the problems and typical solutions.

6.1 *N*-point Correlation Functions

Hierarchical Algorithms

Finding (joint) averages like $\langle \delta_1 \dots \delta_N \rangle$ of a random field δ, where the N-points are taken at a particular configuration, can be done through summation of N-tuples. For the proper averages, the same algorithm is used for the data and a random set describing the geometry of the survey. For the Monte Carlo estimators an ensemble of mixed N-tuples need to be counted, but again the same algorithm applies without further complexity. The algorithm for $N = 2$ is described next; higher orders are exactly analogous, although more tedious to describe. For details see [21].

The spatial configuration of the two-point function is characterized by the distance r between the two points: this should lie within a bin, i.e. $b_1 < r \le b_2$.

An algorithm capable of calculating the sum $\sum_{r_{12} \leq b} \delta_1 \delta_2$ for any b yields the answer for a proper bin $b_1 < r \leq b_2$ as a difference.

The data points or the pixels of the continuous field can be arranged in a tree data structure. At every level, we store the number of elements (or values) below, i.e. sufficient cashed statistics. The sum needed then can be calculated by a recursive double search on the tree starting from the root. If the two nodes are n_1 and n_2 with values δ_1 and δ_2, let us denote with r any distance between points of n_1 and n_2 (the lines between the squares of Fig. 2). If we can prove that (i) for any $r > b \rightarrow$ return, or (ii) for any $r \leq b$ $\rightarrow$ add $\delta_1 \times \delta_2$ to the sum and return (we are done), else (iii) split the tree and continue recursively. In worst case this procedure descends to the leaves of the tree, but typically huge savings are realized when whole chunks of data can be taken into account at coarser levels of the tree.

Algorithms Based on Fourier Transforms

As shown next, pair summation can be reformulated to make use of fast Fourier transforms, one of the fastest algorithms in existence. If $P(x) = a_0 + a_1 x + \cdots + a_{n-1}$ is a polynomial, and $\epsilon = e^{2\pi i/r}$ unit roots, the coefficients of a discrete Fourier transform of the series a_i can be defined as

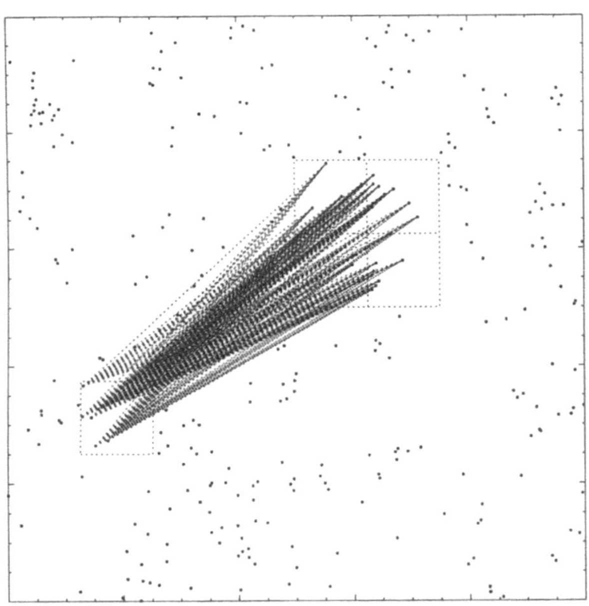

Fig. 2. A particular state of the algorithm is represented. The two squares illustrate two nodes of the tree, which contain the points inside. The number of points is cashed, i.e. available at this level of the tree. The pairs of distances between points are marked with lines. If case (i) or (ii) is true (see text), none of these distances have to be checked, which in turn furnishes the speed-up of the algorithm. The smaller squares within the large square represent the case (iii) when the tree has to be split; then the query is run again recursively

$$\hat{a}_k = P(\epsilon^k) = a_0 + a_1\epsilon^k + \cdots + a_{n-1}\epsilon^{(n-1)k} . \quad (36)$$

Direct calculation confirms that $\sum a_k b_{k+\Delta}$ can be calculated by Fourier transforming the series a_i and b_i, multiplying the resulting $\hat{a}_k \hat{b}^*_k$, and finally inverse Fourier transforming back. This simple observation is the discrete analogue of the Wiener–Khinchin theorem. To obtain correlation one has to work through subtleties involved with the periodic boundary conditions and multidimensionality. The final result is probably a fastest algorithm for calculating correlation functions.

The algorithm in broad terms consists of (i) placing the points in a sufficiently fine grid and storing the value $N_{\mathbf{k}}$, the number of objects at (vector) grid point $\mathbf{k}$ (this step is omitted if the density field is given, such as CMB maps), (ii) calculating fluctuations of the field by $\delta = (N - \langle N \rangle)/\langle N \rangle$, (iii) discrete Fourier transform with a fast FFT engine, (iv) multiplying the coefficients, and (v) Fourier transform back. The same procedure is followed for the mask with zero padding large enough to avoid aliasing effects. The resulting inhomogeneous correlation function is a generalization of the one obtained in the previous subsections; the usual (homogeneous) correlation function can be obtained by summing over spherical shells. Edge effect-corrected power spectrum is obtained with yet another Fourier transform. Measurement of the correlation function on a 768^3 grid takes about 15 minutes on one Opteron processor (cf. Fig. 3).

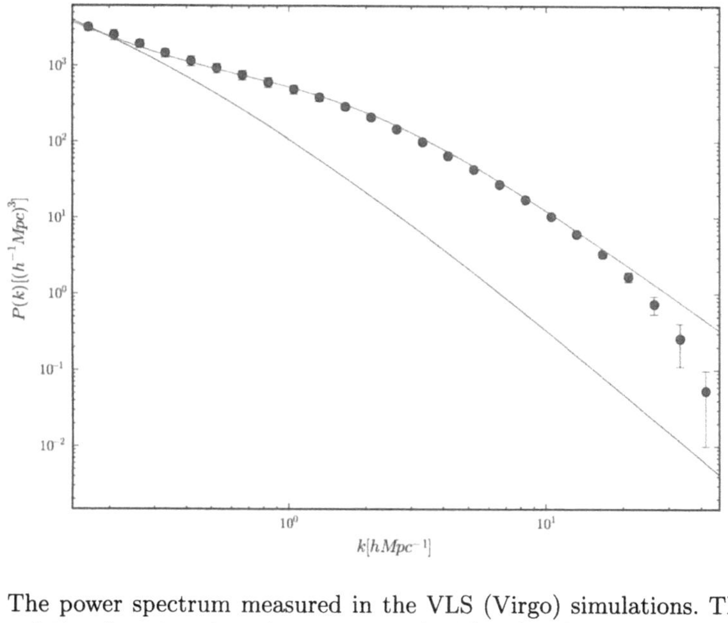

Fig. 3. The power spectrum measured in the VLS (Virgo) simulations. The two-point correlation functions have been measured with a fast hierarchical algorithm, and the edge-corrected power spectrum was obtained from (31). The error bars are calculated from eight sub-cubes from the simulation, and the two curves show the linear and non-linear theoretical power spectra [22]. The measurement of the correlation function with the Fourier algorithm is even faster, and produces identical results

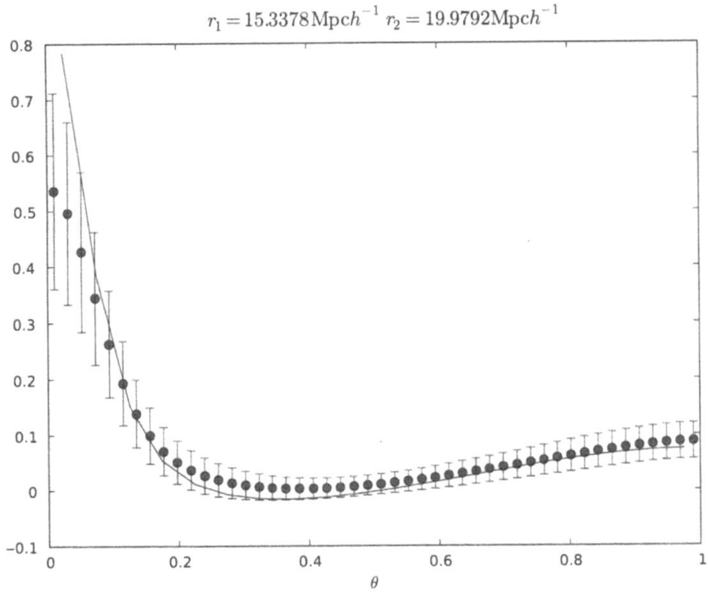

Fig. 4. The three-point correlation function measured with a Fourier-based algorithm. The data used are the same as in Fig. 3. Theory (*solid line*) is calculated through (40). Pure leading order perturbation theory is used, no smoothing effects or non-linearities are taken into account, which probably account for small differences, besides cosmic variance. Note that error bars, estimated from the same eight sub-cubes as in the previous figure, are highly correlated

Pure Fourier algorithms are not practical for edge-corrected three-point statistics. However, it is possible to combine hierarchical and Fourier methods to obtain an algorithm which is faster than either of them. In Fig. 4 I show measurements with such an algorithm in eight realizations of 260 million particle VLS [23] simulations in a 239.5 Mpch cube. The measurement using $\simeq 1400$ bins on a 100^3 grid ($\equiv 10^{15}$ triangles) took less than 3 hours on a 2.4 GHz single CPU.

6.2 Counts-in-Cells Algorithms

I present four different algorithms for counts in cells, because each of them is optimal for different purposes: algorithm I calculates counts in cells with infinite sampling for small scales, II is a massively oversampling algorithm for large scales, III is a massively oversampling algorithm for intermediate/small scales, while IV is a massively oversampling algorithm for large scales and arbitrary cell shapes/sizes. II and IV have very similar performance, although II is more straightforward to generalize for lensing and compensated filters, while IV has more flexibility in terms of cell shape/size. Together I–IV covers the full dynamic range of future surveys, with ample overlap between them for cross-check.

CIC I: Sweep

This algorithm uses the well-known "sweep" paradigm from computational geometry, and it realizes the ideal case of $C = \infty$ number of "random" cells thrown. Here we summarize the basic idea in two dimensions only.

The geometric interpretation of the probability of finding N galaxies in a randomly thrown cell is shown in Fig. 5. There are four points in a rectangular box. Around each object (large dots) a square is drawn, identical to the sampling cell used for counts in cells. The possible centres of random cells all lie within the dashed line, which follows the boundary of the bounding box. Since the square around each point corresponds to the possible centres of (random) cells containing that same point, the question can be reformulated in the following way: let us partition the area of the possible centres of cells according to the overlap properties of the cells drawn around the objects. The ratio of the area with N overlaps to the total area corresponds to P_N.

Imagine a rigid vertical line moving slowly from the left of Fig. 5 towards the right; the boundary can be ignored temporarily. Before the line touches any of the squares, it sweeps through an area contributing to P_0. Therefore at the point of first contact all the swept area contributes to P_0 and can be recorded. After the contact the line is divided into segments sweeping through areas contributing to P_0 and P_1, respectively. The boundaries of these

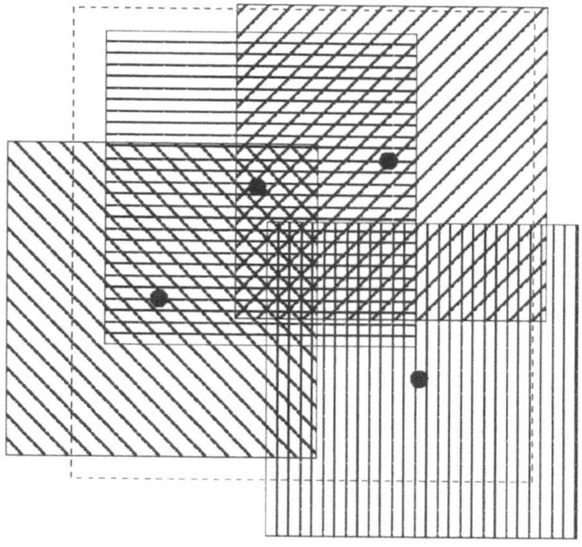

Fig. 5. Illustrates the geometric calculation of counts in cells. There are four points within the solid boundary. The centres of square cells can lie within the dashed boundary. Around each point a square is drawn to represent the possible centres of cells which contain that point. The problem of counts in cells can now be reformulated as calculation of the ratios of all overlap areas (represented with different shadings on the figure) within the dashed boundary

segments can be imagined as two markers on the line, corresponding to the upper and lower corner of the square being touched. As the sweep continues, the results can be recorded at any contact with the side of a square during the movement of the line: the areas swept are assigned according to the markers on the line to different P_Ns. This is done with a one-dimensional sweep on the line counting the two kinds of markers. Then the segmentation of the line is updated. Whenever the line makes contact with the left side of a square, two markers are added; whenever it touches the right hand side of a square, the corresponding markers are dropped. The boundaries and rectangular masks can be trivially taken into account by only starting to record the result of the sweep when entering the area of possible centres. Non-rectangular masks can be converted to rectangular ones by putting them on a grid.

If there are N objects in the plane, the above procedure will finish after $2N$ updating. The algorithm can be trivially generalized for arbitrary rectangles of any dimensions. For instance, in three dimensions the basic sweep is done with a plane, while the plane has to be swept by a line after each contact.

From the definition of the algorithm it follows that the required CPU time scales as $N^D(d/L)^{D(D-1)/2}$ in $D = 2, 3$ dimensions, where N is the number of objects and d/L is the ratio of the scale of the measurement to the characteristic survey length. While for large scales and large D the algorithm becomes impractical, it is clear that for small scales it will be quite fast. It turns out that this is the regime where $C = \infty$ is the most important [18].

CIC II: Successive Convolution

This algorithm is essentially a Fourier (or renormalization) style convolution. It will be explained in one dimension for simplicity; generalization is obvious.

The computations are performed on the largest possible grid with N segments which can be fit into the memory of the computer: this determines the smallest possible scale L/N, where L is the box size and N is the base sampling. A hierarchy of scales are used, with the scale at a given level being

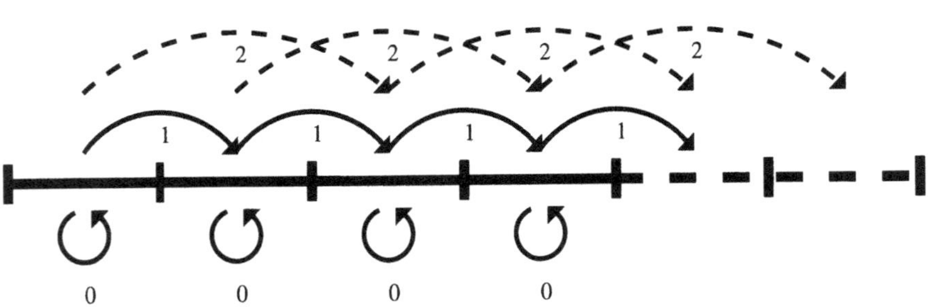

Fig. 6. The successive convolution algorithm is illustrated in one dimension. The solid line represents the grid containing the data, while the dashed line corresponds to auxiliary storage repeating the first two cells. At the 0th level data are placed in the grid, and counts in cells are calculated. At levels 1 and 2 the solid and dashed arrows respectively, represent summations described in the text

twice the scale at one level lower. The results one step lower in the hierarchy are used to keep the number of sampling cells constant even at the largest scales. Counts in cells can be straightforwardly calculated on the resolution scale of the grid, i.e. the smallest scale considered. For the calculation at twice the previous scale the sum of two cells are always stored in one of the cells, for instance in the one with smaller index. Because of the periodic boundary conditions, auxiliary storage is required to calculate the sum of the values in the rightmost cell (if the summations were done left to right), as its right neighbour is the leftmost cell which was overwritten in the first step. After these preparatory steps, counts in cells can again be calculated from the N numbers representing partially overlapping cells. For the next level, twice the previous scale, one needs the sum of four original resolution cells: a calculation simply done by summing every other cell of the previous results into one cell. At this level, two auxiliary storage spaces are needed because of the periodicity. In general, at each level in the hierarchy two cells of the previous results are summed as a preparatory step, and counts in cells are calculated simply by computing the frequency distribution of the N sums stored in the main grid. Auxiliary storage is needed for those rightmost cells, which have the periodic neighbours on the left end.

In D dimensions, 2^D cells are summed in the preparatory step, and the auxiliary storage space enlarges the original hypercube. The scaling of this algorithm is $C \log C \simeq G \log G$ with a preparation step linear in N.

CIC III: Tree

This alternative technique for small scales uses a tree data structure, similar to the algorithm defined for the N-point correlation functions; it is explained in three dimensions for convenience.

The tree data structure can be thought of as an efficient representation of a sparse array, since at small scales most of the cells are empty in a grid spanning the data. The tree is built up recursively by always dividing the particles into two groups based on which half of the volume they belong to. The same function is called on both halves with the corresponding particles until there is no particle in the volume, or the scale becomes smaller than a predetermined value. At each level the scale and the number of particles are known, and when an empty volume is reached, all contained volumes are also empty. These two observations are enough to insert the bookkeeping needed to calculate counts in cells at all scales while the tree is built. The number of sampling cells at each level are 2^l, where l is the level; the original box is represented by $l = 0$. Towards smaller scales the number of cells increases. When $N^3 = 2^l$, where N is the size of the largest grid of the previous algorithm, the two techniques should (and do) give the exact same answers. At larger scales the previous algorithm is superior, since $N > 2^l$, while this algorithm becomes useful at smaller scales. Just as above, this procedure can be further improved by shifting the particles slightly before calculating the tree. However, since this hierarchy of grids has different numbers of cells, random shifts are more

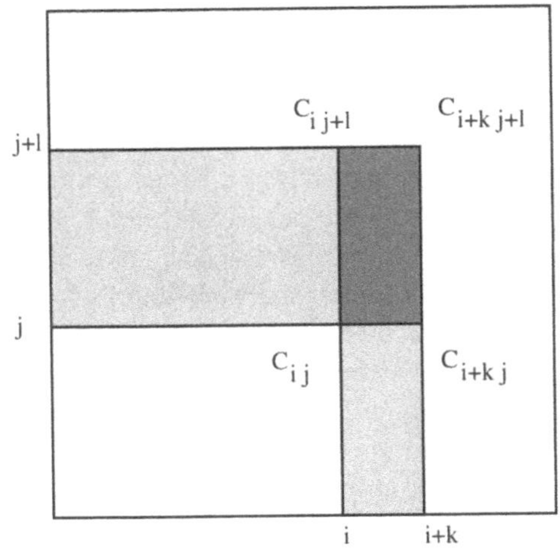

Fig. 7. In the cumulative grid algorithm, each grid point is replaced with a sum of all elements corresponding to rectangles on the figure, with (0,0) as the lower left and the grid point as the upper right coordinate. The dark shaded area is the required sum, which is calculated as $c_{ij} + c_{i+k_1 j+k_2} - c_{i+k_1 j} - c_{ij+k_2}$ (see text)

advantageous. Shifting by a fraction of the smallest scale would not exhaust the possibilities for any larger scale, while shifting by a fraction of the largest grid might not shift the underlying grids at all. With the introduction of random shifts (oversampling grids), the dynamic range of algorithms II and III will develop a substantial overlap, which will be useful for testing. This algorithm also scales as $N \log N$ (with preset depth limiting).

CIC IV: Cumulative Grid

Algorithm II produces counts-in-cells results on scales $2^k l$, where l is the scale associated with the base grid. For calculating counts-in-cells distribution for arbitrary scales $[k_1 \times l, k_2 \times l]$, the following construction will be explained in two dimensions; generalization is obvious.

Let us denote the value of the field n_{ij} at the grid point (i, j). Let us define another grid, with values $c_{ij} = \sum_{p \leq i, q \leq j} n_{pq}$. Then the number of elements in a cell described by $(i, j), (i + k_1, j + k_2)$ can be calculated simply from $c_{ij} + c_{i+k_1 j+k_2} - c_{i+k_1 j} - c_{ij+k_2}$ (a well-known trick in computational geometry). The preprocessing is proportional to N, and counts in cells for arbitrary rectangle that can be calculated is linear with $C \simeq G$.

7 Perturbation Theory

On large scales, where the fluctuations are reasonably small, clustering of cosmic structures can be understood in terms of Eulerian weakly non-linear perturbation theory (PT). An excellent exposition is found [2] with substantial reference list. My goal next is to give an extremely cursory, recipe level introduction to the subject.

PT predicts the behaviour of any statistics from Gaussian initial conditions, assuming an expansion of the density field into first, second, etc. order, $\delta = \delta^{(1)} + \delta^{(2)} + \ldots$. This assumption, when substituted into Euler's equations for the cosmological dark matter "fluid", yields a set of consistent equations for the different orders. The resulting perturbative expansion is most conveniently expressed in Fourier space with kernels F_N as

$$\delta^{(N)}(k) = \int d^3q_1 \ldots d^3q_N F_N(q_1, \ldots, q_N)\delta_D \\ (q_1 + \cdots + q_N - k)\delta^{(1)}(q_1)\ldots\delta^{(1)}(q_N) \quad (37)$$

and an analogous equation for the velocity potential. Euler's equations lead to a simple, albeit "combinatorially exploding" recursion for the kernels [24]. The most important kernel is the first non-trivial F_2

$$F_2(\mathbf{q_1}, \mathbf{q_2}) = 1 + \mu + (\frac{q_1}{q_2} + \frac{q_2}{q_1})\cos(\theta) + (1-\mu)\cos(\theta)^2 , \quad (38)$$

where θ is the angle between the two vectors (in this equation explicitly denoted by bold face) and $\mu = \frac{3}{7}\Omega^{-1/140}$ [25].

The leading order calculation of, e.g. any third-order statistic can be obtained simply by $\langle\delta^3\rangle_c = \langle(\delta^{(1)} + \delta^{(2)} + \ldots)^3\rangle_c = 3\langle(\delta^{(1)})^2\delta^{(2)}\rangle$+higher orders. For Gaussian initial conditions, this leads to formula in terms of $P(k)$, the initial (or linear) power spectrum, and the F_2 kernel.

When the above expansion is executed carefully PT predicts [24, 26, 27, 28] that the bispectrum in the weakly non-linear regime is

$$\left\{\left(\frac{4}{3} + \frac{2}{3}\mu\right)P_0(x) + \left(\frac{k_1}{k_2} + \frac{k_2}{k_1}\right)P_1(x) + \frac{2}{3}(1-\mu)P_2(x)\right\}P(k_1)P(k_2) + \text{perm.} , \quad (39)$$

where x is the cosine of the angle between the two wave vectors, P_l are Legendre polynomials, and $P(k)$ is the linear power spectrum. We have written the F_2 kernel in terms of Legendre polynomials, to make it explicitly clear that the first permutation depends only on terms up to quadrupole.

Together with the formulae in Sect. 5 this fact can be used for the prediction of the first permutation of three-point correlation function in the weakly non-linear regime:

$$\begin{aligned}
\xi^3_0(r_1, r_2) &= \int \tfrac{k_1^2}{2\pi^2}dk_1 \tfrac{k_2^2}{2\pi^2}dk_2 \left(\tfrac{4}{3} + \tfrac{2}{3}\mu\right) P(k_1)P(k_2)j_0(k_1r_1)j_0(k_2r_2) , \\
\xi^3_1(r_1, r_2) &= -\int \tfrac{k_1^2}{2\pi^2}dk_1 \tfrac{k_2^2}{2\pi^2}dk_2 \left(\tfrac{k_1}{k_2} + \tfrac{k_2}{k_1}\right) P(k_1)P(k_2)j_1(k_1r_1)j_1(k_2r_2) , \\
\xi^3_2(r_1, r_2) &= \int \tfrac{k_1^2}{2\pi^2}dk_1 \tfrac{k_2^2}{2\pi^2}dk_2 \tfrac{2}{3}(1-\mu) P(k_1)P(k_2)j_2(k_1r_1)j_2(k_2r_2) . \quad (40)
\end{aligned}$$

I emphasize that the above equation corresponds to the first permutation of the perturbation theory kernel: the other permutations are dealt within the same way, and the results are added. All the integrals factor into one-dimensional ones, which simplifies the calculation in practice. It can be shown with tedious calculations that the above formulae are equivalent to [29]. These equations have been used for the predictions of Fig. 4. One has to be somewhat careful to integrate the Bessel functions with sufficient accuracy.

For completeness we present the analogous formulae for the projected three-point correlation function (see Sect. 9), which is integrated over line of sight coordinates to avoid the effects of redshift distortions. The perturbation theory kernel is simply rewritten in Fourier modes [20]

$$\left\{\left(\frac{3}{2}-\frac{1}{2}\mu\right)+\left(\frac{k_1}{k_2}+\frac{k_2}{k_1}\right)cos(\theta)+\frac{1}{2}(1-\mu)\,cos(2\theta)\right\}P(k_1)P(k_2)+\text{perm.}\,, \tag{41}$$

which yields the equivalent result for the projected three-point function ζ^p as

$$\begin{aligned}
\zeta_0^p(r_1,r_2) &= \int \frac{k_1}{2\pi}dk_1\frac{k_2}{2\pi}dk_2\left(\frac{3}{2}-\frac{1}{2}\mu\right)P(k_1)P(k_2)J_0(k_1r_1)J_0(k_2r_2)\,,\\
\zeta_1^p(r_1,r_2) &= -\int \frac{k_1}{2\pi}dk_1\frac{k_2}{2\pi}dk_2\left(\frac{k_1}{k_2}+\frac{k_2}{k_1}\right)P(k_1)P(k_2)J_1(k_1r_1)J_1(k_2r_2)\,,\\
\zeta_2^p(r_1,r_2) &= \int \frac{k_1}{2\pi}dk_1\frac{k_2}{2\pi}dk_2\frac{1}{2}(1-\mu)\,P(k_1)P(k_2)J_2(k_1r_1)J_2(k_2r_2)\,.
\end{aligned} \tag{42}$$

Here both r_i and k_i are two-dimensional vectors perpendicular to the line of sight, and the other two permutations have to be added up as previously.

8 Bias

Most observations recording the distribution of "light" at some wavelength (except some still singular cases, such as neutrino astronomy or air shower detectors) might have spatial distribution different from that of the underlying dark matter or "mass". This difference means that our estimators might give us a "biased" view of the statistics of the underlying dark matter. This bias is caused by the complex interplay of non-linear gravitational dynamics with dissipative physics. While there have already been advances in ab initio modelling of this complicated process, the most fruitful approach is still phenomenological.

Since galaxy formation is a stochastic process, we can imagine that there is no one-to-one correspondence between dark matter and galaxy density, the former only determines the probability of forming a galaxy (together with its environment and merger history). This general case is called *stochastic* biasing [30, 31], while, if there is a functional dependence between the two fields, it is *deterministic*. In mock catalogues created using semi-analytic galaxy formation models, e.g. [32], most of the stochasticity of the bias is simply due to shot noise [33].

If there is a functional relationship between the dark matter and the galaxy density fields, the simplest possible relation is *linear* and the more general ones are *non-linear*.

Linear bias is described by the equation of $\delta_g = b\delta_{DM}$, where b is a constant. Note that this is an inconsistent equation for $b > 1$ yielding the minimum value of the galaxy density field $-b < -1$, obviously nonsense. Yet, owing to its simplicity, this is the single most used bias model.

A non-linear generalization contains an arbitrary function $\delta_g = f(\delta_{DM})$. This function can either be assumed to be a general function with a few parameters (e.g. exponential) or be expanded. The two most used expansions are Taylor expansion and expansion into Hermite polynomials (and thereby expanding around a Gaussian).

In the weakly non-linear regime, the coefficients of the Taylor expansion are the non-linear bias coefficients;

$$\delta_g = f(\delta_{DM}) = \sum_N \frac{b_N}{N!} \delta_{DM}^N \,. \tag{43}$$

This equation can be used perturbatively to calculate the moments of the biased (galaxy) density field in terms of the moments of the underlying field and bias coefficients [34, 35, 36]. In typical applications, the functional relationship is set up between smoothed fields: it is important to note that smoothing and biasing are two operations which do not commute. Therefore the bias coefficients have a somewhat obscure physical meaning, as they depend not only on the physics of galaxy formation but on the nature of smoothing itself. Note also that the zeroth coefficient b_0 is set to ensure $\langle\delta_g\rangle = 0$, i.e. it is non-zero.

Symmetries can be used to construct estimators which constrain the bias in the weakly non-linear regime. To second order, the biased reduced bispectrum transforms as $q = Q/b + b_2/b^2$ [37], where the lower case denotes the galaxy (measured) and the upper case the dark matter (theory) values. It is clear that b_2 can only effect the monopole term. Thus a simple estimator for the bias can be constructed as [19]

$$\begin{aligned} b &= \frac{Q_1}{q_1} = \frac{Q_2}{q_2} \,, \\ b_2 &= q_0 b^2 - Q_0 b \,. \end{aligned} \tag{44}$$

According to the equations, the quadrupole to dipole ratio does not depend on the bias, thus it serves as a novel, useful test of the underlying assumptions: a quasi-local perturbative, deterministic bias model, and perturbation theory. Figure 8 shows the dipole to quadrupole ratio for BBKS and EH power spectra, respectively. The range of ks to be used for bias extraction can be determined from contrasting the measurements with these predictions. Note that scales where baryon oscillations are prominent are barely accessible with present redshift surveys. On smaller scales non-linear evolution is likely to modify these prediction based purely on leading order perturbation theory [38].

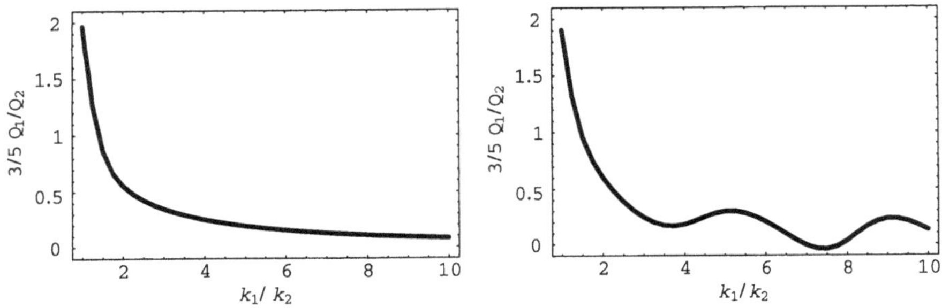

Fig. 8. Left: the dipole to quadrupole ratio $3/5Q_1/Q_2$ is plotted for a BBKS theory (thick solid line) for $k = 0.01$. The right panel shows the same using the EH fit for comparison, featuring baryonic oscillations

The above simple theory illustrates that three-point statistics can constrain the first two bias parameters in the weakly non-linear regime. Similarly four-point statistics constrain the first three parameters, etc.

An alternative to the above simple theory, and possibly more useful on smaller scales, is the halo model. The physical picture is simple [39]: we try to model the distribution by some entities, "halos", which have a relatively simple distribution (from perturbation theory). Large-scale correlations thus follow from the "halo–halo" type terms, while all the small-scale statistics follows from the halo profile ("1-halo term") and the distribution of halo masses. In these notes we cannot give justice to halo models, the interested reader is referred to [6] and references therein. Halo models have considerable success in describing the two-point statistics of the dark matter distributions, and they provide a reasonable approximation (currently at the 20% level) to the three-point statistics. Galaxy bias is then described by the halo occupation probabilities.

For halo model predictions, higher order Bessel integrals are needed [40], but the principle is exactly the same as (40) (see also [41] for a different approach). While halo models are physically motivated, the best fit parameters differ for two- and three-point statistics thus the physical meaning of the parameters is still somewhat nebulous.

9 Redshift Distortions

Redshift distortions, together with biasing, represent the most uncertain aspect of the phenomenology of higher order statistics. Here I only offer a superficial overview of this complicated topic; for a comprehensive review of redshift distortions on two-point statistics, see [42].

The physical idea is fairly simple: in redshift surveys, the radial distance to an object is deduced from its redshift using the Hubble relation in our expanding universe. Therefore, the "peculiar velocity", the velocity of an object with respect to the average Hubble flow, will add to the measured distance. Such a space where position of an object is given in a spherical coordinate system with two angles and a radial distance which contains (a random) velocity is called *redshift space*, as opposed to real space (the one we would really want to measure).

The spatial distribution of objects will no longer be translation invariant, although it will still have rotational symmetry around the observer. The deviation from translational invariance of the observed distribution is called "redshift distortion". Qualitatively, it is mainly due to two distinct effects: on small scales, velocity dispersion of virialized clumps will look elongated along the line of sight, popularly called the "finger of god" effect. On larger scales, infall velocities cause distortions perpendicularly to the line of sight, the Kaiser effect.

The breaking of translational symmetry means that the redshift space quantities depend on larger number of parameters, e.g. the two- and three-point correlation functions will depend on three and six parameters, respectively (two and five in the distant observer approximation).

Redshift distortions can be taken into account in perturbation theory. In the distant observer approximation, when all lines of sights can be taken to be parallel, the effect has been calculated for the three-point function by [43]. The general case needs a complex multipolar expansion; the results are fairly tedious and will be presented elsewhere.

Besides predicting the redshift space quantities using theory or simulation, one can sweep redshift distortions under the rug by introducing the "projected" correlation functions. These assume a distant observer approximation, where one can consider the N-vectors to pointing from the observer to the N vertexes of an N-point configuration parallel. Then, one can introduce $\pi - \sigma$ coordinates [3], with π_i vectors parallel to the line of sight, while the σ_i vectors are perpendicular to it.

The important point about this parametrization is that, approximately, only π coordinates are affected by redshift distortions. A convenient, redshift distortion-free quantity is obtained by integrating over the redshift direction, i.e. π coordinates. The resulting object, the "projected N-point correlation function", integrated over $N-1$ π coordinates is free of redshift distortions. Although it is similar to angular (projected) correlation function, the units of the σ coordinates are still Mpch, and, if properly done, its signal to noise is higher.

Fourier space analogues of the above idea use $k_\perp$ and $k_\parallel$ for perpendicular and parallel k vectors respectively, e.g. the redshift space bispectrum is parametrized by the five parameters $B(k_{1,\perp}, k_{2,\perp}, k_{3,\perp}, k_{1,\parallel}, k_{2,\parallel})$, with $\perp$ denoting transverse, and $\parallel$ parallel quantities with respect to the line of sight in the distant observer approximation. Interestingly, the real space bispectrum can be estimated from taking $k_{1,\parallel} \simeq k_{2,\parallel} \simeq 0$ [19].

10 Example: Conditional Cumulants

In my lectures, I presented a detailed example using a new set of statistics, conditional cumulants, closely following [44]. It illustrates most of the generic features of the theory, baring its strengths and weaknesses.

Conditional cumulants represent an interesting and sensible compromise between N-point correlation functions and cumulants measured from moments of counts in cells. As we will see next, they can be understood as degenerate N-point correlation functions, or integrated monopole moments of the bispectrum, and they are closely related to neighbour counts. They share accurate edge-corrected estimators with N-point correlation functions, yet, they are as straightforward to measure and interpret as counts in cells.

10.1 Basics

The general conditional cumulants, (6), have been defined as the joint connected moment of one unsmoothed and $N-1$ smoothed density fluctuation fields. In the most general case, each top hat window might have a different radius. Further simplification arises if all the top hats are the same, i.e. we define $U_N(r)$ with $r_1 = \ldots = r_{N-1} = r$ as the degenerate conditional cumulant cf. [45]. U_N subtly differs from the usual cumulant of smoothed field $\bar{\xi}_N$ by one less integral over the window function.

The second order, U_2, is equivalent to the (confusingly named) J_3 integral, e.g. [3],

$$U_2(r) = \frac{3}{r^3} J_3(r) = \frac{1}{(2\pi)^3} \int P(k) w(kr) 4\pi k^2 dk \ , \tag{45}$$

where $w(kr) = 3(\sin kr - kr\cos kr)/(kr)^3$ is the Fourier transform of $W_r(s)$ and $P(k)$ is the power spectrum.

For higher orders, we can construct reduced conditional cumulants as

$$R_N(r) = \frac{U_N(r)}{U_2^{N-1}(r)} \ . \tag{46}$$

Both U_N and R_N have deep connection with moments of neighbour counts, e.g. [3] as we show next. Let us define the partition function $Z[J] = \langle \exp(\int iJ\rho) \rangle$ cf. [4], where ρ is the smooth density field. Then we can use the special source function $iJ(x) = W(x)s + \delta_D(x)t$ to obtain the generating function $G(s,t)$. This is related to the generating function of neighbour counts factorial moments as $G(s) = \partial_t G(s,t)|_{t=0}$. Explicitly,

$$G(s) = \sum_{M \geq 0} \frac{(snv)^M}{M!} U_{M+1} \exp \sum_{N \geq 1} \frac{(snv)^N}{N!} \bar{\xi}_N \ , \tag{47}$$

where $nv = \bar{N}$ is the average count of galaxies and $\bar{\xi} = U_1 = 1$ by definition. This generating function can be used to obtain U_Ns and/or R_Ns from neighbour count factorial moments analogously as the generating functions in [46] for obtaining S_Ns from factorial moments of counts in cells. For completeness, the generating function for neighbour counts' distribution is obtained by substituting $s \to s-1$, while the ordinary moment generating function by $s \to e^s - 1$. Expanding $G(e^s - 1)$ recovers the formulae in [3], Sect. 36. The above generating function allows the extraction of U_N from neighbour count statistics for high N. The situation is analogous to the CIC theory presented in [46], and it is fully applicable to neighbour counts statistics with minor and trivial modifications.

So far our discussion has been entirely general; in what follows we will focus on $N = 3$, i.e. the first non-trivial conditional cumulant U_3. $U_3(r_1, r_2)$ is directly related to bispectrum by

$$U_3(r_1,r_2) = \frac{1}{(2\pi)^6} \int B(\mathbf{k}_1,\mathbf{k}_2,\mathbf{k}_3)\delta_D(\mathbf{k}_1+\mathbf{k}_2+\mathbf{k}_3) \\ w(k_1r_1)w(k_2r_2)d^3k_1d^3k_2d^3k_3 \ , \tag{48}$$

where δ_D is the Dirac-delta function. To further elucidate this formula, we use the multi-pole expansion of bispectrum and three-point correlation function of (32) and (33) to find

$$U_3(r_1,r_2) = \frac{4\pi}{V_1V_2}\int_0^{r_1}\int_0^{r_2} \zeta_0(r_1,r_2)r_1^2r_2^2dr_1dr_2 \tag{49}$$

$$= \frac{4\pi}{(2\pi)^6}\int dk_1dk_2\frac{3k_1}{r_1}\frac{3k_2}{r_2}j_1(k_1r_1)j_1(k_2r_2)B_0(k_1,k_2) \ , \tag{50}$$

in which j_1 is the first-order spherical Bessel function. We see that U_3 depends only on the monopole moment of the bispectrum/three-point correlation function. This property significantly simplifies the transformation of this statistic under redshift distortions.

10.2 U_3 in the Weakly Non-linear Perturbation Regime

Using the general machinery of perturbation theory, one can predict the conditional cumulants to leading order as a matter of simple calculation. For third order, the results, as usual, will depend on the F_2 kernel and the linear power spectrum (or correlation function). We leave as an exercise for the reader to show that

$$R_3(r_1,r_2) \equiv \frac{U_3(r_1,r_2)}{U_2(r_1)U_2(r_2)} = \frac{34}{21}\left[1+\frac{\bar{\xi}(r_1,r_2)}{U_2(r_1)}+\frac{\bar{\xi}(r_1,r_2)}{U_2(r_2)}\right] \\ +\frac{1}{3}\frac{\bar{\xi}(r_1,r_2)}{U_2(r_1)}\left[\frac{d\ln U_2(r_2)}{d\ln r_2}+\frac{\partial \ln\bar{\xi}(r_1,r_2)}{\partial\ln r_2}\right] \\ +\frac{1}{3}\frac{\bar{\xi}(r_1,r_2)}{U_2(r_2)}\left[\frac{d\ln U_2(r_1)}{d\ln r_1}+\frac{\partial \ln\bar{\xi}(r_1,r_2)}{\partial\ln r_1}\right] \ , \tag{51}$$

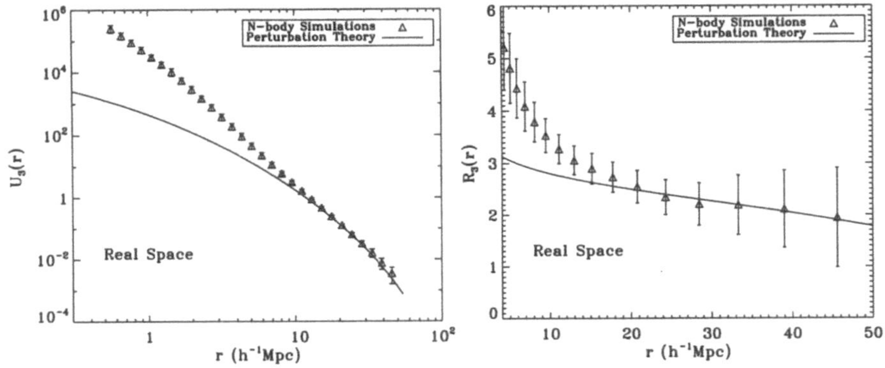

Fig. 9. Predictions of weakly non-linear perturbation theory for third-order conditional cumulant U_3 (*left*); and the reduced statistics R_3 (*right*) in real space (solid line) are compared with measurements in N-body simulations (triangles with error bars) with ΛCDM cosmology. The agreement is remarkable above $\gtrsim 20\,\mathrm{Mpch}$

in which $\bar{\xi}(r_1, r_2) = \frac{1}{2\pi^2}\int k^2 P(k) W(kr_1) W(kr_2) dk$. The special case when $r_1 = r_2 = r$ reads

$$R_3 = \frac{34}{21}\left[1 + 2\frac{\sigma^2}{U_2}\right] + \frac{1}{3}\frac{\sigma^2}{U_2}\left[2\frac{d\ln U_2}{d\ln r} + \frac{d\ln\sigma^2}{d\ln r}\right], \tag{52}$$

where $\sigma^2 = \frac{1}{2\pi^2}\int k^2 P(k) W^2(kr) dk$. Note the similarity of R_3 with the skewness, which is calculated in weakly non-linear perturbation theory as $S_3 = 34/7 - d\ln\sigma^2/d\ln r$ [47, 48].

10.3 Measurements of Conditional Cumulants in Simulations

$U_n(r)$ can be measured similarly to N-point correlation functions. For instance U_2 can be thought of as a two-point correlation function in a bin $[r_{lo}, r_{hi}] \equiv [0, r]$. Taking the lower limit to be a small number instead of 0, one can avoid discreteness effects due to self-counting (this is equivalent to using factorial moments when neighbour counts are calculated directly). Given a set of data and random points, the class of estimators (17) provides an edge (and incompleteness)-corrected technique to measure conditional cumulants. Existing N-point correlation function codes can be used for the estimation; for higher than third order, one also has to take connected moments in the usual way.

While the above suggests a scaling similar to N-point correlation functions, the relation to neighbour count factorial moments outlined in the previous section can be used to realize the estimator using two-point correlation function codes. To develop such an estimator, neighbour count factorial moments need to be collected for each possible combinations, where data and random points play the role of centre and neighbour.

Note that the edge correction of (17) is expected to be less accurate for conditional cumulants than for N-point correlation functions, however, the estimator will be more accurate than CIC estimators. Several alternative ways for correcting edge effects are known, which would be directly applicable to conditional cumulants [49, 50, 51]. In what follows, we use (17) for all results presented.

To test PT of the conditional cumulants, we performed measurements in ΛCDM simulations by the Virgo Supercomputing Consortium [23]. We used outputs of the Virgo simulation and the VLS (very large simulation). Except for box sizes and number of particles, these two simulations have identical cosmological parameters: $\Omega_m = 0.3$, $\Omega_v = 0.7$, $\Gamma = 0.21$, $h = 0.7$, and $\sigma_8 = 0.9$. In order to estimate measurement errors, we divide the VLS simulation into eight independent subsets each with the same size and geometry of the original Virgo simulation. In total, we have used the resulting nine realizations to estimate errors. Note that we corrected for cosmic bias by always taking the average before ratio statistics were formed.

Our measurements of the second- and third-order conditional cumulants are displayed in Figs. 1 and 2, respectively. Results from EPT (52) are denoted with solid lines. The measurements in simulations are in excellent agreement with EPT, especially on large scales $\gtrsim 20\,\mathrm{Mpc}h$.

10.4 Redshift Distortions

As three-dimensional galaxy catalogues are produced inherently in redshift space, understanding effects of redshift distortions on these statistics is crucial before practical applications can follow. In the distant observer approximation, the formula by [52, 53] is expected to provide an excellent approximation for $U_2(r)$. According to Sect. 2, we only need to consider the monopole enhancement

$$U_2(s) = \left(1 + \frac{2}{3}f + \frac{1}{5}f^2\right) U_2(r) , \tag{53}$$

where $f \approx \Omega_m^{0.6}$. This formula essentially predicts a uniform shift of the real space results. To test it, we repeated our measurements in redshift space and found that the above is indeed an excellent approximation in redshift space.

Considering the relatively simple, monopole nature of the statistics, we expect that the overall effect on U_3 should also be a simple shift, similar to the Lagrangian calculations by [28] and the more general Eulerian results by [43]. Specifically, we propose that ratio of R_3 in redshift space to that in real space can be approximated by

$$\frac{5(2520 + 3360f + 1260f^2 + 9f^3 - 14f^4)}{98(15 + 10f + 3f^2)^2} \cdot \frac{7}{4} \, . \tag{54}$$

This is motivated by the notion that the shift from redshift distortions of equilateral triangles should be similar to the corresponding shift for our monopole

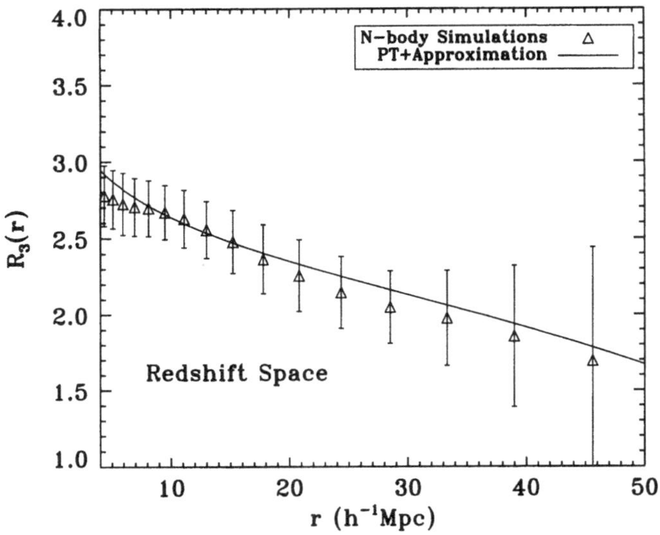

Fig. 10. Same as Fig. 9, right panel, but for R_3 in redshift space. The *solid line* shows phenomenological model based on (54). The theory appears to be a reasonable approximation at the 5% level

statistic. Our simulations results (see Fig. 3) show that this simple idea is indeed a surprisingly good approximation, although the phenomenological theory based on the above formula appears to have $\simeq$ 5% bias on scales $\gtrsim$ 20 Mpch where we expect that weakly non-linear perturbation theory is a good approximation. For practical applications, this bias can be calibrated by N-body or 2LPT [54] simulations.

In addition to the above simple formula, we have calculated the shift due to redshift distortions by angular averaging the bispectrum monopole term in [43]. We have found that the results overpredict redshift distortions, however, they would agree with simulations at the 1–2% level if we halved the terms classified as FOG (finger of god). At the moment there is no justification for such a fudge factor, therefore we opt to use the above phenomenology, which is about 5% accurate. While redshift distortions of third-order statistics are still not fully understood due to the non-perturbative nature of the redshift space mapping (R. Scoccimarro, private communication), detailed calculations taking into account velocity dispersion effects will improve the accuracy of the redshift space theory U_3.

For applications to constrain bias, one has to keep in mind that redshift distortions and non-linear bias do not commute. However, at the level of the above simple theory, it is clear that one can understand the important effects at least for the third-order statistic. There are several ways to apply conditional cumulants for bias determination, either via combination with another statistic [55] or using the configuration dependence of the more general $R_3(r_1, r_2)$. One also has to be careful that in practical applications ratio statistics will contain cosmic bias [56]. We propose that joint estimation with U_2 and U_3 will be more fruitful, even if R_3 is better for plotting purposes. Details of the techniques to constrain bias from these statistics as well as the determination of the bias from wide field redshift surveys are left for future work.

Another way to get around redshift distortions is to adapt conditional cumulants for projected and angular quantities. Such calculations are straightforward and entirely analogous to those performed for S_3 in the past. Further possible generalization of our theory would be to use halo models [6] to extend the range of applicability of the theory well below 20 Mpch. These generalizations are left for subsequent research.

Acknowledgements

This work was supported by NASA through grants AISR NAG5-11996 and ATP NAG5-12101, and by NSF through grants AST02-06243 and ITR 1120201-128440, as well as by Valencia summer school, one of the best organized and most enjoyable ever. The author would like to thank Stephane Colombi, Pablo Fosalba, Jun Pan, Simon Prunet, and Alex Szalay for many discussions and contributions. The simulations used have been carried out by the Virgo Supercomputing Consortium using computers based at Computing Centre of the Max-Planck Society in Garching and at the Edinburgh Parallel Computing Centre.[3]

References

1. Martinez, V.J., Saar, E.: Statistics of the Galaxy Distribution. Chapman and Hall/CRC Press, Boca Raton, USA (2002)
2. Bernardeau, F., Colombi, S., Gaztanaga, E., Scoccimarro, R.: Large-scale structure of the Universe and cosmological perturbation theory., Phys. Rep. **367**, 1–128 (2002)
3. Peebles, P.J.E.: The large-scale structure of the universe, Research supported by the National Science Foundation. Princeton, N.J., Princeton University Press, 1980. 435 p., (1980)
4. Szapudi, I., Szalay, A.S.: Higher order statistics of the galaxy distribution using generating functions, ApJ **408**, 43–56 May (1993)
5. Szapudi, I., Colombi, S., Jenkins, A., Colberg, J.: Experimental cosmic statistics - II. Distribution, MNRAS **313**, 725–733 Apr. (2000)
6. Cooray, A., Sheth, R.: Halo models of large scale structure, Phys. Rep. **372**, 1–129 Dec. (2002)
7. Szapudi, I., Postman, M., Lauer, T.R., Oegerle, W.: Observational Constraints on Higher Order Clustering up to z ˜= 1, ApJ **548**, 114–126 Feb. (2001)

[3] The data are publicly available at `http://www.mpa-garching.mpg.de/Virgo`

8. Szapudi, S., Szalay, A.S.: A New Class of Estimators for the N-Point Correlations, ApJ **494**, L41+ Feb. (1998)
9. Szapudi, I., Szalay, A.S.: The Variance in a New Class of N-point Estimators in Poisson and Binomial Point Processes, Annales de l'I.S.U.P., p. I, (2000)
10. Feldman, H.A., Kaiser, N., Peacock, J.A.: Power-spectrum analysis of three-dimensional redshift surveys, ApJ **426**, 23–37 May (1994)
11. Landy, S.D., Szalay, A.S.: Bias and variance of angular correlation functions, ApJ **412**, 64–71 July (1993)
12. Bond, J.R., Jaffe, A.H., Knox, L.: Estimating the power spectrum of the cosmic microwave background, Phys. Rev. D **57**, 2117–2137 Feb (1998)
13. Tegmark, M., Hamilton, A.J.S., Strauss, M.A., Vogeley, M.S., Szalay, A.S.: Measuring the Galaxy Power Spectrum with Future Redshift Surveys, ApJ **499**, 555–+ May (1998)
14. Oh, S.P., Spergel, D.N., Hinshaw, G.: An Efficient Technique to Determine the Power Spectrum from Cosmic Microwave Background Sky Maps, ApJ **510**, 551–563 Jan. (1999)
15. Hamilton, A.J.S.: Uncorrelated modes of the non-linear power spectrum, MNRAS **312**, 257–284 Feb. (2000)
16. Santos, M.G., Heavens, A., Balbi, A., Borrill, J., Ferreira, P.G., Hanany, S., Jaffe, A.H., Lee, A.T., Rabii, B., Richards, P.L., Smoot, G.F., Stompor, R., Winant, C.D., Wu, J.H.P.: Multiple methods for estimating the bispectrum of the cosmic microwave background with application to the MAXIMA data, MNRAS **341**, 623–643 May (2003)
17. Heavens, A., Santos, M., Ferreira, P.: The bispectrum of MAXIMA, New Astronomy Review **47**, 815–820 Nov. (2003)
18. Szapudi, I., Colombi, S.: Cosmic Error and Statistics of Large-Scale Structure, ApJ **470**, 131–+ Oct. (1996)
19. Szapudi, I.: Three-Point Statistics from a New Perspective, ApJ **605**, L89–L92 Apr. (2004)
20. Zheng, Z.: Projected Three-Point Correlation Functions and Galaxy Bias, ApJ **614**, 527–532 Oct. (2004)
21. Moore, A.W., Connolly, A.J., Genovese, C., Gray, A., Grone, L., Kanidoris, N., Nichol, R.C., Schneider, J., Szalay, A.S., Szapudi, I., Wasserman, L.: Fast Algorithms and Efficient Statistics: N-Point Correlation Functions, in Mining the Sky, 71–+ (2001)
22. Smith, R.E., Peacock, J.A., Jenkins, A., White, S.D.M., Frenk, C.S., Pearce, F.R., Thomas, P.A., Efstathiou, G., Couchman, H.M.P.: Stable clustering, the halo model and non-linear cosmological power spectra, MNRAS **341**, 1311–1332 June (2003)
23. Jenkins, A., Frenk, C.S., Pearce, F.R., Thomas, P.A., Colberg, J.M., White, S.D.M., Couchman, H.M.P., Peacock, J.A., Efstathiou, G., Nelson, A.H.: Evolution of Structure in Cold Dark Matter Universes, ApJ **499**, 20–+ May (1998)
24. Goroff, M.H., Grinstein, B., Rey, S.-J., Wise, M.B.: Coupling of modes of cosmological mass density fluctuations, ApJ **311**, 6–14 Dec. (1986)
25. Kamionkowski, M., Buchalter, A.: Weakly Nonlinear Clustering for Arbitrary Expansion Histories, ApJ **514**, 7–11 Mar. (1999)
26. Fry, J.N.: The Galaxy correlation hierarchy in perturbation theory, ApJ **279**, 499–510 Apr. (1984)
27. Bouchet, F.R., Colombi, S., Hivon, E., Juszkiewicz, R.: Perturbative Lagrangian approach to gravitational instability., A&A **296**, 575–+ Apr. (1995)

28. Hivon, E., Bouchet, F.R., Colombi, S., Juszkiewicz, R.: Redshift distortions of clustering: a Lagrangian approach., A&A **298**, 643–+ June (1995)
29. Jing, Y.P., Boerner, G.: Three-point correlation function in the quasilinear regime., A&A **318**, 667–672 Feb. (1997)
30. Dekel, A., Lahav, O.: Stochastic Nonlinear Galaxy Biasing, ApJ **520**, 24–34 July (1999)
31. Scherrer, R.J., Weinberg, D.H.: Constraints on the Effects of Locally Biased Galaxy Formation, ApJ **504**, 607–+ Sept. (1998)
32. Benson, A.J., Cole, S., Frenk, C.S., Baugh, C.M., Lacey, C.G.: The nature of galaxy bias and clustering, MNRAS **311**, 793–808 Feb. (2000)
33. Szapudi, I., Pan, J.: On Recovering the Nonlinear Bias Function from Counts-in-Cells Measurements, ApJ **602**, 26–37 Feb. (2004)
34. Szapudi, I.: Higher order correlations, Master's Thesis, Eotvos University, Budapest (1990)
35. Fry, J.N., Gaztanaga, E.: Biasing and hierarchical statistics in large-scale structure, ApJ **413**, 447–452 Aug. (1993)
36. Matsubara, T.: Diagrammatic Methods in Statistics and Biasing in the Large-Scale Structure of the Universe, ApJS **101**, 1–+ Nov. (1995)
37. Fry, J.N.: The minimal power spectrum: Higher order contributions, ApJ **421**, 21–26 Jan. (1994)
38. Meiksin, A., White, M., Peacock, J.A.: Baryonic signatures in large-scale structure, MNRAS **304**, 851–864 Apr. (1999)
39. Soneira, R.M., Peebles, P.J.E.: Is there evidence for a spatially homogeneous population of field galaxies, ApJ **211**, 1–15 Jan. (1977)
40. Fosalba, P., Pan, J., Szapudi, I.: Cosmological Three-Point Function: Testing The Halo Model Against Simulations, ApJ **632**, 29 (2005)
41. Takada, M., Jain, B.: The three-point correlation function in cosmology, MNRAS **340**, 580–608 Apr. (2003)
42. Hamilton, A.J.S.: Linear Redshift Distortions: a Review, in ASSL Vol. 231: The Evolving Universe, 185–+ (1998)
43. Scoccimarro, R., Couchman, H.M.P., Frieman, J.A.: The Bispectrum as a Signature of Gravitational Instability in Redshift Space, ApJ **517**, 531–540 June (1999)
44. Pan, J., Szapudi, I.: Conditional Cumulants in Weakly Non-linear Regime, MNRAS **361**, 357 (2005)
45. Bonometto, S.A., Borgani, S., Ghigna, S., Klypin, A., Primack, J.R.: Non-linear clustering in the cold plus hot dark matter model, MNRAS **273**, 101–121 Mar. (1995)
46. Szapudi, I., Szalay, A.S.: Higher order statistics of the galaxy distribution using generating functions, ApJ **408**, 43–56 May (1993)
47. Juszkiewicz, R., Bouchet, F.R., Colombi, S.: Skewness induced by gravity, ApJ **412**, L9–L12 July (1993)
48. Bernardeau, F.: The effects of smoothing on the statistical properties of large-scale cosmic fields, A&A **291**, 697–712 Nov. (1994)
49. Ripley, B.: Statistical inference for spatial processes, University Press,Cambridge (1988)
50. Kerscher, M., Szapudi, I., Szalay, A.S.: A Comparison of Estimators for the Two-Point Correlation Function, ApJ **535**, L13–L16 May (2000)
51. Pan, J., Coles, P.: Boundary corrections in fractal analysis of galaxy surveys, MNRAS **330**, 719–730 Mar. (2002)

52. Kaiser, N.: Clustering in real space and in redshift space, MNRAS **227**, 1–21 July (1987)
53. Lilje, P.B., Efstathiou, G.: Gravitationally induced velocity fields in the universe. I - Correlation functions, MNRAS **236**, 851–864 Feb. (1989)
54. Scoccimarro, R.: Gravitational Clustering from χ^2 Initial Conditions, ApJ **542**, 1–8 Oct. (2000)
55. Szapudi, I.: A New Method for Calculating Counts in Cells, ApJ **497**, 16–+ Apr. (1998)
56. Szapudi, I., Colombi, S., Bernardeau, F.: Cosmic statistics of statistics, MNRAS **310**, 428–444 Dec. (1999)

Phase Correlations and Topological Measures of Large-Scale Structure

P. Coles

School of Physics & Astronomy, University of Nottingham, University Park, Nottingham NG7 2RD, UK
Peter.Coles@Nottingham.ac.uk

1 Introduction

The process of gravitational instability initiated by small primordial density perturbations is a vital ingredient of cosmological models that attempt to explain how galaxies and large-scale structure formed in the Universe. In the standard picture (the "concordance" model), a period of accelerated expansion ("inflation") generated density fluctuations with simple statistical properties through quantum processes (Starobinsky [82], [83], [84]; Guth [39]; Guth & Pi [40]; Albrecht & Steinhardt [2]; Linde [55]). In this scenario the primordial density field is assumed to form a statistically homogeneous and isotropic Gaussian random field (GRF). Over years of observational scrutiny this paradigm has strengthened its hold in the minds of cosmologists and has survived many tests, culminating in those furnished by the Wilkinson Microwave Anisotropy Probe (WMAP; Bennett et al. [7]; Hinshaw et al. [45].

Gaussian random fields are the simplest fully defined stochastic processes (Adler [1]; Bardeen et al. [4]), which make their analysis relatively straightforward. Robust and powerful statistical descriptors can be constructed that have a firm mathematical underpinning and are relatively simple to implement. Second-order statistics such as the ubiquitous power spectrum (e.g. Peacock & Dodds [67]) furnish a complete description of Gaussian fields. They have consequently yielded invaluable insights into the behaviour of large-scale structure in the latest generation of redshift surveys, such as the 2dFGRS (Percival et al. [72]). Important though these methods undoubtedly are, the era of precision cosmology we are now entering requires more thought to be given to methods for both detecting and exploiting departures from Gaussian behaviour.

Even if the primordial density fluctuations were indeed Gaussian, the later stages of gravitational clustering must induce some form of non-linearity. One particular way of looking at this issue is to study the behaviour of

Coles, P.: *Phase Correlations and Topological Measures of Large-Scale Structure.* Lect. Notes Phys. **665**, 493–522 (2009)
DOI 10.1007/978-3-540-44767-2_15

Fourier modes of the cosmological density field. If the hypothesis of primordial Gaussianity is correct then these modes began with random spatial phases. In the early stages of evolution, the plane-wave components of the density evolve independently like linear waves on the surface of deep water. As the structures grow in mass, they interact with other in non-linear ways, more like waves breaking in shallow water. These mode–mode interactions lead to the generation of coupled phases. While the Fourier phases of a Gaussian field contain no information (they are random), non-linearity generates non-random phases that contain much information about the spatial pattern of the fluctuations. Although the significance of phase information in cosmology is still not fully understood, there have been a number of attempts to gain quantitative insight into the behaviour of phases in gravitational systems. Ryden & Gramann [75], Soda & Suto [80] and Jain & Bertschinger [47] concentrated on the evolution of phase shifts for individual modes using perturbation theory and numerical simulations. An alternative approach was adopted by Scherrer et al. [76], who developed a practical method for measuring the phase coupling in random fields that could be applied to real data. Most recently Chiang & Coles [13], Coles & Chiang [19], Chiang [12] and Chiang et al. [14] have explored the evolution of phase information in some detail.

Despite this recent progress, there is still no clear understanding of how the behaviour of the Fourier phases manifests itself in more orthodox statistical descriptors. In particular there is much interest in the usefulness of the simplest possible generalization of the (second-order) power spectrum, i.e. the (third-order) bispectrum (Peebles [69]; Scoccimarro et al. [77]; Scoccimarro, Couchman & Frieman [78]; Verde et al. [89]; [90]; [91]). Since the bispectrum is identically zero for a Gaussian random field, it is generally accepted that the bispectrum encodes some form of phase information but exactly what form of correlation it measures has never been elucidated. Further possible generalizations of the bispectrum are usually called polyspectra; they include the (fourth-order) trispectrum (Verde & Heavens [88]) or a related but simpler statistic called the second-spectrum (Stirling & Peacock [85]). Exploring the connection between polyspectra and non-linearly induced phase association is one of the aims of this paper.

Gravitational instability is expected to generate phase correlations (and non-Gaussianity) even if the primordial fluctuations were Gaussian. The cosmic microwave background (CMB) allows us to probe the fluctuations while they are still in the linear regime and thus test the level of primordial non-Gaussianity without having to worry about non-linear effects. A second aim of this paper is to explain how one can use phase correlations in spherical harmonic expansions of temperature fluctuations in order to detect departures from standard fluctuation statistics.

Finally I discuss the use of topological invariants such as the Euler–Poincaré characteristic of isodensity contours to assess the level of non-Gaussianity in large-scale structure.

2 Basic Statistical Tools

I start by giving some general definitions of concepts which I will later use in relation to the particular case of cosmological density fields. In order to put our results in a clear context, I develop the basic statistical description of cosmological density fields; see also Peebles [69] and Coles & Lucchin [23].

2.1 Fourier Description

I follow standard practice and consider a region of the Universe having volume V_u, for convenience assumed to be a cube of side $L \gg l_s$, where l_s is the maximum scale at which there is significant structure due to the perturbations. The region V_u can be thought of as a "fair sample" of the Universe if this is the case. It is possible to construct, formally, a "realization" of the Universe by dividing it into cells of volume V_u with periodic boundary conditions at the faces of each cube. This device is often convenient, but in any case one often takes the limit $V_u \to \infty$. Let us denote by $\bar{\rho}$ the mean density in a volume V_u and take $\rho(\mathbf{x})$ to be the density at a point in this region specified by the position vector $\mathbf{x}$ with respect to some arbitrary origin. As usual, the fluctuation is defined to be

$$\delta(\mathbf{x}) = [\rho(\mathbf{x}) - \bar{\rho}]/\bar{\rho} \,. \tag{1}$$

We assume this to be expressible as a Fourier series:

$$\delta(\mathbf{x}) = \sum_{\mathbf{k}} \delta_{\mathbf{k}} \ \exp(\mathrm{i}\,\mathbf{k}\cdot\mathbf{x}) = \sum_{\mathbf{k}} \delta^*_{\mathbf{k}} \ \exp(-\mathrm{i}\,\mathbf{k}\cdot\mathbf{x}) \;; \tag{2}$$

the appropriate inverse relationship is of the form

$$\delta_{\mathbf{k}} = \frac{1}{V_u}\int_{V_u} \delta(\mathbf{x})\exp(-\mathrm{i}\,\mathbf{k}\cdot\mathbf{x})d\mathbf{x} \,. \tag{3}$$

The Fourier coefficients $\delta_{\mathbf{k}}$ are complex quantities,

$$\delta_{\mathbf{k}} = |\delta_{\mathbf{k}}|\exp\left(\mathrm{i}\,\phi_{\mathbf{k}}\right) \,, \tag{4}$$

with amplitude $|\delta_{\mathbf{k}}|$ and phase $\phi_{\mathbf{k}}$. The assumption of periodic boundaries results in a discrete $\mathbf{k}$-space representation; the sum is taken from the Nyquist frequency $k_{\mathrm{Ny}} = 2\pi/L$, where $V_u = L^3$, to infinity. Note that as $L \to \infty$, $k_{\mathrm{Ny}} \to 0$. Conservation of mass in V_u implies $\delta_{\mathbf{k}=0} = 0$ and the reality of $\delta(\mathbf{x})$ requires $\delta^*_{\mathbf{k}} = \delta_{-\mathbf{k}}$.

If, instead of the volume V_u, we had chosen a different volume V'_u the perturbation within the new volume would again be represented by a series of the form (2), but with different coefficients $\delta_{\mathbf{k}}$. Now consider a (large) number N of realizations of our periodic volume and label these realizations by V_{u1}, V_{u2}, V_{u3}, ..., V_{uN}. It is meaningful to consider the probability distribution

$\mathcal{P}(\delta_{\mathbf{k}})$ of the relevant coefficients $\delta_{\mathbf{k}}$ from realization to realization across this ensemble. One typically assumes that the distribution is statistically homogeneous and isotropic, in order to satisfy the cosmological principle, and that the real and imaginary parts of $\delta_{\mathbf{k}}$ have a Gaussian distribution and are mutually independent, so that

$$\mathcal{P}(w) = \frac{V_u^{1/2}}{(2\pi\alpha_k^2)^{1/2}} \exp\left(-\frac{w^2 V_u}{2\alpha_k^2}\right) , \tag{5}$$

where w stands for either Re $[\delta_{\mathbf{k}}]$ or Im $[\delta_{\mathbf{k}}]$ and $\alpha_k^2 = \sigma_k^2/2$; σ_k^2 is the spectrum. This is the same as the assumption that the phases $\phi_{\mathbf{k}}$ in (5) are mutually independent and randomly distributed over the interval between $\phi = 0$ and $\phi = 2\pi$. In this case the moduli of the Fourier amplitudes have a Rayleigh distribution:

$$\mathcal{P}(|\delta_{\mathbf{k}}|, \phi_{\mathbf{k}}) d|\delta_{\mathbf{k}}| d\phi_{\mathbf{k}} = \frac{|\delta_{\mathbf{k}}| V_u}{2\pi\sigma_k^2} \exp\left(-\frac{|\delta_{\mathbf{k}}|^2 V_u}{2\sigma_k^2}\right) d|\delta_{\mathbf{k}}| d\phi_{\mathbf{k}} \, . \tag{6}$$

Because of the assumption of statistical homogeneity and isotropy, the quantity $\mathcal{P}(\delta_{\mathbf{k}})$ depends only on the modulus of the wavevector $\mathbf{k}$ and not on its direction. It is fairly simple to show that, if the Fourier quantities $|\delta_{\mathbf{k}}|$ have the Rayleigh distribution, then the probability distribution $\mathcal{P}(\delta)$ of $\delta = \delta(\mathbf{x})$ in real space is Gaussian, so that

$$\mathcal{P}(\delta) d\delta = \frac{1}{(2\pi\sigma^2)^{1/2}} \exp\left(-\frac{\delta^2}{2\sigma^2}\right) d\delta \, , \tag{7}$$

where σ^2 is the variance of the density field $\delta(\mathbf{x})$. This is a strict definition of Gaussianity. However, Gaussian statistics do not always require the distribution (7) for the Fourier component amplitudes. According to its Fourier expansion, $\delta(\mathbf{x})$ is simply a sum over a large number of Fourier modes whose amplitudes are drawn from some distribution. If the phases of each of these modes are random, then the central limit theorem will guarantee that the resulting superposition will be close to a Gaussian if the number of modes is large and the distribution of amplitudes has finite variance. Such fields are called weakly Gaussian.

2.2 Covariance Functions and Probability Densities

I now discuss the real-space statistical properties of spatial perturbations in ρ. The *covariance function* is defined in terms of the density fluctuation by

$$\xi(\mathbf{r}) = \frac{\langle [\rho(\mathbf{x}) - \bar{\rho}][\rho(\mathbf{x}+\mathbf{r}) - \bar{\rho}] \rangle}{\bar{\rho}^2} = \langle \delta(\mathbf{x}) \delta(\mathbf{x}+\mathbf{r}) \rangle \, . \tag{8}$$

The angle brackets in this expression indicate two levels of averaging: first a volume average over a representative patch of the Universe and second an average over different patches within the ensemble, in the manner of Sect. 2.1. Applying the Fourier machinery to (8) one arrives at the *Wiener–Khintchin theorem*, relating the covariance to the spectral density function or power spectrum, $P(k)$:

$$\xi(\mathbf{r}) = \sum_{\mathbf{k}} \langle|\delta_{\mathbf{k}}|^2\rangle \exp(-\mathrm{i}\,\mathbf{k}\cdot\mathbf{r}) \,, \tag{9}$$

which, in passing to the limit $V_u \to \infty$, becomes

$$\xi(\mathbf{r}) = \frac{1}{(2\pi)^3} \int P(k) \exp(-\mathrm{i}\,\mathbf{k}\cdot\mathbf{r}) d\mathbf{k} \,. \tag{10}$$

Averaging (9) over $\mathbf{r}$ gives

$$\langle\xi(\mathbf{r})\rangle_{\mathbf{r}} = \frac{1}{V_u} \sum_{\mathbf{k}} \langle|\delta_{\mathbf{k}}|^2\rangle \int \exp(-\mathrm{i}\,\mathbf{k}\cdot\mathbf{r}) d\mathbf{r} = 0 \,. \tag{11}$$

The function $\xi(\mathbf{r})$ is the *two-point* covariance function. In an analogous manner it is possible to define spatial covariance functions for $N > 2$ points. For example, the three-point covariance function is

$$\zeta(r, s) = \frac{\langle[\rho(\mathbf{x}) - \bar{\rho}][\rho(\mathbf{x}+\mathbf{r}) - \bar{\rho}][\rho(\mathbf{x}+\mathbf{s}) - \bar{\rho}]\rangle}{\bar{\rho}^3} \,, \tag{12}$$

which gives

$$\zeta(\mathbf{r}, \mathbf{s}) = \langle\delta(\mathbf{x})\delta(\mathbf{x}+\mathbf{r})\delta(\mathbf{x}+\mathbf{s})\rangle \,, \tag{13}$$

where the spatial average is taken over all the points $\mathbf{x}$ and over all directions of $\mathbf{r}$ and $\mathbf{s}$ such that $|\mathbf{r} - \mathbf{s}| = t$: in other words, over all points defining a triangle with sides r, s and t. The generalization of (12) to $N > 3$ is obvious.

The covariance functions are related to the moments of the probability distributions of $\delta(\mathbf{x})$. If the fluctuations form a Gaussian random field then the N-variate distributions of the set $\delta_i \equiv \delta(\mathbf{x}_i)$ are just multivariate Gaussians of the form

$$\mathcal{P}_N(\delta_1, ..., \delta_N) = \frac{1}{(2\pi)^{N/2}(\det C)^{1/2}} \exp\Big(-\frac{1}{2}\sum_{i,j} \delta_i \; C_{ij}^{-1} \; \delta_j\Big) \,. \tag{14}$$

The correlation matrix C_{ij} can be expressed in terms of the covariance function

$$C_{ij} = \langle\delta_i\delta_j\rangle = \xi(\mathbf{r}_{ij}) \,. \tag{15}$$

It is convenient to go a stage further and define the N-point *connected* covariance functions as the part of the average $\langle\delta_i ... \delta_N\rangle$ that is not expressible in terms of lower-order functions, e.g.

$$\langle\delta_1\delta_2\delta_3\rangle = \langle\delta_1\rangle_c\langle\delta_2\delta_3\rangle_c + \langle\delta_2\rangle_c\langle\delta_1\delta_3\rangle_c + \langle\delta_3\rangle_c\langle\delta_1\delta_2\rangle_c + \langle\delta_1\rangle_c\langle\delta_2\rangle_c\langle\delta_3\rangle_c + \langle\delta_1\delta_2\delta_3\rangle_c \,, \tag{16}$$

where the connected parts are $\langle\delta_1\delta_2\delta_3\rangle_c$, $\langle\delta_1\delta_2\rangle_c$, etc. Since $\langle\delta\rangle = 0$ by construction, $\langle\delta_1\rangle_c = \langle\delta_1\rangle = 0$. Moreover, $\langle\delta_1\delta_2\rangle_c = \langle\delta_1\delta_2\rangle$ and $\langle\delta_1\delta_2\delta_3\rangle_c = \langle\delta_1\delta_2\delta_3\rangle$. The second- and third-order connected parts are simply the same as the covariance functions. Fourth- and higher-order quantities are different, however. The connected functions are just the multivariate generalization of the cumulants κ_N (Kendall & Stewart [48]). One of the most important properties of Gaussian fields is that all of their N-point connected covariances are zero beyond $N = 2$, so that their statistical properties are fixed once the set of two-point covariances (15) is determined. All large-scale statistical properties are therefore determined by the asymptotic behaviour of $\xi(r)$ as $r \rightarrow \infty$.

3 Phase Coupling

In Sect. 2 we pointed out that a convenient definition of a Gaussian field could be made in terms of its Fourier phases, which should be independent and uniformly distributed on the interval $[0, 2\pi]$. A breakdown of these conditions, such as the correlation of phases of different wavemodes, is a signature that the field has become non-Gaussian. In terms of cosmic large-scale structure formation, non-Gaussian evolution of the density field is symptomatic of the onset of non-linearity in the gravitational collapse process, suggesting that phase evolution and non-linear evolution are closely linked. A relatively simple picture emerges for models where the primordial density fluctuations are Gaussian and the initial phase distribution is uniform. When perturbations remain small, evolution proceeds linearly, individual modes grow independently and the original random phase distribution is preserved. However, as perturbations grow large their evolution becomes non-linear and Fourier modes of different wavenumbers begin to couple together. This gives rise to phase association and consequently to non-Gaussianity. It is clear that phase associations of this type should be related in some way to the existence of the higher-order connected covariance functions, which are traditionally associated with non-linearity and are non-zero only for non-Gaussian fields. In this section such a relationship is explored in detail using an analytical model for the non-linearly evolving density fluctuation field. Phase correlations of a particular form are identified and their connection to the covariance functions is established.

A graphic demonstration of the importance of phases in patterns generally is given in Fig. 1. Since the amplitude of each Fourier mode is unchanged in the phase reshuffling operation, these two pictures have exactly the same power spectrum, $P(k) \propto |\tilde{\delta}(\mathbf{k})|^2$. In fact, they have more than that: they have exactly the same amplitudes for all $\mathbf{k}$. They also have totally different morphology. Further demonstrations of the importance of Fourier phases in defining clustering morphology are given by Chiang [12].

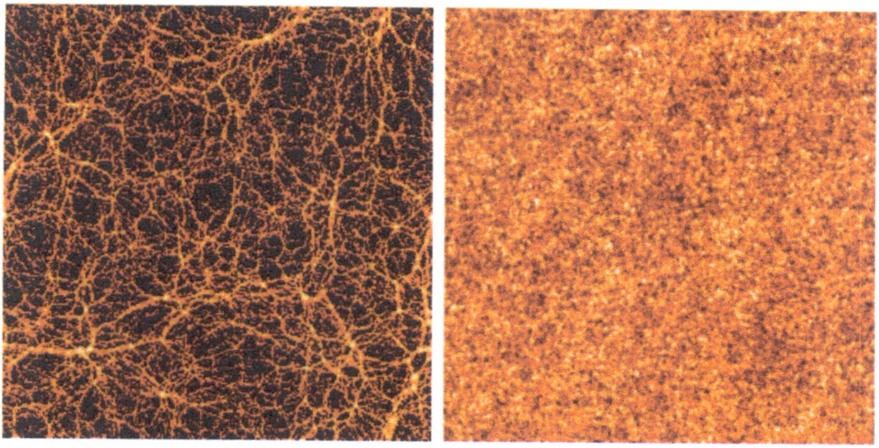

Fig. 1. Numerical simulation of galaxy clustering (*left*) together with a version generated randomly reshuffling the phases between Fourier modes of the original picture (*right*)

3.1 Quadratic Density Fields

It is useful at this stage to select a particular form of non-Gaussian field that serves both as a kind of phenomenological paradigm and as a reasonably realistic model of non-linear evolution from Gaussian initial conditions. The model involves a field which is generated by a simple quadratic transformation of a Gaussian distribution, hence the term *quadratic* non-linearity. Quadratic fields have been discussed before from a number of contexts (e.g. Coles & Barrow [18]; Moscardini et al. [63]; Falk et al. [31]; Luo & Schramm [58]; Luo [57]; Gangui et al. [35]; Koyama et al. [51]; Peebles [70], [71]; Matarrese et al. [59]; Verde et al. [89], [90]; Komatsu & Spergel [50]; Shandarin [79]; Bartolo et al. [6]); for further discussion see below. The motivation is very similar to that of Coles & Jones [22], who introduced the lognormal density field as an illustration of some of the consequences of a more extreme form of non-linearity involving an exponential transformation of the linear density field.

3.2 A Simple Non-linear Model

We adopt a simple perturbative expansion of the form

$$\delta(\mathbf{x}) = \delta_1(\mathbf{x}) + \epsilon\delta_2(\mathbf{x}) \tag{17}$$

to mimic the non-linear evolution of the density field. Although the equivalent transformation in formal Eulerian perturbation theory is a good deal more complicated, the kind of phase associations that we will deal with here is precisely the same in either case. In terms of the Fourier modes, in the continuum limit, we have for the first-order Gaussian term

$$\delta_1(\mathbf{x}) = \int d^3k \; |\delta_{\mathbf{k}}| \exp[\mathrm{i}\,\phi_{\mathbf{k}}] \exp[\mathrm{i}\,\mathbf{k}\cdot\mathbf{x}] \tag{18}$$

and for the second-order perturbation

$$\delta_2(\mathbf{x}) = [\delta_1(\mathbf{x})]^2 = \int d^3k\, d^3k' \; |\delta_{\mathbf{k}}||\delta_{\mathbf{k}'}| \exp[\mathrm{i}\,(\phi_{\mathbf{k}} + \phi_{\mathbf{k}'})] \exp[\mathrm{i}\,(\mathbf{k}+\mathbf{k}')\cdot\boldsymbol{r}] \,. \tag{19}$$

The quadratic field, δ_2, illustrates the idea of mode coupling associated with non-linear evolution. The non-linear field depends on a specific harmonic relationship between the wavenumber and phase of the modes at $\mathbf{k}$ and $\mathbf{k}'$. This relationship between the phases in the non-linear field, i.e.

$$\phi_{\mathbf{k}} + \phi_{\mathbf{k}'} = \phi_{\mathbf{k}+\mathbf{k}'} \,, \tag{20}$$

where the RHS represents the phase of the non-linear field, is termed *quadratic* phase coupling.

3.3 Two-Point Covariance Function

The two-point covariance function can be calculated using the definitions of Sect. 2, namely

$$\xi(r) = \langle \delta(\mathbf{x})\delta(\mathbf{x}+\boldsymbol{r}) \rangle \,. \tag{21}$$

Substituting the non-linear transform for $\delta(\mathbf{x})$ (17) into this expression gives four terms

$$\begin{aligned}\xi(r) = {} & \langle \delta_1(\mathbf{x})\delta_1(\mathbf{x}+\boldsymbol{r})\rangle + \epsilon\langle \delta_1(\mathbf{x})\delta_2(\mathbf{x}+\boldsymbol{r})\rangle + \epsilon\langle \delta_2(\mathbf{x})\delta_1(\mathbf{x}+\boldsymbol{r})\rangle \\ & + \epsilon^2\langle \delta_2(\mathbf{x})\delta_2(\mathbf{x}+\boldsymbol{r})\rangle \,.\end{aligned} \tag{22}$$

The first of these terms is the linear contribution to the covariance function whereas the remaining three give the non-linear corrections. We shall focus on the lowest-order term for now.

As we outlined in Sect. 2, the angle brackets $\langle\rangle$ in these expressions are expectation values, formally denoting an average over the probability distribution of $\delta(\mathbf{x})$. Under the fair sample hypothesis we replace the expectation values in (21) with averages over a selection of independent volumes so that $\langle\rangle \to \langle\rangle_{\mathrm{vol,\ real}}$. The first average is simply a volume integral over a sufficiently large patch of the Universe. The second average is over various realizations of δ_k and ϕ_k in the different patches. Applying these rules to the first term of (22) and performing the volume integration gives

$$\xi_{11}(r) = \int d^3k\, d^3k' \; \langle |\delta_{\mathbf{k}}||\delta_{\mathbf{k}'}| \exp[\mathrm{i}\,(\phi_{\mathbf{k}} + \phi_{\mathbf{k}'})]\rangle_{\mathrm{real}}\, \delta_D(\mathbf{k}+\mathbf{k}') \exp[\mathrm{i}\,\mathbf{k}'\cdot\mathrm{s}] \,, \tag{23}$$

where δ_D is the Dirac delta function. The above expression is simplified given the reality condition

$$\delta_{\mathbf{k}} = \delta^*_{-\mathbf{k}} , \tag{24}$$

from which it is evident that the phases obey

$$\phi_{\mathbf{k}} + \phi_{-\mathbf{k}} = 0 \mod[2\pi] . \tag{25}$$

Integrating (23) one therefore finds that

$$\xi_{11}(r) = \int d^3k \, \langle|\delta_{\mathbf{k}}|^2\rangle_{\mathrm{real}} \exp\left[-\mathrm{i}\,\mathbf{k}\cdot\mathbf{s}\right] , \tag{26}$$

so that the final result is independent of the phases. Indeed this is just the Fourier transform relation between the two-point covariance function and the power spectrum we derived in Sect. 2.1.

3.4 Three-Point Covariance Function

Using the same arguments outlined above it is possible to calculate the three-point connected covariance function, which is defined as

$$\zeta(\boldsymbol{r}, \mathbf{s}) = \langle \delta(\mathbf{x})\delta(\mathbf{x}+\boldsymbol{r})\delta(\mathbf{x}+\mathbf{s})\rangle_c . \tag{27}$$

Making the non-linear transform of (17) one finds the following contributions:

$$\begin{aligned}\zeta(\boldsymbol{r}, \mathbf{s}) = {} & \langle \delta_1(\mathbf{x})\delta_1(\mathbf{x}+\boldsymbol{r})\delta_1(\mathbf{x}+\mathbf{s})\rangle_c + \epsilon\langle \delta_1(\mathbf{x})\delta_1(\mathbf{x}+\boldsymbol{r})\delta_2(\mathbf{x}+\mathbf{s})\rangle_c \\ & + \mathrm{perms}(121, 211) + \epsilon^2\langle \delta_1(\mathbf{x})\delta_2(\mathbf{x}+\boldsymbol{r})\delta_2(\mathbf{x}+\mathbf{s})\rangle_c \\ & + \mathrm{perms}(212, 221) + \epsilon^3\langle \delta_2(\mathbf{x})\delta_2(\mathbf{x}+\boldsymbol{r})\delta(\mathbf{x}+\mathbf{s})\rangle_c . \end{aligned} \tag{28}$$

Again we consider first the lowest-order term. Expanding in terms of the Fourier modes and once again replacing averages as prescribed by the fair sample hypothesis gives

$$\begin{aligned}\zeta_{111}(\boldsymbol{r}, \mathbf{s}) = \int & d^3k\, d^3k'\, d^3k'' \ \langle |\delta_{\mathbf{k}}||\delta_{\mathbf{k}'}||\delta_{\mathbf{k}''}| \exp\left[\mathrm{i}\left(\phi_{\mathbf{k}} + \phi_{\mathbf{k}'} + \phi_{\mathbf{k}''}\right)\right]\rangle_{\mathrm{real}} \\ & \times \delta_D(\mathbf{k}+\mathbf{k}'+\mathbf{k}'') \exp\left[\mathrm{i}\,\mathbf{k}'\cdot\boldsymbol{r}\right] \exp\left[\mathrm{i}\,\mathbf{k}''\cdot\mathbf{s}\right] . \end{aligned} \tag{29}$$

Recall that δ_1 is a Gaussian field so that $\phi_{\mathbf{k}}$, $\phi_{\mathbf{k}'}$ and $\phi_{\mathbf{k}''}$ are independent and uniformly random on the interval $[0, 2\pi]$. Upon integration over one of the wavevectors the phase terms are modified so that its argument contains the sum $(\phi_{\mathbf{k}}+\phi_{\mathbf{k}'}+\phi_{-\mathbf{k}-\mathbf{k}''})$, or a permutation thereof. Whereas the reality condition of (24) implies a relationship between phases of anti-parallel wavevectors, no such conditions hold for modes linked by the triangular constraint imposed by the Dirac delta function. In other words, except for serendipity,

$$\phi_{\mathbf{k}} + \phi_{\mathbf{k}'} + \phi_{-\mathbf{k}-\mathbf{k}''} \neq 0 . \tag{30}$$

In fact due to the circularity of phases, the resulting sum is still just uniformly random on the interval $[0, 2\pi]$ if the phases are random. Upon averaging over sufficient realizations, the phase term will therefore cancel to zero so that the lowest-order contribution to the three-point function vanishes, i.e. $\zeta_{111}(\boldsymbol{r}, \mathbf{s}) = 0$. This is not a new result, but it does explicitly illustrate how the vanishing of the three-point connected covariance function arises in terms of the Fourier phases.

Next consider the first non-linear contribution to the three-point function given by

$$\zeta_{112}(\boldsymbol{r}, \mathbf{s}) = \epsilon\langle\delta_1(\mathbf{x})\delta_1(\mathbf{x}+\boldsymbol{r})\delta_2(\mathbf{x}+\mathbf{s})\rangle \tag{31}$$

or one of its permutations. In this case one of the arguments in the average is the field $\delta_2(\mathbf{x})$, which exhibits quadratic phase coupling of the form (20). Expanding this term to the point of (29) using the definition (19) one obtains

$$\begin{aligned}\zeta_{112}(\boldsymbol{r}, \mathbf{s}) = \int & d^3k\, d^3k'\, d^3k''\, d^3k''' \\ & \langle|\delta_{\mathbf{k}}||\delta_{\mathbf{k}'}||\delta_{\mathbf{k}''}||\delta_{\mathbf{k}'''}| \exp\left[\mathrm{i}\left(\phi_{\mathbf{k}} + \phi_{\mathbf{k}'} + \phi_{\mathbf{k}''} + \phi_{\mathbf{k}'''}\right)\right]\rangle_{\mathrm{real}} \\ & \times\delta_D(\mathbf{k}+\mathbf{k}'+\mathbf{k}''+\mathbf{k}''') \\ & \times\exp\left[\mathrm{i}\,\mathbf{k}'\cdot\boldsymbol{r}\right]\exp\left[\mathrm{i}\left(\mathbf{k}''+\mathbf{k}'''\right)\cdot\mathbf{s}\right].\end{aligned} \tag{32}$$

Once again the Dirac delta function imposes a general constraint upon the configuration of wavevectors. Integrating over one of the $\mathbf{k}$ gives $\mathbf{k}''' = -\mathbf{k} - \mathbf{k}' - \mathbf{k}''$ for example, so that the wavevectors must form a closed loop. This general constraint, however, does not specify a precise shape of loop, instead the remaining integrals run over all of the different possibilities. At this point we may constrain the problem more tightly by noting that most combinations of the $\mathbf{k}$ will contribute zero to $\zeta_{(112)}$. This is because of the circularity property of the phases and (30). Indeed, the only non-zero contributions arise where we are able to apply the phase relation obtained from the reality constraint (25). In other words the properties of the phases dictate that the wavevectors must align in anti-parallel pairs: $\mathbf{k} = -\mathbf{k}'$, $\mathbf{k}'' = -\mathbf{k}'''$ and so forth.

There is a final constraint that must be imposed upon the $\mathbf{k}$ if ζ is the *connected* three-point covariance function. In a graph theoretic sense, the general (unconnected) N-point function $\langle\delta_{l_1}(\mathbf{x}_1)\delta_{l_2}(\mathbf{x}_2)\ \ldots\ \delta_{l_N}(\mathbf{x}_N)\rangle$ can be represented geometrically by a sum of tree diagrams. Each diagram consists of N nodes of order l_i, representing the $\delta_{l_i}(\mathbf{x}_i)$, and a number of linking lines denoting their correlations; see Fry [34] or Bernardeau [8] for more detailed accounts. Every node is made up of l_i internal points, which represent a factor $\delta_{\mathbf{k}} = |\delta_{\mathbf{k}}| \exp(i\phi_{\mathbf{k}})$ in the Fourier expansion. According to the rules for constructing diagrams, linking lines may join one internal point to a single other, either within the same node or in an external node. The *connected* covariance functions are represented specifically by the subset of diagrams for which every node is linked to at least one other, leaving none completely isolated. This constraint implies that certain pairings of wavevectors do not contribute to the connected covariance function. For more details, see Watts & Coles [92].

The above constraints may be inserted into (32) by re-writing the Dirac delta function as a product over delta functions of two arguments, appropriately normalized. There are only two allowed combinations of wavevectors, so we have

$$\delta_D(\mathbf{k}+\mathbf{k}'+\mathbf{k}''+\mathbf{k}''') \rightarrow \frac{1}{2V_u}[\delta_D(\mathbf{k}+\mathbf{k}'')\delta_D(\mathbf{k}'''+\mathbf{k}''')+\delta_D(\mathbf{k}+\mathbf{k}''')\delta_D(\mathbf{k}'+\mathbf{k}'')] \,. \tag{33}$$

Integrating over two of the $\mathbf{k}$ and using (25) eliminates the phase terms and leaves the final result

$$\zeta_{112}(\boldsymbol{r},\mathbf{s}) = \frac{1}{V_u}\int d^3k\, d^3k' \,\langle|\delta_{\mathbf{k}}|^2|\delta_{\mathbf{k}'}|^2\rangle_{\text{real}} \exp\left[\mathrm{i}\,\mathbf{k}'\cdot\boldsymbol{r}\right]\exp\left[-\mathrm{i}\,(\mathbf{k}+\mathbf{k}')\cdot\mathbf{s}\right] . \tag{34}$$

The existence of this quantity has therefore been shown to depend on the quadratic phase coupling of Fourier modes. The relationship between modes and the interpretation of the tree diagrams is also dictated by the properties of the phases.

One may apply the same rules to the higher-order terms in (28). It is immediately clear that the ζ_{122} terms are zero because there is no way to eliminate the phase term $\exp[i(\phi_{\mathbf{k}}+\phi_{\mathbf{k}'}+\phi_{\mathbf{k}''}+\phi_{\mathbf{k}'''}+\phi_{\mathbf{k}''''})]$, a consequence of the property equation (30). Diagrammatically this corresponds to an unpaired *internal* point within one of the nodes of the tree. The final, highest-order contribution to the three-point function is found to be

$$\begin{aligned}\zeta_{222}(\boldsymbol{r},\mathbf{s}) = \frac{1}{V_u^2}\int & d^3k\, d^3k'\, d^3k'' \,\langle|\delta_{\mathbf{k}}|^2|\delta_{\mathbf{k}'}|^2|\delta_{\mathbf{k}''}|^2\rangle_{\text{real}} \\ & \times \exp\left[\mathrm{i}\,(\mathbf{k}-\mathbf{k}')\cdot\boldsymbol{r}\right]\exp\left[\mathrm{i}\,(\mathbf{k}'-\mathbf{k}'')\cdot\mathbf{s}\right] ,\end{aligned} \tag{35}$$

where the phase and geometric constraints allow 12 possible combinations of wavevectors.

3.5 Power Spectrum and Bispectrum

The formal development of the relationship between covariance functions and power spectra developed above suggests the usefulness of higher-order versions of $P(k)$. It is clear from the above arguments that a more convenient notation for the power spectrum than that introduced in Sect. 2.1 is

$$\langle\delta_{\mathbf{k}}\delta_{\mathbf{k}'}\rangle = (2\pi)^3 P(k)\delta_D(\mathbf{k}+\mathbf{k}') \,. \tag{36}$$

The connection between phases and higher-order covariance functions obtained above also suggests defining higher-order polyspectra of the form

$$\langle\delta_{\mathbf{k}}\delta_{\mathbf{k}'}\ldots\delta_{\mathbf{k}^{(n)}}\rangle = (2\pi)^3 P_n(\mathbf{k},\mathbf{k}',\ldots\mathbf{k}^{(n)})\delta_D(\mathbf{k}+\mathbf{k}'+\ldots\mathbf{k}^{(n)}) \,, \tag{37}$$

where the occurrence of the delta function in this expression arises from a generalization of the reality constraint given in (25); see, e.g., Peebles [69]. Conventionally the version of this with $n = 3$ produces the bispectrum, usually called $B(\mathbf{k}, \mathbf{k}', \mathbf{k}'')$ which has found much effective use in recent studies of large-scale structure (Peebles [69]; Scoccimarro et al. [77]; Scoccimarro et al. [78]; Verde et al. [89]; [90]; [91]). It is straightforward to show that the bispectrum is the Fourier transform of the (reduced) three-point covariance function by following similar arguments; see, e.g., Peebles [69].

Note that the delta function constraint requires the bispectrum to be zero except for k-vectors ($\mathbf{k}$, $\mathbf{k}'$, $\mathbf{k}''$) that form a triangle in k-space. It is clear that the bispectrum can only be non-zero when there is a definite relationship between the phases accompanying the modes whose wavevectors form a triangle. Moreover the pattern of phase association necessary to produce a real and non-zero bispectrum is precisely that which is generated by quadratic phase association. This shows, in terms of phases, why it is that the leading order contributions to the bispectrum emerge from second-order fluctuations of a Gaussian random field. The bispectrum measures quadratic phase coupling.

Three-point phase correlations have another interesting property. While the bispectrum is usually taken to be an ensemble-averaged quantity, as defined in (37), it is interesting to consider products of terms $\delta_{\mathbf{k}}\delta_{\mathbf{k}'}\delta_{\mathbf{k}''}$ obtained from an individual realization. According to the fair sample hypothesis discussed above we would hope appropriate averages of such quantities would yield an estimate of the bispectrum. Note that

$$\delta_{\mathbf{k}}\delta_{\mathbf{k}'}\delta_{\mathbf{k}''} = \delta_{\mathbf{k}}\delta_{\mathbf{k}'}\delta_{-\mathbf{k}-\mathbf{k}'} = \delta_{\mathbf{k}}\delta_{\mathbf{k}'}\delta^*_{\mathbf{k}+\mathbf{k}'} \equiv \beta(\mathbf{k}, \mathbf{k}') , \tag{38}$$

using the requirement (25), together with the triangular constraint we discussed above. Each $\beta(\mathbf{k}, \mathbf{k}')$ will carry its own phase, say $\phi_{\mathbf{k},\mathbf{k}'}$, which obeys

$$\phi_{\mathbf{k},\mathbf{k}'} = \phi_{\mathbf{k}} + \phi_{\mathbf{k}'} - \phi_{\mathbf{k}+\mathbf{k}'} \ . \tag{39}$$

It is evident from this that it is possible to recover the complete set of phases $\phi_{\mathbf{k}}$ from the bispectral phases $\phi_{\mathbf{k},\mathbf{k}'}$, up to a constant phase offset corresponding to a global translation of the entire structure (Chiang & Coles [13]). This furnishes a conceptually simple method of recovering missing or contaminated phase information in a consistent way, an idea which has been exploited, for example, in speckle interferometry (Lohmann et al. [56]). In the case of quadratic phase coupling, described by (20), the left-hand side of (39) is identically zero leading to a particularly simple approach to this problem.

4 Phase Correlations in the CMB

Since the release of the first (preliminary) WMAP data set it has been subjected to a number of detailed independent analyses that have revealed some surprising features. Eriksen et al. [30] have pointed out the existence of a north–south asymmetry suggesting that the cosmic microwave background (CMB) revealed by the WMAP data is not statistically homogeneous over the celestial sphere. This is consistent with the results of Coles et al. [21] who found evidence for phase correlations in the WMAP data; see also Hajian & Souradeep [41] and Hajian et al. [42]. The low-order multipoles of the CMB also display some peculiarities (de Oliveira-Costa et al. [25]; Efstathiou [29]). Vielva et al. [87] found significant non-Gaussian behaviour in a wavelet analysis of the same data, as did Chiang et al. [15], Larson & Wandelt [54] and Park [66]. Other analyses of the statistical properties of the WMAP have yielded results consistent with the standard form of fluctuation statistics (Komatsu et al. [49]; Colley & Gott [24]). These unusual properties may well be generated by residual foreground contamination (Banday et al. [3]; Naselsky et al. [65]; de Oliveira-Costa et al. [26]; Dineen & Coles [27]) or other systematic effects, but may also provide the first hints of physics beyond the standard cosmological model.

In order to tap the rich source of information provided by future CMB maps it is important to devise as many independent statistical methods as possible to detect, isolate and diagnose the various possible causes of departures from standard statistics. One particularly fruitful approach is to look at the behaviour of the complex coefficients that arise in a spherical harmonic analysis of CMB maps. Chiang et al. [15], [16] and Coles et al. [21] have focused on the phases of these coefficients on the grounds that a property of a statistically homogeneous and isotropic GRF is that these phases are random. Phases can also be used to test for the presence of primordial magnetic fields (Chen et al. [11]; Naselsky et al. [64]) or evidence of non-trivial topology (Dineen et al. [28]).

4.1 Spherical Harmonics and Gaussian Fluctuations

We can describe the distribution of fluctuations in the microwave background over the celestial sphere using a sum over a set of spherical harmonics:

$$\Delta(\theta,\phi) = \frac{T(\theta,\phi)-\bar{T}}{\bar{T}} = \sum_{l=1}^{\infty}\sum_{m=-l}^{m=+l} a_{l,m} Y_{lm}(\theta,\phi) \,. \tag{40}$$

Here $\Delta(\theta,\phi)$ is the departure of the temperature from the average at angular position (θ,ϕ) on the celestial sphere in some coordinate system, usually galactic. The $Y_{lm}(\theta,\phi)$ are spherical harmonic functions which we define in terms of the Legendre polynomials P_{lm} using

$$Y_{lm}(\theta,\phi) = (-1)^m \sqrt{\frac{(2l+1)(l-m)!}{4\pi(l+m)!}} P_{lm}(\cos\theta) \mathrm{e}^{\mathrm{i}\,m\phi} \,, \tag{41}$$

i.e. we use the Condon–Shortley phase convention. In (1), the $a_{l,m}$ are complex coefficients which can be written as

$$a_{l,m} = x_{l,m} + iy_{l,m} = |a_{l,m}| \exp[\mathrm{i}\,\phi_{l,m}] \,. \tag{42}$$

Note that, since Δ is real, the definitions (40) and (41) require the following relations between the real and imaginary parts of the $a_{l,m}$: if m is odd then

$$\begin{aligned} x_{l,m} &= \Re(a_{l,m}) = -\Re(a_{l,-m}) = -x_{l,-m} \,, \\ y_{l,m} &= \Im(a_{l,m}) = \Im(a_{l,-m}) = y_{l,-m} \,, \end{aligned} \tag{43}$$

while if m is even

$$\begin{aligned} x_{l,m} &= \Re(a_{l,m}) = \Re(a_{l,-m}) = x_{l,-m} \,, \\ y_{l,m} &= \Im(a_{l,m}) = -\Im(a_{l,-m}) = y_{l,-m} \end{aligned} \tag{44}$$

and if m is zero then

$$\Im(a_{l,m}) = y_{l,0} = 0 \,. \tag{45}$$

From this it is clear that the $m = 0$ mode always has zero phase, and there are consequently only l independent phase angles describing the harmonic modes at a given l. Without loss of information we can therefore restrict our analysis to $m \geq 0$.

If the primordial density fluctuations form a Gaussian random field in space the temperature variations induced across the sky form a Gaussian random field over the celestial sphere. This means that

$$\langle a_{l,m} a^*_{l',m'} \rangle = C_l \delta_{ll'} \delta_{mm'} \,, \tag{46}$$

where C_l is the angular power spectrum, the subject of much scrutiny in the context of the cosmic microwave background (e.g. Hinshaw et al. [45]), and $\delta_{xx'}$ is the Kronecker delta function. Since the phases are random, the stochastic properties of a statistically homogeneous and isotropic Gaussian random field are fully specified by the C_l, which determines the variance of the real and imaginary parts of $a_{l,m}$, both of which are Gaussian:

$$\sigma^2(x_{l,m}) = \sigma^2(y_{l,m}) = \sigma_l^2 = \frac{1}{2} C_l \,. \tag{47}$$

4.2 Testing for Phase Correlations

The approach we take is to assume that we have available a set of phases $\phi_{l,m}$ corresponding to a set of spherical harmonic coefficients $a_{l,m}$ obtained from a data set, either real or simulated. We can also form phase differences according to

$$D_m(l) = \phi_{l,m+1} - \phi_{l,m} \,. \tag{48}$$

If the orthodox cosmological interpretation of temperature fluctuations is correct, the phases of the $a_{l,m}$ should be random and so should phase differences of the form $\phi_{l,m+1} - \phi_{l,m}$ and $\phi_{l+1,m} - \phi_{l,m}$. Let us assume, therefore, that we have n generic angles, $\theta_1 \ldots \theta_n$. Under the standard statistical assumption these should be random, apart from the constraints described in the previous section. The first thing we need is a way of testing whether a given set of phase angles is consistent with being drawn from uniform distribution on the unit circle. This is not quite as simple as it seems, particularly if one does not want to assume any particular form for actual distribution of angles, such as a bias in a particular direction; see Fisher [33]. Fortunately, however, there is a fully non-parametric method available, based on the theory of order statistics, and known as Kuiper's statistic (Kuiper [53]).

Kuiper's method revolves around the construction of a statistic, V, obtained from the data via the following prescription. First the angles are sorted into ascending order, to give the set $\{\theta_1, \ldots, \theta_n\}$. It does not matter whether the angles are defined to lie in $[0, 2\pi]$, $[-\pi, +\pi]$ or whatever. Each angle θ_i is divided by 2π to give a set of variables X_i, where $i = 1 \ldots n$. From the set of X_i we derive two values S_n^+ and S_n^-, where

$$S_n^+ = \max\left\{\frac{1}{n} - X_1, \frac{2}{n} - X_2, \ldots, 1 - X_n\right\} \tag{49}$$

and

$$S_n^- = \max\left\{X_1, X_2 - \frac{1}{n}, \ldots; X_n - \frac{n-1}{n}\right\} . \tag{50}$$

Kuiper's statistic, V, is then defined as

$$V = (S_n^+ + S_n^-) \cdot \left(\sqrt{n} + 0.155 + \frac{0.24}{\sqrt{n}}\right) . \tag{51}$$

Anomalously large values of V indicate a distribution that is more clumped than a uniformly random distribution, while low values mean that angles are more regular. The test statistic is normalized by the number of variates, n, in such a way that standard tables can be constructed to determine significance levels for any departure from uniformity; see Fisher [33]. In this context, however, it is more convenient to determine significance levels using Monte Carlo simulations of the "null" hypothesis of random phases. This is partly because of the large number of samples available for test, and partly because we can use them to make the test more general.

The first point to mention is that a given set of phases, say belonging to the modes at fixed l, is not strictly speaking random anyway, because of the constraints noted in the previous section. One could deal with this by discarding the conjugate phases, thus reducing the number of data points, but there is no need to do this when one can instead build the required symmetries into the Monte Carlo generator.

In addition, suppose the phases of the temperature field over the celestial sphere were indeed random, but observations were available only over a part of the sky, such as when a galactic cut is applied to remove parts of the map contaminated by foregrounds. In this case the mask may introduce phase correlations into the observations so the correct null hypothesis would be more complicated than simple uniform randomness. As long as any such selection effect is known, it could be built into the Monte Carlo simulation. One would then need to determine whether V from an observed sky is consistent with having been drawn from the set of values of V generated over the Monte Carlo ensemble.

There is also a more fundamental problem in applying this test to spherical harmonic phases. This is that a given set of $a_{l,m}$ depends on the choice of a particular coordinate axis. A given sky could actually generate an infinite number of different sets of $\phi_{l,m}$ because the phase angles are not rotationally invariant. One has to be sure to take different choices of z-axis into consideration when assessing significance levels, as a random phase distribution has no preferred axis while systematic artifacts may. A positive detection of non-randomness may result from a chance alignment of features with a particular coordinate axis in the real sky unless this is factored into the Monte Carlo simulations. For both the real sky and the Monte Carlo skies we therefore need not a single value of V but a distribution of V-values obtained by rotating the sky over all possible angles. A similar approach is taken by Hansen et al. [44]. The method may seem somewhat clumsy, but a test that has to be sensitive to departures from statistical homogeinity should not be based on measures that are rotationally invariant, such as those suggested by Ferreira et al. [32], as these involve averaging over the very fluctuations one is trying to detect.

4.3 Rotating the $a_{l,m}$

In view of the preceding discussion we need to know how to transform a given set of $a_{l,m}$ into a new set when the coordinate system is rotated into a different orientation. The method is fairly standard, but we outline it here to facilitate implementation of our approach.

Any rotation of the Cartesian coordinate system $S\{x,y,z\} \mapsto S'\{x,y,z\}$ can be described using a set of three Euler angles α, β, γ, which define the magnitude of successive rotations about the coordinate axes. In terms of a rotation operator $\hat{D}(\alpha,\beta,\gamma)$, defined so that a field $f(r,\theta,\phi)$ transforms according to

$$\hat{D}(\alpha,\beta,\gamma)f(r,\theta,\phi) = f'(r,\theta,\phi) = f(r,\theta',\phi') \,, \tag{52}$$

a vector $\mathbf{r}$ is transformed as

$$\mathbf{r}' = \mathbf{D}(0,0,\gamma)\mathbf{D}(0,\beta,0)\mathbf{D}(\alpha,0,0)\mathbf{r} \equiv \mathbf{D}(\alpha,\beta,\gamma)\mathbf{r} \,. \tag{53}$$

Here $\mathbf{D}$ is a matrix representing the operator $\hat{D}$, i.e.

$$\mathbf{D}(\alpha,\beta,\gamma) = \begin{pmatrix} \cos\gamma & \sin\gamma & 0 \\ -\sin\gamma & \cos\gamma & 0 \\ 0 & 0 & 1 \end{pmatrix} \begin{pmatrix} \cos\beta & 0 & -\sin\beta \\ 0 & 1 & 0 \\ \sin\beta & 0 & \cos\beta \end{pmatrix} \begin{pmatrix} \cos\alpha & \sin\alpha & 0 \\ -\sin\alpha & \cos\alpha & 0 \\ 0 & 0 & 1 \end{pmatrix} . \tag{54}$$

The Wigner D functions describe the rotation operator used to realize the transformations of covariant components of tensors with arbitrary rank l. The functions, written as $D^l_{m,m'}$, transform a tensor from $S\{x,y,z\}$ to $S'\{x',y',z'\}$. Consider a tensor $Y_{l,m}(\theta,\phi)$ defined under the coordinate system S and on applying the rotation operator we get

$$\hat{D}(\alpha,\beta,\gamma)Y_{l,m}(\theta,\phi) = Y_{l,m'}(\theta',\phi') = \sum_m Y_{l,m}(\theta,\phi)D^l_{m,m'}(\theta,\phi)\,. \tag{55}$$

This means that the transformation of the tensor under the rotation of the coordinate system can be represented as a matrix multiplication. Finding the rotated coefficients therefore requires a simple matrix multiplication once the appropriate D function is known. To apply this in practice one needs a fast and accurate way of generating the matrix elements $D^l_{m,m'}$ for the rotation matrix. There are $(2l+1)^2$ elements needed to describe the rotation of each mode and the value of each element depends upon the particular values of (α,β,γ) used for that rotation. Details of how to implement this are given in Coles et al. [21].

In order to apply these ideas to make a test of CMB fluctuations, we first need a temperature map from which we can obtain a measured set of $a_{l,m}$. Employing the above transformations with some choice of Euler angles yields a rotated set of the $a_{l,m}$. It is straightforward to choose a set of angles such that random orientations of the coordinate axis can be generated. Once a rotated set has been obtained, Kuiper's statistic is calculated from the relevant transformed set of phases. For example, Coles et al. [21] generated 3000 rotated sets of each CMB map using this kind of resampling of the original data, producing 3000 values of $V_{\rm cmb}$. The values of the statistic were then binned to form a measured (resampled) distribution of $V_{\rm cmb}$. The same procedure is applied to the 1000 Monte Carlo sets of $a_{l,m}$ drawn from a uniformly random distribution, i.e. each set was rotated 3000 times and a distribution of $V_{\rm MC}$ under the null hypothesis is produced. These realizations were then binned to create an overall global average distribution under the null hypothesis.

In order to determine whether the distribution of $V_{\rm cmb}$ is compatible with a distribution drawn from a sky with random phases, we use a simple χ^2 test, using

$$\chi^2 = \sum_i \frac{(f_i - \overline{f_i})^2}{\overline{f_i}}\,, \tag{56}$$

where the summation is over all the bins and $\overline{f_i}$ is the number expected in the ith bin from the overall average distribution. The larger the value of χ^2 the less likely the distribution functions are to be drawn from the same parent distribution. Values of $\chi^2_{\rm MC}$ are calculated for the 1000 Monte Carlo distributions and $\chi^2_{\rm cmb}$ is calculated from the distribution of $V_{\rm cmb}$. If the value of $\chi^2_{\rm cmb}$ is greater than a fraction p of the values of $\chi^2_{\rm MC}$, then the phases depart from a uniform distribution at significance level p. We have chosen 95 per cent as an appropriate level for the level at which the data are said to

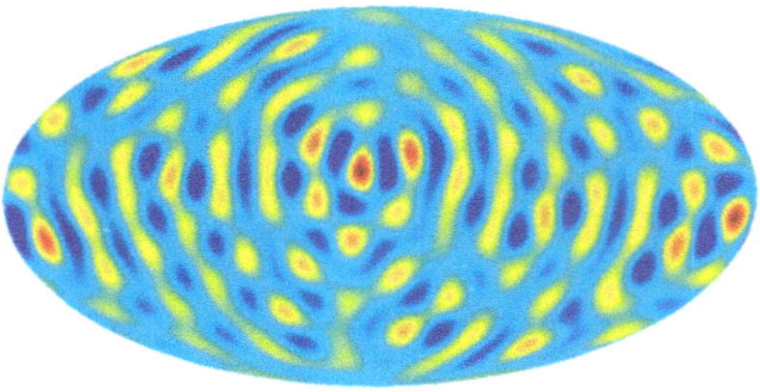

Fig. 2. A reconstruction of the WMAP ILC made using the spherical harmonic mode amplitudes $a_{l,m}$ for $l = 16$ only. Our analysis method (Coles et al. [21]) shows that these modes at different m have correlated phases in harmonic space, and the reconstructed sky shows this is aligned with the galactic plane

display signatures that are not characteristic of a statistically homogeneous Gaussian random field.

Application of this relatively straightforward method to the WMAP first-year data shows the existence of phase correlations, as demonstrated in Fig. 2.

4.4 Random Walks in Harmonic Space

To begin with, we concentrate on a simple measure based on the distribution of total displacements. Consider a particular value of l. The set of values $\{\mathbf{a_{l,m}}\}$ can be thought of as steps in a random walk in the complex plane, a structure which can be easily visualized and which has well-known statistical properties.

The simplest statistic one can think of to describe the set $\{\mathbf{a_{l,m}}\}$ is the net displacement of a random walk corresponding to the spherical harmonic mode l, i.e.

$$\mathbf{R_l} = \sum_{m>0} \mathbf{a_{l,m}} , \tag{57}$$

where the vector $\mathbf{a_{l,m}} \equiv (x_{l,m}, y_{l,m})$ and the random walk has an origin at $a_{l,0}$ (which is always on the x-axis). The length of each step $a_{l,m} = |\mathbf{a_{l,m}}|$ is the usual spherical harmonic coefficient described in the previous section and defined by (1). If the initial fluctuations are Gaussian then the two components of each displacement are independently normal with zero mean and the same variance (8). Each step then has a Rayleigh distribution so that the probability density for $a_{l,m}$ to be in the range $(a, a + da)$ is

$$p(a) = \frac{a}{\sigma_l^2} \exp\left(-\frac{a^2}{2\sigma_l^2}\right) . \tag{58}$$

This is a particularly simple example of a random walk (McCrea & Whipple [60]; Chandrasekhar [10]; Hughes [46]). Since the displacements in x and y are independently Gaussian the next displacement after l steps is itself Gaussian with variance $l\sigma_l^2$. Then the probability density of $|R_l|$ to be in the range $(r, r+dr)$ is itself a Rayleigh distribution of the form

$$p_l(r) = \frac{r}{l\sigma_l^2} \exp\left(-\frac{r^2}{2l\sigma_l^2}\right) . \tag{59}$$

This requires an estimate of σ_l^2. This can be made either using the same data or by assuming a given form for C_l, in which case the resulting test would be of a composite hypothesis that the fluctuations constitute a Gaussian random field with a particular spectrum. For large l this can be done straightforwardly, but for smaller values the sampling distribution of R_l will differ significantly from (59) because of the uncertainty in population variance from a small sample of a_{lm}. This is the so-called "cosmic variance" problem.

So far we have concentrated on fixed l with a random walk as a function of m. We could instead have fixed m and considered a random walk as a function of l. Or indeed randomly selected N values of l and m. In either case the results above still stand except with σ_l^2 replaced by an average over all the modes considered:

$$\sigma^2 = \frac{1}{N} \sum_{l,m} \sigma_{l,m}^2 . \tag{60}$$

We do not consider this case any further in this chapter.

The result (59) is obtained only if the steps of the random walk are independent and Gaussian. If the distribution of the individual steps is non-Gaussian, but the steps are independent, then the result (59) will be true for large l by virtue of the central limit theorem. Exact results for finite l, for example non-Gaussian distributions, are given by Hughes [46]. In such cases the overall 2D random walk comprises two independent 1D random walks in x and y. The Gaussianity of the individual step components can be tested using their empirical distributions via a Kolmogorov–Smirnov (K–S) or similar approach. Lack of independence of step size or step direction (i.e. phase correlations) would appear as anisotropy of their joint distribution which could be quantified by direct measures of cross-correlation or by testing the bivariate distribution using an appropriate 2D K–S test. The latter task is harder, especially if the number of modes available is small. Using the net displacement in 1D or 2D corresponds to using the sum of a sample of n variables to test the parent distribution. This is not necessarily powerful, but is robust and has well-defined properties. The true advantage of the random-walk representation is that it encapsulates the behaviour of the set $\{\mathbf{a_{l,m}}\}$ in a graphical fashion which is ideal for data exploration.

A slightly different approach is to keep each step length constant. The simplest way of doing this is to define

$$\hat{\mathbf{R}}_\mathbf{l} = \sum_{m>0} \frac{\mathbf{a_{l,m}}}{|\mathbf{a_{l,m}}|} , \tag{61}$$

so that each step is of unit length but in a random direction. This is precisely the problem posed in a famous (1905) letter by Pearson [68] and answered 1 week later by Rayleigh [73]. In the limit of large numbers of steps the result maps into the previous result (59) with $\sigma_l^2 = 1$ by virtue of the central limit theorem. For finite values of l there is also an exact result which can be derived in integral form using a method based on characteristic functions (Hughes [46]). The result is that the probability density for $\hat{\mathbf{R}}_\mathbf{l}$ to be in the range $r, r + dr$ is

$$q_l(r) = r \int_0^\infty u J_0(ur)[J_0(u)]^l du \,. \tag{62}$$

The integral is only convergent for $l > 2$ but for $l = 1$ or $l = 2$ straightforward alternative expressions are available (Hughes [46]). One can use this distribution to test for randomness of the phase angles without regard to the amplitudes.

A simple test of the hypothesis that the fluctuations are drawn from a statistically homogeneous and isotropic Gaussian random field on the sky could be furnished by comparing the empirical distribution of harmonic random flights with the form (59). As we explained above, however, the net displacement of the random walk is a simple but rather crude indication of the properties of the $\{\mathbf{a}_{\mathbf{l},\mathbf{m}}\}$, as it does not take into account the ordering of the individual steps. The possible non-Gaussian behaviour of the set $\{\mathbf{a}_{\mathbf{l},\mathbf{m}}\}$ is encoded not so much in the net displacement but in the *shape* of the random walk. To put this another way, there are many possible paths with the same net displacement, and these will have different shapes depending on the correlations between step size and direction. Long runs of directed steps or regular features in the observed structure could be manifestations of phase correlation (Coles et al. [21]). The graphical representation of the set $\{\mathbf{a}_{\mathbf{l},\mathbf{m}}\}$ in the form illustrated by Fig. 3 provides an elegant way of visualizing the behaviour of the harmonic modes and identifying any oddities. These could be quantified using a variety of statistical shape measures: moment of inertia (Rudnick et al. [74]), fractal dimension, first-passage statistics, shape statistics (e.g. Kuhn & Uson [52]) or any of the methods used to quantify the shape of minimal spanning trees (Barrow et al. [5]). Specific examples of correlated random walks are given in Hughes [46].

In practice the most convenient way to assess the significance of departures from the relevant distribution would be to perform Monte Carlo experiments of the null hypothesis. For statistical measures more complicated than the net displacement, the best way to set up a statistical test is to use Monte Carlo re-orderings of the individual steps to establish the confidence level of any departure from Gaussianity. This also enables one to incorporate such complications as galactic cuts.

The WMAP team released an internal linear combination (ILC) map that combined five original frequency band maps in such a way as to maintain unit response to the CMB whilst minimizing foreground contamination. The construction of this map is described in detail in Bennett et al. [7]. The weighted

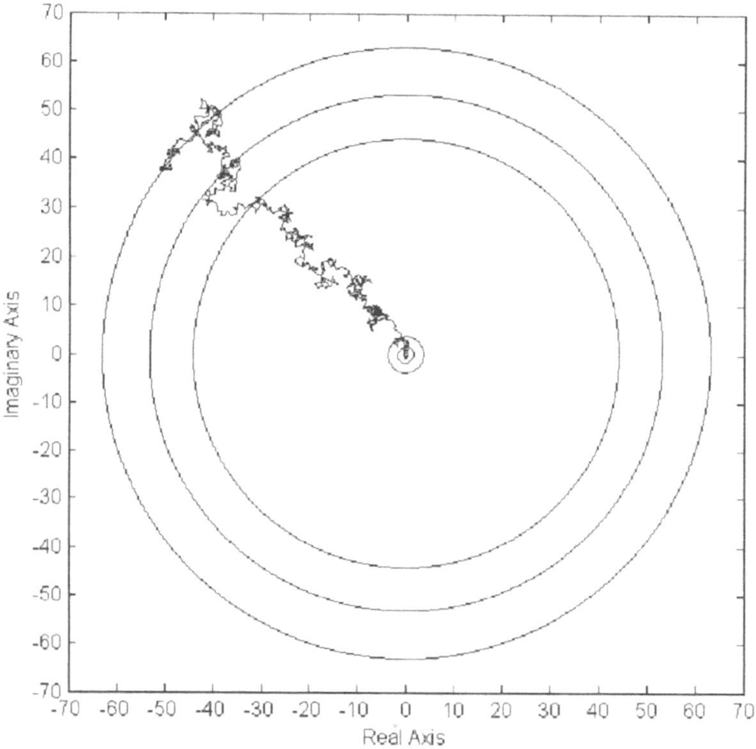

Fig. 3. The random walk performed by the spherical harmonic coefficients for $l = 532$ in the WMAP ILC data, statistically the mode that displays the greatest departure from that expected under the null hypothesis. The *outer circles* correspond to 99.9, 99 and 95 per cent upper confidence limits (from *outer* to *inner*); the *inner circles* are the corresponding lower limits, though the 99.9 per cent lower limit is too small to see

map is produced by minimizing the variance of the temperature scale such that the weights add to one. To further improve the result, the inner galactic plane is divided into 11 separate regions and weights determined separately. This takes account of the spatial variations in the foreground properties. Thus, the final combined map does not rely on models of foreground emission and therefore any systematic or calibration errors of other experiments do not enter the problem. The final map covers the full sky and the idea is that it should represent only the CMB signal. Following the release of the WMAP 1-year data Tegmark et al. ([86]; TOH) produced a cleaned CMB map. They argued that their version contained less contamination outside the galactic plane compared with the ILC map produced by the WMAP team.

The ILC map is not intended for statistical analysis but in any case represents a useful "straw man" for testing statistical techniques for robustness. To this end, we analysed the behaviour of the random walks representing spherical harmonic from $l = 1$ to $l = 600$ in the WMAP ILC. Similar results are obtained for the TOH map so we do not discuss the TOH map here. For both variable-length (57) and unit-length (61) versions of the random walk we generated 100,000 Monte Carlo skies assuming Gaussian statistics. These were used to form a distribution of $|\mathbf{R}_l|$ (or $|\hat{\mathbf{R}}_l|$) over the ensemble of randomly generated skies. A rejection of the null hypothesis (of stationary Gaussianity) at the α per cent level occurs when the measured value of the test statistic lies outside the range occupied by α per cent of the random skies.

Application of this simple test to the WMAP data (Stannard & Coles [81]) does not strongly falsify the null hypothesis, which is not surprising given the simplicity of the measure we have used. The number of modes outside the accepted range is close to that which would be expected if the null hypothesis were true. Notice that slightly more modes show up in the unit-length case than in the other, perhaps indicating that the phase correlations that are known to exist in these data (Chiang et al. [15]) are masked if amplitude information is also included. The most discrepant mode turns out to be $l = 532$ in both cases. For interested readers, a plot of the random walk for this case is included as Fig. 3.

5 Topological Measures of Large-Scale Structure

The application of phase analysis is obviously performed in harmonic space (whether Fourier-harmonic or spherical harmonic). But what does the presence of phase correlations mean for the morphology of large-scale structure? What is the real-space morphology of a fluctuation field with random phases? In studying morphology, one is typically interested in the question of how the individual filaments, sheets and voids join up and intersect to form the global pattern shown in Fig. 1. Is the pattern cellular, having isolated voids surrounded by high-density sheets, or is it more like a sponge in which under- and overdense regions interlock?

Looking at "slice" surveys gives the strong visual impression that we are dealing with bubbles; pencil beams (deep galaxy redshift surveys with a narrow field of view, in which the volume sampled therefore resembles a very narrow cone or "pencil") reinforce this impression by suggesting that a line-of-sight intersects at more or less regular intervals with walls of a cellular pattern. One must be careful of such impressions, however, because of elementary topology. Any closed curve in two dimensions must have an inside and an outside, so that a slice through a sponge-like distribution will appear to exhibit isolated voids just like a slice through a cellular pattern. It is important therefore that we quantify this kind of property using well-defined topological descriptors.

In an influential series of papers, Gott and collaborators have developed a method for doing just this (Gott et al. [36]; Hamilton et al. [43]; Gott et al. [37, 38]; Melott 1990 [62]; Coles et al. [20]). Briefly, the method makes use of a topological invariant known as the *genus*, related to the *Euler–Poincaré characteristic*, of the isodensity surfaces of the distribution. To extract this from a sample, one must first smooth the galaxy distribution with a filter (usually a Gaussian is used; see Sect. 14.3) to remove the discrete nature of the distribution and produce a continuous density field. By defining a threshold level on the continuous field, one can construct excursion sets (sets where the field exceeds the threshold level) for various density levels. An excursion set will typically consist of a number of regions, some of which will be simply connected, e.g. a deformed sphere, and others which will be multiply connected, e.g. a deformed torus is doubly connected. If the density threshold is labelled by ν, the number of standard deviations of the density away from the mean, then one can construct a graph of the genus of the excursion sets at ν as a function of ν: we call this function $G(\nu)$. The genus can be formally expressed as an integral over the intrinsic curvature K of the excursion set surfaces, S_ν, by means of the Gauss–Bonnet theorem.

The general form of this theorem applies to any two-dimensional manifold $\mathcal{M}$ with any (one-dimensional) boundary $\partial\mathcal{M}$ which is piecewise smooth. This latter condition implies that there are a finite number n vertexes in the boundary at which points it is not differentiable. The Gauss–Bonnet theorem states that

$$\sum_{i=1}^{n}(\pi - \alpha_i) + \int_{\partial\mathcal{M}} k_{\mathrm{g}} ds + \int_{\mathcal{M}} k dA = 2\pi\chi_E(\mathcal{M}) \, , \tag{63}$$

where the α_i are the angle deficits at the vertexes (the n interior angles at points where the boundary is not differentiable), k_{g} is the geodesic curvature of the boundary in between the vertexes and k is the Gaussian curvature of the manifold itself. Clearly ds is an element of length taken along the boundary and dA is an area element within the manifold $\mathcal{M}$. The right-hand side of this equation is the Euler–Poincaré characteristic, χ_E, of the manifold.

This probably seems very abstract but the definition above allows us to construct useful quantities for both two- and three-dimensional examples. If we have an excursion set as described above in three dimensions then its surface can be taken to define such a manifold. The boundary is just where the excursion sets intersect the limits of the survey and it will be taken to be smooth. Ignoring this, we see that the Euler–Poincaré characteristic is just the integral of the Gaussian curvature over the all compact bits of the surface of the excursion set. Hence, in this case,

$$2\pi\chi_E = \int_{S_\nu} K dS = 4\pi \left[1 - G(\nu)\right] \, . \tag{64}$$

Roughly speaking, the quantity G is the genus, which for a single surface is the number of "handles" the surface possesses; a sphere has no handles and has zero genus, a torus has one and therefore has a genus of one. For technical reasons to do with the effect of boundaries, it has become conventional not to use G but $G_S = G - 1$. In terms of this definition, multiply connected surfaces have $G_S \geq 0$ and simply connected surfaces have $G_S < 0$. One usually divides the total genus G_S by the volume of the sample to produce g_S, the genus per unit volume.

One of the great advantages of using the genus measure to study large-scale structure, aside from its robustness to errors in the sample, is that all Gaussian density fields have the same form of $g_S(\nu)$:

$$g_S(\nu) = A\left(1 - \nu^2\right) \exp\left(-\nu^2/2\right) , \tag{65}$$

where A is a spectrum-dependent normalization constant. This means that, if one smooths the field enough to remove the effect of non-linear displacements of galaxy positions, the genus curve should look Gaussian for any model evolved from Gaussian initial conditions, regardless of the form of the initial power spectrum which only enters through the normalization factor A. This makes it a potentially powerful test of non-Gaussian initial fluctuations or of models which invoke non-gravitational physics to form large-scale structure. The observations support the interpretation that the initial conditions were Gaussian, although the distribution looks non-Gaussian on smaller scales. The nomenclature for the non-Gaussian distortion one sees is a 'meatball shift': non-linear clustering tends to produce an excess of high-density simply connected regions, compared with the Gaussian curve. The opposite tendency, usually called 'Swiss cheese', is to have an excess of low-density simply connected regions in a high-density background, which is what one might expect to see if cosmic explosions or bubbles formed the large-scale structure. What one would expect to see in the standard picture of gravitational instability from Gaussian initial conditions is a 'meatball' topology when the smoothing scale is small, changing to a sponge as the smoothing scale is increased. This is indeed what seems to be seen in the observations so there is no evidence of bubbles; an example is shown in Fig. 4.

The smoothing required also poses a problem, however, because present redshift surveys sample space only rather sparsely and one needs to smooth rather heavily to construct a continuous field. A smoothing on scales much larger than the scale at which correlations are significant will tend to produce a Gaussian distribution by virtue of the central limit theorem. The power of this method is therefore limited by the smoothing required, which, in turn, depends on the space density of galaxies. An example is given in the figure, which shows the genus curve for the PSCz survey of IRAS galaxies.

Topological information can also be obtained from two-dimensional data sets, whether these are simply projected galaxy positions on the sky (such as the Lick map or the APM survey) or "slices" (such as the various CfA compilations). Here the excursion sets one deals with are just regions of the plane where the (surface) density exceeds some threshold. This method can also be applied to CMB temperature fluctuations where one looks at the

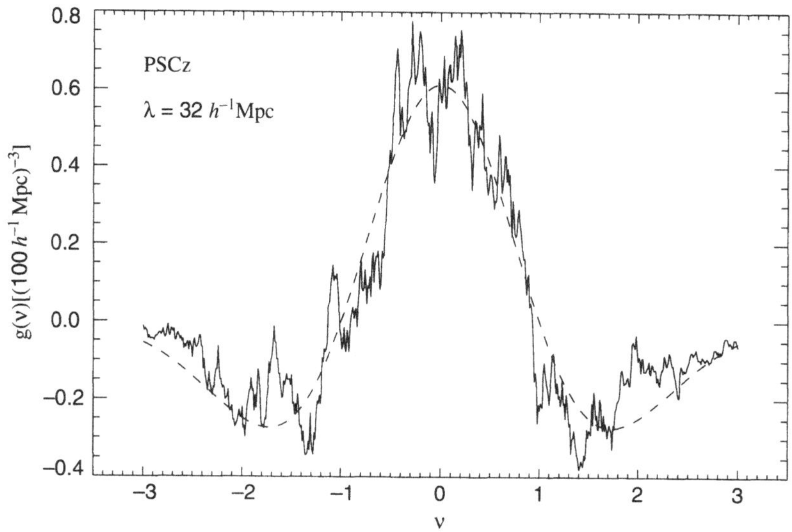

Fig. 4. Genus curve for galaxies in the IRAS PSCz survey. The *noisy curve* is the smoothed galaxy distribution while the *solid line* is the best-fitting curve for a Gaussian field; from Canavezes et al. [9]

topology of regions bounded by lines of constant temperature (Coles [17]; Gott et al. [38]; Colley & Gott [24]; Komatsu et al. [49]).

In such case we imagine the manifold referred to in the statement of the Gauss–Bonnet theorem to be not the surface of the excursion set but the surface upon which the set is defined (i.e. the sky). For reasonably small angles this can be taken to be a flat plane so that the Gaussian curvature of $\mathcal{M}$ is everywhere zero. (The generalization to large angles is trivial; it just adds a constant curvature term.) The Euler characteristic is then simply an integral of the line curvature around the boundaries of the excursion set:

$$2\pi\chi_E = \int k_g ds. \tag{66}$$

In this case the Euler–Poincaré characteristic is simply the number of isolated regions in the excursion set minus the number of holes in such regions.

This is analogous to the genus, but has the interesting property that it is an odd function of ν for a two-dimensional Gaussian random field, unlike $G(\nu)$ which is even. In fact the mean value of χ per unit area on the sky takes the form

$$\chi(\nu) = B\nu \exp\left(-\nu^2/2\right), \tag{67}$$

where B is a constant which depends only on the (two-dimensional) power spectrum of the random field. Notice that $\chi < 0$ for $\nu < 0$ and $\chi > 0$ for $\nu > 0$. A curve shifted to the left with respect to this would be a meatball topology, and to the right would be a swiss cheese.

There are some subtleties with this. First, as discussed above, two-dimensional topology does not really distinguish between 'sponge' and 'Swiss cheese' alternatives. Indeed, there is no two-dimensional equivalent of a sponge topology: a slice through a sponge is topologically equivalent to a slice through Swiss cheese. Nevertheless, it is possible to assess whether, for example, the mean density level ($\nu = 0$) is dominated by underdense or overdense regions so that one can distinguish Swiss cheese and meatball alternatives to some extent. The most obviously useful application of this method is to look at projected catalogues, the main problem being that, if the catalogue is very deep, each line of sight contains a superposition of many three-dimensional structures. This projection acts to suppress departures from Gaussian statistics by virtue of the central limit theorem. Nevertheless, useful information is obtainable from projected data simply because of the size of the data sets available; as is the case with three-dimensional studies, the analysis reveals a clear meatball shift which is what one expects in the gravitational instability picture. The methods used for the study of two-dimensional galaxy clustering can also be used to analyse the pattern of fluctuations on the sky seen in the cosmic microwave background.

More recently, this approach has been generalized to include not just the Euler–Poincaré distribution but all possible topological invariants. This means all quantities that satisfy the requirement that they be additive, continuous, translation invariant and rotation invariant. For an excursion set defined in d dimensions there are $d+1$ such quantities that can be regarded as independent. Any characteristic satisfying these invariance properties can be expressed in terms of linear combinations of these four independent quantities. These are usually called *Minkowski functionals*. Their use in the analysis of galaxy clustering studies was advocated by Mecke et al. [61] and has become widespread since then.

In three dimensions there are four Minkowski functionals. One of these is the integrated Gaussian curvature (equivalent to the genus we discussed above). Another is the mean curvature, H, defined by

$$H = \frac{1}{2}\int\left(\frac{1}{R_1} + \frac{1}{R_2}\right) dA\,. \tag{68}$$

In this expression R_1 and R_2 are the principal radii of curvature at any point in the surface; the Gaussian curvature is $1/(R_1 R_2)$ in terms of these variables. The other two Minkowski functionals are more straightforward. They are the surface area of the set and its volume. These four quantities give a "complete" topological description of the excursion sets.

6 Discussion

In this chapter I have tried to explain how phase correlations, arising from primordial non-Gaussianity, non-linear evolution (or indeed systematic error) can be measured and used to test cosmological models. The use of direct phase information is relatively new in cosmology, so I concentrated on basic properties and explained in some detail how phases relate to more familiar descriptors such as the bispectrum and three-point covariance functions. The magnitude of these statistical descriptors is of course related to the amplitude of the Fourier modes, but the factor that determines whether they are zero or non-zero is the arrangement of the phases of these modes.

The connection between polyspectra and phase information is an important one and it opens up many lines of future research, such as how phase correlations relate to redshift distortion and bias. Using small volumes of course leads to sampling uncertainties which are quite straightforward to deal with in the case of the power spectra but more problematic for higher-order spectra like the bispectrum. Understanding the fluctuations about ensemble averages in terms of phases could also lead to important insights. On the other hand, the application of phase methods to galaxy clustering studies is complicated by the non-linear evolution of perturbations as they collapse and form bound structures. Structures which are highly localized in real space are highly dispersed in Fourier space, so it is quite difficult to disentangle any primordial phase correlations from artifacts of non-linear evolution.

The CMB is a much more promising arena for the application of these methods. Late-time non-linear effects should be small (at least on large angular scales) so any phase correlations will almost certainly arise from either primordial effects or residual foreground contamination. The preliminary analysis we have performed using the WMAP data shows that there are indeed phase correlations, but Fig. 2 suggests that the likely interpretation is that it relates to the galaxy. As the constraints on early Universe physics get stronger, the importance of identifying low-amplitude foregrounds becomes all the more important. The next era of CMB physics will probably be dominated by polarization studies where the effects of foregrounds are likely to be even more complicated. There remains a great deal to learn about how to fully characterize the polarization maps that will soon be obtained. We can be certain, however, that phase information (in one way or another) will help us understand what is going on.

References

1. Adler, R.J.: The Geometry of Random Fields. John Wiley & Sons, New York (1981)
2. Albrecht, A., Steinhardt, P.J.: Phys. Rev. Lett. **48**, 1220 (1982)
3. Banday, A.J., Dickinson, C., Davies, R.D., Davies, R.J., Górski, K.M.: MNRAS **345**, 897 (2003)
4. Bardeen, J.M., Bond, J.R., Kaiser, N., Szalay, A.S.: ApJ **304**, 15 (1986)
5. Barrow, J.D., Bhavsar, S.P., Sonoda, D.H.: MNRAS **216**, 17 (1985)

6. Bartolo, N., Matarrese, S., Riotto, A.: Phys. Rev. D. **65**, 103505 (2002)
7. Bennett, C., et al.: ApJS **148**, 1 (2003)
8. Bernardeau, F.: ApJ **392**, 1 (1992)
9. Canavezes, A. et al., MNRAS **297**, 777 (1998)
10. Chandrasekhar, S.: Rev. Mod. Phys. **15**, 1 (1943)
11. Chen, G., Mukherjee, P., Kahniashvili, T., Ratra, B., Wang, Y.: ApJ **611**, 655 (2004)
12. Chiang, L.-Y.: MNRAS **325**, 405 (2001)
13. Chiang, L.-Y., Coles, P.: MNRAS **311**, 809 (2000)
14. Chiang, L.-Y., Coles, P., Naselsky, P.: MNRAS **337**, 488 (2002)
15. Chiang, L.-Y., Naselsky, P.D., Verkhodanov, O.V., Way, M.J.: ApJ **590**, L65 (2003)
16. Chiang, L.-Y., Naselsky, P.D., Coles, P.: ApJ **602**, L1 (2004)
17. Coles, P.: MNRAS **238**, 509 (1988)
18. Coles, P., Barrow, J.D.: MNRAS, **228**, 407 (1987)
19. Coles, P., Chiang, L.-Y.: Nature **406**, 376 (2000)
20. Coles, P., Davies, A.G., Pearson, R.C.: MNRAS **281**, 1375 (1996)
21. Coles, P., Dineen, P., Earl, J., Wright, D.: MNRAS **350**, 989 (2004)
22. Coles, P., Jones, B.J.T.: MNRAS **248**, 1 (1991)
23. Coles, P., Lucchin, F.: Cosmology: The Origin and Evolution of Cosmic Structure, 2nd edn. John Wiley & Sons, Chichester (2002)
24. Colley, W.N., Gott, J.R.: MNRAS **344**, 686 (2003)
25. de Oliveira-Costa, A., Tegmark, M., Zaldarriaga, M., Hamilton, A.J.S.: Phys. Rev. D. **69**, 063516 (2004)
26. de Oliveira-Costa, A., Tegmark, M., Davies, R.D., Gutierrez, C.M., Lasenby, A.N., Rebolo, R., Watson, R.A.: ApJ **606**, L89 (2004)
27. Dineen, P., Coles, P.: MNRAS **348**, 52 (2004)
28. Dineen, P., Rocha, G., Coles, P.: MNRAS **358**, 1285 (2005)
29. Efstathiou, G.: MNRAS **348**, 885 (2004)
30. Eriksen, H.K., Hansen, F.K., Banday, A.J., Gorski, K.M., Lilje, P.B.: ApJ **605**, 14 (2004)
31. Falk, T., Rangarajan, R., Srednicki, M.: ApJ **403**, L1 (1993)
32. Ferreira, P.G., Magueijo, J., Gorski, K.M.: ApJ **503**, L1 (1998)
33. Fisher, N.I.: Statistical Analysis of Circular Data. Cambridge University Press, Cambridge (1993)
34. Fry, J.N.: ApJ **279**, 499 (1984)
35. Gangui, A., Lucchin, F., Matarrese, S., Mollerach, S.: ApJ **430**, 447 (1994)
36. Gott, J.R., Melott, A.L., Dickinson, M.L.: ApJ **341**, (1986)
37. Gott, J.R. et al.: ApJ **340**, 625 (1989)
38. Gott, J.R., Park, C., Juszkiewicz, R., Bies, W.E., Bennet, D.P., Stebbins, A.: ApJ **352**, 1 (1990)
39. Guth, A.H.: Phys. Rev. D. **23**, 347 (1981)
40. Guth, A.H., Pi, S.-Y.: Phys. Rev. Lett. **49**, 1110 (1982)
41. Hajian, A., Souradeep, T.: ApJ **597**, L5 (2003)
42. Hajian, A., Souradeep, T., Cornish, N.: ApJ **618**, L63 (2004)
43. Hamilton, A.J.S., Gott, J.R., Weinberg, D.H.: ApJ **309**, 1 (1986)
44. Hansen, F.K., Marinucci, D., Vittorio, N.: Phys. Rev. D. **67**, 123004 (2003)
45. Hinshaw, G. et al.: ApJ **148** (2003)
46. Hughes, B.D.: Random Walks and Random Environments. Volume 1: Random Walks. Oxford University Press, Oxford (2004)

47. Jain, B., Bertschinger, E.: ApJ **509**, 517 (1998)
48. Kendall, M., Stuart, A.: The Advanced Theory of Statistics, vol. 1. Griffin & Co, London (1977)
49. Komatsu, E., et al.: ApJ **148**, 119 (2003)
50. Komatsu, E., Spergel, D.N.: Phys. Rev. D. **63**, 063002 (2001)
51. Koyama, K., Soda, J., Taruya, A.: MNRAS **310**, 1111 (1999)
52. Kuhn, J.R., Uson, J.M.: ApJ **263**, L47 (1982)
53. Kuiper, N.H.: Koninklijke Nederlandse Akademie Van Wetenschappen, Proc. Ser. A, LXIII, pp. 38–49 (1960)
54. Larson, D.L., Wandelt, B.D.: ApJ **613**, L85 (2004)
55. Linde, A.D.: Phys. Lett. B. **108**, 389 (1982)
56. Lohmann, A.W., Weigelt, G., Wirnitzer, B.: Appl. Optics **22**, 4028 (1983)
57. Luo, X.: ApJ **427**, L71 (1994)
58. Luo, X., Schramm, D.N.: ApJ **408**, 33 (1993)
59. Matarrese, S., Verde, L., Jimenez, R.: ApJ **541**, 10 (2000)
60. McCrea, W.H., Whipple, F.J.W.: Proc. R. Soc. Edin. **60**, 281 (1940)
61. Mecke, K.R., Buchert, T., Wagner, H.: A&A **288**, 697 (1994)
62. Melott, A.L.: Phys. Rep., **193**, 1 (1990)
63. Moscardini, L., Matarrese, S., Lucchin, F., Messina, A.: MNRAS **248**, 424 (1991)
64. Naselsky, P.D., Chiang, L.-Y., Olesen, P., Verkhodanov, O.: ApJ **615**, 45 (2004)
65. Naselsky, P.D., Doroshkevich, A.G., Verkhodanov, O.: ApJ **599**, L53 (2003)
66. Park, C.-G.: MNRAS **349**, 313 (2004)
67. Peacock, J.A., Dodds, S.: MNRAS **267**, 1020 (1996)
68. Pearson, K.: Nature **72**, 294 (1905)
69. Peebles, P.J.E.: The Large-Scale Structure of the Universe. Princeton University Press, Princeton NJ (1905)
70. Peebles, P.J.E.: ApJ **510**, 523 (1999)
71. Peebles, P.J.E.: ApJ **510**, 531 (1999)
72. Percival, W.J., et al.: MNRAS **327**, 1297 (2001)
73. Rayleigh, L.: Nature **72**, 318 (1905)
74. Rudnick, J., Beldjenna, A., Gaspari, G.: J. Math. Phys. A. **20**, 971 (1987)
75. Ryden, B.S., Gramann, M.: ApJ **383**, L33 (1991)
76. Scherrer, R.J., Melott, A.L., Shandarin, S.F.: ApJ **377**, 29 (1991)
77. Scoccimarro, R., Colombi, S., Fry, J.N., Frieman, J.A., Hivon, E., Melott, A.L.: ApJ **496**, 586 (1998)
78. Scoccimarro, R., Couchman, H.M.P., Frieman, J.A.: ApJ **517**, 531 (1999)
79. Shandarin, S.F.: MNRAS **331**, 865 (2002)
80. Soda, J., Suto, Y.: ApJ **396**, 379 (1992)
81. Stannard, A.J., Coles, P.: MNRAS **364**, 929 (2005)
82. Starobinsky, A.A.: Pis'ma Zh. Eksp. Teor. Fiz. **30**, 719 (1979)
83. Starobinsky, A.A.: Phys. Lett. B. **91**, 99 (1980)
84. Starobinsky, A.A.: Phys. Lett. B. **117**, 175 (1982)
85. Stirling, A.J., Peacock, J.A.: MNRAS **283**, L99 (1996)
86. Tegmark, M., de Oliveira-Costa, A., Hamilton, A.J.S.: Phys. Rev. D. **68**, 123523 (2003)
87. Vielva, P., Martinez-Gonzalez, E., Barreiro, R.B., Sanz, J.L., Cayon, L.: ApJ **609**, 22 (2004)
88. Verde, L., Heavens, A.F.: ApJ **553**, 14 (2001)
89. Verde, L., Jimenez, R., Kamionkowski, M., Matarrese, S.: MNRAS **325**, 412 (2001)

90. Verde, L., Wang, L., Heavens, A.F., Kamionkowski, M.: MNRAS **313**, 141 (2000)
91. Verde, L. et al.: MNRAS **335**, 432 (2002)
92. Watts, P.I.R., Coles, P.: MNRAS **338**, 806 (2002)

Multiscale Methods

E. Saar[1,2]

[1] Tartu Observatoorium, Tõravere, 61602 Tartumaa, Estonia,
saar@aai.ee

[2] Observatori Astronòmic, Universitat de València, Apartat de Correus 22085, E-46071 València, Spain,
Enn.Saar@uv.es

Bernard Jones discussed multiresolution analysis in his wavelet lectures. This is a pretty well-formalized and self-contained area of wavelet analysis. Multiscale methods form a wider and less well-defined area of tools and approaches; the term is used, e.g., in numerical analysis, but the main range of multiscale methods is based on application of wavelets.

In this lecture I shall explain how to carry out multiscale morphological analysis of cosmological density fields. For that, we have to use wavelets to decompose the data into different frequency bands, to calculate densities, and to describe morphology.

Let us start with wavelet transforms.

1 Wavelet Transforms

The most popular wavelet transforms are the orthogonal fast transforms, described by Bernard Jones. For morphological analysis, we need different transforms. The easiest way to understand wavelets is to start with continuous wavelet transforms.

1.1 Continuous Wavelet Transform

The basics of wavelets are most easily understood in the case of one-dimensional signals (time series or data along a line). The most commonly used decomposition of such a signal ($f(x)$) into contributions from different scales is the Fourier decomposition:

$$\hat{f}(k) = \int_{-\infty}^{\infty} f(x) \exp(-\mathrm{i}\, kx)\, dx \,.$$

The Fourier amplitudes $f(k)$ describe the frequency content of a signal. They are not very intuitive, however, as they depend on the behaviour of a signal

Saar, E.: *Multiscale Methods*. Lect. Notes Phys. **665**, 523–563 (2009)
DOI 10.1007/978-3-540-44767-2_16

as a whole; e.g. if the signal is a density distribution along a line, then all the regions of the universe where this line passes through contribute to it. Fourier modes are homeless.

For analysing the texture of images and fields, both scales and positions are important. The right tools for that are continuous wavelets. A wavelet transform of our signal $f(x)$ is

$$W(a,b) = \frac{1}{\sqrt{a}} \int_{-\infty}^{\infty} f(x) \psi \left(\frac{x-b}{a} \right) dx \,,$$

where $\psi(x;a,b)$ is a wavelet profile. Here the argument b ties the wavelet to a particular location, and a is the scale factor.

In order to be interesting (and different from the Fourier modes), typical wavelet profiles have a compact support. Two popular wavelets are shown in Fig. 1 – the Morlet wavelet and the Mexican Hat wavelet (see the formulae in Bernard Jones' lecture). The Morlet wavelet is a complex wavelet.

Continuous wavelets are good for finding sharp edges and singularities of functions. An example of that is given in Fig. 2. The brightness-coded wavelet amplitudes (Mexican Hat wavelet) for the upper-panel skyline show features at all scales.

This example shows us also the information explosion inherent in continuous wavelets – a function $f(x)$ gives rise to a two-argument wavelet amplitude $W(a,b)$; the collection of continuous wavelet amplitudes is heavily redundant. If the wavelet is well behaved (2), the wavelet amplitudes can be used to restore the original function:

$$f(x) = \frac{1}{C_\psi} \int_0^{\infty} \int_{-\infty}^{\infty} \frac{1}{\sqrt{a}} W(a,b) \psi^\star \left(\frac{x-b}{a} \right) \frac{da\, db}{a^2} \,, \tag{1}$$

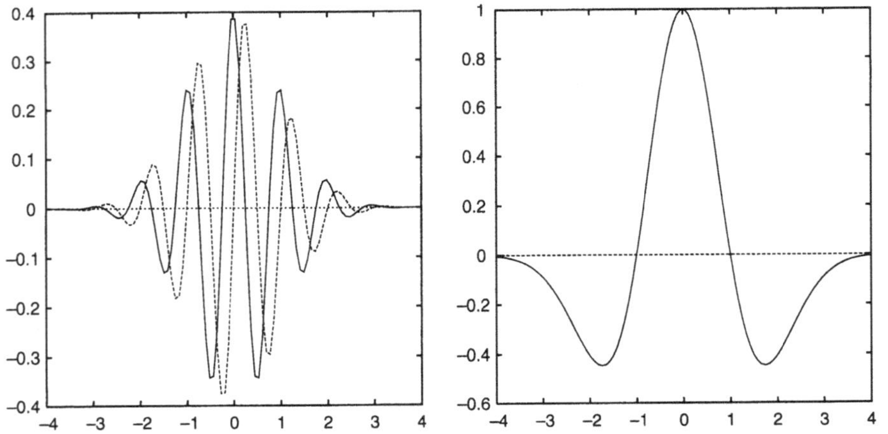

Fig. 1. *Left* – Morlet wavelet (*solid line* shows its real part, *dashed line* – the imaginary part), *right* – (a real-valued) Mexican Hat wavelet

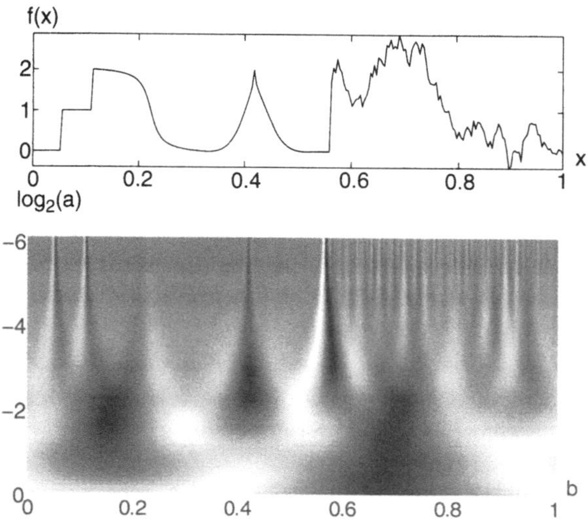

Fig. 2. Example (continuous Mexican Hat wavelet). Figure from S. Mallat [7]

where

$$C_\psi = \int_0^\infty \frac{\hat{\psi}^\star(\nu)\hat{\psi}(\nu)}{\nu} d\nu$$

($\psi^\star$ is a complex conjugate of ψ, and $\hat{\psi}$ is the Fourier transform of ψ). This constant exists ($C_\psi < \infty$), when

$$\hat{\psi}(0) = 0 \Rightarrow \int_{-\infty}^{\infty} \psi(x)\, dx = 0 \,. \tag{2}$$

This is the only requirement that a continuous-transform wavelet has to satisfy. Equation (1) shows that we need to know the wavelet amplitudes of all scales to reconstruct the original function. We also see that large-scale wavelet amplitudes are heavily downweighted in this reconstruction.

1.2 Dyadic Wavelet Transform

It is clear that the information explosion inherent in application of continuous wavelets has to be restrained. The obvious way to proceed is to consider computational restrictions.

First, computations need to be carried out on a discrete coordinate grid b_i. Second, if we look at Fig. 2 we see that the wavelet amplitudes change, in general, smoothly with scale. This suggests that a discrete grid of scales could suffice to analyse the data. While the obvious choice for the coordinate grid is uniform, the scale grid is usually chosen logarithmic, in hope to better catch the scale-dependent behaviour. This discretization generates much less data than the continuous transform did, only $N \times J$, where N is the size of the original data (the number of grid points), and J is the number of different scales. The most popular choice of the scale grid is where neighbouring scales differ by a factor of two. Other choices are possible, of course, but this choice is useful in several respects, as we shall see below.

Such a wavelet transform is called dyadic. As we saw above, any compact function with the zero mean could be used as an analysing wavelet. In the case of a discrete wavelet transform, we, however, have a problem – can we restore the original signal, given all the wavelet amplitudes obtained? It is not clear at first sight, as the restoration integral for a continuous wavelet (1) contains contributions of all wavelet scales.

The answer to that is yes, given that the frequency axis is completely covered by the wavelets of dyadic scales (the wavelets form a so-called frame). This requirement is referred to as the perfect reconstruction condition; I shall show a specific example below.

Let us rewrite now the wavelet transform, for finite grids. As we have heard about the multiresolution analysis already, we shall try to introduce both the smooth and the detail part of the transform. Let the initial data be $a_0(m)$ (the index 0 shows the transform order, and the argument m – the grid point). The smoothing operation and the wavelet transform proper are convolutions, so they can be written as

$$a_{j+1}(m) = \sum_k h_j(k) a_j(m+k) \, , \qquad d_{j+1}(m) = \sum_k g_j(k) a_j(m+k) \, , \tag{3}$$

where the filters $h_j(k)$ and $g_j(k)$ are different from zero for a few values of k around 0 only (the wavelet transform is local). Applying recursively this rule, we can find wavelet transforms of all orders, starting from the data ($a_0(\cdot)$).

Now, we should like to be able to restore the signal, knowing the $a_j(\cdot)$ and $d_j(\cdot)$. As the previous filters were linear, restoration should also be linear, and we demand

$$a_j(m) = \sum_k \tilde{h}_j(k) a_{j+1}(m+k) + \sum_k \tilde{g}_j(k) d_{j+1}(m+k) \, . \tag{4}$$

It can be proved that these rules will work for any dyadic wavelet that satisfies the so-called unit gain condition if the filters are upsampled (see next section).

If you recall (or look up) Bernard Jones' lecture, you will notice that the rules (3), (4) look the same as the rules for bi-orthogonal wavelets. The only difference is that for bi-orthogonal wavelets the filters have to satisfy an extra, so-called aliasing cancellation condition, arising because of the downsampling of the grid.

1.3 À Trous Transform

This downsampling leads us to the next problem. The change of the frequency (scale) between wavelet orders when bi-orthogonal or orthogonal wavelet transforms are used is achieved by downsampling – choosing every other data point. Applying the same wavelet filter on the downsampled data set is equivalent to using twice as wide filter on the original data. Now, in our case the grid is not diluted, and all points participate for all wavelet orders. The obvious solution here is to upsample the filter – to introduce zeros between the filter points. This doubles the filter width, and it is also useful for the computational point of view, as the operations count for the convolutions does not depend now on the wavelet order. Because of these zeros (holes), such a dyadic transform is called '*à trous*' ('with holes' in French; the transform was introduced by French mathematicians, Holschneider et al. [6]).

In Fourier language, such an upsampling halves the frequency:

$$\hat{h}_1(\omega) = \hat{h}_0(2\omega) , \qquad \hat{h}_j(\omega) = \hat{h}(2^j\omega) ,$$

where ω stands for the frequency, and $h(k) \equiv h_0(k), \hat{h}(\omega) \equiv \hat{h}_0(\omega)$. Although we describe localized transforms, it is useful to use Fourier transforms, occasionally – convolutions are frequent in wavelet transforms, and convolutions are converted to simple multiplications in Fourier space. Now, in coordinate space, the explicit expression for the upsampled filter $h(\cdot)$ is

$$h_j(k') = h(k)\delta(k' - k \cdot 2^j) ,$$

saying simply that the only non-zero components of the smoothing filter for the order j are those that have the index $k \cdot 2^j$, and they are determined by the original filter values $h(k) \equiv h_0(k)$. Let us write now the smoothing sum again:

$$a_{j+1}(m) = \sum_{k'} h_j(k')a_j(m+k') = \sum_k h(k)a_j(m+2^jk) , \tag{5}$$

where we retained only the non-zero terms in the last sum. The last equality is of the form usually used in *à trous* transforms; the holes are returned to the data space again. However, the procedure is different from the bi-orthogonal transform, as we find the scaling and wavelet amplitudes for every grid point m, not for the downsampled sets only. The wavelet rule with holes reads

$$d_{j+1}(m) = \sum_k g(k)a_j(m+2^jk) , \tag{6}$$

where, obviously, $g(k) \equiv g_0(k)$.

Let us now construct a particular *à trous* transform, starting backwards, from the reconstruction rule (4). In multiresolution language, this rule tells us that the approximation (sub)space V_j, where the 'smooth functions' live, is a direct sum of two orthogonal subspaces of the next order (the formula is for projections, but it follows from this fact). Let us take it in a much simpler way, literally, and demand

$$a_j(m) = a_{j+1}(m) + d_{j+1}(m) , \tag{7}$$

or $\tilde{h}(k) = \tilde{g}(k) = \delta_{0k}$. This is a very good choice, as applying it recursively we get

$$a_0(m) = a_J(m) + \sum_{j=J}^{j=1} d_j(m) \,, \tag{8}$$

meaning that the data is decomposed into a simple sum of contributions of different details (wavelet orders) and the most smooth picture. As there are no extra weights, these detail spaces have a direct physical meaning, representing the life in the full data space at a given resolution.

Condition (7) gives us at once the formula for the wavelet transform

$$d_{j+1}(m) = a_j(m) - a_{j+1}(m) = \sum_k \left[\delta_{0k} - h(k)\right] a_j(m+k) \,, \tag{9}$$

or $g(k) = \delta_{0k} - h(k)$.

So far, so good. Now we have only to choose the filter $h(k)$ to specify the transform. As the filter is defined by the scaling function $\phi(x)$ via the two-scale equation,

$$\phi(x/2) = 2 \sum_k h(k) \phi(x-k) \,, \tag{10}$$

meaning simply that the scaling function of the next order (note that $f(x/2)$ is only half as fast as $f(x)$) has to obey the smoothing rule in (5) exactly (the space where it lives is a subspace of the lower order space). Different normalizations are used; the coefficient 2 appears here if we omit extra coefficients for the convolution (10) in the Fourier space:

$$\hat{\phi}(2\omega) = \hat{h}(\omega)\hat{\phi}(\omega)$$

(recall that the coordinate space counterpart of $\hat{f}(2\omega)$ is $f(x/2)/2$). Obviously, not all functions satisfy the two-scale equation (10), but a useful class of functions that do are box splines.

1.4 Box Splines

Box splines are easy to obtain – an nth degree box spline $B_n(x)$ is the result of $n+1$ convolutions of the box profile $B_0(x) = 1, \quad x \in [0,1]$ with itself. Some authors like to shift it, some do not; we adopt the condition that the convolution result is centred at 0 when n is odd and at $x = 1/2$ when n is even. This convention gives a simple expression for the Fourier transform of the spline:

$$\hat{B}_n(\omega) = \left(\frac{\sin(\omega/2)}{\omega/2}\right) \exp\left(-\mathrm{i}\,\varepsilon\omega/2\right) \,, \tag{11}$$

where $\varepsilon = 1$, if n is even, and 0 otherwise. You can easily derive the formula yourselves, recalling that the Fourier transform of a $[-1,1]$ box is the sinc$(\cdot)$ function, and using the rule for the argument shifts.

Box splines have, just to start, several very useful properties. First, they are compact; in fact, they are the most compact polynomials for a given degree ($B_n(x)$ is a polynomial of degree n). Second, they are interpolating,

$$\sum_k B_j(x-k) = 1 \; , \tag{12}$$

a necessary condition for a scaling function. And box splines satisfy the two-scale equation (10), with

$$h(k) = 2^{-(n+1)} \binom{n}{k} \; , \tag{13}$$

where n is the degree of the spline (see de Boor [3]). This formula is written for unshifted box splines, and here the index k ranges from 0 to n. It is easy to modify (13) for centred splines; e.g. for centred box splines of an odd degree n the index k ranges from $-(n+1)/2$ to $(n+1)/2$, and we have to replace k at the right-hand side of (13) by $k+(n+1)2$.

I do not intend to be original and shall choose the B_3 box spline for the scaling function. This is the most beloved box spline in astronomical community; see, e.g. the monograph by Jean-Luc Starck and Fion Murtagh [17] for many examples. As any spline, this can be given by different polynomials in different coordinate intervals; fortunately, a compact expression exists:

$$\phi(x) = B_3(x) = \frac{1}{12}\left(|x-2|^3 - 4|x-1|^3 + 6|x|^3 - 4|x+1|^3 + |x+2|^3\right) \; . \tag{14}$$

This function is identically zero outside the interval $[-2,2]$. Formula (13) gives us the filter $h(k)$:

$$h(k) = (1/16, 1/4, 3/8, 1/4, 1/16) \; , \qquad k \in [-2,2] \; . \tag{15}$$

In order to obtain the associated wavelet $\psi(\cdot)$, we have to return to our recipe for calculating the wavelet coefficients (9). These coefficients are, in principle, convolutions (multiplications in Fourier space):

$$\hat{d}_{i+1}(\omega) = \hat{\psi}(\omega)\hat{a}_i(\omega) \; , \tag{16}$$

So, (9) gives us

$$\hat{\psi}(\omega)\hat{a}_i(\omega) = \hat{\phi}(\omega/2)\hat{a}_i(\omega) - \hat{\phi}(\omega)\hat{a}_i(\omega) \; ,$$

or

$$\hat{\psi}(\omega) = \hat{\phi}(\omega/2) - \hat{\phi}(\omega) \; . \tag{17}$$

For coordinate space, the above expression transforms to

$$\psi(x) = 2\phi(2x) - \phi(x) \; . \tag{18}$$

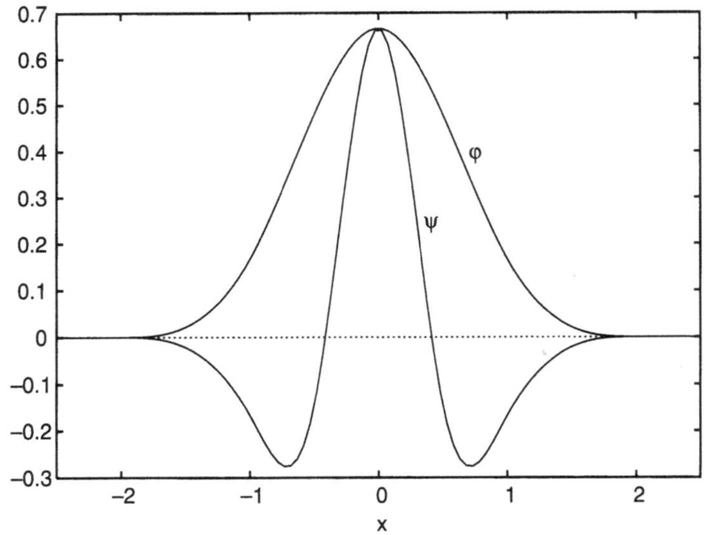

Fig. 3. The B_3 scaling function (ϕ) and its associated wavelet (ψ)

Both the B_3 box splines (the scaling function $\phi(x)$ and its associated wavelet $\phi(x)$) are shown in Fig. 3.

I cannot refrain from noting that the last exercise was, strictly speaking, unnecessary. We could have proceeded with our wavelet transform after obtaining the filter $h(k)$ (15). But it is nice to know the wavelet by face and to get a feeling what our algorithms really do.

Now I can also show you the Fourier transform of the wavelet (Fig. 4), to reassure you that such a wavelet can be built and that it does not leave gaps in the frequency axis. We see, first, that the filter peaks at $\omega \approx \pi$, giving for its characteristic wavelength $\lambda = 2\pi/\omega = 2$ (grid units). Second, we see that neighbouring wavelet orders overlap in frequency. The reason for that is that our wavelets are not orthogonal. So we loose a bit in frequency separation, but gain in spatial resolution.

Two points more: first, about normalization. Although our scaling function $\phi(x)$ is normalized in the right way (its integral is 1), the coefficient in the two-scale equation (10) is different from the conventional one, and, as a result, the filter coefficients h_k sum to 1, not to $\sqrt{2}$. The integral over the wavelet profile is zero (this ensures that $\psi(x)$ is really a wavelet), but its norm $\int \psi^2(x)\,dx \approx 0.2345$; normally it is chosen to be 1. What matters, really, is that it is not zero and does not diverge; welcome to the wavelet world of free normalization.

Second, about initial data. We can recursively apply the transformation rules (5, 6) only if we assume that the data $a_0(m)$ belongs to the class of smooth functions (those obtained by convolution with the scaling function).

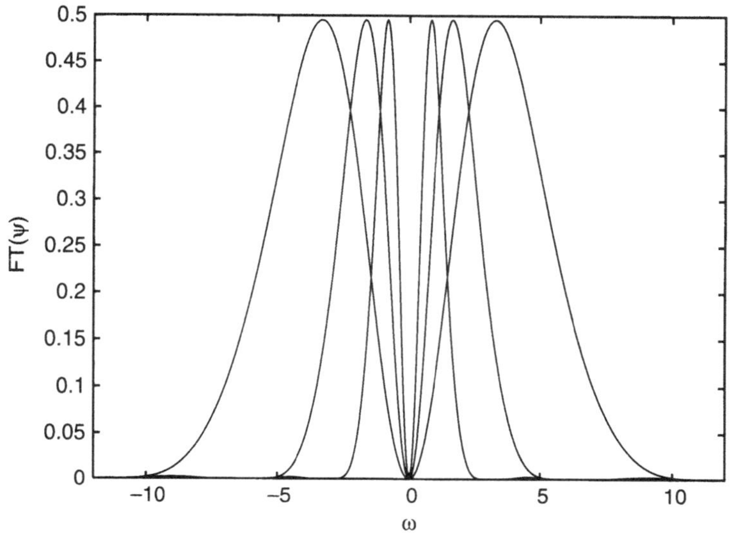

Fig. 4. The Fourier transforms of the three subsequent orders of the $B_3(\cdot)$-associated wavelet. The transforms fully cover the frequency axis, but the overlap between different orders is substantial

So, if the raw data comes in ticking at grid points (regular times), we should smooth it once before starting our transform chain. If the raw data is given at points x that do not coincide with the grid, the right solution is to distribute it to the grid m with the scaling weights $\phi(x-m)$. This procedure was christened 'extirpolation' (inverse interpolation) by Press et al. [11]; a strange fact is that N-body people extirpolate all the time in their codes, but nobody wants to use the term.

So, we have specified all the recipes needed to perform the *à trous* transform. Before we do that, we have to answer the question – why? Bernard Jones demonstrated how well orthogonal wavelet transforms work. An orthogonal wavelet transform changes a signal (picture) of N pixels into exactly N wavelet amplitudes, while an *à trous* transform expands it into $N \times J$ pictures; why bother?

The reason is called 'translational invariance'. As many of the wavelet amplitudes of an orthogonal transform do not have an exact home, when shifting the data, these amplitudes change in strange ways. Sure, we can always use them to reconstruct the shifted picture, but it makes no sense to compare the wavelet amplitudes of the original and shifted pictures. All the *à trous* transforms, on the contrary, keep their amplitudes; these move together with the grid. This is 'translational invariance', and it is important in texture studies, where we want to see different scales of a picture at exactly the same grid point. And cosmic texture is the main subject of this lecture.

A point to note – Fourier transforms are not translation invariant, too.

1.5 Multidimensional À Trous

All the above discussion was devoted to one-dimensional wavelets. This is customary in wavelet literature, as the step into multidimensions is simple – we form direct products of independent one-dimensional wavelets, one for every coordinate. This has been the main approach up to now, although it does not work well everywhere. An important example is a sphere, where special spherical wavelets have to be constructed (I suppose that these will be explained in the CMB lectures).

So, two-dimensional wavelets are introduced by defining the two-dimensional scaling function

$$\phi(x,y) = \phi(x)\phi(y) , \tag{19}$$

and three-dimensional wavelets – by the three-dimensional scaling function

$$\phi(x,y,z) = \phi(x)\phi(y)\phi(z) . \tag{20}$$

A little bit extra care has to be taken to define wavelets; we have to step into Fourier space for a while for that. Recalling (17), we have to write for two dimensions

$$\hat{\psi}(\omega_1,\omega_2) = \hat{\phi}(\omega_1/2)\hat{\phi}(\omega_2/2) - \hat{\phi}(\omega_1)\hat{\phi}(\omega_2)$$

(the direct products (19, 20) look exactly the same in the Fourier space). For coordinate space, it gives

$$\psi(x,y) = 4\phi(2x)\phi(2y) - \phi(x)\phi(y) ,$$

and for three dimensions, respectively,

$$\psi(x,y,x) = 8\phi(2x)\phi(2y)\phi(2z) - \phi(x)\phi(y)\phi(z) ,$$

I show the B_3-associated wavelets in Figs. 5 and 6. Because of their definition, the wavelet profiles are symmetric, but not isotropic, right? A big surprise is that both the B_3 scaling functions and the wavelets are practically isotropic, as the figures hint at. Let us define the isotropic part of the two-dimensional wavelet as

$$\overline{\psi}(r) = \frac{1}{2\pi}\int_0^{2\pi} \psi(r\cos\alpha, r\sin\alpha)\, d\alpha$$

and estimate the deviation from isotropy by

$$\epsilon = \int_{-2}^{2}\int_{-2}^{2} |\psi(x,y) - \overline{\psi}(\sqrt{x^2+y^2})|\, dx\, dy .$$

Comparing ϵ with the integral over the absolute value of our wavelet itself (about 4/9), we find that the difference is about 2%. For three dimensions, the difference is a bit larger, up to 5%.

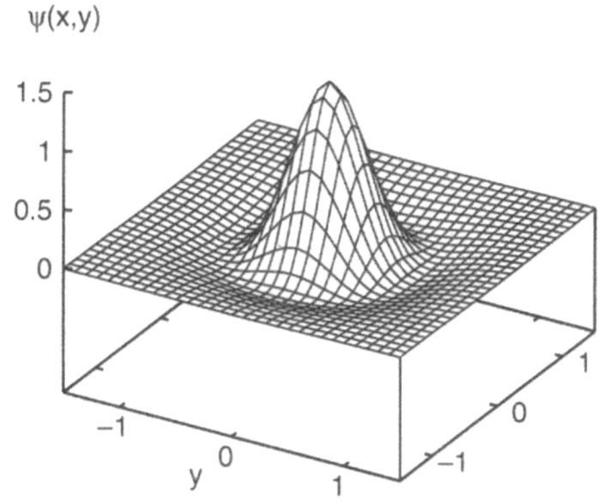

Fig. 5. Two-dimensional B_3-associated wavelet

This isotropy is important for practical applications; it means that our choice of specific coordinate directions does not influence the results we get.

And, as an example, I show a sequence of transforms in Fig. 7. Those are slices of a three-dimensional B_3-associated *à trous* transform sequence for the gravitational potential image of a N-body simulation. The data slice is at the upper left; the left column shows the scaling solutions, higher orders up. The right column shows the wavelet amplitudes; the data can be restored by taking the lower left image, and by adding to it all the wavelet images from the right column. Simple, is it?

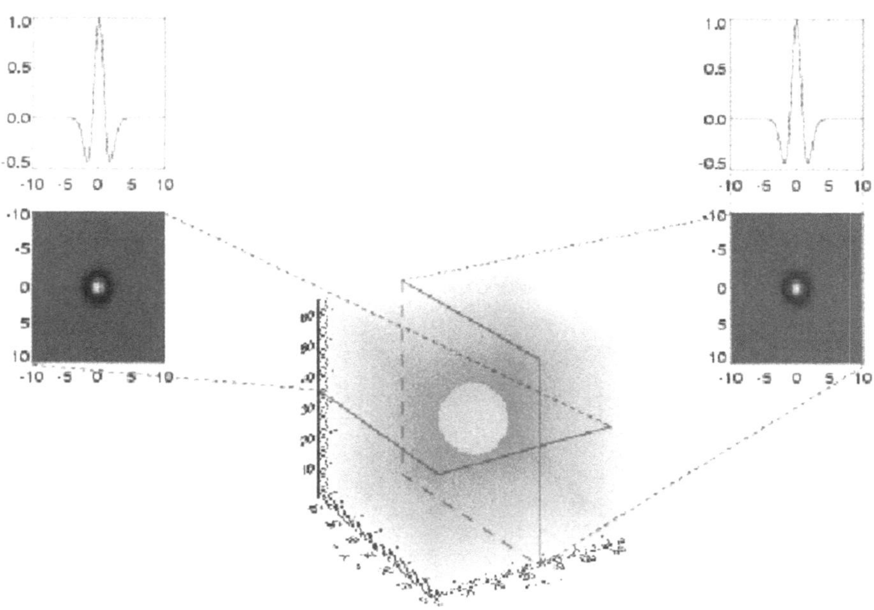

Fig. 6. Three-dimensional B_3-associated wavelet. Figure from J.-L. Starck et al. [16]

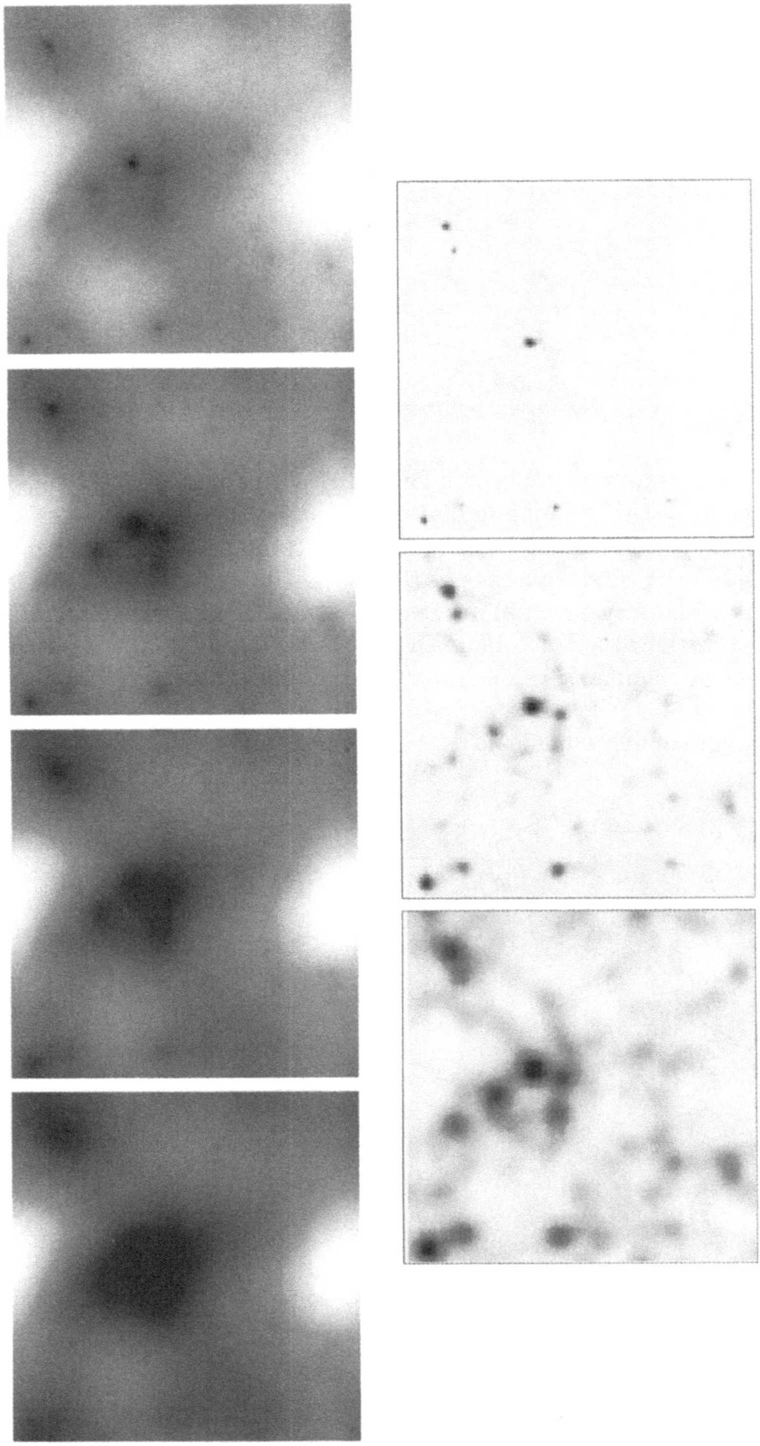

Fig. 7. Á trous potential (*slices*) for a N-body model (*linear scale*)

2 Cosmological Densities

In cosmology, the matter density distribution is a very important quantity that, first, determines the future dynamics of structure, and, second, may carry traces of the very early universe (initial conditions). We, however, cannot observe it. The data we get, after enormous effort, gives us galaxy positions in (redshift) space; even if we learn to associate a proper piece of dark matter with a particular galaxy; we have a point process, not a continuous density field.

There are several ways to deal with that. First we have to introduce a coordinate grid, an extra discrete entity, but necessary when using continuous fields in practical computations.

2.1 N-Body Densities

A similar density estimation problem arises in N-body calculations. There, the fundamental building blocks tare mass points. In order to find gravitational forces, the mass density on the underlying grid has to be calculated, using the positions of the points.

The simplest density assignment scheme is called NGP (nearest grid point), where the grid coordinate i is found by rounding the point coordinate x (to keep formulae simple, consider one-dimensional case and grid step one; generalization is trivial). Thus, the NGP assignment law is $i = \text{floor}(x + 0.5)$. This scheme is pretty rough, but I have seen people using an even worse scheme, $i = \text{floor}(x)$.

A bit more complex scheme is CIC (cloud in cell) where the mass point is dressed in a cubic cloud of the size of the grid cell, and vertexes are dressed in a similar region of influence. The part of the cloud that intersects this region of influence is assigned to the vertex. Sounds complicated, but as the mass weights are obtained by integration over the constant-density cloud, it is, in fact, only linear extirpolation (in three dimensions, three linear).

The most complex single-cloud scheme used is TSC (triangular-shaped cloud), where the density of the cloud changes linearly from a maximum in the centre to zero at the borders. The mass integration needed ensures that this scheme is quadratic extirpolation.

Note that the last two mass assignment schemes are, if fact, centred B-spline assignments – the vertex region is B_0, the CIC density is B_0, too, and the TSC cloud is B_1. The weights are obtained by convolution of those density profiles with the vertex profile, so, finally, NGP is a B_0 extirpolation scheme (no density law to be convolved with), CIC becomes a B_1 extirpolation scheme and TSC a B_2 scheme.

Today's mass assignment schemes are mostly adaptive variations of those listed above, where more dense grids are built in regions of higher density. But the N-body mass assignment schemes share one property – mass conservation. No matter what scheme is used, the total mass assigned to all vertexes of the grid is equal to the total mass of the mass points. No surprise in that – box splines are interpolating (12).

2.2 Statistical Densities

This class of density assignments (estimators) does not care about mass conservation. The underlying assumption is that we observe a sample of events that are governed by an underlying probability density and have to estimate this density. In cosmology, there really is no difference between the two densities, spatial mass density, and probability density.

The basic model of galaxy distribution adopted by cosmological statistics is that of the Cox point process. It says that, first, the universe is defined by a realization of a random process that ascribes a probability density $\lambda(\mathbf{x})$ in space. Then, a Poisson point process gets to work, populating the universe with galaxies, where the probability to have a galaxy at $\mathbf{x}$ is given by the Poisson law with the parameter $\lambda(\mathbf{x})$ fixed by the initial random-field realization. Neat, right? And as we are dealing with random processes; no conservation is required.

Statisticians have worked seriously on probability density estimation problem (see Silverman [15] for a review). The most popular density estimators are the kernel estimators:

$$\varrho_i = \sum_n K(x_n - i) \tag{21}$$

(recall that i are the grid coordinates and x_n are the galaxy coordinates). The kernel K is a symmetric distribution:

$$\int K(x)dx = 1 , \qquad \int xK(x) = 0 .$$

Much work has been done on the choice of kernels, with the result that the exact shape of the kernel does not matter much, but its width does. The best kernel is said to be the Epanechnikov kernel:

$$K_E(\mathbf{x}) = A(1 - \mathbf{x}^2/R^2) , \quad \mathbf{x}^2 \leq R^2, \quad 0 \text{ otherwise} . \tag{22}$$

(I wrote it for multidimensional case to stress that this is not a direct product, but an isotropic kernel; A is, of course, a normalization constant.)

The Gaussian kernel comes close behind:

$$K_G(x) = \frac{1}{\sqrt{2\pi}\sigma} \exp\left(-x^2/2\sigma^2\right) . \tag{23}$$

This is the only kernel where the direct product is isotropic, too. The ranking of kernels is done by deciding how close the estimated probability density $\tilde{f}(x)$ is to the true density $f(x)$, by measuring the MSE (mean standard error):

$$\text{MSE} = E\left[\tilde{f}(x) - f(x)\right]^2 = \text{Var}\left(\tilde{f}(x)\right) + \left[\text{Bias}\left(\tilde{f}(x)\right)\right]^2 . \tag{24}$$

Note that statisticians minimize the MSE, not just the variance, as cosmologists frequently tend to do. As usual in statistics, the results are asymptotic, true for a very big number of galaxies N. Also, the grid we use has to resolve the kernel well (the kernel has to cover a large number of grid vertexes), otherwise the sum of the kernel values over the vertexes will be considerably smaller than 1, and density estimates will be biased low – we will lose galaxies. This concerns especially the Epanechnikov kernel. In this respect, the B_3 kernel we used above is a very good candidate for a density estimation kernel. It is smooth (meaning its Fourier transform decays fast), and it is compact, without wide wings (as the Gaussian kernel has). And it is interpolating – formula (12) guarantees that not a single galaxy is lost.

Kernel density estimators allow a natural generalization for the case of extremely different density amplitudes and scales, as seen in cosmology. Constant-width kernels tend to over-smooth the sharp peaks of the density, if these exist. The solution is using adaptive kernels, by varying the kernel width $h(\cdot)$ from place to place. There are, basically, two different ways to do that. The balloon or scatter estimators are

$$\varrho_i = \sum_n K\left(\frac{x_n - i}{h(i)}\right) ; \tag{25}$$

here the kernels sit on the grid points i. The second type of estimators is called sample point, sandbox, or gather estimators:

$$\varrho_i = \sum_n K\left(\frac{x_n - i}{h(x_n)}\right) . \tag{26}$$

Here the kernel width depends on the sample point. The most difficult problem for adaptive kernels is how to choose the right kernel widths. The usual way is to estimate the density with a constant kernel first, and to select the adaptive kernel widths proportional to some fractional power of the local density obtained in the first pass ($-1/5$ is a recommended choice).

Both estimators are used in cosmology (the terms scatter and gather come from the SPH cosmological hydrodynamics codes). The lore says that the balloon estimators (25) work best in low-probability regions (voids in cosmology), and the sandbox estimators – where densities are high.

2.3 Equal-Mass Densities

A popular density estimator is based on k-d trees. These trees are formed by recursive division of the sample space into two equal-probability halves (having the same number of galaxies). It is a spatial version of adaptive histograms (an equal number of events per bin). Of course, k-d trees give more than just density estimates; they also imprint a tree structure on (or reveal the structure of the geometry of) the density field. An application of k-d trees for estimating densities appeared in `astro-ph` during the school and has already been published by the time of writeup of the lecture (Ascasibar & Binney [2]).

Another popular equal-mass density estimators are kNN (k nearest neighbours) kernels. The name speaks for itself – the local kernel size is chosen to include k particles in the kernel volume. This estimator uses isotropic kernels.

The SPH gather algorithm uses, in fact, the kNN ideology. There is a separate free density estimation tool based on that algorithm ('`smooth`'), written by Joachim Stadel and available from the Washington University N-body shop.[1] Try it; the only problems are that you have to present the input data in the 'tipsy' format and that you get the densities at particle positions, not on a grid. Should be easy to modify, if necessary.

2.4 Wavelet Denoising

Wavelet denoising is a popular image processing methodology. The basic assumption is that noise in an image is present at all scales. Once we accept that assumption, the way to proceed is clear: decompose the image into separate scales (wavelet orders; orthogonal wavelet transforms are the best here), estimate the noise at each wavelet order, eliminate it somehow, and reconstruct the image.

This course of action includes two difficult points – first, estimating the noise. The properties of the basically unknown noise are, ahem, unknown, and we have to make assumptions about them. Gaussian and Poisson noise are the most popular assumptions; this leaves us with the problem of relative noise amplitudes (variances) between different wavelet orders. A popular method is to model the noise. Modelling is started by assuming that all the first-order wavelet data is noise (interesting, is it?) and processing that for the noise variance. After that, noise of that variance is modelled, wavelet transformed, and its properties found for every wavelet order. After that, we face a common decision theory problem, at which p-value should we set the noise limit? If we cut the noise at too low amplitude, we leave much of it in the final image, and if we take the cut too high, we eliminate part of the real signal, too. Once we have selected that level, we can quantify it in the terms of the limiting wavelet amplitude $k\sigma_j$, where σ_j^2 is the modelled noise variance for the level j.

The second problem is how to suppress the noisy amplitudes. The first approach is called 'hard thresholding' and it is simple: the processed wavelet amplitudes $\tilde{w}_j$ are

$$\tilde{w}_j = w_j \;, \quad \text{if} \quad |w_j| > k\sigma_j \;, \quad 0 \quad \text{otherwise} \;. \tag{27}$$

This thresholding leaves an empty trench around 0 in the wavelet amplitude distribution.

[1] `http://www-hpcc.astro.washington.edu/`

Another approach is 'soft thresholding':

$$\tilde{w}_j = w_j - \mathrm{sgn}(w_j)\, k\sigma_j \;, \quad if \quad |w_j| > k\sigma_j \;, \quad 0 \;\; \text{otherwise} \,. \tag{28}$$

This thresholding takes out the same trench, first, but fills it up then, diminishing all the remaining amplitudes.

David Donoho, who was the first to introduce soft thresholding, has also proposed an universal formula for the threshold level:

$$k\sigma_j = \sqrt{2\log(n)}\,\sigma_j \;,$$

where n is the number of pixels in the image. This level corresponds to 3σ for $n = 90$ and to 4σ for $n = 3000$. Of course, astronomers complain – 3000 pixels is a very small size for an astronomical image, but 4σ is a very high cut-off level; we can cut off much of the information in the image. Astronomical information is hard to obtain, and we do not want to waste even a bit. So we better keep our pictures noisy? Fortunately, more approaches to thresholding have appeared in recent years; consult the new edition of the Starck and Murtagh book [17].

Anyway, wavelet denoising has met with resounding success in image processing, no doubts about it. And image processing is an industry these days, so the algorithms that are used are being tested in practice every day. Now, image processing is two-dimensional business; wavelet denoising of a three-dimensional (spatial) density is a completely different story. The difference is that the density contrasts are much bigger in three dimensions than in two dimensions – there simply is more space for the signal to crowd in. And as wavelets follow the details, they might easily over-amplify the contrasts. I know, we have spent the last year trying to develop a decent wavelet-denoising algorithm for the galaxy distributions (well, 'we' means mainly Jean-Luc Starck). Figure 8 shows how the denoising might go awry (right panel), but shows at the same time that good recipes are also possible (left panel). The denoising procedure for the right panel has over-amplified the contrasts and has generated deep black (zero density) holes close to white high-density peaks. So, in order to do a decent denoising job, one has to be careful.

The details of the algorithm we used are too tedious to describe in full; they can be found in Starck and Murtagh [17]. The main points are the following:

1. We used the *à trous* algorithm, not an orthogonal one. The reason for that is that we needed to discriminate between the positions of significant wavelet amplitudes (the multiresolution support) and non-significant amplitudes at the last stage, and when speaking of positions, orthogonal wavelet transforms cannot be used.
2. We hard-thresholded the solution and iterated it, reconstructing and transforming again, to obtain a situation where the final significant wavelet coefficients would cover exactly the original multiresolution support.

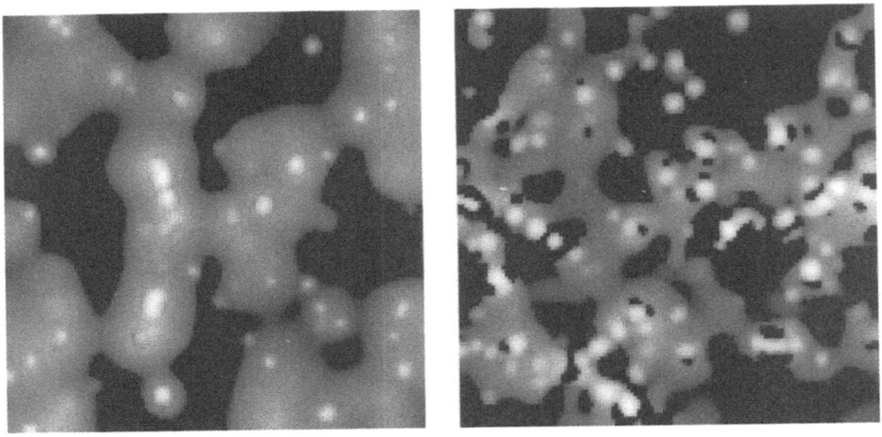

Fig. 8. Wavelet denoising of a (model) galaxy distribution (*left* – a successful attempt, *right* – over-denoising)

3. Finally, we smooth the solution by imposing a smoothness constraint, requiring that the sum of the absolute values of wavelet amplitudes of a given order would be minimal, while keeping the corrected amplitudes themselves within a given tolerance ($\sigma_j/2$) of the original noisy ones. We consider here only the amplitudes in the multiresolution support; this point required using a translation invariant wavelet transform.

Figure 9 compares the results of our three-dimensional wavelet denoising. We started with a three-dimensional galaxy distribution, applied our algorithm to it, and, for comparison, built two Gaussian-smoothed density distributions. It is clearly seen that the typical details, in the case of Gaussian smoothing, are of uniform size, while the wavelet-denoised density distribution is adaptive, showing details of different scales.

2.5 Multiscale Densities

We have made an implicit assumption in this section, namely that a true density field exists. Is that so certain? Our everyday experience tells us that it is. But look at the numbers that stand behind this experience: even one cubic centimetre of air has $6 \times 10^{23}/22.4 \times 10^3 \approx 3 \times 10^{19}$ particles. In our surveys, one gigantic cosmological 'cubic centimetre' of 10 Mpc size contains about ten galaxies. Can we speak about their true spatial density?

One answer is that we can, but there are regions where the density estimates are extremely uncertain; statistics can tell us what the expected variances are. Another answer is that even if there is a true density, it is not always a useful physical quantity, especially for the largest scales we study. One of the reasons we measure the cosmological density field is to find its state

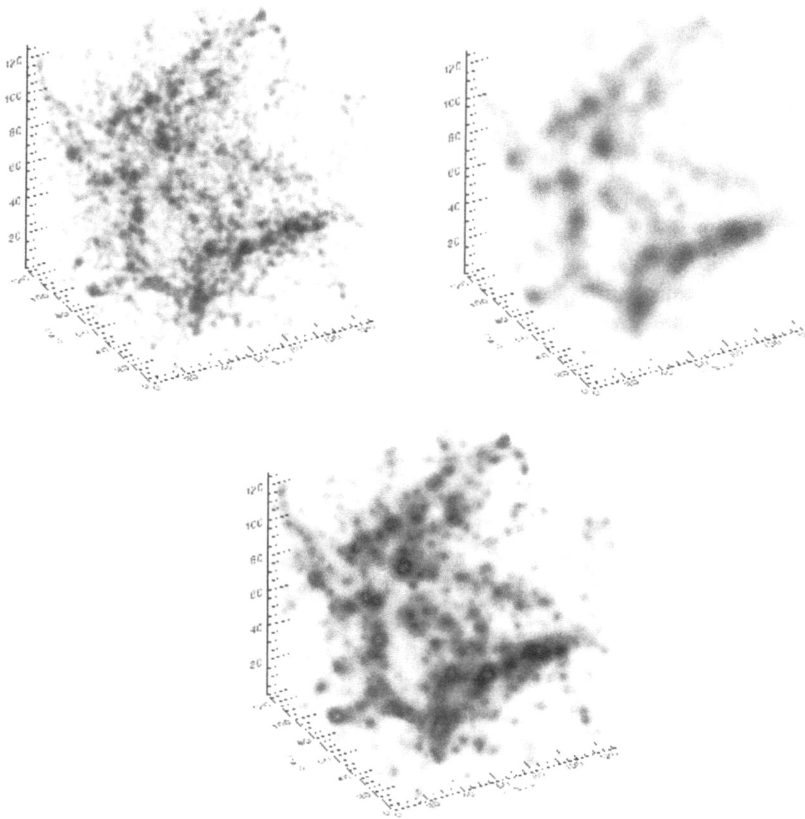

Fig. 9. Density fields for a model galaxy distribution. *Left* – Gaussian $\sigma = 1\,\mathrm{Mpc/h}$ smoothing, middle – wavelet denoising, *right* – Gaussian $\sigma = 3\,\mathrm{Mpc/h}$ smoothing

of evolution and traces of initial conditions in it. The theory of the dynamics of perturbations in an expanding universe predicts that structure evolves at different rates, slowly at large scales and much more rapidly at galactic scales. Observations show that cosmological fields are multiscale objects; the recently determined power spectra span scales from about 600 Mpc/h ($k = 0.01\,\mathrm{h/Mpc}$ to 10 Mpc/h ($k = 0.6\,\mathrm{h/Mpc}$). Thus, should not these fields be studied in multiscale fashion, scale by scale? A true adaptive density mixes effects from different scales and scale separation could give us a cleaner look at the dynamics of large-scale structure.

In case of our everyday densities, this separation of scales can be done safely later, analysing the true density. For galaxy distributions, it is wiser to prescribe a scale (range) and to obtain that density directly from the observed galaxy positions. One advantage is in accuracy, another in that there are places (voids), where, e.g., small-scale densities simply do not exist.

And this is the point where the first part of the lecture (wavelets) connects with the second part (density fields). The representation of the observed density fields by a sum of the densities of different characteristic scales (7) is just what we are looking for. True, there is a pretty large frequency overlap between the neighbouring bands (Fig. 10 shows you their power spectra), but that is possibly the best we can do, while keeping translational invariance. The figure shows also the power spectra of Gaussians of the same scales that are sometimes used to select different scales. As we see, Gaussian frequency bands are heavily correlated; the overlap of the smaller frequency band with that of the higher one is total, and that is natural – smoothing destroys signals of high frequency, but it does not separate frequency bands. So Gaussians should not be used in this business, but, alas, they frequently are.

Figures 11 and 12 show the *à trous* density slices for a Gaussian cube. This is a 256^3 realization of a Gaussian random field with a power spectrum approximating that of our universe. In real universe, the size of the cube would be 256 Mpc/h. The slices are taken from the same height; The images are in grey coding, black shows the densest regions. As in Fig. 7, the scaling solutions form the left column and the wavelets the right column; transform orders grow downwards. In height, the wavelet orders are placed between the scaling order that produced them. The scaling solution of order three is repeated in Fig. 12 to keep the scaling–wavelet alignment. Enjoy.

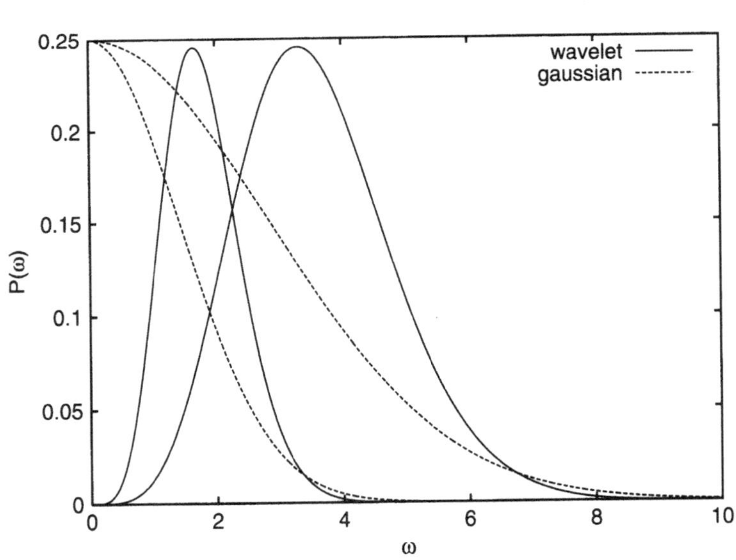

Fig. 10. Power spectra for two neighbouring scales, the B_3-associated *à trous* bands (*solid lines*), and for Gaussian smoothing of the same scales (*dotted lines*). Only the positive half on the frequency axis is shown

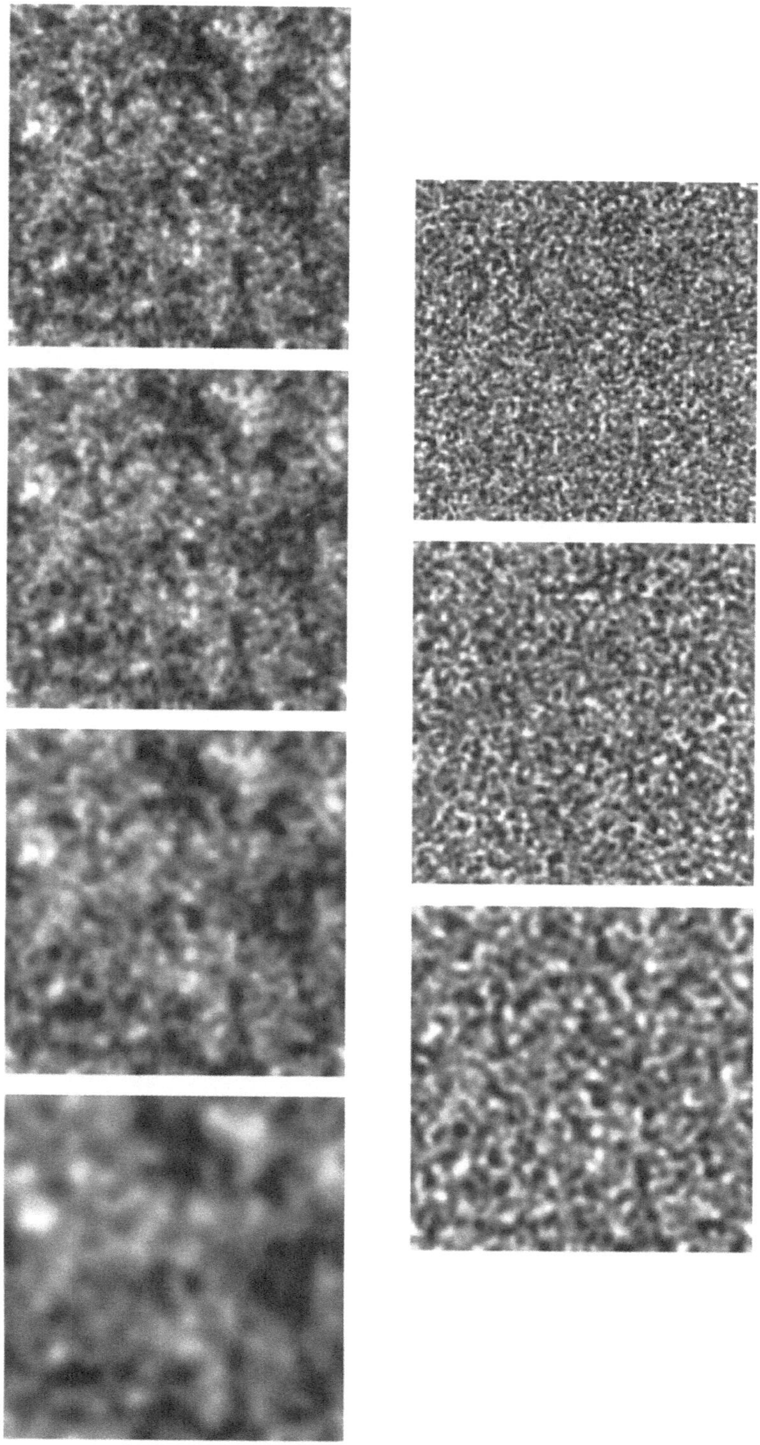

Fig. 11. Á trous density (*slices*) of a Gaussian density field (*linear scale*). The first orders (0–3 for the scaling solutions, 1–3 for the wavelets)

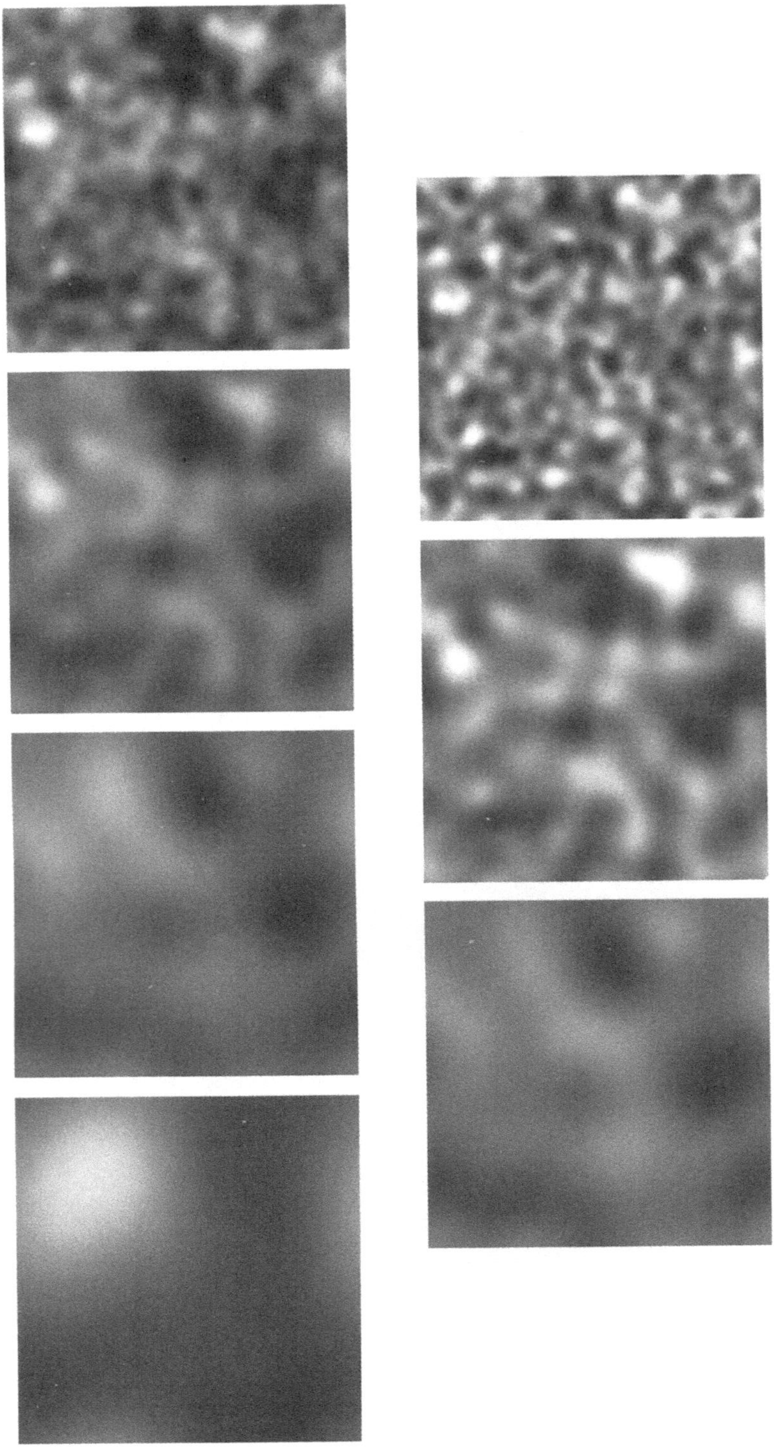

Fig. 12. Á trous density (*slices*) of a Gaussian density field (*linear scale*). The large-scale orders (3–6 for the scaling solutions, 4–6 for the wavelets)

3 Minkowski Functionals

Peter Coles explains in his lecture why it is useful to study morphology of cosmological fields. In short, it is useful because it is sort of a perpendicular approach to the usual moment methods. Our present cosmological paradigm says that the initial perturbation field was a realization of a Gaussian random field. The most direct test of that would be to measure all n-point joint amplitude distributions, starting from the 1-point distribution. Well, we know that even this is not Gaussian, but we know why (gravitational dynamics of a positive density field inevitably skews this distribution), and we can model it. As cosmological densities are pretty uncertain, the more uncertain are their many-point joint distributions. So this direct check does not work, at least presently.

Another possibility to check for Gaussianity is to estimate higher order correlation functions and spectra. For Gaussian realizations, odd-order correlations and power spectra should be zero, and even-order moments should be directly expressible via the second-order moments, the usual two-point correlation function, and the power spectrum. Their dynamical distortions can also be modelled, and this is an active area of research.

Morphological studies provide an independent check of Gaussianity. Morphology of (density) fields depends on all correlation functions at once, is scale dependent, but local, and can also be predicted and its change caused by dynamical evolution can be modelled. This lecture is, finally, about measuring morphology of cosmological fields.

An elegant description of morphological characteristics of density fields is given by Minkowski functionals [10]. These functionals provide a complete family of morphological measures – all additive, motion invariant, and conditionally continuous[2] functionals defined for any hypersurface are linear combinations of its Minkowski functionals.

The Minkowski functionals (MF for short) describe the morphology of isodensity surfaces and depend thus on the specific density level.[3] Of course, when the original data are galaxy positions, the procedure chosen to calculate densities (smoothing) will also determine the result. The usual procedure used in this business is to calculate kernel densities with wide Gaussian kernels; the recipes say that the width of the kernel (standard deviation) should be either the mean distance between galaxies or their correlation length, whichever is

---

[2] We submitted recently a paper on multiscale Minkowski functionals, and the referee wondered what does 'conditionally continuous' mean. So, now I know – they are continuous if the hypersurfaces are compact and convex, and we can approximate any decent hypersurface by unions of such.

[3] In fact, Minkowski functionals depend on a surface; that is why they are called functionals (functions of functions). When we specify the family of isodensity surfaces, the functionals will depend, suddenly, only on a number, the value of the density level, and are downgraded to simple functions, at least in cosmological applications.

larger. Although this produces nice smooth densities, the recipe is bad; it destroys the texture of the density distribution; I shall show it later.

We shall use wavelets to produce densities, and shall look first at the texture of a true (wavelet-denoised) density, and then at the scale-dependent multiscale texture of the galaxy density distribution. We could also start directly from galaxies themselves, as Minkowski functionals can be defined for a point process, decorating the points with spheres of the same radius, and studying the morphology of the resulting surface. This approach does not refer to a density and we do not use it here. Although it is beautiful, too, the basic model that it describes is a (constant-density) Poisson process; a theory for that case exists, and analytical expressions for Minkowski functionals are known. Alas, as the galaxy distribution is strongly correlated, this reference model does not help us much. The continuous density case has a reference model, too, and that is a Gaussian random field, so this is more useful.

For a d-dimensional space, one can find $d+1$ different Minkowski functionals. We shall concentrate on usual three-dimensional space; for that, the Minkowski functionals are defined as follows. Consider an excursion set F_{ϕ_0} of a field $\phi(\mathbf{x})$ in three dimensions (the set of all points where density is larger than a given limit, $\phi(\mathbf{x} \geq \phi_0)$). Then, the first Minkowski functional (the volume functional) is the volume of this region (the excursion set):

$$V_0(\phi_0) = \int_{F_{\phi_0}} d^3x \,. \tag{29}$$

The second MF is proportional to the surface area of the boundary δF_ϕ of the excursion set:

$$V_1(\phi_0) = \frac{1}{6} \int_{\delta F_{\phi_0}} dS(\mathbf{x}) \tag{30}$$

(but not the area itself, notice the constant). The third MF is proportional to the integrated mean curvature of the boundary:

$$V_2(\phi_0) = \frac{1}{6\pi} \int_{\delta F_{\phi_0}} \left(\frac{1}{R_1(\mathbf{x})} + \frac{1}{R_2(\mathbf{x})} \right) dS(\mathbf{x}) \,, \tag{31}$$

where $R_1(\mathbf{x})$ and $R_2(\mathbf{x})$ are the principal radii of curvature of the boundary. The fourth Minkowski functional is proportional to the integrated Gaussian curvature (the Euler characteristic) of the boundary:

$$V_3(\phi_0) = \frac{1}{4\pi} \int_{\delta F_{\phi_0}} \frac{1}{R_1(\mathbf{x}) R_2(\mathbf{x})} dS(\mathbf{x}) \,. \tag{32}$$

The last MF is simply related to the genus that was the first morphological measure used in cosmology; all these papers bear titles containing the word 'topology'. Well, the topological Euler characteristic χ for a surface in three dimensions can be written as

$$\chi = \frac{1}{2\pi} \int_S \kappa \, dS \,, \tag{33}$$

where κ is the Gaussian curvature, so

$$V_3 = \frac{1}{2}\chi \,. \tag{34}$$

Bear in mind, though, that the Euler characteristic (33) describes the topology of a given isodensity surface, not of the three-dimensional region hosting the density distribution; the topology of the latter is, hopefully, trivial.

The first topology papers concentrated on the genus G that is similar to V_3:

$$\chi = 2(1 - G) \,, \qquad V_3 = 1 - G \,. \tag{35}$$

The functional V_3 is a bit more comfortable to use – it is additive, while G is not, and in the case our surface breaks up into several isolated balls, V_3 is equal to the number of balls. If the excursion set contains only a few isolated empty regions (bubbles), V_3 gives their number. In a general case

$$V_3 = \text{\#-of-balls} + \text{\#-of-bubbles} - \text{\#-of-tunnels} \,,$$

where only these tunnels that are open at both ends count.

I have to warn you about a possible confusion with the genus relations (33),(34), and (35) – the coefficient 2 (or 1/2) occupies frequently a wrong position. The confusion is due to a fact that two topological characteristics can be defined for an excursion set – one for its surface, another for the set itself. The relation between these depends on the dimensionality of the space; for three dimensions the topological characteristic for the excursion set is half of that for the surface, and if we mix them up, our formulae become wrong. I know, we have published a wrong formula, too (even twice), but the formulae are right in our book [8]. So, bear in mind that the Minkowski functionals are calculated for surfaces and use only the relations above (33),(34), and (35). When in doubt, consult the classical paper by Mecke et al. [10] and use the Crofton's formula below (44) for a single cubic cell – it gives you $V_3 = 1 \Rightarrow G = 0$.

Figure 13 shows a Gaussian cube (a realization of a Gaussian random process) for two different smoothing widths (the left pair and the right pair of columns, respectively), and for three volume fractions. You can see that the solid figures inside the isodensity surface are awash with handles, especially at the middle 50% density level. Of course, the larger the smoothing, the less the number of handles. You can also see that Gaussian patterns are symmetric – the filled regions are exact lookalikes of the empty regions, for a symmetric change of volume fractions.

Galaxy densities are more asymmetrical, as seen in Fig. 14. This figure shows a model galaxy distribution from a N-body simulation, in a smaller cube. The 50% density volumes differ, showing asymmetry in the density distribution, and the 5%–95% symmetry, evident for the Gaussian cube, is not so perfect any more.

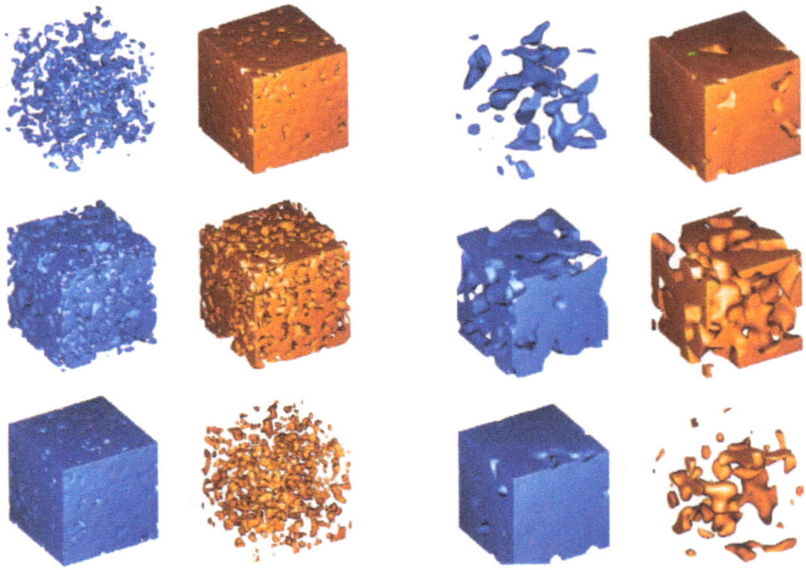

Fig. 13. A Gaussian cube of 256^3 pixels for different Gaussian smoothing filters. The left two columns show isodensity surfaces for $\sigma = 3$ pixels, the right two columns – for $\sigma = 8$ pixels. To better delineate isodensity surfaces, we show two sides of the surface in column pairs, where the left column shows high-density regions, and the right column low-density regions for the same isodensity surface. The rows are for constant volume fractions (7%, 50%, and 93%), starting from below. Figure from Martiínez et al. [9]

Instead of the functionals, their spatial densities V_i are frequently used:

$$v_i(f) = V_i(f)/V \ , \quad i = 0, \ldots, 3 \ ,$$

where V is the total sample volume. The densities allow us to compare the morphology of different data samples.

3.1 Labelling the Isodensity Surfaces

The original argument of the functionals, the density level ϱ_0, can have different amplitudes for different fields, and the functionals are difficult to compare. Because of that, normalized arguments are usually used; the simplest one is the volume fraction f_v, the ratio of the volume *outside* of the excursion set to the total volume of the region where the density is defined (the higher the density level, the closer this ratio is to 1). Another, similar argument is the mass fraction f_m, which is very useful for real, positive density fields, but is cumbersome to apply for realizations of Gaussian fields, where the density may be negative. But when we describe the morphology of single objects (superclusters, say), the mass fraction is the most natural argument. It is

Fig. 14. A galaxy sample (60^3 pixels) for a 3-pixel smoothing scale. The left column shows high-density regions and the right column low-density regions for the same isodensity surface. The rows are for constant volume fractions (7%, 50%, and 93%), starting from below. Figure from Martiínez et al. [9]

also defined to approach 1 for the highest density levels (and for the smallest masses inside the isodensity surface).

The most widely used argument in this business is the Gaussianized volume fraction ν, defined as

$$f_v = \frac{1}{\sqrt{2\pi}} \int_{\nu}^{\infty} \exp(-t^2/2)\, dt \,. \tag{36}$$

This argument was introduced already in the first topology paper by Gott [5], in order to eliminate the first trivial effect of gravitational clustering, the deviation of the 1-point pdf from the (supposedly) Gaussian initial pdf. Notice that using this argument, the first Minkowski functional is trivially Gaussian by definition.

For a Gaussian random field, ν is the density deviation from the mean, divided by the standard deviation. We can define a similar argument for any field:

$$\nu_\sigma = \frac{\varrho - \overline{\varrho}}{\sigma(\varrho)} \,.$$

I show different Minkowski functionals versus different arguments in Figs. 15 and 16. They are calculated for the model galaxy density distribution shown in the previous figure (Fig. 14). Note how much the shape of the same function(al)s depends on the arguments used.

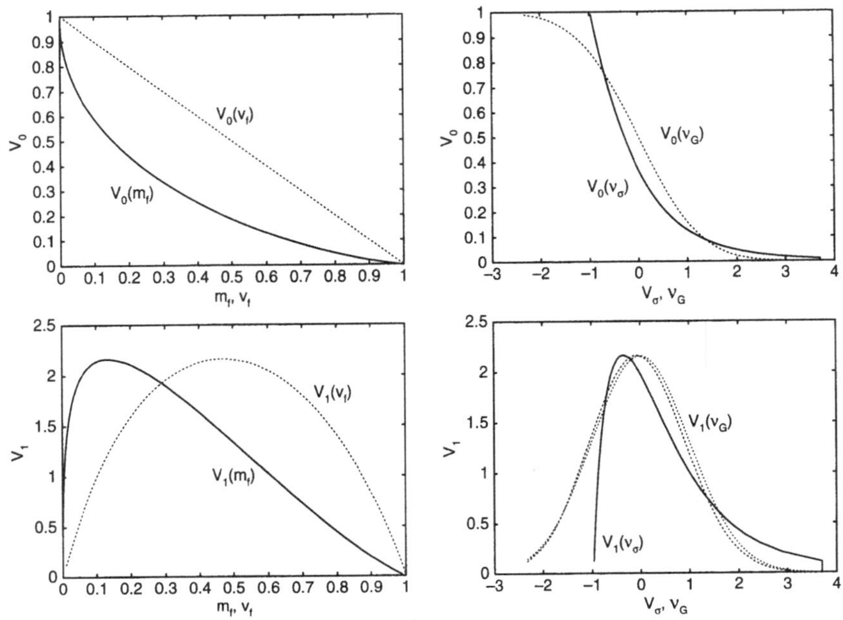

Fig. 15. The first two Minkowski functionals for N-body model galaxies. Here v_f is the volume fraction, m_f the mass fraction, ν_σ the normalized volume fraction, and $\nu_G \equiv \nu$ the Gaussianized volume fraction from (36). The dotted line in the right panels shows the predicted Minkowski functionals for a Gaussian random-field MF

3.2 Gaussian Densities

All the Minkowski functionals have analytic expressions for isodensity slices of realizations of Gaussian random fields. For three-dimensional space they are

$$v_0 = \frac{1}{2} - \frac{1}{2}\Phi\left(\frac{\nu}{\sqrt{2}}\right), \tag{37}$$

$$v_1 = \frac{2}{3}\frac{\lambda}{\sqrt{2\pi}}\exp\left(-\frac{\nu^2}{2}\right), \tag{38}$$

$$v_2 = \frac{2}{3}\frac{\lambda^2}{\sqrt{2\pi}}\nu\exp\left(-\frac{\nu^2}{2}\right), \tag{39}$$

$$v_3 = \frac{\lambda^3}{\sqrt{2\pi}}(\nu^2-1)\exp\left(-\frac{\nu^2}{2}\right), \tag{40}$$

where $\Phi(\cdot)$ is the Gaussian error integral and λ is determined by the correlation function $\xi(r)$ of the field:

$$\lambda^2 = \frac{1}{2\pi}\frac{\xi''(0)}{\xi(0)}. \tag{41}$$

The dimension of λ is inverse length.

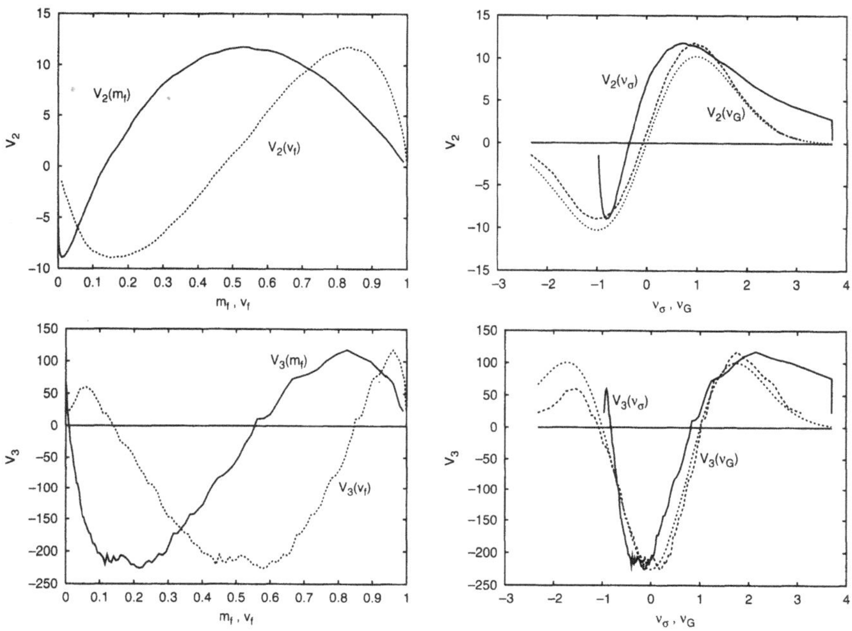

Fig. 16. The last two Minkowski functionals for N-body model galaxies. Here v_f is the volume fraction, m_f the mass fraction, ν_σ the normalized volume fraction, and $\nu_G \equiv \nu$ the Gaussianized volume fraction from (36). The dotted line in the right panels shows the predicted Minkowski functionals for a Gaussian random-field MF

This expression allows to predict all the Minkowski functionals for a known correlation function (or power spectrum). We can also take a more empirical approach and determine λ^2 on the basis of the observed density field itself, using the relations $\xi(0) = \langle \varrho^2 \rangle$ and $\xi''(0) = \langle \varrho_{,i}^2 \rangle$, where $\varrho_{,i}$ is the density derivative in one coordinate direction.

The expected form of these functionals is shown in Fig. 17.

In practice, it is easy to obtain good estimates of the Minkowski functionals for periodic fields. The real data, however, is always spatially limited, and the limiting surfaces cut the isodensity surface. An extremely valuable property of Minkowski functionals is that such cuts can be corrected for – the data volume mask is another body, and Minkowski functionals of intersecting bodies can be calculated. Moreover, if we can assume homogeneity and isotropy for the pattern, we can correct for border effects of large surveys. This is too technical for the lecture, so I refer to our forthcoming paper [12].

3.3 Numerical Algorithms

Several algorithms are used to calculate the Minkowski functionals for a given density field and a given density threshold. We can either try to follow exactly the geometry of the isodensity surface, e.g. using triangulation, or to

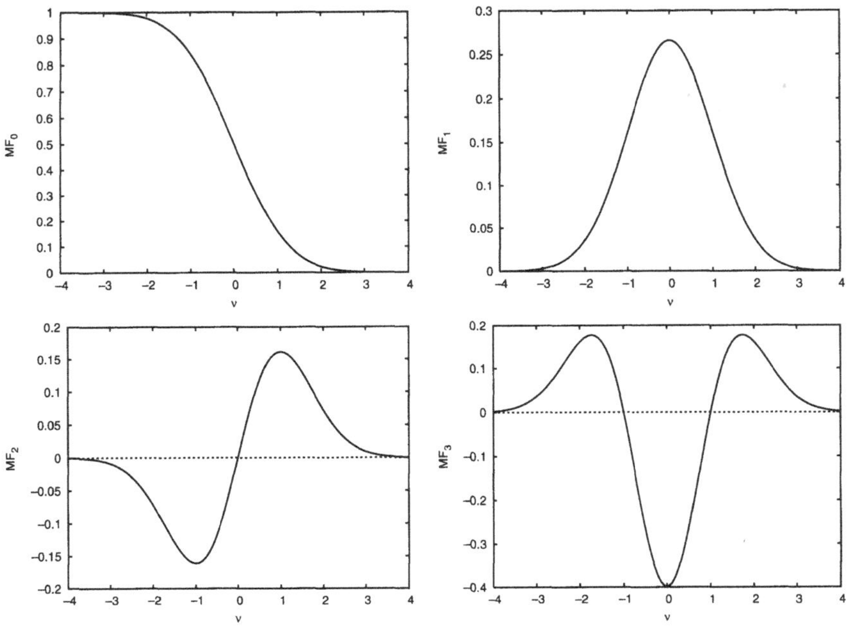

Fig. 17. Gaussian predictions for Minkowski functionals ($\lambda = 1$)

approximate the excursion set on a simple cubic grid. The first algorithms were designed to calculate the genus only; the algorithm that was proposed by Gott, Dickinson, and Melott [5] uses a decomposition of the field into filled and empty cells, and another class of algorithms (Coles, Davies, and Pearson [4]) uses a grid-valued density distribution. The grid-based algorithms are simpler and faster, but are believed to be not as accurate as the triangulation codes.

I advocate a simple grid-based (Crofton) algorithm. To start with, we find the density thresholds for given filling fractions f by sorting the grid densities. This is a common step for all grid-based algorithms. Vertexes with higher densities than the threshold form the excursion set. To calculate the Minkowski functionals for this excursion set (let us denote it by K), we shall use two remarkable results of integral geometry. The first of these is called the Crofton's intersection formula (see Schmalzing and Buchert [14]). Crofton's original result explained how to measure the length of a curve by counting its intersections by random lines (recall the Buffon's needle). The generalization for higher dimensions reads

$$v_k^{(d)}(K) = \frac{\omega_d}{\omega_{d-k}\omega_k} \int_{\mathcal{E}_k^{(d)}} \chi^{(k)}(K \cap E) d\mu_k(E). \tag{42}$$

Here K is the body we study (e.g. an overdensity region), in d dimensions, and E is a k-dimensional hyperplane; $v_k^{(d)}(K)$ is the density of the Minkowski functional V_k for the body K, and the quantity $\chi(K \cap E)^{(k)}$ is the Euler characteristic of the intersection of the body with the hyperplane. The space of all possible hyperplanes is denoted as $\varepsilon_k^{(d)}$, and μ_k is the measure in that space. The normalizing factor is composed of the volumes of unit-radius spheres w_k of dimension k. This formula shows that we can measure lengths, areas and curvatures by simply counting (properly generated) events. Introducing a cubic lattice in the d-dimensional space, the integral in (42) can be reduced to a sum of Euler characteristics.

Another remarkable formula is that for the Euler characteristic in any dimensions, due to Adler [1]:

$$\chi^{(k)}(K) = \sum_{j=0}^{k} (-1)^j N_j(K). \tag{43}$$

Here the body K is assumed to be covered by a cubic grid, and N_i are the numbers of the basic sets of different dimensions within the body. The number N_0 counts points (vertexes), N_1 counts edges formed by two neighbouring vertexes, N_2 counts squares (faces) formed by four edges, and N_3 counts cubes formed by six faces. For morphological studies, this formula was used first by Coles [4] to calculate early genus curves.

Combining these two formulae (42) and (43) and assuming isotropical orientation of the surface elements, Schmalzing and Buchert [14] derived the following rules:

$$\begin{aligned} V_0 &= a^3 N_3 \, , \\ V_1 &= a^2 \left(\frac{2}{9} N_2 - \frac{2}{3} N_3 \right) \, , \\ V_2 &= a \left(\frac{2}{9} N_1 - \frac{4}{9} N_2 + \frac{2}{3} N_3 \right) \, , \\ V_3 &= N_0 - N_1 + N_2 - N_3 \, . \end{aligned} \tag{44}$$

Here a is the grid step. These formulae are really surprising – for example, the second formula says that one can measure the surface area of a body by counting the number of grid cells and faces inside it!

This algorithm is simple to program, and it is very fast, allowing the use of Monte Carlo simulations for error estimation.[4]

The first, and most used, algorithm was 'CONTOUR' and its derivations; this algorithm was written by David Weinberg and was probably one of the first cosmological public-domain algorithms [18]. However, there is a noticeable difference between the results that our grid algorithm produces and those of 'CONTOUR'. You can see it yourself, comparing the Minkowski functional figures in this lecture with Fig. 4 in Peter Coles' chapter. All genus curves ever found

[4] A thorough analysis of the algorithm and its application to galaxy distributions will be available by the time this book will be published [12].

by 'CONTOUR' look like that there, very jaggy. How much I have tried, using Gaussian smoothing and changing smoothing lengths, I have never been able to reproduce these jaggies by our algorithm. At the same time, the calculated genus curves always follow the Gaussian prediction, so there is no bias in either algorithm. As the genus (V_3) counts objects (balls, holes, handles), it should, in principle, have mainly unit jumps, and only occasionally a larger jump. That is what we see in our genus (V_3) graphs. The only conclusion that I can derive is that the algorithm we use is more stable, with a much smaller variance.

3.4 Shapefinders

As Minkowski functionals give a complete morphological description of surfaces, it should be possible to use them to numerically describe the shapes of objects, right? For example, to differentiate between fat (spherical) objects and thin (cylindrical objects), banana-like superclusters and spiky superclusters. This hope has never died, and different shape descriptors have been proposed (a selection of them is listed in our book [8]). The set of shape descriptors that use Minkowski functionals was proposed by Sahni, Sathyaprakash, and Shandarin [13], and is called 'shapefinders'. Now, shapefinder definitions have a habit of changing from paper to paper and it is not easy to follow the changes, so I shall give here a careful derivation of the last version of shapefinders. Shapefinders are defined as properly normalized ratios of Minkowski functionals. In three dimensions, the chain of Minkowski functionals V_i, from V_0 to V_3, has gradually diminishing dimensions, from L^3 to L^0. So, the ratios of neighbours in this chain have a dimension of length; these ratios make up the first set of shapefinders. Take a surface that is delimiting (shaping) a three-dimensional object (e.g. a supercluster). The definitions of the first three Minkowski functionals (29),(30) and (31) can be rewritten as

$$
\begin{aligned}
V &= V_0 = \frac{4\pi}{3} R^3 \,, \\
S &= 6V_1 = 4\pi R^2 \,, \\
C &= 3\pi V_2 = R \,,
\end{aligned}
$$

where the second equalities stand for a sphere of a radius R, V is the volume, S the surface, and C the mean integrated radius of curvature:

$$
C = \int_S \frac{1}{2} \left(\frac{1}{R_1(\mathbf{x})} + \frac{1}{R_2(\mathbf{x})} \right) dS \,. \tag{45}
$$

You can see various definitions of shapefinders in the literature, differing by normalization; the most commonly used shapefinders are normalized to a sphere (for a sphere of a radius R: $H_i \equiv R$, $i = \{1, 2, 3\}$). This normalization leads to

$$H_1 = \frac{3V}{S} = \frac{1}{2}\frac{V_0}{V_1}, \quad \text{thickness}, \tag{46}$$

$$H_2 = \frac{S}{C} = \frac{2}{\pi}\frac{V_1}{V_2}, \quad \text{breadth}, \tag{47}$$

$$H_3 = \frac{C}{4\pi} = \frac{3}{4}V_2, \quad \text{length}. \tag{48}$$

The descriptive names in the formulae above can be easily justified. Consider a brick of the size $L \times W \times H$, where the sizes are really different, $L >> W >> H$. The Minkowski functionals for this brick are easy to calculate:

$$V_0 = LWH, \tag{49}$$

$$V_1 = \frac{1}{3}(LW + LH + WH), \tag{50}$$

$$V_2 = \frac{1}{3}(L + W + H), \tag{51}$$

$$V_3 = 1. \tag{52}$$

A couple of notes: first, in order to calculate the mean curvature use the fact that for a cylinder it does not depend on the radius and an edge of a brick is a quarter of an infinitesimally thin cylinder. Second, check that the formulae (44) give almost the right result, although the orientation of our brick is far from isotropic.

Now, let us normalize, temporarily, the shapefinders to a cube (let them all equal the cube size L). This gives

$$H_1 = V_0/V_1; \quad H_2 = V_1/V_2; \quad H_3 = V_3. \tag{53}$$

Using the Minkowski functionals from (49), the first shapefinder becomes

$$H_1 = \frac{3LWH}{LW + LH + WH}, \quad \text{or} \quad \frac{1}{H_1} = \frac{1}{3}\left(\frac{1}{L} + \frac{1}{W} + \frac{1}{H}\right);$$

it is the harmonic mean of the three sizes. When the sizes are really different, $H_1 \approx 3H$; this justifies the name 'thickness'.

The second shapefinder is a cleverly weighted mean:

$$H_2 = \frac{LW + LH + WH}{L + W + H} \approx \frac{LW}{L} = W$$

(the width); the last equalities assume substantially different sizes.

And as we see from the third equation of (49), the third shapefinder is the mean of the three sizes, or if the sizes are really different, $H_3 \approx L/3$, determined mainly by the length of the brick.

There is a fourth shapefinder – the genus has been given the honour to stand for it, directly. As genus counts 'minus' things (minus the number of isolated objects, minus the number of holes), the fourth Minkowski functional V_3 should be a better candidate.

The first set of shapefinders is accompanied by the 'second-order' shapefinders

$$K_1 = \frac{H_2 - H_1}{H_2 + H_1}, \quad \text{planarity}, \tag{54}$$

$$K_2 = \frac{H_3 - H_2}{H_3 + H_2}, \quad \text{filamentarity}. \tag{55}$$

These five (six, if you count the genus) numbers describe pretty well the shape of smooth (ellipsoidal) surfaces. In this case, the ratios K_1 and K_2 vary nicely from 0 to 1, and another frequently used ratio, K_1/K_2, has definite trends with respect to the shape of the object. But things get much more interesting when you start calculating the shapefinders of real superclusters; as shapes get complex, simple meanings disappear. Also, as shapefinders are defined as ratios of observationally estimated numbers (or ratios of quadratic combinations, as H_i work out in terms of V_i), they are extremely noisy. So, in case of serious use, a procedure should be developed to estimate the shapefinders directly, not by the ratio rules (46), (54); such a procedure does not exist yet. But, for the moment, the shapefinders are probably the best shape description tool (for cosmology) we have.

3.5 Morphology of Wavelet-Denoised Density

So much for the preliminaries. Now we have all our tools (wavelets, densities, Minkowski functionals). Let us use them and see what is the morphology of the real galaxy density distribution.

This question has been asked and answered about a hundred times, starting from the first topology paper by Gott et al. [5]. The first data set was a cube from the CfA I sample and contained 123 galaxies, if I remember right. (Imagine estimating a spatial density on the basis of a hundred points; this is the moment that Landau's definition of a cosmologist is appropriate: 'Cosmologists are often wrong, but never in doubt'.) The answer was – Gaussian!; ten points for courage. The same optimistic answer has been heard about a hundred times since (in all the papers published), with slight corrections in later times, as data is getting better. These corrections have been explained by different observational and dynamical effects, and peace reigns.

But all these studies have carefully smoothed the galaxy catalogues by nice wide Gaussian kernels to get a proper density. Doubts have been expressed that one Gaussianity could lead to another (by Peter Coles, for example), and the nice Gaussian behaviour we get is exactly that we have built in in the density field. So, let us wavelet denoise the density field (a complex recipe, but without any Gaussians) and estimate the Minkowski functionals. The next two figures are from our recent work [9].

The data are the 2dFGRS Northern galaxies; we chose a maximum-volume one-magnitude interval volume-limited (constant-density) sample and cut a maximum-volume brick from it (interesting, every time I say 'maximum', the sample gets smaller). Constant density is necessary, otherwise typical density scales will change with distance, and a brick was cut in order to avoid border corrections (these can bring additional difficulties for wavelet denoising). We carefully wavelet denoised the density; Fig. 18 shows its fourth Minkowski functional.

First, we see that our wavelet-denoised density is never close to Gaussian. Second, the specially built N-body catalogues (mocks) do not have Gaussian morphology, too; although they deviate from the real data, they are closer to it than to the example Gaussian. We see also that the same galaxies, smoothed by a yet very narrow Gaussian kernel ($\sigma = 2\,\mathrm{Mpc/h}$), show an almost Gaussian morphology. Thus, the clear message of this figure is that the morphology of a good (we hope the best) adaptive galaxy density is far from Gaussian; and the Gaussianity-confirming results obtained so far are all the consequence of the Gaussianity input by hand – Gaussian smoothing. This figure exhibits a

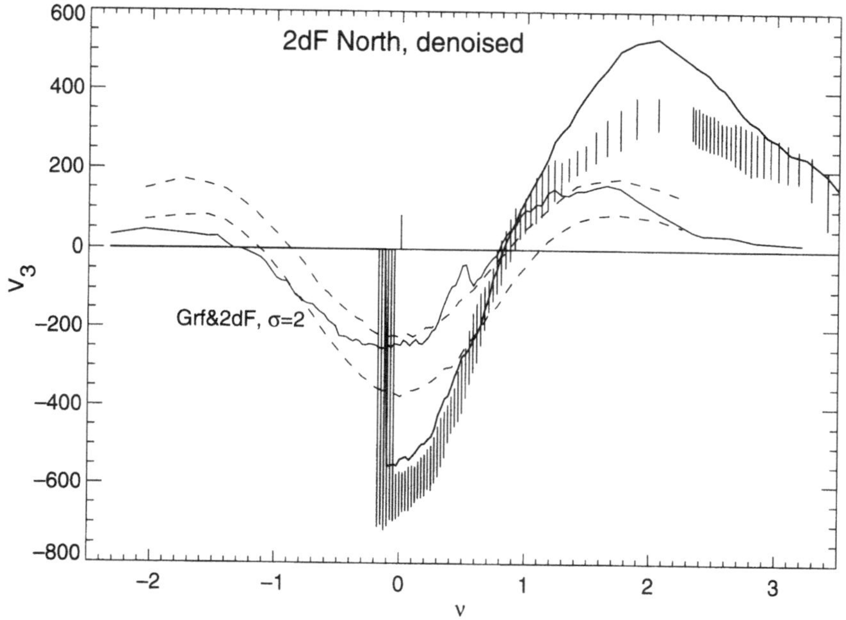

Fig. 18. The Minkowski functional V_3 for the 2dF brick, for the wavelet-denoised data set (*thick solid line*). The variability range of the wavelet-denoised mocks is shown with bars. We show also the 95% confidence limits for 1300 realizations of theoretical Gaussian density fields (*dashed lines*), and the V_3 data curve (*thin solid line*), all obtained for the Gaussian $\sigma = 2\,\mathrm{Mpc/h}$ smoothing. Figure from Martiínez et al. [9]

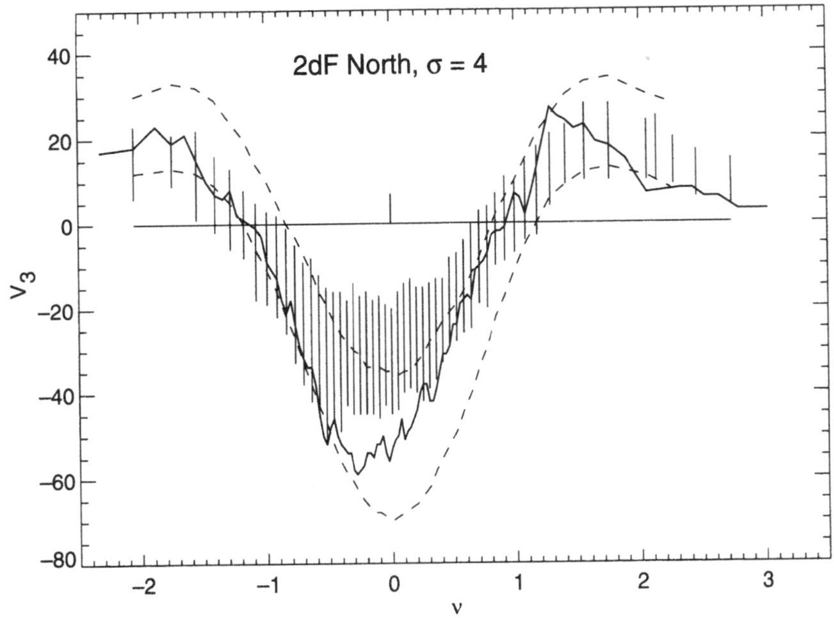

Fig. 19. The Minkowski functional V_3 for the 2dF brick, Gaussian-smoothed with $\sigma = 4\,\mathrm{Mpc/h}$ (*thick solid line*). The variability range of the mocks is shown with bars. We show also the 95% confidence limits for 1300 realizations of theoretical Gaussian density fields (*dashed lines*). Figure from Martiínez et al. [9]

little non-Gaussianity yet; the next one (Fig. 19) almost does not. The filter used there is still narrow ($\sigma = 4\,\mathrm{Mpc/h}$); the filter widths used in most papers are around twice that value.

I am pretty sure that Gaussian smoothing is the culprit here. We built completely non-Gaussian density distributions (Voronoi walls and networks), Gaussian-smoothed them using popular recipes about the kernel size, and obtained perfectly Gaussian Minkowski functionals.

One reason for that, as Vicent Martínez has proposed, is that the severe smoothing used changes a density field into Poissonian, practically. Try it – smooth a density field with a Gaussian of $\sigma = r_0$, where r_0 is the correlation length, and you get a density field with a very flat low-amplitude correlation function, almost Poissonian. And the Minkowski functionals of Poissonian density fields[5] are Gaussian; we tested that.

Another reason that turns Minkowski functionals Gaussian even for small σ must be the extended wings of Gaussian kernels. Although they drop pretty fast, they are big enough to add to a small extra ripple on the main density

[5] To be more exact, here is the recipe – take a Poissonian point process of N points in a volume V and smooth it with a Gaussian kernel of $\sigma = d$, where $d = (V/N)^{1/3}$ is the mean distance between the points.

field. As Minkowski functionals are extremely sensitive to small density variations, all they see is that ripple. This is especially well seen in initially empty regions. Usually, one uses a FFT-based procedure for Gaussian smoothing, as that is at least a hundred times faster (for present catalogue volumes) that direct convolution in real space. This procedure generates a wildly fluctuating small-amplitude density field in empty regions, and we did not realize for a long time at first where those giant-amplitude ghost MF-s came from. Gaussian smoothing and FFT, that was their address.

3.6 Multiscale Morphology

So, the galaxy density, similar to that we have accustomed to find in our everyday experience, is decidedly non-Gaussian. But is that a problem? Cosmological dynamics tells us that structure evolves at different rates at different scales; a true density mixes these scales all together and is not the best object to search for elusive traces of initial conditions. Good. Multiscale densities to the rescue.

The results that will end this chapter did not exist at the time of the summer school. But we live in the present, and time is short, so I will include them. A detailed account is already accepted for publication [12]. I shall show only a collection of Minkowski functionals here.

As our basic data set, we took the 2dFGRS volume-limited samples for the $[-20, -19]$ magnitude interval; they have the highest mean density among similar one-magnitude interval samples. We did not cut bricks this time, but corrected for sample boundaries; we have learnt that by now. We wavelet-decomposed the galaxy density fields and found the Minkowski functionals; as simple as that. Although wavelet decompositions are linear and should not add anything to the morphology of the fields, we checked that on simulated Gaussian density fields. Right, they do not add any extra morphological signal.

The results (for the 2dFGRS North) are shown in Figs. 20, 21, and 22. As the amplitudes of the (densities of) Minkowski functionals vary in a large interval, we use a sign-aware logarithmic mapping:

$$\text{logn}(x; a) = \text{sgn}(x) \log(1 + |x|/a) .$$

This mapping accepts arguments of both signs ($\log(x)$ does not), is linear for $|x| << a$, and logarithmic for $|x| >> a$. As the figures show, for the scales possible to study, the morphology of the galaxy density distribution is decidedly not Gaussian. The deviations are not too large for the second Minkowski functional v_2 (watch how the maxima shift around), but are clearly seen for the other two. The maximum wavelet order here is 3 for a $\sqrt{2}\,\text{Mpc/h}$ grid that corresponds to a characteristic scale of $2^3\sqrt{2} \approx 11.3\,\text{Mpc/h}$. As the mean thickness of the 2dFGR North slice is about 40–50 Mpc/h, we cannot go much further – the higher order wavelet slices would be practically two dimensional. The 2dFGRS Southern data set has similar size limitations. So, as 10 Mpc/h is

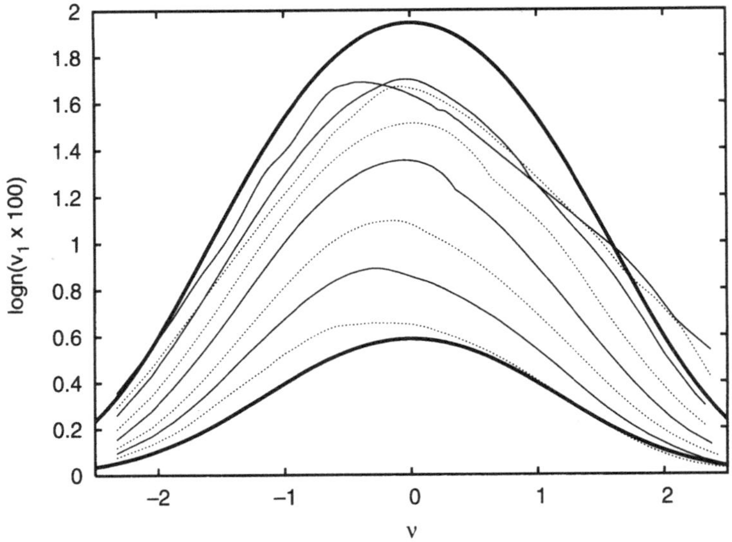

Fig. 20. Summary of the densities of the second MF v_1 for the data and all wavelet orders for the 2dFN19 sample, in the logn mapping. Thick lines show reference Gaussian predictions. Full lines stand for the 1 Mpc/h grid, dotted lines for the $\sqrt{2}$ Mpc/h grid. Figure from Saar et al. [12]

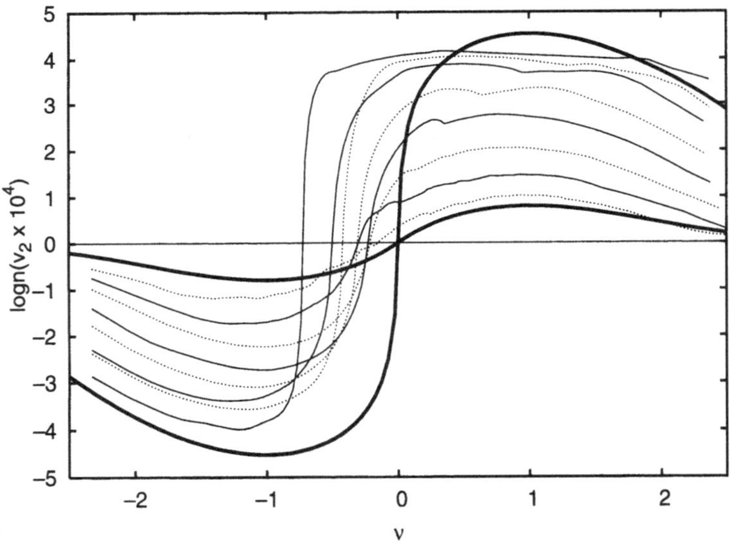

Fig. 21. Summary of the densities of the third MF v_2 for the data and all wavelet orders for the 2dFN19 sample, in the logn mapping. Thick lines show reference Gaussian predictions. Full lines stand for the 1 Mpc/h grid, dotted lines for the $\sqrt{2}$ Mpc/h grid. Figure from Saar et al. [12]

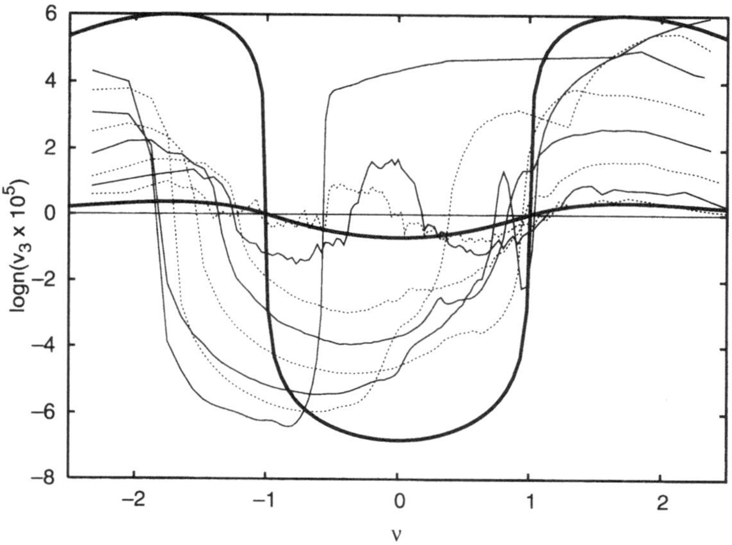

Fig. 22. Summary of the densities of the fourth MF v_3 for the data and all wavelet orders for the 2dFN19 sample, in the logn mapping. Thick lines show reference Gaussian predictions. Full lines stand for the 1 Mpc/h grid, dotted lines for the $\sqrt{2}$ Mpc/h grid. Figure from Saar et al. [12]

a scale where cosmological dynamics might have slight morphological effects, the question of whether the original morphology of the cosmological density field was Gaussian is not answered yet. But we shall find it out soon, when the SDSS will finally fill its full planned volume.

These are the results, for the moment. The morphology of the galaxy density field is far from Gaussian, in contrary to practically every earlier result. Have all these results only confirmed that if we take a serious smoothing effort, we are able to smooth any density field to Poissonian? An indecent thought.

But do not be worried; results come and go; this is the nature of research. Methods, on the contrary, stay a little longer, and this school was all about methods.

Recommended Reading

I was surprised that Bernard Jones recommended a wavelet bookshelf that is almost completely different from mine (the only common book is that of I. Daubechies). So, read those of Bernard's choice, and add mine:

- five books:
  1. Stephane Mallat, A Wavelet Tour of Signal Processing, 2nd ed., Academic Press, London, 1999,

2. C. Sidney Burrus, Ramesh A. Gopinath, Haitao Gao, Introduction to Wavelets and Wavelet Transforms, Prentice Hall, NJ, 1997,
3. Ingrid Daubechies, Ten lectures on wavelets, SIAM, Philadelphia, 2002,
4. Jean-Luc Starck, Fionn Murtagh, "Astronomical Image and Data Analysis", 2nd ed., Springer, 2006 (application of wavelets and many other wonderful image processing methods in astronomy),
5. Bernard W. Silverman, Density Estimation for Statistics and Data Analysis, Chapman & Hall / CRC Press, Boca Raton, 1986 (the classical text on density estimation),

- three articles:
  1. Mark J. Shensa, The Discrete wavelet Transform: Wedding the À Trous and Mallat Algorithms, IEEE Transactions on Signal Processing, 40, 2464–2482, 1992 (the title explains it all),
  2. K.R. Mecke, T. Buchert, H. Wagner, Robust morphological measures for large-scale structure in the Universe, Astron. Astrophys. 288, 697–704, 1994 (introducing Minkowski functionals in cosmology),
  3. Jens Schmalzing, Thomas Buchert, Beyond Genus Statistics: A Unified Approach to the Morphology of Cosmic Structure, Ap. J. Letts 482, L1, 1997, (presentation of two grid algorithms).
- two web pages:
  1. a wavelet tutorial by Jean-Luc Starck at `http://jstarck.free.fr`,
  2. the wavelet pages by David Donoho at `http://www-stat.stanford.edu/~donoho/` (look at lectures and reports).

Acknowledgements

I was introduced to wavelets in about 1990 by Ivar Suisalu (he was my PhD student then). As that happened in NORDITA, I asked advice from Bernard Jones soon, and started on a wavelet road, together with Vicent Martínez and Silvestre Paredes; these early wavelets were continuous. Much later, I have returned to wavelets and have learnt much from Bernard, who is using wavelets in the real world, and from Vicent, Jean-Luc Starck, and David Donoho, the members of our multiscale morphology group. I thank them all for pleasant collaboration and knowledge shared. All the results presented here belong to our morphology group.

My present favourites are the *à trous* wavelets, as you have noticed.

My research has been supported in Estonia by the Estonian Science Foundation grant 6104 and by the Estonian Ministry of Education research project TO-0060058S98. In Spain, I have been supported by the University of Valencia via a visiting professorship and by the Spanish MCyT project AYA2003-08739-C02-01 (including FEDER).

References

1. Adler, R.: The Geometry of Random Fields, Wiley, New York (1981)
2. Ascasibar, Y., Binney J.: MNRAS **356**, 872–882 (2005)
3. de Boor, C.: A Practical Guide to Splines. Springer-Verlag, New York (1978)
4. Coles, P., Davies, A.G., Pearson, R.C.: MNRAS **281**, 1375 (1996)
5. Gott, J.R., Dickinson, M., Melott, A.L.: ApJ **306**, 341 (1986)
6. Holtschneider, M., Kronland-Martinet, R., Morlet, J., Tchamitchian. P.: In: Wavelets, Time-Frequency Methods and Phase Space, pp. 289–297. Springer-Verlag, Berlin (1989)
7. Mallat, S.: A Wavelet Tour of Signal Processing, 2nd ed. Academic Press, London, (1999)
8. Martínez, V.J., Saar, E.: Statistics of the Galaxy Distribution, Chapman & Hall/CRC Press, Boca Raton (2002)
9. Martínez, V.J., Starck, J.-L., Saar, E., Donoho, D.L., Reynolds, S.C., de la Cruz, P., Paredes, S.: ApJ. **634**, 744 (2005)
10. Mecke, K.R., Buchert, T., Wagner, H.: A& A **288**, 697–704 (1994)
11. Press, W.H., Flannery, B.P., Teukolsky, S.A., Vetterling, W.T.: Numerical Recipes in C: The Art of Scientific Computing, CUP, Cambridge (1992)
12. Saar, E., Martínez, V.J., Starck, J.-L., Donoho, D.L.: MNRAS **374**, 1030 (2007)
13. Sahni, V., Sathyaprakash, B.S., Shandarin, S.F.: ApJ. Lett. **495**, L5–L8 (1998)
14. Schmalzing, J., Buchert, T.: ApJ. Lett. **482**, L1 (1997)
15. Silverman, B.W.: Density Estimation for Statistics and Data Analysis, Chapman & Hall/CRC Press, Boca Raton (1986)
16. Starck, J.-L., Martínez, V.J., Donoho, D.L., Levi, P. Querre, E. Saar, EURASIP, J. Appl. Signal Analy. 2455–2469 (2005)
17. Starck, J.-L., Murtagh, F.: Astronomical Image and Data Analysis, 2nd ed., Springer, NY (2006)
18. Weinberg, D.H., PASP **100**, 1373–1385 (1988)

Gaussian Fields and Constrained Simulations of the Large-Scale Structure

Y. Hoffman

Racah Inst. of Physics, Hebrew University, Jerusalem 91904, Israel
hoffman@huji.ac.il

Abstract. The formalism of the Wiener filter (WF) and constrained realizations (CRs) is reviewed here within the cosmological context of the primordial perturbations constituting a random Gaussian field. It is shown that in such a context the WF/CRs methodology is the optimal tool of reconstructing the (linear) large-scale structure of the universe from noisy, sparse and incomplete data. The emphasis of the chapter is on reviewing the WF and CR algorithm as applied to the linear regime of the gravitational instability and Gaussian random fields. Applications to the reconstruction of the cosmography from redshift and peculiar surveys are briefly reviewed.

The main application reviewed here is that of constrained simulations (CSs), namely cosmological simulations in which the initial conditions are Gaussian random fields constrained by actual observational data. CSs of the 'local' universe reproduce, by construction, the large-scale features that are imposed by the observational data. They provide a platform for simulating the small-scale structure and galaxy formation in the 'local' universe within the actual observed large scales. One of the main virtues of the CSs is the reduction o the cosmic variance in simulations aimed at reproducing the observed 'local' universe and thereby improving and making tighter comparison between simulations and observations.

1 Introduction

The mapping of the distribution of galaxies and their velocity field plays a major role in modern cosmology, and it provides the observational basis of the study of the formation of the large-scale structure of the universe (LSS). The observed large-scale density and velocity fields offer a probe of the early universe and can serve as constraints on the value of the cosmological parameters (for a review see Padmanabhan [23]). Another very important window to the early universe is given by the observations of the temperature anisotropies of the cosmic microwave background (CMB) radiation, which probe almost directly the primordial perturbation field at the epoch of decoupling.

Astronomical observations, as any other experiment in the physical sciences, provide a noisy and incomplete information on the universe. In particular,

Hoffman, Y.: *Gaussian Fields and Constrained Simulations of the Large-Scale Structure.* Lect. Notes Phys. **665**, 565–583 (2009)
DOI 10.1007/978-3-540-44767-2_17

cosmological observations are often characterized by a rather low signal-to-noise signal that is subject to incomplete sky coverage. This implies that the theoretical analysis of these observations and their cosmological implications depends heavily on the statistical analysis of the observations. The development of statistical algorithms for parameter estimation and the reconstruction of the underlying dynamical fields constitutes one of the main problems of the current cosmological research (cf. the excellent review in Martínez & Saar [22]).

This chapter focuses on the subject of the estimation and prediction of the LSS from noisy, incomplete and sparse data bases. In particular, the chapter is centred on the reconstruction of the LSS from redshift and peculiar velocity surveys. Much of what is covered here is applicable to the field of the CMB temperature anisotropies, although the technical derivations are within the context of the linear gravitational instability and its application to the LSS. A general introduction to Gaussian random field is given in Sect. 2. The estimation and prediction of the LSS from cosmological surveys is presented in Sect. 3. Section4 presents a short review of the application of the reconstruction methods. Cosmological simulations, based on initial conditions constrained by actual observational data, are introduced in Sect. 5. A brief summary and discussion of future prospects conclude the chapter (Sect. 6).

2 Gaussian Random Fields

The Gaussian random field plays in probability theory and statistical physics the same role the harmonic oscillator plays in classical and quantum mechanics. As in mechanics, where the harmonic oscillator is the second (and lowest) order approximation of any arbitrary potential around a state of equilibrium, the Gaussian random field describes the statistical fluctuations around the most probable field. Given the central role Gaussian fields play in the standard model of cosmology, a general introduction and formulation of the properties of such fields is given here, following Landau and Lifshitz [20]. The formalism presented here assumes physical systems with discrete number of degrees of freedom (hereafter *d.o.f.*); however, this can be easily generalized to the continuous case (see Bertschinger [2]). The term 'field' applies here to discrete systems as well.

Consider a statistical system described by a set of N random *d.o.f.*, $\mathbf{s}' = s'_1, ..., s'_N$, for which a Gibbs entropy functional $\mathcal{S} = S(\mathbf{s})$ is defined. The PDF of the system is given by

$$P(\mathbf{s}') \propto \exp[\, S(\mathbf{s}')\,] \, . \qquad (1)$$

The state of statistical equilibrium $\mathbf{s}^{(0)}$ corresponds to a local maximum of the entropy, and for a global maximum of $\mathcal{S}$ the configuration $\mathbf{s}^{(0)}$ is the most probable state. The entropy is stationary with respect to $\mathbf{s}$ at $\mathbf{s}^{(0)}$ and therefore the PDF of small fluctuations from the state of equilibrium is given by

$$P(\mathbf{s}) \propto \exp\left[-\frac{1}{2} s_i \mathcal{S}_{ij} s_j \right] , \tag{2}$$

where $\mathbf{s} = \mathbf{s}' - \mathbf{s}^{(0)}$, $\mathcal{S}_{ij} = -\frac{\partial^2 \mathcal{S}}{\partial s_i \partial s_j}$ evaluated at $\mathbf{s}^{(0)}$ and summation over repeated indices is assumed throughout chapter paper.

The statistical properties of the system near the point of equilibrium is readily given by (2). The ensemble mean value of s_i vanishes and the covariance is readily evaluated:

$$\langle s_i s_j \rangle = \frac{\int s_i s_j P(\mathbf{s}) \mathrm{d}\mathbf{s}}{\int P(\mathbf{s}) \mathrm{d}\mathbf{s}} = \mathcal{S}_{ij}^{-1} . \tag{3}$$

Thus the matrix of the second derivatives with respect to field *d.o.f.*, evaluated at the point of equilibrium, equals the inverse of the covariance matrix of small fluctuations around this state of equilibrium. It is easy to show that the statistical properties of such a system are determined by the covariance matrix $\langle s_i s_j \rangle$.

The derivation given here applies to general systems whose statistical behaviour is determined by an entropy functional. However, no entropy formulation exists for the cosmological case where an infinite self-gravitating system is considered. Yet in the inflationary model, quantum fluctuations of the field that drives the inflation give rise to a Gaussian random perturbation field. The Gaussian nature of the field is a generic feature of the inflationary model. Some of the general aspects of the estimation and prediction presented here are independent of the nature of the PDF of the perturbation field; however, to the extent that the assumption of Gaussianity is made, it is independent of the mechanism that gives rise to it.

3 Estimation and Prediction

3.1 Inversion Problem

Consider the following problem of observing a physical system described by N *d.o.f.*, $\mathbf{s} = s_1, ..., s_N$, yielding a data set of M observations, $\mathbf{d} = d_1, ..., d_M$. It is assumed here that the act of observations can be modelled as a linear convolution on the field underlying *d.o.f.*:

$$\mathbf{d} = \mathbf{R}\mathbf{s} + \epsilon , \tag{4}$$

where $\mathbf{R}$ is the linear operator that relates the data to the underlying field, also known as point spread function or response function, and the statistical observational errors are given by $\epsilon = \epsilon_1, ..., \epsilon_{\mathbf{M}}$. Here it is assumed that the data $\mathbf{d}$ is corrected for the systematic errors and therefore it is statistically uncorrelated with the field $\mathbf{s}$. The operator R represents the response of the measuring device to the underlying field, such as the blurring introduced by the finite resolution of telescopes, or the linear procedure by which the raw data is analysed to obtain the data $\mathbf{d}$.

There are two major sources of errors that often occur in the analysis of the LSS and CMB. One is the finite accuracy of observations, such as the errors in the determination of the peculiar velocities (cf. Dekel [7]). The other is the shot noise which is introduced by sampling a continuous field by discrete events. This is the case with the map making of the CMB sky out of time-ordered data where the shot noise is determined by the photon counting rate. Shot noise errors arise also in the determination of the continuous density field from the finite and discrete galaxy distribution in redshift surveys. In the case of errors due to the finite accuracy their distribution is often taken to be normal, and the errors are usually assumed to be uncorrelated. Shot noise errors are more complicated. The distribution of the shot noise errors is Poisson, which in the limit of large numbers tends to a normal distribution. In all applications presented here, the large number is assumed and the error distribution function is taken to be Gaussian. The procedure of counting is based on dividing the observational domain into pixels, and each observed event is assigned to one or few pixels. In the case of overlapping pixels, such as in the case of Gaussian smoothing of the galaxy distribution often used in the analysis of redshift surveys, the shot noise errors are correlated. In design or determination of the pixel size one should consider the following trade-off. By increasing the pixel size, the mean number of events per pixel increases and therefore the shot noise level decreases; however, this leads to a loss in spatial resolution and reduces the effective number of observed *d.o.f.* The optimal size is usually obtained for a mean number of events (per pixel) less than ten. The size, shape and weighting of the data are mathematically modelled by the point spread function $\mathbf{R}$.

The problem of the reconstruction of the underlying field $\mathbf{s}$ from the data set $\mathbf{d}$ is essentially a problem of inversion or deconvolution. In the absence of noise, ϵ, the field $\mathbf{s}$ can be simply obtained by applying the inverse of $\mathbf{R}$ on the data vector $\mathbf{d}$. However, in practice two major problems are encountered. First, all cosmological observations of the CMB and LSS are characterized by a quite low signal-to-noise ratio, and the noise cannot be neglected. Second, in all relevant cosmological case the number of *d.o.f.* of the underlying field is much larger than the number of data points, $N \gg M$. In such a case, the data base does not have enough information to constrain all the field *d.o.f.*, and the problem is ill-posed. However, even in the case where $N \approx M$ the inversion would amplify the errors resulting in an unstable deconvolution which is noise dominated. Some additional information or operation is needed to stabilize the inversion.

The problem of reconstruction can be classified into two major classes, estimation and prediction. Equation (4) is rewritten here as $\mathbf{d} = \mathbf{s}^{\rm obs} + \epsilon$, where $\mathbf{s}^{\rm obs}$ is a vector of length M, $\mathbf{s}^{\rm obs} = \mathbf{Rs}$. The vector $\mathbf{s}^{\rm obs}$ corresponds to dynamical variables that are directly observed and are given by linear combinations of the field *d.o.f.*, $\mathbf{s}$. The reconstruction of these is referred to here as an estimation, corresponding to an estimate of the signal from a noisy measurement. The reconstruction of the other *d.o.f.* not directly observed is referred to as prediction. The following example would clarify the difference between the two. Consider the observations of peculiar velocities, where only the radial (line-of-sight) component is measured with a finite accuracy. An estimation of the actual radial velocities is based on some procedure by which the radial velocities are 'cleaned' from the noise. The reconstruction of the other two components of the velocities, and the extrapolation from regions where velocities are measured to unobserved regions, is termed here as prediction. It is clear that while estimation can be performed based on some general statistical algorithms, predictions should be made based only on some physical assumptions or model made on the nature of the particular system at hand that have to be made a priori. This set of assumptions is defined here as the *prior* model. The aim of the present chapter is to present a coherent and systematic approach to the problem of estimation and prediction from noisy and incomplete data. Throughout the chapter no distinction will be made between prediction and estimation and be termed as estimation.

3.2 Wiener Filter

The canonical model of cosmology assumed here is that structure has evolved from a Gaussian random field, whose statistical properties are determined by the covariance matrix,

$$\mathbf{S} = \left\langle \mathbf{s}\mathbf{s}^{\dagger} \right\rangle = \left\{ \left\langle s_i s_j \right\rangle \right\} . \tag{5}$$

Here the explicit assumption about the nature of the PDF is relaxed and the *prior* model consists only of the knowledge of the covariance (or correlation) matrix. The errors covariance matrix is also assumed to be known:

$$\mathbf{N} = \left\langle \epsilon \, \epsilon^{\dagger} \right\rangle . \tag{6}$$

The data auto-covariance matrix is given by

$$\mathbf{D} = \left\langle \mathbf{d} \; \mathbf{d}^{\dagger} \right\rangle = \mathbf{R}\mathbf{S}\mathbf{R}^{\dagger} + \mathbf{N} , \tag{7}$$

where the last equality follows from our assumption on the lack of correlation of the errors and the underlying field. The simplest possible estimator is one that depends linearly on the data,

$$\mathrm{s}^{\mathrm{MV}} = \mathbf{F}\mathbf{d} . \tag{8}$$

The operator $\mathbf{F}$ is an $N \times M$ matrix that minimizes the variance of the residual $\mathbf{r}$,

$$\left\langle \mathbf{r} \; \mathbf{r}^{\dagger} \right\rangle = \left\langle \left(\mathbf{s} - \mathbf{s}^{\mathrm{MV}} \right) \left(\mathbf{s}^{\dagger} - \mathbf{s}^{\mathrm{MV}\dagger} \right) \right\rangle . \tag{9}$$

Minimizing the variance of the residual (9) with respect to $\mathbf{F}$ yields

$$\mathbf{F} = \left\langle \mathbf{s}\mathbf{d}^\dagger \right\rangle \left\langle \mathbf{d}\mathbf{d}^\dagger \right\rangle^{-1} . \tag{10}$$

The $N \times M$ data-field cross-correlation matrix is given by

$$\left\langle \mathbf{s}\mathbf{d}^\dagger \right\rangle = \left\langle \mathbf{s}\mathbf{s}^\dagger \right\rangle \mathbf{R}^\dagger = \mathbf{S}\mathbf{R}^\dagger . \tag{11}$$

The minimal variance estimator is

$$\mathbf{s}^{\mathrm{MV}} = \mathbf{S}\mathbf{R}^\dagger \left(\mathbf{R}\mathbf{S}\mathbf{R}^\dagger + \mathbf{N} \right)^{-1} \mathbf{d} . \tag{12}$$

The variance of the residual from the minimal variance estimator is given by

$$\left\langle \mathbf{r}\,\mathbf{r}^\dagger \right\rangle = \mathbf{S} - \mathbf{S}\mathbf{R}^\dagger (\mathbf{R}\mathbf{S}\mathbf{R}^\dagger + \mathbf{N})^{-1} \mathbf{R}\mathbf{S} . \tag{13}$$

The minimal variance estimator is also known as the WF (Wiener [39]) and it has been recently reintroduced into astrophysics by Rybicki and Press [27]. In the field of image processing and data analysis the WF is considered to be a rather 'naive' and elementary tool, inferior to the more advanced tools of analysis such as maximum entropy (cf. Skilling [30] and references therein). However, it will be shown below that for the case of the analysis and reconstruction of the CMB and the LSS, where the underlying field is assumed to be Gaussian or close to it, the WF is indeed the optimal and very powerful tool of reconstruction.

3.3 Bayes' Theorem and Conditional Probability

The Bayesian approach to estimation and reconstruction is based on Bayes' Theorem which states that given the PDF of the underlying field and the observational random errors, the *posterior* PDF of the underlying field given the data is

$$P\left[\mathbf{s}|\mathbf{d}\right] = \frac{P\left[\mathbf{s},\mathbf{d}\right]}{P\left[\mathbf{d}\right]} \propto P\left[\mathbf{d}|\mathbf{s}\right] P\left[\mathbf{s}\right] , \tag{14}$$

where $P\left[\mathbf{d}\right]$ serves as a mere normalization constant. The reader is referred to the general probability theory literature for a general discussion on the merits of the Bayesian approach, and more specific applications in the field of cosmology are reviewed in Kaiser [13]. The general approach adopted here is essentially Bayesian, where a *prior* model is assumed and the various algorithms of estimation and reconstruction are formulated within its framework.

In the cosmological context of Gaussian random fields the Bayesian approach is extremely powerful. The PDF of the underlying field is a multivariate Gaussian:

$$P\left[\mathbf{s}\right] = \frac{1}{\left[(2\pi)^N \det(\mathbf{S})\right]^{1/2}} \exp\left(-\frac{1}{2} \mathbf{s}^\dagger\, \mathbf{S}^{-1} \mathbf{s} \right) . \tag{15}$$

The conditional probability of the data given the field $\mathbf{s}$ is

$$P\left[\mathbf{d}|\mathbf{s}\right] \propto \exp\left[-\frac{1}{2} (\mathbf{d} - \mathbf{R}\mathbf{s})^\dagger \mathbf{N}^{-1} (\mathbf{d} - \mathbf{R}\mathbf{s}) \right] . \tag{16}$$

Combining the last two equations and after some algebraic manipulations (Bertschinger [2]) the *posterior* PDF is found to be a shifted Gaussian:

$$P\big[\mathbf{s}|\mathbf{d}\big] \propto \exp\Big[-\frac{1}{2}(\mathbf{d}-\mathbf{SR}^\dagger(\mathbf{RSR}^\dagger+\mathbf{N})^{-1}\mathbf{d})^\dagger((\mathbf{RSR}^\dagger)^{-1}+\mathbf{N}^{-1})$$
$$\times(\mathbf{d}-\mathbf{SR}^\dagger(\mathbf{RSR}^\dagger+\mathbf{N})^{-1}\mathbf{d})\Big]\,. \tag{17}$$

Given the *posterior* PDF there are two possible estimators of the field. One is the conditional mean field given the data,

$$\mathbf{s}^{\text{mean}} = \frac{\int \mathbf{s}P\big[\mathbf{s}|\mathbf{d}\big]\,\mathrm{d}\mathbf{s}}{\int P\big[\mathbf{s}|\mathbf{d}\big]\,\mathrm{d}\mathbf{s}}\,. \tag{18}$$

The other estimator is the maximum a posteriori (MAP) estimate of the field, namely the one that maximizes $P\big[\mathbf{s}|\mathbf{d}\big]$. For the particular case of Gaussian random field these two estimates and the WF do coincide. Not only do the various estimators coincide but also the variance of the residual calculated from the *posterior* PDF,

$$\langle \mathbf{r}\mathbf{r}^\dagger\rangle = \Big(\mathbf{S}^{-1}+\mathbf{R}^\dagger\mathbf{N}^{-1}\mathbf{R}\Big)^{-1}\,, \tag{19}$$

which equals the expression of (13).

An alternative definition of an estimator of the underlying field is given by the maximum entropy algorithm, where the entropy defined in Sect. 2 as second-order expansion around a state of local equilibrium is used to stabilize the inversion of the χ^2 distribution of the data given an underlying field (i.e. model). The estimator is defined by the MAP estimate of the field, with respect to $P\big[\text{entropy}\big]P\big[\chi^2\big]$, which can be easily shown to coincide with $P\big[\mathbf{s}|\mathbf{d}\big]$. Thus, for Gaussian fields also the maximum entropy estimators coincide with the WF.

3.4 Constrained Realizations

The algorithm of estimation and prediction from noisy and incomplete data provides one with an estimator of the underlying field, whose nature depends on the properties of the random field. It provides also a way for evaluating the variance of the residual from the estimator and this can be used to estimate the scatter. In all practical cases, the variance of the residual is not negligible and consequently the estimated field is smoother and less fluctuating than the actual underlying field. It follows that even for the case where the data used is typical to the problem of hand (i.e. $\chi^2/d.o.f. \approx 1$) the estimated field does not resemble typical realization of such a field. Such an estimator cannot be used to set initial conditions for N-body simulations, for example. This defines the need for a method for generating typical realizations of the underlying field that are consistent with the field's PDF and are constrained to obey a set of constraints, the so-called constrained realizations (CRs). The statistical properties of constrained Gaussian fields were thoroughly studied in the cosmological context by Bardeen et al.[1], who focused mainly on the ensemble average aspect. Bertschinger [2] extended this work and investigated the conditional PDF of the field given constraints that are linear functionals of the field. Bertschinger was the first to recognize the need for CRs and suggested an algorithm for the construction of these. His method was based on iterative simulated annealing for properly sampling the conditional PDF. However, simulated annealing methods are well known for their slow convergence, and therefore Bertschinger's method could be practically used to set only a very few constraints. The method was improved later by Binney and Quinn [3] who showed that the method can be solved exactly in a non-iterative way in the spherical Bessel/spherical harmonics representation and for a localized set of constraints. Another improvement on Bertschinger's [2] method was developed by Hoffman and Ribak [11], whose iterative method was more efficient than the original one.

The problem of constructing CRs of Gaussian fields subject to linear constraints has been completely solved by Hoffman and Ribak [11], who suggested the simplest and most efficient method for generating CRs. The key point of this method relies on the fact that the conditional PDF of the residual from the mean is independent of the actual numerical values of the constraints. This can be easily shown from (17), where the data vector $\mathbf{d}$ is taken as the set of constraints. The derivation here generalizes the original treatments of Bertschinger (1987) and Hoffman and Ribak [11] to enable constraints whose exact values are uncertain, i.e. noisy data. The conditional PDF of the residual is given by

$$P\big[\mathbf{r}|\mathbf{d}\big] \propto \exp\Big[-\frac{1}{2}\mathbf{r}^{\dagger}\big((\mathbf{RSR}^{\dagger})^{-1}+\mathbf{N}^{-1}\big)\mathbf{r}\Big] , \tag{20}$$

where the residual is defined by $\mathbf{r} = \mathbf{s} - \mathbf{s}^{\mathrm{MV}} = \mathbf{s} - \mathbf{SR}^{\dagger}(\mathbf{RSR}^{\dagger}+\mathbf{N})^{-1}\mathbf{d}$. The variance of the residual is given by $\big((\mathbf{RSR}^{\dagger})^{-1}+\mathbf{N}^{-1}\big)^{-1}$. It is the property of Gaussian fields that the conditional PDF depends on the nature of the constraints and is independent of the actual numerical values. Thus the statistical properties of the residual depend on whether the constraints are imposed on the density, velocity or potential fields and where exactly in configuration space these are imposed, but the residual is independent of the numerical values of the constraints.

The CR is written as the sum of the deterministic (given the *prior* and the data) mean field and the random residual,

$$\mathbf{s}^{\mathrm{CR}} = \mathbf{s}^{\mathrm{MV}} + \mathbf{r} . \tag{21}$$

The problem of generating CRs is reduced to the problem of generating the residual $\mathbf{r}$. Consider now an unconstrained realization of the underlying field, $\tilde{\mathbf{s}}$, and the errors, $\tilde{\epsilon}$. This realization is 'observed' in the same way as the actual field to yield a 'fake' data set:

$$\tilde{\mathbf{d}} = \mathbf{R}\tilde{\mathbf{s}} + \tilde{\epsilon} \,. \tag{22}$$

A realization of the residual is obtained by taking the difference between the field $\tilde{\mathbf{s}}$ and its estimator,

$$\mathbf{r} = \tilde{\mathbf{s}} - \mathbf{SR}^{\dagger}(\mathbf{RSR}^{\dagger} + \mathbf{N})^{-1}\tilde{\mathbf{d}} \,. \tag{23}$$

This is used to construct the desired CR,

$$\mathbf{s}^{\mathrm{CR}} = \tilde{\mathbf{s}} + \mathbf{SR}^{\dagger}(\mathbf{RSR}^{\dagger} + \mathbf{N})^{-1}(\mathbf{d} - \tilde{\mathbf{d}}) \,. \tag{24}$$

The algorithm described here is optimal because it correctly samples the conditional PDF and it involves only one realization of the unconstrained field and the calculation of the mean field by the application of the WF. To see best the statistical properties of the CRs, the case considered here is that of $\mathbf{R} = \mathbf{I}$, where the constraints are imposed on the field itself, and the noise or measurement errors are neglected. Assuming the Gaussian field to be reconstructed is some $f(\mathbf{r})$, whose auto-correlation function $\xi(\mathbf{r}_1, \mathbf{r}_2) = \langle f(\mathbf{r}_1) f(\mathbf{r}_2) \rangle$ is known. Consider a set of constraints $F_i = f(\mathbf{r}_i)$ is imposed on the desired realization, i.e. at the locations $\{\mathbf{r}_i\}$ the value of the field should be equal to the specified values, $\{F_i\}$. In these new notations (24) is rewritten as

$$f^{\mathrm{CR}}(\mathbf{r}) = \tilde{f}(\mathbf{r}) + \xi_i(\mathbf{r})\xi_{i,j}^{-1}\left(F_j - \tilde{f}(\mathbf{r}_j)\right) \,, \tag{25}$$

where $\xi_i(\mathbf{r}) = \xi(\mathbf{r}_i, \mathbf{r})$ and $\xi_{i,j} = \xi(\mathbf{r}_i, \mathbf{r}_j)$. It is easy to show that at the point where a constraint is imposed, namely $\mathbf{r} = \mathbf{r}_i$, the field indeed obeys the constraint, $F_i = f(\mathbf{r}_i)$. In the limit where the distance between $\mathbf{r}$ and all the points of constraints is larger than the correlation length, $\xi_i(\mathbf{r}) \approx 0$, the realization becomes essentially unconstrained, $f(\mathbf{r})^{\mathrm{CR}} \approx \tilde{f}(\mathbf{r})$. In the case where the errors are characterized by a non-vanishing error covariance matrix, $\mathbf{N}_{ij}$, the expression for the CR is given by

$$f^{\mathrm{CR}}(\mathbf{r}) = \tilde{f}(\mathbf{r}) + \xi_i(\mathbf{r})\Big(\xi_{i,j} + \mathbf{N}_{ij}\Big)^{-1}\left(F_j - \tilde{F}(\mathbf{r}_j)\right) \,, \tag{26}$$

where $\tilde{F}(\mathbf{r}_j)$ is the sum of the random realization of the field and the noise. Note that in the case of noise-dominated constraints or in the limit where the field at question is not correlated with the constraints, the WF mean field vanishes and one obtains an unconstrained realization.

A different presentation of the Hoffman & Ribak [11] methodology was presented by van de Weygaert & Bertschinger [36], who suggested a somewhat different practical implementation of the method. The WF/CR formalism can be extended to the case of log-normal random fields (Sheth [29]). CRs of the log of the density field are completely analog to CRs of normal random fields, yet the statistical interpretation of the CRs is less obvious. For Gaussian fields the mean and the most probable fields coincide, a property that is not obeyed by log-normal fields. It follows that the minimal variance, most probable and conditional mean estimator do not coincide in the log-normal case and, one needs to judge the merits of one estimator over the others (cf. Sheth [29]).

A common criticism of the WF estimator is that in the above limit of noisy dominated data and away from the region where the constraints are imposed the resulting estimator tends to zero. Indeed, in the absence of any relevant information the null field is the most probable one. The CRs provide a possible remedy to this shortcoming. In regions dominated by good data, the CRs are mainly dominated by the mean field and the random residual is negligible; however, in regions of noisy or no data at all the mean field vanishes and the random residual dominates the CRs. This provides one with an efficient method for generating initial conditions for N-body simulations which sample correctly the field PDF, and at the same time obey the constraints taken from the observed universe.

3.5 Dynamical Reconstruction

So far the WF has been applied in the case of the statistical reconstruction of an underlying field from observations that sample that same field. In particular we have focused on the density field which is sampled by the galaxy distribution. Here, the reconstruction method is extended further to do dynamical reconstruction, namely using observational data that sample one field to reconstruct a different field that is dynamically related. Immediate applications of that approach include the estimation of the underlying density field from observed radial peculiar velocities, or going the other way round to use density data to construct the peculiar velocity field. This dynamical approach depends on the availability of a theoretical model which relates the two different fields via a cross-correlation function. Most of the discussion here is based on the linear theory of gravitational instability; however, it will be shown how the formalism can be extended beyond the linear regime.

The velocity–density relation plays a crucial role in the study of LSS. An indirect density-from-velocity reconstruction was done before by Kaiser and Stebbins [14] and Stebbins [31], who have defined the problem within the framework of Bayesian statistics. Their approach is equivalent to the WF method used by Zaroubi et al. [43] in their analysis of the MARK III peculiar velocities survey. The WF dynamical reconstruction, $\delta^{\rm WF}(\mathbf{r})$, is derived by cross-correlating the radial velocity, $u(\mathbf{r}) = \mathbf{v}(\mathbf{r}) \cdot \hat{r}$ with $\delta(\mathbf{r})$:

$$\delta^{\rm WF}(\mathbf{r}) = \Big\langle \delta(\mathbf{r}) u_i \Big\rangle \Big\langle u_i u_j + \epsilon_i \epsilon_j \Big\rangle^{-1} U_j \ . \tag{27}$$

Here the $U_i = u_i + \epsilon_i$ is the observed radial velocity and for simplicity we assume a diagonal error covariance matrix. Given an assumed power spectrum the auto- and cross-correlation functions are readily calculated. Note that within the linear theory the velocity and density fields are related by a simple convolution, and therefore the problem of density reconstruction from observed velocity is equivalent to a deconvolution in the presence of noise (with possible predictions to unobserved regions). The inverse problem of estimating the velocity field from the measured galaxy distribution is very similar to the above velocity reconstruction. The velocity–density relation is analogous to (27), i.e.

$$\mathbf{v}^{\mathrm{WF}}(\mathbf{r}) = \left\langle \mathbf{v}(\mathbf{r})\delta(\mathbf{r}_i)\right\rangle \left\langle \delta(\mathbf{r}_i)\delta(\mathbf{r}_j) + \epsilon_i\epsilon_j \right\rangle^{-1} \Delta(\mathbf{r}_j) \,, \tag{28}$$

where $\Delta(\mathbf{r}_i) = \delta(\mathbf{r}_i) + \epsilon_i$ is the observed overdensity at $\mathbf{r}_i$.

A major problem in analysing redshift surveys is the transformation of the galaxies from redshift to real space. Within the limitation of the linear theory a WF can be used to remove noise, transform from redshift to real space, deconvolve smoothing and extrapolate across unobserved regions (cf. Fisher et al. [10]). In general one can write

$$\delta^{\mathrm{WF}}(\mathbf{r}) = \left\langle \delta(\mathbf{r})\delta_S(\mathbf{s}_i) + \epsilon_i\epsilon_j \right\rangle \left\langle \delta_S(\mathbf{s}_i)\delta_S(\mathbf{s}_j)\right\rangle^{-1} \Delta_S(\mathbf{s}_j) \,, \tag{29}$$

and here $\mathbf{s}$ is the position vector in redshift space, subscript S denotes quantities evaluated in redshift space and Δ_S is the observationally determined density (contrast) as evaluated in redshift space. The auto- and cross-correlation matrices which relate the redshift space and real space densities are readily calculated in the linear theory (cf. Zaroubi and Hoffman [42]). In the linear theory the redshift space density depends linearly on the actual (real space) densities, and therefore one can write a linear transformation to relate the two and the WF approach amounts to a matrix inversion in the presence of noise. Indeed, the redshift to real space transformation in the spherical harmonics/Bessel functions representation is given in terms of a regularized inversion of a distortion matrix (Fisher et al. [10]). The problem of the WF reconstruction from redshift to real space gets simplified considerably when applied to a survey that extends a small angular scale on the observer's sky. The 'distant observer' approximation of Kaiser [12] is applicable in such a case and the problem at hand simplifies considerably (Erdoğdu et al. [9]).

4 Cosmography

The WF/CR methodology has been used extensively to reconstruct the local cosmography and the CMB sky map from a variety of redshift and peculiar surveys and from observations of the CMB temperature anisotropies. No attempt is made here to describe all the work done in this field and only a brief and incomplete review is presented.

Lahav et al. [19] and Fisher et al. [10] analysed the IRAS redshift survey by means of the WF. The Fisher et al. analysis was based on the spherical harmonic/Bessel function expansion in which the redshift distortions are naturally accounted for. Following that approach Schmoldt et al. [28] analysed the PSCz redshift survey. The study of deeper redshift surveys which are not full sky, such as the 2dF Galaxy Redshift Survey, requires different approach. The spherical harmonic/Bessel functional basis is not adequate for surveys with partial sky coverage. Erdoğdu et al. [9] used the Cartesian presentation in their analysis of the 2dF survey. The limited angular extent of the survey makes the 'distant observer' approximation (Kaiser [12]) viable and the treatment of redshift distortions simplifies considerably. Indeed, Erdoğdu et al. [9] used the galaxy distribution in redshift space to recover the density and velocity fields in real space.

Large-scale surveys of peculiar velocities have played a major role in the study of the local cosmography. Peculiar velocities are directly related to the density field in the linear regime, and hence provide an attractive venue for the reconstruction of the total mass density field. However, the determination of the peculiar velocities is plagued by a variety of statistical and systematic uncertainties. Peculiar velocity surveys constitute a very noisy and sparse data base in which the individual data points are highly correlated. Wiener filtering is the ideal tool for analysing such surveys. Indeed, Zaroubi et al. [43] analysed the MARK III catalogue of peculiar velocity by means of the WF and CRs and reconstructed the density and velocity fields out to $\approx 80h^{-1}$Mpc. This analysis was extended to the analysis of the ENEAR survey (Zaroubi et al. [40]) and the combined SEcat catalogue (made of the ENEAR and SFI surveys) (Zaroubi et al. [41]). The latter study compared the SEcat-recovered density and velocity fields with the ones derived from the PSCz redshift survey and found a good agreement.

The WF/CR methodology plays a major role in the study of the CMB temperature anisotropies. The WF is the optimal tool for making maps of the CMB sky and the CRs provide an efficient way of making Monte Carlo realizations and mock surveys. Bunn et al. [4] were the first to make a WF map of the CMB sky out of the COBE first-year data. The data itself was noisy and direct estimation of the sky map was very unreliable. The full data of COBE later on confirmed and was found to be consistent with the WF reconstructed map, based on the first-year data. This analysis was later extended by Bunn et al. [5]. Tegmark [33] showed that the WF is an optimal tool for making two-dimensional CMB maps from the raw time-ordered data. Tegmark et al. [34] used the WF to make a high-resolution map of the CMB around the North Celestial Pole from the Saskatoon experiment. Stompor et al. [32] used CRs to create mock data bases and realistic realizations of the errors in their analysis of the MAXIMA CMB experiment. Tegmark et al. [35] and Patanchon et al. [24] used the WF to make CMB map from the WMAP data.

5 Constrained Simulations

5.1 Motivation

Over the last three decades cosmological simulations have proved to be an invaluable tool in testing theoretical models in the highly non-linear regime. However, comparisons of simulations with observational data are typically done only in a statistical sense. The standard approach is to assume a cosmological model and to use the appropriate power spectrum of the primordial perturbations to construct a random realization of the density field within a given simulation volume. The evolution of the initial density field is then followed using a numerical code and results are compared with observations. A wide variety of statistical measures have been used for comparisons, including the two-point correlation function, the power spectrum, the mass function, different shape statistics.

The statistical approach is well suited to cases in which there is a statistically well-represented sample of objects, selected in a controlled way both from the observed universe and in simulations. Such cases include the various two-point statistics of galaxies, such as the two-point correlation function and the pair-wise velocity dispersion. Assuming that the surveys at question and the simulations cover many independent uncorrelated volumes, the statistical analysis is undoubtedly the optimal way to compare simulations and observations and impose constraints on theoretical methods. The statistical approach fails when dealing with unique individual objects and in particular in the modelling of the very local structure, such as the Local Group (LG) or the Local Supercluster (LSC). The traditional approach has been to choose 'observers' or regions in the simulations that resemble the desired configuration as closely as possible. The results of numerical simulations aimed at identifying structures similar to those observed locally in random realizations were generally inconclusive because one is never sure that the selection of objects in simulations is right.

An alternative approach is to design the initial conditions of the simulations so as to reproduce the desired objects or environments by construction. This can be easily done by means of CRs of Gaussian fields which are used to set the initial conditions. The basic idea is to impose either observational data or 'synthetic' data that capture the essential features of the desired objects and impose these as constraints on the Gaussian random field that constitutes the initial conditions. The resulting simulations recover the assumed constraints and are consistent with the power spectrum of the initial conditions and the assumed cosmological model. The particular choice followed here consists of using surveys of the nearby LSS as the constraining data. This alternative approach is dubbed here as a constrained simulation (CS).

5.2 Choice of Constraints

The algorithm of CR of Gaussian fields enables one to impose a large number of constraints on the initial conditions of the simulations. Thus, in principle one can use more than a few thousands of observational data points and use them to constrain the initial conditions. However, this can be done subject to a few conditions. The constraining data should correspond to linearized variables, namely dynamical variables that are linear in its deviations from an unperturbed Friedman model. Also, the statistical nature of the errors in the observed data should be well understood and be easily modelled. Given these, one can model the observed data within the linear theory and impose it as linear constraints on the random realizations of the primordial perturbation field.

The two natural sources of observational data to be used as constraints are redshift surveys of the galaxy distribution and surveys of peculiar velocities. Indeed the IRAS and its extension the PSCz surveys were used by Kolatt et al.(1996) [17], Bistolas & Hoffman [6] and Mathis et al. [21]. The MARK III peculiar velocities were used as constrains by van de Weygaert & Hoffman [37], [38], Kravtsov et al. [18], Klypin et al. [15], Romano-Diaz [26] and Dover et al. [8].

Full sky redshift surveys are the most natural choice of data for CSs of the local universe. The main advantage of using redshift surveys are the well-understood selection of galaxies and the sampling errors. The main disadvantage of redshift surveys lies with the very indirect relation between the observed non-linear galaxy density field and the underlying primordial matter density field. Two physical processes make this relation non-trivial: the poorly known biasing relation between the two fields and the non-linear dynamical evolution. Non-linearities change the statistical distribution of the evolved density field and break the one-to-one relation between the linear and non-linear fields. This implies that even in the case of an ideal redshift survey, deep with full sky coverage and with negligible errors, the reconstruction of the primordial perturbation field is severely limited by these two effects. Peculiar velocity surveys provide almost an orthogonal alternative to redshift surveys. The measurement of peculiar velocities of galaxies poses a formidable challenge to observational astronomy. The difficulty of that measurement lies with the complexities of distance measurement and it suffers from a whole range of statistical and in particular systematic uncertainties (cf. Dekel [7]). Typical statistical errors in the determination of the peculiar velocity of a galaxy amount to (20%–25%) of the Hubble velocity of the galaxy, which amounts to an error larger than the peculiar velocity for a galaxy at a distance of a few tens of Mpc away. There are two main advantages for using velocities as constraints. One is the simple relation of the peculiar velocities to the linear density perturbation. The other is the 'linear' property of the peculiar velocity, namely velocities are more 'linear' than the densities at a given dynamical stage of evolution. Closely related to this is the peculiar velocities PDF which remains Gaussian at the time when the densities become non-linear and obey a log-normal distribution (Kofman et al. [16]). Another useful feature of using velocities as constraints is the much larger coherence scale of the velocities compared with the densities, namely the velocity–velocity correlation length is much larger than the density–density length. This long-range coherence partially compensates for the sparseness of the surveys and the large errors in the determination of the peculiar velocity of individual galaxies.

A different kind of constraints is provided by clusters of galaxies. Consider a cluster of known virial parameters, such as radius, velocity dispersion and mass. Assuming that the cluster dynamics can be approximated by the spherical top-hat model, one can translate the present-day parameters to the linear density contrast on the scale of the cluster. The main shortcoming of cluster constraints is the use of the rather simplified top-hat model to relate the initial and final structure of a given cluster. The CSs of Klypin et al. [16] and Dover et al. [15] are based on the MARK III peculiar velocities and the nearby X-ray selected clusters of Reiprich & Böhringer [25]).

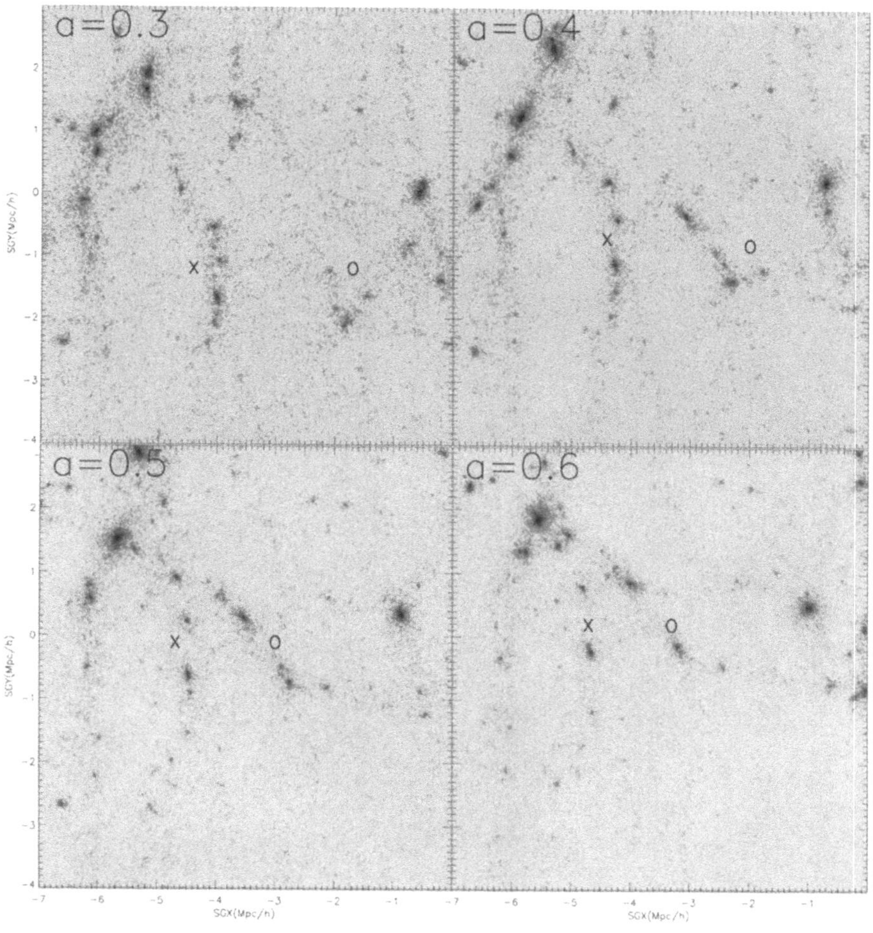

Fig. 1. Constrained simulation of the formation of the Local Group: The cosmological evolution of the density field in a thin slice centred on the supergalactic plane. The different panels present the density field at different values of the expansion parameter a ranging from 0.3 to 0.6. The 'X' and 'O' mark the Milky Way and Andromeda candidates of the simulated Local Group. The location of the simulated objects deviates by a few Mpc from their actual location

5.3 Local Supercluster

CSs have been used to study the LSC, focusing on its general structure and dynamics (Klypin et al., [15]), the structure and observational signature of the LSC warm–hot intergalactic medium (Kravtsov et al. [18]) and the coldness of the local flow (Dover et al. [8]). Future work will focus on conducting N-body and hydrodynamical simulations aiming at studying galaxy formation at the local neighbourhood, and in particular the formation of the LG.

The Dover et al. [8] paper reports the first step in a project of simulating the LG by means of CSs. A very brief introduction to these simulations is presented here. The constraints consist of peculiar velocities and X-ray selected clusters. The large-scale features of the simulations faithfully

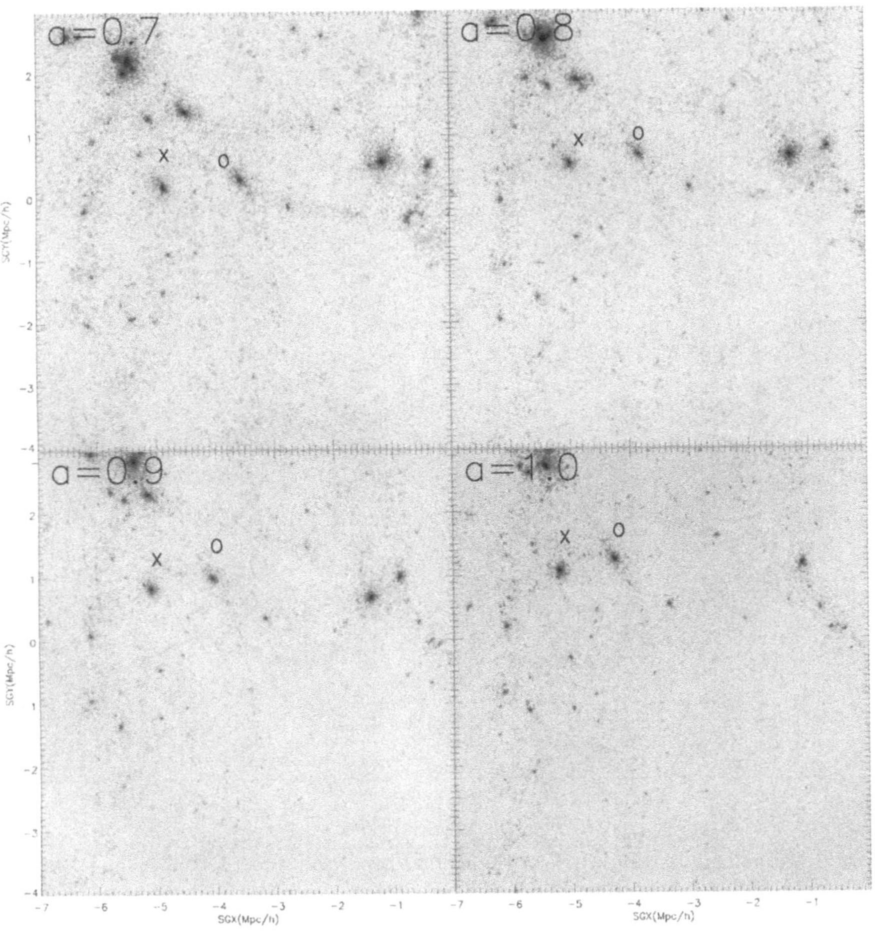

Fig. 2. Same as Fig. 1: panels correspond to expansion parameter ranging from 0.7 to 1.0

reconstruct the LSC and in particular the Virgo cluster. The filamentary structure that connects to the Virgo is reproduced as well, in particular the one that contains the LG. The velocity constraints do not directly constrain the LG system of its two massive galactic-scale halos. Yet the resulting simulated structure makes the emergence of LG-like objects to be very probable. An interesting possibility that emerges is that the Milky Way and M31 formed in two separate filaments that are converging into one. This is shown in the sequence of the time evolution of the density field in the LG neighbourhood (Figs. 1 and 2). The interesting feature that follows is that the LG does not evolve as a quasi-two-body system, but rather the Milky Way and M31 are evolving within their own filaments and their observed infall is driven by the convergence of the filaments. The nature of this process and its observational consequence are to be the subject of a very intensive study.

6 Summary and Future Prospects

The general methodology of the WF and CRs has proved to be very powerful when applied to random Gaussian fields, hence also to the analysis of the LSS. The WF is the optimal tool for the reconstruction of the CMB temperature anisotropies map and the LSS in its linear regime from noisy, sparse and incomplete data basis. The WF is optimal in the Gaussian case in the sense that it coincides with many different estimators such as the minimal variance solution, the Bayesian most probable, the conditional mean field and the maximum entropy estimators. CRs provide a self-consistent way for performing Monte Carlo simulations of the residual field from the WF estimated field. CRs can be used to set mock catalogues of LSS and CMB surveys. Also, CRs properly map the residual of the actual field from the WF estimator and thus can be used to estimate confidence levels around the WF reconstructed field. It follows that WF and CRs should be the method of choice in the analysis of future surveys of the linear regime of the LSS and the CMB sky.

A very different application of the CRs is that of numerical cosmological simulations constrained by observations, namely CSs. Realizations of the primordial perturbation fields constrained by redshift and/or peculiar velocity surveys are used to set up the initial conditions of cosmological simulations. The resulting simulations reproduce, by construction, the main features of the nearby LSS, such as the LSC and local clusters and voids. Smaller scales are not directly constrained by the data and therefore the small-scale structure is simulated within the framework of the observed large-scale structure of the 'local' universe. This enables a detailed study of the formation of nearby objects, the LG in particular, within the framework of a given cosmological model, with most of the cosmic scatter being removed by the imposed constraints. CSs complement the standard cosmological simulations and provide a valuable tool for studying and simulating the nearby universe.

Acknowledgements

This research has been supported by ISF-143/02 and the Sheinborn Foundation at the Hebrew University. The author acknowledges many useful discussions with his collaborators on the constrained simulation of the local universe project, S. Gottlöber, A. Klypin and G. Yepes, and for their generous permission to present some yet unpublished results. E. Romano-Diaz is gratefully acknowledged for his careful reading of the manuscript.

References

1. Bardeen, J.M., Bond, J.R., Kaiser, N., Szalay, A.S.: ApJ. **304**, 28 (1986)
2. Bertschinger, E.: ApJ. Lett. **323**, L103 (1987)
3. Binney, J. Quinn, T.: MNRAS **249**, 678 (1991)
4. Bunn, E.F., Fisher, K.B., Hoffman, Y., Lahav, O. Silk, J., Zaroubi, S.: ApJ. Lett. **432**, L75 (1994)
5. Bunn, E., Hoffman, Y., Silk, J.: ApJ. Lett. **464**, 1 (1996)
6. Bistolas, V., Hoffman, Y. ApJ **492**, 439 (1998)
7. Dekel A.: ARA&A **32**, 371 (1994)
8. Dover, Y., Romano-Diaz, E., Sivan, M., Hoffman, Y., Yepes, G., Klypin, A., Gottlöber, S. (preprint) (2006)
9. Erdǒgdu P. et al.: MNRAS **352**, 939 (2004)
10. Fisher, K.B., Lahav, O., Hoffman, Y., Lynden-Bell, D., Zaroubi, S.: MNRAS **272**, 885 (1994)
11. Hoffman, Y., Ribak, E.: ApJ. Lett. **380**, L5 (1991)
12. Kaiser, N.: MNRAS **227**, 1 (1987)
13. Kaiser, N.: MNRAS **231**, 149 (1988)
14. Kaiser, N., Stebbins, A.: in Large Scale Structure and Peculiar Motions in the Universe, eds. D.W. Latham, L.N. DaCosta (ASP Conference Series), p. 111 (1991)
15. Klypin, A., Hoffman, Y., Kravtsov, A.V., Gottlöber, S.: ApJ. **596**, 19 (2003)
16. Kofman, L., Bertschinger, E., Gelb, J.M., Nusser, A., Dekel, A.: ApJ. **420**, 44 (1994)
17. Kolatt, T., Dekel, A., Ganon, G., Willick, J. A.: ApJ., **457**, 460 (1996)
18. Kravtsov, A.V., Klypin, A., Hoffman, Y.: ApJ **571**, 563 (2002)
19. Lahav, O., Fisher, K.B., Hoffman, Y., Scharf, C.A., Zaroubi, S.: ApJ. Lett. **423**, L93 (1994)
20. Landau, L.D., Lifshitz. E.M. Statistical Physics, Part 1, 3rd edn, Pergamon Press, UK (1982)
21. Mathis, H., Lemson, G., Springel, V., Kauffmann, G., White, S.D.M., Eldar, A., Dekel, A.: MNRAS **333**, 739 (2002)
22. Martínez, V.J. Saar, E.: Statistics of the Galaxy Distribution, Chapman & Hall/CRC, Boca Raton (2002)
23. Padmanabhan, T.: Structure Formation in the Universe, Cambridge University Press, Cambridge, UK (1993)
24. Patanchon, G., Cardoso, J.-F., Delabrouille, J., Vielva, P.: MNRAS **364**, 1185 (2005)

25. Reiprich, T.H., Boehringer, H.: ApJ. **567**, 716 (2002)
26. Romano-Diaz, E.: Probing Cosmic Velocity Flows in the Local Universe, PhD Thesis, University of Groningen, Groningen (2004)
27. Rybicki, G.B., Press, W.H.: ApJ. **398**, 169 (1992)
28. Schmoldt et al.: Astron. J. **118**, 1146 (1999)
29. Sheth, R.K.: MNRAS **277**, 933 (1995)
30. Skilling, J.: Proceedings of the 8th Workshop on Maximum Entropy and Bayesian Methods, ed. Skilling, J., Kluwer, Dordrecht (1989)
31. Stebbins, A.: Proc. of the 9th IAP Conference on Cosmic Velocity Fields, eds. F. Bouchet and M. Lachiéze-Rey, (Gif-sur-Yvette Cedex: Editions Frontiéres) p. 253, (1994)
32. Stompor, R., et al.: Phys. Rev. D. **65**, 022003 (2002)
33. Tegmark, M.: ApJ. Lett. **480**, L87 (1997)
34. Tegmark, M., de Oliveira-Costa, A., Devlin, M.J., Netterfields, C.B., Page, L., Wollack, E.J.: ApJ. Lett. **474**, L77 (1997)
35. Tegmark, M., de Oliveira-Costa, A., Hamilton, A.J.: Phys. Rev. D. **k68**, 123523 (2003)
36. van de Weygaert, R., Bertschinger, E.: MNRAS **281**, 84 (1996)
37. van deWeygaert, R., Hoffman, Y.: Evolution of Large Scale Structure: From Recombination to Garching, Banday, A.J., Sheth, R.K., Costa, L.N. (eds), (1999)
38. van de Weygaert, R., Hoffman, Y. in ASP Conf. Ser. 201, Cosmic Flows 1999: Towards an Understanding of Large-Scale Structures, Courteau, S., Strauss, M.A., Willick, J.A. (eds) (2000)
39. Wiener, N.: Extrapolation and Smoothing of Stationary Time Series. Wiley, New York (1949)
40. Zaroubi, S., Bernardi, M., da Costa, L. N., Hoffman, Y., Alonso, M. V.,Wegner, G., Willmer, C. N. A., Pellegrini, P.S.: MNRAS **326**. 375 (2001)
41. Zaroubi, S., Branchini, E., Hoffman, Y., da Costa, L. N.: MNRAS **336**, 1234 (2002)
42. Zaroubi, S., Hoffman, Y.: ApJ. **462**, 25 (1996)
43. Zaroubi, S., Hoffman, Y., Dekel, A.: ApJ. **520**, 413 (1999)
44. Zaroubi, S., Hoffman, Y., Fisher, K.B., Lahav, O.: ApJ. **449**, 446 (1995)

Weak Gravitational Lensing

A. Heavens

Institute for Astronomy, University of Edinburgh, Blackford Hill,
Edinburgh EH9 3HJ
afh@roe.ac.uk

1 Introduction

Gravitational lensing is strong if the distortions to the images are substantial, and weak if the distortions are small. Weak lensing of background images can be used to learn about the mass distribution in individual objects, such as galaxy clusters, but I will concentrate on weak lensing on a cosmological scale, which is usually analysed in a statistical way. Recent reviews of this area include [4, 26, 32].

The basic effect of weak lensing is for the clumpy matter distribution to perturb slightly the trajectories of photon paths. By considering how nearby light paths are perturbed, one finds that the shapes of distant objects are changed slightly. Associated with the size change is a change in the brightness of the source. The size and magnitude changes can, in principle, be used to constrain the properties of the matter distribution along the line of sight (and cosmological parameters as well), but it is the change in shape of background sources which has almost exclusively been used in cosmological weak lensing studies. The reason for this is simply that the signal to noise is better. These notes will concentrate on shear (=shape changes), but the magnification and amplification of sources can also be used and will probably be used in future when the surveys are larger. The great promise of lensing is that it acts as a direct probe of the matter distribution (whether dark or not) and avoids the use of objects which are assumed to trace the mass distribution in some way, such as galaxies in large-scale structure studies. Theoretically, lensing is very appealing, as the physics is very simple, and very robust, direct connections can be made between weak lensing observables and the statistical properties of the matter distribution. These statistical properties are dependent on cosmological parameters in a known way, so weak lensing can be employed as a cosmological tool. The main uncertainties in lensing are observational – it is very challenging to make images of the necessary quality. Lensing can, of course, be used to investigate the mass distribution of a discrete lens lying along the line of sight, but I will concentrate here on the weak effects on

Heavens, A.: *Weak Gravitational Lensing*. Lect. Notes Phys. **665**, 585–600 (2009)
DOI 10.1007/978-3-540-44767-2_18

a cosmological scale of the non-uniform distribution of matter all the way between the source and observer, an effect often referred to as *cosmic shear.*

2 Distortion of Light Bundles

The distortion of a light bundle has to be treated with general relativity, but if one is prepared to accept one modification to Newtonian physics, one can do without GR.

In an expanding universe, it is usual to define a *comoving coordinate* $\mathbf{x}$, such that 'fundamental observers' retain the same coordinate. Fundamental observers are characterized by the property of seeing the universe as isotropic; the Earth is not (quite) a fundamental observer, as from here the cosmic microwave background looks slightly anisotropic. The equation of motion for the transverse coordinates (about some fiducial direction) of a photon in a flat universe is

$$\frac{d^2 x_i}{d\eta^2} = -\frac{2}{c^2}\frac{\partial \Phi}{\partial x_i}, \qquad i = 1,2, \tag{1}$$

which can be derived using general relativity (see Appendix for details). We will use $x_i; i = 1, 2$, for coordinates transverse to the line of sight, and r to indicate the radial coordinate. η is the conformal time related to the coordinate t by $d\eta = cdt/R(t)$ and $R(t)$ is the cosmic scale factor equal to R_0 at the present time. $\Phi(x_i, r)$ is the peculiar gravitational potential related to the matter overdensity field $\delta \equiv \delta\rho/\rho$ by Poisson's equation

$$\nabla^2 \Phi = \frac{3H_0^2 \Omega_m}{2a(t)} \delta, \tag{2}$$

where H_0 is the Hubble constant, Ω_m is the present matter density parameter and $a(t) = R(t)/R_0 = (1+z)^{-1}$, where z is redshift.

The equation of motion is derived in the Appendix from general relativity, in a (nearly flat) metric given by $ds^2 = (1 + 2\Phi/c^2)c^2 dt^2 - (1 - 2\Phi/c^2)R^2(t)(dr^2 + r^2 d\theta^2 + r^2 \sin^2\theta d\varphi^2)$. From a Newtonian point of view, equation (1) is understandable if we note that time is replaced by η (which arises because we are using comoving coordinates), and there is a factor 2 which does not appear in Newtonian physics. This same factor of two gives rise to the famous result that in GR the angle of light bending round the Sun is double that of Newtonian theory.

The coordinates $\mathbf{x}_i$ are related to the (small) angles of the photon to the fiducial direction $\boldsymbol{\Theta} = (\theta_x, \theta_y)$ by $\mathbf{x}_i = r\boldsymbol{\Theta}_i$.

3 Lensing Potential

The solution to (1) is obtained by first noting that the zero-order ray has $ds^2 = 0 \Rightarrow dr = -d\eta$, where we take the negative root because the light ray is incoming. Integrating twice and reversing the order of integration gives

$$x_i = r\theta_i - \frac{2}{c^2}\int_0^r dr' \frac{\partial \Phi}{\partial x'_i}(r - r') \,. \tag{3}$$

We now perform a Taylor expansion of $\partial\Phi/\partial x'_i$ and find that the deviation of two nearby light rays is

$$\Delta x_i = r\Delta\theta_i - \frac{2}{c^2}\Delta\theta_j \int_0^r dr' r'(r - r')\frac{\partial^2 \Phi}{\partial x'_i \partial x'_j} \,, \tag{4}$$

which we may write as

$$\Delta x_i = r\Delta\theta_j(\delta_{ij} - \phi_{ij}) \,, \tag{5}$$

where δ_{ij} is the Kronecker delta $(i = 1, 2)$ and we define

$$\phi_{ij}(\mathbf{r}) \equiv \frac{2}{c^2}\int_0^r dr' \frac{(r - r')}{rr'}\frac{\partial^2 \Phi(\mathbf{r}')}{\partial \theta_i \partial \theta_j} \,. \tag{6}$$

The integral is understood to be along a radial line (i.e. $\mathbf{r} \parallel \mathbf{r}'$); this is the *Born approximation*, which is a very good approximation for weak lensing [5, 25, 31]. In reality the light path is not quite radial.

It is convenient to introduce the *lensing potential*, which controls the distortion of the ray bundle:

$$\phi(\mathbf{r}) \equiv \frac{2}{c^2}\int_0^r dr' \frac{(r - r')}{rr'}\Phi(\mathbf{r}') \,. \tag{7}$$

Note that $\phi_{ij}(\mathbf{r}) = \partial^2\phi(\mathbf{r})/\partial\theta_i\partial\theta_j$.

3.1 Averaging over a Distribution of Sources

If we consider the distortion averaged over a distribution of sources with a radial distribution $p(r)$ (normalized such that $\int dr\, p(r) = 1$), the average distortion is again obtained by reversing the order of integration:

$$\Delta x_i = r\Delta\theta_j \left(\delta_{ij} - \frac{2}{c^2}\int_0^r \frac{dr'}{r'} g(r') \frac{\partial^2 \Phi(\mathbf{r}')}{\partial \theta_i \partial \theta_j}\right) \,, \tag{8}$$

where

$$g(r) \equiv \int_r^\infty dr'\, p(r') \frac{r' - r}{r'} \,. \tag{9}$$

In order to estimate $p(r)$, surveys began to estimate distances to source galaxies using photometric redshifts. This has opened up the prospect of a full 3D analysis of the shear field, which we will discuss briefly later in this article.

3.2 Convergence and Shear

If we define the *amplification matrix* $A_{ij} \equiv \delta_{ij} - \phi_{ij}$, then we see that A is symmetric, and it can be decomposed into an isotropic expansion term, and a shear. A general amplification matrix also includes a rotation term (the final degree of freedom being the rotation angle), but we see that weak lensing does not introduce rotation of the image and has only three degrees of freedom, rather than the four possible in a 2×2 matrix. We decompose the amplification matrix as follows:

$$A_{ij} = \begin{pmatrix} 1-\kappa & 0 \\ 0 & 1-\kappa \end{pmatrix} + \begin{pmatrix} -\gamma_1 & -\gamma_2 \\ -\gamma_2 & \gamma_1 \end{pmatrix} , \tag{10}$$

where κ is called the *convergence* and $\gamma = \gamma_1 + i\gamma_2$ is the *complex shear*. For weak lensing, both $|\kappa|$ and $|\gamma_i|$ are $\ll 1$. A non-zero κ represents an isotropic expansion or contraction of a source; $\gamma_1 > 0$ represents an elongation of the image along the x-axis and contraction along y. $\gamma_1 < 0$ stretches along y and contracts along x. $\gamma_2 \neq 0$ represents stretching along $x = \pm y$ directions.

Making the decomposition, we find that

$$\begin{aligned} \kappa &= \frac{1}{2}\left(\phi_{11} + \phi_{22}\right) , \\ \gamma_1 &= \frac{1}{2}\left(\phi_{11} - \phi_{22}\right) , \\ \gamma_2 &= \phi_{12} . \end{aligned} \tag{11}$$

Note that $\kappa > 0$ corresponds to *magnification* of the image. Lensing preserves surface brightness, so this also amounts to *amplification* of the source flux.

It is worth noting that the amplification matrix may be written as

$$A_{ij} = (1-\kappa)\begin{pmatrix} 1-g_1 & -g_2 \\ -g_2 & 1+g_1 \end{pmatrix} , \tag{12}$$

where $g \equiv g/(1-\kappa)$ is called the *reduced shear*. Since the $1-\kappa$ multiplier affects only the overall size (and hence brightness, as lensing preserves surface brightness) of the source, but not its shape, we see that shear measurements can determine only the reduced shear, and not the shear itself. For weak lensing, $\kappa \ll 1$, so the two coincide to linear order.

3.3 Relationship to Matter Density Field

The gravitational potential Φ is related to the matter overdensity field $\delta \equiv \delta\rho/\rho$ by Poisson's equation (2). The convergence is then

$$\kappa(\mathbf{r}) = \frac{3H_0^2\Omega_m}{2c^2}\int_0^r dr' \frac{r'(r-r')}{r}\frac{\delta(\mathbf{r}')}{a(r')} . \tag{13}$$

Note that there is an extra term $\partial^2\Phi/\partial r'^2$ in ∇^2 which integrates to zero to the order to which we are working.

3.4 Convergence Power Spectrum and Shear Correlation Function

The average shear is zero, so the most common statistics to use for cosmology are two-point statistics, quadratic in the shear. These may be in 'configuration' ('real') space, or in transform space (using Fourier coefficients or similar). I will focus on two quadratic measures, the convergence power spectrum and the shear–shear correlation function.

To find the expectation value of a quadratic quantity, it is convenient to make use of the matter density power spectrum, $P(k)$, defined by the following relation between the overdensity Fourier coefficients:

$$\langle \delta_{\mathbf{k}} \delta^*_{\mathbf{k}'} \rangle = (2\pi)^3 \delta^D(\mathbf{k} - \mathbf{k}') P(k) \,, \tag{14}$$

where δ^D is the Dirac delta function. $P(k)$ is evolving, so we write it as $P(k;r)$ in future, where r and t are related through the lookback time. (This r-dependence may look strange; there is a subtlety: (14) holds if the field is homogeneous and isotropic, which the field on the past light cone is not, since it evolves. In the radial integrals, one has to consider the homogeneous field at the same cosmic time as the time of emission of the source.) The trick is to get the desired quadratic quantity into a form which includes $P(k;r)$.

For the convergence power spectrum, we first transform the convergence in a 2D Fourier transform on the sky, where ℓ is a 2D dimensionless wavenumber:

$$\kappa_\ell = \int d^2\boldsymbol{\Theta}\, \kappa(\boldsymbol{\Theta}) e^{-i\boldsymbol{\ell}.\boldsymbol{\Theta}} \tag{15}$$

$$= A \int_0^\infty dr\, r\, \frac{g(r)}{a(r)} \int d^2\boldsymbol{\Theta}\, \delta(r\boldsymbol{\Theta}, r) e^{-i\boldsymbol{\ell}.\boldsymbol{\Theta}} \,, \tag{16}$$

where $A \equiv 3H_0^2\Omega_m/2c^2$. We expand the overdensity field in a Fourier transform,

$$\delta(r\boldsymbol{\Theta}, r) = \int \frac{d^3\mathbf{k}}{(2\pi)^3}\, \delta_{\mathbf{k}} e^{ik_\parallel r} e^{i\mathbf{k}_\perp . r\boldsymbol{\Theta}} \,, \tag{17}$$

and substitute into (16). We form the quantity $\langle \kappa_\ell \kappa^*_{\ell'} \rangle$, which, by analogy with (14), is related to the (2D) *convergence power spectrum* by

$$\langle \kappa_\ell \kappa^*_{\ell'} \rangle = (2\pi)^2 \delta^D(\boldsymbol{\ell} - \boldsymbol{\ell}') P_\kappa(|\ell|) \,. \tag{18}$$

Straightforwardly,

$$\langle \kappa_\ell \kappa^*_{\ell'} \rangle = A^2 \int_0^\infty dr\, G(r) \int_0^\infty dr'\, G(r') \int d^2\boldsymbol{\Theta} d^2\boldsymbol{\Theta}' \frac{d^3\mathbf{k}}{(2\pi)^3} \frac{d^3\mathbf{k}'}{(2\pi)^3} \tag{19}$$
$$\langle \delta_{\mathbf{k}} \delta^*_{\mathbf{k}'} \rangle \exp(ik_\parallel r - ik_\parallel{}' r') \exp(-i\boldsymbol{\ell}.\boldsymbol{\Theta} + i\boldsymbol{\ell}'.\boldsymbol{\Theta}') \,,$$

where $G(r) \equiv rg(r)/a(r)$. Using (14) we remove the $\mathbf{k}'$ integration, introducing the power spectrum $P(k) = P\left(\sqrt{k_\parallel^2 + |\mathbf{k}_\perp|^2}\right)$. For small-angle surveys, most of the signal comes from short wavelengths, and the $k_\parallel$ is negligible, so $P(k) \simeq P(|\mathbf{k}_\perp|)$. The only $k_\parallel$ term remaining is the exponential, which integrates to $(2\pi)\delta^D(r - r')$. The integrals over $\boldsymbol{\Theta}$ and $\boldsymbol{\Theta}'$ give $(2\pi)^2\delta^D(\boldsymbol{\ell} - r\mathbf{k}_\perp)$ and $(2\pi)^2\delta^D(\boldsymbol{\ell}' - r\mathbf{k}_\perp{}')$, respectively, so the whole lot simplifies to give the convergence power spectrum as

$$P_\kappa(\ell) = \left(\frac{3H_0^2\Omega_m}{2c^2}\right)^2 \int_0^\infty dr \left[\frac{g(r)}{a(r)}\right]^2 P(\ell/r; r) \,. \tag{20}$$

An exercise for the reader is to show that the power spectrum for γ is the same: $P_\gamma = P_\kappa$. The shear correlation function for points separated by an angle θ is

$$\begin{aligned} \langle\gamma\gamma^*\rangle_\theta &= \int \frac{d^2\ell}{(2\pi)^2} P_\gamma(\ell) \exp\left({}^{(1)}\, \boldsymbol{\ell}.\boldsymbol{\Theta}\right) \\ &= \int \frac{\ell d\ell}{(2\pi)^2} P_\kappa(\ell) \exp({}^{(1)}\, \ell\theta \cos\varphi) d\varphi \\ &= \int \frac{d\ell}{2\pi} \ell P_\kappa(\ell) J_0(\ell\theta) \,, \end{aligned} \tag{21}$$

where we have used polar coordinates, with φ the angle between $\boldsymbol{\ell}$ and $\boldsymbol{\Theta}$, and we have exploited the isotropy (P_κ depends only on the modulus of $\boldsymbol{\ell}$). J_0 is a Bessel function.

Other quadratic quantities (examples are shear variances on different scales, aperture mass (squared)) can be written similarly as integrals over the power spectrum, with different kernel functions.

4 Matter Power Spectrum

As we have seen, the two-point statistics of the shear and convergence fields depend on the power spectrum of the matter, $P(k; t)$. The power spectrum grows in a simple way when the perturbations in the overdensity are small, $|\delta| \ll 1$, when the power spectrum grows in amplitude whilst keeping the same shape as a function of k. However, gravitational lensing can still be weak, even if the overdensity field is nonlinear. Poisson's equation still holds provided we are in the weak-field limit as far as general relativity is concerned, and this essentially always holds for cases of practical interest. In order to get as much statistical power out of lensing, one must probe the nonlinear regime, so it is necessary for parameter estimation to know how the power spectrum grows. Through the extensive use of numerical simulations, the growth of dark matter clustering is well understood down to quite small scales, where uncertainties in modelling, or uncertain physics, such as the influence of baryons on the dark matter [34], make the predictions unreliable. Accurate fits for the nonlinear power spectrum have been found [27] up to $k > 10\,\mathrm{h\,Mpc}^{-1}$, which is far

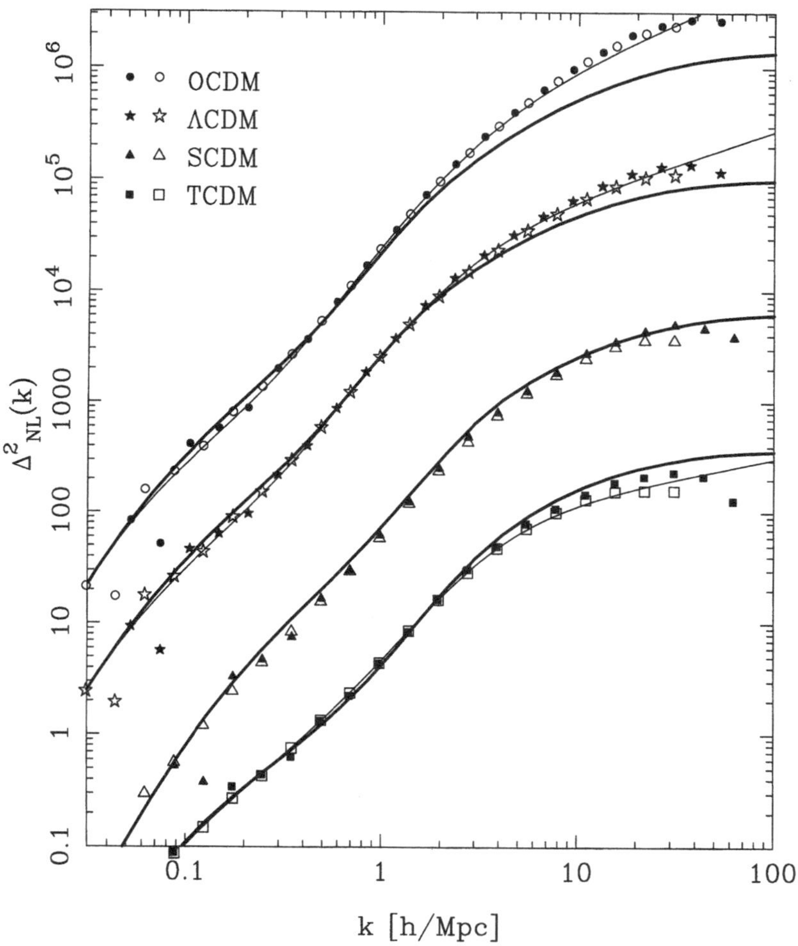

Fig. 1. The nonlinear power spectrum from numerical simulations, along with fitting functions (from Smith et al. [27])

beyond the linear/nonlinear transition $k \sim 0.2\,\mathrm{h\,Mpc}^{-1}$. Figure 1 shows fits for a number of CDM models. For precision use, one must make sure that the statistics do not have substantial contributions from the high-k end where the nonlinear power spectrum is uncertain. This can be explored by looking at the kernel functions implicit in the quantities such as the shear correlation function (22).

5 Estimating the Shear Field

We need to have some estimator for the shear field. The most obvious is to use the shape of a galaxy somehow. We need some measure of shape, and we require to know how it is altered by cosmic shear. I will describe the most common shape measurement used, but note that there are new techniques (shapelets, polar shapelets, etc. [6, 10]) which are coming into use.

The method I describe is commonly referred to as KSB (Kaiser, Squires & Broadhurst 1995 [22]). The shape measurement part of KSB uses moments of the surface brightness distribution of the source galaxy (here assumed to be centred on the origin):

$$Q_{ij} \equiv \int d^2\boldsymbol{\Theta}\theta_i\theta_j I(\boldsymbol{\Theta})W(\boldsymbol{\Theta}) , \tag{22}$$

where I is the surface brightness of the galaxy on the sky, and W is an optional weight function. The shape and orientation of the galaxy are encapsulated in the two *ellipticities*

$$e_1 \equiv \left(\frac{Q_{11} - Q_{22}}{Q_{11} + Q_{22}}\right) , \qquad e_2 = \frac{2Q_{12}}{Q_{11} + Q_{22}} , \tag{23}$$

and it is sometimes convenient to define a *complex ellipticity* $e \equiv e_1 + ie_2$.

It is straightforward to show that the source moments Q^s_{ij} are related to the observed moments by

$$Q^s_{ij} = A_{il}Q_{lm}A_{jm} . \tag{24}$$

A straightforward but tedious calculation relates the complex source and image ellipticities:

$$e = \frac{e + 2\gamma + g^2 e^{s*}}{1 + |g|^2 + 2Re(ge^{s*})} , \tag{25}$$

where $g \equiv \gamma/(1-\kappa)$. To estimate the shear correlation function $\langle\gamma\gamma^*\rangle$, one can use $\langle ee^*\rangle$, provided the terms $\langle e^s e^{s*}\rangle$ and $\langle e^s g^*\rangle$ are zero. The first term corresponds to intrinsic alignments of galaxies, which may be non-zero due to tidal effects. We will look at this later. Even the cross-term may be non-zero, but this term has received little attention to date [18].

The estimate of the shear will be noisy, as galaxies are not round and have an intrinsic ellipticity dispersion of around $\sigma_e \simeq 0.3$. With many (N) sources, the error on the average ellipticity can be reduced to $\sigma_e/\sqrt{N}$, and progress can be made.

5.1 Point Spread Function

The main practical difficulty of lensing experiments is that the atmosphere and telescope affect the shape of the images. These modifications to the shape may arise due to such things as the point spread function, or poor tracking of the telescope. The former needs to be treated with great care. Stars (whose images should be round) can be used to remove image distortions to very high accuracy, although a possibly fundamental limitation may arise because of the finite number of stars in an image. Interpolation of the anisotropy of the PSF needs to be done carefully, and examples of how this can be done in an optimal way are given in [33].

6 Intrinsic Alignments

The main signature of weak lensing is a small alignment of the images, at the level of a correlation of ellipticities of $\sim 10^{-4}$. One might be concerned that physical processes might also induce an alignment of the galaxies themselves. In the traditional lensing observations, the distances of individual galaxies are ignored, and one simply uses the alignment on the sky of galaxies, and one might hope that the galaxies will typically be at such large separations along the line of sight that any physical interactions would be rare and can be ignored. However, the lensing signal is very small, so the assumption that intrinsic alignment effects are sufficiently small needs to be tested. This was first done in a series of papers by a number of groups in 2000–2001, e.g. [9, 11, 12, 16], and the answer is that the effects may not be negligible. The contamination by intrinsic alignments is highly depth dependent. This is easy to see, since at fixed angular separation galaxies in a shallow survey will be physically closer together in space, and hence more likely to experience tidal interactions which might align the galaxies. In addition to this, the shallower the survey, the smaller the lensing signal. In a pioneering study, the alignments of nearby galaxies in the SuperCOSMOS survey were investigated [7]. This survey is so shallow (median redshift ~ 0.1) that the expected lensing signal is tiny. A non-zero alignment was found, which agrees with at least some of the theoretical estimates of the effect. The main exception is the numerical study of Jing [21], which predicts a contamination so high that it could dominate even deep surveys. For deep surveys, the effect is expected to be rather small, but if one wants to use weak lensing as a probe of subtle effects such as the effects of altering the equation of state of dark energy, then one has to do something. There are essentially two options – either one tries to calculate the intrinsic alignment signal and subtract it or one tries to remove it altogether. The former approach is not practical, as, although there is some agreement as to the general level of the contamination, the details are not accurately enough known. The latter approach is becoming possible, as lensing surveys are now obtaining estimates of the distance to each galaxy, via photometric redshifts (spectroscopic redshifts are difficult to obtain, because one needs a rather deep sample, with median redshift at least 0.6 or so, and large numbers, to reduce shot noise due to the random orientations of ellipticities). With photometric redshifts, one can remove physically close galaxies from the pair statistics (such as the shear correlation function) [14, 24]. Thus one removes a systematic error in favour of a slightly increased statistical error. The analysis in [13] is the only study which has explicitly removed close pairs.

6.1 E- and B-Modes

The shear field is an example of a spin-weight 2 field. Spin weight is defined by how an object changes if the coordinate system is rotated by an angle ψ. Spin-weight s objects change $f \to f\, e^{si\psi}$. Thus a scalar field (such as the lensing potential), which does not change at all, has spin weight 0. The shear field γ, on the other hand, changes according to $\gamma \to \gamma e^{2i\psi}$, so has spin weight 2. This can be seen easily since a 45° rotation changes pure γ_1 into pure γ_2, so changes real γ into imaginary γ, and vice versa.

The weak lensing shear is derivable from the lensing potential with

$$\gamma = \frac{1}{2}\left[-\left(\frac{\partial}{\partial x} + i\frac{\partial}{\partial y}\right)\right]\left[-\left(\frac{\partial}{\partial x} + i\frac{\partial}{\partial y}\right)\right]\phi\,. \tag{26}$$

This is the small-area limit of a more general operation (for more details of this sort of description, see[8])

$$\gamma = \frac{1}{2}\tilde{\partial}\,\tilde{\partial}\phi\,, \tag{27}$$

where $\tilde{\partial}$ (pronounced 'edth' and usually written with the tilde crossing the ∂) is defined on the full sky. Note that the lensing potential ϕ is real. A general spin-weight 2 object can be written as the second derivative of a complex field

$$\gamma = \frac{1}{2}\tilde{\partial}\,\tilde{\partial}\,(\phi_E + i\phi_B)\,, \tag{28}$$

so we see that weak lensing has $\phi_E = \phi$ and $\phi_B = 0$. This decomposition of the field into 'E' and 'B' modes has an analogue in polarization; they are sometimes called grad and curl modes as well. Examples of patterns from E modes and B modes are shown in Fig. 2.

The presence of significant B-modes in the data indicates some problem, and B-mode analysis is a useful check on systematics. Note that of course the absence of B-modes is a necessary, but not sufficient, requirement for absence of systematics.

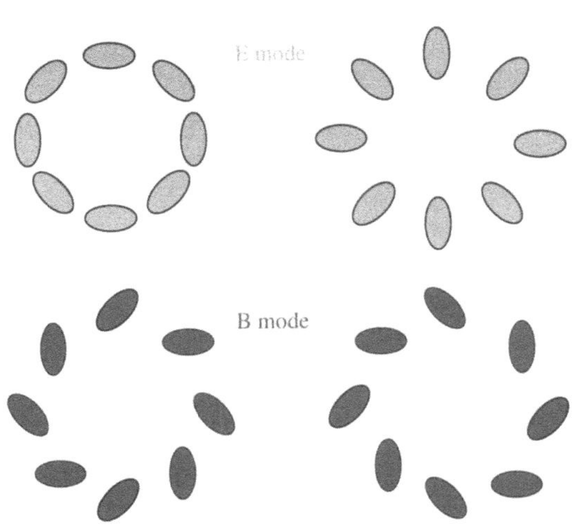

Fig. 2. Example patterns from E-mode and B-mode fields (from [32]). Weak lensing only produces E-modes at any significant level, so the presence of B-modes can indicate systematic errors

7 Results

The first results from cosmic shear were published in 2000 [1, 23, 30, 35], so as an observational science, cosmological weak lensing is very young. To date, the surveys have been able to show clear detections of the effect, and reasonably accurate determination of some cosmological parameters, usually the amplitude of the dark matter perturbations (measured by the rms fractional fluctuations in an $8\,h^{-1}$ Mpc sphere and denoted by σ_8), and the matter density parameter Ω_m. Current surveys cannot lift a near-degeneracy between these two, and usually a combination (typically $\sigma_8\Omega_m^{0.5}$) is quoted. This makes sense – it is difficult, but certainly not impossible, to distinguish between a highly clumped low-density universe and a modestly clumped high-density universe. There is no question that the surveys do not yet have the size nor the careful control of systematics required to compete with the microwave background and other techniques used for cosmological parameter estimation.

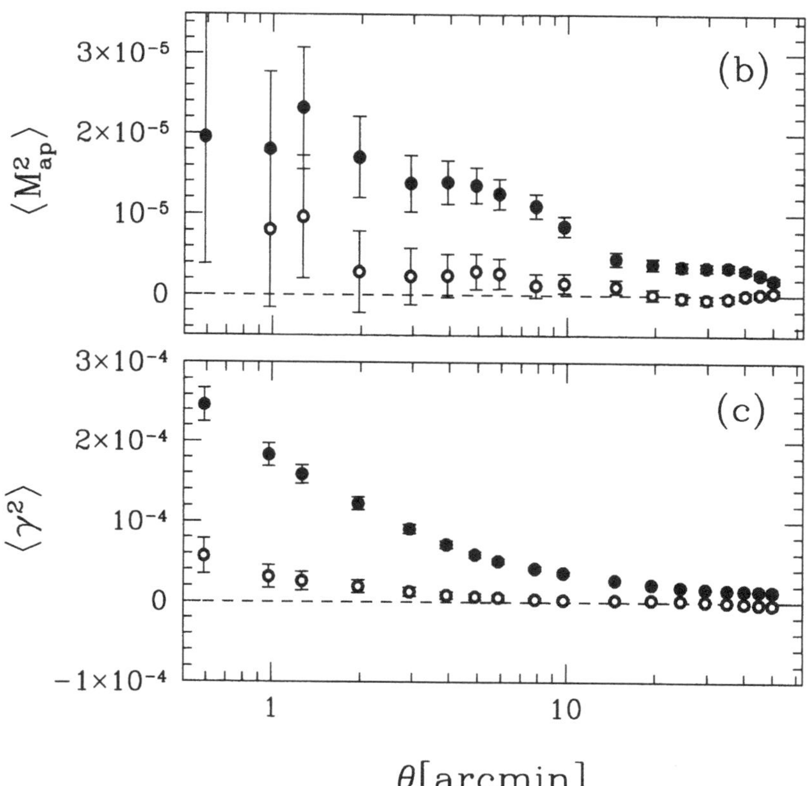

Fig. 3. E- and B-modes from a reanalysis of the VIRMOS-DESCART survey [33]. Top panel shows the aperture mass statistic (a weighted quadratic measure) and the lower panel the shear variance on different scales. Top points are the E-modes and bottom the B-modes

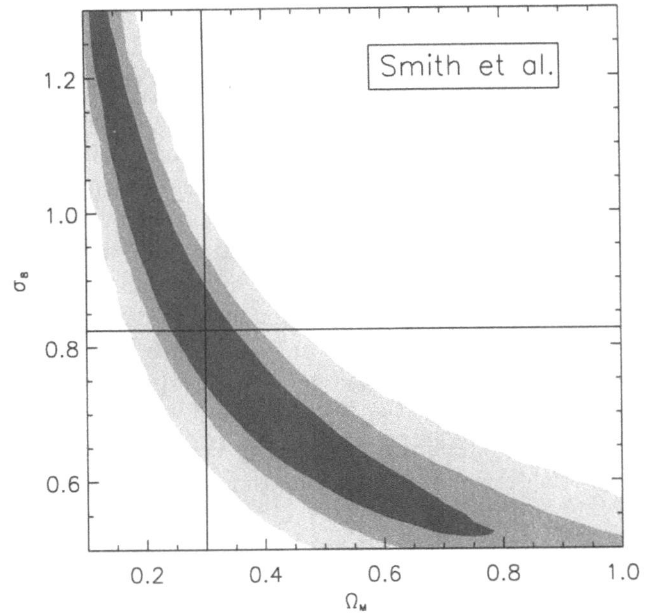

Fig. 4. Cosmological parameters from a reanalysis of the VIRMOS-DESCART survey [33]. Contours are 68%, 95% and 99.9% joint confidence levels. The vertical line is at $\Omega_m = 0.3$, roughly what is favoured by microwave background and large-scale structure studies

However, this situation is changing fast, particularly with the CFHT Legacy Survey, which is underway, and the upcoming VST weak lensing survey. There are other more ambitious surveys which are in some stage of planning or funding, such as Pan-STARRS, the Dark Energy Survey and VISTA. A summary of recent results is in the review by van Waerbeke and Mellier [32], and I will not list the results here. I mention only one recent study, a reanalysis of the VIRMOS-DESCART survey [33], as this one uses a neat trick to deal with varying point spread function and quotes a small error ($\sigma_8 = (0.83 \pm 0.07)(\Omega_m/0.3)^{-0.49}$. The technique employed was to make use of the fact that lensing does not induce a significant B-mode. As discussed earlier, one effectively needs to interpolate the shape of the point spread function between the positions of the stars in the image, and this can be done in a number of ways. van Waerbeke et al. chose the interpolation scheme which minimized the B-mode. Although not foolproof in principle, the results are impressive, and shown in Fig. 3. The cosmological parameters found from this analysis are shown in Fig. 4.

8 Future

The main promise of weak lensing in the future will come from larger surveys with optics designed for excellent image quality. Currently, the CFHTLS is underway, designed to cover 170 square degrees to a median redshift in excess of one. In future Pan-STARRS, VST and VISTA promise very small PSF distortions and large areal coverage, and in the far future LSST on the ground, and satellites such as SNAP or DUNE may deliver extremely potent lensing surveys. In parallel with these developments, the acquisition of photometric redshifts for the sources has opened up the exciting possibility of analysing weak lensing surveys in 3D. Each source represents a noisy estimate of the shear field at a location in 3D space, and this extra information turns out to be extremely valuable, increasing substantially the statistical power of lensing surveys. In particular, it can lift the degeneracy between σ_8 and Ω_m, measure directly the growth of dark matter clustering [3] and, more excitingly still, represents a powerful method to measure the equation of state of dark energy [8, 15, 17, 20] – surely one of the most important remaining questions in cosmology. In addition, photometric redshifts allow the possibility of direct 3D dark matter mapping [2, 28, 29], thus addressing another of the unsolved problems (Fig. 4).

Appendix 1: Propagation of Light Through a Weakly Perturbed Universe

Geodesic Equation

The geodesic equation governs the worldline x^λ ($\lambda = 0, 1, 2, 3$) of a particle, and is readily found in textbooks on general relativity. It is

$$\frac{d^2x^\lambda}{dp^2} + \Gamma^\lambda{}_{\mu\nu}\frac{dx^\mu}{dp}\frac{dx^\nu}{dp} = 0\,, \tag{29}$$

where p is an affine parameter, and Γ^μ_ν is the affine connection, which can be written in terms of the metric tensor $g_{\mu\nu}$ as

$$\Gamma^\lambda{}_{\mu\nu} = \frac{1}{2}g^{\sigma\lambda}\left\{\frac{\partial g_{\mu\nu}}{\partial x^\sigma} + \frac{\partial g_{\sigma\nu}}{\partial x^\mu} - \frac{\partial g_{\mu\sigma}}{\partial x^\nu}\right\}\,. \tag{30}$$

For weak fields, the interval is given by

$$ds^2 = \left(1 + \frac{2\Phi}{c^2}\right)c^2dt^2 - \left(1 - \frac{2\Phi}{c^2}\right)R^2(t)\left[dr^2 + S_k^2(r)d\psi^2\right]\,, \tag{31}$$

where Φ is the peculiar gravitational potential, $R(t)$ is the scale factor of the universe and r, θ, φ are the usual comoving spherical coordinates. The angle $d\psi = \sqrt{d\theta^2 + \sin^2\theta d\varphi^2}$. $S_k(r)$ depends on the geometry of the universe, being given by

$$S_k(r) = \begin{cases} \sinh r, & \text{if } k < 0\,; \\ r, & \text{if } k = 0\,; \\ \sin r, & \text{if } k > 0\,. \end{cases}$$

The curvature $k = -1, 0, 1$ corresponds to open, flat and closed universes, respectively. We are interested in the distortion of a small light bundle, so we can concentrate on a small patch of sky. If we choose the polar axis of the coordinate system to be along the centre of the light bundle, we can define angles $\theta_x \equiv \theta \cos\varphi$ and $\theta_y \equiv \theta \sin\varphi$. For convenience we also use the *conformal time*, defined by $d\eta = cdt/R(t)$, in place of the usual time coordinate. With these definitions the interval is more simply written as

$$ds^2 = R^2(t)\left\{\left(1+\frac{2\Phi}{c^2}\right)d\eta^2 - \left(1-\frac{2\Phi}{c^2}\right)\left[dr^2 + S_k^2(r)(d\theta_x^2 + d\theta_y^2)\right]\right\} .$$

The metric tensor for weakly perturbed flat Friedman–Robertson–Walker metric is then

$$R^2(t)\begin{pmatrix} 1+2\Phi/c^2 & 0 & 0 & 0 \\ 0 & -\left(1-2\Phi/c^2\right) & 0 & 0 \\ 0 & 0 & -r^2\left(1-2\Phi/c^2\right) & 0 \\ 0 & 0 & 0 & -r^2\left(1-2\Phi/c^2\right) \end{pmatrix} .$$

We are interested in how the angles of the ray, (θ_x, θ_y), change as the photon moves along its path, responding to the varying gravitational potential. The unperturbed, radial, path is set by $0 = ds^2 \simeq d\eta^2 - dr^2$, i.e. For a radial incoming ray,

$$\frac{dr}{d\eta} = -1 .$$

With $g^{\mu\nu}$ defined as the inverse of $g_{\mu\nu}$ (so $g_{\mu\nu}g^{\nu\alpha} = \delta_\mu{}^\alpha$), the affine connections are readily computed.

The parametrized equation for η is required only to zero order in Φ and reduces to

$$\frac{d^2\eta}{dp^2} = -2\frac{\dot{R}}{R}\dot{\eta} ,$$

where a dot here denotes d/dp. By choosing the unit of p appropriately, we find

$$\frac{d\eta}{dp} = \frac{1}{R^2} .$$

We can also relate the radial coordinate to the conformal time, again to zero-order in Φ: The first-order equations governing θ_x and θ_y are obtained from the geodesic equation or by the variational methods (see, for example, d'Inverno (1992), section 7.6 [19])

$$\frac{\partial L^2}{\partial x^\mu} - \frac{d}{dp}\left(\frac{\partial L^2}{\partial \dot{x}^\mu}\right) = 0 ,$$

where $L^2 \equiv (ds/dp)^2$. With $x^\mu = \theta_x$,

$$R^2\frac{2}{c^2}\frac{\partial\Phi}{\partial\theta_x}\dot{\eta}^2 + \frac{2}{c^2}R^2\frac{\partial\Phi}{\partial\theta_x}\left(\dot{r}^2 + r^2\theta_x^2 + r^2\theta_y^2\right) - \frac{d}{dp}\left[-2R^2r^2\left(1-\frac{2\Phi}{c^2}\right)\dot{\theta}_x\right] = 0 .$$

With the zero-order solutions for $d\eta/dp$ and $dr/d\eta$, to first order this reduces simply to

$$\frac{d^2\theta_x}{d\eta^2} - \frac{2}{r}\frac{d\theta_x}{d\eta} = -\frac{2}{c^2 r^2}\frac{\partial\Phi}{\partial\theta_x} .$$

It is convenient to write this as an equation for the comoving displacement of the ray from a fiducial direction,

$$x_i \equiv r\theta_i , \qquad i = 1, 2 ,$$

and the equation for θ_x and a similar one for θ_y simplify to

$$\frac{d^2\mathbf{x}}{d\eta^2} = -\frac{2}{c^2}\nabla\Phi ,$$

where ∇ here is a comoving transverse gradient operator (∂_x, ∂_y).

We see that the propagation equation for the displacement looks similar to what one would guess from a Newtonian point of view; the presence of η (instead of t) in the acceleration term on the left is a result of the expansion of the universe and the choice of comoving coordinates. The right-hand side looks like the gradient of the potential, but is larger than the naive gradient by a factor of two. This is the same factor of two which leads to the classic result of general relativity, famously tested by Eddington's 1919 solar eclipse observations, that the angle of light bending by the Sun is double what Newtonian theory predicted.

Appendix 2: Exercises

1. Generalize the result for the propagation of light to a non-flat universe. Show that the perturbed equation for the photon trajectory is

$$\frac{d^2\mathbf{x}}{d\eta^2} + k\mathbf{x} = -\frac{2}{c^2}\nabla\Phi .$$

References

1. Bacon, D.J., Refregier, A., Ellis, R.S.: MNRAS **318**, 625 (2000)
2. Bacon, D.J., Taylor, A.N.: MNRAS **344**, 1307 (2003)
3. Bacon, D.J., et al.: MNRAS **363**, 723 (2005)
4. Bartelmann, M., Schneider, P.: Weak Gravitational Lensing, Phys. Rep. **340**, 291–472; astro-ph/9912508 (2001)
5. Bernardeau, F., van Waerbeke, L., Mellier, Y.: A&A **322**, 1 (1997)
6. Bernstein, G., Jarvis, M.: AJ **123**, 583 (2002)
7. Brown, M. et al.: MNRAS **333**, 501 (2002)
8. Castro, P.G., Heavens, A.F., Kitching, T.: Phys. Rev. D. **72**, 023516 (2005)
9. Catelan, P., Kamionkowski, M., Blandford, R.D.: MNRAS **320**, L7 (2001)

10. Chang, T.-C., Refregier, A.: ApJ **570**, 447 (2002)
11. Crittenden, R., Natarajan, P., Pen, U.-L., Theuns, T.: ApJ **559**, 552 (2001)
12. Croft, R.A.C., Metzler, C.A.: ApJ **545**, 561 (2000)
13. Heymans, C.E.C. et al.: MNRAS **347**, 895 (2004)
14. Heymans, C.E.C., Heavens, A.F., MNRAS **337**, 711 (2003)
15. Heavens, A.F., MNRAS **343**, 1327 (2003)
16. Heavens, A.F., Refregier, A., Heymans, C.E.C.: MNRAS **319**, 649 (2000)
17. Heavens, A.F., Kitching, T.: MNRAS **373**, 105 (2006)
18. Hirata, C., Seljak, U.: Phys. Rev. **D70**, 063526 (2004)
19. d'Inverno, R.: Introducing Einstein's Relativity, Oxford University Press, Oxford (1992)
20. Jain, B., Taylor, A.N.: PRL **91**, 141302 (2003)
21. Jing, Y.P.: MNRAS **335**, L89 (2002)
22. Kaiser, N., Squires, G., Broadhurst, T.: ApJ **449**, 460(1995)
23. Kaiser, N., Wilson, G., Luppino, G., astro-ph/0003338 (2000)
24. King, L., Schneider, P.: A&A **396**, 411 (2003)
25. Schneider, P. et al.: MNRAS **296**, 893 (1998)
26. Schneider, P., Gravitational lensing as a probe of structure, to appear in the proceedings of the XIV Canary Islands Winter School of Astrophysics, Dark Matter and Dark Energy in the Universe, Tenerife, astro-ph/0306465 (2003)
27. Smith, R.E. et al.: MNRAS **341**, 1311 (2003)
28. Taylor, A.N.: astro-ph/0111605 (2001)
29. Taylor, A.N., et al.: MNRAS **353**, 1176 (2004)
30. van Waerbeke, L., et al.: A& A **358**, 30 (2000)
31. van Waerbeke, L. et al.: A&A **393**, 369 (2002)
32. van Waerbeke, L., Mellier, Y., Gravitational Lensing by Large Scale Structures: A Review. Proceedings of Aussois Winter School, astro-ph/0305089 (2003)
33. van Waerbeke, L., Mellier, Y., Hoekstra, H., A& A **429**, 75 (2005)
34. White, M., Astroparticle Physics **22**, 211–217 (2004)
35. Wittman, D. et al.: Nature **405**, 143 (2000)

Mass Reconstruction from Lensing

J.M. Diego

MIT Center for Space Research, 70 Vassar Street, Cambridge, MA 02138, USA
jdiego@space.mit.edu

One of the most striking consequences of Einstein's general relativity is the ability of a large concentrated mass to deflect photons as they pass near a massive object. An example of this is a galaxy or a galaxy cluster. Because of this quality, massive objects are usually referred as *gravitational lenses.* The level of deflection differs depending on the minimum distance the photon approaches the deflecting object and the total mass of the lens. The *weak lensing* regime occurs when the deflection angle is small, normally occurring when the photon is at a large distance from the gravitational lens. The deflection angle in this regime decreases approximately linearly as the distance to the lens increases. When the distance is small we are in the *strong lensing* regime (hereafter SL). The SL effect creates a more complex distortion of the background sources (normally galaxies or quasars). The deflection angle becomes less linear as the photons get closer to the lens. As a result, the effect can create multiple images of the source. The number and distribution of these multiple images will depend on the geometry of the problem (observer–lens–source) as well as on the internal distribution of matter in the source. A beautiful example is shown in Fig. 1 where large arcs can be seen around the centre of gravity of the cluster.

While weak lensing distortions appear as a stretched version of the background sky, strong lensing distortions are rich in complexity, making evident the non-linear nature of the deflection angle. This complex structure of arcs appearing in the strong lensing regime can be used effectively to reveal the internal gravitational structure of the lens. Solving the problem of inverting this rich data to get the gravitational potential responsible for it is not a trivial task. The unknowns of the problem are the distribution of matter in the lens and the position and shape of the source (or sources). These two unknowns have to be modelled and adjusted in order to reproduce the observed arcs. Usually, degeneracies between the model (lens and sources) and the data (position and shape of the arcs) appear.

The range of possible models explaining the data (arcs) reduces as the number of available arcs increases. Problems arise when the observed arcs are

Diego, J.M.: *Mass Reconstruction from Lensing.* Lect. Notes Phys. **665**, 601–627 (2009)
DOI 10.1007/978-3-540-44767-2_19

Fig. 1. Multiple arcs in A2218 captured by the Hubble telescope

low. In this case a parametrization of the model involving few parameters is needed. The parametrization of the model is done over the mass distribution but assuming that the lens is a superposition of haloes, each one described by a small set of parameters. The parameters are then varied until a fair reproduction of the observed data (i.e. arcs) is achieved. The methods following this type of approach are called parametric methods.

In opposition to this, if the number of arcs is sufficiently large, or the arcs extend over a significant fraction of the image (large arcs), one can attempt to invert the data without recalling any parametrization of the mass but allowing instead the mass to be distributed in any way. The mass distribution will then be varied over the lens plane until it reproduces the observed data. This second approach is known as non-parametric inversion. It is obvious that the requirement of having a large number of arcs will apply only in the case of strong lensing in galaxy clusters where a sufficiently large number of arcs can be found. In the case of galaxy–galaxy lensing, non-parametric methods are of little help since the number of arcs is usually limited to a small number. In these cases parametric methods are the only way to solve the problem. In this chapter we will focus only on the case of galaxy clusters where both parametric and non-parametric algorithms can be used.

Both parametric and non-parametric methods aim at recovering the internal distribution of dark matter in deep gravitational potentials. Knowing the internal structure, for instance galaxy clusters, is fundamental to answer questions relative to the nature of dark matter or to understand the formation of galaxy clusters as well as their evolution. An accurate estimate of the mass in galaxy clusters is also needed in order to use their mass function in cosmological studies. The internal distribution of dark matter is also needed to understand the dynamics of the plasma as seen in X-rays and more recently through the Sunyaev–Zel'dovich effect.

In this lesson we will briefly discuss parametric methods and will focus our attention more on non-parametric ones but before we will review some fundamental relations used in strong lensing. More detailed information can be found in the abundant bibliography. Here we recommend the comprehensive review of Narayan & Bartelmann [15]. Other reviews can be found in Blandford & Narayan [3], Schneider, Ehlers & Falco [20], Wambsganss [23], and Kneib [12].

A more detailed description of some of the algorithms described here can also be found in Diego et al. [7].

1 Basics of Strong Lensing

The fundamental problem in lens modelling is the following: Given the positions of lensed images, $\boldsymbol{\theta}$, what are the positions of the corresponding background galaxies, $\boldsymbol{\beta}$, and the mass distribution of the lens, $M(\boldsymbol{\theta})$. Mathematically this entails inverting the lens equation

$$\boldsymbol{\beta} = \boldsymbol{\theta} - \hat{\boldsymbol{\alpha}}(\boldsymbol{\theta}, M(\boldsymbol{\theta})) , \tag{1}$$

where $\hat{\boldsymbol{\alpha}}(\boldsymbol{\theta})$ is the deflection angle created by the lens which depends on the observed positions, $\boldsymbol{\theta}$. From now on we will omit the vector notation unless otherwise noted.

The deflection angle, α, at the position θ is found by integrating the contributions from the whole mass distribution:

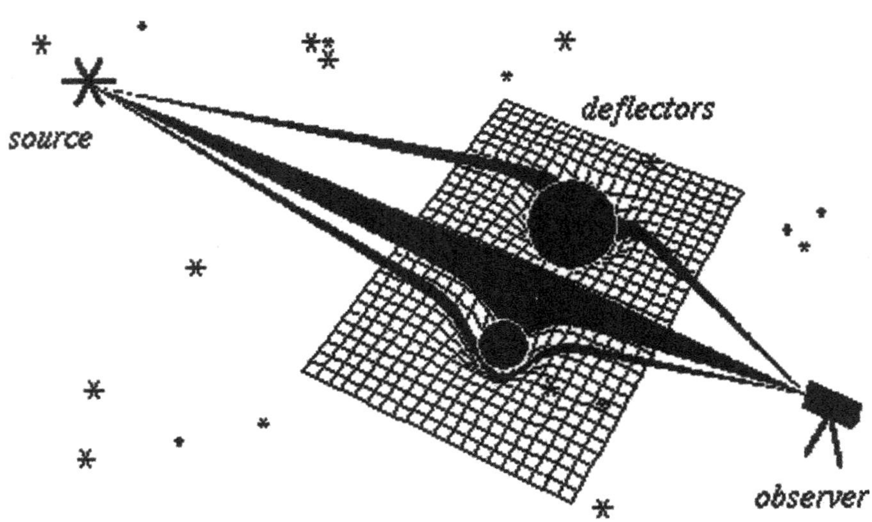

Fig. 2. Light can be deflected in complex ways when passing near a strong gravitational potential. In the above figure, the observer sees three *ghost* images of the original source

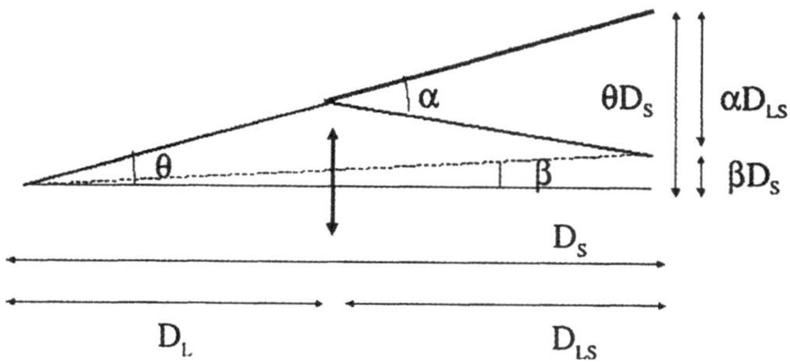

Fig. 3. Plot showing the fundamental relation between the different angles from which the lens equation is derived; $\theta D_S = \alpha D_{LS} + \beta D_S$

$$\hat{\alpha}(\theta) = \frac{D_{LS}}{D_S}\alpha(\theta) = \frac{4G}{c^2}\frac{D_{LS}}{D_S D_L}\int M(\theta')\frac{\theta - \theta'}{|\theta - \theta'|^2}d\theta' \ , \qquad (2)$$

where D_{ls}, D_l, and D_s are the angular distances from the lens to the source galaxy, the distance from the observer to the lens, and the distance from the observer to the source galaxy, respectively. In (2) we have made the usual thin lens approximation so the mass $M(\theta')$ is the projected mass along the line of sight θ'.

Due to the (non-linear) dependency of the deflection angle, α, on the position in the sky, θ, this problem is usually regarded as a typical example of a non-linear problem. We will see that this is only partially true.

At any position in the sky, θ, the deflection angle, α, can be computed for any redshift of the source by integrating (2). If after subtracting α from the position θ, the resulting angle, β, coincides with the position of the source at redshift z_s, that means, a lensed image of the source at the position β will appear in the position θ. The lensed image will be magnified (or demagnified) by a factor μ. The magnification is given by the Jacobian of the lens mapping (e.g. Young et al. [26]):

$$\mu(\theta)^{-1} = \frac{\beta}{\theta}\frac{d\beta}{d\theta} = 1 - \nabla\hat{\alpha} + \frac{d\hat{\alpha}_x}{dx}\frac{d\hat{\alpha}_y}{dy} - \frac{d\hat{\alpha}_x}{dy}\frac{d\hat{\alpha}_y}{dx} \ . \qquad (3)$$

The magnification, μ, tends to infinity in certain regions called the critical curves. Typically one finds two different kinds of critical curves. In the outer region is the *tangential* critical curve or *Einstein ring* (for lenses with symmetry). Near the radial critical curve is where the giant arcs appear. The background sources suffer the largest magnifications and distortions in this area. The diameter of the radial curve is proportional to the enclosed mass within it. Interior to the tangential critical curve is the *radial* critical curve. This is the area where multiple images merge or split into two as they move in the source plane (or the lens changes its distribution or relative position to the source). The images near this curve are distributed in a radial form normally pointing towards the centre of gravity of the potential. The relative

distance between the two critical curves gives an idea of how steep the mass profile is in the central regions.

2 Parametric Methods

Parametric methods have been the traditional approach to analysing strong lensing data, and serve a purpose when strong lensing data contain far less data than the impressive systems we see today. In these cases, parametric methods are the only way to go since the amount of data is not enough to constrain the mass distribution without the help of a parametrization. Parametric methods rely on assumptions or priors about the mass distribution. The priors are well motivated both theoretically and observationally. The most usual prior is to assume that the mass follows the luminosity profile of the cluster.

The basic idea of parametric methods is to decompose the mass in the lens plane as a superposition of *smooth* components or haloes following a parametrized form, usually NFW profiles or isothermal spheres. This approach is motivated by hierarchical models of structure formation. N-body simulations also motivate the use of NFW profiles as the parametrized form to describe the individual haloes. Each of the individual haloes is described by a set of parameters, usually the location of the centre, a concentration parameter, and a total mass, although additional parameters can be incorporated such as ellipticity.

The best model is determined by using the arcs as constraints to determine the parameters. Normally only the position of the arcs is used as constraints although more elaborate analysis can also include their extension or shape. The process is intrinsically slow since for each generated model a back-forward calculation has to be done. In the back step the observed arcs are *back-projected* into the source plane so the predicted position of the source is found; in a second step this predicted position has to be *moved forward* into the image plane to find other possible arcs or to make it coincide with existing images. The forward step is by far the slowest one since in this case all the sky has to be checked for possible images (or solutions of the non-linear problem) while the back-projection can be sped up significantly since in this case the problem can be linearized as we will see below.

Parametric methods have been applied successfully in a wide variety of images (Kochanek & Blandford [13], Kneib et al. [9, 10, 11], Broadhurst et al. [5], Koopmans et al. [14], Sand et al. [19], Gavazzi et al. [19]). More recently, Broadhurst et al. [6] have applied a parametric algorithm to the best strong lensing data set ever with satisfactory results. This data set contains more than a 100 arcs with photometric redshift on the cluster A1689. A comparison of the results using parametric and non-parametric algorithms on this data will be presented at the end of this lesson.

3 Non-parametric Methods

If the number of arcs is large enough, the model mass distribution can be constrained without (almost) ambiguity. In this context it is interesting to use non-parametric methods which give a fair treatment to the data in the sense that they do not rely on assumptions. Non-parametric methods are also the way to do *sanity* tests by comparing their results with those derived from parametric methods. Another very interesting feature of non-parametric methods is that, as we will see below, they can work with a linearized version of the problem. This will allow to speed up the estimation of a solution by orders of magnitude.

Among the non-parametric methods available, we would like to emphasize the pixelization methods of Saha et al. [17, 18], Abdelsalam et al. [1, 2], and also the multipole approach of Kochanek & Blandford [13] and Trotter et al. [22].

The pixelization method presented in Saha et al. [17] (and subsequent papers) divides the lens plane into a grid of cells. The mass in each cell is estimated after minimizing a quadratic quantity which involves an estimate of the luminosity of the cluster. As we will see later, this *prior* information can be avoided by alternative pixelization algorithms. The multipole approach decomposes the gravitational potential of the lens using a Taylor expansion. This allows to generalize the potential in a way that is less subject to any preconception about the mass distribution.

In the next sections we will describe a different type of algorithm based on the pixelization of the lens plane. The main feature of that approach is that the problem can be formulated in a linear form and then be solved using standard techniques.

4 Non-parametric Methods: The Problem Formulated in Its Basic Linear Form

The key approximation we make is to split the lens plane into N_c small regions (hereafter cells) over which the projected mass is more or less constant. We can then rewrite (2) as

$$\hat{\alpha}(\theta) = \frac{4G}{c^2}\frac{D_{ls}}{D_s D_l}\sum_{N_c} m_i \frac{\theta - \theta_i}{|\theta - \theta_i|^2} \,. \tag{4}$$

The first point we want to make here is that the deflection angle α may be thought of as the net contribution of many small masses m_i in the positions θ_i, each one pulling the deflection in the direction of $(\theta - \theta_i)$ and with a magnitude which is proportional to $m_i/(\theta - \theta_i)$. If we divide the lens plane into a grid with N_c cells, the masses m_i can be considered as the mass contained in the cell i

($i = 1, ..., N_c$). If the cells are sufficiently small then the above pixelization of the mass plane will give a good approximation to the real mass distribution.

Our second point is that the problem is non-linear in one direction but linear in the other. That is, given a position in the sky θ (and given a lens) there is only one β which satisfies (1) but given a position of the background galaxy β, there may be more than one position in the sky (θ) satisfying (1) or equivalently, the source galaxy may appear lensed in more than one position in the sky. The linear nature of the problem is evident when one realizes that the only non-linear variable, namely the θ positions, is fixed by the observation and that the problem depends linearly on the unknowns (β positions and masses, m_i, in the cells).

Let us now assume that we have a data set consisting of a system of radial and tangential strong lensing arcs which are spread over N_θ pixels in the image. We will also assume that we know which arcs originate from the same sources. Since both the data and the mass distribution have been discretized, we can rewrite (1) as a simple matrix equation:

$$\beta = \theta - \Upsilon M \ , \tag{5}$$

where θ and β are now $2N_\theta$ element vectors containing x and y values of the observed positions and the (unknown) source positions, respectively. M is the mass vector containing all N_c mass cells and Υ is the $(2N_\theta \times N_c)$ matrix casting the mass vector into a vector of displacement angles. A more technical account of the make-up of the Υ matrix can be found in the appendix. Equation (5) clearly demonstrates the linear nature of the problem when formulated in this manner. The problem has now been reduced to a set of $2N_\theta$ linear equations with $2N_\theta + N_c$ unknowns (lens masses and source galaxy positions). Notice that when the problem is formulated in this form, there are more unknowns than equations which means it is an underdetermined system with an infinite number of solutions. In order to identify a suitable solution for such a system, we need to add extra information or impose constraints.

One way of doing this is by reducing the number of unknowns, for instance by removing the source positions from the unknown category. This can be achieved by minimizing the dispersion in the source plane, i.e. demanding that the pixels in the source plane be as concentrated as possible for each source. In this case we are left with only N_c unknowns.

Another way of constraining the system is by assuming that the N_s sources are well approximated by point sources, which reduce the number of unknowns to $N_c + 2N_s$. This means effectively demanding that all observed θs for arcs corresponding to the same source can be traced back through the lens to N_s single points. With this assumption we can rewrite the lens equation in the compact form of

$$\theta = \Gamma X \ . \tag{6}$$

Γ is now a matrix of dimension $2N_\theta \times (N_c + 2N_s)$ and X is the vector of dimension $(N_c + 2N_s)$ containing all the unknowns in our problem (see Appendix A), the mass elements and the $2N_s$ central (x and y) coordinates of the N_s sources. Now the system is overdetermined and has a unique *point source solution.* (As long as the number of mass cells is reasonable.)

The linearization of the problem means that it is in principle solvable by both matrix inversion and simple linear programming routines. In practice, the problem quickly becomes ill conditioned and too large for direct matrix inversion, and approximate numerical methods are more suitable. The main problem with the linearization is that we do not know if the obtained linearized solution creates artificial tangential or radial arcs. Checking this requires forward solving of the lens equation which is non-linear due to the complicated dependence of the deflection angle on θ.

We suggest a novel approach to this problem by using all the available information in the images, i.e. the information inherent in the dark areas as well as the observed arcs. By pixelization of the dark areas, tracing these pixels back through the lens and imposing that they fall outside the sources it is possible to find the true solution without overpredicting arcs. This use of the null space is to our knowledge unprecedented and in principle allows for a complete, linear solution to a problem usually considered non-linear.

5 Inversion Methods

In this section we will describe some inversion methods which can be applied to solve the problem. Most of these methods can be found in popular books like *Numerical Recipes* (Press et al. [16]). All these algorithms are being implemented in a package SLAP (Strong Lensing Analysis Package).[1]

To test the different algorithms we will use simulated data where all the unknowns (mass distribution of the cluster and positions of the sources) are known a priori.

Rather than taking a uniform grid, it is better to construct a *dynamical* or *multi-resolution* grid. By sampling dense regions more heavily, it is possible to reduce drastically the number of cells needed to accurately reproduce the lensing properties of the cluster. In other words we choose an adaptive grid which samples the dense cluster centre better than the outer regions. The multi-resolution grid is obtained through an iterative procedure. Given a mass estimate (a first mass estimate can be obtained with a coarse regular grid), a new solution for the mass is found. This new solution is used again to define a new grid. Typically the mass estimate will improve with each iterative step as this dynamical grid allows for the relevant regions of the cluster to become resolved. An example of a dynamical grid is shown in Fig. 5. Having the sources and mass distribution, we can calculate the arcs produced by this combination. The result can be seen in Fig. 6 where we compare the exact solution with the approximation obtained using the gridded mass distribution. The problem now reduces to finding the mass distribution of Fig. 4 and the

[1] Available at `http://darwin.cfa.harvard.edu/SLAP/`

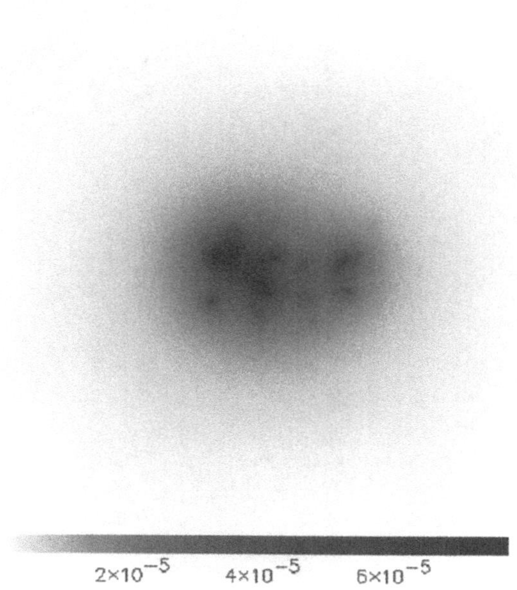

Fig. 4. Original simulated mass profile. The total mass is $1.119 \times 10^{15}\mathrm{h}^{-1}\,\mathrm{M}_{\odot}$. The field of view is 0.1 degrees across and the cluster is at $z = 0.18$. The units in the grey scale map are $10^{15}\mathrm{h}^{-1}\,\mathrm{M}_{\odot}/pixel$

position of the sources using the observed distribution of the arcs of Fig. 6 (left).

Once we have the mass divided into cells, we can formulate the problem in its linear form with all the unknowns on one side (see (6)). In these circumstances, it is tempting to try a direct inversion of (6). Although the Γ matrix is not square, one can find its inverse, Γ^{-1}, by decomposing Γ into orthogonal matrices. This is similar to finding the *eigenvalues* and *eigenvectors* of Γ. This approximation has its advantages as well as its drawbacks and we will explore this possibility later. However, we anticipate that degeneracies between neighbouring pixels in the arcs as well as neighbouring cells in the lens plane (not to mention the compact nature of the sources) will result in a system of linear equations which is not well behaved. The rank of the matrix Γ will be normally smaller than its smaller dimension. Calculating the inverse in this situation is not a trivial task.

A second approach is *rotating* our system of linear equations using a transformation which is just Γ^T. This transforms Γ into a square, symmetric, and positive definite matrix of dimension $(2N_s + N_c) \times (2N_s + N_c)$, $A = \Gamma^T\Gamma$ which is better behaved than the original Γ matrix. However, the rank of A is in general smaller than its dimension and its inverse does not exist. The

Fig. 5. Same mass of Fig. 4 but divided into cells of a dynamical grid

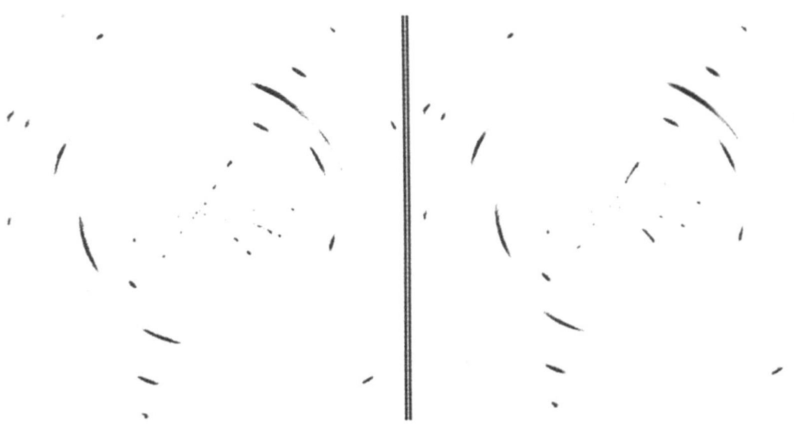

Fig. 6. Thirteen sources lensed by the mass distribution of Fig. 4. The right image is the lensed image using 325 cells in the dynamical grid. The left image is the exact solution using no griding. Note the differences in the radial arcs

hope in this case must therefore be to find an approximate rather than exact solution to the system.

The third approach is the simplest and will be explored first. We assume we know nothing about the sources other than their redshift and that they are much smaller than the strong lensing arcs. This is the same as saying that the lens has to be such that it *focuses* the arcs into compact sources at the desired (known) redshift. This simple argument alone will turn out to be powerful enough to get a quick but good first estimate of the mass distribution in the lens plane.

We explore all these approaches below in reverse order.

5.1 First Approach: Minimizing Dispersion in the Source Plane

In this subsection we will discuss the simplest (although effective) method to get a fast estimation of the mass using no prior information on the lens nor the sources. The problem then contains two sets of unknowns: the mass vector we want to determine, M, and the β positions of the sources. In general, for a finite number of observed arcs, there are several combinations of β and M which can reproduce the observations. The most obvious unphysical solution is the null solution, where the mass is zero and the sources identical to the observed arcs. The easiest way to avoid such unphysical solutions is to minimize the variance of the β positions. This is equivalent to imposing that the vector M really acts as a true lens: We require big arcs with large magnifications and multiple images separated by arcsec or even arcmin to *focus* into a rather compact region in the source plane. This minimization process assumes that we are able to associate the multiple lensed images with a particular source. This can be achieved with high-quality spectroscopy.

To minimize the variance in the source plane it is illustrative to follow the steepest descent path although other more effective minimization algorithms can be used (see below). Given an initial guess for the mass vector, one can calculate the derivative of the variance in the source plane as a function of the mass and minimize in the direction of the derivative. Once a minimum is found, we calculate the derivative in the new mass position and minimize again in an iterative procedure.

The quantity to be minimized is

$$f(M) = \sum_s \sigma_s^2 , \tag{7}$$

where the sum is over the number of identified sources and σ_s^2 is the variance of the source s in the source plane. That is,

$$\sigma_s^2 = <\beta^2>_s - <\beta>_s^2 , \tag{8}$$

where the βs are calculated from (5) and the average is over the βs corresponding to source s. By combining equations (5) and (8) it is easy to compute the derivative of σ_s^2 with respect to M:

$$\frac{\partial \sigma_s^2}{\partial M_j} = 2 < \beta >_s < \Upsilon_j >_s - 2 < \beta \Upsilon_j >_s , \tag{9}$$

where Υ_j is the column j of the Υ matrix and the average is made only on the elements associated with the source s. We should note that all equations involving the vectors β, θ or $\alpha = \Upsilon M$ have two components, x and y, so there will be in fact two equations like (9): one for the x component of β and the other one for the y component. At the end, the quantity we aim to minimize is $\sigma^2 = \sigma_x^2 + \sigma_y^2$. As we already mentioned, the minimization can be done following the path of steepest descent given by (9). This path will end in a minimum at the new mass M^j. The process can be repeated by evaluating the new path at the new mass position until the variance is smaller than certain ϵ. A good choice for ϵ is to take a few times the expected variance for a population of N_s galaxies at the measured N_s redshifts. The minimization of the variance is a powerful and robust method for finding a first guess for the mass vector without making assumptions about the sources. The final result is shown in Fig. 7. Since the source positions can be estimated from (5), the minimization of the variance also provides us with an initial guess for these. The recovered

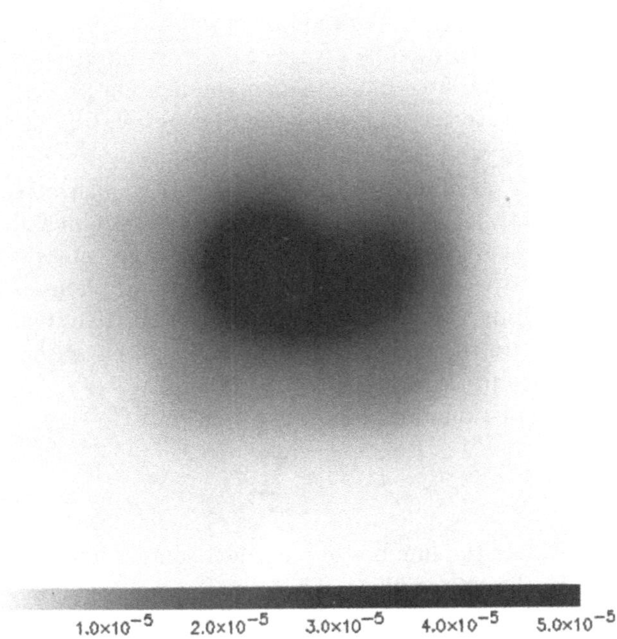

Fig. 7. Smooth version of the recovered mass after minimizing the variance. The total recovered mass is $1.01 \times 10^{15} h^{-1} M_\odot$. Compare this mass with the original one in Fig. 4

positions are found to be very close to the real position of the sources. The drawback of this algorithm is its slow convergence. A typical minimization may take several hours on a 1 GHz processor. In the next subsection we will go a step further and will include the β positions in the minimization as well as speed up the convergence by orders of magnitude.

5.2 Biconjugate Gradient

Inversion of linear systems where the matrix dimensions are of order 10^3 is a numerically trivial problem for today's computers provided the matrix is well behaved. If the matrix has null or negative eigenvalues, a direct inversion is not feasible and one has to aim to solve for some approximated solution. Our system of linear equations is a good example of an ill-conditioned one. Direct inversion of the matrix is not possible due to negative eigenvalues. However, there is another important reason why we do not want to solve exactly (or invert) the system of equations. An exact solution means that we will recover a mass distribution which puts the arcs into delta function sources. As we will see later, this solution will be unphysical. Instead, we are interested in an approximated solution which does not solve exactly the system of equations and which has a residual. This residual will have the physical meaning of the extension of the sources or the difference between the point-like sources and the real, extended ones. The biconjugate gradient will be a useful way to *regularize* our problem.

The biconjugate gradient algorithm is one of the fastest and most powerful algorithms to solve for systems of linear equations. It is also extremely useful for finding approximate solutions for systems where no exact solutions exist or where the exact solution is not the one we are interested in. The latter will be our case. Given a system of linear equations

$$Ax = b\ , \tag{10}$$

a solution of this system can be found by minimizing the following function:

$$f(x) = c - bx + \frac{1}{2}x^t Ax, \tag{11}$$

where c is a constant. When the function $f(x)$ is minimized, its gradient is zero:

$$\nabla f(x) = Ax - b = 0\ . \tag{12}$$

That is, at the position of the minimum of the function $f(x)$ we find the solution of (10). In most cases, finding the minimum of (11) is much easier than finding the solution of the system in (10) especially when no exact solution exists for (10) or A does not have an inverse.

The biconjugate gradient finds the minimum of (11) (or equivalently, the solution of (10)) by following an iterative process which minimizes the function $f(x)$ in a series of steps no longer than the dimension of the problem. The beauty of the algorithm is that the successive minimizations are carried out on a series of orthogonal conjugate directions, p_k, with respect to the metric A. That is,

$$p_i A p_j = 0, \quad j < i \,. \tag{13}$$

This condition is useful when minimizing in a multidimensional space since it guarantees that successive minimizations do not spoil the minimizations in previous steps. Let us now turn to the system we want to solve, namely (6). The biconjugate gradient method assumes that the matrix Γ (matrix A in (10)) is square. For our case this does not hold since we typically have $N_\theta >> (N_c + N_s)$. Instead we build a new quantity, called the square of the residual, R^2:

$$\begin{aligned} R^2 &= (\theta - \Gamma X)^T(\theta - \Gamma X) \\ &= 2(\frac{1}{2}\theta^T\theta - \Gamma^T\theta X + \frac{1}{2}X^T\Gamma^T\Gamma X) \,. \end{aligned} \tag{14}$$

By comparing (14) and (11) it is easy to identify the terms, $c = \frac{1}{2}\theta^T\theta$, $b = \Gamma^T\theta$, and $A = \Gamma^T\Gamma$. Minimizing the quantity R^2 is equivalent to solving (6). To see this we only have to realize that

$$b - AX = \Gamma^T(\theta - \Gamma X) = \Gamma^T R \,. \tag{15}$$

If an exact solution for (6) does not exist, the minimum of R^2 will give the better approximated solution to the system. The minimum can be now found easily with the biconjugate gradient (Press et al. [16]). For the case of symmetric matrices A, the algorithm constructs two sequences of vectors r_k and p_k and two constants, α_k and β_k:

$$\alpha_k = \frac{r_k^T r_k}{p_k^T A p_k} \,, \tag{16}$$

$$r_{k+1} = r_k - \alpha_k A p_k \,, \tag{17}$$

$$\beta_k = \frac{r_{k+1}^T r_{k+1}}{r_k^T r_k} \,, \tag{18}$$

$$p_{k+1} = r_{k+1} + \beta_k p_k \,. \tag{19}$$

At every iteration, an improved estimate of the solution is found by

$$X_{k+1} = X_k + \alpha_k p_k \,. \tag{20}$$

The algorithm starts with an initial guess for the solution, X_1, and chooses the residual and search direction in the first iteration to be

$$r_1 = p_1 = b - AX_1 \,. \tag{21}$$

Note that p_1 is nothing but ∇R^2. Thus the algorithm chooses as a first minimization direction the gradient of the function to be minimized at the position of the first guess. Then it minimizes in directions which are conjugate to the previous ones until it reaches the minimum or the square of the residual R^2 is smaller than certain value ϵ. The method has one potential pathological behaviour when applied to our problem. One cannot choose ϵ to be arbitrarily small. If one chooses a very small ϵ the algorithm will try to find a solution which focuses the arcs in N_s sources which are delta functions. This is not surprising as we are assuming that all the $2N_\theta$ unknown βs are reduced to just $2N_s$ βs, i.e. the *point source solution.* The mass distribution which accomplishes this is usually very much biased compared to the right one. It shows a lot of substructure and it has large fluctuations in the lens plane. One therefore has to choose ϵ with a wise criteria. Since the algorithm will stop when $R^2 < \epsilon$ we should choose ϵ to be an estimate of the expected dispersion of the sources at the specified redshifts. This is the only prior which has to be given to the method. However, we will see later how the specific value of ϵ is not very critical as long as it is within a factor of a few of the right source dispersion. Instead of defining ϵ in terms of R^2 it is better to define it in terms of the residual of the conjugate gradient algorithm, r_k^2. This will speed the minimization process significantly since we do not need to calculate the real dispersion at each step but to use the already estimated r_k. Both residuals are connected by the relation,

$$r_k = \Gamma^T R \, . \tag{22}$$

Imposing a prior on the sizes of the sources means that we expect the residual of the lens equation, R, to take typical values of the order of the expected dispersion of the sources at the measured redshifts. Hence we can define an R_{prior} of the form

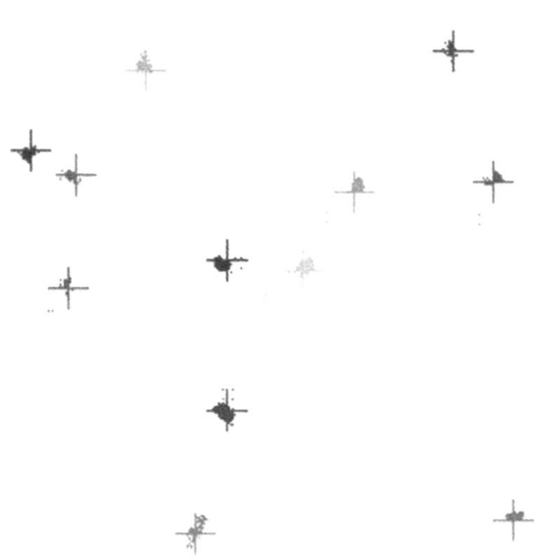

Fig. 8. Recovered βs after minimizing R^2. Crosses represent the true position of the sources. The recovered β_o falls in the middle of its corresponding cloud β points

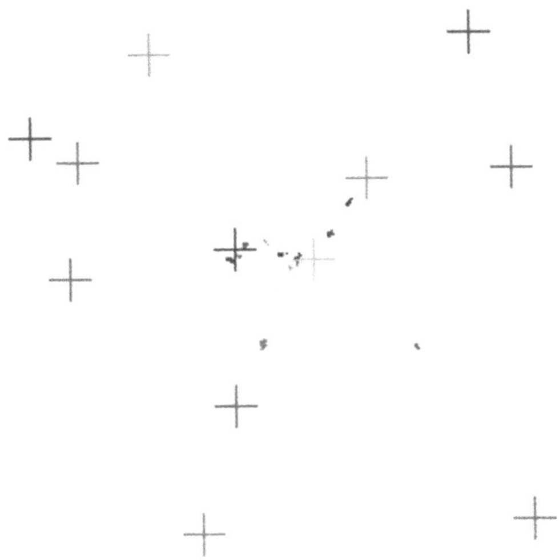

Fig. 9. Recovered βs after minimizing R^2 for the *point source solution*. The real source positions are shown as crosses

$$R^i_{prior} = \sigma_i * RND \,, \tag{23}$$

where the index i runs from 1 to N_θ and σ_i is the dispersion (prior) assumed for the source associated to pixel i and RND is a random number uniformly distributed over -1 and 1. Then, we can estimate ϵ as

$$\epsilon = r_k^T r_k = R^T_{prior} \Gamma \Gamma^T R_{prior} \,. \tag{24}$$

If we choose as a prior the sources that are Gaussians with a $\sigma = 40\,\mathrm{kpc}$ located at the measured redshift, this renders $\epsilon \approx 2 \times 10^{-10}$. The reader will note that the chosen σ is a few times larger than the one we would assign to a typical galaxy. The reader can argue that a more clever way of including this prior information in the algorithm is by perturbing the β elements in (35) (see Appendix). This is done by adding some noise to the 1s in the two $\hat{1}$ matrices in (35). One could for instance add Gaussian noise with a dispersion similar to the expected dispersion of the source at redshift z. The reality, however, is that the quadratic nature of R^2 cancels out any symmetric perturbation added to the elements of Γ. Thus, the result is similar if we perturb Γ or not and we still have to include the prior and fix ϵ to be large enough so we do not recover the *point source solution*. This also tells us that this method is not very promising if one wants to include parity information in the recovery of the mass and sources. In the next subsection we will show a different approach which can include this parity information.

5.3 Singular Value Decomposition

The Singular Value Decomposition (hereafter SVD) algorithm allows for decomposition of a generic $m \times n$ (with $m >= n$) matrix A into the product of

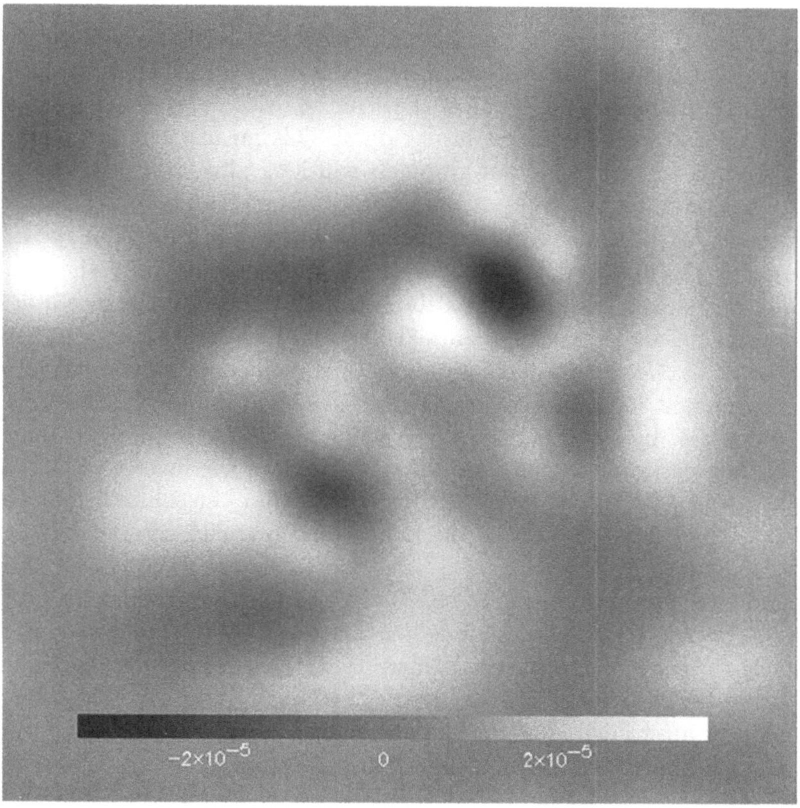

Fig. 10. Recovered mass after minimizing R^2 using the biconjugate gradient algorithm and with a very small ϵ (*point source solution*). The total recovered mass is $2.43 \times 10^{15} \mathrm{h}^{-1}\,\mathrm{M}_{\odot}$, but there are also regions with negative masses

three matrices, two orthogonal and one diagonal:

$$A = UWV^T, \tag{25}$$

where U is an $m \times n$ orthogonal matrix, W is an $n \times n$ diagonal matrix whose elements in the diagonal are the *singular values* and V^T is the transpose of an $n \times n$ orthogonal matrix. When A is symmetric, the SVD reduces to finding the *eigenvectors* and *eigenvalues* of A.

The advantage of this decomposition is that the inverse of A is given by

$$A^{-1} = VW^{-1}U^T\,, \tag{26}$$

where W^{-1} is another diagonal matrix whose elements are just the inverse of the elements of W, that is, $W_{jj}^{-1} = 1.0/W_{jj}$. The proof $A^{-1}A = I$ follows from the property that U and V^T are orthogonal ($U^TU = VV^T = I$).

Although the SVD allows us to invert the problem by calculating Γ^{-1}, its full power lies in its ability to solve a system approximately. The level of approximation can be controlled by setting a threshold in the matrix W^{-1}. In our problem, there will be many equations which are strongly correlated in the sense that most of the θ positions in a single arc will come from the same source (that is, they will have almost the same β). Also we have to keep in mind that we are using all the pixels in our data. This means that two equations corresponding to two neighbouring pixels will look almost exactly the same. When one computes the SVD of the matrix Γ, these two facts translate into a matrix W with elements in the diagonal which are 0 or close to 0. The inverse of W would be dominated by these small numbers and the solution will look very noisy. The good news about using SVD is that the most relevant information in the Γ matrix is packed into the first few values (the largest values in W) while the small single values in W will contain little information or the redundant information coming from neighbouring pixels. One can just approximate W by another matrix W' where all the elements in its diagonal smaller than certain threshold are set to 0. Also, in the inverse of W', these elements are set to 0. The magic of this trick is that the solution found with this approximation will contain the main trend or main components of the mass distribution.

Another advantage of using the SVD algorithm is that in this case no prior on the extension of the sources is needed. The degree of accuracy is controlled by setting the threshold in the singular values of the matrix W. Those elements in W below the threshold are set to 0 and the same in its inverse. The threshold is usually set after looking at the singular values. The first ones will normally stay in some kind of *plateau* and after that the singular values will decrease rapidly. The threshold should be normally set immediately after the *plateau*. After decomposing Γ in its SVD decomposition and calculating its inverse with (26), the result is very similar to the one obtained with the previous algorithms.

Like the two previous algorithms, the SVD has its own pathologies. Using standard subroutines to find the SVD of Γ usually return a *no convergence* error. This error comes from the nearly degenerate nature of Γ. One has then to increase the number of maximum iterations on these subroutines or use a coarse version of Γ where only a small fraction of the θ positions (or equivalently, Γ rows) are considered. Another solution is inverting the preconditioned system of equations where we previously multiply for Γ^T. This allows to use all the θ pixels and find the SVD of $\Gamma^T\Gamma$ in a small number of iterations. However, using the SVD of Γ instead of $\Gamma^T\Gamma$ has a very interesting feature. It allows to introduce parity information in an effective way.

Contrary to the quadratic cases of minimization of the variance or the square of the residual R^2, the SVD allows us to include parity information in the Γ matrix which will not disappear when we look for the best solution. Since no $\Gamma^T\theta$ or $\Gamma^T\Gamma$ operations are involved, parity information will not cancel out. SVD could be an interesting way of fine-tuning the solution by including the extra information coming from the parity of the arcs.

6 Incorporating the Null Space

So far we have made use of the information contained in the observed strong lensing arcs. This gives us a solution which explains the data in the sense that it predicts arcs in the right positions but it could well happen that the solution overpredicts arcs. An example of this can be seen by comparing the reproduced arcs in Fig. 13 with the true arcs in Fig. 6. Some of the reproduced arcs are slightly larger and span a larger area than the true ones, although the sources are perfectly reproduced in their true positions and the recovered mass is very close to the true mass distribution.

To avoid this we propose for the first time to include information from the null space, $\tilde{\theta}$. This space tell us where an arc should not appear but it does not tell us where the hypothetical $\tilde{\beta}$ should be. The only fair thing we can do is to impose that none of the pixels in the null space fall into the estimated β of our solution. By achieving this we will have a solution which predicts the right arcs while not overpredicting arcs. The solution will be then fully consistent with the observed data.

The null space is connected to the solution X by

$$\tilde{\beta} = \tilde{\theta} - \tilde{\Upsilon} M \; . \tag{27}$$

It is evident that we want the new solution X (M and β_o) to be such that the new $\tilde{\beta}$ does not fall within a circle of radius $p(k)$ centred in each one of the β_o where $p(k)$ is the prior with information on how extended the sources are. The null space will perturb the solution X in such a way that the new solution, $X' = X + \Delta\tilde{X}$, is an approximated solution of (35) and satisfies all the constraints of the form $\tilde{\beta} \ni \beta$.

The way we incorporate the constraints is by adopting an approach commonly used in penalized quadratic programming. We will minimize the new function

$$\phi(X) = R^2 + \lambda \sum_k g(f_k) \; , \tag{28}$$

where λ is a constant which guarantees that in the first iterations the second term in (28) does not dominate the first and g_k is a function which will penalize those models predicting $\tilde{\beta}$ falling near the β positions which minimize R^2. As a penalizing function we will choose an asymptotically divergent Gaussian

$$g(f_k) = \frac{1}{e^{f_k} - \mu} \tag{29}$$

with

$$f_k = \frac{(\tilde{\beta}(k)_x - \beta^o_x)^2 + (\tilde{\beta}(k)_y - \beta^o_y)^2}{\sigma_k^2} \; , \tag{30}$$

where $\tilde{\beta}(k)_x$ is the x component of $\tilde{\beta}$ for the pixel k in the null space. β^o_x is our estimated value of β (x component) and similarly for the y component.

There are as many constraints of the form f_k as there are pixels, $\tilde{\theta}$, in the null space.

The parameter μ controls the degree of divergence of the Gaussian function. When $\mu = 0$ we recover the classical Gaussian but as μ approaches 1 the Gaussian becomes more and more sharp without increasing its dispersion. For $\mu = 1$ the Gaussian is infinite at $f_k = 0$ (see Fig. 11). By minimizing the function $\phi(X)$ with increasing values of μ we will find that in the limit $\mu \to 1$ the solution will push away those $\tilde{\beta}$ which originally were falling in the region defined by the set of β.

Ideally we want to include all the *dark* or empty pixels in the $\tilde{\theta}$ space but this is, in general, found to be a waste of memory and computing time. The fact is that only the $\tilde{\theta}$ pixels which are hitting (or close to hit) one of the N_s-estimated β positions of the sources will have some impact on the solution. Most of the $\tilde{\theta}$ pixels in the null space already satisfy all the constraints for the actual solution X. For this reason we will include only those $\tilde{\theta}$ for which the solution X predicts their corresponding $\tilde{\beta}$ are close to hitting (or actually hitting) a source. The $\tilde{\theta}$ space will include the observed θs as a subspace. We have to exclude this subspace from $\tilde{\beta}$ before minimizing (28).

After the minimization process is finished, the new solution will have the arcs falling in compact regions around β_0 while the extra arcs produced by the previous solution will fall in regions outside areas around β_o.

From our simulation we have seen that addition of the null space induces small changes in the mass plane and it tends to stabilize the solution in the sense that it makes the recovered mass profile independent of the threshold ϵ. This is in fact an interesting bonus which comes with the addition of the null space. The new function to be minimized, ϕ, can be minimized until the true minimum is found. In this case, there is no equivalent of a point source solution. Since some of the $\tilde{\theta}$ are in fact very close to the observed θ, the solution which minimizes ϕ will *focus* those $\tilde{\theta}$ and θ in neighbouring regions

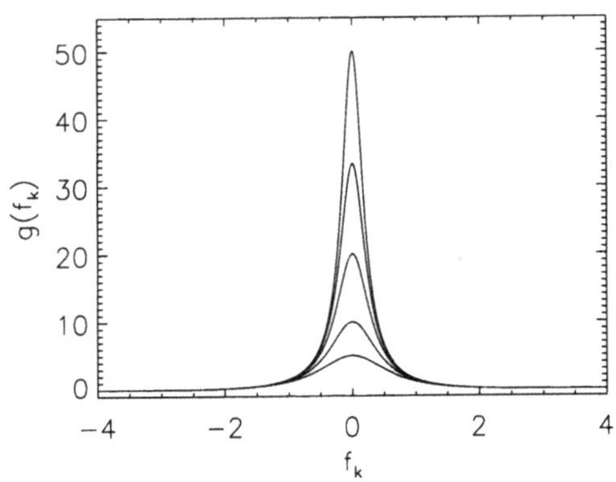

Fig. 11. Penalizing function showing values of $\mu = 0.8, 0.9, 0.95, 0.97$, and 0.98 and $\sigma = 1$. Note that the width of the curve does not change when increasing μ

in the source plane. When minimizing ϕ there will be two competing effects. One will tend to increase the mass so it minimizes R^2 (point source solution). The other will tend to reduce the mass so the $\tilde{\beta}$ will be pushed away from the β positions. The outcome will be a balanced situation between the β trying to collapse into compact sources and the $\tilde{\beta}$ trying to escape the wells in the *beta* positions. We show in Fig. 12 the recovered mass after imposing the constraints in the $\tilde{\theta}$ space. The total mass is now only 1.2% lower than the true mass. The new mass also contains more structure and starts showing the internal distribution of the main components of the cluster.

In Fig. 13 we show the predicted position for the arcs after combining the best mass together with the best estimate for the position of the sources. By comparing Figs. 13 and 6 (left) we see that the predicted arcs match very well the observed (simulated) data with the exception of some of the arcs near the centre of the image.

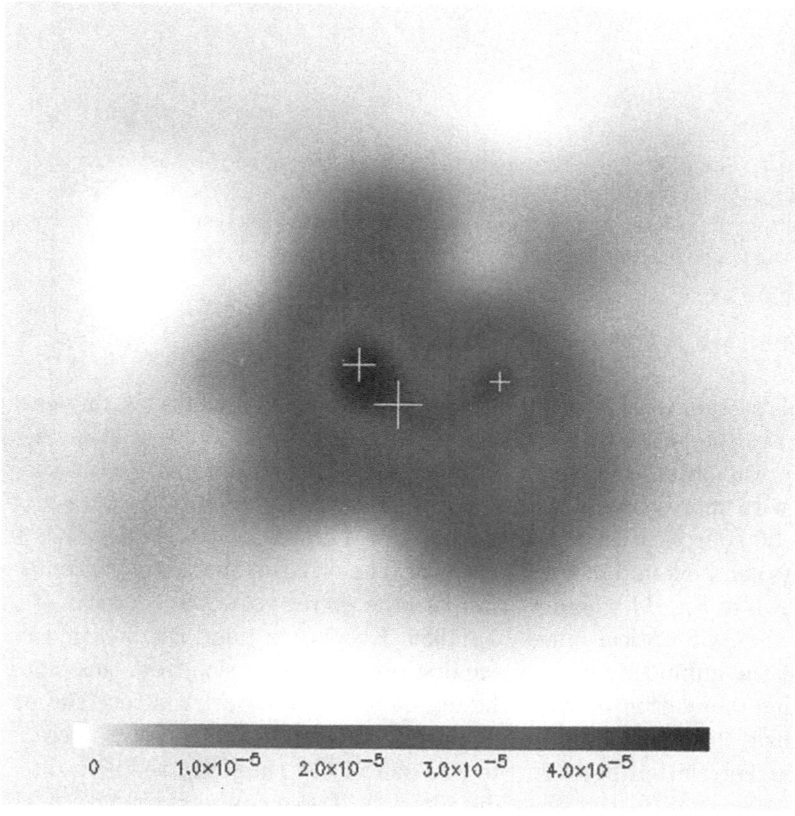

Fig. 12. Recovered mass (smoothed) after minimizing ϕ. The crosses show the position of the three main haloes in the cluster. The size of the cross is proportional to the mass of the halo. The total recovered mass is $1.1055 \times 10^{15}\,\mathrm{h}^{-1}\,\mathrm{M}_{\odot}$

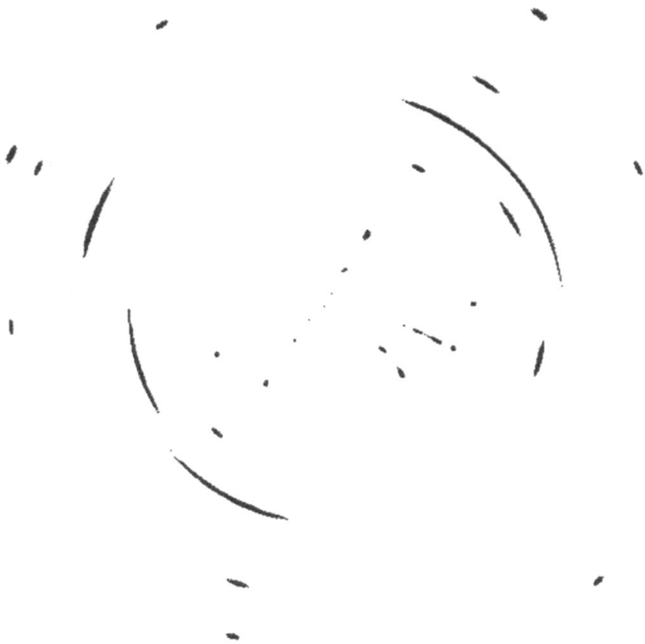

Fig. 13. The plot shows the predicted arcs according to the best mass and source solution. We have assumed that the sources are circles with radius $15\,\mathrm{h}^{-1}\mathrm{kpc}$ centred in the source position and at the measured redshifts. This result should be compared with the true arcs seen in Fig. 6. The match is almost perfect

7 Case of A1689

In this section we will apply one of the previous algorithms (biconjugate gradient) to the best data set on strong lensing ever (Broadhurst et al. [6]). The image was obtained with the Hubble-ACS camera and contains hundreds of arcs with more than 100 arcs having measured photo-z. The arcs originate from 30 sources with redshifts from $z = 1$ up to $z = 6$. Full details of the results can be found in Diego et al. [8]. The result of the minimization process is shown in Fig. 14 where we compare the average of 1000 recovered solutions with the ACS optical image of A1689. Keeping in mind that *no* information about the luminosity is used, the first obvious conclusion from this plot is the existing correlation between the luminous and the dark matter. The peak of the mass distribution falls very close to the central cD galaxy. There is also a clear correlation between the position of the subgroup to the right and a secondary peak in the mass distribution. The small subgroup at $\approx 30''$ to the south of the cD seems to be sitting close to the top of other overdensity. Comparing the result derived from the non-parametric algorithm with results using parametric methods (Broadhurst et al. [6]), the agreement is good in the region where the data are more sensitive to the mass distribution (see

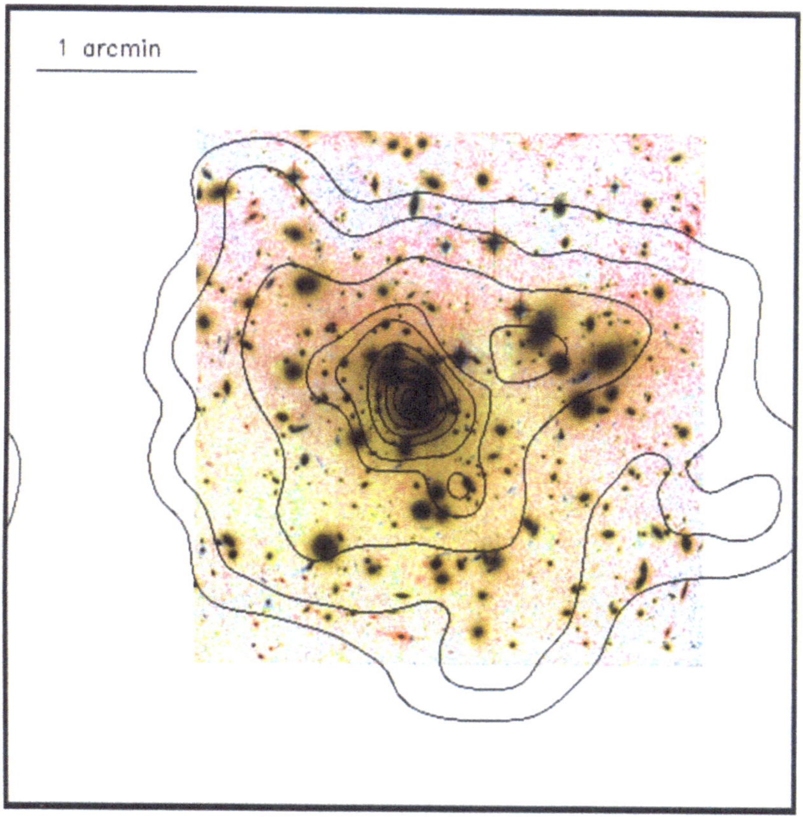

Fig. 14. Mean recovered mass (contours) compared with true ACS image. The mass is the average of 1,000 minimizations of the lens equation where at each minimization we change the grid, the initial conditions, X_o, and the redshifts of the sources. Contours go from 0.1 to 0.97 times the maximum mass density in intervals of 0.1 (0.097 last interval). Total mass in the field of view is about $5.2 \times 10^{14} h^{-1} M_{\odot}$. The field of view is $5 \times 5\,\mathrm{arcmin}^2$

Fig. 15). This is a very interesting conclusion considering that it confirms that dark matter models with a NFW profile are consistent with the best fitting model obtained by both parametric and non-parametric models. It also adds robustness to the derived mass profile since this one seems to be insensitive to the assumptions made in the parametric method or the intrinsic biases in the non-parametric one (Diego et al. [8]).

8 Concluding Remarks

Previous works have suggested some discrepancy between mass estimates derived from different data (Wu & Fang [25]). Strong lensing analyses were predicted to yield interesting cosmological constraints, but due to the

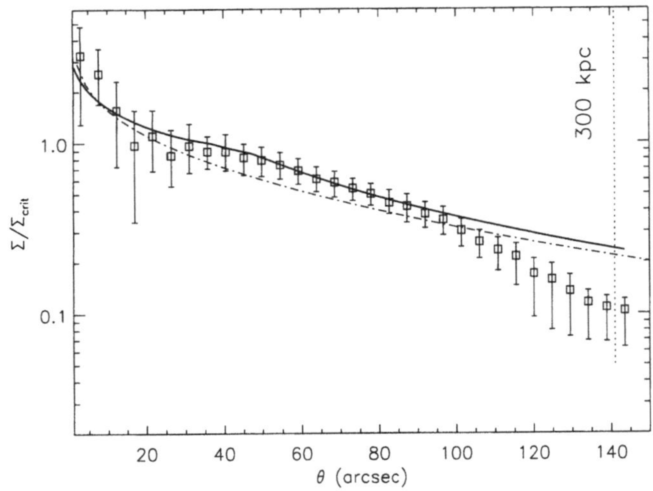

Fig. 15. The plot shows the mean value (squares) and the 99% confidence region of the 1D profiles for the 1,000 minimizations. The dot-dashed line is the best fitting NFW profile found in Broadhurst et al. [6]. The density has been rescaled by the critical density, Σ_{crit}. The thick solid line is a very similar NFW profile plus an excess

uncertainties in our understanding of galaxy clusters they have yet to live up to these predictions. A fast, per cent level determination of cluster masses from lensing observations could allow for sufficient statistical sampling to provide information about cosmological parameters.

However, the methods presented here should not be applied indiscriminately. For instance, if the reconstructed mass is systematically biased towards recovering less cuspy central regions this reduces the possibility of making conclusions about mechanisms for dark matter annihilation (Spergel & Steinhardt [21], Wyithe et al. [24], and Boehm et al. [4]).

The conclusions derived from analyses based on strong lensing data have to be checked carefully for possible systematics. A way to gain confidence in the results is by deriving the mass distribution using different, independent algorithms, that is, parametric and non-parametric ones. If the results agree, we may be more confident in our conclusions. If on the other hand they disagree, this opens the door to interesting debates to understand the discrepancy. The above comparison using the ACS image of A1689 shows that the agreement is possible.

Acknowledgement

This work was supported by NSF CAREER grant AST-0134999, NASA grant NAG5-11099, the David and Lucile Packard Foundation, and the Cottrell Foundation.

Appendix: Υ and Γ Matrices

The Υ matrix contains the information of how each mass element j affects the ith deflection angle. That is,

$$\hat{\alpha}_i = \Upsilon_{ij} M_j. \tag{31}$$

Precisely how the matrix is constructed is a matter of convenience. Our specific choice is to put both x and y components of all the arc pixels into vectors with $2N_\theta$ elements. The resulting Υ is then a $2N_\theta \times N_{cells}$ matrix.

In fact the structure of the Υ matrix and the vectors are irrelevant as long as they combine to correctly represent the lens equation as

$$\beta_i = \theta_i - \Upsilon_{ij} M_j. \tag{32}$$

Each element in the Υ matrix is computed as follows:

$$\Upsilon_x(i,j) = \lambda[1 - \exp(-\delta/2\sigma^2)]\frac{\delta_x}{\delta^2}, \tag{33}$$

where

$$\lambda = 10^{15} M_\odot \frac{4G}{c^2} \frac{D_{ls}}{D_l D_s}. \tag{34}$$

The index i runs over the N_θ-observed θ pixels and index j runs over the N_c elements in the mass vector, M. The factor δ_x in (33) is just the difference (in radians) between the x position in the arc (or x of pixel θ_i) and the x position of the cell j in the mass grid ($\delta_x = \theta_x(i) - \theta'_x(j)$). Similarly we can define $\delta_y = \theta_y(i) - \theta'_y(j)$ and $\delta = \sqrt{\delta_x^2 + \delta_y^2}$. Also, for Υ_y we only have to change δ_x by δ_y. Since we include the factor $10^{15}\mathrm{M}_\odot$ in λ (see (34)), the mass vector M in (5) will be given in 10^{15} h^{-1} $\mathrm{M}_\odot$ units. The h^{-1} dependency comes from the fact that in λ we have the ratio $D_{ls}/(D_l D_s)$ which goes as h. Also we calculate Υ_x and Υ_y separately, the final Υ matrix entering in (5) contains both Υ_x and Υ_y (the same holds for the vectors β and θ). One can rearrange the x and y components in any order.

The structure of the Γ matrix is identical to the matrix Υ but with the difference that it has $2N_s$ additional columns (the location of the extra columns is irrelevant as long as it is consistent with the location of the $2N_s$ β_o unknowns in the X vector). It is easy to see that each one of these extra columns (with dimension $2 * N_\theta$) corresponds to one of the N_s sources. Since the Γ matrix has to contain both the x and y components, the first/second half of each one of the extra columns will be all 0s depending on whether it corresponds to the y/x component of B_o. The other half will be full of 0s and 1s, the 1s being in the positions associated with that particular source, the 0s elsewhere. That is, the lens equation can be written explicitly as ($\hat{a}$ denotes matrix and $\boldsymbol{a}$ denotes vector)

$$\begin{pmatrix} \theta_x \\ \theta_y \end{pmatrix} = \begin{pmatrix} \hat{\Upsilon}_x & \hat{1} & \hat{0} \\ \hat{\Upsilon}_y & \hat{0} & \hat{1} \end{pmatrix} \begin{pmatrix} M \\ \beta_o^x \\ \beta_o^y \end{pmatrix}, \tag{35}$$

where again, θ_x and θ_y are two N_θ-dimensional vectors containing the x and y positions, respectively, of the pixels in the observed arcs. The two $(N_\theta \times N_c)$ matrices $\hat{\Upsilon}_x$ and $\hat{\Upsilon}_y$ contain the x and y lensing effect of the cell j on the θ pixel i. The $N_\theta \times N_s$ dimensional matrices $\hat{1}$ and $\hat{0}$ are full of 0s (the $\hat{0}$ matrix) or contain 1s (the $\hat{1}$ matrix) in the i positions ($i \in [1, N_\theta]$) where the ith θ pixel comes from the j source ($j \in [1, N_s]$) and 0 elsewhere. The vector $\boldsymbol{M}$ contains the N_c-gridded masses we want to estimate. β_o^x contains the central x positions of the N_s sources. Similarly, β_o^y will contain the central y positions.

References

1. Abdelsalam, H.M., Saha, P., Williams, L.L.R.: MNRAS **294**, 734 (1998)
2. Abdelsalam, H.M, Saha, P., Williams, L.L.R.: AJ **116**, 1541 (1998)
3. Blandford, R.D., Narayan, R.: ARA&A **30**, 311 (1992)
4. Boehm, C., Hooper, D., Silk, J., Casse, M., Paul, J.: PhRvL **92**, 1301 (2004)
5. Broadhurst, T.J., Taylor, A.N., Peacock, J.A.: ApJ **438**, 49 (1995)
6. Broadhurst, T., Benítez, N., Coe, D., Sharon, K., Zekser, K., White, R., Ford, H., Bouwens, R.: ApJ **621**, 53 (2005)
7. Diego, J.M., Protopapas, P., Sandvik, H.B., Tegmark, M.: MNRAS **360**, 477 (2004)
8. Diego, J.M., Sandvik, H.B., Protopapas, P., Tegmark, M., Benitez, N., Broadhurst, T.: MNRAS **362**, 1247 (2004)
9. Kneib, J.-P., Mellier, Y., Fort, B., Mathez, G.: A&A **273**, 367 (1993)
10. Kneib, J.-P., Mellier, Y., Pello, R., Miralda-Escudé, J., Le Borgne, J.-F., Boehringer, H., Picat, J.-P.: A&A **303**, 27 (1995)
11. Kneib, J.-P., Ellis, R.S., Smail, I.R., Couch, W., Sharples, R.: ApJ **471**, 643 (1996)
12. Kneib, J.-P.: The shapes of galaxies and their dark halos, In: Natarajan, P. (ed) Proceedings of the Yale Cosmology Workshop: The Shapes of Galaxies and their Dark Matter Halos, New Haven, Connecticut, USA, 28–30 May 2001. p. 50. World Scientific: Singapore 2002
13. Kochanek, C.S., Blandford, R.D.: ApJ **375**, 492 (1991)
14. Koopmans, L.V.E., Bruy, A.G., Jackson, N.: MNRAS **295**, 534 (1998)
15. Narayan, R., Bartelmann, M.: Formation of Structure in the Universe, Dekel, A., Ostriker, J.P., Cambridge Univ. Press, Cambridge 360; on-line version at http://www.ita.uni-heidelberg.de/~msb/Lectures/EinfSeminar/JeruLect.pdf (1999)
16. Press, W.H., Teukolsky, S.A., Vetterling, W.T., Flannery, B.P.: Numerical Recipes in Fortran 77. Cambridge University Press, Cambridge (1997)
17. Saha, P., Williams, L.L.R.: MNRAS **292**, 148 (1997)
18. Saha, P.: AJ **120**, 1654 (2000)
19. Sand, D.J., Treu, T. Ellis, R.S.: ApJ **574**, 129 (2002)
20. Schneider, P., Ehlers, J., Falco, E.E.: Book Review: Gravitational lenses. Springer-Verlag (1993)

21. Spergel, D.N., Steinhardt, P.J.: PhRvL **84**, 3760 (2000)
22. Trotter, C.S., Winn J.N., Hewitt J.N.: ApJ **535**, 671 (2000)
23. Wambsganss, J.: Living Reviews in Relativity **1**, 12 (1998)
24. Wyithe, J.S.B., Turner, E.L., Spergel, D.N.: ApJ **555**, 504 (2001)
25. Wu, X.-P., Fang, L.-Z.: ApJ **483**, 62 (1997)
26. Young, P., Gunn, J.E., Oke, J.B., Westphal, J.A., Kristian, J.: ApJ **244**, 736 (1981)

Index

Kolibri 360 GmbH
Ferdinand-Jühlke-Straße 7
99095 Erfurt, Deutschland
produktsicherheit@kolibri360.de